3rd International Conference on Manufacturing and Industrial Technologies 2016 (ICMIT 2016)

MATEC Web of Conferences Volume 70 (2016)

Instanbul, Turkey
25-27 May 2016

Editors:

T. Deaconescu **A. Deaconescu**

ISBN: 978-1-5108-3015-8

Printed from e-media with permission by:

Curran Associates, Inc.
57 Morehouse Lane
Red Hook, NY 12571

Some format issues inherent in the e-media version may also appear in this print version.

Printed by Curran Associates, Inc. (2016)

For additional information, please contact EDP Sciences – Web of Conferences at the address below.

EDP Sciences – Web of Conferences
17, Avenue du Hoggar
Parc d'Activité de Courtabœuf
BP 112
F-91944 Les Ulis Cedex A
France

Phone: +33 (0) 1 69 18 75 75
Fax: +33 (0) 1 69 28 84 91

contact-edps@webofconferences.org

Additional copies of this publication are available from:

Curran Associates, Inc.
57 Morehouse Lane
Red Hook, NY 12571 USA
Phone: 845-758-0400
Fax: 845-758-2633
Email: curran@proceedings.com
Web: www.proceedings.com

TABLE OF CONTENTS

Committee

Conference Chair

Prof. Dr. Osman Adiguzel, Firat University, Department of Physics, Turkey
Prof. Wenzhi Cui, Chongqing University, China

Program Chair

Dr. Mário S. Ming Kong, University Lisbon, Portugal
Prof. Andrea Deaconescu, Transilvania University of Brasov, Romania
Prof. İlhan Asiltürk, Selcuk University, Turkey
Dr. Serhat İkizoğlu, Istanbul Technical University, Turkey
Prof Fuqiang Gao, Chongqing University, China

Technical Committee

Dr. Abdelkebir Charkaoui,University Hassan 1st - FST Settat, Morocco
Prof. Engku Muhammad Nazri Engku Abu Bakar, Universiti Utara Malaysia
Dr. Mehdi Jafari Vardanjani,University of Tehran, Iran
Dr. Balhassn S. M. Ali, Department of Mechanical Engineering,Bursa Orhangazi University, Turkey
Prof. Yahya Bozkurt, Marmara University, Turkey
Prof. Tudor Deaconescu, Transilvania University of Brasov, Romania
Dr. Mahdi Alajmi, College of Technological Studies, PAAET, Kuwait
Prof. Mehmet Savsar, Kuwait University, Kuwait
Prof. Supakorn Pukird, Ubon Ratchathani University, Thailand
Prof. Vladimir Strezov,Macquarie University,Australia
Dr.Vladimir Konstantinov Kotev, Bulgarian Academy of Sciences (BAS), Bulgaria
Dr. Hasan AL Dabbas, Philadelphia University, Jordan
Dr. Sabuj Mallik, University of Greenwich, United Kingdom
Prof. Boran Sekeroglu, Near East University, Turkey
Dr. K. K. Soundrapandian, Pdpm-Indian Institute Of Information Technology Design And Manufacturing, India
Dr. Wan Ahmad Yusmawiza Wan Yusoff, Hail University, Kingdom of Saudi Arabia

Preface

In modern society, manufacturing and industrial technologies always plays an important role in people's daily life. And it has been changing all aspects of our lives in such diverse areas and makes our lives easier. In light of the fast-paced advancements in manufacturing and industrial technologies taking place all over the world, it is of interest to keep an eye on the state-of-the-art research and development and to facilitate collaboration in multidisciplinary research areas. With this end in view, 2016 3th International Conference on Manufacturing and Industrial Technologies was held once again this year in Istanbul, Turkey on May 25-27, 2016.

The objective of the conference was to address and deliberate on the latest technical status and recent trends in the research and applications of manufacturing and industrial technologies. This conference has been so designed with the view to provide an opportunity for the scientists, engineers, industrialists, students and other professionals from all over the world to interact and exchange their new ideas and research outcomes for future collaboration. The conference was also aimed at motivating the next generation of researchers to promote their interests in manufacturing and industrial technologies and their applications.

This volume consists of 69 selected papers from the conference which are presented orally at the conference or via poster sessions, and provide up-to-date, comprehensive and worldwide state-of-the art knowledge in this field. Each contributed paper was strictly peer-reviewed by international reviewers who were drawn from the organizing committee members and other experts in the related field from all over the world. The proceedings are divided into 11 chapters as follows: material properties and processing, mechanical design-manufacture and automation, mechanical design-manufacture and automation, mechanical design-manufacture and automation, mechanical design-manufacture and automation, supply chain management, fluid mechanics, control system and application, mechatronics, electronic and power system, environmental science and engineering, environmental science and engineering.

We would like to thank the program chairs, reviewers and the members of the conference committee for their hard work. We are grateful to all those who have contributed to the success of ICMIT 2016. We hope that all participants and other interested readers benefit scientifically from the proceedings.

Prof. Dr. Osman Adiguzel, Firat University, Department of Physics, Turkey
Prof. Wenzhi Cui, Chongqing University, China

June 26, 2016

ISSN (electronic edition): 2261-236X

Statement of Peer review

In submitting conference proceedings to MATEC Web of Conferences, I certify to the Publisher that I adhere to the **Policy on Publishing Integrity** of the journal in order to safeguard good scientific practice in publishing.

1. All articles have been subjected to peer review administered by the proceedings editors.
2. Reviews have been conducted by expert referees, who have been requested to provide unbiased and constructive comments aimed, whenever possible, at improving the work.
3. Proceedings editors have taken all reasonable steps to ensure the quality of the materials they publish and their decision to accept or reject a paper for publication has been based only on the merits of the work and the relevance to the journal.

Title, date and place of the conference

2016 The 3rd International Conference on Manufacturing and Industrial Technologies
Istanbul, Turkey , May 25-27, 2016

Proceedings editor(s):

Tudor Deaconescu, Transilvania University of Brasov, Romania; tdeacon@unitbv.ro

Andrea Deaconescu, Transilvania University of Brasov, Romania; deacon@unitbv.ro

Date and editor's signature

Measuring Dilution and Wear for Laser Cladding of Stellite 6 Produced on a P91 Steel Substrate using Two Different Heat Inputs

Alain Kusmoko, Druce Dunne, Huijun Li

Faculty of Engineering, University of Wollongong, NSW 2522, Australia

Abstract. Stellite 6 was deposited by laser cladding on a P91 substrate with energy inputs of 1 kW (P91-1) and 1.8 kW (P91-1.8). The chemical compositions, microstructures and surface roughnesses of these coatings were characterized by atomic absorption spectroscopy, scanning electron microscopy and atomic force microscopy. The microhardness of the coatings was measured and the wear mechanism of the coatings was evaluated using a pin-on-plate (reciprocating) wear testing machine. The results showed less cracking and pore development for Stellite 6 coatings applied to the P91 steel substrate with the lower heat input (P91-1). Further, the Stellite coating for P91-1 was significantly harder than that obtained for P91-1.8. The wear test results indicated that the weight loss for P91-1 was much lower than for P91-1.8. The surface topography data indicated that the surface roughness for P91-1 was much lower than for P91-1.8. The measurements of dilution and carbon content showed that P91-1 has lower dilution and higher concentration of carbon than P91-1.8. It is concluded that the lower hardness of the coating for P91-1.8, together with the softer underlying substrate structure, markedly reduced the wear resistance of the Stellite 6 coating and the lower hardness of the coating for P91-1.8 was due to higher level of dilution and lower concentration of carbon.

1 Introduction

P91 steels are high strength alloys that normally transform completely to martensite during air cooling. This means that the specification of preheating and post weld heat treating conditions is an important concern. One important issue is specification of maximum tempering temperatures which are limited by the lower critical temperature for the ferrite-to-austenite phase transformation, A1. This steel is an important steam generator material for thermal and nuclear power plants. P91 steel displays three distinct temperature regimes for the variations of tensile strength, average work hardening rate and ductility. It has the advantages of low thermal expansion, high thermal conductivity, good steam corrosion resistance and excellent creep resistance [1]. These properties improve the resistance of P91 to thermally-induced stresses and they provide a better match of properties with the lower alloy steel used throughout other components of the boilers. The higher thermal conductivity also improves heat transfer through tubing walls into cooling water. Also, P91 steels have relatively good oxidation resistance at intermediate temperatures and can exhibit high strengths up to about 600°C. P91 is relatively resistant to tempering, and higher PWHT temperatures are generally needed to reduce hardness in weld heat-affected zones and to develop the necessary toughness in weld deposits [2].

Stellite 6 is a very versatile material that is used for hardfacing of various component parts for applications requiring wear resistance. Stellite alloys have been considered as candidate materials for surface hardening of P91 steel. The microstructure of Stellite 6 contains hard M_7C_3 carbides in interdendritic regions in both as-cast and as welded conditions. Stellite alloys also contain a hard Laves phase in a softer matrix of eutectic or solid solution, which is useful for unlubricated wear conditions [3]. A little works have been done for Stellite 6 deposited on P91 steel substrate.

Surface alloying is usually done by advanced techniques such as laser cladding due to significant advantages like fast processing speed, relative cleanliness, a very high heating/cooling rate (105 K/s) and high solidification velocity (up to a maximum of 30 m/s) [4]. As a result, the process has a low energy input and causes less distortion of the component than hand or arc welding. Steen and Bruck [5] have reviewed laser cladding processes. In the coaxial laser cladding process, metal powder is injected through a nozzle, which is coaxial with the laser beam. The powder absorbs laser energy and become partially melted before reaching the substrate. Part of the laser energy is also absorbed by the substrate to cause surface melting, forming a strong metallurgical bond between the substrate and the clad layer. Laser clad layers can be produced that are defect-free and result in low dilution [5].This technique has not been extensively studied for the development of P91 steel with different energy input.

The purpose of this study was to evaluate the sliding wear characteristics and measuring the dilution and carbon content of Stellite 6 coating materials produced by laser cladding of a P91 steel substrate with low (1 kW) and high (1.8 kW) energy inputs The sliding wear tests were carried out on flat samples in an unlubricated (dry) condition using a reciprocating wear tester with a tool steel ball. This paper discussed further on dilution measurements and estimation of carbon content of Stellite 6 coatings and the role of carbon in forming Cr_7C_3 particles to harden the deposit and contribute the increase wear resistance.

2 Experimental methods

2.1 Laser cladding deposition

Table 1. Nominal compositions (wt%) of the P91 steel and Stellite 6 alloy [3].

(%)	**Co**	**Cr**	**Fe**	**W**	**Ni**	**C**	**Si**	**Mn**
S 6	60	27	2.5	5	2.5	1	1	1
P91		9.08			0.08	0.09	0.45	0.46
(%)	**P**	**S**	**Mo**	**V**	**Al**	**Ti**	**Nb**	B
S 6								
P91			0.96	0.19	0.08			

The laser cladding process of P91 steel substrates with Stellite 6 was carried out by a commercial coating manufacturing company using 1 kW and 1.8 kW energy input. The initial coating thickness as received was about 0.35 mm for both energy inputs. Table 1 shows the nominal compositions of the P91 steel and the Stellite 6 alloy.

2.2 Characterisation of stellite coated samples

The microhardness measurements were made at intervals of 0.05 mm through the coating thickness using a Leco M-400-H1 hardness testing machine with a load of 300 g. The samples were then etched in a mixed acid solution to reveal the microstructure of the Stellite 6 coating. Subsequently, coatings were studied using a Leica DMRM optical microscope.

2.3 Wear testing

Wear testing was carried out using a pin-on-plate (reciprocating) mode with a 6 mm tool steel ball as the pin. During operation, the ball remained stationary while the flat specimen moved in a linear, back and forth sliding motion, under a prescribed set of conditions.

Since the aim of the work was to examine the wear of Stellite 6 coating materials, it was necessary to grind and polish the flat specimens (coatings) to the required surface finish for the wear test. The coatings were about 0.3 – 0.4 mm thick and approximately 0.05 mm of the coating was removed.

Prior to carrying out the wear tests, the test specimens were weighed to an accuracy of 0.0001 g. The flat specimen was then screwed firmly in place on the base of the wear tester. After the test was complete, wear debris was removed from the sample, which was then washed in alcohol, dried, and reweighed.

The tool steel ball was also washed in alcohol, dried and weighed to an accuracy of 0.0001 g at the start of each test and at the same time as the flat specimen. The ball was re-weighed after testing but, as the weight of the steel ball did not change significantly, it was not considered in assessing the wear damage.

The test speed, number of cycles and test duration were held constant: 50 rpm, 10,000 cycles and 200 minutes. The various tests conducted are: Test# 1, P91-1 with applied load of 2 kg; Test# 2, P91-1.8 with applied load of 2 kg; Test# 3, P91-1 with applied load of 5 kg; Test# 4, P91-1.8 with applied load of 5 kg.

2.4 Examination of wear damage and topography.

In order to study the effect of laser energy input and the applied load during wear testing on the wear track, the surfaces of the samples from Tests# 1-4 were examined after testing using a S440 scanning electron microscope (SEM) operating at 20 kV and using Atomic Force Microscopy (Veeco) with a scan size of 50 μ and scan rate of 1 Hz.

2.5 Measurements of dilution and estimation C content

A measured compositions of Stellite 6 coatings has been conducted using AAS (Atomic Absorption Spectroscopy) to determine the chemical analyses present using two different laser power (1 kW and 1.8 kW). However, the C content was not detected. Therefore it was necessary to estimate the C content by calculating dilution. The method of calculation is shown in Fig. 1.

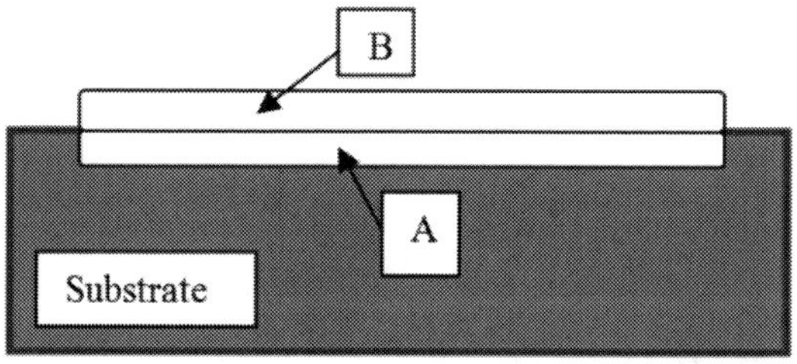

Figure 1. Schematic diagram showing the clad layer, which consists of two parts: added Stellite 6 alloy (region B) and melted base plate (region A). The dilution D of the Stellite alloy is given by A/(A+B) [6].

If the C content (wt%) of the substrate is given by $[C]_{BM}$ and the C content (wt%) of the Stellite 6 is $[C]_S$ then the estimated C content of the weld deposit $[C]_{WD}$ is given as follows:

$$[C]_{WD} = D \times [C]_{BM} + (1-D) \times [C]_S$$

The nominal carbon content $[C]_S$ of the Stellite 6 alloy is 1.0 wt% C. Therefore, $[C]_{WD} = D \times [C]_{BM} + (1-D)$.

3 Results

3.1 Coating compositions

The compositions of the Stellite 6 coatings were determined by AAS (Atomic Absorption Spectroscopy), see Table 2. Table 2 shows that the two chemical analyses (P91-1 and P91-1.8) of the coatings were similar, but there were some differences in alloy content of the coatings. The most significant differences were the lower Fe and Mn contents of the coating for P91-1, consistent with a lower level of dilution of the Stellite coating by the substrate.

3.2 Scanning Electron Microscopy (SEM) of deposit coating cross-sections

The coatings on the P91 steel substrate had a cellular-dendritic appearance. The higher heat input of 1.8 kW produced a coarser cellular-dendritic structure, see Fig. 2 (a-b).

(a) (b)

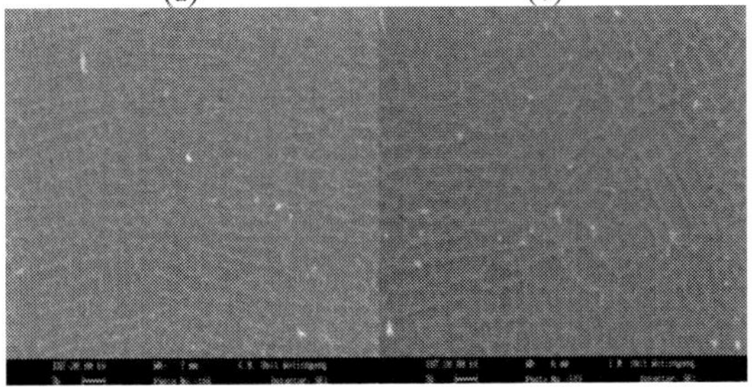

Figure 2. SEM micrograph of cross-sections of the Stellite 6 layer deposited on (a) P91-1, (b) P91-1.8.

3.3 Microhardness testing of coating cross-sections

Microhardness profiles for the Stellite 6 weld samples is shown in Fig. 3. For the coating deposited at 1 kW, the coating hardness was about 550 HV compared with 500 HV for 1.8 kW. The HAZ hardness was generally lower than that of the coating, but higher than the substrate. The hardness of the unaffected substrate was about 225 HV.

3.4 Wear testing

Tests# 1-2 were conducted using an applied load of 2 kg. It was found that the deposit for P91-1 wore substantially less, while the deposit for P91-1.8 showed significant wear with deep grooves. The effect of a higher load (5 kg), Test# 3-4, was generally similar to that for Tests# 1-2, see Fig. 4 (a-b).

3.5 Mass loss

Table 3 shows the weight loss measurements for the Stellite coated samples. It can be seen in Table 3 (Tests# 1-4) that the weight loss increased with load and was higher for P91-1.8.

Table 2. Measured compositions of the Stellite 6 coatings.

(wt%)	P91-1	P91-1.8
P	0.23	0.25
Mn	0.33	0.40
Si	0.57	0.59
Ni	2.40	2.00
Cr	29.25	28.40
Mo	0.23	0.13
Cu	0.011	0.039
Nb	0.02	<0.01
Ti	0.02	0.02
V	0.028	0.024
Fe	6.2	8.2
W	4.1	3.9
Co	54.0	53.1

Table 3. Weight loss for Stellite coatings deposited at a power input of 1 kW and 1.8 kW.

Loads (kg)	P91-1 (g)	P91-1.8 (g)
2	0.00488	0.00496
5	0.01962	0.02736

3.6 Characterisation of wear

In order to study the effect of load on the wear track, Stellite coated samples were examined at the completion of the wear test by scanning electron microscopy and atomic force microscopy to establish the nature of wear.

The worn surface of P91-1, Fig. 5 (a), is smooth compared to the P91-1.8 surface which was more porous and showed greater surface roughness. The effect of a higher load (5 kg) at 1.8 kW heat input is illustrated by Fig. 5 (b).

The surface roughness of P91-1, Fig. 6 (a), has R_a of 74.284 nm, while for the higher heat input, P91-1.8 has a value of 89.237 nm, Fig. 6 (b).

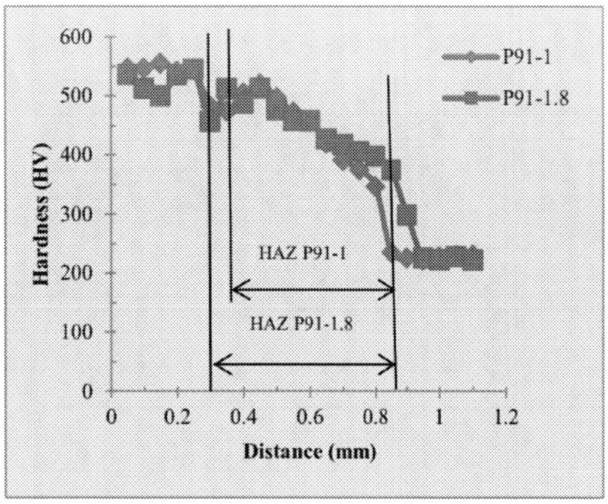

Figure 3. Graph of hardness profiles with distance from the coating surface for Stellite 6 deposited on P91 steel.

(a) (b)

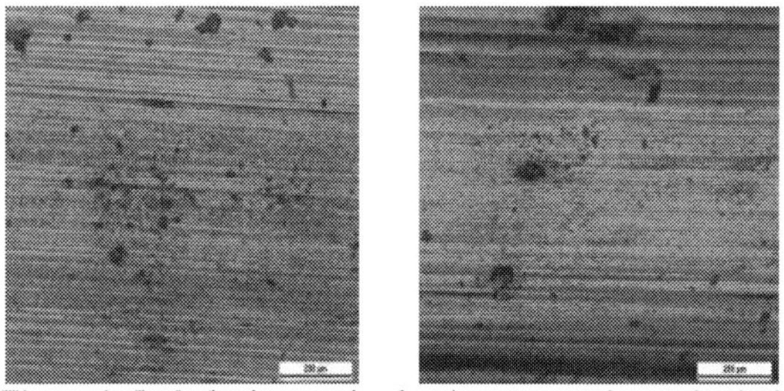

Figure 4. Optical micrographs showing wear tracks at a load of (a) 2 kg for P91-1, (b) 5 kg for P91-1.8.

(a) (b)

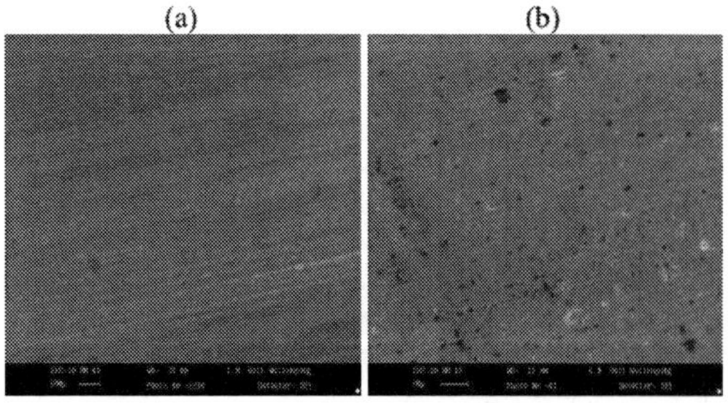

Figure 5. SEM micrograph of worn surface tested at a load of (a) 2 kg for P91-1, (b) 5 kg for P91-1.8.

3,7 Dilution measurements and estimation of carbon content

In order to estimate the carbon content of the Stellite 6 coatings as the carbon was not determined in Table 2, it was necessary to measure the dilution shown in Fig. 1, subsequently the C content was estimated. Tables 4 and 5 show the estimation of dilution and carbon content of Stellite 6 coatings.

(a) (b)

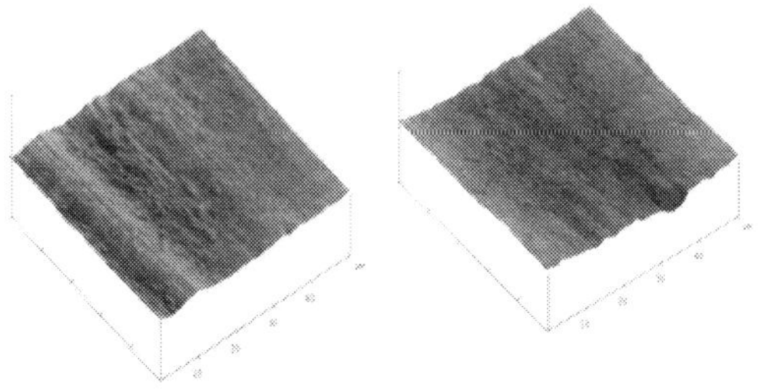

Figure 6. AFM micrographs of surface roughness tested at a load of 2 kg for (a) P91-1 and (b) P91-1.8.

Table 4. Measurements of dilution of the Stellite 6 coatings.

	Area of A (μm^2)	Area of B (μm^2)	Dilution D (wt%)
P91-1	69332.749	805707.663	0.1292
P91-1.8	115010.680	1157493.983	0.1904

Table 5. Estimation of C content of the Stellite 6 coatings.

	$[C]_{BM}$ (wt%)	$[C]_S$ (wt%)	Dilution D (wt%)	$[C]_{WD}$ (wt%)
P91-1	0.09	1.00	0.1292	0.882
P91-1.8	0.09	1.00	0.1904	0.827

4 Discussion

The comparative tests conducted for laser clad P91 steel substrate showed that the weight loss was lower for coated samples deposited at a heat input of 1 kW (P91-1). The amount of wear (mass loss) of the Stellite coated samples was greater for the tests conducted on coatings deposited with 1.8 kW than for those deposited at 1 kW, as shown in Table 3.

For deposits produced at 1 kW, the weight loss increased by a factor of about 4 with increasing test load up to 5 kg, but for the higher heat input the rate of weight loss strongly increased by a factor of about 7 with increasing load. It is likely that the greater incidence of microcracks and porosity observed after wear testing of P91-1.8 samples is due in part to the substrate, particularly to the HAZ, being softer and less rigid [7].

The higher wear rate for the P91-1.8 Stellite coated samples is also consistent with the lower surface hardness of approximately 500 HV compared with 550 HV for the P91-1 Stellite coated samples, as shown in Fig. 3. Acceleration of the wear rate is therefore likely for the Stellite coating P91-1.8 as the wear grooves penetrate the coating [7].

As Table 2 describes, the Stellite composition was modified by the substrate. This change occurred by melting of the substrate and mixing with the deposited alloy (dilution). Because of substrate dilution, the coating produced on the P91 steel substrate showed higher Cr, Fe, Mo, V and Nb contents than those of the nominal Stellite 6 composition given in Table 1, while the Co, Ni, Mn, Si and W contents were reduced (Table 2).

Although the compositions of the two coatings are similar, there are significantly lower Fe and Mn contents for the coating deposited at 1 kW, consistent with a lower level of dilution, hence promoting the formation of Cr_7C_3 particles and hardening of the deposit [8]. The deposit on P91-1 was about 50 HV points higher than for the coating on P91-1.8 (Fig. 3), this difference would be expected to substantially contribute the increase wear resistance [8]. The underlying HAZ was also marginally harder for the lower heat input (Fig. 3), providing a more rigid layer beneath the coating [8].

The higher level of dilution for P91-1.8 Stellite coated samples, Table 4 are consistent with higher wear rate, Table 3 and lower surface hardness, Fig. 3 of approximately 500 HV compared with 550 HV for the P91-1 Stellite coated samples, indicating that P91 has strong carbide formers such as Cr, V and Nb to produce higher hardness on P91-1 Stellite coated samples [7, 8].

As Table 5 shows, the carbon content of weld deposit for the P91-1 Stellite coated samples was higher by a factor about 1.0665% than P91-1.8 Stellite coated samples, suggesting the higher of C content consistent with a lower level of dilution, Table 4, hence promoting the carbide formation of Cr_7C_3 particles and to harden the deposit, Fig. 3, thus producing higher wear resistance [9].

5 Summary and conclusions

The present study compared the wear behaviour of Stellite 6 under reciprocating wear conditions as laser clad deposits on two different heat inputs (1.0 kW and 1.8 kW) of P91 steel. The coating composition was slightly different in the two cases because of differential dilution by the substrate. The compositional differences combined with different cooling rates after deposition resulted in substantially different coating hardnesses. The coating on P91-1 had a hardness of approximately 550 HV, whereas the coating on P91-1.8 had a hardness of approximately 500 HV. The tests were carried out unlubricated, using loads of 2 and 5 kg and a speed of 50 rpm for 10000 revolutions.

The results described that the rate of weight loss and the total weight loss were higher for the higher load and also for the higher heat input. The wear rate was lower for P91-1 coated samples, with less cracking and pore development in the Stellite 6 coatings. The measurements of dilution showed that the level of dilution was higher for the higher heat input and the carbon content estimation also indicated that carbon content was higher for the lower heat input.

It is concluded that the deposit obtained at 1 kW was harder because of compositional, microstructural, surface topography and level of dilution differences; and the harder coating resulted in the higher wear resistance due to higher C content of weld deposit to form carbide (Cr_7C_3 particles) to harden the deposit than the Stellite 6 coating deposited at the higher heat input of 1.8 kW.

Acknowledgements

The authors would like to thank to Laser Bond and Jarvie Engineering companies who have provided the laser cladding.process for this research.

References

1. M. El-Azim, O. El-Desoky, H. Ruoff, F. Kauffmann, and E. Roos, Creep Fracture Mechanism in Welded Joints of P91 Steel, J. Ins. Mat. Min. and Mig., **29**, 1027-1033 (2013)
2. E. Samuel and B. Choudhury, Tensile Work Hardening Behaviour of P91 Steel, J. Mat. Sci. and Eng. A, **528**, 7827-7830 (2011)
3. Tribology T-900 *Technical Data*, Stellite Coating Division, Deloro Stellite Company, Goshen, Indiana (2002)
4. J. D. Majumdar, A. Kumar, and L. Li, Direct Laser Cladding of SIC Disperse AISI 316L Stainless Steel, J. Trib. Int, **42**, 750-753 (2009)
5. G. J. Bruck, High Power Laser Beam Cladding, J. Min. Met. & Mat Sco, **39**, 10-23 (1987)
6. F. Meriaudeau, F. Truchetet, D. Grevey and A. B. Vannes, Laser Cladding Process and Image Processing, J. Laser in Eng, **6**, 161-187 (1997)
7. T. Lolla, J. Siefert, S. S. Babu and D. Gand, Delamination Failures of Stellite Hardfacing in Power Plants : A Microstructural Characterisation Study, J. Man. Onl, **19**, 476-486 (2014)
8. A. Kusmoko, D. Dunne, H. Li and D. Nolan, 'Effect of Two Different Energy Inputs for Laser Cladding of Stellite 6 on P91 and P22 Steel Substrates, J. Proc. Mat. Sci, **6**, 26-36 (2014)
9. A. Kusmoko, The Effect of Heat Input for Stellite 6 Deposited on High Strength Steel Substrates by Laser Cladding, *PhD Thesis*, The University of Wolongong (2016)

MATEC Web of Conferences 70, 01002 (2016) DOI: 10.1051/matecconf/20167001002
ICMIT 2016

Effect of Graphite Electrode to Surface's Characteristic of EDM

Apiwat Muttamara[1] , Warunee Borwornkiatkaew[2], Anuprong Pronpijit[2] and Songkran Nuanchom[2]

[1]*Faculty of Engineering, Thammasat University, Pathum Thani 12120, Thailand*
[2]*National Metal and Materials Technology Center (MTEC), Pathum Thani 12120, Thailand*

Abstract. Electrical discharge machining process (EDM) is a process for removing material by the thermal of electrical discharge. Some of the melted and all of the evaporated material is then quenched and flushed away by dielectric liquid and the remaining melt recast on the finished surface. The recast layer is called as white layer. Beneath the recast layer, a heat affected zone is formed. The quality of an EDM product is usually evaluated in terms of its surface integrity, which is characterized by the surface roughness, existence of surface cracks and residual stresses. This paper presents a study of surface's characteristics by EDM in de-ionized water due to decarbonisation. The machining tests were conducted on mild steel JIS grade SS400 with copper and graphite electrodes. The workpiece surfaces are analyzed by scanning electron microscope and XRD technique. The carbon transfers from graphite electrode to the white layer relating to martensitic phrase of recast layer.

1 Introduction

Electrical Discharge Machining (EDM) is one of the most effective methods for removing material from electrical conductive materials. When machining a workpiece with EDM, a recast layer is formed at the surface of the workpiece [1]. The recast layer shows white color after etching so it is called the white layer. The white layer is the result of the solidification of a melted zone and is known to exhibit high hardness, good adherence to the bulk and good resistance to corrosion. However, it containes a lot of microcracks, which causes a problem for certain applications. The microstructure and the properties of the melted zone after solidification are quite different with respect to those of the workpieces and tool electrodes used in de-ionized water [2]-[5]. EDM process can be performed in decarbonization for example EDM process in deionized water. Graphite is widely used as electrode material in EDM, this is due to its low density and good electrical along with its machinability [5]-[6]. This research was studied the effect of electrode to EDM process in deionized water.

2 Experimental

The machine used for this study is a FORM 2-LC of Charmilles Technologies. From the previous studies of Lee et al. [7], they used the Taguchi approach and found that the characteristics of an EDMed surface, such as surface roughness and white layer thickness are mainly influenced by current and discharge duration.

Decarbonisation is interesting for the experiment in order to observe carbon in electrode material affect to the surface's characteristics. Thus this study was carried out in de-ionized water and also used the EDM conditions for examining the characteristics of an EDMed surface. The workpieces was mild steel (SS400). Electrodes are copper and graphite, diameter of 12 mm. The properties are shown in Table 1. The parameters are shown in Table 2. The formation of surface cracks was checked by scanning electron microscope and the structural phrases were analysed by X-ray diffraction (XRD).

Graphite is the preferred electrode material because light weight and easy for machining. Graphite was made by compressing the powder/binder mixture in only one direction, resulting in grain relate to impact force direction. Graphite has significantly lower mechanical strength properties than metallic electrode ex. copper. It is neither as hard as strong, nor ductile as copper electrode. Due to the significant differences between copper and graphite electrode, the study was carried out in water di-electric fluid.

Table 1. Properties materials.

Properties	Value
Thermal Conductivity (W/m°C)	53.66
Density (g/cm^3)	7.85
Melting point (°C)	1,495
Young's Modulus (GPa)	400

Table 2. EDM conditions.

Properties	Value
Polarity	Negative
Open Load Voltage (Volts)	250
Current (A)	5-30
Discharge duration (μs)	300
Duty Factor (%)	50

3 Results and Discussions

The machined workpiece surfaces were observed by scanning electron microscope (SEM) as shown in Fig. 1 The machinied surface were treated with copper (Fig.1(a)) and graphite electrode (Fig.1(b)). Some micro-cracks were found on the EDMed surface when using a copper electrode, but no crack was observed when using a graphite electrode. This may explain that powder from graphite electrode easily drop to recast layer during discharge process. Consequently, the recast layer contains many holes and channels caused by the bubble air that generated during EDM process. The copper electrode gives more small pores and hole compared to graphite electrode. The recast layer also contained some microcracks. The heat from EDM process melts and evaporates the workpiece material. The treated surface is covered with debris and voids because of the high heat energy released by the electrical discharges, followed by rapid cooling. It seems that the copper electrode generated more bubble air than that of the graphite electrode [8]. However, graphite is a powder / sintering material by compressing. The graphite does not melt absolutely, but sublimes from an electrode during discharge.

a) SS400 with copper electrode

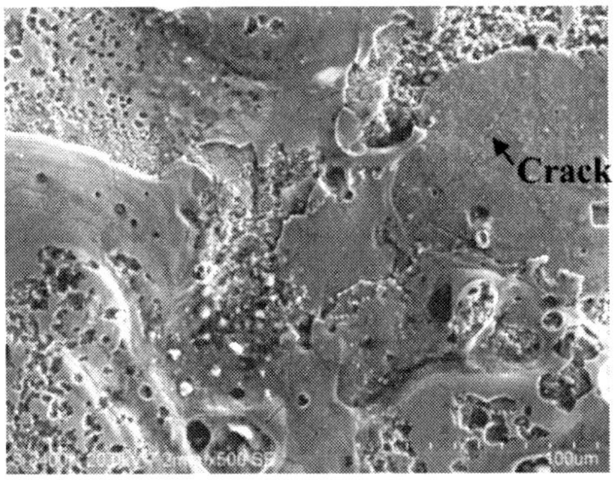

b) SS400 with graphite electrode

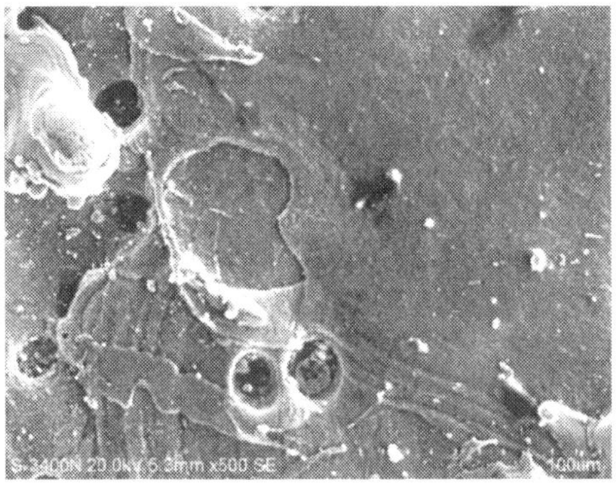

Figure 1. SEM image of EDMed surface machined on SS400 with (a) copper and (b) graphite electrode.

Fig. 2 (a) shows a cross-sectional image of etched SS400 with copper electrode. Fig. 2 (b) shows a cross-sectional etched SS400 with graphite electrode. During the solidification, as the temperature drops near the eutectic temperature it tends to transform into ferrite, carbide and retained austenite.

a) SS400 with copper electrode

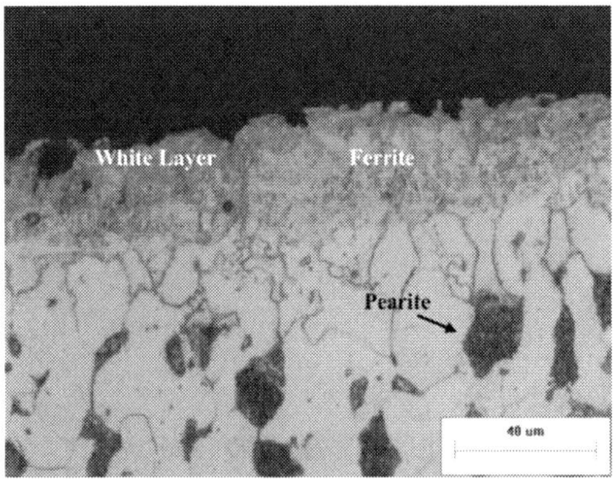

b) SS400 with graphite electrode

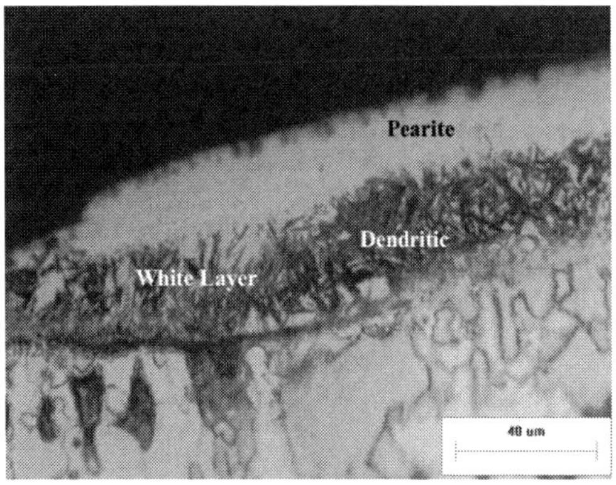

Figure 2. Cross-sectional of EDMed surface for SS400 with (a) copper and (b) graphite electrode (I=25 A, t_{on}=300 μs)

Now we can determine that quench is rapid enough to precipitate carbides in martensite structure formed in the white layer in Fig. 2 (b).A thin columnar dendritic sublayer is observed on the recast layer. This reveals the phases consisting of martensite and residual austenite with dendritic structure [9-12]. Beneath the recast layer is a sublayer that is not melted called the heat affected zone (HAZ). The matrix of the workpiece is composed of pearlite and ferrite.

Fig. 8 displays XRD patterns of SS400 with copper and graphite electrode. The result shows iron oxides and α-Fe_in the recast layer in Fig. 8(a). FeO came from chemical reaction between the workpiece and elemental oxygen decomposed from the deionized water during sparking [13]. Fig. 8(b) shows the XRD pattern with graphite electrode. Some carbon was found in the recast layer. However, we did not detect the presence of Fe_3C or any composed carbon with iron, so their contents in the recast layer were low.

a) SS400 with copper electrode

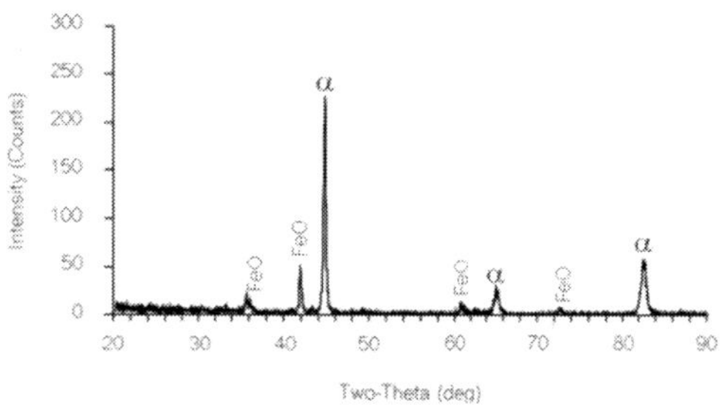

b) SS400 with graphite electrode

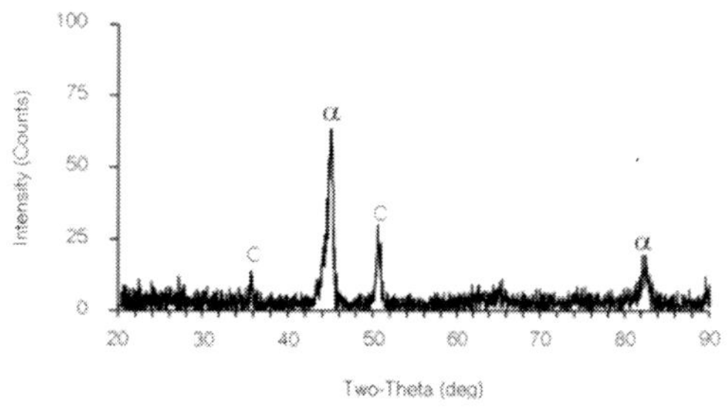

Figure 3. XRD patterns of surfaces treated by EDM with a) copper and b) graphite electrodes.

To study the characteristics of the material phases, the microhardness of the recast layer of each sample was also measured at different distances from the top layer. The comparison shown in Fig.4 denotes compares the microhardness of the cross section of the recast layers of different electrode materials: graphite and copper. A low load of 10 gf was used for the hardness measurements. The machined samples contained the recast layer on mild steel, with an average thickness of 40 μm. The hardness of the matrix is 184-220 HV. In addition, the hardness of the recast layer machined with graphite electrode was higher than that of the recast layer machined with copper electrode. The microhardness of the recast layer with graphite electrode reached 420 HV, which is a little higher than that of machined with copper electrode (320HV). This is attributed to the higher carbon content on the cast layer. The dendritic which has a needle structure can be seen in Fig. 2(b). In this case, carbon is transferred from the graphite electrode and the white layer freeze to martensitic structure [14].

The relationship between the surface roughness of EDMed surface treated by graphite and copper electrode are presented in Fig 5. The surface roughness that produced with graphite electrode was higher than that of copper electrode. This phenomenon may explain that graphite powder from graphite electrode make pores to the layer and make rougher than that EDMed with copper.

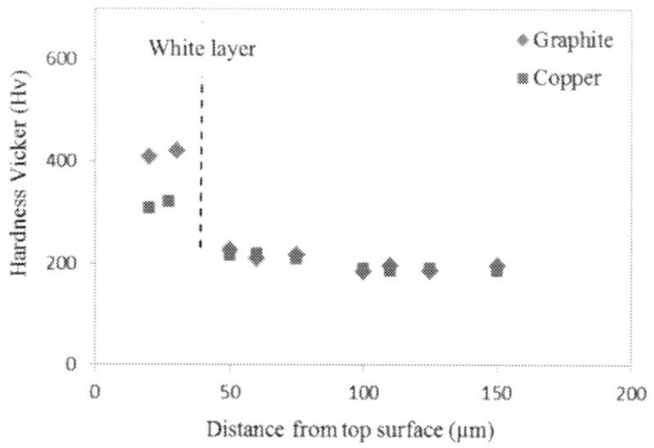

Figure 4. Microhardness distribution of the cross section of layers treated by EDM.

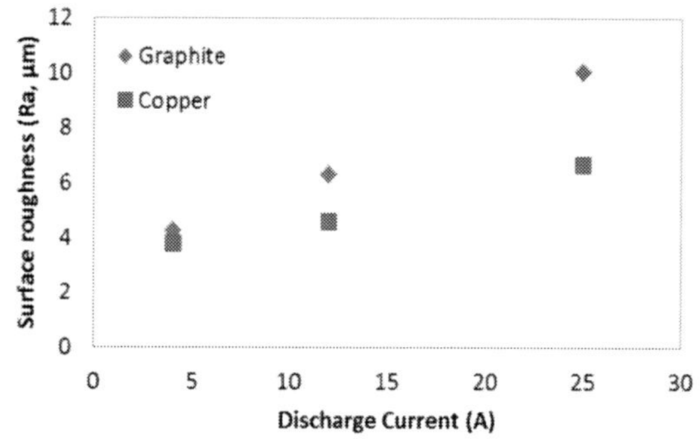

Figure 5. Surface roughness of gray iron and mild steel surfaces treated by EDM in different media.

4 Conclusion

1.There is no crack on white layer when EDM with graphite electrode.

2.Carbon from electrode material regenerate to the white layer.

3. Carbon can be found on EDMed surface with graphite electrode. However, it is not enough to generate to Fe_3C.

4. Graphite electrode make more surface roughness than that made by copper electrode.

Acknowledgements

The authors acknowledge the support from the Thammasat University research's fund, the Commission on Higher Education of Thailand (the National Research University Project).

References

1. Y. Zhang, Y. Liu, Y. Shen, R. Ji, Z. Li, C. Zheng, J. Mater, Process. Technol., **214**, 5, 1052–1061 (2014)
2. M. Barash, M. G. Sri-Ram, Proceedings of Third International Machine Tool Design and Research Conference, 85–91, Birmingham (1962)
3. J. C. Rebelo, A. M. Diaz, D. Kremer, J. L. Lebrun, J. Mater. Process. Technol., **84**, 1–3, 90–99 (1998)
4. B. Ekmekci, Metall. Mater. Trans. B, Process., Sci. **40**, 70–81 (2009)
5. J. P. Kruth, L. Stevens, L. Froyen, B. Lauwers, K. U. Leuven, Ann. CIRP, **44**, 169–172 (1995)
6. F. L. Amorim, W. L. Weingaertner, J. Braz., Soc. Mech. Sci. & Eng., **29**, 4 (2007)
7. H. T. Lee, T. Y. Tai, J. Mater. Process. Technol., **142**, 3, 676–683(2003)
8. S. L. Chen, B. H. Yan, F. Y. Huang, J. Mater. Process. Technol., **87**, 107–111 (1999)
9. G. Cusanelli, A. Hessler-Wyser, F. Bobard, R. Demellayer, R. Perez, R. Flükiger, J. Mater. Process. Technol., **149**, 289–295 (2004)
10. F. J. O'Brien, D. Taylor, T. C. Lee, J. Biomech., **36**, 973–980 (2003)
11. F. Ghanem, C. Braham, H. Sidhom, J. Mater. Process. Technol., **142**, 163–173 (2003).
12. J. P. Kruth, J. V. Humbeeck, L. Stevens, Proceedings of the ISEM XI, 849–862 (1995)
13. E. D. Cabanillas, J. Desimoni, G. Punte, R. C. Mercader, Mater. Sci. Eng. A, 276, 133–140 (2000)
14. M. Barash, Metals Eng., **5**, 48–51 (1965)

The Potential of a Water-in-Diesel Emulsion for Increased Engine Performance and as an Environmentally Friendly Fuel

N. H. Abdurahman[1], Y. M. Rosli[1], N. H. Azhari[2] and A. Abdul Adam[3]

[1]*Facuty of Chemical and Natural Resources Engineering, University of Malaysia Pahang-UMP, Malaysia*
[2]*Faculty of Industrial Sciences and Technology, University of Malaysia Pahang-UMP, Malaysia*
[3]*Faculty of Mechanical Engineering, University of Malaysia Pahang-UMP, Malaysia*

Abstract. With increasing energy prices and the drive to reduce CO_2 emissions, universities and industries face the challenge of finding new technologies to reduce energy consumption, to meet legal emission requirements and to reduce cost while increasing quality. In this study, an experiment was conducted to investigate the effect of water-in-diesel (WiDE) emulsions on the combustion performance and emulsion characteristics of a direct injection diesel engine under 0 to 50% engine loads at 0 to 300 psi with engine operating speeds of 1200 to 2500 rpm. The five types of emulsified diesel fuel used in this study consisted of WiDE-5%, WiDE-10%, WiDE-20%, WiDE-30% and WiDE-40% with 3% of the surfactants tested. The results show that 15-85% emulsion diesel fuel is comparable and compatible with pure diesel. The power curve shows the emulsion diesel to be relatively similar to pure diesel, but with slightly higher results at 0% (0 psi). The torque curve at 0% also shows the emulsion diesel to be similar to pure diesel where the 15 - 85% torque is higher than that of pure diesel. The engine torque, power and brake efficiency increase as the water percentage in the emulsion increases. The NO and CO_2 emission results of the 15-85% fuel were better than that of pure diesel; however, the CO emissions were higher, especially at higher engine speeds.

1 Introduction

Emulsions of water in diesel for use in internal combustion engines are easily applicable alternative fuels for existing vehicle fleets. Emulsion fuel is defined as an emulsion of water in standard diesel fuel with specific additives and surfactants to stabilize the system. There is growing interest in the use of diesel emulsions, and the environmental aspects are the main driving force behind the attention. The presence of water has a significant effect on several emission constituents, including exhaust gases, such as nitrogen oxides (both NO and NO_2, which are collectively referred to as NOx) and carbon monoxide (CO), as well as black smoke and particulate matter. There may be a certain loss in engine performance due to the presence of water, but the diesel consumption is often reduced. The typical water content of the diesel emulsions is between 10 and 20%. Both the global and local effects of environmental emissions are of concern when using alternative vehicle fuels. The global effect due to the gradual increase of atmospheric CO_2 was the decisive factor behind the 1997 Kyoto Protocol. In the transport sector, the use of renewable fuels is the most effective way to limit the net increase of atmospheric CO_2. In addition, the replacement of diesel and gasoline by natural gas will reduce CO_2 emissions because of the lower carbon-to-hydrogen ratio in natural gas; however, the gain is much less than that of renewable fuels. The primary focus of the effect of local emissions is placed on the health hazardous NO_x and the particulate matter emissions. For new engines, these emissions are strictly regulated by legislation. In Euro 4 (2005) and in Euro 5 (2008), the emissions of NO_x, particulate matter, hydrocarbons and carbon monoxide are clearly regulated; for existing vehicles, other rules apply.

There is growing interest in the use of diesel emulsions, and the environmental aspects are the main driving force behind the attention. The presence of water has a significant effect on several emission constituents, including exhaust gases, such as nitrogen oxides (both NO and NO_2, which are collectively referred to as NOx) and carbon monoxide (CO), as well as black smoke and particulate matter. There may be a certain loss in engine performance due to the presence of water, but the diesel consumption is often reduced. The typical water content of the diesel emulsions is between 10 and 20%. Both the global and local effects of environmental emissions are of concern when using alternative vehicle fuels. The global effect due to the gradual increase of atmospheric CO_2 was the decisive factor behind the 1997 Kyoto Protocol. In the transport sector, the use of renewable fuels is the most effective way to limit the net increase of atmospheric CO_2. In addition, the replacement of diesel and gasoline by natural gas will reduce CO_2 emissions because of the lower carbon-to-hydrogen ratio in natural gas; however, the gain is much less than that of renewable fuels. The primary focus of the effect of local emissions is placed on the health hazardous NO_x and the particulate matter emissions. For new engines, these emissions are strictly regulated by legislation. In Euro 4 (2005) and in Euro 5 (2008), the emissions of NO_x, particulate matter, hydrocarbons and carbon monoxide are clearly regulated; for existing vehicles, other rules apply.

The interest in water-in-diesel emulsions derives from the fact that water in the form of micrometre-sized droplets exerts various positive effects on fuel combustion. Water-in-oil emulsions have been formulated and evaluated for most types of fuels, ranging from light hydrocarbons to triglycerides. However, the main interest lies particularly in water-in-diesel emulsions because the high combustion temperature and high pressure that exist in diesel engines are particularly suitable for the concept. This idea was experimentally discovered [1] and theoretically verified [2] nearly 30 years ago. The use of diesel emulsions provides several interesting effects, including reduced nitrogen oxide (NOx) emissions, lower soot and particulate contents in the exhaust, and improved combustion efficiency. The presence of water in diesel fuels brings about a considerable reduction in NOx and particulate matters (PM) emissions [3-8]. The effect on emissions is the main reason for the large interest in diesel emulsions today. It has been reported that 15% water in diesel fuel can reduce nitrogen oxide (NOx) emissions up to 35% under regular conditions [9]. Other authors have reported such reduced levels of NOx [5, 10-13]; however, these results relate to diesel and other lighter hydrocarbon fractions with very low nitrogen contents. For fuels with high nitrogen contents, such as some residual oils, the NOx in the exhaust mainly comes from the oxidation of the nitrogen-containing fuel components. In such cases, the emulsified water only has a small effect on the NOx level [14]. For diesel fuels, there may be a concomitant increase in the emission of carbon monoxide (CO) as well as in hydrocarbons [9]. The amount of PM in the exhaust from fuels with high nitrogen contents is normally significantly reduced [14]. Experimental studies have been performed on the effect of water in the form of a water-in-diesel emulsion on the formation of polycyclic aromatic hydrocarbons, some of which are known to be mutagenic [15]. Emulsion fuels produced lower amounts of polycyclic aromatic hydrocarbons in flames and reduced atmospheric emissions. A more specific analysis showed a decrease of two substances, pyrene and fluorene, when emulsion fuel was used. It is believed that micro-explosions promote the oxidation of hydrocarbons, possibly involving an attack by hydroxyl radicals. The reduced amount of the formed polycyclic aromatic hydrocarbons is likely to be one of the reasons for the lower amount of soot obtained with emulsion fuels because these hydrocarbons can be observed as precursors to soot particles.

2 Materials and methods

2.1 Materials

The oil samples used in this study were from commercial diesel. A detailed procedure for the preparation of water-in-diesel (W/D) emulsions and procedures, including the formation of the W/D emulsion, their characteristics and preparation methods, are thoroughly described in previous research by Abdurahman [16]. This paper merely describes the main experimental steps. The characteristics of the prepared emulsion are specified in Table 1. For preparation of water-in-diesel (W/D) emulsions, the agent-in-oil method was implemented. The emulsifying agent dissolved in the continuous phase (diesel) and water was gradually added to the mixture (diesel + emulsifying agent). The emulsions were vigorously agitated using a standard three-blade propeller at room temperature (25 - 30). The prepared emulsions were used to check for w/o or o/w emulsions. All of the emulsions investigated were a type of water-in-oil emulsion (oil continuous phase). The properties of the water-in-diesel emulsions are listed in Table 2. The surfactant used in this study was Triton X-100 (polyethylene glycol octylphenyl ether), which has a chemical formula of $C_{33}H_{60}O_{10}$. This surfactant is a non-ionic hydrophilic surfactant that is suitable for use in the production of both W/O and O/W emulsions.

Table 1. Physical properties of the diesel fuel.

Density at 20 °C, (kg/m^3)	840
Dynamic viscosity at 20 °C, mPas	6.5
Surface tension at 20 °C (mN/m)	29.4
Flash point, °C	78
Initial boiling point, °C	174.5
Final boiling point, °C	380.3

Table 2. Density of the diesel and WiDE emulsions (5%, 10%, 20%, 30% & 40%).

Diesel	**Kg/m^3 at 20 °C 840**	**Kg/m^3 at 30 °C 835**	**Kg/m^3 at 40 °C 828**	**Kg/m^3 at 50 °C 820**
WiDE-5%	848	842	835	828
WiDE-10%	854	849	842	834
WiDE-20%	870	865	858	850
WiDE-30%	877	872	865	857
WiDE-40%	930	925	918	910

2.2 Equipment and conditions

In this study, the results of the formulation and evaluation of the 15-85% water-in-diesel emulsions with pure diesel were determined on a TF120M Yanmar engine. The engine sound was normal with no knocking. The experiments were conducted at 0% (0 psi) and 50% (300 psi), while the engine speed ranged from 1200 to 2500 rpm. The power curve, torque curve, NO, CO2, and CO results of WiDE were compared with traditional diesel fuel results.

3 Results and discussion

3.1 Emulsion diesel for fuel properties

The stability of WiDE was investigated under the influence of the water concentration, surfactant concentration, mixing speed, viscosity, density and time. The most important factor of the emulsion preparation is the selection of a suitable surfactant or blend of surfactants that can satisfactorily emulsify the chosen ingredients at a specific temperature. Emulsification is greatly influenced by the hydrophilic–lipophilic balance (HLB) of any surfactant/emulsifier [16]. The percentage of water in the emulsion, stirring intensity, stirring duration, emulsifying temperature and operational pressure are other parameters that affect the stability of an emulsion. The WiDE prepared for the current study was observed for more than 65 days because it was emulsified and found to be stable for that period. It was noted that the larger the water content in the emulsion, the brighter and milkier the produced emulsion.

In this study, a wide range of water concentrations was investigated, from 0% to 40% by volume. Figure 1 shows the WiDE emulsion water content percentages versus the elapsed time for three samples with water concentrations of 5%, 20% and 40%. These samples were prepared with no surfactant added at a mixing speed of 3000 rpm for 20 minutes. Figure 1 shows that the percentage of water content within the three emulsions dropped very significantly over 15 minutes. The amount of separated water reached 27%, 50% and 80% for WiDE-5%, WiDE-20% and WiDE-40%, respectively. Try to ensure that lines are no thinner than 0.25 point.

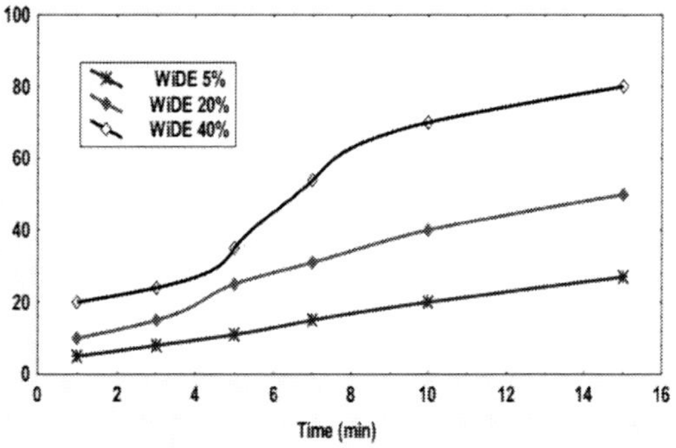

Figure 1. Emulsion behaviour without surfactant.

These samples show the strong unstable behaviour of water within WiDE emulsions, attributed to the absence of surfactants within the emulsion. Therefore, this preliminary study indicates that the presence of surfactant is crucial for WiDE emulsion stability. Figure 2 shows the effect of added water on the emulsion stability in the presence of a 3% Triton-X100 concentration.

The emulsion was prepared using a mixing speed of 3000 rpm and was mixed for 20 minutes. Figure 2 shows that the WiDE-5% and WiDE-20% remained stable for 60 days before any water was separated. The WiDE-40 remained stable for 40 days before any water was separated.

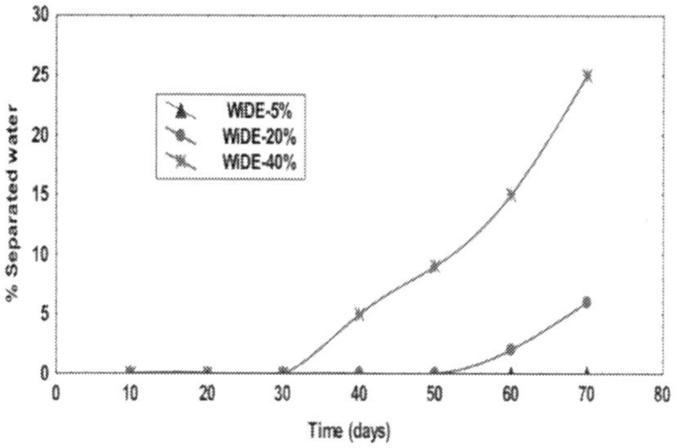

Figure 2. Stability profile for different WiDE emulsions.

3.2 Diesel and engine performance

The interest in water-in-diesel emulsions derives from the fact that water in the form of micrometre-sized droplets exerts some positive effects on fuel combustion. In this study, 15-85% water-in-diesel oil emulsions were formulated and evaluated using a TF120M Yanmar engine. The engine sound was normal with no knocking. The results show that the presence of water had a positive effect on the combustion efficiency of the engine. Figures 3 & 4 show how 15-85% water in the form of an emulsion affects the engine torque output for various speeds (0 psi and 300 psi). A comparison was made between 15-85% WiDE and pure diesel performance on the TF120M Yanmar engine. Figure 3 illustrates the results with a 0 % load (0 psi); Figure 4 shows the performance of a 50% load (300 psi). In both figures, the torque increases as the engine speed increases. Interestingly, the engine with emulsion fuel is able to ingest a full charge of air compared to when traditional diesel fuel is used. This conclusion was supported by Abu-Zaid [17], who used the engine speed over a range from 1200 to 3300 rpm. However, Abu-Zaid found that at low speed, the torque increases as the engine speed increases. At approximately 1500 rpm, the torque begins to decrease because the engine is unable to ingest a full charge of air; these results are also in agreement with research by Harbach and Agosta [18].

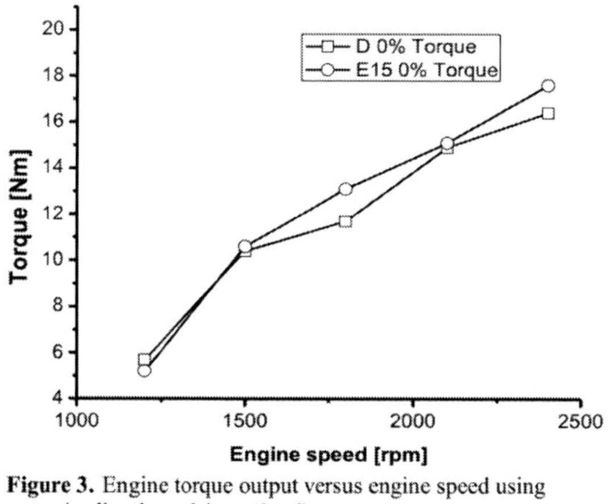

Figure 3. Engine torque output versus engine speed using water-in-diesel emulsions (0 psi).

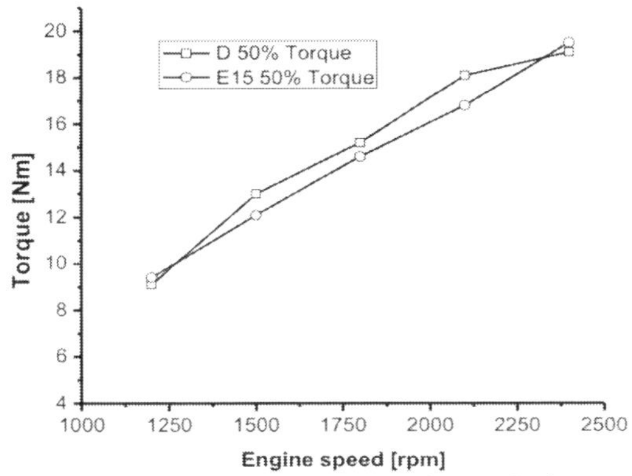

Figure 4. Engine torque output versus engine speed using water-in-diesel emulsions (300 psi).

It can be clearly observed from Figures 3 & 4 that the addition of water in the form of an emulsion has a positive effect on the combustion efficiency compared with the use of pure diesel, especially at 0 psi (Fig. 3).

The torque increases with the water content over the entire rpm range (1200 - 2500 rpm). When the charge is fired in the cylinder, the water turns into steam at high pressure. The combustion efficiency is also improved because of the presence of water and, in particular, the presence of the oil-water interface with a very low interfacial tension that leads to a finer atomization of the fuel during injection Walavalkar [19].

3.3 Carbon dioxide emissions

The formation of carbon dioxide (CO_2) reflects the complete combustion of the fuel; incomplete combustion typically reduces CO_2 but increases the CO and hydrocarbon (HC) contents. The variation of CO_2 emissions with the engine speed for the 15-85% emulsion fuel and pure diesel at (0 psi and 300 psi) is presented in Fig. 5. As shown in the figure, the CO_2 content increases with engine speed and decreases with an increasing water content (15%). The combustion of the emulsified fuels produces lower CO_2 emissions compared to the combustion of pure diesel because the burning of the emulsified fuels generates a large number of micro-explosions, leading to a greater mixing of the reactant mixture, Lin. et al [20].

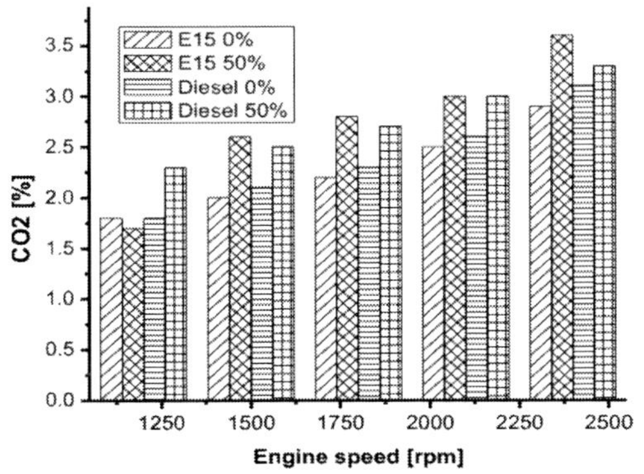

Figure 5. CO_2 emissions at different engine.

3.4 Nitrogen monoxide emission

The variation of NO with engine speed for the addition of the 15% water emulsion and pure diesel is presented in Fig. 6. The engine speed has a minor effect on NO production compared with the water content in the fuel. The combustion of the emulsified fuel produced significantly less amounts of NO compared to pure diesel, possibly due to the finely dispersed water of the emulsion causing a phenomenon known as heat sink that reduces the local adiabatic flame temperature. These results also may due to the heat absorption by the water vaporization causing a reduction in the peak ignition temperature that further reduces NO formation, Kannan et al, [21].

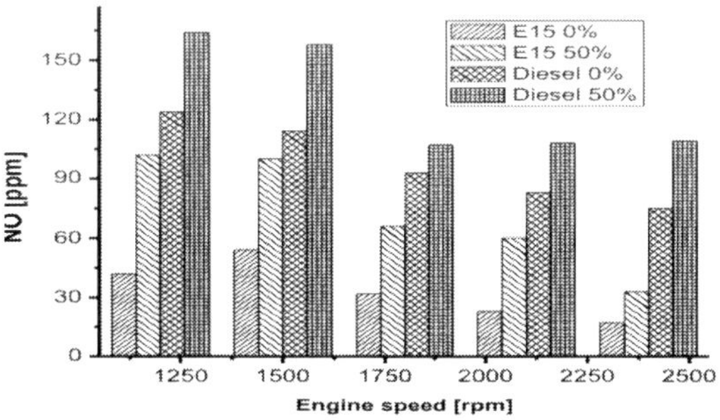

Figure 6. NO variation for pure diesel and 15% water addition.

3.5 Engine power output

The effect of the emulsions on the engine power is shown in Figures 7 & 8. The power increases as the engine speed increases, thus there is clearly no reduction of power within the experimental engine speeds (1200 - 2500 rpm). These results may be attributed to the lack of friction losses with increased engine speed, which is not a dominant factor at high speeds. Observation has shown that the 15-85% emulsion fuel was very compatible with pure diesel, and in some cases, it is even better than pure diesel. The maximum power occurred at 2500 rpm, as illustrated in Figures 7 & 8, and the maximum torque (20 Nm) occurred at a maximum engine speed of 2500 rpm, as presented Figs. 3 & 4. The power output increase for the emulsions resulted due to the influences of the water in the emulsion on fuel combustion. Harbach and Agosta [18] found that the introduction of water in diesel for a single cylinder engine prolongs the ignition delay.

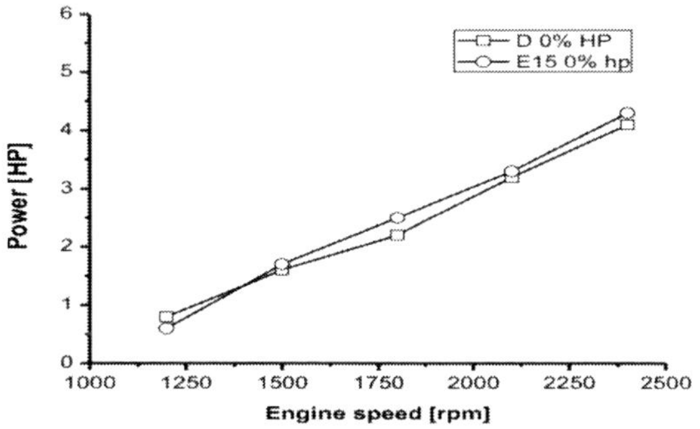

Figure 7. Engine power output versus engine speed using diesel-water emulsions (0 % load).

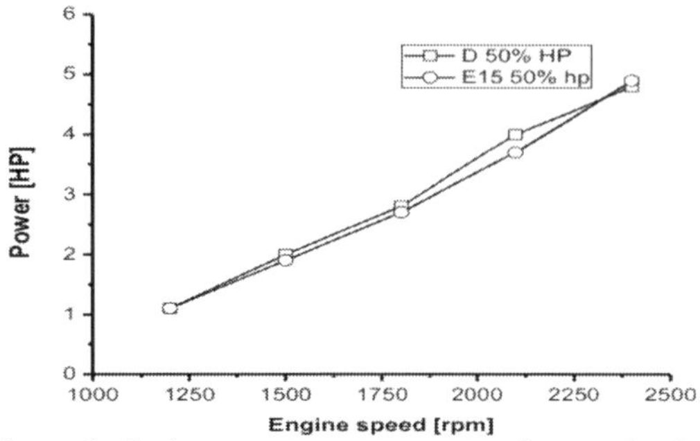

Figure 8. Engine power output versus engine speed using diesel-water emulsions (50 % loads).

4 Conclusions

In this study, an experimental investigation was conducted to examine the effect of water diesel emulsions on the performance of a TF120M Yanmar engine. The findings are summarized as follows:

- The addition of water in the form of an emulsion improves the combustion efficiency in the diesel engine, thus increasing the performance of the engine.
- Within a limited amount of running time during the engine tests (85 hrs.), no apparent wear, knocking or corrosion were observed.
- NO emissions were significantly reduced by using diesel water emulsions.
- CO emissions decreased with an increase in water percentage in the fuel.
- The water-in-diesel emulsion fuel is the best alternative fuel to substitute for diesel fuel in both transport and stationary CI engines because of the reduction in both NO_x and PM.

References

1. D. H. Cook, C. K Law, Combust Sci Technol **18**, 217 (1978)
2. C. K. Law, Combust Sci Technol **17**, 29 (1977)
3. J. M. Ballester, N. Fueyo, C. Dopazo, Fuel **75**, 695 (1996)
4. R. W. Gunnerman, R. L. Russel, SAE Tech Pap (1997) 972099.
5. G. E. Andrews, S. W. Bartle, S. W. Pang, A. M Nurein, P. T. Williams, Proceedings of the International Centre for Heat and Mass Transfer, (Sept 1987). p. 613, (New York, NY, USA: Hemisphere Publ Corp; 1989)
6. W. Bernhardt, K. Weidmann, Erdol Kohle Erdgas Petrochem 40, 86 (1987)
7. D. Hall, C. Thorne, S. Goodier, SP-1802, 153 (SAE Spec Publ 2003)
8. M. Tsukahara, Y. Yoshimoto, International Congress and Exposition, Detroit 71, (Feb 24–28 1992)
9. A. Lawson, Vergeer EC, Mitchell EW, Dainty ED. Heavy-Duty Diesel Emission Control: A Review of Technology, vol. 36. Montreal, Quebec, Canada: CIM; 1986. p. 238.
10. J. W. Park, K. Y. Huh, J. H Lee, Proc Inst Mech Eng Part D J Automob Eng **215**, 83 (2001)
11. E. M. Afify, Report ARO194353EG (DTIC No ADA-161-652). Raleigh, NC, USA: 106 (North Carolina State Univ., 1985)
12. F. Barnaud, P. Schmelzle, P. Schulz, SP-1551:95 (SAE Spec Publ 2000)
13. M. Fromager, P. Schmelzle, P. Schulz, 2nd International Symposium on Fuels and Lubricants, Symposium Papers, New Delhi, **2**, 435 (Mar 10–12, 2000)
14. R. E. Hall, Trans ASME J Eng Power **98**, 6425 (1976)
15. J. A. Kozinski, Combust Flame **96**, 249 (1994)
16. H. N. Abdurahman, M. Y. Rosli, H. N. Azhari, Stabilization of crude oil Emulsions, Journal of Applied Sciences, 350-357 (2006).
17. Abu Zaid, M. Energy Convers Manag, **45**, 697 (2004).
18. J. Harbach, V. Agosta, Effects of emulsified fuel on combustion in a four-stroke Diesel engines. J Ship Res, **35 (4)**, 356-63 (1991)
19. A. Y. Walavalkar, PhD Thesis, Pennsylvania State University (2001)
20. C. Y. Lin, K. H. Wang. The fuel properties of three-phase emulsions as an alternative fuel for diesel engines. Fuel, **82**, 1367-75 (2003)
21. T. K. Kannan, Thevetia, Peruviana. Biodiesel emulsion used as a fuel in a single cylinder diesel engine reduces NO and smoke, Thermal science. **15**, 1185-1191 (2011)

Robust Design Based Optimisation of Waterjet Cutting

Tudor Deaconescu

Transilvania University of Brasov, Industrial Engineering and Management Department, Faculty of Technological Engineering and Industrial Management, Romania

Abstract. The most important input quantities of waterjet cutting are the jet pressure, feed speed, stand-off distance, abrasive graining, mass flow, etc. Other quantities contributing to machining efficiency are the type of utilized abrasive or the tilt of the jet. Each of these quantities can be assigned different set points. The roughness of the machined surfaces and the thickness of the cut part are output quantities of the system, their values depending on the input parameters and the influence of various disturbing factors (noises). The paper discusses surface roughness obtained consequently to abrasive jet cutting. Optimisation of the machining system was achieved by intervening on five selected input quantities (factors), with two set points considered for each. Upon applying Taguchi methods, eventually the combination of factor set points was determined that ensures robust behaviour of the system.

1 Introduction

Waterjet cutting has proved its utility due to its versatility and easy deployment. The benefits that ensure waterjet abrasive machining its top position among special technologies are the diversity of machinable materials at low cost, reduced environmental impact, high flexibility, easy programming, rapid machining, and conservation of the material's properties due to the low processing temparatures. All these advantages call for in-depth research and swift solving of the problems posed by waterjet cutting: relatively reduced dimensional accuracy of the products (about 0.1 mm) and the occurrence of certain defects (striations, mini-craters, high surface roughness).

The relatively low machining cost as well as the minimum impact on the environment render waterjet erosion a technology of the future. Emerged and refined over the last years, abrasive waterjet microprocessing has succeeded in improving processing precision to 0.02 mm, while surface roughness can be optimised by means of adequately configuring the influential input factors of the process.

One of the methods of abrasive erosion of parts is waterjet machining or waterjet cutting, where erosion is initiated by the kinetic energy of a moving fluid. Waterjet as an erosive agent (without added abrasive) dates back as far as the 1900, when it was used in mining for cutting coals at pressure not exceeding 100 MPa.

The abrasive jet machining technology entails introducing a certain quantity of abrasive grains into the waterjet, thus improving its coherence, amplifying its mechanical effect and increasing cutting speed by about 30%. In waterjet machining erosion is activated by the kinetic energy of a fluid in motion.

Waterjet machining has become possible in the 1970s with the launching of dedicated pumps [1], [2]. The first cutting installations were developed allowing increased performance of machining. Further milestones in the evolution of this technology are the abrasive mixing tubes made by ROCTEC, a ceramic composite of tungsten carbides (1980), the fluid motion control systems (1990), the tilting system of the cutting head (2000) and the CNC software implemented on waterjet cutting machines in the first decade of the 21st century [3].

Due to its numerous advantages, in many industrial countries this process has been employed intensively. Thus, for example in Germany abrasive jets were employed for the cutting of aluminium (41%), steel (29%), glass (22%), granite, marble and rock (21%), non-ferrous metals (7%), wood (7%), rubber (5%), titanium (5%), etc. Abrasive jet cutting devices currently hold about 35% of the total cutting machines worldwide, what proves once again the efficiency and advantages of this processing method.

Due to their benefits, waterjet machining procedures are continuously expanding. The Global Waterjet Machine Market report for 2015-2019 highlights the waterjet machining technology sales forecast for the American market [4]. Thus, starting from sales of USD 781.16 million in 2014, an increase up to USD 1221.12 million is envisaged until 2019. The main beneficiaries are the aerospace, the automotive and the rock and ore extractive industries.

Figure 1 shows the current market shares of the most important manufacturers of waterjet machines as established in the mentioned report [4].

The mentioned report also estimates a number of about 20,000 waterjet machines deployed worldwide in 2014 and an increasing trend of 5-6% yearly.

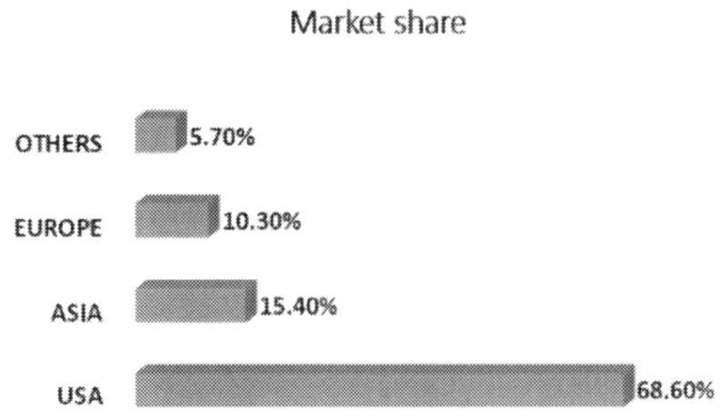

Figure 1. Current market shares of the top waterjet machine manufacturers.

The same report presents the increase of the market due to the benefits of the technology and forecasts sales exceeding a billion dollars starting 2018 (Figure 2).

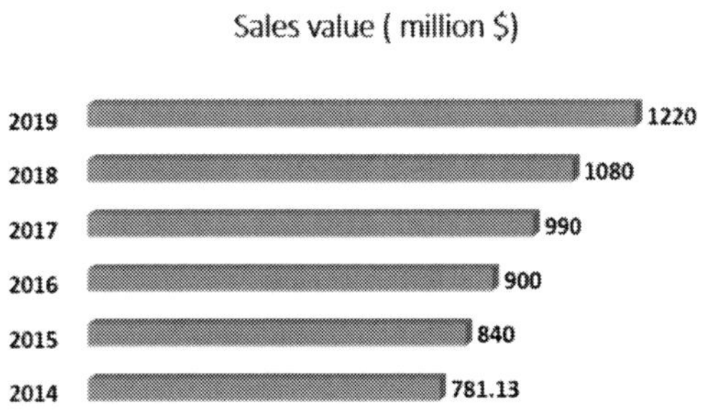

Figure 2. Annual sales values of waterjet machines

Pressure waterjet machining yields a number of significant benefits, including:

- special quality of the edges resulted upon cutting, eliminating the need for subsequent deburring;
- the cut contour can have complex shapes; zig-zag lines can be machined, as well as narrow bridges and interior contours;
- depending on the diameter of the jet, a very narrow cutting slot is generated in the part, thus achieving important material savings;
- tool sharpening and re-sharpening operations are eliminated;
- no preliminary contour drawing on the surface of the part is necessary;
- the machining process lends itself to automation;
- high productivity of cutting;
- minimum processing expenditure, given the reduced cost of water, abrasives, etc.

In relation to the latter advantage, Table 1 features the hourly costs of an abrasive jet cutting installation.

Table 1. Costs generated by abrasive jet erosion.

	[%]
Labour	35
Machine (amortisation in 5 years - 10^4 hours)	25
Abrasives	15
Nozzle	10
Pump maintenance	10
Energy	5
Total	100

A comparison between waterjet cutting and cutting with welding equipment, with plasma, laser, abrasive stones or a mechanical saw highlights the benefits of the former, as cutting occurs at normal, even low temperatures and no alteration of the material structure are caused by melting or internal stress. The relatively low price and minimum impact on the environment render waterjet cutting a technology of the future. Emerged and developed over the last years waterjet micromachining has improved precision to 0.02 mm and allows optimisation of surface roughness by configuring the process input factors by robust design [5].

An exact analysis of the mechanism of pressure abrasive fluid jet erosion is difficult, due to the multitude of individual processes which take place at the contact between the incident beam and the processed surface. The complexity of the process results from the effects of impact pressures, compression and shearing stresses, erosion and cavitation, all combined with the various properties of the processed materials.

Abrasive jet erosion is not "cutting" in the sense of carving, but a chip removal process, not unlike saw cutting. Analysis of material removed by an abrasive jet reveals chips of up to 5 μm, comparable to those generated in cutting with a diamond disk.

Machining by pressure abrasive fluid jets entails, in a first stage calibration of the fluid in the nozzle. In the second stage the fluid-abrasive mix is formed, which is sent towards the surface of the workpiece. It is extremely difficult to observe the forming process of the abrasive mix, because of the high speed of the grains (200...800 m/s) and their small dimensions. Figure 3 illustrates the stages of abrasive jet forming, and presents the main quantities contributing to the erosive process.

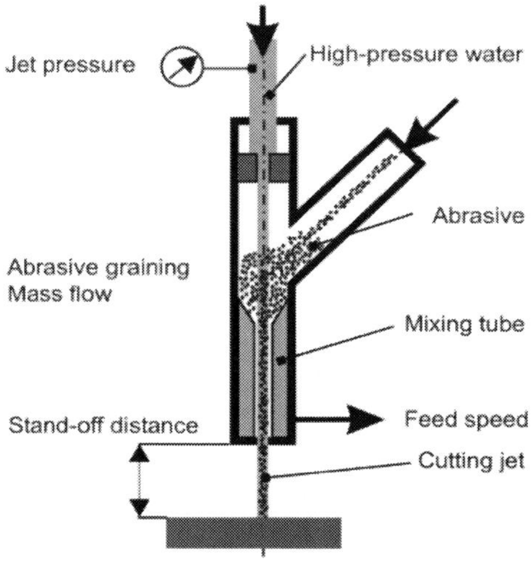

Figure 3. Principle of abrasive jet cutting.

The above figure shows that the abrasive grains are transported to the mixing tube by aspiration, via the water jet. Once in the mixing tube, the grains repeatedly collide with its walls and the water jet, being eventually included in the liquid jet. Because of these repeated collisions, the abrasive grains are continuously crushed, so that they have significantly smaller dimensions when leaving the mixing tube and hitting the workpiece.

Particularly significant is the form of the grains utilized in abrasive jet machining. Grain shape should be as compact and close to spherical as possible, as illustrated in Figure 4.

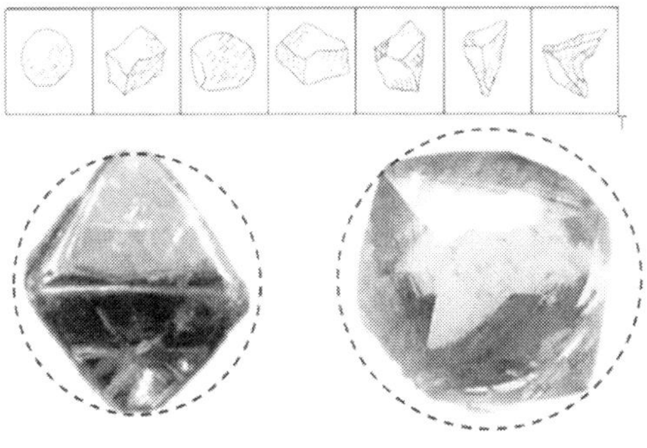

Figure 4. Recommended abrasive grain shapes.

The abrasive grains used in waterjet machining are selected by a form factor calculated by equation:

$$F_f = \frac{d_{min}}{d_{max}} \tag{1}$$

where d_{min} is the largest diameter of a circle that can be inscribed in the abrasive grain, while d_{max} is the smallest diameter of a circle circumscribed to the grain. The usual values of the form factor range between 0.64 and 0.70.

When leaving the mixing tube, the abrasive jet is characterised by four important properties: speed, coherence, density and symmetry. Speed is the most important characteristic of a jet, as it determines its cutting capacity. A higher speed improves capability of the jet to perform as a cutting tool.

Jet speed depends also on the stand-off distance; evidently jet speed decreases as the workpiece surface is located at a greater distance. Figure 5 presents jet speed variation versus stand-off distance, for a pressure of 200 MPa and an initial jet diameter of 0.2 mm [6].

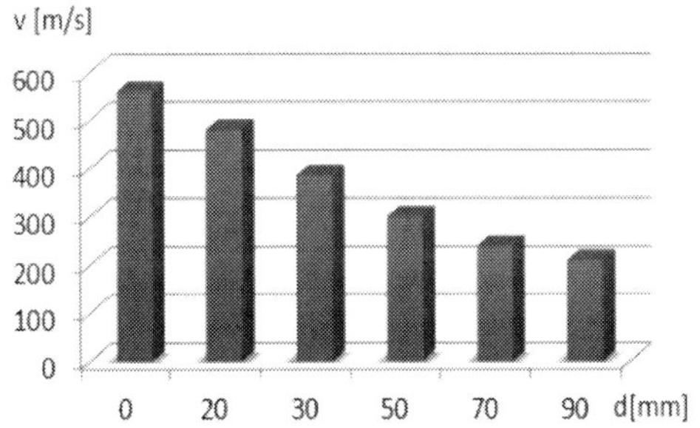

Figure 5. Jet speed versus stand-off distance.

As the jet penetrates the material, its kinetic energy decreases continuously, causing a curved trajectory of the cut (Figure 6). This disadvantage can be compensated by increasing jet speed v, or by decreasing the feed speed w_a. In general, a cut of high quality requires adequate correlation of jet speed, workpiece thickness and material.

Jet feed speed w_a is particularly important for the quality of the cut. A higher feed speed reduces machining quality, and in some situations, when certain critical values are exceeded, the jet is deviated and cutting does not occur.

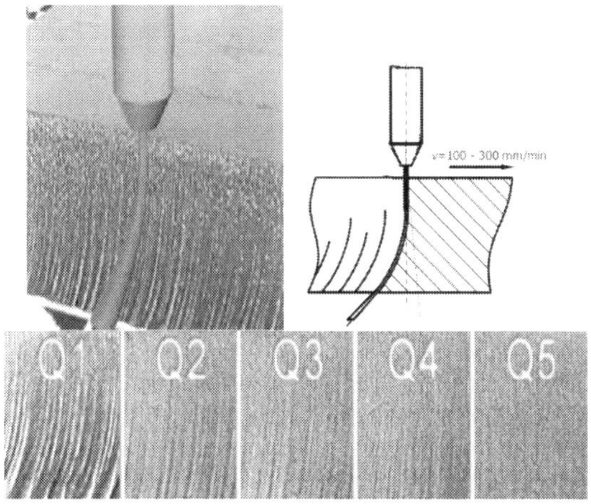

Figure 6. Curved trajectory of the cut.

In the above figure the quality of the cut is classified into five categories. While Q1 stands for a low quality cut yielding high surface roughness, Q5 indicates a cut achieved upon adequately setting the working parameters.

The mass flow of the abrasive grains is a further significant parameter for assessing the cutting capacity of the abrasive jet. Figure 7 presents the dependency of cut material thickness on the abrasive mass flow.

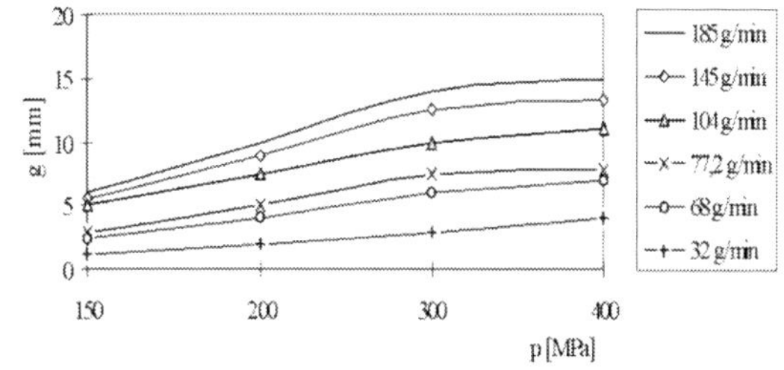

Figure 7. Influence of abrasive mass flow on cut thickness.

Erosion cutting by abrasive jets is achieved consequently to the cumulated effect of two phenomena:

- cutting and scraping carried out by each abrasive grain, during the first part of the machining process, as long as each grain still has sharp tips and cutting edges;
- plastic deformation of the material during the second part of the machining process, once the grains have lost their cutting capability. In this phase the remaining kinetic energy of the abrasive is still great enough to exceed to flow limit of the material, and thus in time to cause the removal of the tooling allowance.

Because of these two successive phenomena the quality of the cut surfaces is not uniform along the entire machined depth. A first area with numerous narrow striations can be observed, followed by an area with wider scratches (Figure 8). The depth of the first area, with a better roughness, depends directly on the speed of the abrasive jet.

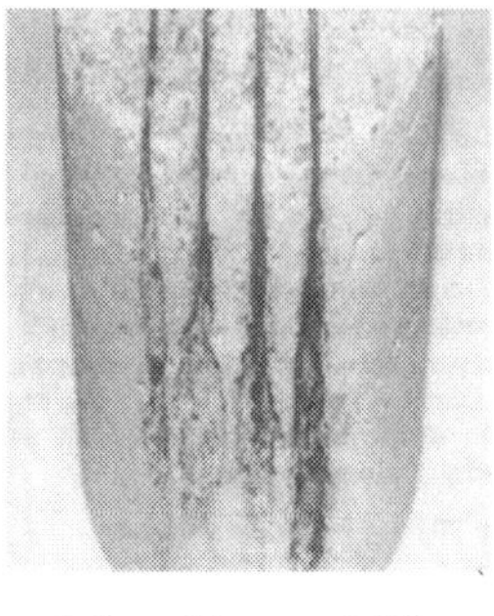

Figure 8. Width variation of the trace left by an abrasive grain.

It is known that as each abrasive grain penetrates deeper into the machined material, the quality of the cut decreases. Hence it is recommended to select the set points for the working parameters (pressure, graining, mass flow, feed speed) consequently to thorough experimental study. The thickness of the material that can be cut while ensuring acceptable quality needs to be determined by experiment-based multi-criteria optimisation.

Following analysis of the abrasive jet erosion process and knowing the main factors that contribute to obtaining surfaces of acceptable quality, further on a novel approach to ensuring machined surface quality is presented, based on the concept of robust design of these technological systems. This approach allows implementation of quality by conducting specially designed experiments of short duration, at low costs and with minimum material consumption.

The fundamental difference to classical R&D of a system consists in the pragmatism of robust design that does not aim at identifying and eliminating the causes responsible for non-quality, considered "noises", but only render the system insensitive to their action.

2 Experimental set-up

Experimental studies for the robust optimisation of pressure abrasive jet machining were conducted on a Maxiem 1530 machine. Such a piece of equipment allows the cutting of workpieces of up to 3099 mm x 1575 mm, at a maximum available pressure of 3450 bar and a maximum feed speed of 9 m/min. The dimensional precision ensured by this machine is of 0.05 mm.

The main parameters that contribute to attaining the goals of abrasive jet cutting are the pressure provided by the equipment, feed speed, stand-off distance, abrasive graining and mass flow. Other quantities impacting on machining efficiency are the type of the utilized abrasive and the tilting angle of the jet. All enumerated parameters represent input quantities of the system, each having a range of possible values.

Experimental research consisted of cutting workpieces made from an aluminium alloy with the following composition: Al: 92.02%; Cu: 1.65%; Mg: 2%; Cr: 0.23%; Zn: 4%; Mn: 0.1%. The roughness (R_a) of the surface was measured after testing.

The data sets obtained upon experimenting and measuring were processed by dedicated software, *LappMaster*, which yielded the optimum set-points of the tested controllable input factors.

3 Experimental design

Robust design of manufacturing systems is a major contribution brought by Japanese quality expert Genichi Taguchi to the set of instruments available in industry for quality analysis and improvement. Robustly designed processes ensure a constant quality, by reducing the scattering of the values measured for the quality characteristic. Significantly for this method, this desideratum is achieved without eliminating the causes responsible for variability, namely by rendering the process insensitive to the influence of such uncontrollable factors (noise factors).

Robust design is a method to reduce the interval of variability of the quality characteristics, by providing the methodology for systematically obtaining solutions that render manufacturing processes less sensitive to disturbing causes. This type of design can be utilized for optimising product development, as well as for conceiving new manufacturing processes.

According to Taguchi's strategy non-quality can be eliminated without eliminating the disturbing factors from the system, but by adequately combining the set-points of the input factors.

The method used to optimize the pressure abrasive jet cutting process was the method of fractioned orthogonal factorial arrays of experiments devised by G. Taguchi. This method entails a planned approach, aimed at studying and establishing the links between the controllable input factors of a system and the obtained effects. The method identifies the combination of set-points of the input parameters of the system that reduce the effects of parasite causes, without however addressing these directly.

From the numerous input quantities affecting the pressure abrasive jet cutting process five factors were selected considered to have a major contribution to the machining process. Table 2 presents these factors, each being assigned two set-points during experiments.

Table 2. Selected factors and associated set-points.

Set-point	Factors				
	A Feed speed	B Jet pressure	C Stand-off distance	D Abrasive graining	E Mass flow
	[mm/min]	[bar]	[mm]	[μm]	[g/s]
1	50	1250	1	60	0.5
2	100	1750	1.5	90	2

If a full array of experiments were used, the total number of tests would be $2^5 = 32$. Such an array entails the experimental study of all possible combinations of the set-points of the considered input factors, influencing system operation. Not all tests have, however, the same practical significance, some of the runs providing more information than others. Starting from this observation, G. Taguchi has devised a set of fractioned orthogonal factorial arrays of experiments. These arrays allow considerable reduction of the number of test runs to be conducted, without diminishing relevance and accuracy of the yielded results.

The experiments were conducted according to an array of experiments requiring only 8 runs, each with a different combination input factor set-points.

Table 3 presents this array of experiments, indicating in brackets, next to the set-point for each factor, the respective concrete values.

Table 3. The utilized array of experiments.

Run	Factors				
	A	B	C	D	E
1	1 (50)	1 (1250)	1 (1)	1 (60)	1 (0.5)
2	1 (50)	1 (1250)	1 (1)	2 (90)	2 (2)
3	1 (50)	2 (1750)	2 (1.5)	1 (60)	1 (0.5)
4	1 (50)	2 (1750)	2 (1.5)	2 (90)	2 (2)
5	2 (100)	1 (1250)	2 (1.5)	1 (60)	2 (2)
6	2 (100)	1 (1250)	2 (1.5)	2 (90)	1 (0.5)
7	2 (100)	2 (1750)	1 (1)	1 (60)	2 (2)
8	2 (100)	2 (1750)	1 (1)	2 (90)	1 (0.5)

Each run was repeated five times, and roughness R_a was measured in different areas of the machined surface. Further arithmetic mean, standard deviation and signal-to-noise ratio were computed for the measured results.

The signal-to-noise ratio (S/N) is the performance indicator that simultaneously considers the target value of the analysed quality criterion and the variability around the target. The desired value is defined as signal (expressed by $\bar{y}$ = the arithmetic mean of the measured values), while its undesired variability, to be diminished, is defined as noise, (expressed by the standard deviation s).

In certain practical applications merely determining the results of measurements may not be sufficient. In the case of pressure abrasive jet cutting, one of the aims is to machine surfaces such as to achieve a roughness as small of possible. This means imposing a sense of this quantity's evolution, in other words considering an optimisation criterion of "*smaller is better*" type. The computational relationship of the S/N ratio for a quality characteristic whose value, in order to be optimum, should tend towards a minimum, as is the case of surface roughness is:

$$S/N_m = -10 \cdot \log(s^2 + \bar{y}^2) \quad (2)$$

Obviously, the higher the algebraic value of the signal-to-noise ratio is (maximised signal – to minimized noise), the better the performance of the machining system will be. Consequently, based on the results obtained by running the *LappMaster* programme the optimum set-points for each analysed input factor could be established. Table 4 presents these results.

Table 4. Optimum combination of factor set-points.

Factor	Factor description	Optimum set-point	Value of optimum set-point
A	Feed speed	1	50 mm/min
B	Jet pressure	1	1250 bar
C	Stand-off distance	2	1.5 mm
D	Abrasive graining	1	60 μm
E	Mass flow	1	0.5 g/s

By setting the values of the five factors to the determined points, the expected resulting signal-to-noise ratio is of -6.4276 dB, which value would generate a robust abrasive jet cutting process. This value of resulting S/N ratio was computed according to equation (3):

$$\frac{S}{N_{res}} = \frac{\bar{S}}{N} + \sum \textit{the effects with the highest alg. value} \quad (3)$$

4 Conclusion

The paper presents a study of high pressure abrasive jet cutting, optimised by the method of robust parameter design. This approach is aimed at rendering the analysed system insensitive to the action of noise factors, by an adequate combination of the input factor set-points.

In the first stage of applying the method of fractioned orthogonal factorial arrays of experiments devised by G. Taguchi, five controllable input factors were selected, and two set-points were considered for each. Within the ensuing phase, a set of experiments were run, with a set-point assigned to each factor as provided by Taguchi's fractioned orthogonal factorial array; eventually, upon mathematical processing of the measured results a combination of optimum set-points of the analysed factors resulted. By assigning these optimum set-points to the input quantities the machining system becomes robust, and consequently the roughness of the machined surfaces will not deviate from the desired and predicted value.

References

1. P.W. Bridgman, *The Physics of High Pressure*, (Dover Publications Inc., New York., 1970)
2. A.W. Momber, R. Kovacevic, *Principles of abrasive waterjet machining*, (Springer Verlag, London, 1998)
3. B. Barabaş, *Current State of Research on Waterjet Machining of Surfaces*, (Transilvania University of Braşov, 2015)
4. Report Buyer - Global Waterjet Machine Market 2015-2019, accessed November 2015, https://www.reportbuyer.com/product/869256/global-waterjet-machine-market-2015-2019.html
5. T. Deaconescu, Contributions to the abrasive waterjet machining of parts, Nonconventional Technologies Rev., **4** (2015)
6. C. Birtu, *Contributions to Materials Machining by Hydro-Abrasive Jets*, (Doctoral Thesis, Bucureşti, 2000)

Improvement of the Critical Speed of the Ball Screw Mechanism by Using Flexible Supported Nut

Songwei Li, Min Wang

Beijing Municipal Key Lab of Advanced Manufacturing Technology,Beijing Municipal Key Lab of EDM Technology, College of Mechanical Engineering and Applied Electronics Technology, Beijing University of Technology, Beijing ,100124,China

Abstract. The ball screw with a long stroke has small rigidity in radial direction and is very liable to resonate dramatically during operation at high speed. For improving the critical speed of the ball screw, this paper investigates the influence of the supporting stiffness of the screw shaft at the nut on the resonance frequency of the screw shaft in the radial direction. The screw shaft is modeled as an Euler-Bernoulli beam with elastic supports at both ends and the position of the nut. According to the established model, the relationship between the natural frequencies and the supporting stiffness of the shaft at the nut can be obtained. The simulation results show that the flexible support at the position of the nut can effectively increase the minimum natural frequencies and reduce the variation range of the natural frequencies over the travel range of the nut. It is conducive to the vibration suppression and the critical speed improvement of the ball strew.

1 Introduction

The ball screw is originated in the late 19th century, which has a high precision, high efficiency, high rigidity, so it is widely used in the mechanical engineering, aviation, aerospace and many other fields [1]. Generally, in the ball screw mechanism, the motor drives the screw shaft to rotate, then the rotary motion of the motor is converted into a liner reciprocation of the nut. The worktable rigidly connected to the screw shaft through the nut, which moves with the travel range of the nut. As an important functional component in high precision machine tools, the performance parameters of the ball strew affect the positioning and repetitive positioning accuracy during the procession, which affect the quality of the workpiece ultimately.

As a linear drive execution unit in the servo feed system of the CNC machine tool, the moving stability and vibration resistance of the ball screw mechanism are important factors affecting the accuracy of the machine tool. Due to the limit of the resonance frequency of the screw shaft, its critical speed is rather difficult to improve. The elongated shaft structure and the large aspect ratio of the ball screw can easily lead to lateral, axial and torsional vibration in the transmission process. During which, the lateral vibration is a common phenomenon. When the variation range of the natural frequencies of the controlled structure is too large, it is difficult to suppress the vibration [2]. In contrast, when the variation range of the natural frequencies of the controlled structure is narrow, the tuned mass dampers or other methods can be used to reduce the vibration, and the effect is also obvious. The mass tuned damper has a simple structure and good stability, which is widely used in the construction, machinery, vehicles and many other industries [3-5].

The nut is continuous moving in the whole stroke, which leads to the alteration of the natural frequencies and mode shapes of the screw shaft, so it is not likely to achieve the vibration control [6]. In terms of the ball screw feed system, scholars have used different methods to build the modeling of the ball screw [7-9], which providing references for the dynamics study of the screw shaft. In previous, most studies did not consider the influence of the travel range of the nut and the supporting stiffness of the screw shaft at the nut on resonance frequency of the screw shaft in the radial direction. This paper mainly studies the lateral vibration of the ball screw, and then builds a dynamic model of the screw shaft, which is elastic supported at both ends and the position of the nut. According to the established model, the relationship between the natural frequencies and the supporting stiffness of the shaft at the nut can be obtained. By optimizing the supporting stiffness of screw shaft at the nut in the radial direction, it is dramatic to increase the minimum natural frequencies and reduce the variation range of the natural frequencies over the travel range of the nut. Based on the optimization results, the flexible supported nut is proposed, which can be used to adjust the supporting stiffness of the screw shaft at the position of the nut and increase the critical rotation speed and vibration resistance of the ball screw.

2 The function of the lateral vibration mode shapes of the screw shaft

In the ball screw feed system, the worktable is connected to the screw shaft with nut, and both ends of the screw shaft are supported by bearings. Among them, there is generally rigid connection between the worktable and the nut.

The screw shaft is modeled as an Euler - Bernoulli beam with elastic supports at both ends and the position of the nut. The length of the screw shaft is L, the position of the nut is L_1, the fulcrum of both ends are O_1 and O_2, the origin of coordinates is O_1, the worktable contacts O point through the nut and the screw shaft, k_1, k_2 represent the radial supporting stiffness of bearings at both ends, k represents the support stiffness of the screw shaft at the nut, the simplified model of the screw shaft is shown in Figure 1.

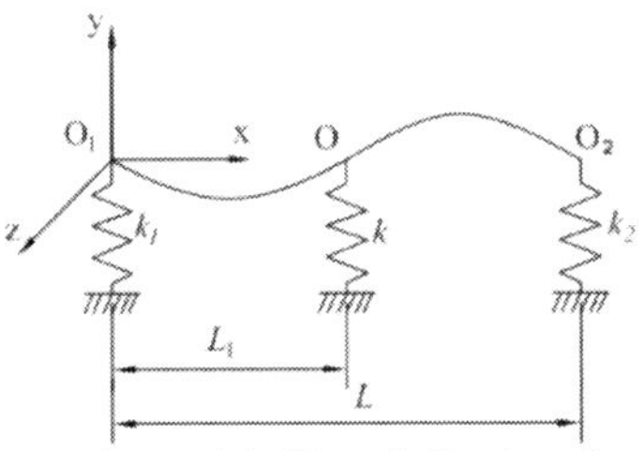

Figure 1. Supporting model of lateral vibration of screw shaft.

According to kinetic equilibrium equations of the beam, the lateral bending vibration mode function is given as follows [10]:

$$\begin{aligned}\varphi(x) = & D_1\left[\cosh(\gamma x)+\cos(\gamma x)\right] \\ & + D_2\left[\sinh(\gamma x)+\sin(\gamma x)\right] \\ & + D_3\left[\cosh(\gamma x)-\cos(\gamma x)\right] \\ & + D_4\left[\sinh(\gamma x)-\sin(\gamma x)\right]\end{aligned} \tag{1-a}$$

$$A = \pi d^2 / 4 \tag{1-b}$$

$$I = \pi d^4 / 64 \tag{1-c}$$

$$\gamma^4 = A\omega^2\rho / (EI) \tag{1-d}$$

where, D_1-D_4 are modal shape modulus, ω is the natural frequency of the screw shaft, ρ is the density of the screw shaft, A is the cross-sectional area of the screw shaft, d is the diameter of the screw shaft, E is the elastic modulus of the screw shaft, I is the moments of inertia of the screw shaft's cross-section.

Through the formula (1-a) to (1-d), the vibration mode function of O_1O segment is expressed as:

$$\begin{aligned}\varphi_1(x) = & C_1\left[\cosh(\gamma x)+\cos(\gamma x)\right] \\ & + C_2\left[\sinh(\gamma x)+\sin(\gamma x)\right] \\ & + C_3\left[\cosh(\gamma x)-\cos(\gamma x)\right] \\ & + C_4\left[\sinh(\gamma x)-\sin(\gamma x)\right]\end{aligned} \tag{2}$$

The vibration mode function of OO_2 segment is written as:

$$\begin{aligned}\varphi_2(x) = & C_5\left[\cosh(\gamma x)+\cos(\gamma x)\right] \\ & + C_6\left[\sinh(\gamma x)+\sin(\gamma x)\right] \\ & + C_7\left[\cosh(\gamma x)-\cos(\gamma x)\right] \\ & + C_8\left[\sinh(\gamma x)-\sin(\gamma x)\right]\end{aligned} \tag{3}$$

According to the support conditions of the screw shaft at O_1 and O_2, the relationship between the bending moment and shear, the following expressions can be obtained.

$$\varphi_1''(0) = 0 \tag{4}$$

$$EI\varphi_1'''(0) = -k_1\varphi_1(0) \tag{5}$$

$$\varphi_2''(L) = 0 \tag{6}$$

$$EI\varphi_2'''(L) = k_2\varphi_2(L) \tag{7}$$

Considering the screw shaft is continuous at O point, the continuous conditions include displacement, rotation, bending moment and shear, which can be expressed as:

$$\varphi_1(L_1) = \varphi_2(L_1) \tag{8}$$

$$\varphi_1'(L_1) = \varphi_2'(L_1) \tag{9}$$

$$\varphi_1''(L_1) = \varphi_2''(L_1) \tag{10}$$

$$EI\varphi_1'''(L_1) + k\varphi_1(L_1) = EI\varphi_2'''(L_1) \tag{11}$$

By the formula (1) to (11), it is easy to get the screw shaft's natural frequencies ω, modal modulus C_1~C_8, and the vibration mode shapes function. In order to make it convenient to calculate, the modal mass of the screw shaft can be normalized, the n-th modal quality of the screw shaft is given as:

$$m_n = \int_0^{L_1} \rho A\varphi_{n,1}^2(x)d_x + \int_{L_1}^{L} \rho A\varphi_{n,2}^2(x)d_x \tag{12}$$

The vibration mode function of the screw shaft which is quality normalized can be expressed as follows:

$$\varphi_{n,1}(x) = \frac{1}{\sqrt{m_n}}\left\{\begin{aligned} & C_{n,1}\left[\cosh(\gamma_n x)+\cos(\gamma_n x)\right] \\ & +C_{n,2}\left[\sinh(\gamma_n x)+\sin(\gamma_n x)\right] \\ & +C_{n,3}\left[\cosh(\gamma_n x)-\cos(\gamma_n x)\right] \\ & +C_{n,4}\left[\sinh(\gamma_n x)-\sin(\gamma_n x)\right]\end{aligned}\right\} \tag{13-a}$$

$$\varphi_{n,2}(x) = \frac{1}{\sqrt{m_n}}\left\{\begin{aligned} & C_{n,5}\left[\cosh(\gamma_n x)+\cos(\gamma_n x)\right] \\ & +C_{n,6}\left[\sinh(\gamma_n x)+\sin(\gamma_n x)\right] \\ & +C_{n,7}\left[\cosh(\gamma_n x)-\cos(\gamma_n x)\right] \\ & +C_{n,8}\left[\sinh(\gamma_n x)-\sin(\gamma_n x)\right]\end{aligned}\right\} \tag{13-b}$$

where, $\varphi_{n,1}(x)$, $\varphi_{n,2}(x)$ respectively represent the n-th vibration mode function of the screw shaft's O_1O and OO_2 segment, $C_{n,m}$(m=1,2...,8)is the modal modulus of the n-th vibration mode function.

3 The analysis of the natural frequencies of the screw shaft

In the ball screw feed system, the screw shaft is driven by a motor, and then the worktable makes a linear motion accompanied with the screw shaft's rotation. During the whole stroke, the worktable is constantly moving with the travel range of the nut, which lead to the change of the screw shaft's natural frequencies and mode shapes.

In this paper, the screw shaft adopts angular contact ball bearings. The screw shaft's length L= 0.808m, the screw shaft's diameter d=0.025m, the screw shaft's material density ρ= 7.85 × 10^3kg / m^3, and the screw shaft's elastic modulus E =2.07×10^{11}Pa. According to the calculation formula of the stiffness of angular contact ball bearings, the bearing's supporting stiffness k_1 = k_2= 10^7N / m.

3.1 The influence of the travel range of the nut on the natural frequencies of the screw shaft

Take the supporting stiffness of the screw shaft at the nut in the radial direction k = 10^7N / m. Considering the worktable is moving with the travel range of the nut, the nut's position from 0.1L to 0.9 L is selected to calculate. Thus, the natural frequencies over the travel range of the nut are obtained, the results are shown in Figure 2.

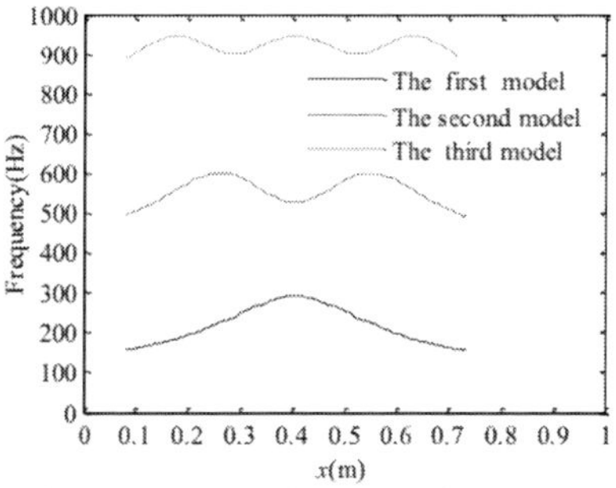

Figure 2. The natural frequencies at the different positions of the nut.

According to Figure 2, During the whole stroke of the nut, the natural frequencies of the screw shaft in radial direction continuously changes within a certain range. With the increase of the vibration models, the variation range of the natural frequencies reduces gradually. The natural frequencies are symmetric with respect to the neutral position over the travel range of the nut. The natural frequency arrives at a peak value in the neutral position. The maximum and minimum values of the second and third natural frequency appear alternately.

3.2 The effect of the supporting stiffness of the screw shaft at the nut on the screw shaft's natural frequencies

Due to the limitation of screw shaft's natural frequencies, its critical speed is difficult to be improved. It is also necessary to improve the natural frequencies to improve the vibration resistance of the screw shaft. In order to ensure the stability control of the worktable's vibration in the whole stroke, it is necessary to optimize the natural frequencies of the screw shaft. Changing the supporting stiffness of the screw shaft at the nut, the relationship between the natural frequencies and the

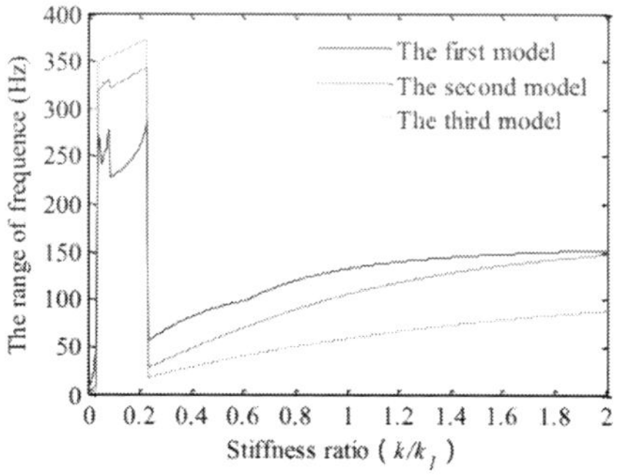

Figure 3. The variation range of the natural frequencies with different supporting stiffness of the screw shaft at the nut

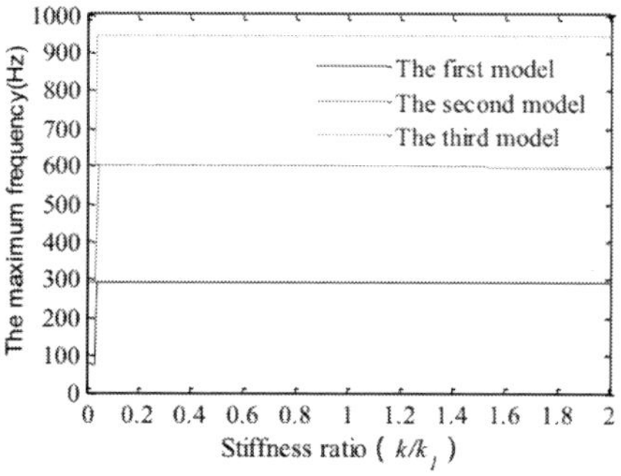

Figure 4. The maximum natural frequencies with different supporting stiffness of the screw shaft at the nut

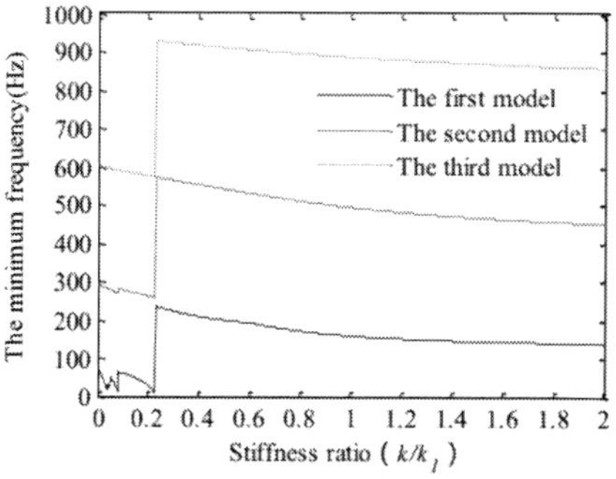

Figure 5. The minimum natural frequencies with different supporting stiffness of the screw shaft at the nut supporting stiffness of the screw shaft at the nut can be obtained.

Considering the travel range of the nut, this paper choose the nut at the positions of 0.1L, 0.2L, 0.3L, 0.4L and 0.5L to calculate. Changing the supporting stiffness of the screw shaft at the nut k, and the natural frequencies at the different positions of the nut can be calculated. Finally, the variation range of the natural frequencies, the minimum and maximum natural frequencies over the travel range of the nut are obtained. Results are shown in Figure 3-Figure 5.

According to Figure 3, during the whole stroke of the nut, the trend of the variation range of the natural frequency is divided into two stages. At the beginning, it increases rapidly with the stiffness ratio varies, which reaches the maximum value ultimately. When the stiffness ratio is about 0.23, it is reduced to a minimum value quickly. At this time, the natural frequency difference is 53*Hz*, the second-order natural frequency difference is 27*Hz*, and the third-order natural frequency difference is 18*Hz*. When the stiffness ratio is greater than 0.23, it is gradually reduced, and tends to be stable finally. Figure 4 shows that, the maximum natural frequencies increases rapidly with the stiffness ratio varies, and then it reaches a steady state. Figure 5 shows that, with the increasing of the stiffness ratio, the minimum natural frequencies decreases slowly, and then they reach a peak value suddenly, and finally they decreases to a stable value. When the stiffness ratio is about 0.23, the minimum natural frequencies arrive at the maximum value.

The results show that optimizing the supporting stiffness of the screw shaft at the nut, the minimum natural frequencies of the shaft over the whole stroke of the nut can be improved obviously, and the variation range of the natural frequencies can be reduced at the same time. According to the optimization results, it is necessary to design a flexible supported nut which can effectively adjust the supporting stiffness of the screw shaft at the position of the nut, and increase the critical rotation speed and vibration resistance of the ball screw mechanism. When designing the flexible supported nut, the stiffness ratio is required to be strict controlled. If the stiffness ratio is relatively too small, the variation range of the natural frequencies is large in the whole stroke, so it is difficult to suppress vibration. Conversely, if the stiffness ratio is relatively too large, the variation range of the natural frequencies is stable, the maximum and the minimum natural frequencies reach a steady state, so there is no use to improve the natural frequencies and suppress vibration with great stiffness ratio. In terms of the screw shaft above, a flexible supported nut whose stiffness ratio is slightly greater than 0.23 can be designed, which can increase the minimum natural frequencies and reduce the variation range of the natural frequencies over the travel range of the nut.

4 The analysis of the lateral vibration mode shapes of the screw shaft

Along with the moving of the nut in the range of the whole stroke, the vibration mode shapes of the screw shaft change continuously, which resulting in a negative effect on the running stability of the screw shaft. Mostly, the vibration of the screw shaft occurs in low order natural frequencies, so it is sensible to consider the change of the first vibration mode shape of the screw shaft. Changing the supporting stiffness of the screw shaft at the nut, the vibration mode shape of the screw shaft at the different nut positions is obtained. In order to ensure the modal shape of the screw shaft is stable in the whole stroke of the nut, it is benefit to select the nut at the positions of the 0.1*L*, 0.3*L*, 0.5*L* to calculate, the results are shown in Figure 6-8.

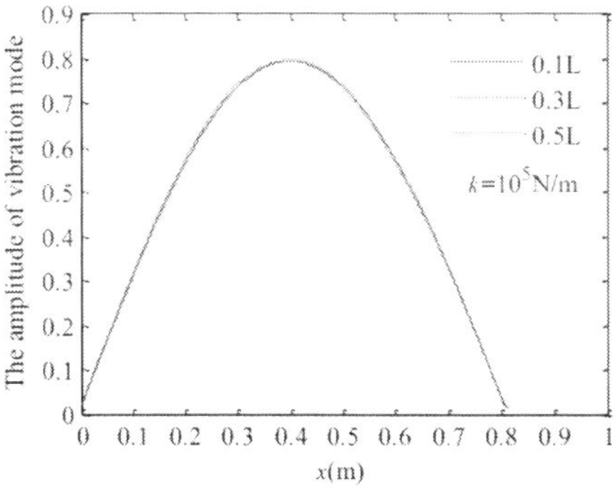

Figure 6. The lateral vibration mode shapes of the screw shaft.

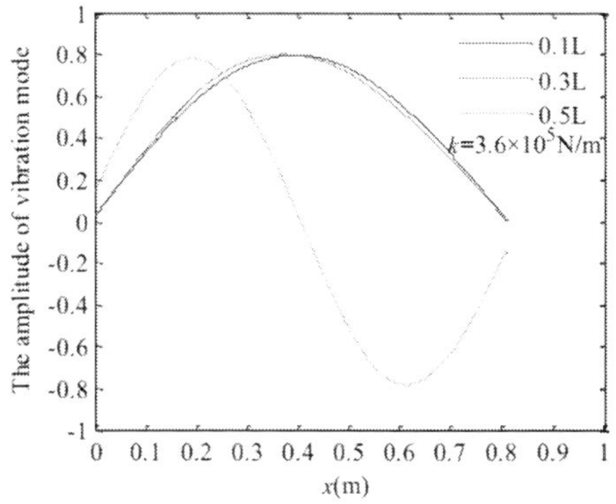

Figure 7. The lateral vibration mode shapes of the screw shaft.

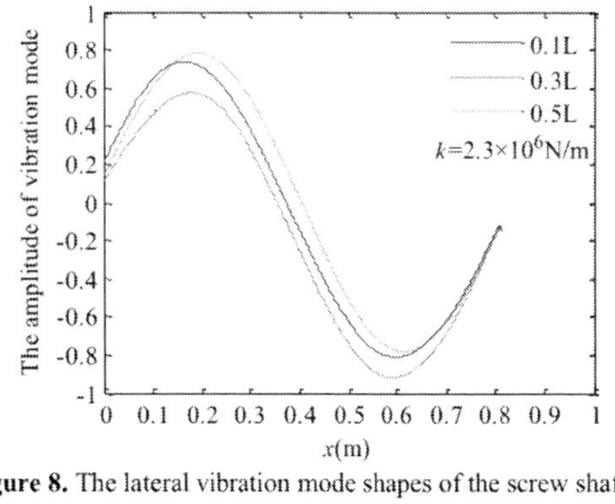

Figure 8. The lateral vibration mode shapes of the screw shaft.

From Figure 6 to Figure 8, when the supporting stiffness of the screw shaft at the nut is changing, the vibration mode shapes is varying. In Figure 6, as the supporting stiffness of the screw shaft at the nut is small, that is to say the screw shaft is similar to the state of no support, so the mode shapes of the screw shaft in the three different nut's position are almost the same. In Figure 7, when the nut at the position of 0.5*L* and the supporting stiffness of the screw shaft at the nut is

3.6×10^5N/m, the vibration mode shape changes suddenly, indicating that the mode shapes of the screw shaft in the whole stroke of the nut is not stable, resulting in the variation range of the natural frequencies change bigger abruptly in Figure 3. When the supporting stiffness of the screw shaft at the nut is 2.3×10^7 N/m, the mode shape of the nut at the three different positions are similarity and maintain stable, resulting in the variation range of natural frequencies changes smaller abruptly in Figure 3.

Simulation results show that, when the supporting stiffness of the screw shaft at the nut changes within a certain range, the mode shape of the screw shaft at different nut's positions are varying. The change of the vibration mode shapes of the screw shaft reflect that the natural frequencies of the screw shaft have a big change, resulting in the variation range of natural frequencies change larger in the whole stroke of the nut. By optimizing the supporting stiffness of the screw shaft at the nut, the vibration mode shapes of the screw shaft can remain stable, meanwhile, the variation range of the natural frequencies are the smallest in the whole stroke of the nut. A flexible supported nut can be used to adjust the supporting stiffness of the screw shaft at the nut, it maintains the stability of the vibration mode shapes and reduces the variation range of the natural frequencies of the screw shaft in the whole stroke of the nut effectively, meanwhile, it is also benefit for the smooth operation of the ball screw mechanism.

5 Conclusions

By modeling the ball screw mechanism considering the influence of the moving nut to the dynamics of the screw shaft, we analyze the effect of supporting stiffness on the natural frequencies and mode shapes of the screw shaft. Aiming to control the lateral vibration of the screw shaft, several conclusions can be obtained as follows:

(1) During the whole stroke of the nut, the natural frequencies of the screw shaft in the radial direction continuously changes within a certain range. With the increase of the vibration modes, the variation range of the natural frequencies of the screw shaft in radial direction reduces.

(2) By optimizing the supporting stiffness of the screw shaft at the nut, the minimum natural frequencies of the shaft over the whole stroke of the nut can be improved obviously, and the variation range of the natural frequencies can be reduced at the same time.

(3) According to the optimization results, the flexible supported nut can be used to adjust the supporting stiffness of the shaft at the position of the nut, increase the critical rotation speed and vibration resistance of the ball screw.

References

1. Z. Xiao, Mach. Tool. Bib , **9** (2001)
2. J. Zhang, Struc. Envi. Engi, **31**, 3 (2004)
3. Z. Wang, W.Ren, J. Su, etal. Engi. Mech, **22**, 5(2005)
4. Y. Yang, Q. Liu, J. Vibr. Engi, **23**, 4 (2010)
5. C. Tong, X. Zhang, J. Vibr. Meas. Diag. **27**, 6 (2007)
6. M.Wang, F. Li, T. Zan, X. Gao, J. Vibr. Shock, **34**, 10 (2015).
7. H. Zhang, J. Sun , J. Changchun . Univ , **21**, 2 (2011)
8. M.C. Lin, B. Ravani,S.A. Velinsky. J. Mech. Design, **116**, 3 (1994)
9. G. Holroyd, C. Pislaru, D.G. Ford.Tran. Engi. Sci, 2003
10. D. Zhu, Y. Xing, *Engineering Vibration Foundation*, (Beihang University Press, 2004)

Time-Frequency Enhancement Technique for Bevel Gear Fault Diagnosis

Dennis Hartono[1], Dunant Halim[1], Achmad Widodo[2] and Gethin Wyn Roberts[3]

[1]*Department of Mechanical, Materials and Manufacturing Engineering, The University of Nottingham Ningbo China, China 315100*
[2]*Department of Mechanical Engineering, Diponegoro University, Semarang, Indonesia 50275*
[3]*Department of Civil Engineering, The University of Nottingham Ningbo China, China 315100*

Abstract. Gear failure is one of the most common causes of breakdown in rotating machineries. It is well known that vibration signals from machineries can be effectively used to detect certain gear faults. Yet it is still not an easy task to find a symptom that reflects a particular fault from vibration signals. This paper presents an advanced time-frequency signal processing technique for extracting important gear fault information from the vibration signal that is heavily corrupted by measurement noise. Experiments were performed on a bevel gearbox test rig using vibration measurements. The Time Synchronous Average (TSA) was initially utilized to eliminate all asynchronous component of vibration signal obtained from the gear. The Continuous Wavelet Transform (CWT) method was then used to capture the non-stationary behaviour of the impulse signal generated from the broken bevel gear tooth. It was shown that the diagnosis method using the Continuous Wavelet Transform combined with Time Synchronous Averaging outperformed the conventional spectral analysis, capable of identifying the angular location of broken teeth in the gear.

1 Introduction

Diagnostic methods using vibration from geared systems have provided a significant contribution in preventing major failures in rotating machineries. However, most signal processing algorithms that have been developed for vibration diagnostic process rely on the assumption that the vibration signal is stationary. The complexity of vibration signal, comprising both transient and stationary components, can lead to insufficiency of these diagnostic methods to accurately detect gearbox faults from vibration signal.

It is well known that the main component in gearbox vibration signals is the gearmesh frequency, its harmonics and their sidebands around the gearmesh frequency. The modulation sidebands that are present around the fundamental frequency or harmonics of the tooth meshing frequency can provide useful information of gear defect locations [1]. Moreover, the amplitude of these sidebands can offer a quantitative measure of the severity of the gear fault [2]. However when the defects need to be identified in term of their angular locations most conventional methods, such as the Fast Fourier Transform (FFT) and Cepstrum, are not capable to provide such information accurately. Moreover, FFT assumes the underlying signal is a stationary signal, whereas the impulse generated by the gear fault is typically a transient signal. In this case, the use of time-frequency analysis has been proved to be able to detect the transient phenomena [3, 4], e.g. the impulse response generated from faulty gear. There is the need of using more sophisticated signal processing methods to provide such useful diagnostics information.

The objective of this work is to compare several advanced time-frequency techniques such as Short Time Fourier Transform and Continuous Wavelet Transform to capture the transient vibration signals generated by the gear fault accurately. Initially, the Time Synchronous Averaging technique was utilized to extract the gear fault feature from the vibrations signals that were heavily corrupted by measurement noise. The Fast Fourier Transform was utilized to extract the sidebands around the gearmesh frequencies that indicated which particular shaft the faulty gear was attached. Finally, STFT and CWT methods were compared to offer an accurate identification of the faulty gear angular.

2 Theoretical background

2.1 The time synchronous averaging

The Time Synchronous Averaging is the process of ensemble averaging by sampling of the vibration signal according to the rotation of a particular gear of interest. In practice it will contain only components which are synchronous with the revolution of the gear in question. This method requires a rotational reference signal to determine angular position of the required gear. The meshing vibration x(*t*) is determined by the signal average of one of the gears with T teeth, which contains the fundamental frequency and its harmonics associated with the tooth meshing frequency ($T*f_s$) as formulated below:

$$x(t) = \sum_{m=0}^{M} X_m \cos(2\pi m T f_s t + \emptyset_m) \quad (1)$$

where X_m is the amplitude at the m-th harmonic frequency, f_s is the rotational frequency of the gear and $\emptyset_m$ is the initial phase. When a local defect occur in a gear, the changes in vibration signal will be observed because of the stiffness between neighbouring teeth will be influenced by the fault [5, 6]. As a result the amplitude and phase modulation will occur in the meshing vibration $x(t)$ as [5-8]:

$$g(t) = \sum_{m=0}^{M} X_m\big(1 + a_m(t)\big)\cos(2\pi m f_s t + \emptyset_m + b_m(t)) \quad (2)$$

where

$$a_m(t) = \sum_{n=0}^{N} A_{mn} \cos(2\pi n f_s t + \alpha_{mn}) \quad (3)$$

$$b_m(t) = \sum_{n=0}^{N} B_{mn} \cos(2\pi n f_s t + \beta_{mn}) \quad (4)$$

here, a_m and b_m are the amplitude and phase modulation functions caused by the changes in vibration signals due to the defect. α_{mn} and β_{mn} are the phase angle of amplitude and phase modulation functions respectively. A_{mn} and B_{mn} are the amplitude of each harmonic in the amplitude and phase modulation function respectively. The index m indicates that the modulation function may differ within each harmonic m. The index n represents the harmonic index of the amplitude and phase modulation functions with respect to gear shaft rotation frequency.

2.2 The Short Time Fourier Transform (STFT)

The Short Time Fourier Transform is achieved by breaking down the signal into several overlapping usual Fourier Transform analysis. It is assumed that the underlying segmented signal will behave as a stationary signal so that the FFT assumption can be validated in this particular segment. STFT is formulated as [3, 9]

$$X(t,f) = \int_{-\infty}^{\infty} x(\tau) w(t-\tau) e^{-j\pi f \tau}\, d\tau \quad (5)$$

where $w(t\text{-}\tau)$ is the window function and τ is the variable that slides the window across the underlying waveform.

2.3 The Continuous Wavelet Transform (CWT)

The continuous wavelet transform $WT_x(t,s)$ of a signal $x(t)$ is defined as the convolution integral of $x(t)$ with a translated and dilated version of mother wavelet function $\psi(t)$,

$$WT_x(t,s) = \int_{-\infty}^{\infty} x(\tau)\sqrt{s}\,\psi\big(s(t-\tau)\big)d\tau \quad (6)$$

where s and t are the scale and time parameters that produce the respective dilation and translation of the mother wavelet function [10]. One of the important considerations in CWT is the proper choice of the mother wavelet function that is determined by the characteristics of the underlying signals. It is well known that the morlet wavelet is suitable for gearbox fault diagnosis [2, 4, 11-13] because it has a similiar shape to that of the impulse signal. There are several types of morlet wavelet used in the existing literatures however, and the morlet wavelet used in this paper is formulated as [14]:

$$\psi(x) = \frac{1}{\sqrt{\pi F b}} e^{2i\pi F_c x} e^{\frac{-x^2}{Fb}} \quad (7)$$

where F_b is the positive bandwith parameter and F_c is the central frequency of the morlet wavelet.

3 Gearbox experimental setup

Vibration measurements from two pairs of a bevel gear unit from a gearbox test rig were performed. One of the pairs was with a broken tooth and another was without it. The gearbox test-rig from the Machinery Fault Simulator-Spectra Quest located at Diponegoro University was used to conduct the experiment. A magnetic brake was incorporated as a load. Three accelerometers sensor were used to capture the vibration signals from the test rig, although only the axial vibration measurement will be analysed in this work. The picture of the gearbox test-rig and accelerometers attached on it are shown in Fig. 1.

Figure 1. The gearbox test rig.

In this work the tooth breakage was introduced and the results from the faulty gear were compared with the healthy gear. The figure of the broken gear that is used in this work is shown in Fig. 2.

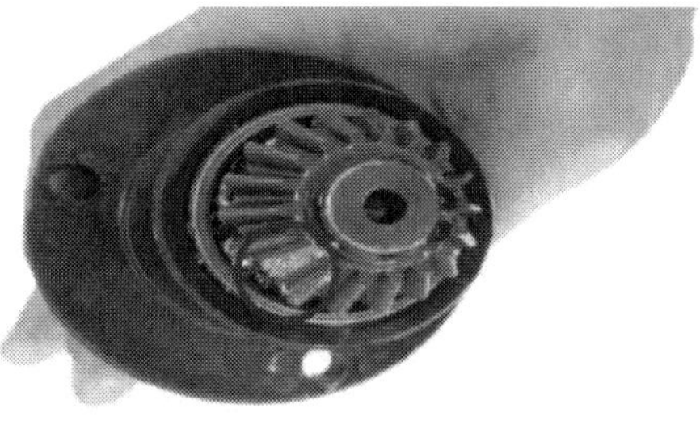

Figure 2. The gear with faulty tooth.

The sampling rate used to sample the vibration signals was 5.12 kHz. Eight seconds length data was gathered for each measurement performed in this experiment. Moreover, a tachometer sensor from Lattice Instruments was also used to provide the phase information from the gear. This information will be used for pre-processing the vibrations signal by TSA method to eliminate the asynchronous component from the faulty gear under consideration.

4 Results and discussions

4.1 Vibration measurements from the test rig

There were two sets of vibrations signal acquired from the gearbox test rig. The vibration signals (in terms of axial acceleration output from the accelerometer) from the healthy gearbox and the faulty gearbox are presented in Figs. 3 and 4, respectively.

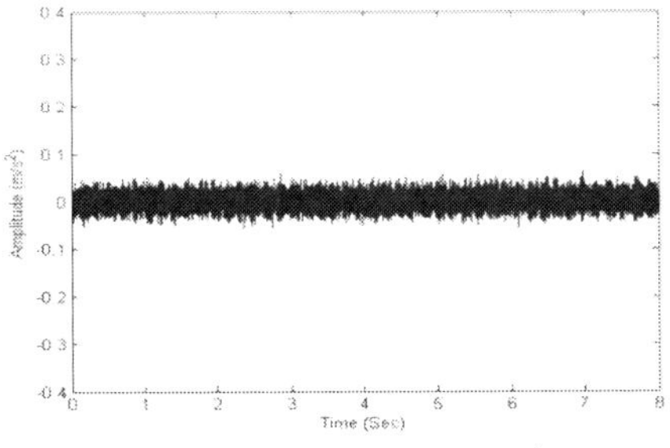

Figure 3. Vibration signal from a healthy gearbox.

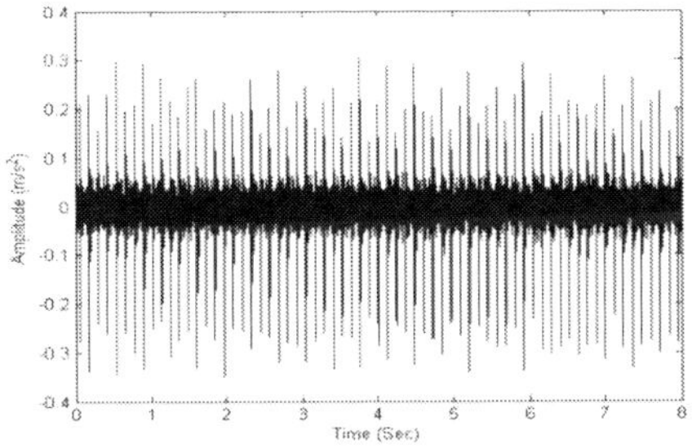

Figure 4. Vibration signal from a faulty gearbox.

As can be clearly seen from Figs. 3 and 4, there are a slight increase in the amplitude of the vibration signal from the broken gear. An impulse-like response (spikes) can occur when the faulty gear tooth makes a contact with the other gear [1]. However, it was difficult to identify the periodicity of this particular spike, which can correspond to the faulty gear tooth [15], solely by a direct visual inspection of the plot. Moreover this information is not enough to provide location the defect gears in terms of its angular location in the gear. Instead of the change in amplitude of the vibration signal, there is no further useful information can be gathered from these two plots.

4.2 Pre-Processing by TSA

Based on the obtained vibration, TSA can then be utilized to extract the periodic component that gets buried in the noise and eliminate all nonsynchronous vibration components. It represents an estimate of the meshing vibration over one revolution of a gear of interest. The TSA signal of a healthy gear is presented in Fig. 5, which clearly shows a regular amplitude pattern. In general, the TSA of a healthy gear is usually composed of gear meshing waveforms modulated by some low shaft order (first or second) functions, where the shaft order is an integer multiple of the rotational speed of the shaft.

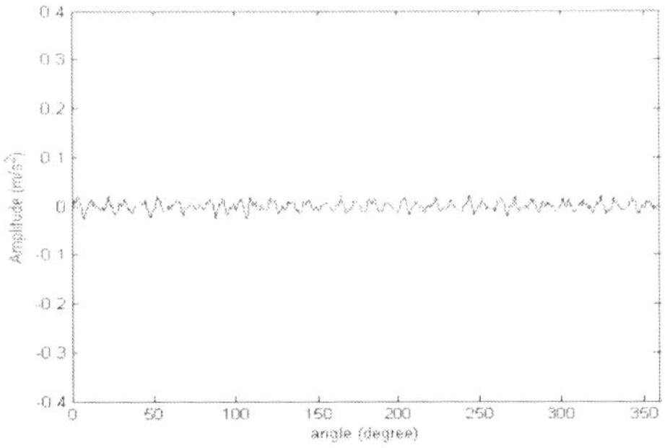

Figure 5. TSA vibration signal from a healthy gearbox.

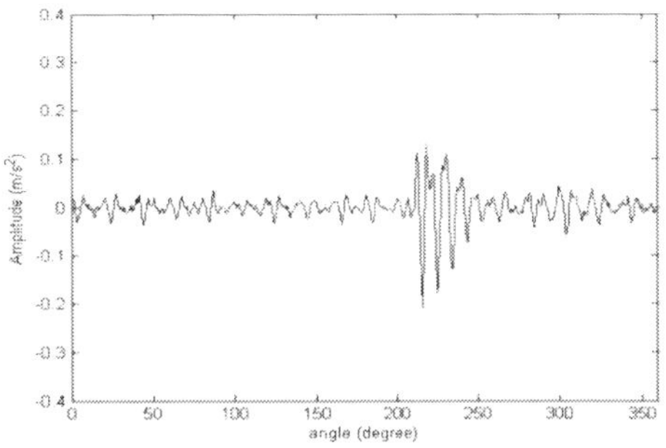

Figure 6. TSA vibration signal from a faulty gearbox.

On the other hand, the TSA of the faulty gear reveals a significant change in vibration amplitude at rotational angle of around 220^0, caused by a tooth breakage as shown in Fig. 6. This impulse-like signal was generated by the faulty tooth when it meshed with other gears, which was observable in the TSA signal. These results demonstrate the ability of TSA to offer a precise location of the faulty tooth, which is not observable in the raw vibration signal.

Another way of interpretation of the vibration signal is in frequency domain by the use of FFT. FFT plot offers diagnostic information of the health of gearbox, particularly indicated through the sidebands that appear around the gearmesh frequency and its harmonics. For healthy gears, the FFT plot of the TSA signal is only composed of the fundamental gearmesh frequency and the low shaft order peaks (1-10 orders). When the fault occurs, the FFT of TSA signal will produce higher order of sidebands. As previously mentioned in the introduction, when a fault occurs, the amplitude of the sidebands increases and its spacing from the corresponding gearmesh harmonics corresponds to the location of the fault. Figs. 7 and 8 showed the FFT plots of the healthy gear and broken gear respectively. It is obvious that when the fault occurs there are more high amplitude sidebands that appear around second, third and fourth gearmesh frequencies compared to the normal FFT of healthy gear. However, several spurious peaks can still be observed appearing in FFT plot. This is due to spectral leakage and smearing as inherent drawbacks of analyzing non-stationary signal with FFT. It is happened because when the fault occurs, the speed and load applied to the gearbox system are not constant anymore at that particular instant of time, so the stationary assumption that FFT relies on does not hold anymore.

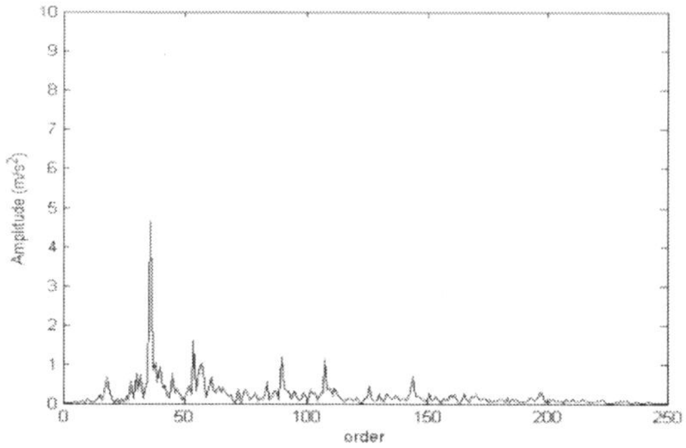

Figure 7. FFT of TSA vibration signal from a healthy gearbox.

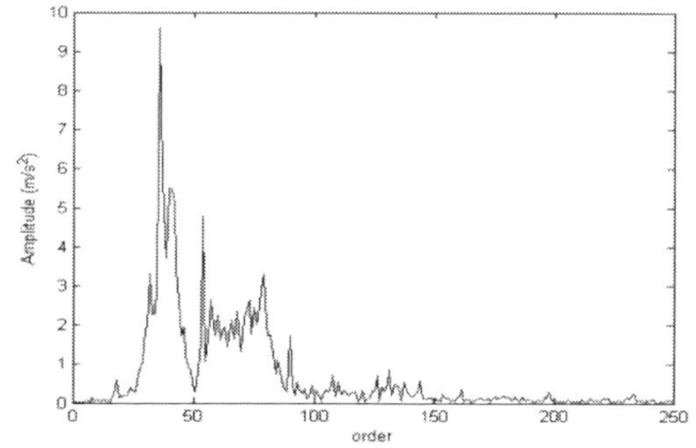

Figure 8. FFT of TSA axial vibration signal from a faulty gear.

4.3 Time-Frequency Method

From the drawback of the FFT, this work proposes a time-frequency technique to analyse the transient phenomena generated by the impulse signal in TSA. STFT plots of the healthy and broken gear are respectively shown in Fig. 9 and 10.

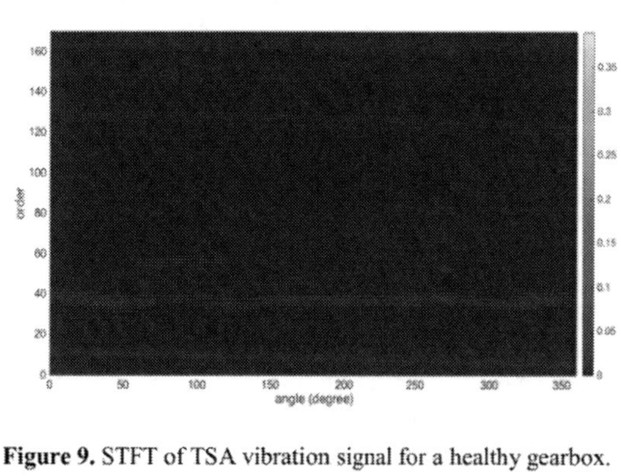

Figure 9. STFT of TSA vibration signal for a healthy gearbox.

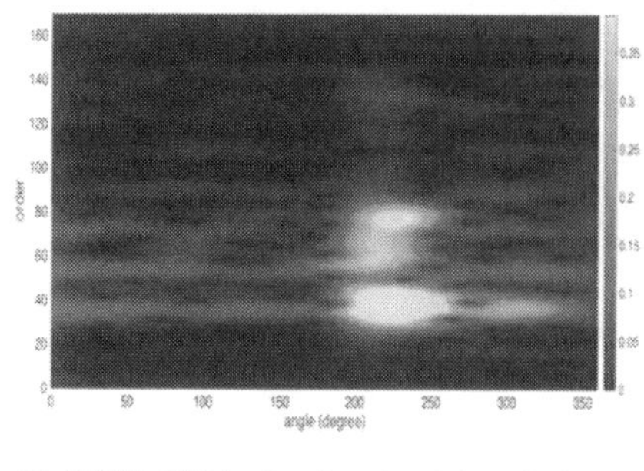

Figure 10. STFT of TSA vibration signal for a faulty gearbox.

Fig. 10 clearly indicated that STFT was able to capture the transient phenomena of the impulse signal when the faulty tooth was meshed with its neighbour. This is not possible to achieve using the usual Fourier Transform because the angle information is lost. Moreover, it also can distinguish between the healthy and broken gear TSA signals by showing the higher amplitude intensity observed from the STFT plot for the broken gear, compared to STFT plot for the healthy gear. The appearance of more sidebands and 4th harmonics (72nd order) of the gearmesh frequency that do not appear in the results of the healthy gearbox can be seen from Fig. 10. STFT plot, however still provides an absurd interpretation by showing a significant intensity at an angle larger than 250^0 which does not correspond to any fault in the gear.

To address the drawback of STFT, the continuous wavelet transform of the TSA signal was investigated. Figs. 11 and 12 show the CWT plot of TSA vibration signal from the healthy and faulty gearboxes, respectively.

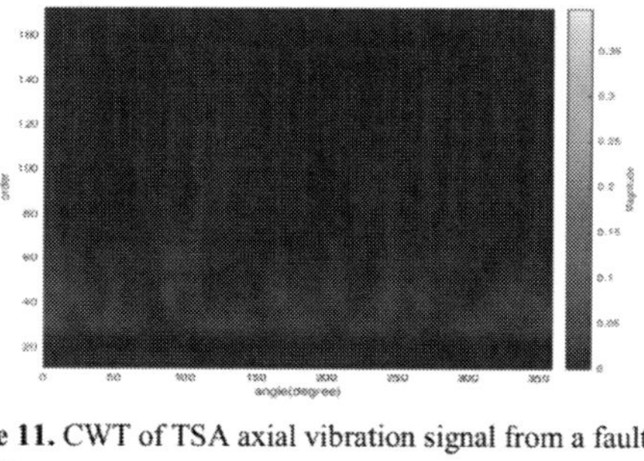

Figure 11. CWT of TSA axial vibration signal from a faulty gearbox.

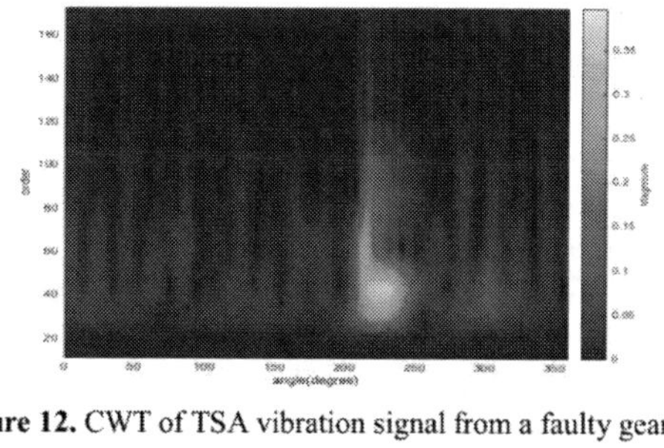

Figure 12. CWT of TSA vibration signal from a faulty gearbox.

It can clearly be seen that the contrast of the intensity between the vibration signals from the healthy and faulty gearboxes from Figs. 11 and 12. It is demonstrated that

CWT not only able to distinguish the normal gear and broken gear vibration signals but also offer a better time resolution compared to STFT. CWT can better indicate that the fault occurs at the angle between 200^0 and 250^0. It is thus shown that the joint angle-frequency representation is quite beneficial to identify the location of the gear fault, in terms of angular location. Such information cannot be obtained using the conventional spectral analysis like FFT, which can only offer the frequency content of the signal at the expense of the time information.

5. Conclusions

This study has demonstrated the usefulness of time-frequency signal processing of vibration signal for diagnosing the gearbox health based on experimental results from a bevel gearbox test rig. It has been shown that the TSA method had the ability to identify the angular location of the gear fault, which was not possible to obtain by using the raw vibration signal alone. FFT was used to provide a different representation of vibration signal in frequency domain. It was found that the information about angular location of gear fault was lost once the FFT analysis was performed. Moreover, FFT analysis was not a suitable method to analyse transient behaviour generated by broken tooth. In this work, STFT and CWT were utilized to provide a better interpretation of the impulse signal obtained from the TSA method in the joint time frequency domain. Although STFT of TSA signals were able to distinguish the difference between the normal and faulty gearboxes, it was still not good enough to offer an accurate angular location of the faulty gear tooth due to the need to choose the appropriate parameter for the window function. On the other hand, CWT was not only able to distinguish TSA signals from healthy and faulty gearboxes, but can also provide a better time resolution of the TSA vibration signal compared to STFT. The experimental results in this work have demonstrated that the combination of TSA and CWT is a reliable technique for enhancing gearbox fault diagnosis by offering an accurate angular location of the gear fault.

Acknowledgment

The authors gratefully acknowledge the support provided by Ningbo Science and Technology Bureau-International Cooperation Programme (Project: 2012D10029), China.

References

1. R. B. Randall, J. Mech. Des **104**, 259-267 (1982)
2. G. Dalpiaz, A. Rivola, R. Rubini, Mech. Sys. and Sig. Proc, **84**, 387-412 (2000)
3. W. J. Wang, P. D. McFadden, Mech. Sys and Sig. Proc, **7**, 193-203 (1993)
4. W. J. Staszewski, G. R. Tomlinson, Mech. Sys and Sig. Proc, **8**, 289-307 (1994)
5. P. D. McFadden, J. Vib. Aco. Strss. Real. **108**, 165-170 (1986)
6. P. D. McFadden, Mech. Sys and Sig. Proc, **2**, 403-409 (1988)
7. R. B. Randall, J. Mech, Des **104**, 259-267 (1982)
8. P. D. McFadden, J. D. Smith, Proocedings of the Institution of Mechanical Engineers **199**, 287-292 (1985)
9. L. Cohen, Proceedings of the IEEE **77**, 941-981 (1989)
10. W. J. Wang, P.D. McFadden, J. Snd. Vib **192**, 927-939 (1996)
11. D. Bouhlabal, M. F. Golnaraghi, F. Ismail, Mech. Sys and Sig. Proc, **13**, 423-436 (1999)
12. W. Q. Wang, F. Ismail, M. F. Golnaraghi, Mech. Sys. Sig. Proc, **15**, 905-922 (2001)
13. H. Zheng, Z. Li, X. Chen, Mech. Sys. Sig. Proc, **16**, 447-457 (2002)
14. A. Teolis, *Computational Signal Processing with Wavelets*. 65 (1998)
15. C. K. Sung, H. M. Tai, C. W. Chen, Mech. Mach. Thry **35**, 1169-1182 (2000)

Vibration-based Fault Diagnostic of a Spur Gearbox

Dennis Hartono[1], Dunant Halim[1], Gethin Wyn Roberts[2] and Qianyu Liu[1]

[1]Department of Mechanical, Materials and Manufacturing Engineering The University of Nottingham Ningbo China, China 315100
[2]Department of Civil Engineering, The University of Nottingham Ningbo China, China 315100

Abstract. This paper presents comparative studies of Fast Fourier Transform (FFT), Short Time Fourier Transform (STFT) and Continuous Wavelet Transform (CWT) as several advanced time-frequency analysis methods for diagnosing an early stage of spur gear tooth failure. An incipient fault of a chipped tooth was investigated in this work using vibration measurements from a spur gearbox test rig. Time Synchronous Averaging was implemented for the analysis to enhance the clarity of fault feature from the gear of interest. Based on the experimental results and analysis, it was shown that FFT method could identify the location of the faulty gear with sufficient accuracy. On the other hand, Short Time Fourier Transform method could not provide the angular location information of the faulty gear. It was found that the Continuous Wavelet Transform method offered the best representation of angle-frequency representation. It was not only able to distinguish the difference between the normal and faulty gearboxes from the joint angle-frequency results but could also provide an accurate angular location of the faulty gear tooth in the gearbox.

1 Introduction

Vibration signals generated from a gearbox contain abundant information about the operating condition and health of the gearbox. However, the underlying vibration signal is generally very complex and specialized signal treatments using certain signal processing methods will be needed. It is well known that impulse-like vibration signal is generated when a fault occurs in a gearbox. Impulse is a short duration event that excites vibration over a wide range of frequencies. It can be considered as a transient and a non-stationary signal. However, most signal processing methods used to analyze vibration signal are holding on the assumption that the signal is stationary, For example, Cepstral analysis, FFT, and the amplitude or frequency demodulation have been applied for gearbox fault diagnosis [1, 2]. Thus, more sophisticated signal processing methods such as beyond the classical Fourier Transform approach are needed to achieve reliable and accurate gearbox diagnosis. A relatively small number of advanced time-frequency methods have been used to analyse non-stationary signals such as The Short Time Fourier Transform, Continuous Wavelet Transform and Wigner Ville Distribution, although they have not been implemented as rigorously as stationary signal processing methods in the condition monitoring industry [3-5].

In this work, the Time Synchronous Averaging (TSA) technique was initially implemented to extract the fault feature from the vibrations signals that are heavily corrupted by measurement noise. TSA has the capability of presenting the vibration information in terms of angular rotation of gear, enabling the comparison between the healthy and faulty gears. FFT was then implemented to the TSA signal to extract the sidebands around gearmesh frequencies that indicate the fault location [6]. However, FFT only could provide the frequency content of the TSA signals and the angular information of the faulty gear is lost. To overcome this drawback, the use of STFT and CWT were proposed in this work to further analyze the TSA signal for accurate diagnostic purposes. This analysis can reveal both the frequency content of the TSA signals and also indicate how the frequency content evolves with respect to the angular location of the gear. Such information can also be used to determine the gear fault location in the angular domain.

This paper is organized as follows. Section 2 provides a brief description about the background theory of TSA, STFT and CWT methods, while Section 3 presents the experimental set-up using a spur gearbox test rig developed at the University of Nottingham Ningbo china. Section 4 discusses the experimental results based on the vibration measurement from the gearbox, and finally Section 5 offers the main conclusion of this work.

2 Theoretical background

2.1 Time synchronous averaging

Time Synchronous Averaging (TSA) is commonly used to extract the vibration of a single gear from the overall vibration of the gearbox system, which is quite useful for diagnostic purposes. In this situation, the gears are supposed to be operated at constant load and speed, and all the teeth of the gears can be considered to be identical and equally spaced. The meshing vibration $x(t)$ is thus determined by the signal average over one of the gears with T teeth and rotates at f_s Hz as formulated as:

$$x(t) = \sum_{m=0}^{M} X_m \cos(2\pi m T f_s t + \emptyset_m) \quad (1)$$

where X_m is the amplitude of the meshing vibration, f_s is the rotational frequency of the gear, $\emptyset_m$ is the inital phase. If one of the teeth has a local defect such as a chipped tooth, it will cause amplitude and phase modulations. The resulting modulated gear meshing vibration is formulated as [7]:

$$g(t) = \sum_{m=0}^{M} X_m\big(1 + a_m(t)\big)\cos(2\pi m f_s t + \emptyset_m + b_m(t)) \quad (2)$$

where the amplitude modulation function and the frequency modulation function, $a_m(t)$ and $b_m(t)$, can be formulated as:

$$a_m(t) = \sum_{n=0}^{N} A_{mn}\cos(2\pi n f_s t + \alpha_{mn}) \quad (3)$$

$$b_m(t) = \sum_{n=0}^{N} B_{mn}\cos(2\pi n f_s t + \beta_{mn}) \quad (4)$$

Here, a_m and b_m are the amplitude and phase modulation caused by the changes in vibration signals due to the defect. α_{mn} and β_{mn} are the phase angle of amplitude and phase modulation functions respectively. A_{mn} and B_{mn} are the amplitude of each harmonic in the amplitude and phase modulation function respectively. The index m indicates that the modulation function may differ for each harmonics m. The index n represents the harmonic index of the amplitude and phase modulation functions with respect to the gear shaft rotation frequency (f_s).

2.2 Short time fourier transform

The basic idea of STFT is slicing the signal into different segments (with possible overlaps). Assuming that within each slice/segment the underlying signal will behave as a stationary signal, then each slice can be subjected to Fourier Transform analysis. It can be formulated as [8]:

$$X(t,f) = \int_{-\infty}^{\infty} x(\tau)w(t-\tau)e^{-j\pi f\tau}\,d\tau \quad (5)$$

where $w(t\text{-}\tau)$ is the window function used and τ is the variable that slides the window across the underlying waveform.

2.3 Continuous wavelet transform

Continuous Wavelet Transform is another type of linear transformation of a signal based on decomposition of the signal in terms of a basis of elementary functions. It is formulated as [9, 10]:

$$W_g(a;b) = \frac{1}{\sqrt{a}}\int_{-\infty}^{\infty} x(t)g^*(\tfrac{t-b}{a})\,dt \quad (6)$$

where $g(t)$ is the analysing wavelet that is translated and dilated by parameter b and a respectively

$$g_{a,b}(t) = \frac{1}{\sqrt{a}}h^*(\tfrac{t-b}{a}) \quad (7)$$

There are a number of different real and complex valued functions for mother wavelet. Morlet wavelet has been proven to be the most suitable for gearbox fault diagnosis [11]. It can be formulated as [12]:

$$\psi(x) = \frac{1}{\sqrt{\pi Fb}}e^{2i\pi F_c x}e^{\frac{-x^2}{Fb}} \quad (8)$$

where F_b is the bandwidth parameter and F_c is the centre of the morlet wavelet.

3 Gearbox experimental setup

A spur gearbox test rig was used for this work as presented in Fig. 1. During the experiment, a 1 HP DC motor was used as the driver of the system. The rotational speed of the input was set at 500 RPM, corresponding to the frequency of 8.37 Hz. Two torque and speed sensors located at the input and output of the gearbox, were incorporated to ensure the constant operational speed during the experiment. There was no load applied to the gearbox for this experiment although additional load could be provided using a magnetic brake attached to the test rig. The whole experiment was conducted on the Nexus Optical Table to isolate vibration from external environment.

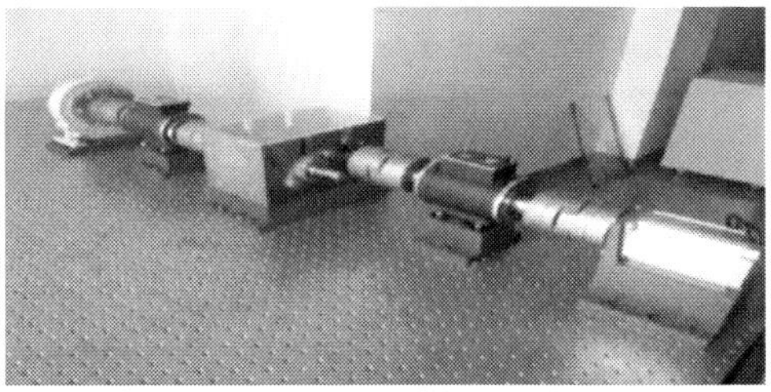

Figure 1. The gearbox test rig used for experiments.

The experimental data were obtained from a spur gearbox comprising a pinion gear of 18 teeth in mesh with another gear of 27 teeth, driven by the DC motor. The faulty tooth considered in this work was a chipped tooth as shown in Fig. 2, where it can be seen that only a slight part of the gear has been taken out. This was done in purpose to test whether the proposed signal processing algorithm was sensitive enough to detect such a small incipient failure in gearbox.

Figure 2. The gear with faulty tooth.

In this work, three accelerometers were placed on a bearing housing as depicted in Fig. 3. However, only vibration measurement in the radial direction was analyzed further in this work. NI CompactRio was used for data acquisition with 9234 module and 5.120 kHz sampling rate was used to obtain 200000 data samples from vibration measurement. In addition, a tachometer sensor was also used to process the raw vibration data using Time Synchronous Averaging.

Figure 3. Accelerometers location on the bearing housing.

4 Results and discussion

Vibration signal, in terms of radial acceleration, from a healthy and faulty gearboxes are presented in Figs. 4 and 5. As observed from the figures, it is difficult to differentiate vibration signals obtained from the healthy or faulty gearboxes based solely on evaluating the raw vibration signals. The only thing that can be observed is that there is a slight increase of the vibration amplitude of the faulty gearbox compared to that of the healthy gearbox. This can be explained by the fact that at the early stage of the gear fault, relatively small impulse responses are generated. This makes the amplitude of vibration signal from the faulty gearbox almost indistinguishable compared to that of the healthy gearbox.

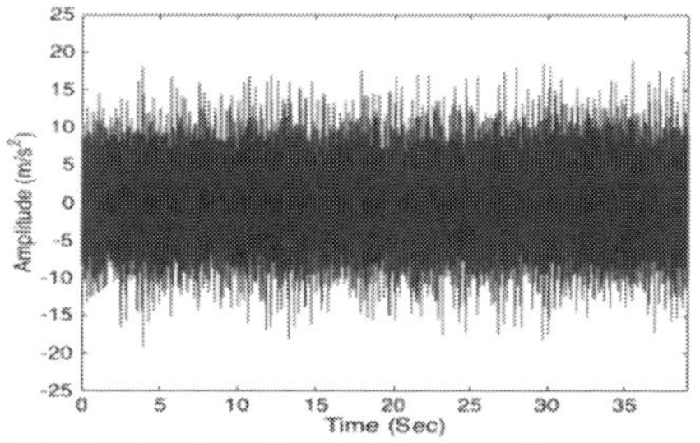

Figure 4. Vibration signal from a healthy gearbox

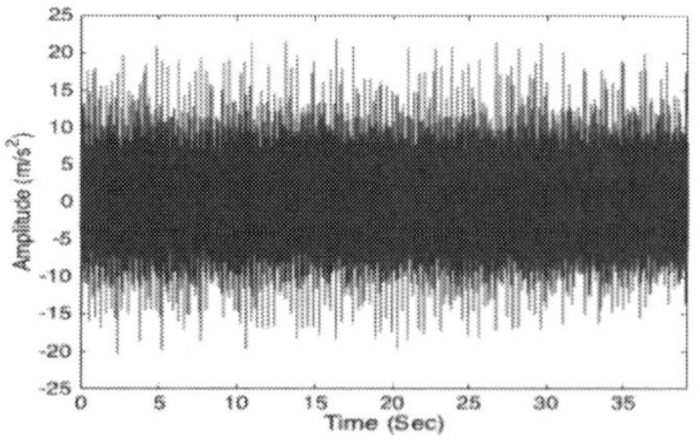

Figure 5. Vibration signal from a faulty gearbox

From the results, a large number of spikes were also observed in the vibration signals from both healthy and faulty gearboxes. The spikes that came from the faulty gear could be used to identify which gear contained the fault by identifying the periodicity patterns of signal [7]. However, it was found that it was not possible to detect the periodicity from the time domain plot itself. Not only that the background noise that made it difficult to observe the signal periodicity, but also the the nature of the gearbox vibration signal itself that was rather complex due to vibration contributed by various components, such as gearbox housing. Therefore, an advanced pre-processing tool was required to filter out the signal from the asynchronous signal components and background noise. In this work, TSA was proposed as a pre-processing tool to eliminate the asynchronous signal component and enhance the signal to noise ratio of the vibration signals.

TSA vibration signal from a healthy gearbox is presented in Fig. 6. From the results, it is clear how the background noise in vibration signal has been diminished significantly, compared with the raw vibration signals shown in Fig. 4. It was shown that TSA was not only able to reduce the background noise that corrupted the vibration measurement, but it could also eliminate asynchronous signal components captured by the accelerometer, so that it could identify the fault feature more accurately. It was also observed that the amplitude of the TSA signal showed a regular pattern that could be analysed further.

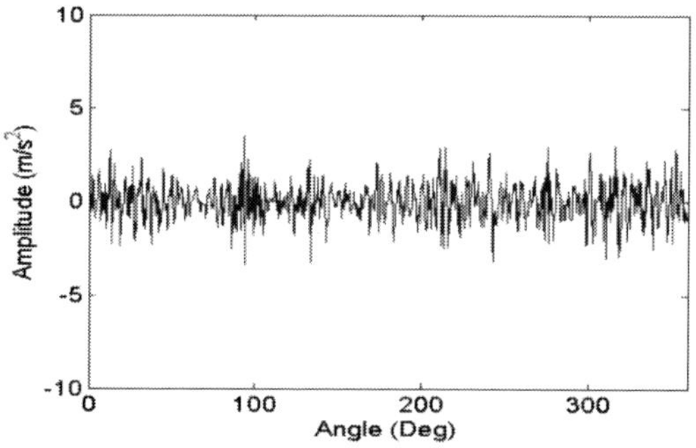

Figure 6. TSA vibration signal from a healthy gearbox

Moreover, based on the performed analysis, it was found that the TSA processing was able to extract the weak impulse signal generated from a faulty gearbox that was not possible to observe from the raw vibration signal, as shown Fig. 7. The results indicate that there is a relatively large spike observed in the signal from a faulty gearbox, which was located at a particular gear angle of around 270^0 as observed in Fig. 7. This spike was associated with the impulse response generated from the faulty gear located at angle of 270^0.

The results demonstrated that TSA was able to identify the angular location of the faulty gear tooth with sufficient accuracy. This was done by determining the angular location of the impulse response generated by the faulty gear. However, the method could not be used to identify which particular gear had the faulty tooth In this case, spectral analysis method could be used. FFT of TSA signal from a healthy gearbox is presented in Fig. 6. As expected from a normal gearbox, several peaks were observed at the gearmesh frequency and its harmonics with low amplitude sidebands around them. These sidebands were the fault indicator that would be compared with the FFT plot from the faulty gearbox.

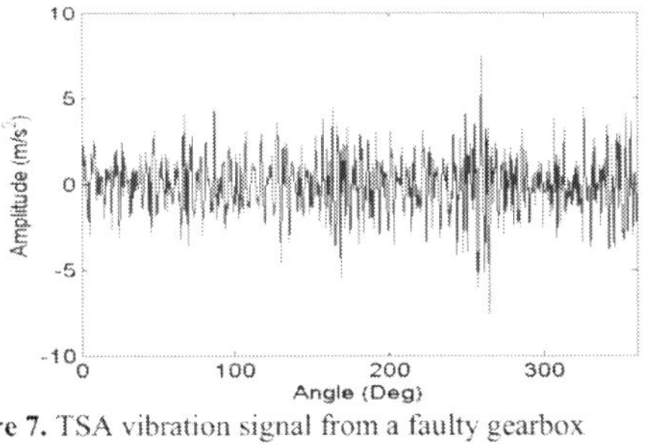

Figure 7. TSA vibration signal from a faulty gearbox

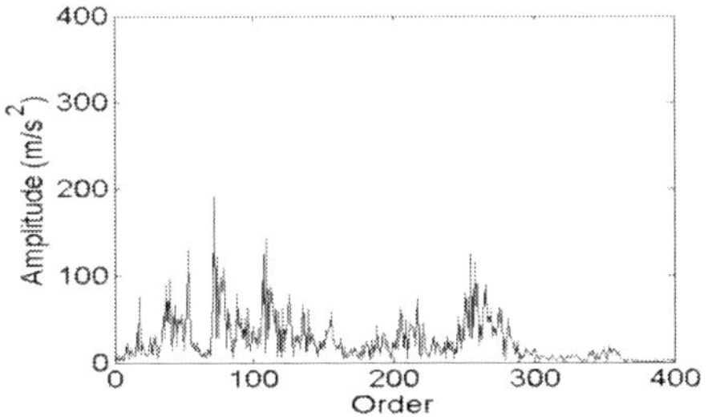

Figure 8. FFT vibration signal from healthy gearbox

The FFT of the faulty gearbox is presented in Fig. 8. Compared to the FFT of the healthy gearbox, an increase of amplitude in several gearmesh frequencies and sidebands around them could be observed. Sidebands are considered as a useful indicator to provide information about the location of the faulty tooth from its spacing from the gearmesh frequency [6]. However, FFT is not suitable to analyze the transient signal such as the impulse vibration signal generated from a faulty gearbox, due to its stationary assumption of the underlying signal. From the experimental results, it was quite difficult to see the sidebands spacing because they were corrupted by the background noise. In addition, there is an inherent drawback of the FFT due to the spectral leakage that will smear the spectral peaks. Furthermore FFT is only able to present the information in frequency domain and cannot provide the time domain useful for fault diagnostic purposes.

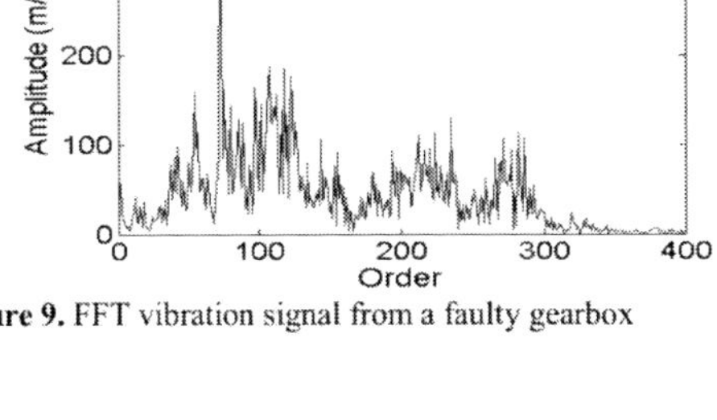

Figure 9. FFT vibration signal from a faulty gearbox

Thus, another time-frequency signal processing method was investigated in this work. STFT of TSA signal from healthy and faulty gearboxes are presented in Figs. 6 and 7. From the results, it was clearly observed the contrast of intensity between TSA signals from the healthy and faulty gearboxes. However, the results were not specific enough to provide the angular location of the faulty tooth accurately. As observed in Fig. 10, STFT provided a fault indication by indicating strong impulse signals around 160^0 and 340^0, whereas the actual faulty tooth only occurred at approximately 270^0. This result can be explained by the fact that STFT requires an appropriate selection of parameter for its window function. Once it is chosen, it will be used in the entire analysis since STFT has a constant time-frequency resolution in the entire time-frequency plot. Therefore it is not suitable for analyzing gearbox vibration signals that contain multi-scale components [13].

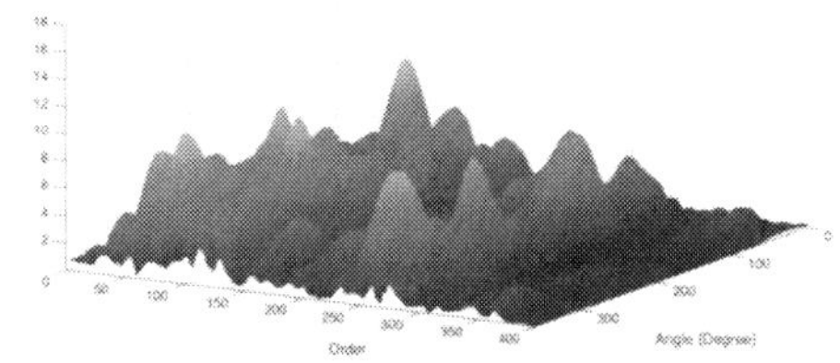

Figure 10. STFT of TSA vibration signal from a healthy gearbox

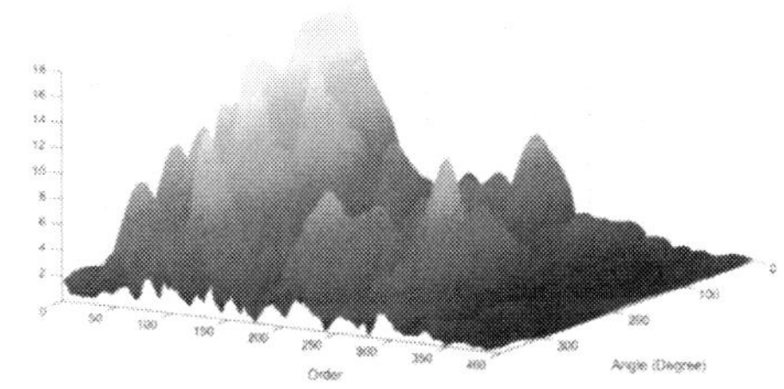

Figure 11. STFT of TSA vibration signal from a faulty gearbox

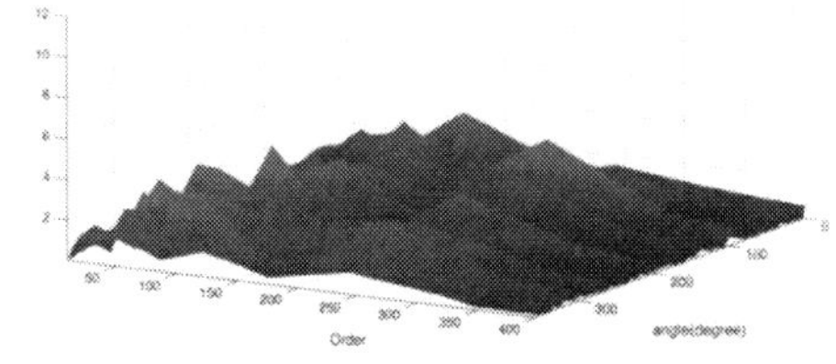

Figure 12. CWT of TSA vibration signal from a healthy gearbox

From the drawback of STFT method, an alternative CWT method was later implemented in this work. The plot of the CWT from healthy and faulty TSA signals from the gearbox are presented in Figs. 12 and 13. Based on the results, it was demonstrated that CWT was not only able to distinguish the TSA vibration signal from

 DOI: 10.1051/matecconf/20167002004

healthy and faulty gearboxes, but also provided an accurate angular location where the fault occured compared to STFT. CWT was able to pinpoint the strong frequency component occurred around 270^0, as also correctly identified from the TSA results in Fig. 7.

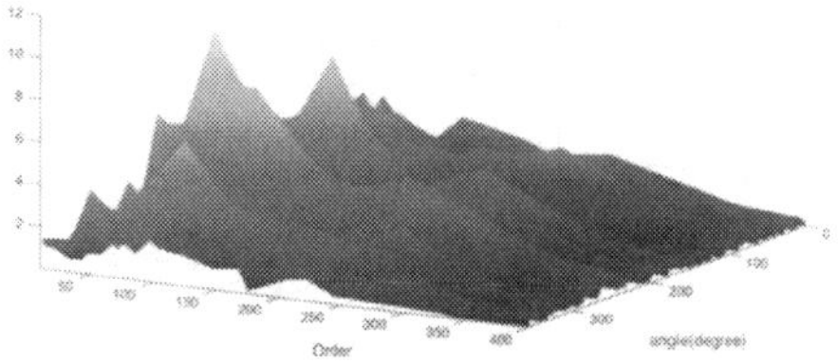

Figure 13. CWT of TSA vibration signal from a faulty gearbox

5 Conclusions

Vibration-based fault diagnostic of a gearbox has been presented in this work using experimental measurements from a spur gearbox test rig. The work compared FFT, STFT and CWT as several advanced signal processing methods for analyzing the non-stationary signal associated with an incipient gear fault in the form of a chipped tooth. TSA was initially implemented to enhance the meshing vibration of the gear of interest. It was shown in this work that FFT was able to show the sidebands around the gearmesh frequencies. However, the background noise could cover make it difficult to observe the frequency contents of vibration signal accurate, potentially leading to inaccurate diagnostic. STFT did not perform well in indicating the location of the faulty gear although it was still able to distinguish between the normal and faulty gearbox from its time-frequency representation. However, STFT could be used as a fault indicator when there was no fault in the gear. Finally, CWT combined with the TSA could provide satisfactory results by not only be able to distinguish the fault pattern from faulty gearbox compared to a healthy gearbox but could also provide a reasonably accurately angular location of the faulty tooth at around 270^0.

Acknowledgment

The authors gratefully acknowledge the support provided by Ningbo Science and Technology Bureau-International Cooperation Programme (Project: 2012D10029), China.

References

1. J. Bao, B. Tang, J. Zhao, Msrmnt **83**, 57-71 (2016)
2. B. Liang, S. D. Iwnicki, Y. Zhao, Mech. Sys. Sig. Proc **39**, 342-360 (2013)
3. G. He, K. Ding, W. Li. X. Jiao, Ren. Engy **87**, 364-375 (2016)
4. Y. Qin, B. Tang, Y. Mao, Sig. Proc **120**, 480-494 (2016)
5. Y, Yang, X. J. Dong, Z. K. Peng, W. M. Zhang, G. Meng, J. Snd. Vib **335**, 350-366 (2015)
6. G. Dalpiaz, A. Rivola, R. Rubini, Mech. Sys. Sig. Proc **14**, 387-412 (2000)
7. G. He, K. Ding, H. Lin, Mech. Sys. Sig. Proc **66-67**, 597-611 (2016)
8. N. H. Chandra, A. S. Sekhar, Mech. Sys. Sig. Proc **72-73**, 105-133 (2016)
9. J. Chen, Z. Li, J. Pan, G. Chen, Y. Zi, J. Yuan, B. Chen, Z. He, Mech. Sys and Sig. Proc **70-71**, 1-35 (2016)
10. W. He, Q. Miao, M. Azarian, M. Pecht, Mech. Sys. Sig. Proc **64-65**, 149-161 (2015)
11. W. Liu, B. Tang, Exp. Sys. Appl **38**, 7575-7581 (2011)
12. P. He, P. Li, H. Sun, Proc. Eng **15**, 464-468 (2011)
13. L. Jedlinski, J. Jonak, Appl. Sft. Cmpt **30**, 636-641 (2015)

Dynamic Characteristics Analysis and Structure Optimization Study of Glaze Spraying Manipulator

Hao Feng, Yong-jun Zhan and Xiao-ping Wang

School of Mechanical and Electronic Engineering, Jingdezhen Ceramic Institute, Jiang xi, Jingdezhen, 333403, China

Abstract. A new type of glaze spray manipulator is studied with a three-dimensional entity modal. The dynamic characteristics of its structure have been analysis by finite element method. The first ten other modes of vibration and their natural frequencies were obtained. It is found out that the frequency peak appears at 10Hz and the max deformation occurs in sixth order vibration mode by the observation of the end of spray gun on the manipulator. Furthermore, an improvement structure is provided to reinforce the stiffness and the thickness of rotation support at the manipulator's wrist (the end joint). By test verification, it is proved that the max deformation is reduced after structure improvements.

1 Introduction

The technology of spraying manipulator was developed in the early 1990s and widely adopted in the auto industry [1] etc. There are many advantages such as reducing labour intensity, making the spraying process constant, improving spraying efficiency and so on. It was quite obvious in the production with a little change of the size. For many varieties and small quantity production, a new type manipulator was designed whose structure is a suspension type with a movable base [2]. For eliminating the bob-weight, a linkage mechanism was applied to the movement of the third arm [3]. As a result, the nature frequency is reduced with the weight saving. The precision should be decreased in high-speed status too. So it is necessary to analysis its dynamic characteristics and study the optimum its structure for improving the accuracy of its end movement.

2 Pre-processing Finite Element Model (FEM)

2.1Three-dimensional model simplification

The glaze spraying manipulator is composed of base, pillar, roof spport rotation mechanism, arm rotation mechanism, wrist rotation mechanism, spray gun etc.The structure of the study is huge and complex, thus it is necessary to simplify the model.According to Sain Vaint Principle, the model is simplified by deletingfillets, chamfers, undercuts, threaded holes and other small structures while improving the uniformity of mesh distribution[4]. Then the calculation efficiency and accuracy will utmost improved. The three-dimensional model is shown in Fig. 1.

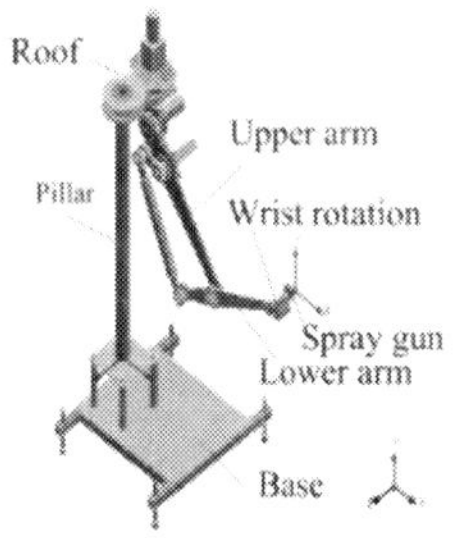

Figure 1. The three-dimensional model and coordinate system.

A movement interference check of moving parts for the model is processed by Solidworks Motion [5] based on the simplified model.The structure space layout interference collision and motion mechanism collision will be found by motion interference checking. If the interference occurred, the exact location will be found out and displayed in the window which is shown in Figure2. Then, the interference parts will be modified accordingly.

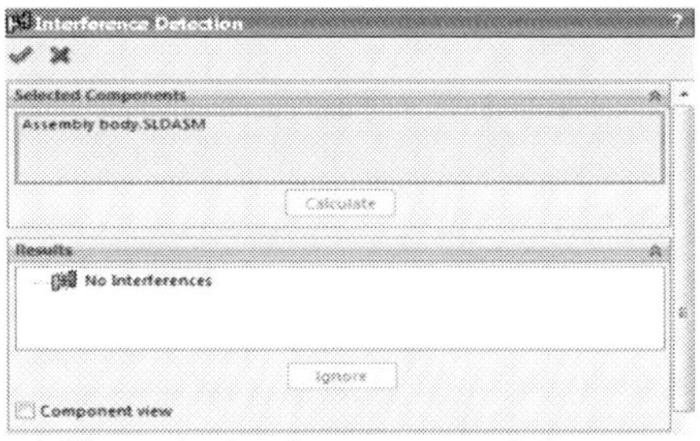

Figure 2. The motion interference prompt window.

The operating procedure is that selecting the "IntelliMotion Builder" button in drop-down menu and setting motion parameters of each joint in Motion plugin. Click the "Interferences" and " Calculate" button in turn, so that the motion interference checkwill be carried out. In this paper, there is no motion interference of the model after analysis.Therefore, it provides a reasonable and effective model for the subsequent finite element analysis.

Import the model into ANSYS Workbench and create a global coordinate system where the circle center is at the center of uprights bottom [6]. The global coordinate systems is shown in Fig. 1. The X axis is parallel the upper and lower arm plane, the Y axis is parallel the pillar axis and the Z axis is vertical the upper and lower arm. Building a local coordinate system with the same directions of global coordinate at the end of spraying gun and let the circle center at the center of the plane.

2.2 Material definition and constraints addition

In order to reduce the rotational inertia of manipulator, an operation that cutting weight from upper arm to the end of spray gun has launched. Therefore, the upper and lower arm of the manipulator are replaced by aluminum alloy (ρ = 2.77×10-6kg/mm3, E= 7.1×1010N/mm2, v= 0.33) and other parts are steel. According to the actual situation, the constraints of bond are employed to connect rotary joint and the basis with fix support restraints that make it cannot move and rotate in any direction.In harmonic response analysis, 5N harmonic force is applied at the end of spray gun in the direction of vertical the plane.

2.3 Mesh designing and quality controlling

Table 1. Mesh quality evaluation parameters.

NO.	Evaluation Parameters	Evaluation Standard and Value Range	Averages Value
1	element quality	ranges from 0 to 1, Optimum value is 1	0.8032
2	aspect ratio	Optimum value is 1, Warning value is 20	1.9973
3	Jacobian Ratio	Optimum value is 1, Warning value is 40	1.1799
4	warping factor	Optimum value is 0, Shell unit limit is 1, Three-dimensional unit limit is 7	3.6752
5	parallel deviation	Optimum value is 0, Warning value is 70°	8.6005
6	maximum corner angle	Optimum value for a triangle is 60°, Optimum value for Quadrilateral is 90°, Warning value is 155°	100.6°
7	skewness	ranges from 0 to 1, Optimum value is 0	0.2468
8	Orthogonal	ranges from 0 to 1, Optimum value is 1	0.8595

Meshing is a major preprocessing step in finite element analysis. The mesh quality will influence the accuracy of calculation result directly and high-quality mesh will make the result more accurate [7]. The method combined with global and local meshing is used to achieve a high quality mesh, thus the model is divided into 352,962 nodes and 66,388 units.The mesh quality evaluation parameter of ANSYS Workbench is shown in Table 1.

The element quality and skewness are the most important parameters to evaluate the mesh quality. Form table 1, the element quality value is 0.8032 and skewness is 0.2468. To the best of our knowledge, the mesh quality is excellent when the skewness value less than 0.25. Therefore, the mesh quality achieves a good foundation for the model calculation.

3. Modal and harmonic response analysis

A modal analysis is done for the manipulator to obtain the first ten orders natural frequency and vibration mode, which create an optimized condition for gaining a better dynamic characteristic[8]-[9]. A steady-state response is obtained by harmonic response analysis of a certain linear structure that under a harmonic load and then the frequency amplitude response curve is acquired [10].The dynamic characteristic of manipulator is calculated out after modal and harmonic response analysis, and then provided a basis for further research.

3.1 Theoretical foundations of modal and harmonic response analysis

According to the classical mechanics theory, the general kinetic equation is shown in formula (1):

$$M\{x''\} + C\{x'\} + K\{x\} = \{f(t)\} \qquad (1)$$

where M is mass matrix, C is damping matrix, K is stiffness matrix, $\{x\}$ is displacement vector, $\{f(t)\}$ is force vector, $\{x'\}$ is velocity vector and $\{x''\}$ is acceleration vector.

The classical eigenvalue problem is solved with undamped modal analysis and the correspond kinetic equation is shown in formula (2):

$$M\{x''\} + K\{x'\} = \{0\} \qquad (2)$$

The free vibration mode of the structure is harmonic, so the displacement is a sine function

$$x = A\sin(\omega t) \qquad (3)$$

Combined formula (3) and formula (2) that the result is shown in formula (4):

$$(K - \omega^2 M) = \{0\} \qquad (4)$$

A classic eigenvalue problem is represented in formula (4), where ω_i^2 is eigenvalues and ω_i is natural circular frequency. In harmonic response analysis, the general kinetic equation of the manipulator is shown in formula (1). However, the right side of the equation is $F(t)=F_0\cos(\omega t)$.

3.2 Simulation results analysis

According the vibration theory, the lower order modes have a huge impact on vibration, thus the first ten orders modal is calculated [11].The natural frequency and the vibration trend of first ten orders are shown in Table 2 and the parts of modal vibration modes are shown in Fig. 3.

Form Table 2, the deformation mainly concentrated in the parts from upper arm to the end of spray gun and the natural frequency varied from 3.014Hz to 35.483Hz.According to the vibration trend, the vibration part is limited to the local position with the increase of modal. Fig. 3 shows that the maximum deformation position of first nine orders modal is at the end of the spray gun and the tenth order is at wrist. The maximum amplitude in sixth modal is 29.351mm while the value of other modals decreased more than10mm.According to above-mentioned analysis, the deformation of the rotating support is the main reason for the manipulator shape.

Table 2. The natural frequency and vibration mode.

NO.	Frequency (Hz)	Modal description
1	3.014	The whole swing back and forth in Y direction
2	3.744	Whole panning in X direction
3	3.8673	Upper and lower arm swing in X direction
4	5.0374	upper and lower arm swing in X direction and torsion
5	9.9721	Wrist and the spray gun bobbing in X direction
6	11.431	Upper and lower arm swing and torsion
7	12.752	Whole bobbing around the Z axis and The upper and lower arm swing in X direction
8	25.905	Whole torsion around the Y axis and The upper and lower arm swing
9	29.917	Whole torsion around the Y axis and lower arm and wrist swing
10	35.483	Upper and lower arm torsion in the X direction

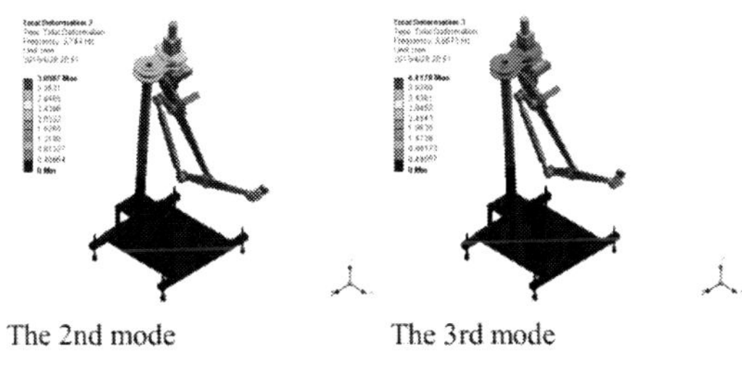

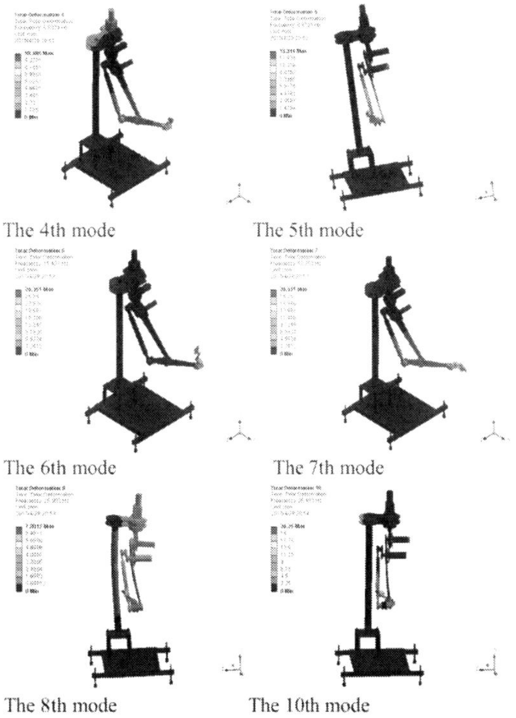

Figure 3. Parts of the modal shapes.

In harmonic response analysis,the constraint is as same as modal analysis and the frequency range is set to 0-34Hz.Therefore, the result of harmonic response analysis is shown in Fig. 4:

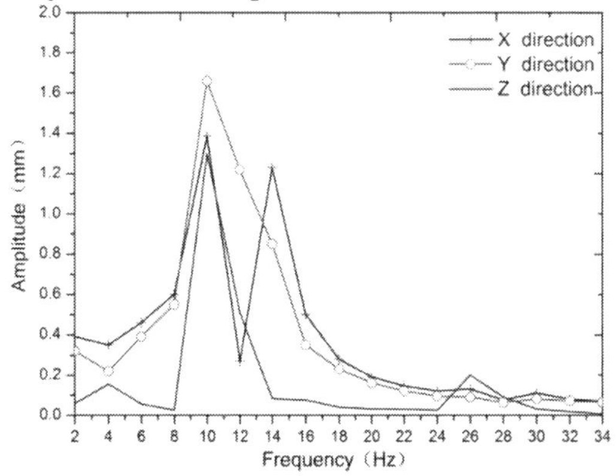

Figure 4. The frequency amplitude response curve.

Fig. 4 shows that the maximum amplitude is generated in 10Hz in X,Y,Z directions and the corresponding value is 1.3861mm, 1.6567mm and 1.3056mm. Moreover, the second peak appears in14Hz in X direction.However, all the peak frequencies are too low and the amplitudes are too large, thus it will affect the manipulator stability and generate an additional dynamic load. Besides, the maximum amplitude frequency is close to the frequency of fifth and sixth order modal thus it will cause the resonance of themanipulator. Thus the structure of manipulator needs redesigned to raise the resonant frequency and stability.

4. Structure optimization

A structure improvement method is proposed in this section.According to the previous analysis,the deformation of rotating support is the main factor which lead to poor dynamic performance of the manipulator. we increase the thickness of rotating support from 2 mm to 4 mm and add two ribs to achieve a more stable performance.

Reconstructing and calculating the model, the results of modified and non-modified model are shown in Table 3 and the harmonic response analysis result is shown in Fig. 5.

Table 3. The natural frequency and maximum deformation before and after optimization.

NO.	Non-modified Model		Modified Model	
	Frequency (Hz)	Maximum deformation (mm)	Frequency (Hz)	Maximum deformation (mm)
1	3.014	3.6597	3.927	17.589
2	3.744	4.4178	4.6008	5.2506
3	3.8673	10.485	5.8411	5.0431
4	5.0374	13.314	6.386	16.7
5	9.9721	29.351	10.115	13.573
6	11.431	20.531	14.484	22.415
7	12.752	7.2012	31.683	15.694
8	25.905	10.494	35.672	13.45
9	29.917	20.25	39.743	13.603
10	35.483	10.68	47.524	5.5207

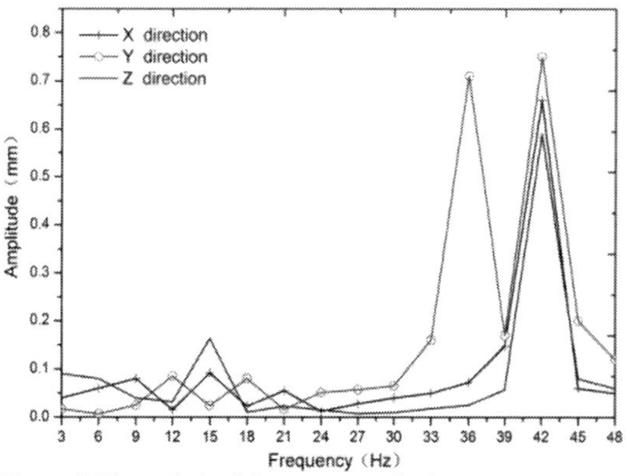

Figure 5. The optimized frequency amplitude response curve.

From Table 3, we can see that the vibration trend is similar to non-modified model, but the natural frequency is raised in different degrees. In particular, the maximum deformation of first order increased 13.485mm, the fifth order decreased 13.657mm, the maximum deformation descend form 29.351mm to 22.415mml, and the maximum natural frequency is raised to 54Hz in modified mode. From Fig. 5, compared with Fig. 3, the maximum amplitude response frequencies ascend from 10Hz to 42Hz in harmonic response analysis. Furthermore, the second peak frequency is presented to 36Hz in Y direction. The maximum amplitude reduce 0.7255mm (from 1.3861mm to 0.6606mm) in X direction, reduce 0.8991mm (from 1.6567mm to 0.7567 mm) in Y direction and reduce 0.71mm (from 1.3056mm to 0.5956mm) in Z direction. Through the preceding analysis, the natural frequencies and resonance frequencies are advanced meanwhile the tendency of maximum frequency amplitude is opposite.

The improved-structure based manipulator is manufactured to further verify the dynamic characteristic and the physical picture is shown in Fig. 6. Subsequently, the dynamic performance experiments will be carried out later.

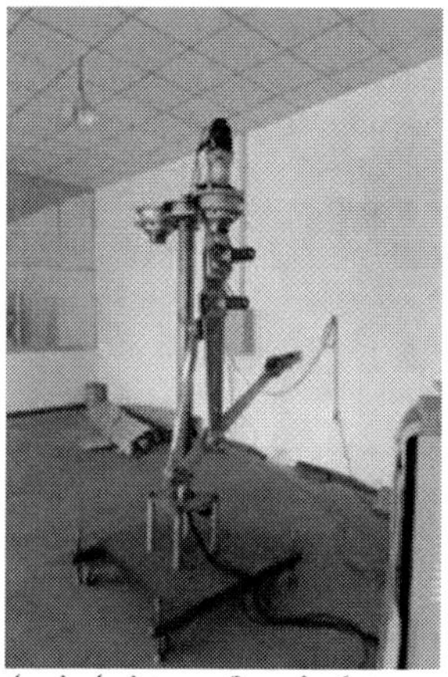

Figure 6. The physical picture of manipulator.

5. Conclusion

This article carried on the finite element analysis of a new type of glaze spraying manipulator and improved its structure to achieve a better dynamic characteristics. The calculation results shows that the rotary support has great influence on the deformation of the manipulator. Then an improved scheme is proposed to increase the thickness of rotation support and add ribs in both sides. The natural frequency of the modified structure is grown by 33.9%. Moreover, the maximum deformation of the manipulator was reduced by 52.3%, 54.3%, 54.4% in three directions. The dynamic performance of the modified structure is better than non-modified. In addition, the study could lay a theoretical foundation for kinematics, dynamics and deviation analysis, etc.

Acknowledgment

This work is supported by the National Natural Science Foundation of China ((51165011), the Natural Science Foundation of Jiangxi Province (2010GZC0090), Science and Technology Support Program of Jiangxi Province (20122BBE500039) and Technology Landed Plan of Jiangxi Province for Institutions of higher learning.

References

1. Wang Feng-chao, Wang Li-ping, Liang Xin-cheng, Zhao Jin-shan, Qiu Yun-jie. Research Review of the Intelligent Spraying Robots. The sixth youth science and technology forum of china aviation academy, pp. 675-679, (2014)
2. Wang Tian-miao, Tao Yong, Research Status and Industrialization Development Strategy of Chinese Industrial Robot. *Journal of Mechanical Engineering*, **9**, pp. 1-13, (2014)
3. Zhu Lei-lei, Chen Jun. A Review of Wheeled Mobile Robot. MACHINETOOL & HYDRAULICS, 242-247, (2009).
4. Song Shao-yun, Yin Fang, Sain Vaint Principle of Meshing in Finite Element Method and its Application. Mechanical Design and Manufacturing, **8**, pp. 63-65, (2012)
5. Li Da-lei, Ding Tian-tao, Cheng Jian-min, LiNing.Motion analysis for space swinging institutions based on SolidWorks Motion. Manufacturing Automation, **11**, pp. 70-71, (2011)
6. Wang Ze-peng, Hu Ren-xi, Kang Shi-ting, Finite element analysis from the entry to the master based on ANSYS Workbench14.5. Mechanical Industry Press, pp. 195-220, (2014)
7. Dong Xu, Li Zhi-jie, Xu Jin-ming, Six degrees freedom industrial robot dynamic characteristics analysis based on ANSYS Workbench. Precision Manufacturing and Automation, **3**,pp. 10-13, (2014)
8. Liu Tao, Wang Wei-wu, Lu Fei, Yuan Ren-wu, The model and harmonic response analysis of trunk telescopic boom by Workbench. Manufacturing Automation, **2**,pp. 80-82, (2015)
9. Tokhi M O, Mohamed Z, Modeling of a flexible robot manipulator using finite methods: a symbolic approach.Journal of Low Frequency Noise Vibration and Active Control, **18**(2), pp. 63-76, (1999).
10. Wang Shu-fen, Zhao Xu-yang, Li Yu-guang, Dynamic Characteristics Analysis of an Automotive Seat Frame. Mechanical Design and Manufacturing, **6**, pp. 145-147, (2015)
11. Ji Wen-mei, Fang Tong, Chen Qi-song, Mechanical Vibration. Science Press, **6**, 1985.

Pneumatic Muscle Actuated Compliant Gripper Systems

Andrea Deaconescu

Transilvania University of Brasov, Industrial Engineering and Management Department, Faculty of Technological Engineering and Industrial Management, Romania

Abstract. The paper presents the stages of developing new, light, eco-friendly and bionic gripper systems. Gripping is achieved by means of original, self adaptive, bio-inspired systems, with a pneumatic muscle as motion generator. The method underlying the development of these new gripping systems is based on the creation of concepts by analogy, an instrument aimed at widening the inspiration horizon in designing by using models from nature.

1 Introduction

Amongst the numerous functions of the human hand the most important one is grippping. While the grasping capacity of the extremities of the limbs can also be found in animals (*e.g.* lobster claws), it is in humans that it has reached its highest functionality. Thus, various combinations of the independent motions of the finger segments (phalanges) of the human hand allow gripping, holding, moving and releasing of various bodies.

The gripping function in humans entails a sequence of phases, starting with choosing the grasping method, followed by the actual seizing action and ending with controlling the manipulated object. Gripping is a consequence of desire generated internally or externally, that causes reactions at visual and brain level. Hence the hand is positioned close to the envisaged object, seizes it and carries out the programmed task. Figure 1 shows the phases entailed by gripping and the links between these [1, 2].

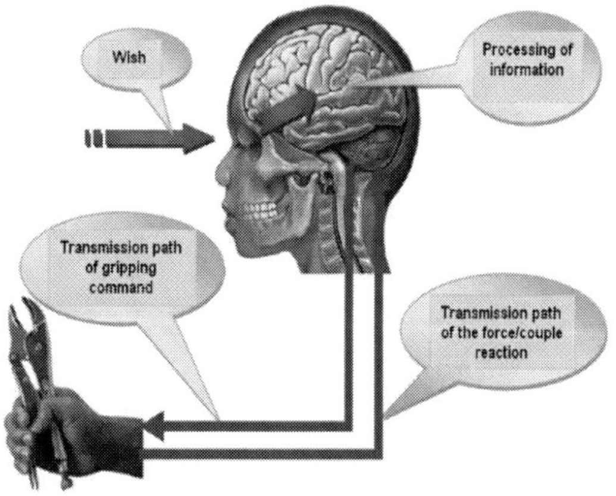

Figure 1. Command flows of a human hand gripping action.

In the field of robotics gripping entails the contact between the effector element of a robot (the gripper) and a body in view of its manipulation. The grippers are those components of robot systems that facilitate the temporary contact with the manipulated object, ensuring its position and orientation during transport and assembly.

The main functions of gripping systems depend on the specific applications they are part of and include [3]:

- temporary maintaining of a final position and orientation of a body in relation to the manipulation equipment;
- firm holding of the gripped object under static conditions (und the action of its own weight), dynamic conditions (in motion, during acceleration and braking) or when subjected to the action of external forces (cutting forces, torques);
- determining and modifying the position and orientation of the object depending on the concrete requirements of the application, etc.

In order to carry out the above functions, a gripping system needs to include the following main components:

- an energy system, meant to provide the type and quantity of energy required by the gripper;
- a mechanical execution system that carries out the actual gripping task;
- a system of sensors and transducers, meant to collect information (forces, displacements, accelerations etc.) and to transmit these to the control system;
- an information processing and control system that receives the signals coming from the sensors and transducers, processes these and consequently issues the necessary commands to the energy system.

The studies discussed in this paper are meant to present novel, light and eco-friendly gripper systems of bionic type. Gripping is achieved by means of original, self adaptive, bio-inspired systems, with a pneumatic muscle as motion generator. The proposed solutions of gripper systems are of non-anthropomorphic type designed for industrial or medical robotics applications.

The method underlying the development of these new gripping systems is based on design by analogy, and is a powerful working instrument aimed at widening the inspiration horizon in designing, at increasing the number of proposed variants, and consequently at generating a data base of solutions for gripper systems usable in various applications.

2 Design by analogy. Biomimetics

Design by analogy represents a frequently utilized method for the generation of new products and/or technologies. Numerous scientific papers address the issue of systematic knowledge transfer from natural sciences, from biology to engineering, and argue the enormous potential of bionic design (biomimetics) for developing new products and technologies [4].

The functional morphology of living organisms regarded from an engineering perspective represents a permanent source of inspiration for innovative solutions of high-tech constructions. In this respect since 1960 bionics (biomimetics), a novel scientific branch has developed, combining knowledge from biology, mathematics, medicine and engineering. Bionics draws upon biological intuition and engineering pragmatism to adapt nature's projects to the requirements of modern technics. Nature is but the starting point for innovations, it only offers clues as to what is useful in a mechanism. Starting from such clues it is the task of engineers to develop and refine the analyzed system. Thus bionics represents systematic learning from nature, as opposed to "inspiration from (copying) nature" [5, 6].

The motive for that bionics has not yet reached maturity from an engineering point of view, is that nature is an unbounded and impossible to control complex. Nature does not "design" a structure to attain a final target, as would an engineer; it encompasses billions of random experiments conducted over thousands of generations yielding beings whose aim it is to survive long enough such as to produce the next generation and launch the next round of random experiments.

Globally bionics is increasingly active, some of its achievements being, for example, the studies completed in England and the USA concerning the protuberances of the front side of the humpback whale tail with applications in the geometry of the airplane wing, and the studies conducted in Germany on the finger-shaped feathers of raptor birds that inspire engineers in creating airplane wings with geometries variable during flying, such as to reduce air resistance and fuel consumption. Medical research in Japan was aimed at reducing the pain caused by an injection by means of serrated edge hypodermic needles similar to a mosquito's proboscis, such as to minimise nerve stimulation.

Among the objectives of current bionic research, special attention comes to the study of effector organs. Their study, as well as that of the transmission processes of commands to them represents an essential part of bionics. In this respect terrestrial animals, birds and fish represent the source of inspiration for the development of various motion generating systems. Such an example is the study of the lizard's climbing motion and the subsequent transfer of the obtained results to the project of a climbing robot (Fig. 2a). Another example is that of an aeroplane wing, a constructive solution inspired from the wing of a bat (Fig. 2b).

Figure 2. Examples of biomimetics.

3 Gripper systems with bionic actuation

An essential problem related to the design of a novel gripper system concerns the functional characteristics this has to fulfil. Requirements like the generated force, rigidity/compliance, dexterity, the number of degrees of mobility depend on the application the gripper is designed for.

Natural gripper systems represent an inexhaustible source of inspiration for engineers, and in time they have underlain numerous practical applications. Thus figure 3 presents a mechatronic system bio-inspired from the seizing of an egg by a person using two fingers and based on this the accordingly developed bio-inspired mechatronic system [2].

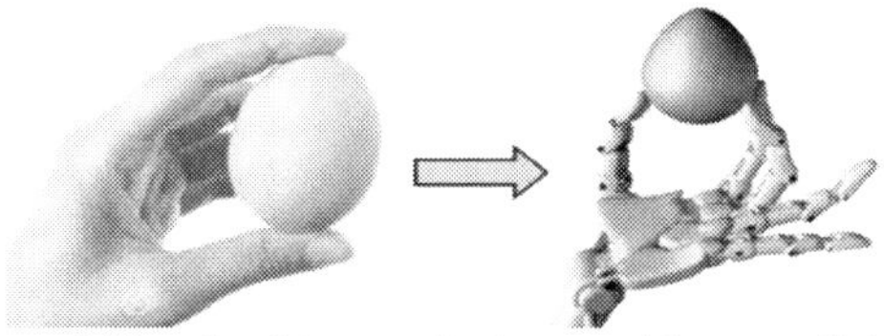

Figure 3. Transfer of the natural gripper model to an artificial system.

An artificial gripper system needs to ensure not only the seizing of the object, but also its safe manipulation. Thus, in the case of the egg of the above example, the magnitude of the applied force needs to be controlled, such as to not deteriorate the seized object. This is the case on compliant behaviour (Figure 4).

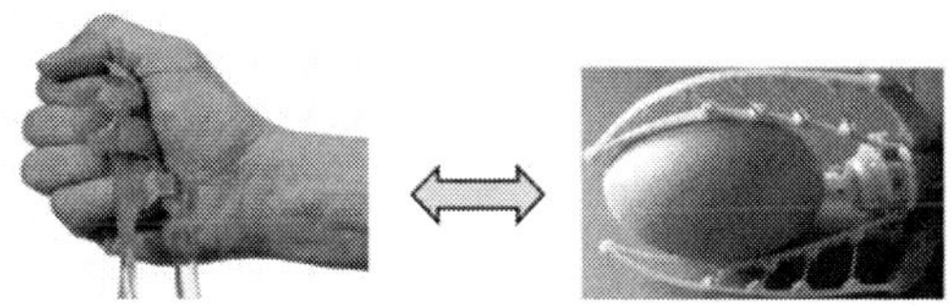

Figure 4. Seizing the gripped object with an uncontrolled vs. controlled force.

Compliance, the inverse of rigidity, represents one of the most important properties required from a gripper system. A compliant gripper system allows the safe manipulation of objects, without causing their deformation, by applying of forces of adaptable magnitude. Such behaviour can be ensured by deploying adjustable compliant actuators – ACAs, like pneumatic muscles.

Figure 5 presents some of the main phases of the gripping process, namely approaching the object to be seized, initiation of the contact and its firm gripping between the two jaws by the application of an increasing force. For these phases the classic profile of the jaws' velocities is presents, as well as the desired and necessary form of the curve that describes the compliance of the entire system.

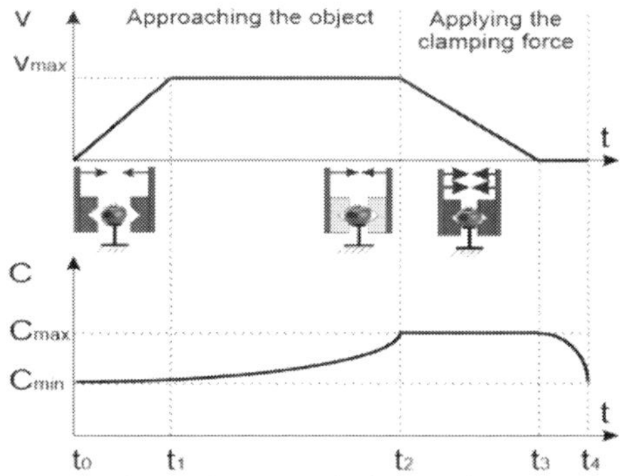

Figure 5. Variation of the jaw velocity and of the compliance of a gripper versus time.

The two graphs suggest that in the absence of contact between the jaws and the object, between moments t_0 and t_2 compliance can have smaller values, what allows a better positioning precision given by higher rigidity. The form of the curve that describes the variation of compliance needs to be concave, what allows a greater increase of compliance towards the moment of jaw-object contact initiation.

Between moments t_2 and t_3, when contact has been achieved, compliance needs to have the greatest values, thus ensuring the safe gripping of the object without its deformation/destruction.

An adjustable compliant actuator, like the pneumatic muscle, can adapt its functional behaviour between two limits. Thus its behaviour can vary from very rigid – required for a good positioning precision – to compliant, when the main requirement is the safety of the motion.

From the above follows that novel, innovative, bio-inspired solutions of gripper systems require light, flexibly structured equipment, with a high useful load-to-eigen-weight ratio, with integrated position adjustment and force control at an affordable price. Currently no gripper systems are available that would satisfy the majority of these requirements.

The paper aims at presenting innovative bio-inspired gripper systems the novelty of which consists in their actuation, namely with pneumatic muscles. In this respect the principles of design by analogy are applied via a top-down bionic approach. Such approach entails that subsequently to stating the problem, natural systems are to be identified able to provide ideas for adequate solutions.

Gripping is achieved by means of mechanical contact forces. While in natural systems these forces are generated by a muscle, in artificial grippers forces are generated by motors. In the novel gripper systems proposed in this paper the deployed motor is a pneumatic muscle, a bio-inspired system analogous to human and animal muscles.

Between the motor and the jaws (the effectors) typically a mechanism intervenes, consisting of rigid elements. In this case the proposed gripper systems are based on gears.

The pneumatic muscle copies by biomimetics the functioning of the human muscle fibre (Figure 6), and has a number of characteristics like shock absorbing capacity and shock resistance, low weight, small overall dimensions and mass per unit of power and elasticity [7]. These characteristics render pneumatic muscles optimum constructive elements for applications in robotics, for both orientation and gripper mechanisms.

Transition from the human to the pneumatic muscle

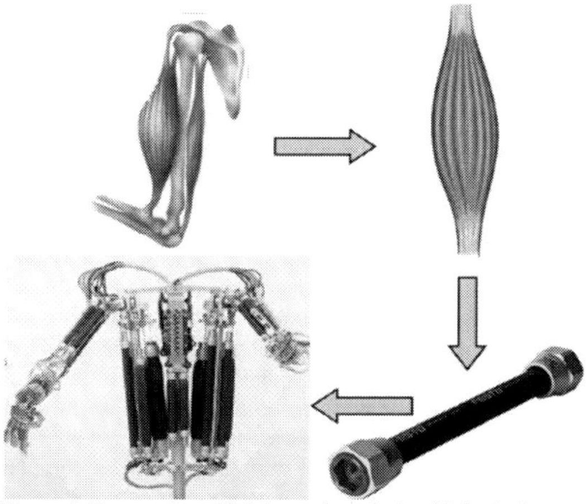

Figure 6. Pneumatic artificial muscles mimic a biological muscle.

Further on three variants of a gripper system are proposed, with a pneumatic muscle as the motor element linked to a gear mechanism for power transmission.

The pneumatic muscle actuated gripper systems were dimensioned based on the following input data: mass of the object m = 0.7 kg; acceleration of the motion of the system consisting of the gripper and the object: $a = 5\ m/s^2$; gravitational acceleration: $g = 9.81\ m/s^2$; emergency stop deceleration: $a_S = 10\ m/s^2$; friction coefficient: $\mu = 0.2$; safety coefficient: S = 2.5. Considering the imposed requirements related to functioning and overall dimensions, the models proposed in this paper are based on the utilization of the smallest pneumatic muscle produced by Festo, Germany, (MAS-10-45N-AA-MC-O-ER-EG), with the following characteristics:

- muscle diameter: 10 mm;
- length of the active part: 45 mm;
- maximum working pressure: 6 bar.

The maximum forces that have to be developed by this pneumatic muscle occurr for the highest values of the accelerations, namely in sudden braking. In such a case the necessary gripping force the be exerted by one jaw is computed by equation (1):

$$F_G = \frac{m \cdot (g + a_S) \cdot S}{\mu \cdot n} = \frac{0.7 \cdot (9.81 + 10) \cdot 2.5}{0.2 \cdot 2} = 86.67\ N \qquad (1)$$

The proposed gripper systems have two jaws, thus requiring the selected pneumatic muscle to develop a force of minimum 2 · 86.67 N = 173.34 N.

Figure 7 presents the graph describing the force developed by the selected pneumatic muscle versus pressure and stroke. The hatched rectangle marks the optimum deployment area of the pneumatic muscle, so that for a 4 mm maximum stroke of the jaws the developed force exceeds 173.34 N.

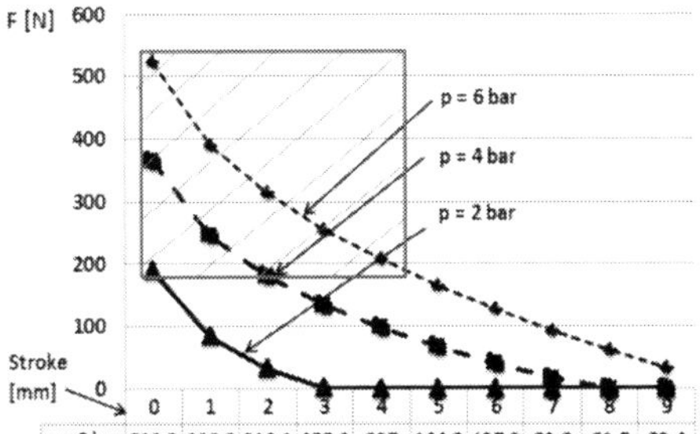

Stroke [mm]	0	1	2	3	4	5	6	7	8	9
p = 6 bar	522.5	390.8	313.1	255.2	207	164.9	127.2	92.9	61.5	32.4
p = 4 bar	364.3	245.5	180	134.1	97.8	67.3	40.8	17.3	0	0
p = 2 bar	187.9	83.2	32.3	1.2	0	0	0	0	0	0

Figure 7. Variation of the developed force versus pressure and stroke.

Figures 8 to 10 show the schematics of the three variants of gripper systems. All the three proposed variants of gripper systems benefit from compliant behaviour, thus ensuring the safe manipulation of fragile objects, without deforming or destroying them. The utilization of a pneumatic muscle as an adjustable compliant actuator allows deviations from the position of equilibrium, their magnitude depending on the level of straining external forces.

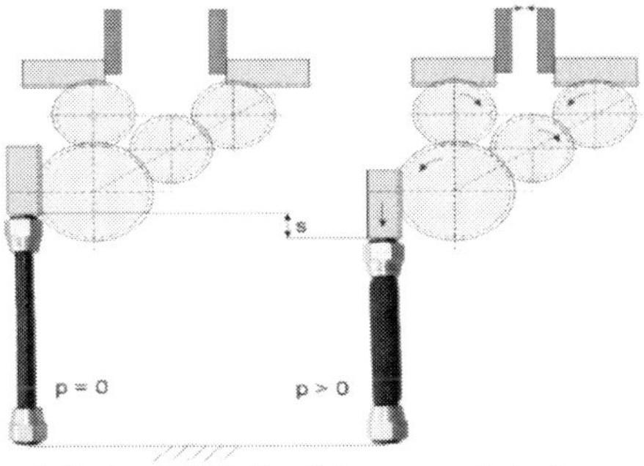

Figure 8. Variant 1 (V1): Parallel, asymmetrical gripper system with two mobile jaws.

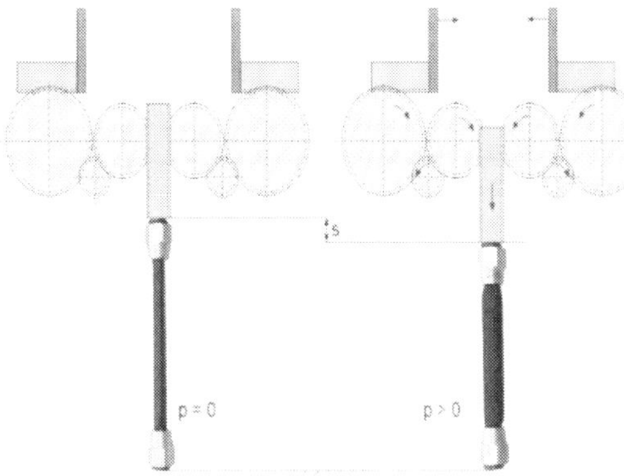

Figure 9. Variant 2 (V2): Parallel, symmetrical gripper system with two mobile jaws.

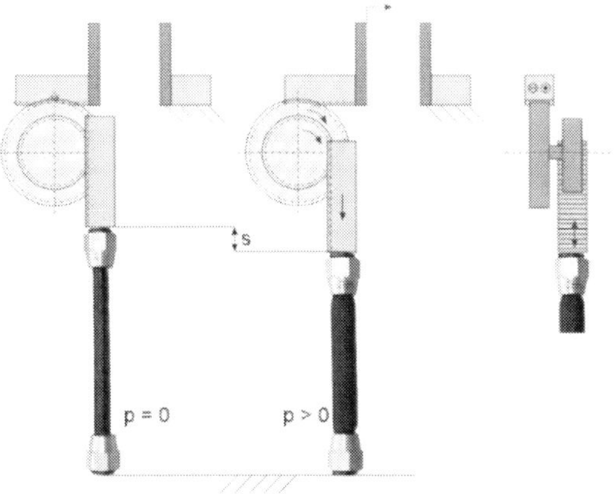

Figure 10. Variant 3 (V3): Parallel, asymmetrical gripper system with one mobile and one fixed jaw.

From the three proposed variants the asymmetrical one with two mobile jaws was built as a prototype and tested. Figure 11 shows the experimental model.

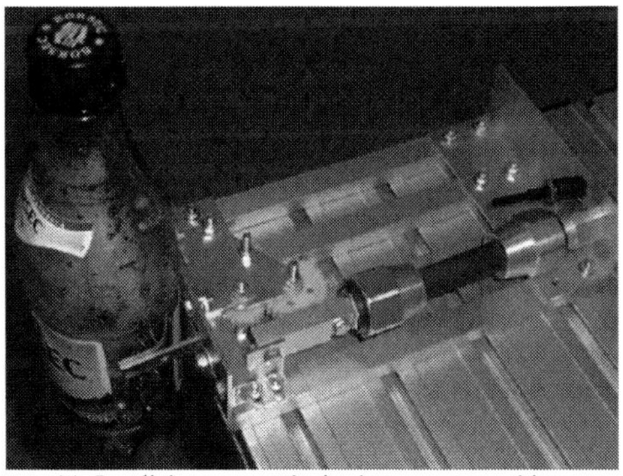

Figure 11. Parallel, asymmetrical gripper system with two mobile jaws.

The power transmission mechanism consists of gears and racks that ensure the final motion of the two gripping jaws. The pneumatic muscle is positioned asymmetrically, thus allowing the metal structure of the gripper system to include the proportional pressure regulator that controls the motion of the linear motor. The displacement of the jaws is achieved by means of an MPPES-3-1/4-6-010 proportional pressure regulator (manufactured by Festo) that can be controlled by a MPZ-1-24DC-SGH-6-SW reference module. This actuation concept allows the slow and uniform charging of the pneumatic muscle, while avoiding possible shocks being introduce into the system.

Of particular interest within the conducted experiments was determining the compliant and adjustable character of the gripper system. Adjustable compliance is obtained only for a non-linear dependency of the force developed by one jaw on its displacement. An adjustable compliant actuator, as is the pneumatic muscle, can adapt its behaviour between two limits. Thus its behaviour can vary from very rigid (required for a good positioning accuracy) to compliant, when the main requirement is ensuring motion saftey.

Figure 12 presents the variation of the force developed by one jaw versus its stroke, the pressure varying from 0 to 6 bar. It can be observed that at the beginning of the motion the value of the force is at its maximum, and as the jaw progresses towards its limit, the force decreases.

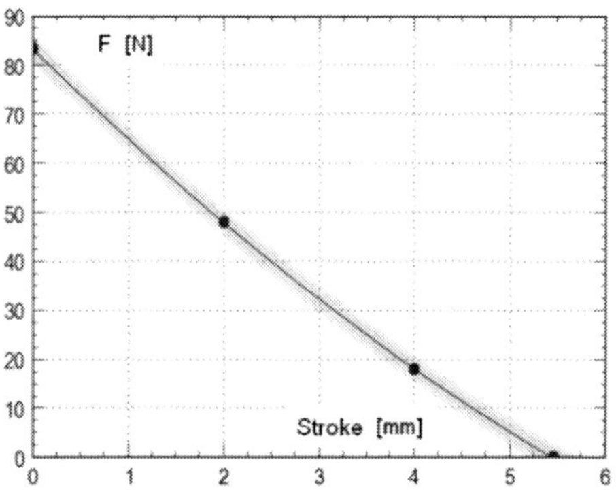

Figure 12. Variation of the force developed by one jaw vs its stroke

The regression function corresponding to the previous curve is (2):

$$F_{jaw} = 83.36 - 19.146 \cdot c + 0.7088 \cdot c^2 \quad (2)$$

The compliance C of the gripper system is computed as the inverse of rigidity (k):

$$C = k^{-1} = \left(-\frac{dF}{dc}\right)^{-1} = \frac{1}{19.146 - 1.4176 \cdot c} \quad (3)$$

Figure 13 shows the evolution of compliance versus the stroke of a jaw.

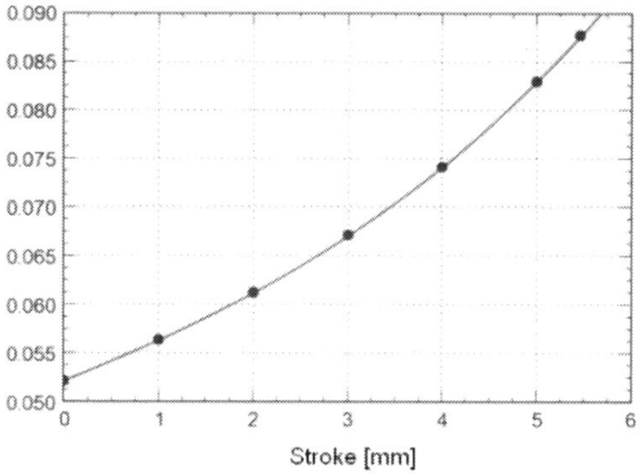

Figure 13. Compliance of the gripper system.

The above graph reveals the concave shape of the compliance curve, what facilitates the accelerated increase of this quantity once the jaw-object contact has been achieved.

4 Conclusions

The paper has presented several innovative constructive solutions of non-anthropomorphic gripper systems that ensure the safe seizing and manipulation of fragile objects. The idea of utilizing pneumatic muscles as actuators was taken from nature, by biomimetics, based on the functioning of the human muscle fibre.

The utilization of the linear pneumatic muscle as the actuator of the gripper system entails a light construction of significantly compliant behaviour, which corresponds to the safety requirements concerning the interaction with humans.

References

1. S. Nof (editor), *Handbook of Industrial Robotics*. (John Wiley and Sons Inc. New York, (1985)
2. D. Negrea - *Research concerning the gripper systems of industrial robots actuated by pneumatic muscles*. (PhD Thesis Universitatea Transilvania din Braşov, 2014)
3. G.J. Monkman, S. Hesse, R. Steinmann, H. Schunk, *Robot Grippers*. (WILEY-VCH Verlag GmbH & Co. KGaA, Weinheim, 2007)
4. B.T. Christensen, C.D. Schunn, The Relationship of Analogical Distance to Analogical Function and Pre-inventive Structure: The Case of Engineering Design. Memory & Cognition, **35(1)** (2007)
5. G.E. Dieter, *Engineering Design: A Material and processing Approach*. (McGraw-Hill, 2000)
6. K.T. Ulrich, S.D. Eppinger, *Product Design and Development*. (Boston McGraw–Hill/Irwin, 2004)
7. Festo. Precisely to the point: membrane technology. https://www.festo.com/cms/en_corp/14047.htm Accessed 2015

Experimental Research on Liquid Desiccant Air-conditioning Unit

Yueyan Feng, Yinglin Li

Nanjing Normal University, School of Energy and Mechanical Engineering, 210042, China

Abstract. An experimental device of liquid desiccant air conditioning system is established. Experimental tests about the temperature difference between diluted solution of inlet and concentrated solution of exit in the solution heat exchanger are carried on, and $CaCl_2$ solution is used as desiccant. Results show that: the fluctuation range in the day at different times of the basic difference of the measured temperature does not exceed 1 ℃, and the temperature difference between diluted solution of inlet and concentrated solution of exit in solution heat exchanger appears the minimum value of 2.7℃ and the maximum value of 10.2℃. Also, the percent of the additional load and the ratio of additional load to the evaporator load are analyzed.

1. Introduction

At present, the researches on the liquid desiccant system mainly concentrate in the padding, dehumidifiers, regenerators and other components, and few researches focus on the solution heat exchanger. Solution heat exchanger can recover thermal energy and increase thermal coefficient of unit [1]. On one hand, solution heat exchanger of the liquid desiccant unit increases the temperature of dilute solution and decreases the temperature of concentrated solution; on the other hand, after going through dehumidifier, concentrated solution becomes thin and cannot absorb moisture, so dilute solution needs to be concentrated and regenerated, the concentration of solution after going through heat exchanger is just to meet the requirements. Although the solution heat exchanger has above advantages, in terms of liquid desiccant system, there is a certain difference between the temperature of the concentrated solution and the dilute solution after going through solution heat exchanger. Thus this adds an additional consumption of the evaporator load.

In this paper the temperature difference between the diluted solution and concentrated solution in the solution heat exchanger is deeply explored based on experimental research, also the analysis about the additional consumption of load and the corresponding percentage of additional load is presented.

2. The principle of liquid dehumidification air-conditioning system and test rig

2.1 The principle of liquid dehumidification air-conditioning system

The temperature difference between the air and the solution is the driving force of heat transfer, the difference of the partial pressure of water vapor between in the air and in the surface of the solution is the driving force of mass transfer [2]. A liquid desiccant diagram is shown in Fig. 1. During the process of dehumidification, partial pressure of water vapor in the air is greater than partial pressure of water vapor in the surface of the solution, moisture in the air can be absorbed by the solution, and the moisture is passed to the solution, the air humidity decreases to achieve the purpose of dehumidification. During the absorbing process, the concentration of concentrated solution decreases and the capacity of moisture absorption gradually declines [3].

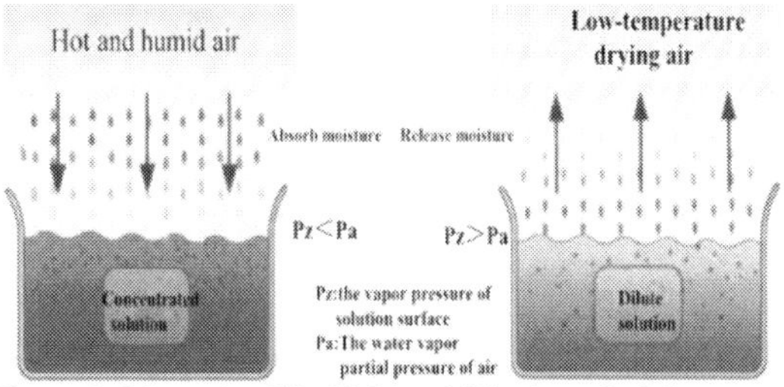

Figure 1. Schematic of liquid dehumidification principle

Regeneration process is the reverse process of dehumidification process, namely, the partial pressure of water vapor in the air is less than in the surface of the solution, so that the moisture transfers from the solution to the air.

The partial pressure of saturated water vapor in the surface of the solution increases with the increase of the solution temperature, the temperature of desiccant solution increases, and the water vapor pressure of the surface of the solution is greater than the partial pressure of the regenerative water vapor in the air, moisture transfers from the solution to the air, so that to achieve the purpose of the regeneration of the solution [4].

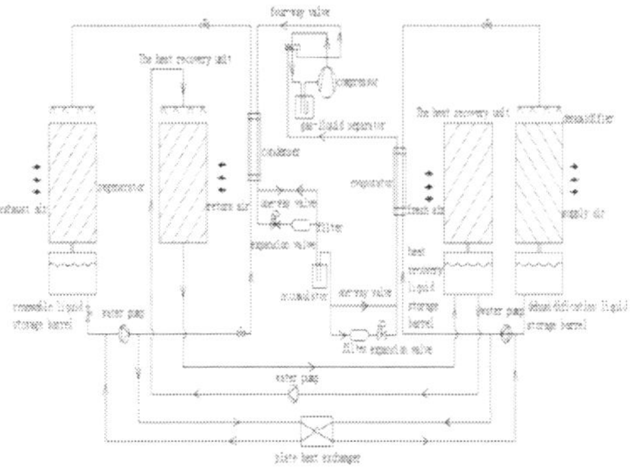

Figure 2. Diagram of heat pump heat recovery solution dehumidification unit

Figure 3. Test rig picture of pump type heat recovery solution dehumidification unit

Table 1. Main parameters of testing component in the test rig

Component	Parameters
Heat recovery module	Celdex，980mm×500mm×300mm
Dehumidification module	Celdex，980mm×500mm×300mm
Regeneration module	Celdex，980mm×500mm×300mm
Solution tank	Plastic, diameter of 315 mm, height of 450 mm
Solution heat exchanger	Plate heat exchanger
Water pump1	7.5 m^3/h，Lift of 12.5 m
Water pump2	7.2 m^3/h，Lift of 14.6 m
Water pump3	15.6 m^3/h，Lift of 4 m
Fan in dehumidification side	Centrifugal fan RDK2.8A, 3000 m^3/h
Fan in regeneration side	DKNB 500-MOT90, 2280 m^3/h

2.2 Test rig of heat pump heat recovery solution dehumidification unit

Fig. 2 presents a heat pump heat recovery solution dehumidification unit; it mainly composes of the heat recovery module, dehumidification module, regeneration module, the solution heat exchanger, solution pump, solution tank, compressor, condenser, evaporator, expansion valves, the solution control valves, solution pipeline and other components. $CaCl_2$ solution is chosen as the absorbent. Fig. 3 is the picture of the experimental rig. And the main parameters of relevant apparatus in the system are shown in Table 1.

3. Analysis of temperature difference of dilute solution and concentrated solution

In the experiment, the plate heat exchanger is chosen as solution heat exchanger in the heat pump heat recovery solution dehumidification unit. The structure of plate heat exchanger is small size and light weight with the feature of higher heat transfer coefficient.

The calculation formula is given below for the specific heat capacity of Calcium chloride solution, enthalpy of solution, the percent of the additional load, the percent of the evaporator load and the ratio of additional load to the evaporator load .

1) The specific heat capacity of $CaCl_2$ solution $C_{Psol}(T,\xi)$[5]

$$C_{Psol}(T,\xi)=C_{PH2O}(\theta)\times(1-f_1(\xi)\times f_2(\theta)) \quad (1)$$

$$C_{PH2O}(\theta)=A+B\theta^{0.02}+C\theta^{0.04}+D\theta^{0.06}+E\theta^{1.8}+F\theta^{8} \quad (2)$$

$$\theta=T/228-1 \quad (3)$$

Type of each parameter are shown in Table 2

When $\varepsilon\leq0.31$,then $f_1(\xi)=A\xi+B\xi^2+C\xi^3$ (4)

When $\varepsilon>0.31$,then $f_1(\xi)=D+E\xi$ (5)

$$f_2(\theta)=F\theta^{0.02}+G\theta^{0.04}+H\theta^{0.06} \quad (6)$$

ε—concentration representation

Type of each parameter are shown in table 3

2) Enthalpy of solution h_L

$$h_L=C_{Psol}(T,\xi)\times t_L \quad (7)$$

3) The percent of the additional load Δh

$$\Delta h=C_{c,out}t_{c,out}-C_{d,in}t_{d,in} \quad (8)$$

4) The percent of the evaporator load q_0

$$q_0=C_{in}t_{l,in}-C_{out}t_{l,out} \quad (9)$$

5) The ratio of additional load to the evaporator load φ

$$\varphi=\Delta h/q_0 \quad (10)$$

Table 2. The value of various parameters from formula(1)to formula(3)

	A	B	C	D	E	F
T≤0℃	830.54602	-1247.52013	-68.6035	491.2765	-1.80692	-137.51511
T>0℃	88.7891	-120.1958	-16.9264	52.4654	0.10826	0.46988

Table 3. The value of various parameters from formula(4)to formula (6)

A	B	C	D	E	F	G	H
1.63799	-1.69002	1.05124	0	0	58.5225	-105.6343	47.7948

Experimental data of inlet and outlet temperature of the heat exchanger is selected from seven time points: 2: 30, 2: 45, 3:00, 3: 15, 3: 30, 3: 45, 4: 00 on April 20, April 25, April 30, May 5, May 10, which is plotted in Fig.4. Temperature experimental data is the average of data selected from seven periods: 1: 30-1: 55, 2: 00-2: 25, 2: 30-2: 55, 3: 00-3: 25, 3; 30-3: 55, 4: 00-4: 25, 4: 30-4: 55 on April 30, it is plotted in Fig.5.

The temperature difference in Fig. 4 and 5 represents the temperature difference between the concentrated solution in the exit and diluted solution in the inlet of solution heat exchanger.

As shown in Fig. 4, it can be seen that the value of temperature difference on April 20, April 25 and April 30 is almost in a straight line, the value of temperature difference on May 5 and May 10 fluctuates slightly, the line is gentle as a whole. The maximum value of temperature difference is 10.2 ℃ at 15:30 on May 5th, the minimum value of temperature difference is 2.7 ℃ at 16:00 on April 20.

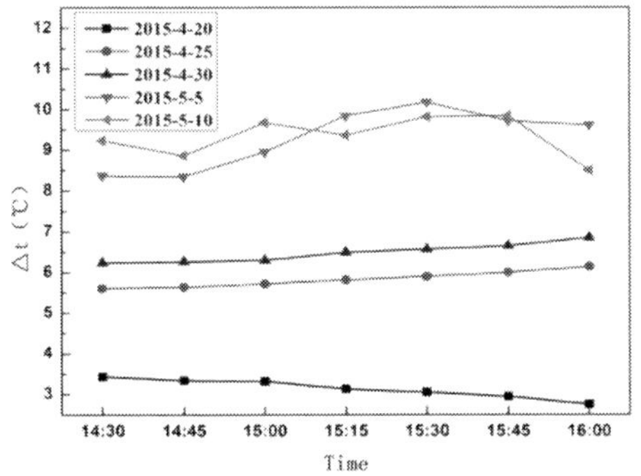

Figure 4. Temperature difference at different time points on five days

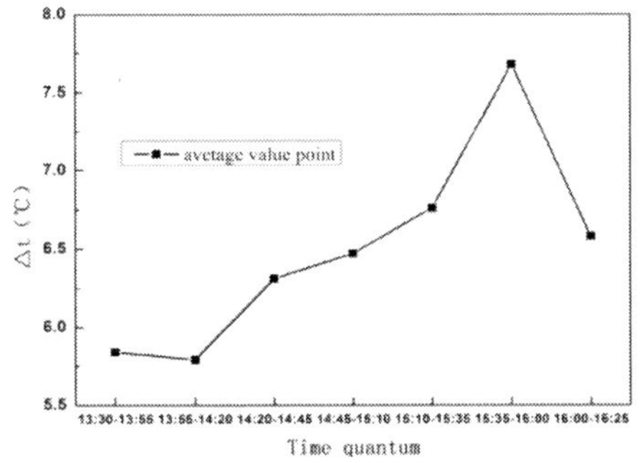

Figure 5. Temperature difference in different periods during one day

As can be seen from Fig. 5, the value of temperature difference fluctuates between 5.5 ℃ and 7.7 ℃ and has a peak in the period of 15: 30-16: 00, the fluctuating range is about 2 ℃. Fig. 6 and Fig. 7 are corresponding percent of additional load of Fig.4 and 5.

As presented in Fig. 6, the value of temperature difference on April 20, April 25 and April 30 is also in a straight line, the percent of additional load on May 5 and May 10 fluctuates slightly and the line is gentle on the whole. The minimum value of percent of additional load is 7.42 kJ/kg at 16:00 on April 20 and the maximum value is 28.5 kJ/kg at 15:30 on May 5.

As can be seen from Fig. 7, the value of per unit mass of additional load on April 30 fluctuates between 15.8 kJ/kg and 22 kJ/kg and has a peak during the period of 15: 30-16: 00, fluctuating value is about 6.2 kJ/kg.

Percentage of per unit mass of additional load accounts for evaporator load indicates the degree of influence that per unit mass of additional load has on evaporator load. Calculated percentage at different time points during five days and during one day are respectively plotted in Fig. 8 and 9.

According to Fig. 8, we can see percentage that per unit mass of additional load accounts for evaporator load on April 20 and April 30 presents almost a straight line and fluctuates slightly. The percentage has a valley value at 15:15 on April 25 and other values have less extent fluctuation. Fluctuating values on May 5 and May 10 are larger than other three days.

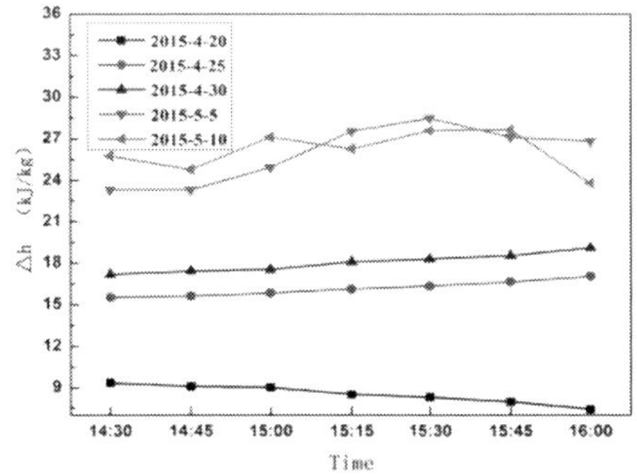

Figure 6. Percent of the additional load at different time points during five days

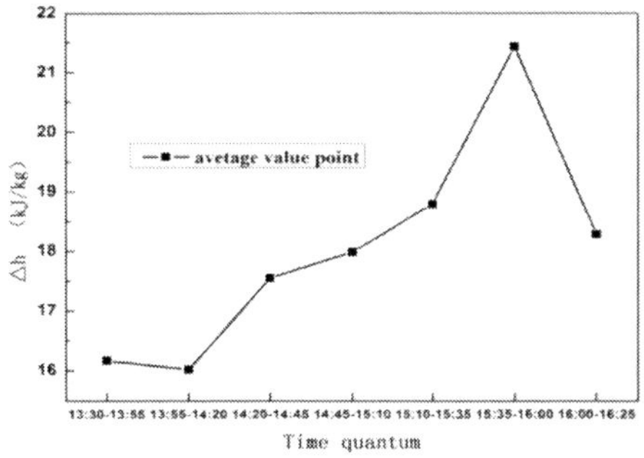

Figure 7. per unit mass of additional load diagram during different periods a day

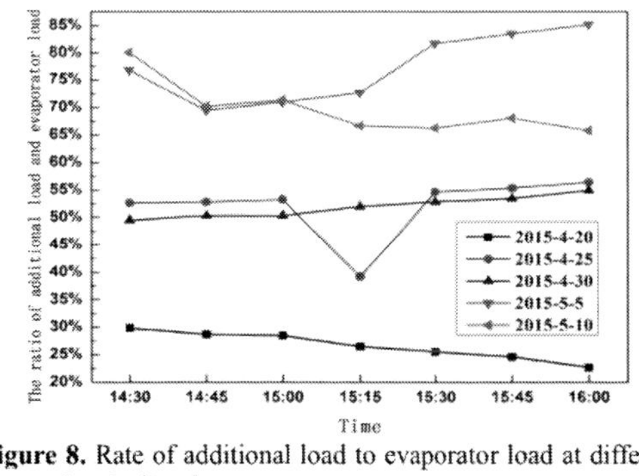

Figure 8. Rate of additional load to evaporator load at different time points in five days

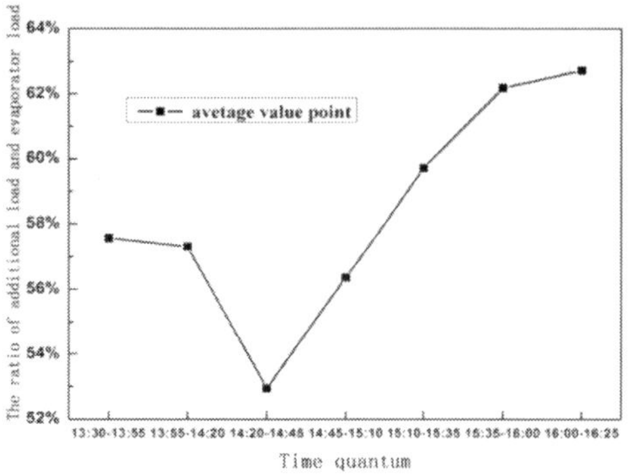

Figure 9. Rate of additional load to evaporator load at different time points

The variation trend of Fig. 5 and Fig. 7 is different; it has a valley value 52.9% during 14: 20-14: 45. There is no rule from the whole variation.

4. Conclusion

1）The fluctuation range of temperature difference measured at different time a day is mainly within 1℃;Experimental data measured on different dates are different

2）On April 30, the value of temperature difference stabilizes between 5.5 ℃ and 7.7 ℃ and fluctuates about 2 ℃, the data may be affected by a particular value.

3）The temperature difference of concentrated solution on exit and dilute solution on entry of solution heat exchanger has the minimum value 2.75℃ and the maximum value 10.18℃, the minimum value and the maximum value of per unit mass of additional cooling load respectively are 7.42kJ/kg and 28.48kJ/kg.

4）we calculate the percentage that additional load accounts for evaporator load and the number of the value indicates the degree of influence that dilute solution and concentrated solution has on evaporator after heat exchanging. The minimum value is 52.94% and the maximum value is 85.14%, which has great impact on the evaporator but what kind of actions should be taken to decrease additional load need further study.

References

1. L. Jian, Study on heat transfer performance of solution heat exchanger about lithium bromide absorption refrigeration system, Refrigeration and Air Conditioning, **3**, 68-70 (2006).
2. Z. W. Lian, *Heat and mass transfer principle and equipment*, China Building Industry Press, (2011).
3. SinoRefine Air Conditioning Technology *Co., Ltd*,. Liquid desiccant type air handling units comprehensive manual[M] (Internal Information), First Edition, (2010).
4. J. W. Tang, Investigation of dehumidification process on falling-film of LiCl solution[D], Beijing University of Chemical Technology, (2007).
5. M. R. Conde-Petit, Aqueous solutions of lithium and calcium chlorides: Property formulations for use in air conditioning equipment design, M. CONDE ENGINEERING, (2009).

Design of Batch Pressing Mechanism for Bedded Standard Samples

Jianguang Li, Yongyan Wang, Nan Qin, Huihua Zhu, Yanchun Wang

College of Electromechanical Engineering, Qingdao University of Science and Technology, Qingdao 266061, China

Abstract. In stability analysis of rock soil engineering, before establishing similar physical simulation model, many match experiments on similar material are required, the current standard sample forming processes are difficult to satisfy the uniformity and compactness in the same sample set often, and have obvious defects in the molding process, molding efficiency and bedding processing. A pressing mechanism consists of adjustable concave-convex composite mould, multi-function trough, drive screw and framework is designed, the mould is pressed nto material to fill directly. Each layered material, dip angle, thickness and position can be controlled, forming number for each lot is adjustable, uniform of the same lot is good, all parts are detachable, operation is also simple. This mechanism is suitable for small batch forming, and has obvious advantages in bedding controlling, forming efficiency, forming quality and operation convenience, which can work well for match experiments before large-scale physical similar simulation experiment, geotechnical experiment, composite rock mass experiment, also can be used to make micro physical model with engineering structure.

1 Introduction

In some design for large engineering, such as slope, caverns, storage and so on, stability analysis usually are done to forecast the possible deformation and damage of rock mass in engineering. Common stability analyses include physical model similar simulation and numerical simulation. In the physical model, large-scale geomechanical model is always made, according to the geological conditions, rock layers and geometrical size of current engineering, model materials for similar materials follow a similar ratio and proportion, which is related with the rock mechanical properties, engineering scale in actual geological environment. Due to discrete and uncertainty of underground rock mass, in each new project, the geomechanical model need a lot of preparation of similar material standard samples and conventional mechanical test. Usually, the test goal is getting the elastic modulus, compression strength and other parameters of similar materials at a certain similar ratio, and it is required a series of sample group with sufficient number and consistent performance, to ensure the reliability of data. In practice, different experimental project require different sample number, which often are done manually or by simple mold, and can meet the requirement of uniformity partly. For example, Ma zhen-guo [1] et al make sample by pressing mechanism and matching die (similar method is widely used, most of them is operated manually, the pressing force is small and the forming number is small, see Fig. 1), but there are obvious disadvantages in forming efficiency and quality.

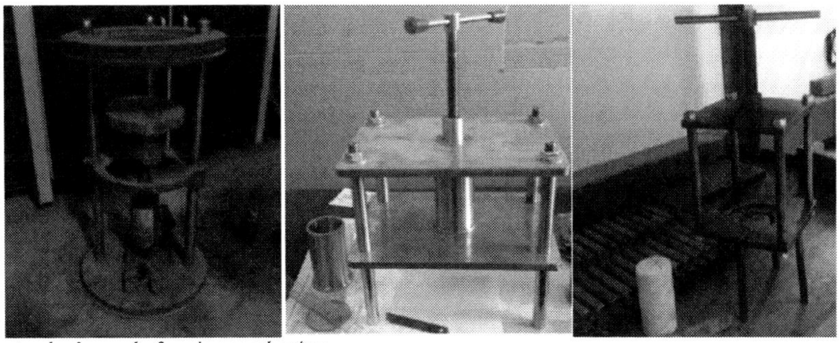

Figure 1. Common standard sample forming mechanism.

If the sample contain bedding and have special requirements for angle and materials, it is difficult to ensure the consistency of the same batch of samples by making manually, and depend more on the special machinery and tooling. In essence, layered rock mass is composed by different material, thickness, angle, adhesion.etc, the production process or equipments, which can achieve any combination of above variables, all belong to reasonable forming process. Different researchers use different ideas and designs. Jiang de-yi et al [2] aimed to get the approximation of original rock mass, with the help of general press machine, pressed salt rock powder and interlayer material powder into cylindrical molding mould orderly, and obtained salt rock standard sample with horizontal interlayer; Yong ming tien [3] designed a set of equipment to make standard sample with any layer characteristic, by mixture, adsorption layer by layer, concentration pressure, cutting and drilling etc. Zhang gui-min [4] made further improvement based on Yong ming tien, he designed a set of thin layer adsorption, translation and cutting device, by which, the bedded standard sample with different angle and thickness ratio could be obtained.

The above designs have practicability to a certain degree, but there are also some disadvantages. Based on the comprehensive investigation on the previous research, author design a set of simple, efficient pressing mechanism, by which, standard sample with bedding, uniform density, controllable angle and thickness can be made. The main part of this mechanism consists of an adjustable concave-convex composite mould, multi-function silo, drive screw and framework, can be equipped with force measuring ring, can also be used in combination with ordinary testing machine.

2 Adjustable composite mould

The forming quality of pressing mechanism depends on mould. Generally speaking, the forming principle of split mould is simple, but which require higher machining quality, and there is poor roundness after the mould closing, it is difficult to guarantee the shape accuracy of cylinder and uniformity of sample. The mould group with rigid connection have fixed concave-convex mould, can not be disassembled freely, it is not easy to control forming quantity, but easy to waste material. So the seamless pipes machined are used to assemble removable mould group this design.

8 cold-drawn steel cylinder (2×4) of inner diameter 50mm, height 150mm are machined as concave mould group, of which, one end has a 15 degrees blade, and the other end with whorl is assemble with rectangular concave mould plate through a set of slotted round nut. The screw holes are drilled at corresponding positions on convex mould plate, and convex module group could be formed once pressing head are installed. When using, in order to form or demould, people need place smaller round pie into concave mould cylinder as movable pressing head. According to the requirement of different test project, the number of standard sample made each batch could be controlled to 2, 4, 6 or 8. In addition, all components of this mechanism could be disassembled, cleaned, moved and carried easily.

In practice, due to different adhesion and solidification time of different simulation materials, the surface of sample has certain damage during demoulding, so before the material into the mould, people could daub some mold-release agent on the inner surface of concave mould cylinder, such as paraffin wax, silicone oil and graphite etc, to ensure the final molding quality.

3 Multi-function trough

Because the molding method this design require the molds are pressed into the material, therefore, as far sample with bedding, before the material into molds, worker need set thickness, inclination, position, number of each layer (according to predetermined design, finished in the trough), which means in addition to holding material, the trough space should be adjustable, should have observation window, and could bear certain expansion pressure.

Eventually, trough of the designed by a side plate, a bottom plate and a movable partition board, which baffle according to selection of molding quantity to separating of the feed slot space, control material usage, material saving. Length direction of the side of the side plate reserved installation position, layered loading, fixed thick organic glass, as the observation window, both real-time observation and material clearing export function. As Fig. 2.

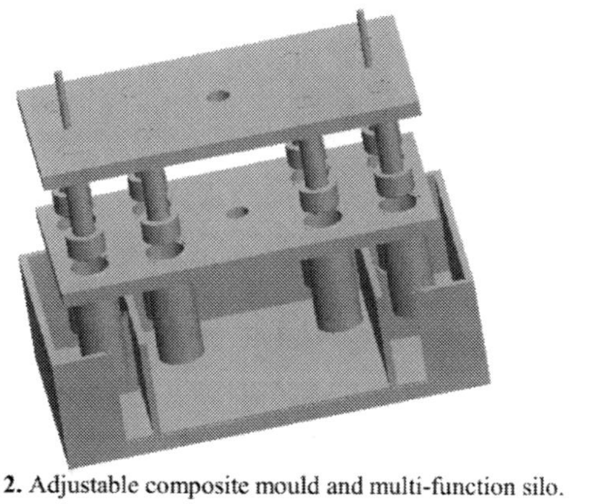

Figure 2. Adjustable composite mould and multi-function silo.

4 Drive screw and frame

The design is different from other existing schemes on molding process, therefore, it is necessary to guarantee a certain rolling force by mechanical means. According to the design, in molding process, different pressure value is needed according to different conditions, the maximum value in the convex die compaction link, a single test sample's maximum pressure molding is designed to 3 kN, then at the maximum efficiency, the maximum load of pressing 8 samples is 24kN; in loading ways, by rotating the handle, drives the single screw propulsion thrust to produce thrust, converting to linear thrust formulas from rotating torque. As Eq. 1.

$$\begin{cases} P = 8P_i \\ T = Fd \\ P = \dfrac{2\pi\eta_1 T}{l} \end{cases} \tag{1}$$

wherein, P, the thrust generated by the screw rod; P_i, the single test piece pressed strength; T, the driving torque; l, the feed screw guide path; η_1, the positive efficiency of the feed screw; F, the handle thrust; d, the handle length.

The straight line thrust designed meets the design requirements and proves manual drive mode is feasible. When the bigger compaction force needed, it could be obtained by several ways like (1) connecting a longer handle; (2) reducing each batch forming quantity (3) using press machine loading through conversion joint (4) changing manual drive to motor drive.

In addition, regarding the central screw bearing main pressure, its size design requires for the stability. Taking into consideration of the overall comfort and reserve force measuring ring space, screw length is set as 550 mm, the end part and the frame are connected by sleeve threads, regarded as articulated. The stability safety factor of the design is n_{st}=3, by a slender rod stability formula. As Eq. 2.

$$\begin{cases} \dfrac{P_{cr}}{P} \geq n_{st} \\ P_{cr} = \dfrac{\pi^2 EI}{(\mu l)^2} \\ I = \dfrac{\pi D^4}{64} \end{cases} \tag{2}$$

To reach the conclusion, under maximum screw load, selecting steel as material, the diameter should be bigger than 22mm, comprehensive consideration of service life, space interference, lead and other issues, a 8.8 grade M27 screw is chosen finally.

The machine frame adopts four column type, connected by two layers beam clapboards, takes into account of the stability problem on material selecting and cross section size designing issues, here no repeat any more.

To quantitative control the pressure, try to ensure the compactness and uniformity of each batch, before each pressure, force measuring ring is placed between central screw and die plate, to get real-time observation pressure data; instead, or the whole mold after loading could be put pressure under pressure machine to get data through online software.

5 Forming process of the equipment

The mechanism has the advantages of simple structure and convenient operation, the whole process could be completed by a single person, and the specific forming process consists of 4 steps:

(a) Material compaction. Setting watch window glass, fixing the position of trough, the material which are mixed well according to a predetermined ratio and stirred well are put into trough successively, observing in real time by lateral organic glass window, making different angles and beddings as needed; rotating the handle, letting central screw contact temporary plate to compact material in batches.

(b) Concave mold pressing. Assembling concave mold combination, adjusting to the horizontal plane and tighten the screw in turn, apply mold releasing agent, overall combination is placed on top of the materials, installing vertical positioning guide rod, rotating the handle, pressing concave mold into the materials until contacting anti-sticky pad.

(c) Convex mold compaction. Putting active press plate into concave mold combination successively; assembling convex mold combination, adjusting to the horizontal plane and tighten the screw in turn, positioned on the top of concave mold combination, installing positioning guide rod; putting convex mold into concave mold at a predetermined pressure, re-compacting materials, standing its shaping until primary solidification, removing the watching window glass, opening the trough from the side, cleaning up the scraps outside the mold, eliminating the adhesion with sample, reducing the next step enhancing resistance, keeping static state.

(d) Demolding and forming. After materials coagulation, lifting concave mold mechanism away from t trough by the handle, ascending to a certain height then fixed; reinstalling convex mold combination by single mold release plan and fixing position by guide rod; dropping the central guide rod, putting convex mold into concave mold at a certain speed, demolding the samples. To avoid sample edge damages caused by falling after demolding, proper amount of sand could be put in the bottom of trough as buffer. To prevent corrosion, disassembling and cleaning the parts sticky with materials timely after each forming, brushing oil for maintenance when stopping use for a long time. Complete forming process is referred as Fig. 3.

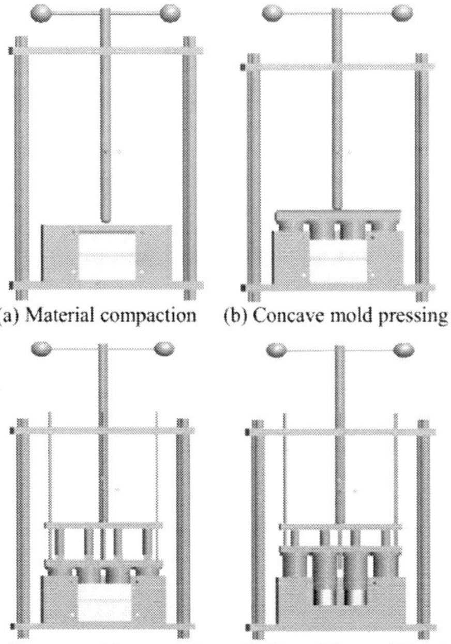

(a) Material compaction (b) Concave mold pressing

(c) Convex mold compaction (d) Demolding and forming

Figure 3. Bedded standard samples forming process.

DOI: 10.1051/matecconf/20167002008

Finally, the physical mechanism and bedded standard samples made by this mechanism are shown in Fig. 4, the mold quality achieves technical requirements. When putting materials, artificial control could be given to each layer for angle, thickness, and position; the quantity of one time forming could be controlled when installing mold combination; pressing force and sample height could be controlled in compaction process, which has outstanding advantages and is better than other type forming mechanisms for bedded samples making. The design got "four column type geotechnical material batch pressing mechanism for bedded standard samples" [5] and "geotechnical experiment standard samples forming combination mold" [6] two national new practical patents authorization.

Figure 4. Physical mechanism and bedded standard samples.

6 Conclusion

A batch pressing mechanism for bedded standard samples is designed by research on existing standard samples forming mechanisms, by putting mold into materials, each layered material, dip angle, thickness and position can be controlled, forming quantity for each lot is adjustable, uniformity of the same lot is also good and all parts are detachable. The mechanism is suitable for small batch forming and has obvious advantages in bedding controlling, forming efficiency, forming quality and operation convenience, etc. which could be used well in similar material simulation experiment, geotechnical experiment, also could be used to make micro physical models with engineering structure.

Acknowledgement

I gratefully acknowledge the financial support from the National Natural Science Fund project (51374134), the Chinese Education Ministry doctoral program project (20133719110005), and the Natural Science Foundation of Shandong Province (ZR2014EL20).

References

1. Z. G. Ma, Study of Simulating Creep Properties of Soft Rock with Latex Cement. Qingdao: Qingdao University of Science and Technology, (2014):43-48.
2. D. Y. Jiang, T. Ren, J. Chen et al. Experimental study of mechanical characteristic of molded salt rock with weak interlayer. *Chinese Journal of Rock Mechanics and Engineering*, (2012), **31**, 9: 1797-1803.
3. Y. M. Tien, M.C. Kuo, C. H. Juang. An experimental investigation of the failure mechanism of simulated transversely isotropic rocks. *International Journal of Rock Mechanics & Mining Sciences*. **43** (2006) 1163-1181.
4. G. M. Zhang, Y. P. Li, X. L. Shi, et al. Research on a model material preparation method for alternate layered rock mass and preliminary experiment. *Rock and Soil Mechanics*, 2011, **32** (supp.2): 284-289.
5. J. G. Li, Y. Y. Wang, W. X. Wang et al. Four column types batch pressing mechanism for geotechnical materials bedded standard samples: China, 201420140161.6 (P). 2014-07-30.
6. Y. Y. Wang, J. G. Li, W. X. Wang et al. Composite mould for geotechnical test standard sample forming: China, 201420140748.7 (P). 2014-07-30.

Study of Micro Grid Hybrid System of Photovoltaic and Diesel Engine

Dwi Novitasari [1], Y. S. Indartono [1,2], J. E. Harjono [3,2], MIrsyad [1,2], Tri D. Rachmildha [4,5] and I. K. Reksowardojo [2,3]

[1] *New and Renewable Energy Research Center, InstitutTeknologi Bandung, Indonesia*
[2] *Faculty of Mechanical and Aerospace Engineering, InstitutTeknologi Bandung, Indonesia*
[3] *Combustion Engine and Propulsion System Laboratory, InstitutTeknologi Bandung, Indonesia*
[4] *Electrical Energy Conversion Research Laboratory, Institut Teknlogi Bandung, Indonesia*
[5] *School of Electrical Engineering and Informatics, Institut Teknlogi Bandung, Indonesia*

Abstract. Indonesia has abundant potentials of new and renewable energy that can be used for electricity generation, especially in rural areas which have no access for grid electricity yet. The energy resources can be from solar, water, biomass or biofuel. Many villagers still use diesel generators to produce electricity in their villages. It is considered expensive because fuel price in rural areas increases 2-3 times than the normal price due to transportation cost. Hybrid system using renewable energy resources is one of the solutions to produce electricity in affordable cost for rural area. The idea is to combine diesel generators and photovoltaic toproduce electricity. Moreover, the diesel engine fuel can be replaced with biofuel. This study will analyze the hybrid system in a small scale which consists of 1kWp photovoltaic and 3 kW diesel engine. Electric load power will vary. The system is controlled by a single bidirectional inverter whichconverts power from DC to AC and vice versa.

1 Introduction

Indonesia has many remote areas and small islands which have yet to get access on electricity. Its geographical condition and profile of demand become the main barriers to develop large electricity grid. In 2014, electrification ratio in Indonesia was at 81.70% [1]. Electrification ratio is defined as the number of households already getting electricity compared to the total number of households in Indonesia. The data show that there were 3,132 of 78,609 villages which have not been electrified yet [2].

On the other hand, Indonesia has various energy sources such as solar, water and bioenergythat can be used for generating electricity in rural areas. In Indonesia, solar energy potential is about 4.8kWh/m^2/day, and wind velocity is about 3-6m/s [3]. The potential of solar energy is distributed in almost all regions but the potential of wind energy is not available in all regions, but it is only in the eastern area of Indonesia such as Papua, Sulawesi and also in coastal areas [3]. Whereasfor bioenergy, Calophyllum inophyllum (Nyamplung), and Jatrophacurcas are some of the plants in Indonesia that can be used as biofuel. The potential yield of these plants to produce bioenergy is about 4,680 and 1,900-2,500kg of oil/ha per year, respectively [4]. Potential area of land for Jatropha planting in Indonesia is 49,5317 million hectares [5].

These renewable energysourcesare highly recommended to generate electricity in remote areas and small islands. Ministry of Energy and Mineral Resources (MEMR) has built 37 solar power generations across Indonesia, particularly in outer islands and border areas in 2014 with a total capacity of 1,470kW for 6,116 households [6]. However, currently most rural communities in Indonesia are generating electricity diesel generators. The problem of using generatorsis the price of diesel fuel and also its impact of air polution. The communities use diesel fuel to operate the generators, while the price of fuel is not cheap in rural areas. It can be 2-3 times more expensive than the price from the government due to the cost of transportation. It causes theelectricity generators able to operate for only 1-3 hours per day, usually in the evening and only for powering the lamps. This is also the reason why using renewable energy, already existing in their areas, is expected to decrease the operational cost and increase the operational time of electricity generation.
Furthermore, the characteristics of renewable energy resources is intermittent in nature, e.g. photovoltaic generatepower depends on the sunlight. When the rainy season comes, there will be only a small energy from sunlight that can be absorbed by photovoltaic. So, the application of renewable energy resources should work together with energy resources that can produce more stable energy. A system which consists of 2 or more resources is called hybrid system. It can consist of all renewable energy or renewable and non renewable energy, such as wind and solar energy, wind, solar and engine using gasoline or biofuel, etc. Considering that

there are generator sets in rural areas, the idea is to produce electricity generation using hybrid system which consists of engine using biofuel and photovoltaic (PV). PV is a good option as renewable energy resource in Indonesia because Indonesia is located in tropical area, so it gets sunlight whole year. That is the background of this research. The experiment will focus on observing and analyzing the characteristics of the hybrid system which consists of 1kWp photovoltaic and 3kW diesel engine in a small scale. The experiment was conducted in the Combustion Engines and Propulsion Systems Laboratory, Institut Teknologi Bandung, Indonesia.

2 Materials

The whole system in this experiment consists of 1kWp photovoltaic, 3kW diesel engine, bidirectional inverter, solar charge controller, battery, load and monitoring system. There are 20 solar panels each with 50Wp. The diesel engine specification is lister type with the power rating of 3000 watts and operates normally at 650 RPM. This engine can be operated using several types of fuel, not only diesel oil but also biofuel which comes from jatropha oil, palm oil, and coconut oil. The inverter type is bidirectional inverter. It does not only convert from DC to AC but also from AC to DC. The specification of inverter is Conext XW+ 7048 by Schneider Electric with continuous power output of 5500W. The specification of solar charge controller is Conext MPPT 60 150 from Schneider Electric as well. The battery capacity is 200Ah consisting of 8 batterieswith 100Ah, 12 VDC, with a configuration of 4 batteries in series to satisfy inverter input voltage, 48 VDC then the batteries connected in parallels so that the capacity become 200Ah. The total load in this experiment is 2500 W consisting of 5 lamps each at 500W. ConextCombox (CC) from Schneider Electric is a monitoring and control system at once. It controls inverter and solar charge controller. Monitoring system using CC can be seen at real time and the data which was captured every minute for 24 hours can be saved in SD card. The hybrid system scheme is shown in Figure 1.

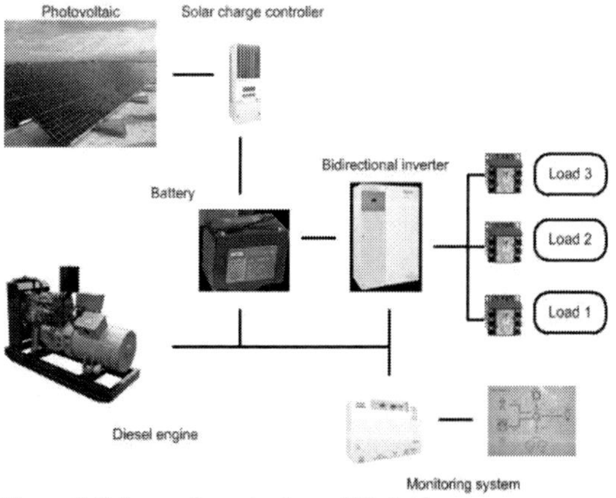

Figure 1. Scheme of smart micro grid hybrid system.

3 Procedure

Hybrid experiment started with the setting of diesel engine supply as the base load. Independent variable to be changed was load power. First, load was set at 1000W, and diesel engine supply of 500W. The load was increased each with 500 watt after 10 minutes, but keeping diesel engine supply constant. So, it would have 4 variations of load when the diesel engine supply was 500W, i.e. 1000W, 1500W, 2000W, 2500W.Then adding the diesel engine supply to become 1000W with 3 variations of load, 1500W, 2000W, 2500W. If the diesel engine produce 1500 watt, the load variations are 2000 W and 2500W and the last was the diesel engine supplywas 2000W which is the maximum real output of diesel engine and the load was 2500 watt. The procedures of experiment are shown in Figure 2. There are some parameters that need to be observed: power, current, voltage and frequency. The entire parameters are saved in a monitoring system and can be analyzed.

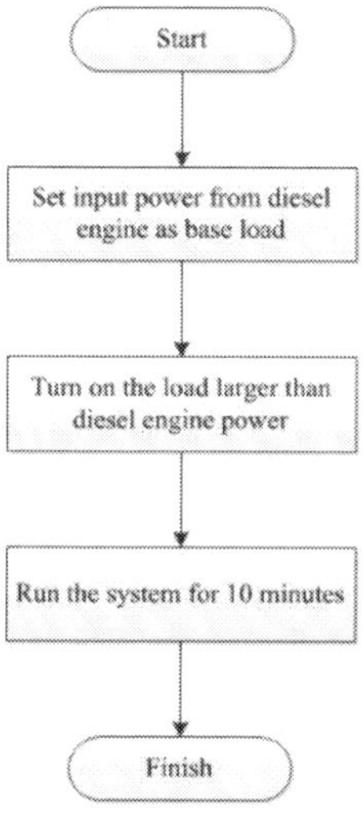

Figure 2. Scheme of experiment.

4 Results and discussion

4.1 PV – Battery - Load

Before starting the experiment, the measurement of solar radiation at the place of experiment is shown in Figure 3. The data was captured in September 4th 2015, about 2 months before the experiment. The data of solar radiation was captured every 5 seconds started from 08:07:15 to 14:39:35. Based on the data, maximum solar radiation was 918.1W/m^2 at 12:00, the minimum value was 88.2W/m^2 at 14:00 and the average was about 683.1W/m^2.

For the experiment, the first configuration used only battery to supply load. There were 2 battery banks parallelly connected in this experiment. The first battery banks supply power to load and second battery bank receives power from photovoltaic. The result of this experiment is shown in Figure 4. The battery will supply more power when the load is higher. It means that battery needs more power to convert from DC to AC. If the high load exists for a long time, the battery will no longer able to supply the demand. The trend of this result match with the convertion efficiency characteristic shown in Figure 5, which shows that for high load,the convertion efficiency will decrease.However, based on the experiment, the efficiency of inverter is slightly lower than its value as shown in characteristic chart.

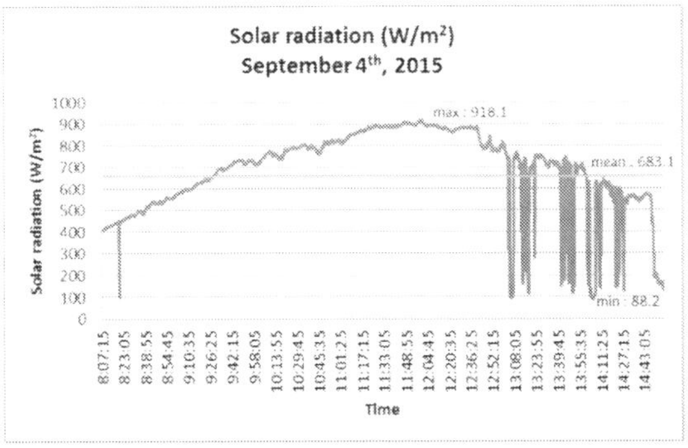

Figure 3. Measured solar radiation in this experiment.

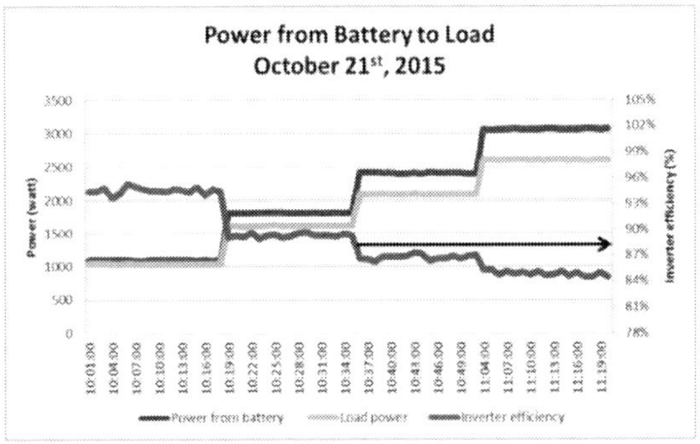

Figure 4. Power supply from battery to load.

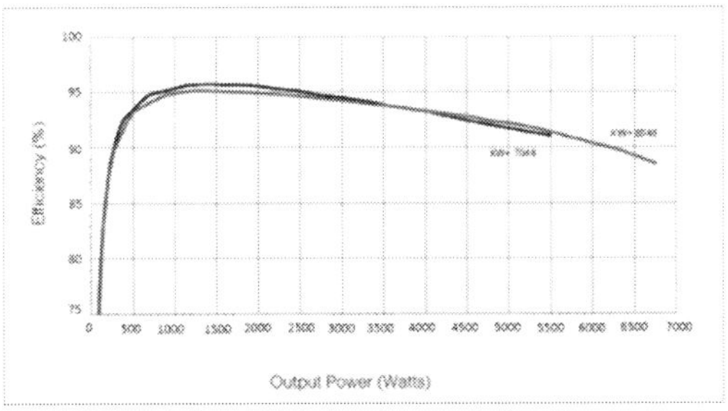

Figure 5. Convertion efficiency characteristic of Inverter Conext XW+ 7048 [7].

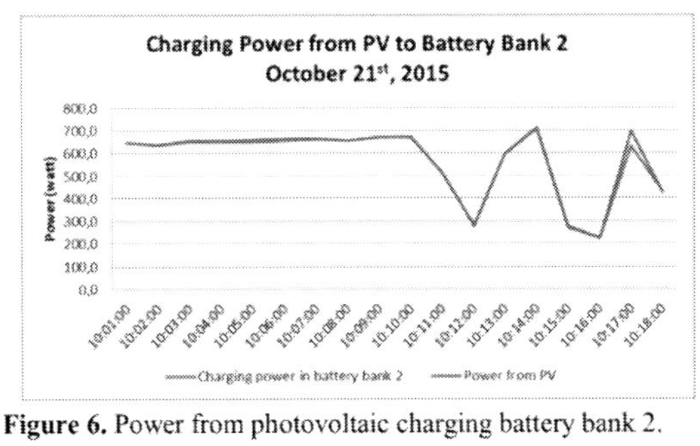

Figure 6. Power from photovoltaic charging battery bank 2.

Figure 6 shows that the power from photovoltaic was saved in battery bank 2. The power from photovoltaic pass through the solar charge controller and bidirectional inverter before it was saved in battery bank 2 in DC mode.

4.2 Diesel engine – Battery – Load

The second configuration in this experimentused diesel engine to supply load.The supply from diesel engine was adjusted from inverter by defining the limit of maximum current which was allowed by inverter.The output voltage from the generator was 230 volt. Equation (1) shown how to calculate the current. I is current (ampere), P is power (watt) and V is voltage (volt).

$$I = \frac{P}{V} \quad (1)$$

Table 1 shows the maximum current that could pass through the inverter. The inverter always searches maximum current value to provide load requirement.

Table 1. Maximum current setting in inverter.

Load (watt)	Maximum current (A)
500	2.5
1000	4.5
1500	6.7
2000	8.7
2500	10.9

In factin this experiment, the current and voltage was not always fixed as shown in Table 1. Based on captured data, for 500W load, the output voltage from generator was about 233-234 volt and the current fixed was 2.3 ampere. The voltage was relatively higher than 230 volt when the load less than 2000W. For 1500W load, the voltage was still 232-233 volt and current was 6.5 ampere, but for the load of 2000W, the voltage was 228-229 volt and current output was 8.2-8.4 ampere. It happened because the RPM of diesel engine did not fixin 650 RPM. The range value for diesel rotation per minutes was 645-655 RPM. Higher or lower than the range value, diesel will not operate in optimum condition.

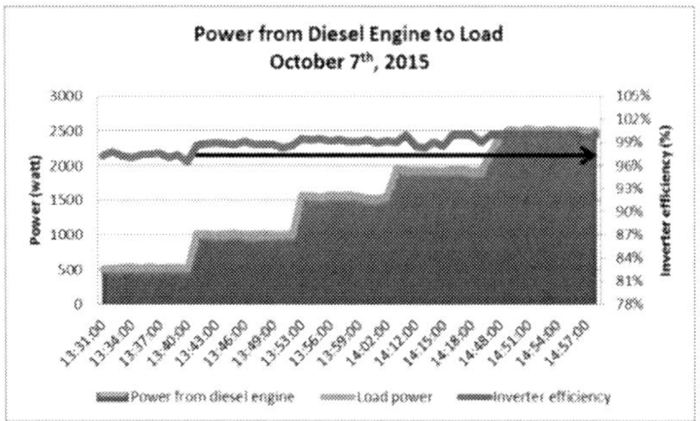

Figure 7. Power supply from diesel engine to load.

Figure 7 shows the results of diesel experiments. The inverter efficiency for diesel engine was different from inverter efficiency for battery. It was due to the converting current from battery. The battery was in

Direct Current (DC) and the load was in Alternating Current (AC) mode, so there was converting efficiency in this process. It is different from diesel engine where the output from generator was AC mode, the same mode with load. So, basically the power only passed through the inverter, without any process. Figure 7 also shows that diesel engine will perform better when the load is high. The diesel engine capacity was 3000 watt, but the experiments did not set until 3000 watt because the power could not pass through the inverter due to frequency drop.

Figure 8 shows that power from diesel engine charge the battery bank 1. The diesel engine supplied power more than load requirements meaning that the power from diesel engine was used for charging the battery bank 1. By using 2 battery banks, it was easier to define the power flow from and to battery. In this case battery bank 1 was charged by diesel engine and battery bank 2 was charged by photovoltaic. When battery bank 2 was charged by photovoltaic, voltage in battery bank 1 will follow the battery bank 2. When battery bank 1 was charged by diesel engine, inverter will automatically prohibit power from PV to charge battery bank 2 and allows diesel engine to charge battery bank 1. The power from PV might be converted as heat.

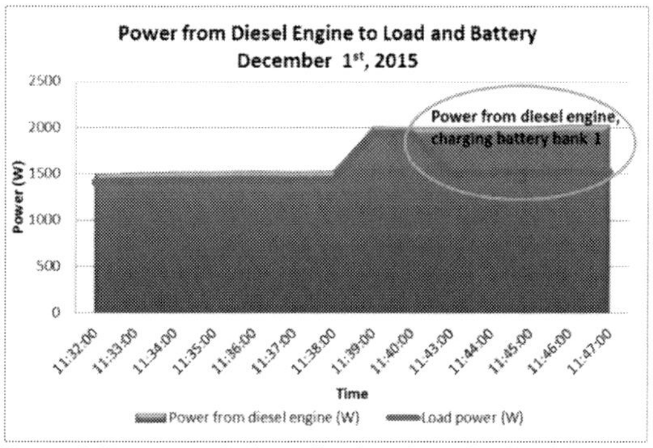

Figure 8. Power supply from diesel engine to load and battery bank 1.

4.3 Diesel engine – Photovoltaic – Battery – Load

The last experiment and the most important part set the system in hybrid configuration. The diesel engine was set below the required load. The diesel engine was run to cover baseload whilebatteries which was charged by PVwould cover the rest of power demand from load. The result of these experiment was shown in Figure 9 and Figure 10.

In Figure 9, for the first 10 minutes diesel engine was set to generate power of 2000W and the load was 2500W. The second condition kept the diesel engine supply of 1500W with variations of loads at 2000W and 2500W. Until the last experiment, the diesel engine supported 500W with variations of loads from 1000W to 2500W.

This experiment started with high supplied power from diesel engine. The power from diesel engine was decrease followed the variations of loads. Based on the experiment, this hybrid system could run well. The load in this experiments used 5 lamps each 500 watt. The supply from diesel engine was controlled by delimit the maximum current. The fuel setting in this diesel engine was controlled manually. It means when there was request from the inverter to supply 500W, and the maximum current was 2.5 ampere, the governor in diesel engine should be controlled how big the aperture of fuel should be opened. The aperture should be controlled in order to keep the rotation of engine in about 650 rotation per minutes (RPM).

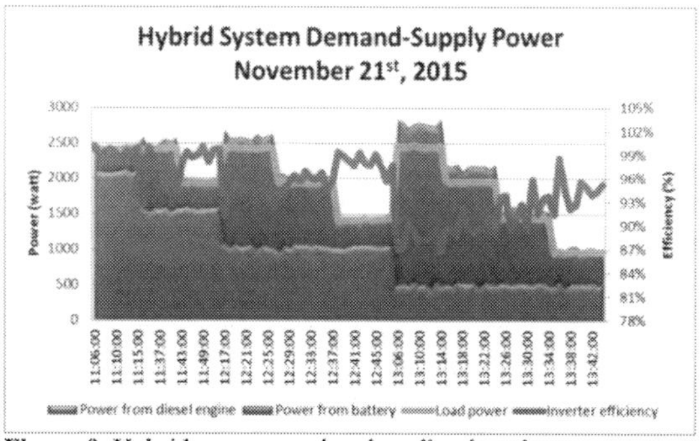

Figure 9. Hybrid power supply when diesel engine power was decrease.

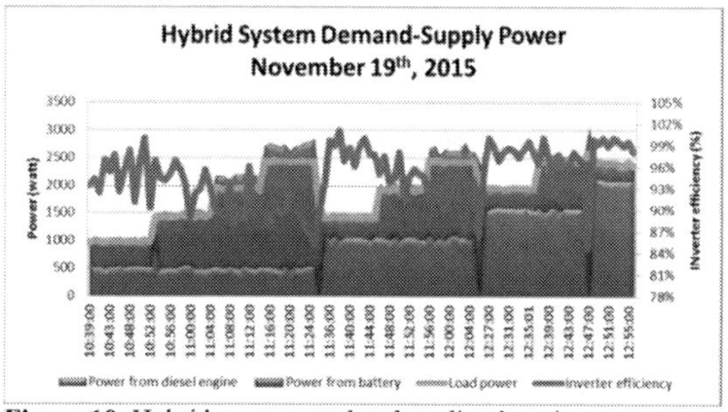

Figure 10. Hybrid power supply when diesel engine power was increase.

Another experiment for hybrid system is shown in Figure 10. The differences between Figure 9 and Figure 10 is power supply from diesel engine. The diesel engine in Figure 9 supplied power from high to low and in Figure 10, supplied power from diesel engine started from low power then increased until its maximum power. Evidently, it has different result. Figure 10 shows that the inverter took time to synchronize with diesel engine when the supply from diesel engine was changed.When it happened, load will be automatically supplied by battery bank 1.

The inverter needs to synchronize with diesel engine because when output power from diesel engine is changed, the RPM of diesel engine will change, which means changing the frequency as well.The inverter has requirement for frequencies from diesel engine. Based on the inverter specification, there are 2 range of frequencies, define as low and high. Low frequency for diesel engine is in range 40-48Hz and high frequency is between 52-68Hz. In Figure 9, the frequency of diesel engine was still in range when the power decreased so that no effect for the load. But in Figure 10, increasing power from diesel engine resulted the diesel engine work harder, so the rotation will be lower then the frequency will be lower than the inverter frequency's range. When it happened, the aperture of fuel should be opened larger than before, so that there would be more fuel go to combustion chamber to keep the rotation of diesel engine at about 650 RPM.

5 Conclusion

The result of experiments shows that: 1) inverter'sefficiency for system consists of photovoltaic-battery-load is about 83%-95% and 2) inverter's efficiency for system consists of diesel engine-load is about 96%-100%. The reasons are by using battery, the inverter must convert the current from DC to AC, while by using diesel engine, the current passes through the inverter without converting current. Based on the experiments, the efficiency of inverter in converting mode (DC to AC mode) is slightly lower from its specification. In hybrid system, the method on operating diesel engine will affect the whole system. If diesel engine started to deliver power from high mode then the power from diesel engine decreased, the system would stable and operate well. But, if diesel engine started to produce low power, then it needed to increase the power production, the inverter needs to synchronize wth diesel engine. In high power produced by diesel engine, time needed to synchronize between diesel engine and inverter was about 1-3 minutes. In this condition, battery bank will supply the load automatically so the load can operate normally.

The use of diesel engine can increase the efficiency of inverter. On the other hand the use of PV in this system is proven reduce the consumption of fuel significantly. Batteries are still necessary in stabilizing system. It is good to use 2 battery banks because it will be easier to capture the power flow from diesel engine or PV. In addition, using 2 battery banks in parallel is good forthe reliability of the whole system than single battery bank.

Acknowledgment

This research received fund from New and Renewable Energy Research Center, Institut Teknologi Bandung and National Strategic Research, Ministry of Research and The Ministry of Higher Education, Indonesia, 2015.

References

1. PT PLN (Persero), Statistik PLN 2014, (2015)
2. PT Perusahaan Listrik Negara (Persero), Rencana Usaha Penyediaan Tenaga Listrik (RUPTL) 2015 - 2024, (2015)
3. Badan Pengkajian dan Penerapan Teknologi, Outlook Energi Indonesia 2014, Jakarta, (2014)
4. M. Irsyad, Y. S. Indartono, A. Suwono, and A. D. Pasek, Thermal Characteristics of Non-edible Oils as Phase Change Materials Candidate to Application of Air Conditioning Chilled Water System, *IOP Conf. Ser. Mater. Sci. Eng.*, **88**:012051, (2015)
5. A. Kuncahyo, Priyohadi Fathallah and Semin, Analisa Prediksi Potensi Bahan Baku Biodiesel Sebagai Suplemen Bahan Bakar Motor Diesel di Indonesia, *J. Tek. Pomits*, **2**(1):1-5, (2013)
6. Ferial, 2014, Pulau Terluar Dan Perbatasan Dapat 37 Unit PLTS, 2015. [Online]. Available: http://ebtke.esdm.go.id/post/2015/01/21/763/2014.pulau.terluar.dan.perbatasan.dapat.37.unit.plts
7. Schneider Electric, *Owner's Guide Conext ™ XW Inverter / Charger*, (2014)

The Performance Evaluation of Horizontal Axis Wind Turbine Torque and Mechanical Power Generation Affected by the Number of Blade

Rodney H.G. Tan[1] and Matthew Y.W. Teow[2]

[1]UCSI University, Faculty of Engineering, Technology and Built Environment, 56000 Kuala Lumpur, Malaysia
[2]KDU University College, School of Engineering, 40150 Shah Alam, Malaysia

Abstract. This paper presents the evaluation of horizontal axis wind turbine torque and mechanical power generation and its relation to the number of blades at a given wind speed. The relationship of wind turbine rotational frequency, tip speed, minimum wind speed, mechanical power and torque related to the number of blades are derived. The purpose of this study is to determine the wind energy extraction efficiency achieved for every increment of blade number. Effective factor is introduced to interpret the effectiveness of the wind turbine extracting wind energy below and above the minimum wind speed for a given number of blades. Improve factor is introduced to indicate the improvement achieved for every increment of blades. The evaluation was performance with wind turbine from 1 to 6 blades. The evaluation results shows that the higher the number of blades the lower the minimum wind speed to achieve unity effective factor. High improve factors are achieved between 1 to 2 and 2 to 3 blades increment. It contributes to better understanding and determination for the choice of the number of blades for wind turbine design.

1 Introduction

Wind energy is one of the top choices of renewable energy resources widely harvest by many countries around the world particularly China, Germany, UK and USA. According to Global Wind Energy Council 2014 statistics report [1], wind energy in the past decade has recorded an average growth rate of 21% and global total installed capacity of 318105MW as end of year 2013. Wind turbines are used to harvest wind energy. It has the advantage over others renewable energy resources harvesting method in term of smaller footprint per MW making it possible to install on uneven land terrain and offshore. If the installed location is strategically favor to wind meteorology, wind energy allows continuous harvesting day and night throughout the years.

Wind energy is extracted using either horizontal or vertical axis wind turbine configuration. Wind turbine relies on its rotating blades to convert wind energy into mechanical energy. There are many literatures on how wind turbine performance affected by the blade efficiency and aerodynamic design [2], [3]. However, there is not many literature reported on how the number of blades affect the performance of a given wind turbine. Even it does, there is no evaluation on the mechanical power and torque generation with respect to the number of blades [4]. This paper presents the evaluation of horizontal axis wind turbine affected by the number of blades in terms of mechanical power and torque output at a given wind speed.

2 Wind turbine model

Wind turbine extracts wind energy when wind flows through the blades perpendicularly resulting in mechanical rotational motion due to aerodynamic force. The blade rotational motion that sweeps across its area converts into mechanical energy which in turn drives the generator to generate electricity. The mechanical energy in watt of a given wind turbine can be modeled using Eq. (1), where ρ is the air density of 1.225 kg/m^3, v is the wind speed in m/s, A is the wind turbine sweep area of πr^2 where r is the blade radius. 2 x 252 mm.

$$P_m = \frac{1}{2}\rho A v^3 C_p(\lambda,\beta) \tag{1}$$

The power coefficient C_p is the function of wind turbine tip speed ratio λ and the blade pitch angle β. The tip speed ratio λ is defined in Eq. (2) as the ratio of the turbine blade tip speed Ts in m/s to the wind speed v in m/s.

$$\lambda = \frac{Ts}{v} \tag{2}$$

The generic wind turbine power coefficient [5] can be modeled using Eq. (3) and Eq. (4), where c_1=0.5176, c_2=116, c_3=0.4, c_4=5, c_5=21, c_6=0.0068.

$$C_p = c_1\left(\frac{c_2}{\lambda_i} - c_3\beta - c_4\right)e^{\frac{-c_5}{\lambda_i}} + c_6\lambda \tag{3}$$

$$\frac{1}{\lambda_i} = \frac{1}{\lambda + 0.08\beta} - \frac{0.035}{\beta^2+1} \tag{4}$$

The power coefficient performance with respect with tip speed ratio for a fixed pitch wind turbine of blade pitch angle β=0 is shown in Figure 1. For this generic wind turbine model, it can be clearly see that the highest power coefficient of 0.4798 can be achieved at tip speed ratio of 8.

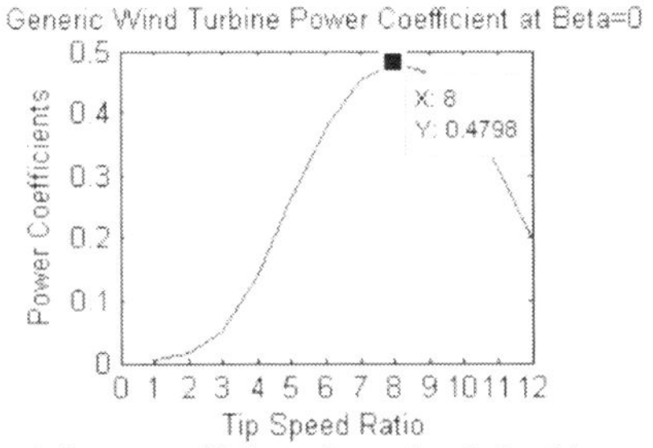

Figure 1. Power coefficient of generic wind turbine model.

The wind turbine torque in newton-meters can be determined using Eq. (5), where P_m is the wind turbine mechanical power in watts, rpm is the wind turbine angular velocity in revolution per minute.

$$P_t = \frac{P_m \cdot 60}{2\pi \cdot rpm} \quad (5)$$

3 Blade wind energy extraction

The basic principle for a wind turbine to effectively extract wind energy is to complete the sweep area with its blades in one second. With refers to Figure 2, for a single blade B1 wind turbine to effectively extract wind energy passing through its sweep area A1 is to complete one revolution in one second. Therefore, the frequency must not be less than 1Hz to effectively extract wind energy without losses. The wind turbine tip speed in m/s is defined in Eq. (6) where ω is the angular velocity of $2\pi f$ and r is the blade radius in meter. Thus, the tip speed require is proportional to the length or radius of the blade.

$$Ts = \omega \cdot r \quad (6)$$

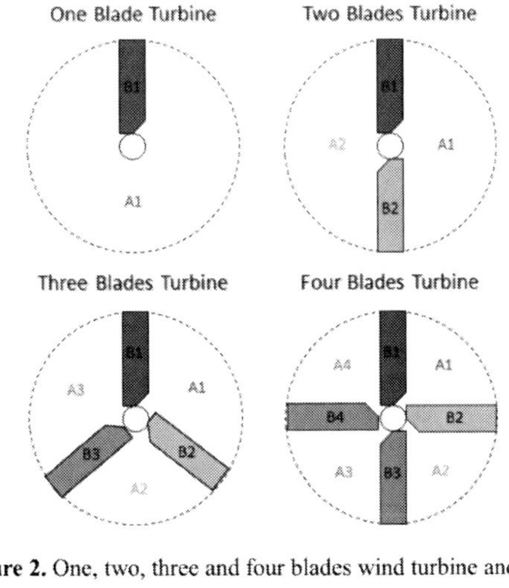

Figure 2. One, two, three and four blades wind turbine and its sweep area.

Since the frequency require to complete one revolution is 1Hz, tip speed became the circumference of the wind turbine in meter, which can be directly translated to the minimum tip speed in m/s to effectively extracts the wind energy.

For two blades wind turbine, the frequency needed to complete the sweep area is half or 0.5Hz, simply because when the first blade B1 sweep the first half area A1 of the wind turbine, at the same time the second blade B2 at its opposite direction sweep the second half area A2 of the wind turbine. Thus, the frequency needed to complete the sweep area is half. For three blades wind turbine, blade B1 sweep area A1, blade B2 sweep area A2 and blade B3 sweep area A3 at the same time, therefore it require only one third of the frequency 0.33Hz to complete the turbine sweep area. The same principle applies to any number of blade as defined in Eq. (7), where f is the frequency needed to complete the turbine sweep area and n is the number of blade.

$$f = \frac{1}{n} \quad (7)$$

In practice, the wind turbine rotational speed is expressed in revolution per minute (rpm), as it is used in relation to drive the electric generator. Eq. (8) shows the wind turbine *rpm* required to complete the sweep area in relation to the number of blade n.

$$rpm = \frac{60}{n} \quad (8)$$

In summary the wind turbine rotational frequency and rpm is the reciprocal function of the number of blades. The higher the number of blades and slower the tip speed required to complete the wind turbine sweep area. With Eq. (6) and Eq. (7) in place, we can rearrange Eq. (2) to determine the minimum wind speed Eq. (9) required for the wind turbine to complete the wind turbine sweep area and effectively extract wind energy.

$$v_{\min} = \frac{2\pi f r}{\lambda n} \quad (9)$$

If the wind speed v experienced by a wind turbine for a given n number of blades is greater than the minimum wind speed $v_{\min}$, then the wind turbine effective factor η is 1. On the other hand, if the wind speed v experienced by the wind turbine is less than $v_{\min}$, then the wind turbine effective factor can be determined in Eq. (10). The reduction of effective factor is simply due to the turbine tip speed drive by the wind speed is not fast enough to complete the sweep area.

$$\eta = \frac{v}{v_{\min}} \qquad \eta = 1 : v \geq v_{\min} \quad (10)$$

In the general wind turbine mechanical power equation the effective factor η is 1. Therefore effective factor η can be neglected in Eq. (1). However, with the wind turbine effective factor η taken into account, the wind turbine mechanical power can now be determined in Eq. (11).

$$P_m = \frac{1}{2}\rho A v^3 C p \eta \quad (11)$$

4 Results and discussion

To demonstrate how the wind turbine mechanical power affected by the number of blades and wind speed. A commercial wind turbine EWT500kW was chosen as reference. The EWT500kW is a three blades wind turbine with blade radius of 26 meter, cut-in speed of 2.5m/s and rated wind speed of 10m/s. The tip speed ratio is assumed to be maintained at 8 from cut-in to rated wind speed based on the generic wind turbine model.

Table 1 shows the Wind turbine mechanical power generation with respect to the number of blades, wind speed and tip speed. It is arranged in such a way that the wind turbine mechanical power highlighted in bold refers to the minimum wind speed for a given number of blades. For example, the minimum wind speed for the reference three blades wind turbine is 6.8 [m/s] and produces mechanical power of 196.9 [kW] in bold. At the wind speed of 10.2 [m/s], the turbine produces 664.6 [kW] mechanical power, which is above the rated 500 [kW] electrical power rating for the turbine. This indicates that the turbine generates more than enough mechanical power to drive the rated 500 [kW] electrical power generation conversion.

At the wind speed of 6.8 [m/s] for every additional increment of blade from 3, 4, 5 and 6 blades to the same turbine, the mechanical power produce is the same, this is because the minimum wind speed for 4, 5 and 6 blades are lower than 6.8 [m/s], therefore the effective factor is 1. On the other hand, at wind speed 6.8 [m/s], if the blades is reduce from 3, 2 and 1 blade, the effective factor less than 1 is taken into effect and the turbine mechanical power reduces from 196.9 [kW] to 131 [kW] and 65.6 [kW] respectively. Figure 3 illustrate Table 1 in graphical plot. The similar characteristics apply to wind turbine torque generation with respect to the number of blades, wind speed and revolution per minute as shown in Table 2.

Table 1. Wind turbine mechanical power generated with respect to wind speed, tip speed and number of blades.

Wind Speed [m/s]	Turbine Blade Tip Speed [m/s]	Number of Blades					
		1	2	3	4	5	6
		Wind Turbine Mechanical Power [kW]					
20.4	163.4	**5295.4**	5295.4	5295.4	5295.4	5295.4	5295.4
10.2	81.7	331	**661.9**	661.9	661.9	661.9	661.9
6.8	54.5	65.4	130.8	**196.1**	196.1	196.1	196.1
5.1	40.8	20.7	41.4	62.1	**82.7**	82.7	82.7
4.1	32.7	8.6	17.3	25.9	34.6	**43**	43
3.4	27.2	4.1	8.2	12.3	16.3	20.4	**24.5**

Table 2. Wind turbine torque generated with respect to wind speed, rpm and number of blades

Wind Speed [m/s]	Turbine Rotational Speed [rpm]	Number of Blades					
		1	2	3	4	5	6
		Wind Turbine Torque [kNm]					
20.4	60	**843.6**	843.6	843.6	843.6	843.6	843.6
10.2	30	105.5	**210.9**	210.9	210.9	210.9	210.9
6.8	20	31.2	62.5	**93.7**	93.7	93.7	93.7
5.1	15	13.2	26.4	39.5	**52.7**	52.7	52.7
4.1	12	6.8	13.7	20.5	27.4	**34.1**	34.1
3.4	10	3.9	7.8	11.7	15.6	19.5	**23.4**

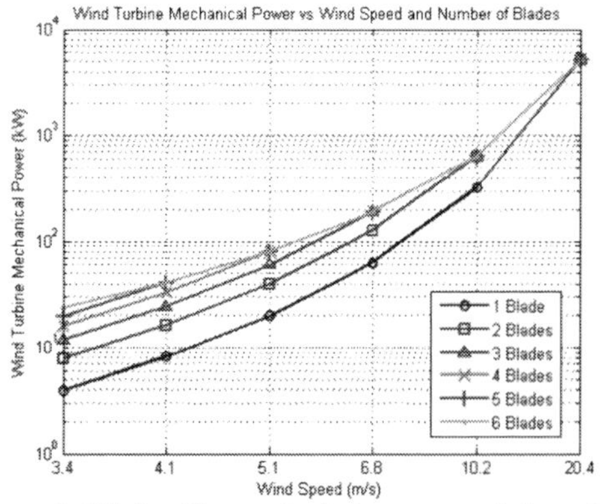

Figure 3. Wind turbine mechanical power vs wind speed and blades.

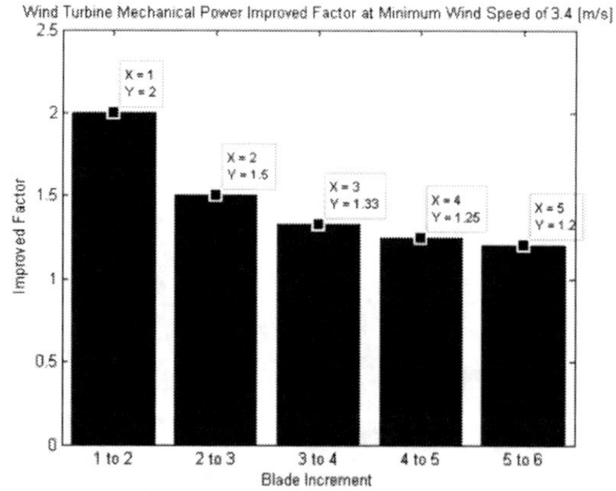

Figure 4. Improve factor for each blade increment.

The improve factor for every incremental number of blades shows in Figure 4 is fundamentally governed by Eq. (12), where n_{ref} is the reference number of blades.

$$\frac{n_{ref}+1}{n_{ref}} \tag{12}$$

In summary, the higher the number of blades, the lower the minimum wind speed and tip speed require to achieve unity effective factor. The improve factor decreases as the number of blades increases. It is observed in Figure 4 that the improve factor from 3 to 4 blades onward is significantly lower as compared to 1 to 2 and 2 to 3 blades increment. This is the main reason why modern large scale wind turbines are designed with only three blades. Secondly, three blades design is more optimum design in term of cost per blade and electric power generation aspect. This leads to modern large scale horizontal axis wind turbine to have three blades standard design. However, for micro scale turbine operates at low wind speed maybe more feasible in term of cost per blade to have more than three blades design to achieve unity effective factor at lower minimum wind speed.

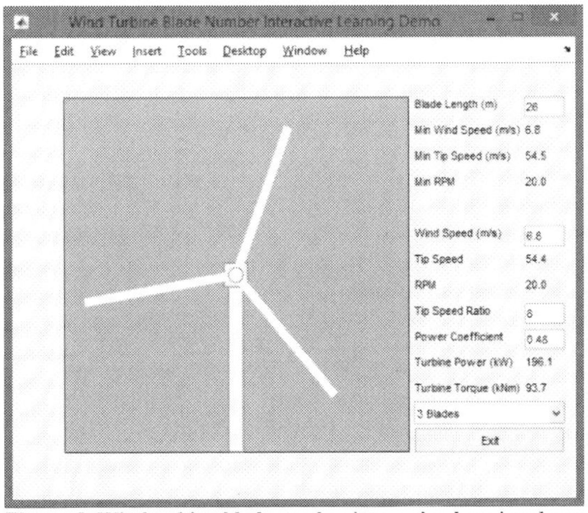

Figure 5. Wind turbine blade number interactive learning demo.

A Wind Turbine Blade Number Interactive Learning Demo was developed in MATLAB as shown in Figure 5. It provides interactive demonstration on how the blade number affect the turbine torque and mechanical power which allows the reader to reproduce the results shown in Table 1 and 2. This interactive demo is made available for reader to download at Mathworks official MATLAB Central File Exchange link below http://www.mathworks.com/matlabcentral/fileexchange/55536-wind-turbine-blade-number-interactive-learning-demo

5 Conclusion

The relationship of the wind turbine mechanical power, torque, wind speed and the number of blades are presented. The evaluation results shows that the higher the number of blades the more effective the wind turbine extracts wind energy at low wind speed. However, high improve factor can only achieve at 1 to 2 and 2 to 3 blades increment. This paper contributes to better understanding and determination for the choice of the number of blades for wind turbine design.

References

1. Global Wind Statistic 2013, Global Wind Energy Council, (2014)
2. A. Sarkar, D. Kumar Behera, International Journal of Scientific and Research Publications, **2**, (2012)
3. P. J. Schubel, R. J. Crossley, Energies, **5**, 3425, (2012)
4. M. Predescu, A. Bejinariu, O.Mitroi, A. Nedelcu, International Conference on Renewable Energies and Power Quality, (2009)
5. A. W. Manyonsge, R. M. Ochieng, F. N. Onyango, Applied Mathematical Sciences, **6**, 4527-4536 (2012)

Numerical Simulation of Combustion Chamber for Button Turbojet Engine

Hongpeng Ma, Shuzhou Fang, Hang Gao, Teng Li and Guanlin Fang

School of Aerospace Science and Engineering, Beijing Institute of Technology, Beijing,100081, China

Abstract. To provide reference data for ultra-micro combustor, a new type button turbojet engine was designed and simulated the combustion's steady-state process. The boundary condition of inlet was calculated using isentropic numerical calculation, taken into turbulent chemical reaction, heat radiation, and so on, getting the combustion chamber's steady-state of the velocity, temperature and component concentration distribution, analysis the fuel/air flow and backflow, combustion efficiency and total pressure recovery coefficient, and compared with the experimental data. The calculation results can accurately reflect the actual combustion. The results show that combustion chamber exit velocity is about 65m/s, outlet temperature is around 1000K, the simulation and experimental data are similar, combustion chamber structure design is reasonable, and this paper will provide a basis for the future improvement of the millimeter scale turbojet engine.

1 Introduction

As a high energy density of tactical powerplant, Micro Turbojet Engine(MTE) can be applied to small UAV (Unmanned Aerial Vehicle), target drone, cruise missiles and other kinds of MAV(Micro Aerial Vehicle). It will play an increasingly important role in the future air war, military reconnaissance, environment monitoring and even anti-terrorism operations, etc, [1]-[3]. However, combustion chamber as a key technology of MTE, due to the requirements of more and more miniaturization (millimeter level), there are many technical difficulties: (1) the structure size is too small to MEMS processing easily; (2) compared to normal engine, bigger surface-to-volume ratio will increase its heat loss; (3) burning resident time is short, easy to appear the incomplete combustion;(4) resistance factors are complicated, many process related to the structure size are diffenrent, such as fule injection, mixing of fluid mechanics, the flow field analysis will become more difficult, [4]-[6].

Up to now, both home and abroad researches are mainly concentrated on the study of micro turbojet engine's traditional structure design and improvement, [7]-[9], the flow field analysis of the micro small engines are very rare, millimeter turbojet engine's three-dimensional flow field researches are basically in the blank stage. This paper discussion is based on the researches at home and abroad [10]-[15], design a round Button Turbojet Engine (BTE), and do some numerical simulations for the combustion chamber. Therefore, not only we got the combustion efficiency factors, characteristic curve and total pressure recovery coefficient, but also obtained the distribution of temperature, velocity and component concentration about the internal combustion chamber. This work will provide the foundation for the future development, processing and production of the micro turbojet engine's combustion chamber.

2 Physical model and calculation method

2.1 Physical model

Reference some of past design experience, restricted to small structure size and MEMS processing, combustion chamber structure was finally confirmed as follows: the combustion chamber is divided into six independent fan-shaped combustion chamber, diffuser blade surface form the outer wall of the combustion chamber, combustion chamber cavity is composed of backflow region wall and the diffuser blade wall, choose the methane as fuel, each entrance of combustion chamber set an intake pipe. The overall vertical view of engine (XOZ plane) and one sixth engine structure as shown in Fig. 1 and Fig. 2 respectively.

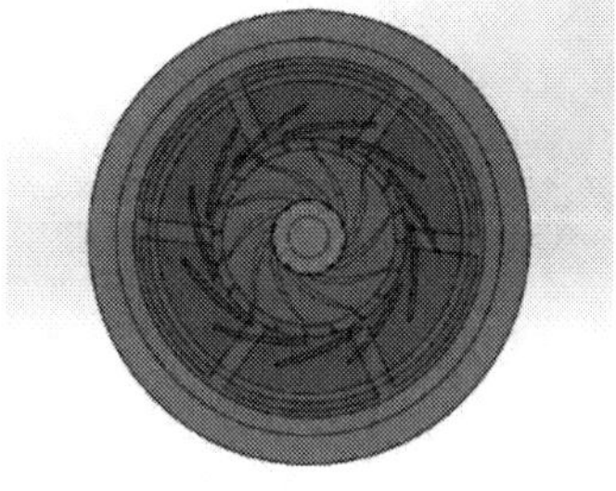

Figure 1. Overall vertical view of engine (XOZ plane)

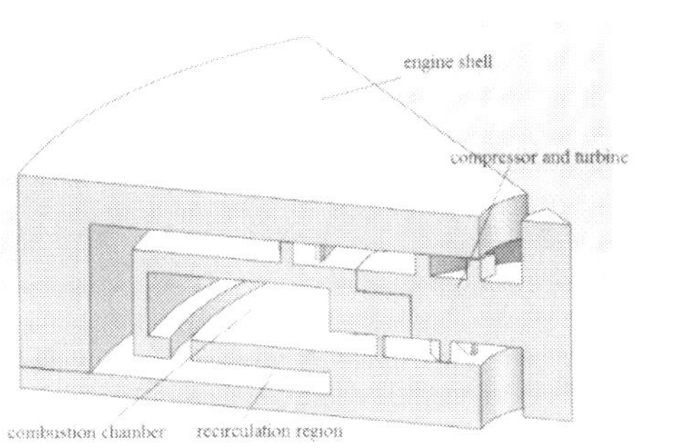

Figure 2. One sixth turbo engine structure

Combustor design dimensions are as follows: outer diameter D_1=16.3mm, inner diameter D_2=15.74mm, combustion chamber height H=1.2mm, each volume of the combustion chamber V_c'=22.692mm^3, total volume V_c=136.152mm^3, outlet temperature of the combustion chamber designed for 1000K, affordable for turbine materials, another innovation is an U-shaped recirculation zone at the entrance of the combustion chamber in order to improve the stability of the combustion flame.

2.2 Combustor thermodynamic calculation

The design of thermodynamic parameters: inlet combustor temperature T_2^* =400K, outlet combustor temperature T_3^* =1000K, fuel initial temperature T_f =350K, fuel mass flow rate m_f =0.002(kg/s), the combustor oil-gas ratio σ=0.0225, methane combustion heat value H_u=50.07(kJ/kg).

2.3 Combustion model

Calculation using the EBU-Arrhenius turbulent combustion model, this model takes into account the turbulence and chemical dynamic factors influence on chemical reaction rate. This model's average response rate is showed as following formula (1):

$$\overline{w} = \min\left\{\overline{w_{st}}, \overline{w_{sv}}\right\} \tag{1}$$

where $\overline{w_{st}} = C_{EBU}\rho g\varepsilon / k$, $\overline{w_{sv}} = A\rho^2 Y_1 Y_2 \exp(-\frac{E}{T})$, $\overline{w_{st}}$ is the response rate controlled by turbulence, C_{EBU} is empirical constant, g is determined by the corresponding transport equation, $\overline{w_{sv}}$ is response rate controlled by laminar flow.

2.4 Radiation model

Common thermal radiation model include Zone Method, Heat Flux, Discrete Transfer Method, Monte-Carlo Method and the Discrete Coordinate Method, etc. [16]. This article choose the Discrete Transfer Method, because it is wider using in engineering application and has higher precision in sloving flow equation.

The radiation heat transfer's integro-differential equations is as following formula(2):

$$s \cdot \frac{\partial I}{\partial x_j} = k_a I_b - (k_a + k_s) I + \frac{K}{4\pi} \int \Phi(\Omega, \overline{\Omega}) I d\Omega \tag{2}$$

where I is radiation intensity, I_b is black-body radiation intensity, s is direction vector, k_a is medium absorbance, k_s is medium scattering coefficient, $\Phi(\Omega,\overline{\Omega})$ is phase function and show the scattering characteristics of spatial distribution.

2.5 Turbulence model

In this paper, using double combustion model to simulate the interior flow field, the Reynolds Number is based on the wall distance, $\text{Re}_y = \frac{\rho\sqrt{k}y}{\mu}$. Where k is turbulence energy, y is the distance between computational grids and the wall, turbulence viscosity ratio and Reynolds number have direct ratio relations, taken the experience value between 1 to 10, turbulence intensity(I) can be calculated by the following formula(3) :

$$I = 0.16 \times (Re)^{(-1/8)} \tag{3}$$

where Re is Reynolds number, given by the study of Alan h. Epstein, et al, [17]. In high Reynolds number turbulence model, compared with standard k-ε model, k-ε / RNG model equation has a new additional generated item, it can improve the simulated result of complex rapid flow. In the calculation software of FLUENT, double-layer model through turbulence model combined with enhanced wall function. In order to guarantee the accuracy of calculation, the wall grid is encrypted, ensure wall boundary layer $y^+ < 5$, other parameters set as default.

2.6 Computational zone, mesh and boundary conditions

Because of the engine combustion chamber is divided into six same size single chamber, refer to Fig. 1. Computational zone is choosen to one sixth combustion chamber, using ICEM divides the unstructured grids, the total number of grids are 2431003, partial encryption, combustion chamber inlet conditions set on the basis of the compressor's integration simulation results, computational mesh as shown in Fig. 3.

Boundary condition setting :

- Adopt speed inlet conditions, the numerical simulation results of cold combustion flow field as the the

DOI: 10.1051/matecconf/20167003003

combustion chamber's import speed border, get the entrance velocity vector distribution, and then adjust inlet velocity according to the residual gas coefficient, flow flux, import area, density and other conditions.

- Set pressure-outlet for outlet conditions.
- Two sides of sector surfaces are spin cycle boundary, the combustion chamber internal walls set as fluid-structure coupled wall.

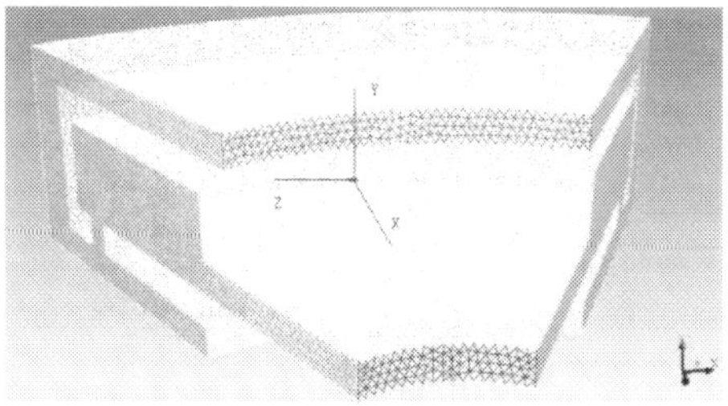

Figure 3. Computational mesh

3 Calculation result

Because the whole combustion chamber around Y axis is symmetrical rotation period, data analysis are mainly selected in X=0mm cross section and Y=0.45~0.6mm cross section whicn can reflect the combustion chamber internal component concentration and chemical reaction.

3.1 Velocity field analysis

Fig. 4 is the velocity distribution of X=0mm cross section which is the combustion chamber's center section. The figure shows that primary combustion zone is full of the whole combustion chamber, outlet velocity is about 65m/s. Fig. 5 shows velocity nephogram marked primary combustion zone and backflow zone. As shown in Fig. 5, backflow zone exists at the entrance of the combustion chamber, which is beneficial to air to enter, improves the gas's burning time and the stability of the flame.

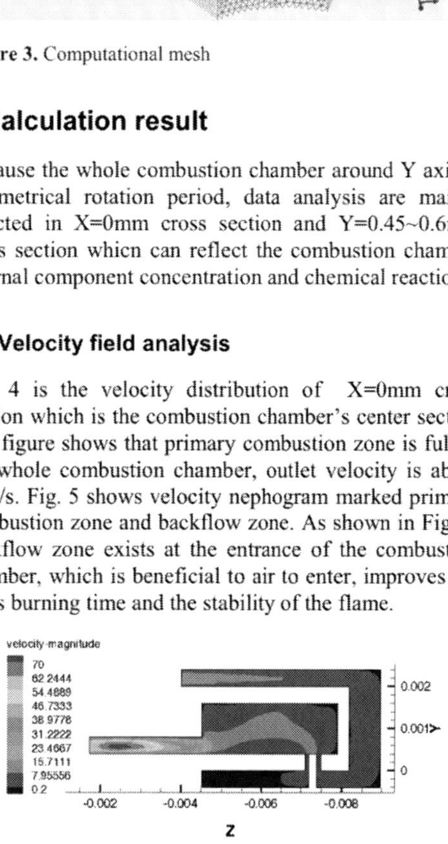

Figure 4. X=0mm cross section velocity nephogram

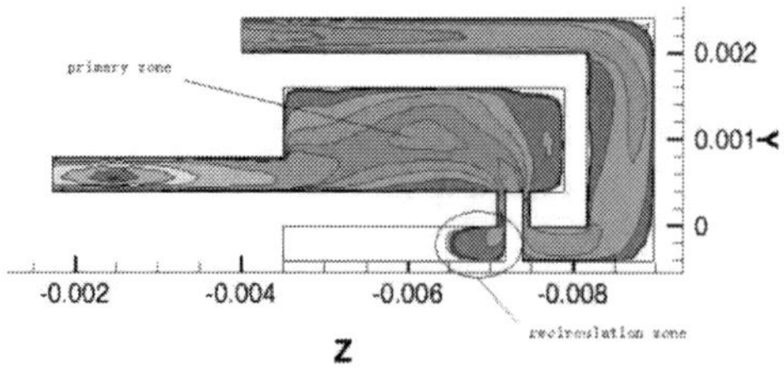

Figure 5. Marked primary zone and backflow zone of X=0mm cross section velocity nephogram

The following Fig. 6 is the velocity nephogram of Y=0.5mm cross section. As seen in Fig. 6, the exit of guide vane changed the originally vertical velocity vector into an angle of oblique direction which is perpendicular to the export turbine rotor blade surface, the high-speed fule gas impact the turbine rotor and get the power for driving the compressor.

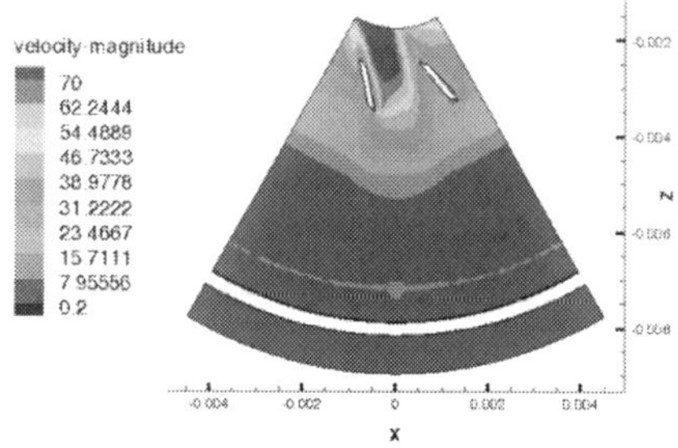

Figure 6. Y=0.5mm cross section velocity nephogram

3.2 Temperature field analysis

In order to analyze the main combustion distribution area, select multiple consecutive cross section and semi-section X=0mm for the whole flow field's temperature analysis. The selected cross sections are Y=0.45mm, Y=0.5mm, Y=0.55mm and Y=0.6mm. Fig. 7~Fig. 12 show the temperature nephogram respectively. As seen in these figures, burning mainly concentrate in the region of the Y=0.4~0.6mm, the mixture of fuel and air combust sufficiently, the highest temperature is 1800K, the average temperature is about 1000K. Burning area presents a fan-shaped distribution, fit with the shape of combustion chamber design, and the wall temperature distribute uniformly, even there is no centralized hot point, all of these prove the rationality of the design of the combustion chamber powerfully.

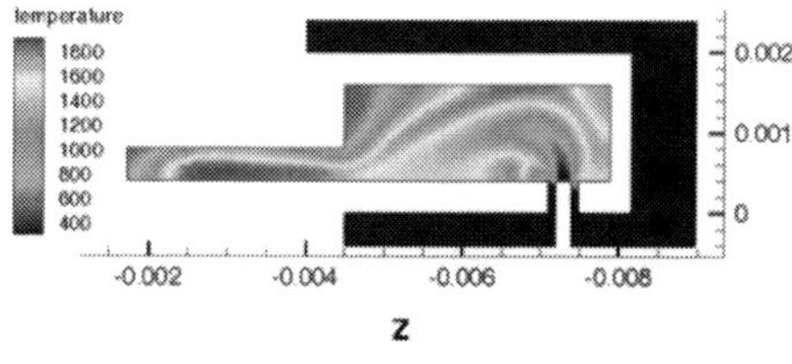

Figure 7. X=0mm cross section temperature nephogram

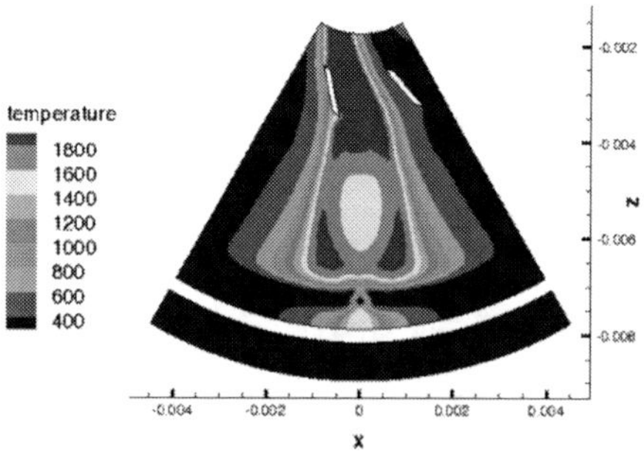

Figure 8. Y=0.45mm cross section temperature nephogram

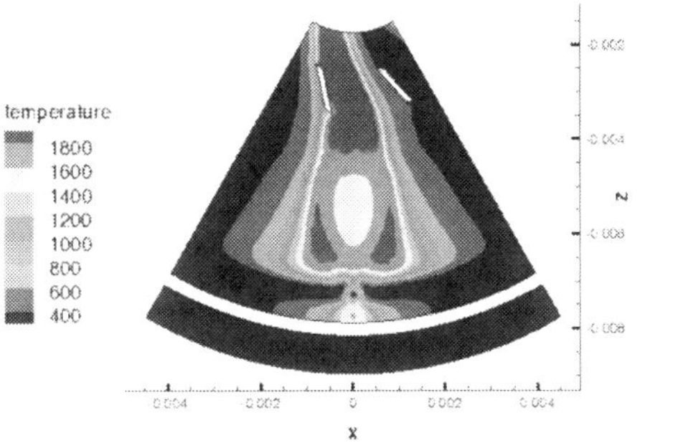

Figure 9. Y=0.5mm cross section temperature nephogram

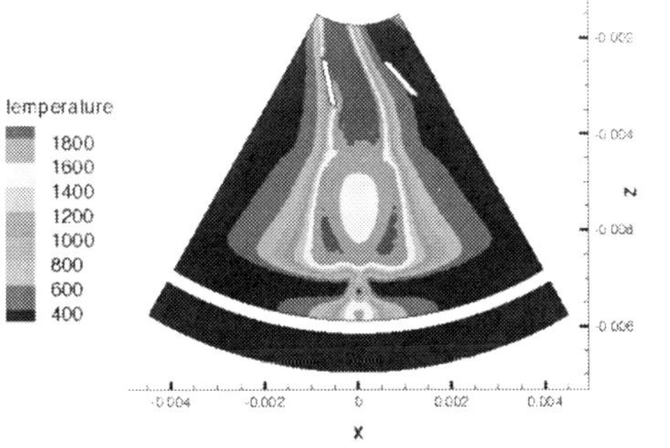

Figure 10. Y=0.55mm cross section temperature nephogram

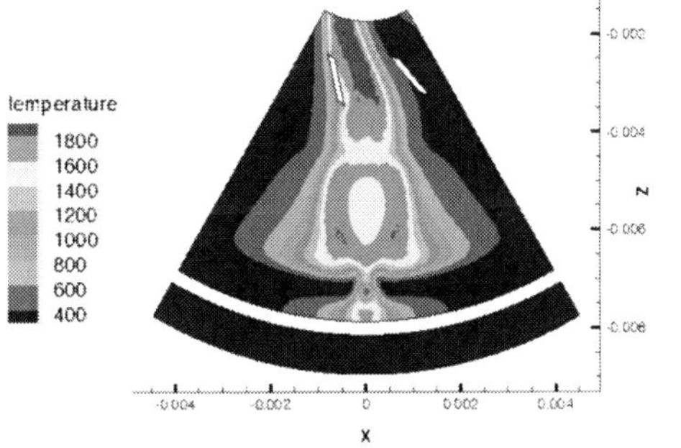

Figure 11. Y=0.6mm cross section temperature nephogram

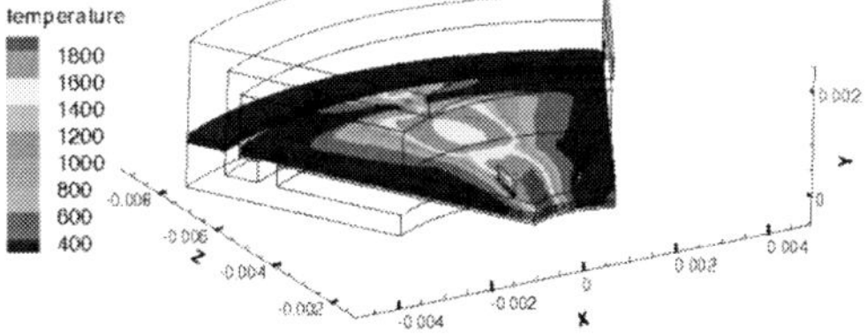

Figure 12. Y=0.45~0.6mm 3D temperature nephogram

3.3 Component concentration field analysis

Take the fuel inlet section X=0mm and Y=0.5mm cross section which can perfectly present the main combustion area to analyze the concentration of components. Fuel components of CH_4 concentration distribution as shown in Fig. 13 and Fig. 14, it is observed that the highest fuel concentration centres on these areas of Y=0.4~0.6mm. Combustion occurs in the primary zone, totally accord with the former analysis data of temperature. As the axis distance increase, burning is ongoing, fuel concentration is getting lower and lower. The outlet fuel mass fraction is very small, fuel combustion efficiency is above 90%, it is proved that the overall design of the combustion chamber is reasonable.

Fig. 15 distribution of O_2 concentration in the section of Y=0.5mm, it can be seen the concentration distribution is in a triangular shaped diffusion, so as to ensure the stability of combustion. Fig. 16 and Fig. 17 are for the concentration distribution of combustion products of CO_2 and H_2O components, the concentration distribution in the diagrams proves once again the position of theoretical main combustion zone.

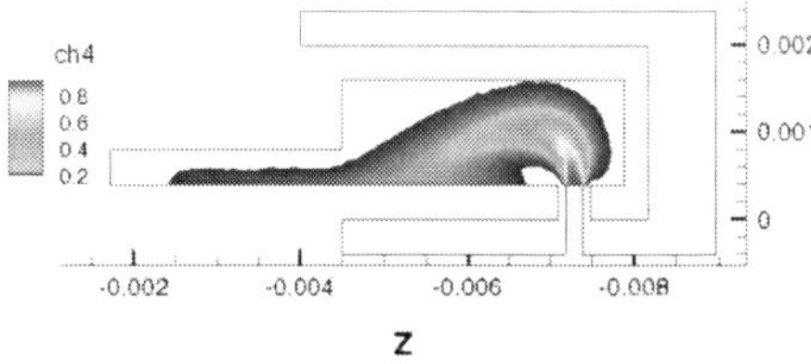

Figure 13. X=0mm cross section CH_4 concentration distribution

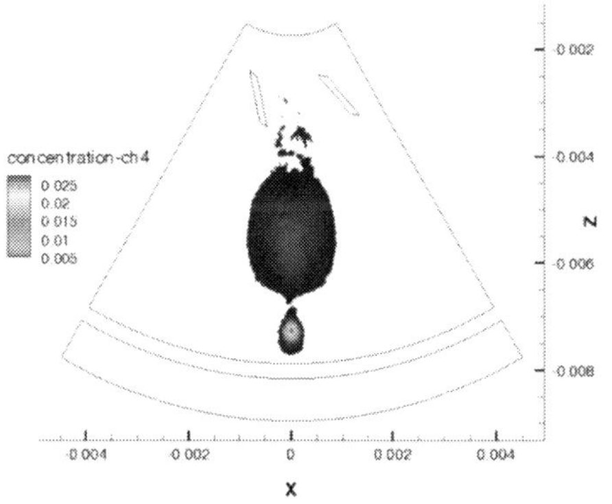

Figure 14. Y=0.5mm cross section CH_4 concentration distribution

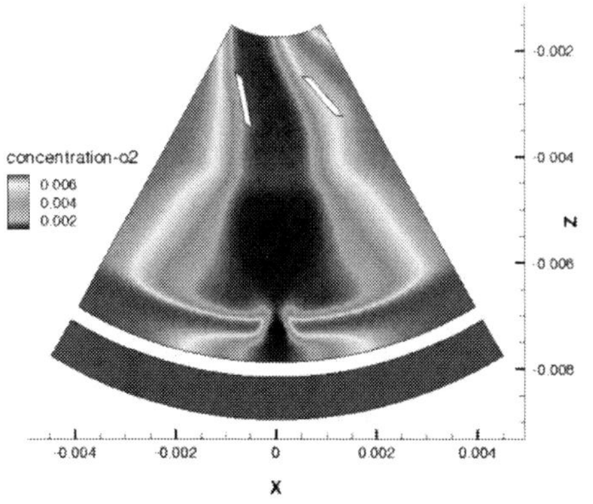

Figure 15. Y=0.5mm cross section O_2 concentration distribution

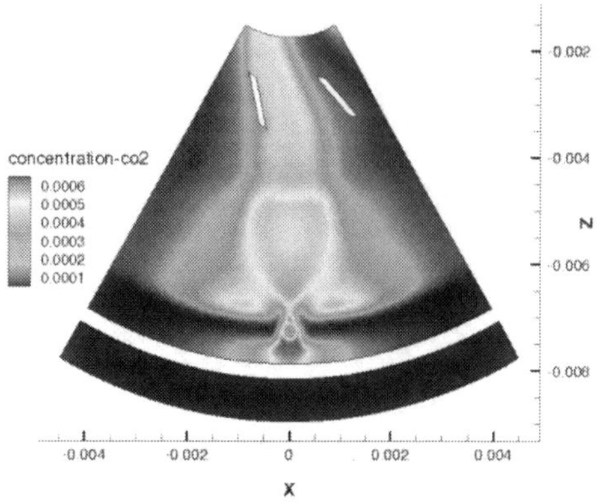

Figure 16. Y=0.5mm cross section CO_2 concentration distribution

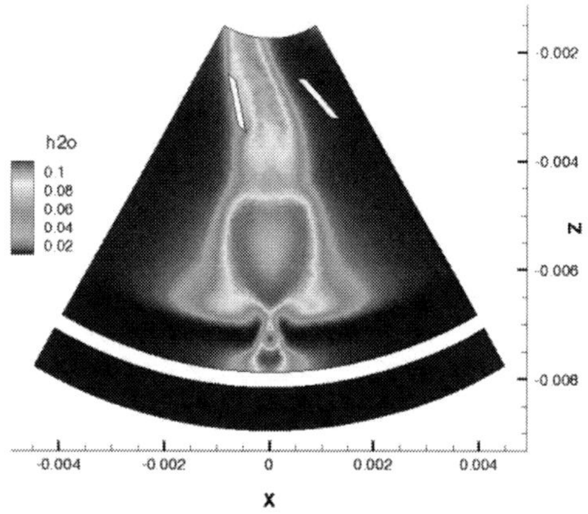

Figure 17. Y=0.5mm cross section H_2O concentration distribution

3.4 Calculation and analysis of performance parameters of combustion chamber

It can be concluded that the main performance parameters of the combustion chamber are as follows:

- Total pressure recovery coefficient σ_c

$$\sigma_c = \frac{P_3^*}{P_2^*} \tag{4}$$

In the formula (4), P_3^* and P_2^* respectively represent the average total outlet-pressure and the average total inlet-pressure. By calculating σ_c=0.9.

-Total volumetric heat intensity Q_{Vc}

$$Q_{Vc} = \frac{3600\eta_c m_f H_u}{P_2^* V_c} = 6238.3 \ kJ/\left(m^3 \cdot Pa \cdot h\right)$$

volumetric heat intensity is bigger than common turbojet engines, This is because the volume of the new type of turbojet engine chamber is smaller, surface-to-volume ratio is bigger, the combustion residence time is much shorter and the flow resistance increases exponentially caused by the smaller size of the combustion chamber. But the change trend of the combustion efficiency characteristic is not very different from the common ones. It should be pointed out, although through the calculation, the combustion efficiency can reach 90%, but in fact, due to the micro-scale viscous resistance and combustion residence time of various factors influence, the actual combustion efficiency inevitably reduce.

3.5 Comparative analysis of experimental data

Because of the limit of the MEMS processing and material cost and other aspects of the conditions, at present home and abroad in terms of this new button type turbojet engine ((combustion chamber diameter in 20mm), there is no actual experiments for comparison, so this paper chooses existing the experimental data of micro turbojet engine (combustion chamber diameter in 50mm to 100mm) and the theoretical study of Massachusetts Institute of Technology in micro turbojet as the contrast, the selected data are as follows: (1) a study on tiny annular combustion chamber by Nanjing University of Science and Technology (NUST)(combustion chamber diameter is 50mm), [18]. (2) a micro turbojet engine study by Beijing Institute of Technology (BIT) (the outer diameter of the combustion chamber is 70mm), [19]. (3) research on micro turbojet based on MEMS processing method by the Massachusetts Institute of Technology (MIT), [17]. The compared data are shown as Table 1.

Table 1. Simulation and experiment contrast

	NUST	BIT	MIT	BTE
σ_c	0.98	0.9	>0.9	0.9
Q_{Vc}	3895	3980	*	6238.3
η	0.93	0.91	>0.9	0.9

From the table, the total pressure recovery coefficient, volumetric heat intensity and combustion efficiency and so on is in the reasonable range, it shows that the design of the button type turbine engine design is reasonable. Limited by the MEMS processing technology and experimental conditions, only do some simulation research work at present, the follow-up will eventually need to go through the test of unity machine working condition. The research of this paper has certain reference value for the design and improvement of the future ultra-micro turbojet engine.

4 Conclusion

In this paper, a new combustion chamber model was designed and simulated based on the conventional micro turbine engine, realizing the integration simulation of working process in atmospheric environment and getting the following conclusions :

(1) The average velocity of the combustion chamber outlet is 65m/s, the combustion chamber outlet

temperature is about 1000K, fuel and air are mixed evenly, the combustion is sufficient.

(2) The combustion chamber is segmentation for six of the same fan combustion chamber, respectively of fuel supply, effectively guarantee the reliability of the engine's thrust. And there is a recirculation zone at the entrance of the combustion chamber, which can effectively promote the stability of the flame. Also the wall temperature is well-distributed, no obvious hot spots.

(3) The simulation data are compared with the experimental and simulation data of the domestic and foreign engine. The results show that the design of the button type engine is reasonable.

(4) One approach to realizing BTE utilizes the micromachining technology known as MEMS, the process impact of these engines will be dependent on the performance levels and the manufacturing costs. It's possible that BTE may be competitive with conventional machines. It will be very useful as power sources for small aero&space vehicles.

References

1. Liang D W, Huang G P. Recent Development and Key Techniques of Micro Turbine in Centimeter Size. *Gas Turbine Experiment and Research*, **17**(2):9-13, (2001).
2. Yuan P Y. The Present Situation and Development Prospect of Micro Turbine Engine. *International Aviation*, 7 1996
3. S.W. Janson, H. Helvajian, MEMS, Micro-engineering and Aeiospace Systems. *AIAA*, 99-3802.
4. Li C, Fang S Z, Zhang P. Numerical Simulation of Annular Combustion Chamber for Micro Turbine Engine. *Journal of Propulsion Technology*, **5**:513-518, (2008)
5. Jacobson S A. Aero-thermal challenges in the design of a micro-fabricated gas turbine engine. *AIAA*, 98-2545.
6. Saburo Y, Kana O. Concept and experiment of a flat-flame micro combustor for ultra micro gas turbine. *AIAA*, 2002-3771.
7. Yuan Y Z, Wang L P, Guan L W. Numerical Simulation and Optimization od A Micro Annular Combustor. *J Tsinghua Univ (Sci&Tech)*, **2**:198-201+209, (2007)
8. Amit, et al. A Six-Wafer Combustion System for a Silicon Micro Gas Turbine Engine. *Journal of Micro-electromechanical Systems*, 9-4, (2000)
9. Stuart A. Jacobson. Aero-thermal challenges in the design of a micro-fabricated gas turbine engine. Albuquerque: *29th AIAA Fluid Dynamics Conference*, 98-2445.
10. A. H. Epstein, S. D. Serturia, *et al.* Micro-heat engines, gas turbines, and rocket engines-The MIT micro-engine project. Snowmass Village: *28th AIAA Fluid Dynamics Conference*, 97-1773.
11. C. M. Spadaccini, *et al.* High Power Density Silicon Combustion Systems for Micro Gas Turbine Engines, *Journal of Engineering for Gas Turbines and Power*,: **125**: 709-719, (2003)
12. P. J. Conelho. Numerical Simulation of a Mild Combustion Burner. *Combustion and Flame*, ,**124**: 503-518, (2001)
13. Denis Veynante, Luc Vervisch. Turbulent combustion modeling, *Progress in Energy and Combustion Science*, **28**:193-266, (2002).
14. Jinsong Hua, Meng Wu, Kurichi Kumar. Numerical Simulation of The Combustion of Hydrogen–Air Mixture in Micro-Scaled Chambers. Part I: Fundamental study, *Chemical Engineering Science*,: **60**:3497-3506, (2005)
15. Jinsong Hua, Meng Wu, Kurichi Kumar. Numerical Simulation of The Combustion of Hydrogen–Air Mixture in Micro-Scaled Chambers. Part II: CFD analysis for a micro-combustor, *Chemical Engineering Science*, **60**:3507-3515, (2005)
16. Wen Z, Shi L Y, Ren Y R. FLUENT Fluid Computing Applications Course, *Tsinghua University Press*, 1 (2009)
17. Alan H. Epstein. Millimeter-Scale, Micro-Electro-Mechanical Systems Gas Turbine Engines, *Journal of Engineering for Gas Turbines and Power*, **126**:205-226, (2004)
18. Wang D. Experimental Study of Micro Turbojet Engine, *Nanjing University of Science and Technology*, (2011)
19. Li C. The Design and Numerical Simulation Performance of Micro Turbojet Engine, *Beijing Institute of Technology*, (2008)

MATEC Web of Conferences 70, 03004 (2016) DOI: 10.1051/matecconf/20167003004
ICMIT 2016

Aeroacoustic Simulation for NASA CC3 Centrifugal Compressor Operating at off Design Condition

Mohamed Alqaradawi[1], Ibrahim Shahin[2], Mohamed Gadala[3] and Osama Badr[4]

[1] *Mechanical and Industrial Engineering Department Collage of Engineering, Qatar University, Doha (2713), Qatar.*
[2] *Mechanical and Industrial Engineering Department, Collage of Engineering, Qatar University, Doha (2713), Qatar.*
[2] *Mechanical Engineering Department, Shoubra faculty of Engineering, Benha University, Egypt.*
[3] *Mechanical Engineering Department, UBC-University of British Columbia. Vancouver, BC, Canada.*
[4] *Mechanical Engineering Department, the French University in Egypt, Al-Shorouk City, New Cairo.*

Abstract. This paper covers the characterization of the acoustic noise and the unsteady flow field of a high speed centrifugal compressor NASA CC3. In order to accurately predict the noise, all analyses are carried out through the use of Large Eddy Simulation and Ffowcs Williams–Hawkings model for noise prediction. The relative effect of hub cavity on flow characteristics and sound levels is investigated, for a compressor stage with a total pressure ratio equal to 4, working from surge to near choke condition. In comparison with the experimental results from literature, the predicted compressor performance and flow field are predicted well. The hub cavity flow effect on the compressor aeroacoustic generated noise is shown in the paper. The unsteady static pressure and sound pressure levels are compared not only at different location but also for design and off design operating points. The internal flow results inside the hub cavity are presented at surge, design and near choke points. The conclusion is that the cavity effect of the centrifugal compressor cannot be ignored in the numerical prediction of aerodynamic generated noise. The impeller back plate of the rotor experiences a strong pressure fluctuation, which is maxima at the impeller outer radius for all operating point, but higher pressure values at the surge point.

1 Introduction

The aero-acoustic analysis of the centrifugal compressor has become urgent not only due to demands requirements for environment-friendly products [1] but also to prevent compressors from mechanical failure [2]. If acoustics resonances occur within centrifugal compressor, the pressure fluctuations may reach very high levels. Even though compressor parts are designed for high cycle fatigue; this would jeopardizing the safe operation of the compressor and the facility. Rotating stall induces large vibratory stresses in the blading of compressors and is therefore undesirable for structural reasons although the compressor may continue to give acceptable performance [3]. Surge consists of large-amplitude oscillations of the flow through the entire compressor which also produces large oscillations in compressor delivery pressure [4].
Extensive studies have been developed on the unsteady flow in centrifugal compressors focusing on rotating surge phenomena. Mckain and Holbrook defined both aerodynamic and manufacturing design details for 4:1 pressure ratio single stage centrifugal compressor [5]. The data published by Skoch, *et al.* [6] included experimental measurements using a Laser-Doppler Anemometer to obtain flow-field velocity measurements in a centrifugal impeller. Wernet, et al. [7] conducted pressure transducer measurements to capture the transient response of a compressor prior to rotating stall. During this investigation, Digital Particle Image Velocimetry was applied to capture transient velocity and pressure measurements during compressor surge. In fact, reviewing existing studies on aero-acoustic interaction indicates that centrifugal compressors have received less care than their axial compressor counterparts and especially in the unsteady aerodynamics flow. Interference generated in axial compressors has received much attention, due to the strict patterns in the aero-engine industry because of its aeronautical use and public demand for aircraft with lower noise [8], [9]. Hanson [10] showed that under certain inlet and exit conditions, an acoustic mode could get trapped between blade rows leading to amplification and higher frequency propagation. Mengle [11] analysed the physical aspects of spinning acoustic modes produced by blade vibration. Besides, the theoretical development is conducted to understand and anticipate the frequency spectra observed in the stationary and rotating reference frames.

Raitor and Neise [12] measured aero acoustic performance and investigated the sources and mechanisms of sound generation of the spectral components governing the overall noise level of centrifugal compressors. The sound generation

mechanisms of centrifugal compressors are well described for low compressor speeds. Recently, Konig et al. [13] presented two case studies on shrouded impeller under high cycle fatigue, one based on a 2D impeller stage with a vaned diffuser, the other on a 3D impeller stage with a vaneless diffuser. The authors presented a coupling mechanism between impeller vibration mode, a Tyler-Sofrin excitation resulting from impeller/stator interaction, and secondary flow cavity acoustic modes.
Petry et al. [14], [15] conducted experimental study with unsteady pressure instrumentation. Acoustic resonances in the side cavities of the centrifugal compressor stage have been characterized. The results showed that a cavity acoustic mode is excited when the Tyler-Sofrin mode and the cavity eigen-mode have the same frequency and circumferential harmonics. Richards et al. [16] conducted experimental and unsteady CFD investigation for complex aero-acoustic interaction between the impeller and the upstream and downstream return channel vanes. However, the CFD study of Richards et al. [16]-[14] conducted only on blade row interactions and not including the hub and shroud cavities acoustic. Comprehensive descriptions and investigations of the flow between rotating and stationary discs, can be found in the work of Tuliszka-Sznitko [17], [18], Gauthier et al. [19], Pellé [20] and Serre [21], but most of these studies neglected the complicated flow out from the impeller region and the unsteady flow during the surge event. Medic et al. [22] performed a numerical study using Large-eddy simulation (LES) with wall-adapting local eddy-viscosity (WALE) subgrid scale model to explain the complex turbulent flow in NACA CC3 centrifugal impeller, obtained with LES, Reynolds Averaged Navier-Stokes (RANS) turbulence models and hybrid RANS/LES approaches. The present paper is a continuation of a previously presented study of Shahin et al. [23] and focuses on the influence of the unsteady pressure and swirling flow in the hub side cavity on the acoustic levels. Since cavity effects are shown to be significant, in this research, a model of compressor with and without the hub cavity is proposed. Centrifugal compressor hub cavity flow simulation has been neglected by a lot of previous works, which discuss the aeroacoustic simulation for centrifugal compressors [16], [24] and [25].

2 Test Case Presentations

For the simulations performed in this work, we consider the well-studied NASA CC3 compressor that was tried over many years at the Small Engine Components Test Facility at NASA Glenn Research Center [5]-[7]. The computational model consists of three parts; the inlet pipe with an inlet bell at its exit, rotating impeller and Vaneless diffuser with impeller hub side cavity followed by shaft seal, shown in Fig. 1. A straight pipe with length 5 times the inlet diameter is attached at the compressor inlet to insure fully developed flow at the inlet. The duct wall is included in the numerical implementation to get very accurate predictions of noise field. The impeller is followed by a Vaneless diffuser with an annular radial-to-axial bend. The axial bend is extended to minimize the effect of outlet boundary conditions on the flow inside the diffuser. Fig. 1 shows the basic geometry and simulation domain for the moving and stationary parts. The computational grid generated for one passage is shown in Fig. 2 and the flow path through impeller, diffuser and hub cavity is shown in Fig. 3. The compressor stage with no cavity is also shown in Fig. 3; the impeller exit flow is completely go through the diffuser.

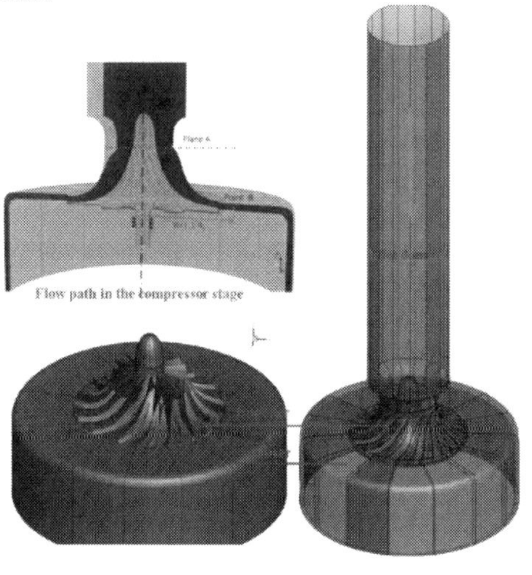

Figure 1. The geometry of simulated parts

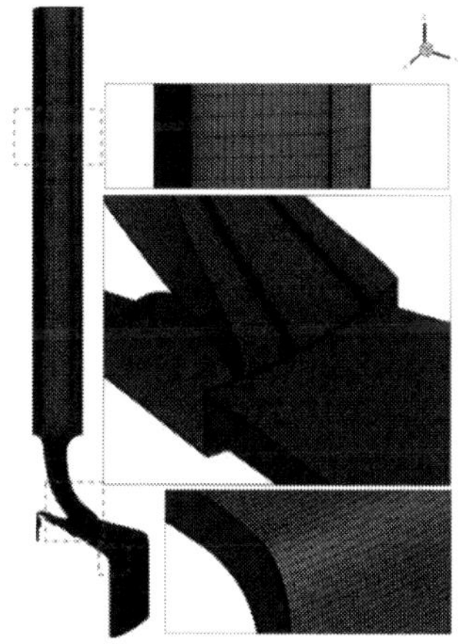

Figure 2. Computational domain grid "3.23 Million of nodes"

3 Numerical procedures

3.1 Mathematical model

Three-dimensional, Navier-Stokes equations are solved for compressible flow. The sliding mesh is used for simulating the impeller rotation, in which the multiple cell zones are connected to each other through interface boundary condition. As the mesh motion is updated in time, the interfaces are likewise updated to reflect the new positions of each zone [26].

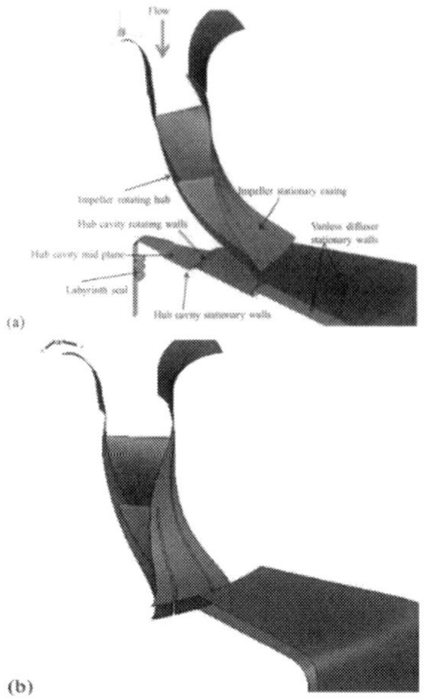

Figure 3. Flow path in the compressor stage for: (a) with cavity, (b) without cavity

The unsteady large scale structures and separation zones are the main characteristics of the flow in centrifugal compressor; therefore the large eddy simulation (LES) approach is used. In LES, large eddies are resolved directly, while small eddies are modelled with sub-grid scale model. A major role of the small scales is to dissipate the turbulent energy that is transferred from the larger scales to the smaller ones through the energy cascade. In the LES approach, the separation between the resolved and unresolved scales is achieved by low-pass spatial filtering of the governing equations. The subgrid-scale stresses resulting from the filtering operation are unknown, and require modelling. In the present case, subgrid Smagorinsky-Lilly model is used [27].

The frequency domain approach of Ffowcs Williams–Hawkings formulation applicable to the computed domain is chosen as the governing equation for the noise prediction [28], which has been used also by [24], [25] for predicting the aerodynamic noise in a centrifugal compressor. The ANSYS-FLUENT program with parallel processing, version 14.5 [26] was used to solve the filtered governing equations, which are discretized by the finite volume method. For the spatial discretization, a second order upwind scheme was used. The second order implicit formulation was used for the temporal discretization. In order to resolve the relevant fluctuations, high grid resolution with enough small time steps were employed to have the Courant number smaller than one to be consistent with the flow properties which, for this case, give a time step of 6.119 e-6 sec. The unsteady calculations are performed with a time-step size so that the impeller mesh turns less than 1° per time step, this time-step is sufficient to accurately resolve pressure spectra within human hearing range. In LES it is desirable to make sure that no physical information can propagate further than one grid cell at one time step. This gives a coupling between the temporal and the spatial discretization of the problem. This condition is expressed in terms the Courant Friedrichs Lewy (CFL) condition, which states that the Courant number must be less than one.

3.2 Boundary conditions

The mass flow inlet boundary condition is used at the inlet of the pipe with atmospheric pressure and 288 K and pressure outlet boundary condition at the domain outlet. The pressure outlet boundary condition is specified at the seal exit which equal to atmospheric pressure. Rotating periodic boundary condition is used. Wall boundary condition with no slip velocity is employed on walls. The rotating cavity face has the same rotational speed of the impeller, while the fixed cavity face has zero speed. The impeller casing wall has zero absolute speed.

Table 1. Operating conditions and unsteady simulation parameters

Condition		Value
Design Inlet pressure	bar	1.013
Design pressure ratio		4:1
Inlet temperature	K	288
Design mass flow rate	lbm/sec	10
Surge mass flow rate	lbm/sec	7.2
Rotational Speed	rpm	21789
Total simulated flow time	sec	0.175
Time step	sec	6. 119e-6
Number of time steps		28600
Number of iteration per time step		20
Hardware operating time per case	day	75
Number of simulated surge cycles		2

The multi-block structured hexahedral mesh created using ICEM CFD 14.5 blocking tools is used for meshing computational domain, which consist of one blade sector. In order to capture the boundary layer, which is important for compressor flows, the mesh is refined at the walls. The blade tip clearance is resolved with 10 cells in the radial direction. The grid is concentrated near the walls and the measured y+ is less than 1. The grid is enlarged by a ratio of 5% as it moved further away from the walls. Operating conditions and unsteady simulation parameters are listed in Table 1.

3.3 Mesh sensitivity study and numerical model validation

In order to ensure the accuracy of present numerical solution, a careful check for present numerical model was conducted. In this section, the computed data are compared with measured data to validate the developed model for both the performance map data and the internal flow velocity in the vaneless diffuser. Careful check for grid independence was also conducted; three numbers of computational nodes were used 3.23, 4.36 and 6.451 million of nodes. The absolute velocities normalized by impeller tip speed at design flow rate with different number of computational nodes are shown in Fig. 4, for Skotch et al. [6] experimental results and present CFD. The present CFD results for mesh 3.23 and 4.36 million of nodes under estimate the velocity profile after the impeller outlet, while the results with 6.451 million nodes agree well with experimental data from [6].

It is clear that the tip leakage flow is well detected, so the grid with 6.451 million nodes was used for all the studies conducted in the present simulation. Overall, results compare favourably, although the present CFD simulation predicts slightly lower velocities near the hub and shroud walls of the vaneless diffuser. These differences may be attributable to the smooth and the adiabatic walls assumption made during the simulation.

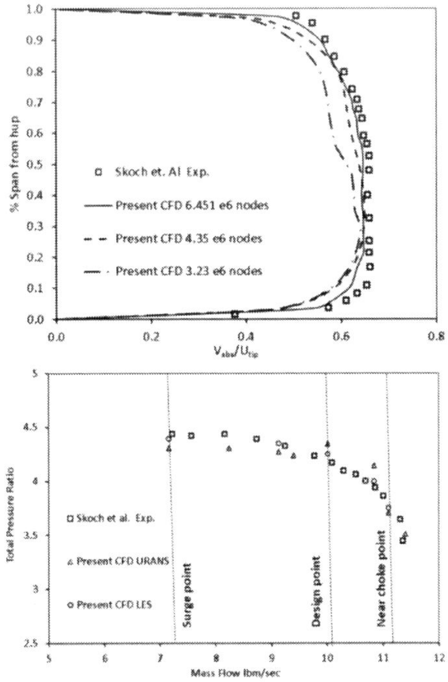

Figure 4. Comparison of measured data [6] and present simulation: (Upper) Velocity in the vaneless diffuser for different computational number of nodes, (Lower) Stage performance map.

4 Results and discussion

In the following sections, the flow physics in the mid span and near tip plans at surge, design and near choke points are presented. Also the flow field in the hub cavity region will be analysed at these flow rates, so as to explain the reason for high SPL for lower flow rates and also for the compressor with cavity case.

4.1 Compressor impeller flow

In this section, the mass flow signals are monitored with the flow time. These signals are acquired at three different operating points; surge, design and near choke conditions. Fig. 5 presents the time resolved mass flow rate at the impeller inlet, and the fluctuation in the mass flow is increased by moving to lower mass flow rate. Near the surge point noticeably higher amplitude fluctuations were observed. Also, the surge event is characterized not only by high mass flow fluctuation at impeller inlet, but also by the flow reversal at the front of the impeller. This can be observed in Fig. 5a in which the mass flow at impeller inlet is monitored with the flow time. On the other hand, no flow reversal was observed in the impeller front when operating at the design flow rate Fig. 5b, but only a small fluctuation in the mass flow is observed with no change in the flow direction. Fig. 6 shows the tip leakage vortex structure, the tip leakage vortices are clearly found rolling up from the leading edge and the corresponding stream lines also indicated that. The tip leakage vortex of the full-blade seems to be moving from the main blade suction side towards but not reaching the next main blade pressure side.

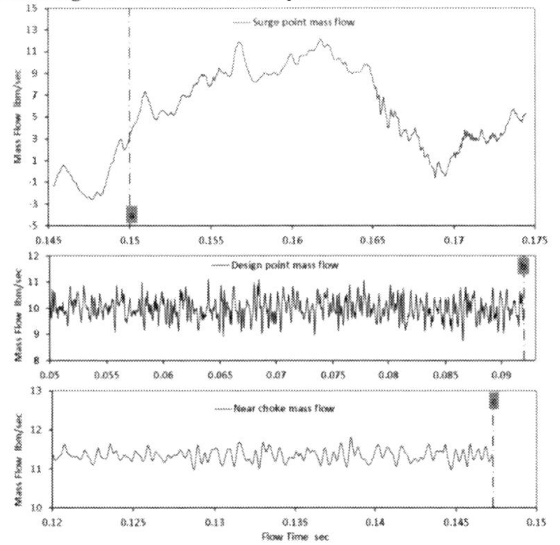

Figure 5. Mass flow with time monitored at impeller inlet

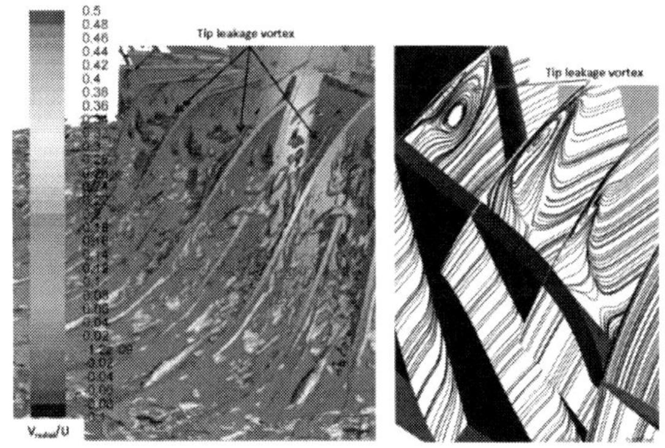

Figure 6. Tip leakage vortex structure by: (Left) Normalized Q-Criterion equal 0.1 colored by radial velocity, (Right) Stream lines at different meridional sections at the impeller inducer.

4.2 Vaneless diffuser flow

Fig. 7a shows the vaneless diffuser core flow during surge condition and it is clear the vaneless diffusers have rotating instability that is similar to rotating stall. The rotating instabilities, characterized by a core of outward radial velocity followed by a zone of inward radial velocity, are quite well identifiable. Fig. 7b shows the absolute iso-velocity values corresponding to the design operating flow conditions. The flow pattern in the diffuser inlet is characterized by the tip leakage vortex and the jet/wake structure, which resulting from the interaction of the impeller secondary flows including the tip leakage flow. Fig. 8 shows the instantaneous static pressure at 0.5 span and 0.95 span planes, for design, surge and choke flow conditions. The pressure values are higher at 0.95 span planes more than that at 0.5 spans due to the tip leakage flow from the impeller outlet. The pressure is increased gradually from the impeller inlet to the diffuser outlet, but the pressure fluctuation is higher for design point more near chock operating point, the highest fluctuation is detected at the surge point. The rotating flow instability in the diffuser and flow reversal in the impeller are the main reasons for higher pressure fluctuation.

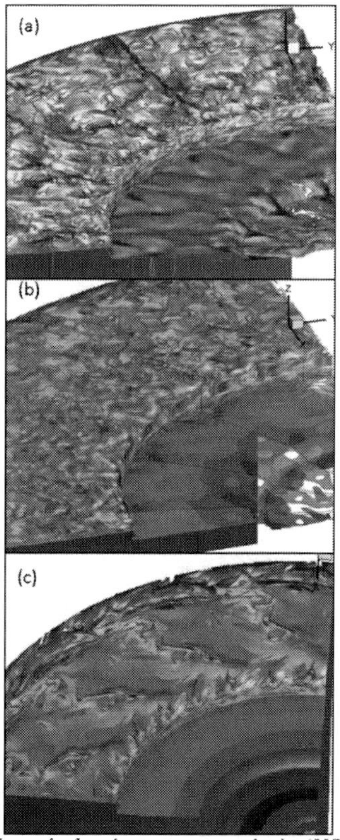

Figure 7. Dimensionless instantaneous velocity 'V/U2' iso-values in diffuser and hub cavity at (a) Surge (b) Design (c) Near choke

The instantaneous radial velocity is shown in Fig. 9 for different operating flow rats. The aerodynamic behaviour in the diffuser is strongly changed by moving from design operating point towards left and right in the compressor performance map. For the surge operating condition, the low momentum zone appears along the diffuser hub side and continues downstream up to the diffuser end. This is due to the impeller outlet fluid removal through the back face cavity, and the flow is aspirated through the cavity. Separation zones near the casing just after the impeller outlet are detected as a result from the impeller tip leakage flow and the flow tends to separate close to the diffuser end due to the diffusion inside the diffuser. Close to choke operating point, the influence of hub cavity leakage flows is much smaller and lower separated regions are detected. The main flow is pushed towards the casing side, as a result from the swirling flow coming to the diffuser domain from the hub cavity. Consequently, a low momentum flow close to the hub appeared due to the removal of the mainstream flow into the hub cavity flow. The radial velocity for the no cavity case is shown in Fig 9.d, it is clear that radial velocity is increase than that in Fig. 9.b as no flow extracted through the cavity, also no flow separation is detected at the diffuser hub due to the absence of the cavity. The absence of flow separation near the cavity will results in low acoustic noise when compared with case with cavity.

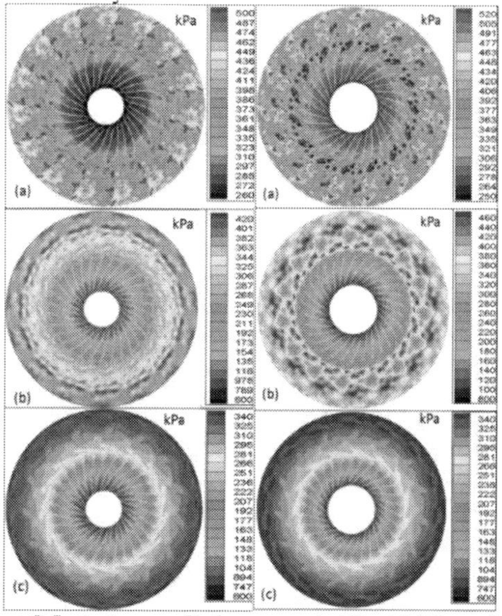

Figure 8. Instantaneous Static pressure in kPa at (a) Surge (b) Design (c)Near choke (Left: 0.5 span, Right: 0.95 span)

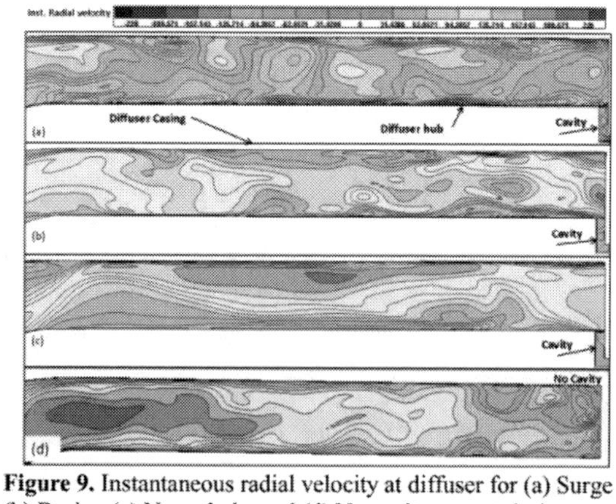

Figure 9. Instantaneous radial velocity at diffuser for (a) Surge (b) Design (c) Near choke and (d) No cavity case at design point

4.3 Hub cavity flow

Fig. 10 shows the leakage flow through the hub cavity as a percentage of the operating mass flow during the simulation time. The amount of mass flow passing through the hub cavity depends on pressure recovery and total pressure losses of downstream stationary parts. Changing the flow from the design operating point towards surge and choke affects the amount of flow through the cavity. The hub cavity flow rate is increased as surge condition, as the pressure loss is increased in the diffuser and flow blockage. Leakage flow in the hub cavity tends to decrease at near choke conditions as the pressure recovery in the diffuser is decreased. The amount of flow passing through the cavities depends, for certain geometry, on the combined effects of pressure recovery and total pressure losses of downstream stationary components for the hub one. Moving from design operating point to lower or higher flow rates point, the amount of flow through the cavity tends to increase and decrease, respectively.

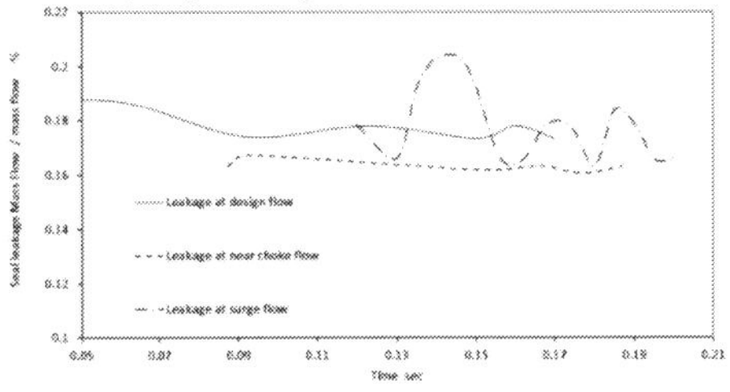

Figure 10. Leakage flow through the hub cavity with time

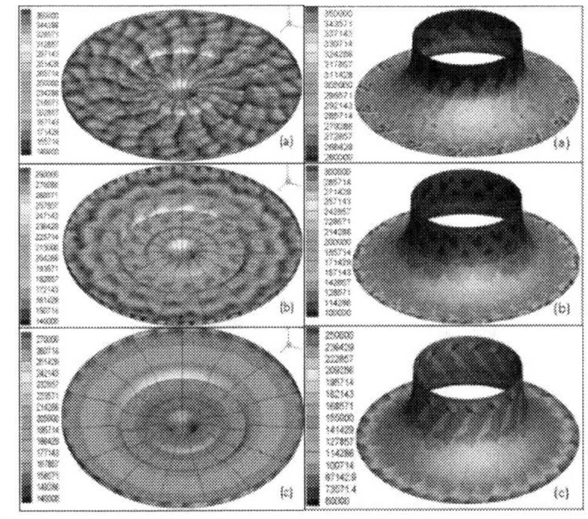

Figure 11. Instantaneous static pressure in Pa on the impeller casing side at (a) Surge (b) Design (c) Near choke. (Left: Impeller back plat, Right: Impeller shroud)

The generation of rotating pressure patterns due to the interaction between rotating and stationary components is shown in Fig. 11. The instantaneous distribution of such pressure pattern is shown for the impeller back plate and the casing. It can be seen that the impeller back plate of the rotor experiences a strong pressure fluctuation. The pressure maxima are located at the impeller outer radius all operating point, but higher pressure values are generated for the surge point. The plots show that instantaneous static pressure is greatest near the impeller trailing edge, with a significant increase in loading values when moving from near choke towards the surge point. The tip leakage vortices at the impeller exit plane is one of reasons causes a noticed increase of noise downstream of the impeller.

The fluctuated flow during the surge point also causes a fluctuation on the cavity flow, which results in rotating flow instability inside the cavity and an increase for the static pressure fluctuation. The unsteady pressures in the cavities are much higher than that on the compressor casing.

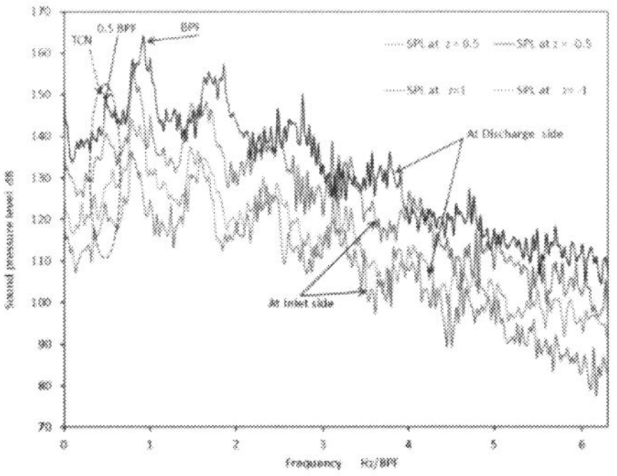

Figure 12. Sound pressure spectra for no cavity case at different locations

4.4 Sound pressure level signal analysis

The acoustic pressure signals were captured at the inlet side and discharge side of the compressor. The SPL is monitored and captured during one complete surge cycles for the surge operating conditions. These signals were computed in time domain and then converted to the frequency domain using the FFT algorithm. Sound pressure receiver locations relative to compressor parts are shown in Fig. 12, All the receivers are located on the compressor rotation axis and z represents the distance on compressor axis measured from the compressor nose which represents the zero value for z, positive values of z corresponds to the inlet side while negative z corresponds to the discharge side. The receivers on the discharge side are located outside the CFD domain, and this is agreed by ANSYS FLUENT [26]. In Fig. 12 the sound pressure spectra monitored during simulation at compressor suction and discharge receivers are plotted. The x axis represents the frequency normalized by the BPF (blade passing frequency, 10895 Hz). The BPF sound level component is seen obviously in the spectrum with high magnitudes. With increasing BPF harmonic order, the levels of the blade tone harmonics become smaller, and they decay at high frequency levels. The receiver position has a big effect on the SPL levels. As the receiver becomes close to the impeller outlet, the more noticeable are the BPF and its harmonics. Other noise component is observed at about half the BPF; this component magnitude is lower than the BPF level. It is produced by the tip leakage flow through the gap between the compressor casing and the impeller blade tips "tip clearance noise" (TCN), as observed by [10]. The sound pressure level spectra monitored on the compressor discharge side is higher than that measured in the inlet side, because noise components generated at the interface between impeller exit and stationary parts " diffuser inlet " could not transferred through the complicated flow fields in the rotor blade channels to the compressor inlet side.

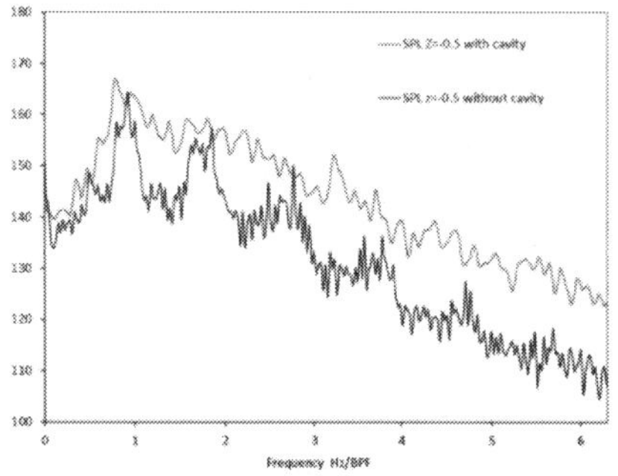

Figure 13. Sound pressure spectra for case with and without cavity z=-0.5, for design point.

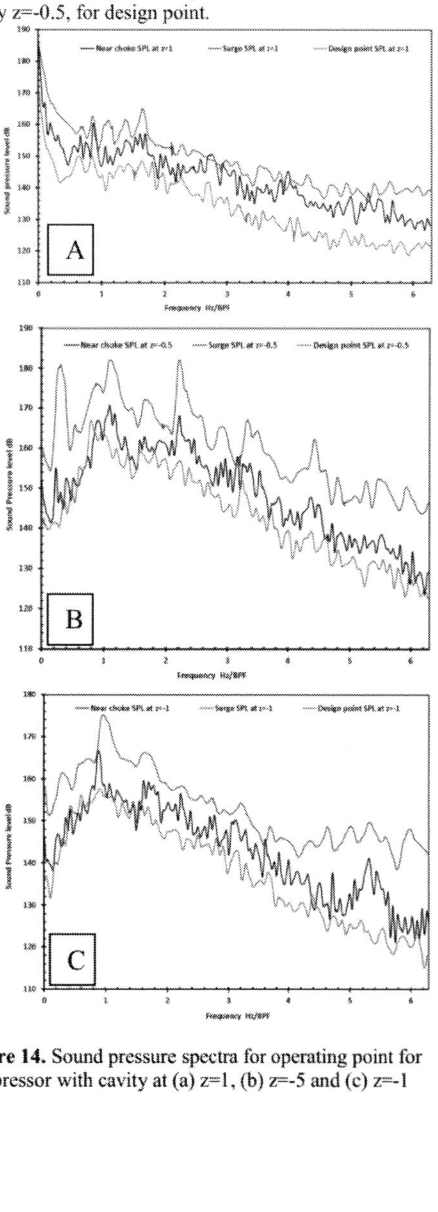

Figure 14. Sound pressure spectra for operating point for compressor with cavity at (a) z=1, (b) z=-5 and (c) z=-1

Sound pressure spectra for the compressor with and without hub cavity are compared in Fig. 13. The SPL for the cavity case is higher than that without cavity. This difference is increased at higher frequencies. The flow swirling and high rotating pressure generated in the hub cavity is the cause of this difference, which has been shown in this paper before (Section 5.3) that the flow in the hub cavity is characterized by high swirling flow. Fig. 14 compares the SPL spectra of the compressor working from surging to near choking conditions under full speed; the compressor produces sound with significant changes as the mass flow varies. The SPL at surge condition is the maximum for all receiver positions.

5 Conclusions

In this paper, we examined the aerodynamic and aero acoustics performance of a transonic centrifugal compressor working from surge to near choke conditions. The effect of hub cavity on the flow characteristics and sound levels is investigated numerically. Three dimensional simulations for the compressor with and without hub cavity are performed. The entire compressible flow inside the centrifugal compressor stage is simulated through the use of LES, and Ffowcs Williams–Hawkings model for noise prediction. The SPL for the cavity case is more eminent than that without cavity, this difference increases at higher frequencies. The flow swirling and high rotating pressure generated in the hub cavity may be the reason of this conflict.

The results also demonstrate that, the vaneless diffuser during surge have a rotating instability that is similar to rotating stall. The hub cavity flow rate is increased at surge condition, as the pressure losses is increased in the diffuser and flow blockage. The impeller back plate of the rotor experiences a strong pressure fluctuation higher than that detected on the shroud, which results in a generation of axial force on the impeller. The pressure maxima are located at the junction of the impeller and diffuser for all operating point, but higher pressure values are brought forth from the surge point. The propagation of rotating pressure patterns in the hub cavity is due to the interaction between rotating and stationary parts. Finally, The research done here gives a details on the hub side cavity effect on the acoustic sources for the centrifugal compressor and shows the importance of simulating the impeller cavities when deal with the compressor acoustics. The cavities of the centrifugal compressor should be included in the mathematical prediction of the noise.

Acknowledgments

This publication was made possible by NPRP grant No. 4-651- 2-242 from the Qatar National Research Fund (a member of Qatar Foundation). The statements made herein are solely the responsibility of the authors.

References

1. Gottlob, D. Regulations for community noise. Noise News International, **3**(4), 223-236, (1995)
2. König, S., Petry, N., & Wagner, N. G. Aeroacoustic Phenomena in High-Pressure Centrifugal Compressors—A Possible Root Cause for Impeller Failures. *In Proceedings of the 38th Turbomachinery Symposium*, pp. 103-121, (2009)
3. Stenning, A. H. Rotating stall and surge. Journal of Fluids Engineering, **102**(1), 14-20, (1980)
4. Emmons, H. W., Pearson, C. E., & Grant, H. P. Compressor surge and stall propagation. *Trans. ASME*, **77**(4), 455-469, (1955)
5. McKain, T. F., & Holbrook, G. J. Coordinates for a high performance 4: 1 pressure ratio centrifugal compressor. *Lewis Research Center*, NASA-23268, (1997)
6. Skoch, G. J., Prahst, P. S., Wernet, M. P., Wood, J. R., & Strazisar, A. J. Laser anemometer measurements of the flow field in a 4: 1 pressure ratio centrifugal impeller. In ASME 1997 International Gas Turbine and Aero engine Congress and Exhibition American Society of Mechanical Engineers, pp. V001T03A049-V001T03A049, (June 1997)
7. Wernet, M. P., Bright, M. M., & Skoch, G. J. An investigation of surge in a high-speed centrifugal compressor using digital PIV. *Journal of Turbomachinery*, **123**(2), 418-428, (2001)
8. Holste, F., & Neise, W. Noise source identification in a propfan model by means of acoustical near field measurements. Journal of Sound and Vibration, **203**(4), 641-665, (1997)
9. Tyler, J. M., & Sofrin, T. G. Axial flow compressor noise studies (No. 620532). SAE Technical Paper, (1962)
10. Hanson, D. B. Mode trapping in coupled 2D cascades—acoustic and aerodynamic results. AIAA paper, (93-4417), (1993)
11. Mengle, V. G. Acoustic spectra and detection of vibrating rotor blades, including row-to-row interference. *In AIAA, Space Programs and Technologies Conference*, vol. 1, (October 1990)
12. Raitor, T., & Neise, W. Sound generation in centrifugal compressors. *Journal of Sound and Vibration*, **314**(3), 738-756, (2008)
13. Konig, S. Acoustic eigenmodes in the side cavities of centrifugal compressors. In ASME Turbo Expo 2009: Power for Land, Sea, and Air American Society of Mechanical Engineers, pp. 547-557, (January 2009)
14. Petry, N., Benra, F. K., & Koenig, S. Experimental study of acoustic resonances in the side cavities of a high-pressure centrifugal compressor excited by rotor/stator interaction. In ASME Turbo Expo 2010: Power for Land, Sea, and Air. American Society of Mechanical Engineers, pp. 2339-2351, (October 2010)
15. Petry, N., König, S., & Benra, F. K. Influence of the Swirling Flow in the Side Cavities of a Coupled Blade Row Interactions. Journal of Turbomachinery, **134**(6), (2013)
16. Richards, S. K., Ramakrishnan, K., Shieh, C. M., Moyroud, F., Picavet, A., Ballarini, V., & Michelassi, V. Unsteady Acoustic Forcing on an Impeller Due to, (2012)
17. Tuliszka-Sznitko, E., & Zieliński, A. DNS/LES of Transitional Flow in Rotating Cavity. International Journal of Transport Phenomena, **10**(3), (2008).
18. Tuliszka-Sznitko, E., Zielinski, A., & Majchrowski, W. Large eddy simulation of non-isothermal flow in rotor/stator cavity. Ichmt Digital Library Online, (2009)
19. Gauthier, G., Gondret, P., Moisy, F., & Rabaud, M. Instabilities in the flow between co-and counter-rotating disks. Journal of Fluid Mechanics, **473**: 1-21, (2002)
20. Pellé, J., & Harmand, S.. Heat transfer measurements in an opened rotor–stator system air-gap. Experimental thermal and fluid science, **31**(3), 165-180, (2007)
21. Serre, E., Crespo Del Arco, E., & Bontoux, P. Annular and spiral patterns in flows between rotating and stationary discs. Journal of Fluid Mechanics, **434**: 65-100, (2001)
22. Medic, G., Feng, J., Chen, L., & Sharma, O.. Towards Large-Eddy Simulation of Turbulent Flow in a Centrifugal Impeller. In ASME 2011 Turbo Expo: Turbine Technical Conference and Exposition, American Society of Mechanical Engineers, pp. 2057-2066, (January 2011)
23. Shahin, I., Gadala, M., Alqaradawi, M., & Badr, O.. Unsteady CFD Simulation for High Speed Centrifugal Compressor Operating Near Surge. In ASME Turbo Expo 2014: Turbine Technical Conference and Exposition (pp. V02DT44A045-V02DT44A045). American Society of Mechanical Engineers, (June 2014)
24. Wang, P., & Zangeneh, M. Aerodynamic and Aeroacoustic Optimization of a Transonic Centrifugal Compressor. In ASME Turbo Expo 2014: Turbine Technical Conference and Exposition (pp. V02AT41A012-V02AT41A012). American Society of Mechanical Engineers, (June 2014)
25. Sun, H., Shin, H., & Lee, S. Analysis and optimization of aerodynamic noise in a centrifugal compressor. Journal of sound and Vibration, **289**(4), 999-1018, (2006)
26. ANSYS, "Academic Research Release 14.5, Theory Manual," ANSYS, Irvine CA. (2012)
27. Meneveau, C., & Katz, J. Scale-invariance and turbulence models for large-eddy simulation. Annual Review of Fluid Mechanics, **32**(1), 1-32, (2000)
28. Ffowcs Williams, J.E. and Hawkings, D.L.. Sound generated by turbulence and surfaces in arbitrary motion. Philos Trans R Soc A, **264**:321-342, (1969)

MATEC Web of Conferences 70, 03005 (2016) DOI: 10.1051/matecconf/20167003005
ICMIT 2016

Multi-load Optimal Design of Burner-inner-liner Under Performance Index Constraint by Second-Order Polynomial Taylor Series Method

TU Gaoqiao, Wong Chun Nam, Zheng Min and Tang Kongzheng

Lanzhou University of Technology, School of Mechanical and Electronical Engineering, Lanzhou, Gansu 730050, China

Abstract. Using maximum expansion pressure of n-decane, the aeroengine burner-inner-liner combustion pressure load is computed. Aerodynamic loads are obtained from internal gas pressure load and gas momentum. Multi-load second-order Taylor series equations are established using multi-variant polynomials and their sensitivities. Optimal designs are carried out using various performance index constraints. When 0.25 to 0.8 rectifications of different design variants are implemented, they converge under 5×10^{-4} d-norm difference ratio.

1 Introduction

According to the turbine combustion technique [1], maintenance costs on turbine and combustion chamber account for 60% of the whole aircraft maintenance. Therefore, high performance on the optimal design, operation reliability and structural safety are demanded on modern aeroengine thermal components. As burner inner liner (BIL) is the major component in combustion chamber, it becomes one of the most significant components of the aeroengine. BIL is a metallic thin-wall cylinder that controls the combustion, mixing and cooling processes. It guides the combustion chamber cylinder and rotors from thermal combustion products [2]. From the maintenance survey [3], BIL accounts for the 63% of the combustion chamber faults. Combustion chamber cylinder is 5% and fuel nozzle is 4%. Since performance indices are determined by BIL mainly, it becomes the research focus on cannular combustion chamber design.

In the operation process of BIL, complicated pressure loads occurs which can be classified as follows: 1) Due to gas combustion, expansion pressure is exerted on its cylinder; 2) Pressure load is generated by the pressure different between the inner and outer cylinder surfaces. Most traditional optimization methods work on unique load where only one disciplinary analysis is involved. To improve the performance of BIL under these loads in different disciplinary, multi-load optimization technique is implemented in this design.

Multi-load optimization technique allows the multiple responses objective functions to be optimized simultaneously under the coupled design variants. This method caters for the complicated interactions among different disciplinary, and the requirements of various performance indices. Inequality constraints are imposed by these indices.

2 Expansion pressure load created in fuel combustion

The changes in reactant, product and composition mole fraction of *n*-decane and aviation fuel premixed combustion flame are basically consistent as indicated by Zeng[4]. Although the aviation fuel contains complicated ingredients, *n*-decane can be used for numerical simulation as alternative of aviation fuel. Details of his proposed reaction mechanism can describe detailed dynamic characteristics of *n*-decane premixed combustion. Because percentages of C and H within hexane hydrogen classes are the same, one can infer its complete combustion product proportions are the same. As a result, *n*-decane, propane and cyclohexane possess same maximum expansion pressure of 0.86MPa.

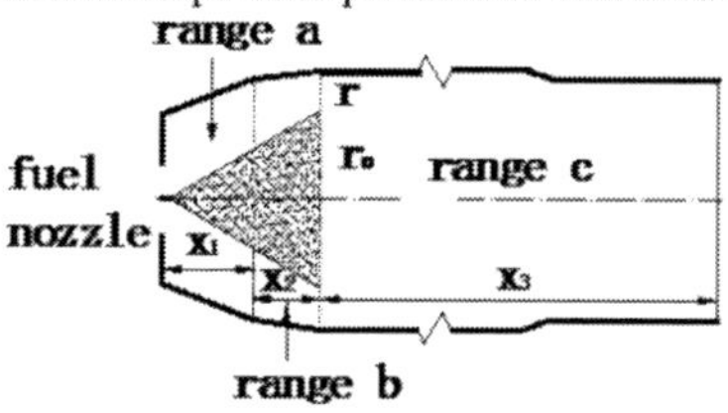

Figure 1. Sectional dimensions in each range.

From this known combustion pressure, BIL inside pressure distribution is obtained using inverse-square law $P_0/P_1 = (r_0+r)^2/r_0^2$ as in Fig.1. In range a, $r_0 + r = 0.364 \cdot x_1 + 0.05 = y_1$ for $0 \le x_1 \le 0.055$. In range b, $r_0 + r = 0.07002 = y_2$ for $0.055 \le x_2 \le 0.095$. In range c, $r_0 + r = \sqrt{(y_2 - 0.095\tan(\pi/6))^2 + (x_3 - 0.095)^2} \Big/ (0.07002 - 0.095\tan(\pi/6))$

for $0.095 \le x_3 \le 0.51$. Using the distributed pressure on each range computed above, its thrust load is computed by equivalent nodal loads

$$\begin{aligned} F^1 &= 12\cdot(F_{11}/2+\sin(\alpha/2+\pi/4)\cdot F0_1+\sin(\beta/2+\pi/4)\cdot F0_2+\sin(\pi/2-\beta/2)\cdot F0_3-\sin(\pi/2-\gamma/2)\cdot F0_8-\sin(\gamma/2+\pi/2)\cdot F0_9) \\ &= 6.982\times10^3\,daN \end{aligned} \tag{1}$$

where α, β, γ are the axial angles of ring 1,2,8 respectively. F_n is the pressure load on ring n. $F0_n$ is the equivalent nodal load of ring n.

3 Aerodynamic load generated by gas pressure and gas momentum

According to the aerodynamic equation, when the gas passes the design model (Fig. 2), total thrust is the gas momentum difference between outlet and inlet. Set the gas pressure load inside as P. Taking the thrust direction as positive, aerodynamic force on BIL is $P - A_0^{'}P_{in} + A_1^{'}P_{out}$. Gas momentum is $G(V_{out} - V_{in})$ with $G = 50.7\ \mathrm{kg}/s$. Thus we have the balance equation $P = G(V_{out} - V_{in}) + A_0^{'}P_{in} - A_1^{'}P_{out}$ with $V_{in} = 150m/s$, $V_{out} = 600m/s$. Using the action and reaction rule, thrust load due to aerodynamic force is $F^2 = -P$. Substituting all computed parts, one has

$$F^2 = G(V_{in} - V_{out}) - P_{in}(A_0 + t(S_{11} + S_3\sin\beta)) + P_{out}A_1 = 3.7907\times10^4\,daN \tag{2}$$

where t is the gas hole area, S_n is surface area of ring n. Therefore the total thrust load F_T is composed of these multi-load as

$$F_T = F^1 + F^2 = 4.489\times10^4 daN \tag{3}$$

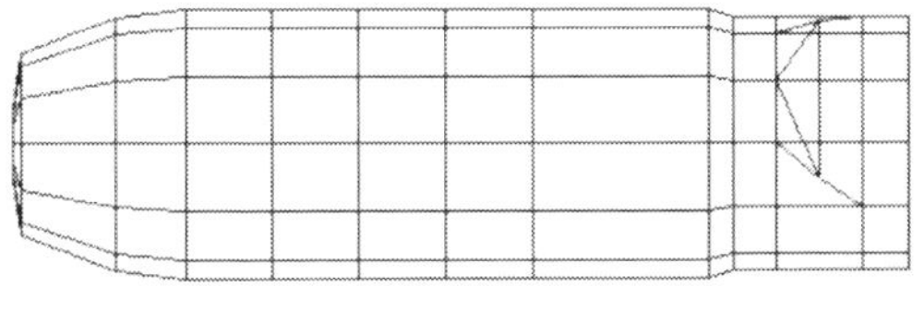

Figure 2. Design model of BIL.

4 Establishment of multi-load optimization principle

In this method, second-order Taylor Series expansion equation [5] is utilized to relate the changes of the multi-load responses with the rectifications of design variants in the design model. By inverse computation, these rectifications in the design variants are estimated from the BFGS algorithm[6]. The optimization process is repeated until specific termination criteria are reached. Let $\overline{Q}_a = (Q_{a1}, Q_{a2}, \cdots, Q_{am})^T$ be the a-th updated design variant vector, where Q_i denotes the fuel nozzle jet angle, gas flow rate, fuel combustion pressure and compressor outlet pressure. Meanwhile $\overline{Q}_d = (Q_{d1}, Q_{d2}, \cdots, Q_{dm})^T$ is the actual design variant vector. The selected sets of n_t thrust response values t^k and n_S thrust response vectors $\overline{S}^k$ corresponding to the updated and actual structures are denoted by $\overline{Y}_A$ and $\overline{Y}_D$ respectively, where $\overline{Y}_A = \left(t_a^1, t_a^2, \cdots, t_a^{n_t}, \overline{S}_a^1, \overline{S}_a^2, \cdots, \overline{S}_a^{n_S}\right)^T$ and $\overline{Y}_D = \left(t_d^1, t_d^2, \cdots, t_d^{n_t}, \overline{S}_d^1, \overline{S}_d^2, \cdots, \overline{S}_d^{n_S}\right)^T$. In the thrust load system equation, t^k are thrust force due to the fuel combustion, flow rate and air pressure. $\overline{S}^k$ are distributed stress due to expansion pressure and distributed air pressure on the cylinder surface.

In general, one can use second-order Taylor series expansion[7] to obtain the change of the k-th thrust response vector due to change in design variant δQ_i ($i = 1,2,...,m$):

$$\overline{S}_d^k - \overline{S}_a^k = \sum_{i=1}^{m}\frac{\partial \overline{S}_a^k}{\partial Q_i}\delta Q_i + \frac{1}{2!}\sum_{i=1}^{m}\sum_{j=1}^{m}\frac{\partial^2 \overline{S}_a^k}{\partial Q_i \partial Q_j}\delta Q_i \delta Q_j + \overline{\varepsilon}_t^{k(2)} \tag{4}$$

where $\overline{\varepsilon}_S^{k(2)}$ is the residual error vector in its expansion. For the p-th order expansion of the k-th thrust response value, one gets

$$t_d^k - t_a^k = \sum_{i=1}^{m}\frac{\partial t_a^k}{\partial Q_i}\delta Q_i + \frac{1}{2!}\sum_{i=1}^{m}\sum_{j=1}^{m}\frac{\partial^2 t_a^k}{\partial Q_i \partial Q_j}\delta Q_i \delta Q_j + \varepsilon_t^{k(2)}, \tag{5}$$

in which $\varepsilon_t^{k(2)}$ is the residual error value in the expansion. Termination criterion is established to control the accuracy of iterative process. For the optimal design of BIL, the objective function is met when the sum of thrust response residual errors drop to the global minimum. As a result, the d-norm $\|\mathrm{d}_a\|$ difference ratio is chosen as the termination indicator, i.e. $\left\|\Delta \mathrm{d}_a\right\| = \left\|\left\|\mathrm{d}_a\right\| - \left\|\mathrm{d}_{a-1}\right\|\right\| / \left\|\mathrm{d}_0\right\| < 5\times 10^{-4}$ where

$$\left\|\mathrm{d}_a\right\| = \sum_{kt}^{n_t}\sum_{ks=1}^{n_s}\left\|\left\{\overline{S}_d^{ks} - \overline{S}_a^{ks}, \frac{t_d^{kt} - t_a^{kt}}{t_d^{kt}}\right\}\right\|. \tag{6}$$

Based on the $k1\times k2\times\cdots\times km$ ordered sets, one can interpolate m-variate thust response vector polynomial function as

$$\overline{S}_a^{k}\left(\overline{Q}\right) = \sum_{k1=1}^{K1}\sum_{k2=1}^{K2}\cdots\sum_{km=1}^{Km}\overline{S}^{k}_{a\langle k1k2\cdots km\rangle} L_1^{\langle k1\rangle}\left(Q_1^{\langle k1\rangle}\right) L_1^{\langle k2\rangle}\left(Q_2^{\langle k2\rangle}\right)\cdots L_m^{\langle km\rangle}\left(Q_m^{\langle km\rangle}\right) \tag{7}$$

where $L_i^{\langle ki\rangle}\left(Q_i^{\langle ki\rangle}\right)$ is the Lagrange factor function of the ith design variant at the kith interpolated stiffness value, given by

$$L_i^{\langle ki\rangle}\left(Q_i^{\langle ki\rangle}\right) = \prod_{\substack{k=1\\ k\neq ki}}^{ki}\left(\frac{Q_i - Q_i^{\langle k\rangle}}{Q_i^{\langle ki\rangle} - Q_i^{\langle k\rangle}}\right)$$

$$(ki = 1,\ 2,\ \cdots, Ki) \tag{8}$$

Meanwhile, its thust response value polynomial function is

$$t_a^{t}\left(\overline{Q}\right) = \sum_{k1=1}^{K1}\sum_{k2=1}^{K2}\cdots\sum_{km=1}^{Km} t^{k}_{a\langle k1k2\cdots km\rangle} L_1^{\langle k1\rangle}\left(Q_1^{\langle k1\rangle}\right) L_2^{\langle k2\rangle}\left(Q_2^{\langle k2\rangle}\right)\cdots L_m^{\langle km\rangle}\left(Q_m^{\langle km\rangle}\right) \tag{9}$$

First-order derivative terms can be obtained by direct derivative on Eqs. (7), (9) with respect to Q_i. For multiple variants rectification, the first-order thust response vector and value derivatives [7] can be expressed as

$$\frac{\partial \overline{S}_a^k}{\partial Q_i} = \sum_{k1=1}^{K}\sum_{k2=1}^{K}\cdots\sum_{km=1}^{Km}\overline{S}^k_{a\langle k1k1\cdots km\rangle} L_1^{\langle k1\rangle}\left(Q_1^{\langle k1\rangle}\right)\cdots\frac{\partial L_i^{\langle ki\rangle}\left(Q_i^{\langle ki\rangle}\right)}{\partial Q_i}\cdots L_m^{\langle km\rangle}\left(Q_m^{\langle km\rangle}\right) \tag{10}$$

$$\frac{\partial t_a^k}{\partial Q_i} = \sum_{k1=1}^{K1}\sum_{k2=1}^{K2}\cdots\sum_{km}^{Km} t^k_{a\langle k1k2\cdots km\rangle} L_1^{\langle k1\rangle}\left(Q_1^{\langle k1\rangle}\right)\cdots\frac{\partial L_i^{\langle ki\rangle}\left(Q_i^{\langle ki\rangle}\right)}{\partial Q_i}\cdots L_m^{\langle km\rangle}\left(Q_m^{\langle km\rangle}\right) \tag{11}$$

where

$$\frac{\partial L_i^{\langle ki\rangle}\left(Q_i^{\langle ki\rangle}\right)}{\partial Q_i} = \sum_{\substack{k=1\\ k\neq ki}}^{Ki}\frac{1}{Q_i^{\langle ki\rangle} - Q_i^{\langle k\rangle}}\prod_{\substack{kj=1\\ kj\neq ki\\ kj\neq k}}^{Ki}\left[\frac{Q_i - Q_i^{\langle kj\rangle}}{Q_i^{\langle ki\rangle} - Q_i^{\langle kj\rangle}}\right].$$

On the other hand, the second-order derivative consists of two terms namely the repeated differential and the unrepeated differential. For the unrepeated term, it is given by the direct first-order derivative of Eqs. 7,9 according to Q_i involved. Special care is given to the repeated terms where direct second-order derivative with respect to Q_i are encountered. For multiple variants, the second-order thust response vector and value derivatives [7] are expressd as

$$\frac{\partial^2 \overline{S}_a^k}{\partial Q_i \partial Q_j} = \begin{cases} \sum_{k1=1}^{K1}\sum_{k2=1}^{K2}\cdots\sum_{km}^{Km}\overline{S}^k_{a\langle k1k1\cdots km\rangle} L_1^{\langle k1\rangle}\left(Q_1^{\langle k1\rangle}\right)\cdots\frac{\partial L_i^{\langle ki\rangle}\left(Q_i^{\langle ki\rangle}\right)}{\partial Q_i}\cdots\frac{\partial L_j^{\langle kj\rangle}\left(Q_j^{\langle kj\rangle}\right)}{\partial Q_j}\cdots L_m^{\langle km\rangle}\left(Q_m^{\langle km\rangle}\right)\cdots i\neq j \\ \sum_{k1=1}^{K1}\sum_{k2=1}^{K2}\cdots\sum_{km}^{Km}\overline{S}^k_{a\langle k1k1\cdots km\rangle} L_1^{\langle k1\rangle}\left(Q_1^{\langle k1\rangle}\right)\cdots\frac{\partial^2 L_i^{\langle ki\rangle}\left(Q_i^{\langle ki\rangle}\right)}{\partial Q_i^2}\cdots L_m^{\langle km\rangle}\left(Q_m^{\langle km\rangle}\right)\cdots\cdots i = j \end{cases} \tag{12}$$

$$\frac{\partial^2 t_a^k}{\partial Q_i \partial Q_j} = \begin{cases} \sum_{k1=1}^{K1}\sum_{k2=1}^{K2}\cdots\sum_{km}^{Km} t^k_{a\langle k1k2\cdots km\rangle} L_1^{\langle k1\rangle}\left(Q_1^{\langle k1\rangle}\right)\cdots\frac{\partial L_i^{\langle ki\rangle}\left(Q_i^{\langle ki\rangle}\right)}{\partial Q_i}\cdots\frac{\partial L_j^{\langle kj\rangle}\left(Q_j^{\langle kj\rangle}\right)}{\partial Q_j}\cdots L_m^{\langle km\rangle}\left(Q_m^{\langle km\rangle}\right)\cdots\cdots i\neq j \\ \sum_{k1=1}^{K1}\sum_{k2=1}^{K2}\cdots\sum_{km}^{Km} t^k_{a\langle k1k2\cdots km\rangle} L_1^{\langle k1\rangle}\left(Q_1^{\langle k1\rangle}\right)\cdots\frac{\partial^2 L_i^{\langle ki\rangle}\left(Q_i^{\langle ki\rangle}\right)}{\partial Q_i^2}\cdots L_m^{\langle km\rangle}\left(Q_m^{\langle km\rangle}\right)\cdots\cdots i = j \end{cases} \tag{13}$$

where

$$\frac{\partial^2 L_i^{(ki)}\left(Q_i^{(ki)}\right)}{\partial Q_i^2}=\sum_{\substack{k=1\\k\neq ki}}^{Ki}\frac{1}{Q_i^{(ki)}-Q_i^{(k)}}\sum_{\substack{km=1\\km\neq ki\\km\neq k}}^{Ki}\frac{1}{Q_i^{(ki)}-Q_i^{(km)}}\prod_{\substack{kj=1\\kj\neq ki\\kj\neq k\\kj\neq km}}^{Ki}\left[\frac{Q_i-Q_i^{(kj)}}{Q_i^{(ki)}-Q_i^{(kj)}}\right]$$

Substituting Eqs. (10), (12) into Eq. (4), second-order Taylor series expansion of thust response vector is furnished. Moreover, second-order thust response value equation of Eq. (5) is completed by Eqs. (11), (13).

5 Normalization of taylor series equation using accuracy number

Here we carry out the normalization of the optimization equations. For the response vectors, we reduce the data precision as small as possible, while not affecting its variation characteristics. We multiply the equations in each discipline by accuracy numbers, so that they are normalized to the same precision level. In this case, the computation problems such as the singularity, accuracy and convergency can be avoided. For example, accuracies of the distributed stress and pressure are 10^{19} and 10^{6} respectively, then the normalization accuracy numbers are $[10^{-19}, 10^{-6}]$. Under this treatment, faster and more accurate convergences are attained.

6 Optimal design of BIL under performance index constraint

In this optimization approach, inequality constraints are imposed using various performance indices. Specific thrust is constrained in the range $60 \le F_s = F_T/G \le 80\text{da}N\cdot\text{s/kg}$; specific fuel consumption is $0.8 \le C_F = 3600f/F_S \le 1.0kg/(h\cdot daN)$ where $f \approx 0.018$; thrust weight ratio is $3.5 \le F_w = F_T/W_g \le 4.0$ and compressor outlet pressure ratio is $1.15 \le P_r = P_{out}/P_{atm} \le 1.35$ where P_{atm} is the atmospheric pressure.

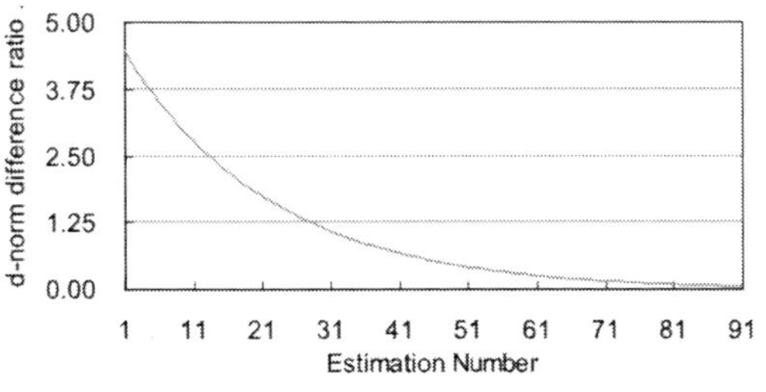

Figure 3. Convergence of $\|\Delta \mathrm{d}_a\|$ under 0.5 rectification of Q_2 and Q_4.

Using different combinations of one to four Q_i, various 0.25 to 0.8 rectification cases are established. When two design variants, gas flow rate Q_2 and compressor outlet pressure Q_4 are rectified in 0.5 level, their optimization and convergence processes are recorded as follows.

From Fig. 3, the minimization of $\|\Delta \mathrm{d}_a\|$ is faster up to estimation 40. It becomes slower up to estimation 80, and stable afterwards. Termination criterion is met at estimation 91 where the $\|\Delta \mathrm{d}_a\|$ drops to 4.8×10^{-4}.

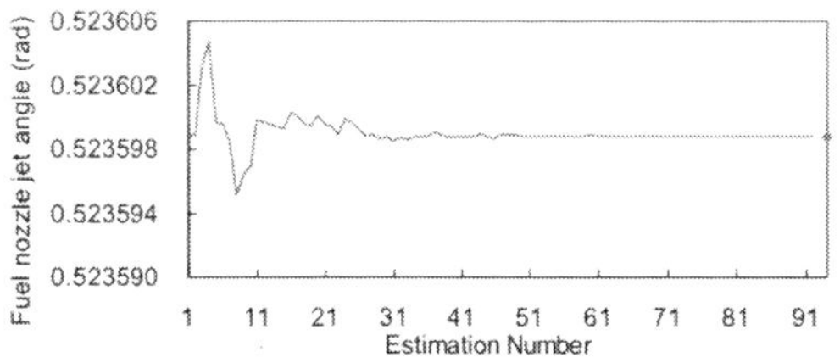

Figure 4. Estimation pattern of Q_1 under 0.5 rectification

From Fig. 4, rapid adjustment of Q_1 occurs up to estimation 40. Then it becomes stable afterwards. As its rectification is relatively small, wave-like pattern having peak and valley is observed.

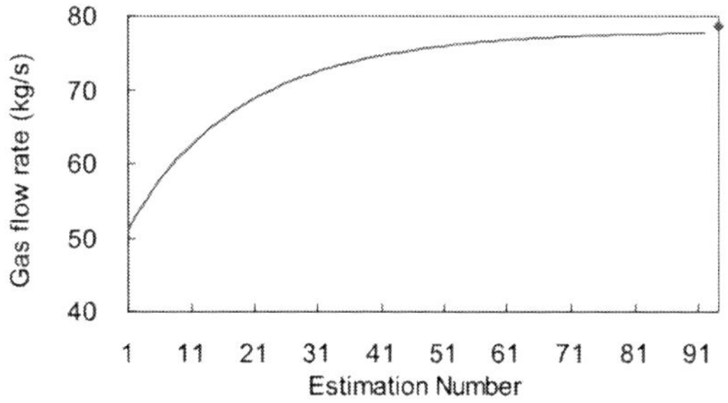

Figure 5. Estimation pattern of Q_2 under 0.5 rectification.

Estimation process of Q_2 is illustrated in Fig.5. Similar to the pattern of $\|\mathrm{d}\|$, it increases rapidly up to estimation 40. Afterwards, it changes gradually until termination criterion is met at estimation 91 where its percentage error is constrained by the performance indices at 1.21%. From the estimation pattern of Q_3, its rapid adjustment occurs up to estimation 30. Then it becomes stable afterwards. Not the same as Q_1, its rectification cycle is shorter leading to less significant peak and valley features.

Clear cut pattern is observed as in Fig. 6. As its nominal value is relatively large among the other variants, its priority is high in the pattern. Thus its optimization is rapid and accurate. Using 0.25 rectification for same design variants, 92 estimations are needed. Thus, the convergence rate is slightly decreased for smaller rectification level.

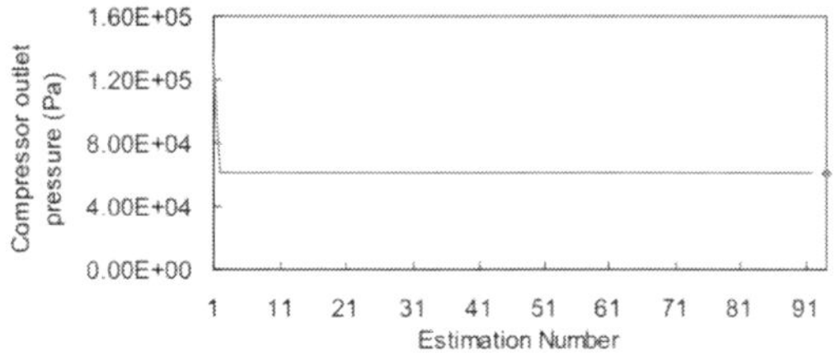

Figure 6. Estimation pattern of Q_4 under 0.5 rectification.

When three design variants Q_1, Q_2, Q_4 (m=3) are rectified in 0.8 level, their optimization and convergence processes are recorded as follows.

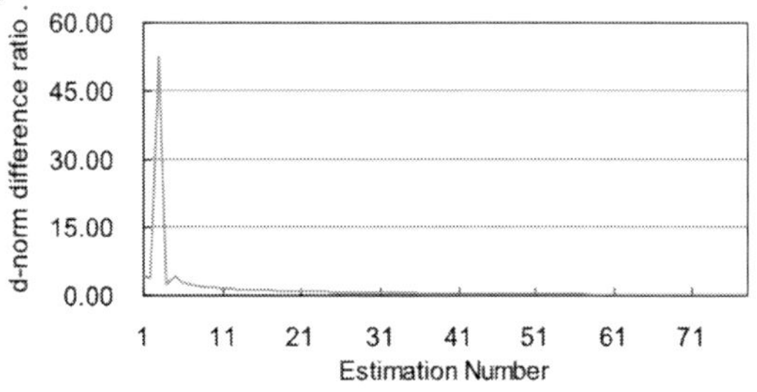

Figure 7. Convergence of $\|\Delta d_a\|$ under 0.8 rectification of Q_1, Q_2, Q_4.

From Fig. 7, $\|\Delta d_a\|$ is ramped up to the peak 52.5 at estimation 3. Then it drops rapidly to 2.26 and minimizes gradually afterwards. Termination criterion is met at estimation 78 where $\|\Delta d_a\|$ drops to 4.8×10^{-4}.

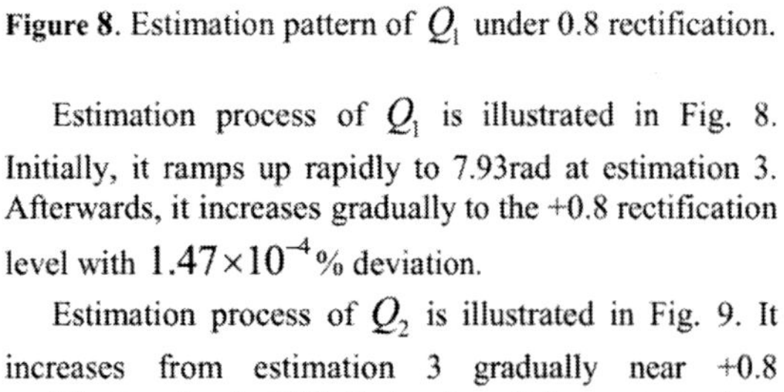

Figure 8. Estimation pattern of Q_1 under 0.8 rectification.

Estimation process of Q_1 is illustrated in Fig. 8. Initially, it ramps up rapidly to 7.93rad at estimation 3. Afterwards, it increases gradually to the +0.8 rectification level with 1.47×10^{-4}% deviation.

Estimation process of Q_2 is illustrated in Fig. 9. It increases from estimation 3 gradually near +0.8 rectification level of 89.4kg/s where it is constrained by the performance indices at 2.08%. As Q_3 rectification is zero, wave-like pattern having large peak and valley of amplitudes up to 7.5×10^6 Pa is observed in the initial stage. Then it drops gradually to the initial level with -6.94×10^{-4}% deviation.

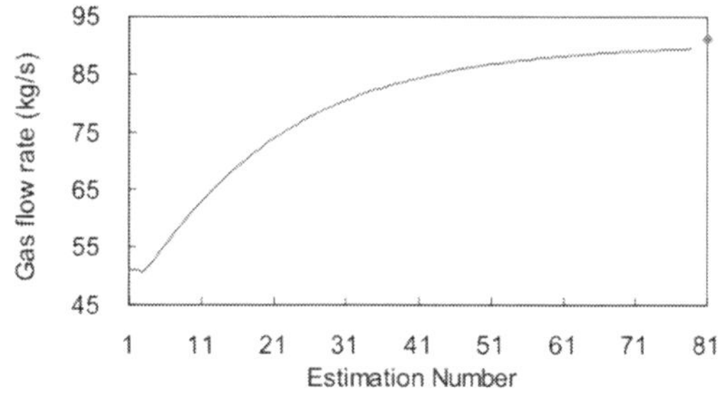

Figure 9. Estimation pattern of Q_2 under 0.8 rectification.

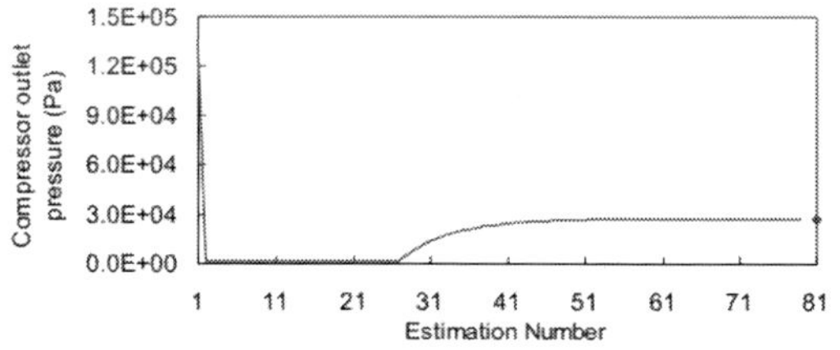

Figure 10. Estimation pattern of Q_4 under 0.8 rectification.

From Fig. 10 Q_4 drops rapidly from 1.36×10^5 Pa to minimum level of 1.36×10^3. It remains stationary until estimation 27, and increases gradually to the -0.8 rectification level with 4.44×10^{-2} % deviation. Using same design variants with 0.25 rectification, 87 estimations are required. Therefore, the convergence rate is lower for larger rectification level. When m=1, 97 estimations are needed in both 0.25 and 0.8 rectification cases. For m=4, there is 69 estimations in 0.25 rectification case. In the 0.8 rectification, 44 estimations are required. When rectification level is larger, convergence rate is decreased. Meanwhile, for the increase in m, the convergence rate is increased. For the illustrated cases, m=2,3, their convergence rates follow these trends.

Summary

BIL combustion pressure load F^1 is calculated to be $6.982\times10^3 daN$ using 0.86MPa n-decane expansion pressure. Aerodynamic load F^2 is obtained as $3.7907\times10^4 daN$. Using developed equations, 0.25 to 0.8 rectifications of one to four design variants constrained by various performance indices converge under 5×10^{-4} $\|\Delta d_a\|$ criterion. In general, larger rectification level leads to slower convergence. This trend is more significant for larger m. When m increases, its convergence becomes faster.

Acknowledgements

This research is supported by the National Natural Science Foundation of China under grant numbers 50905028 and 51275554.

References

1. X. Hou, H. Ji, Q. Liu, *et al.*, High performance aviation gas turbine combustion technique. Peking: Militrary Defence Industrial Press, (2002), pp. 5-11.
2. Y. Wang, Aeroengine Principle. Peking: Beihang University Press, (2009), pp. 116-124.
3. S. Peng, Aeroengine Combustion Chamber Structure. Peking: Militrary Defence Industrial Press, (1978), pp. 3-7.
4. W. Zeng, J. Liu, X. Chen, M. Jie, Detailed reaction kinetic modeling of n-decane premixed combustion, *Journal of Aerospace Power*, **26**(10):2258-2266. (2011)
5. C.N. Wong, A.A. Barhorst, Polynomial interpolated Taylor series method for parameter identification of nonlinear dynamic system, *ASME Journal of Computational and Nonlinear Dynamics*, **1**(3):248-256 (2006).
6. C.N. Wong, J. Xiong, H. Huang, T. Hu, Damage Detection of Space Truss using Second Order Polynomial Method with BFGS Quasi-Newton Optimization, *Proceedings of the ASME 2010 IDETC/CIE, DETC2010-28091*, (2010), pp.753-762.
7. C.N. Wong, J. Xiong, H. Huang, Y.J. Zhao, A polynomial algorithm for model updating of engineering truss, *Mechanics Based Design of Structures and Machines*, **38**:1-24, (2010).

Hydrodynamic Performance Analysis on Different Airfoils of Straight Blade H Type of Vertical Axis Tidal Current Energy Turbine

Kan Kan[1], Yuan Zheng[2, 3], Xiaoling Liang[4], Shifeng Fu[1], Zhongwei He[1] and Bin Sun[1]

[1] College of Water Conservancy and Hydropower Engineering, Hohai University, Nanjing, 210098, P.R. China
[2] National Engineering Research Center of Water Resources Efficient Utilization and Engineering Safety, Nanjing, 210098, P.R. China
[3] College of Energy and Electrical Engineering, Hohai University, Nanjing, 210098, P.R. China
[4] College of Mechanical Engineering, Hohai University Wentian College, Maanshan, 243031, P.R.China

Abstract. Airfoil is a key factor that influences the hydrodynamic performance of vertical axis tidal current energy turbine. In order to explore the influences from the runner's blade airfoil towards its hydrodynamic performance, three-dimensional unsteady flow numerical simulation on four airfoils of straight blade H type of tidal current energy turbines was carried out, under the precondition of same turbine compactness. Through investigating the influences from the four different airfoils of H blade towards the runner in terms of its dynamic torque feature and hydropower utilization coefficient, this research has analyzed the hydrodynamic performance of the tidal current energy turbine. As the research result indicates, the maximum dynamic torque value of the single-blade turbine with NACA0015 airfoil is significantly higher than that of the other three airfoils; while the maximum value of the overall runner torque is determined by the operating conditions such as different stream speed etc. As a whole, hydropower utilization coefficient of the turbine with asymmetric airfoil NACA4415 is better than that of the other three airfoils. This research has provided references for the design and airfoil selection for the new type of tidal current energy turbine.

1 Introduction

With the constant consumption of the non-renewable energy resources like coal and petroleum, environmental pollution issue is increasingly severe and the contradiction between energy supply and demand is becoming more and more intensified. Under such background, the tidal current, as a kind of abundant, renewable and clean energy, has obtained substantial attention [1]. Tidal current energy is rather common in China with a total production as high as 140 million KW. Under contemporary times when energy crisis and maritime rights issue are becoming more and more prominent, reinforcing the development and utilization efforts in tidal current energy has become beneficial to the world's rapid economic development pace.

In recent years, vertical axis tidal current energy turbine has obtained rapid development across the world. In 2002, Italian (PdA) company has successfully launched the floating vertical axis straight blade varying roar turbine in the Mexico strait [2]. U.S. GCK Technology Company has developed the vertical axis tidal current energy turbine (GHT). By adopting spiral blades with certain rotation degree, the vertical axis turbine will operate in a steadier mode and is much easier to launch compared to the one using straight blade [3]. In 2000-2011, Korea applied the American technology and established the tidal power station at the downstream of the Myeongryang Strait Bridge of Jindo County. With two 500KW vertical axis turbines installed, the station is supported by steel jackets with transformer and control room located on the bank. After the resource exploration and site selection at the first phase as well as site construction and pilot run at the second phase, this power station has been officially put into operation in 2011 [4]. In addition, Ocean University of China has developed a floating vertical axis tidal current turbine power generating set [5]. By adopting soft blade, which is similar to the sailing wing, this set can fully utilize the elevating force and resistance effects. In 2008, this vertical axis tidal turbine power generation set has been successfully tested in Zhaitang Island water area of Jiaonan City of Qingdao. In August 2012, "Marine Energy I", which is developed by renewable energy research institute of Harbin Engineering University, has been accepted with total installed capacity of 300KW [6]. Vertical axis water turbine is featured by simple structured blade, no convection device and lower maintenance costs [7]. Most current studies concentrate on one single aspect without covering the comprehensive impacting factors that influence the hydrodynamic performance of the straight blade H vertical axis water turbine. At the present study, through combining the blade element theory and the three-dimensional numerical simulation method, analysis on the influences from different airfoils and stream speeds towards the runner's dynamic torque feature and hydrodynamic utilization was accomplished. Based on the analysis, the optimal blade tip speed ratio providing essential references to the design and airfoil selection for the tidal current energy water turbine.

2 Physical model

2.1 Geometric modeling

In order to analyze the influences from different airfoils towards the hydrodynamic performances of the vertical axis tidal current energy water turbine, this paper has first ensured the precondition of same turbine compactness. And then by employing Gambit software, it has established the three-dimensional model for the turbine runner under different types of airfoils. The airfoils used in this research include NACA4415, NACA0015, NASANLF0416 and S1210 [7], which are shown in Fig. 1.

(a) NACA4415

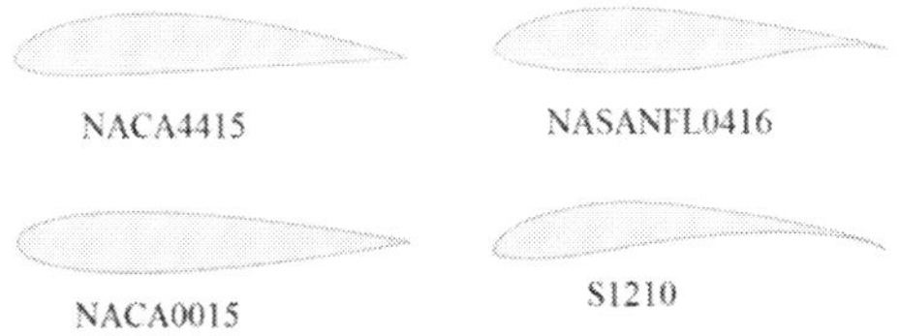

(b) Four types of airfoils

Figure 1. Turbine's runner model under different airfoils

Compactness definition for vertical axis H type of blade:

$$\sigma = \frac{zc}{\pi d} \tag{1}$$

Compactness of the turbine is normally set as 0.15~0.5. This paper has set the value as 0.35 with z as the blade number, c as the chord and d as the runner's diameter.

2.2 Parameters of the turbine model

Basic parameters for straight H type of tidal current energy water turbine are shown in Table 1.

Table 1. Basic parameters for runner of vertical axis tidal energy water turbine

Parameter Name	Straight blade
External diameter of the turbine $D_1\ (m)$	3.5
Hub diameter of the turbine $D_2\ (m)$	0.5
Blade height $h(m)$	4
Incidence angle of the blade $\varphi(°)$	-7
Number of blades Z	3
Airfoil of the support frame	NACA0015
Chord of the main airfoil $C_1(m)$	1.28
Parameter Name	Straight blade

3 Three-dimensional Numerical Simulation Computation

3.1 CFD computation method

During the operation of vertical axis turbine, the external flowing medium water is incompressible viscous fluid and the blade surface is mainly subject to gravity, water pressure and centrifugal force. Thus, this research adopts the Navie-Stokes equations based on Reynolds averaged equations and Spalart-Allmaras turbulence model [8] in performing the CFD numerical simulation on the flowing field at the runner area of vertical axis turbine. The equation is:

$$\frac{\partial u_i}{\partial x_i} = 0 \tag{2}$$

N-S equation:

$$\rho\frac{\partial}{\partial t}(u_i) + \rho u_j \frac{\partial}{\partial x_j}(u_i) = \rho F_i - \frac{\partial p}{\partial x_i} + \mu\frac{\partial^2 u_i}{\partial x_j \partial x_j} \tag{3}$$

Wherein, ρ is the fluid density; u is the fluid velocity vector; F is the mass force; p is the pressure; μ is the eddy viscosity; above computation has adopted the Second Order Upwind format so as to improve the computing accuracy. Besides, it has utilized non-steady implicit solution with time step as 0.01s. Moreover, it has employed SIMPLEC computation method in achieving the coupling between pressure and speed.

3.2 Boundary condition

Boundary condition at the inflow entrance is set as the velocity inlet; the cylinder surface where the turbine locates is set as the glide plane. Under the condition where pressure is resumed to the inflow static pressure, set the boundary condition at the cross section of the stream effluent as the average static pressure outlet, for incompressible fluid, the surface of the runner and support blade as well as the flow passage is set as no slip boundary condition.

3.3 Computational domain and meshing

Actually, tidal current energy turbine is operating in a very large area under the sea. According to the basic parameters of the turbine model, this paper has established a rectangular area as the flowing area around the vertical axis turbine as well as the computational domain. Select the central point under the bottom of turbine's rotation axis as the origin of coordinates with flow direction along the positive direction of x and principle axis direction as z. Then the computational domain size is: X_{max} is 14m, X_{min} is -7m; Y_{max} is 3.5m and Y_{min} is -3.5m; Z_{max} is 4.25m and Z_{min} -0.25m. And the overall computational domain size is: $21m \times 7m \times 4.5m$. To ensure the accuracy of the simulation computation, this research further classifies the computational domain into two parts. Part one is defined as a cylinder area with radius as 2m and height as 4.5m, which closely encircles the vertical axis turbine. Moreover, the meshes in this area has refined and the cylinder surface was defined as the sliding boundary. And the rest part of the rectangular area is defined as the part two. Furthermore, ANSYS ICEM Software mesh on above two areas. And given the runner is normally coated with irregular complex curved surfaces, this research has adopted tetrahedral unstructured mesh in segmentation. At last, there are around 2.35 million meshes in the rotating area and 0.9 million meshes in the external flowing field.

4 Computation result analysis

4.1 Influence from the airfoil towards runner's dynamic torque feature

As shown by Fig. 2, under the operating condition when the velocity inlet for four airfoils of turbines is v=1.5m/s and v=2m/s respectively and the rotating speed is 13.38r/min, the torque change curve graph for single runner blade is presented as below. Indicating that when the velocity inlet is v=1.5m/s or v=2m/s, torque change curve is basically the same for different airfoils of single blades. And with the rising of speed, maximum torque value for turbine blade at the same airfoil is improving; whereas the minimum value is decreasing. Within one cycle, single blade torque fluctuation of airfoil S1210 is significantly large but the torque value generated is relatively smaller.

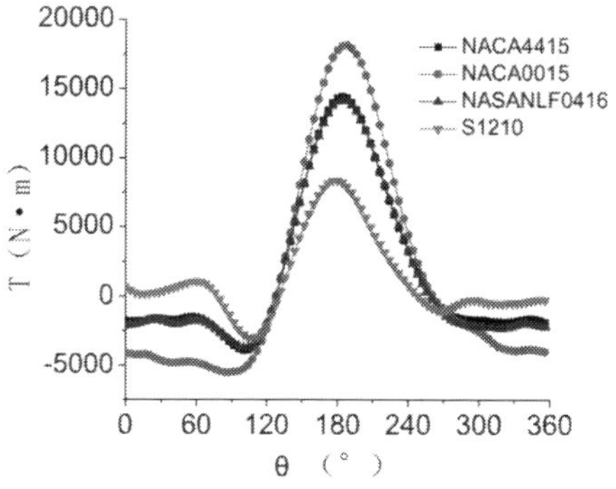

(a) v=1.5m/s

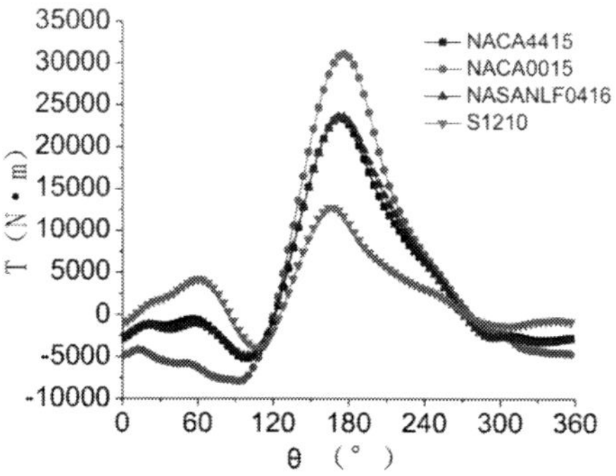

(b) v=2.0m/s

Figure 2. Comparative graph for single blade under different airfoils

Through comparing the torque value of the single blade under different airfoils in Fig. 2, it's detected: maximum torque value of NACA0015 airfoil is significantly higher than that of the other three airfoils and its minimum value is smaller than the other three airfoils as well; torque value generated by NACA4415and NASANLF0416 is close to each other; when the phase degree θ is 120° ~270°, torque value of all airfoils will first increase and then decrease with all of them as positive and the blades doing positive work; whereas, when the phase degree is beyond above range, torque value of single blade for S1210 airfoil is obviously higher than that of the other airfoils; meanwhile, only the blade for this type of airfoil alone is obtaining positive value within a small range, and it's majorly doing negative work within this area.

As shown from Fig. 3: the torque change curve for the overall runner of different airfoils is composed by the torques of the straight blades from three runners. Under the conditions when the rotation speed is 13.38r/min and velocity inlet is v=1.5m/s or v=2.0m/s, overall torque value for runner with airfoil S1210 is the lowest. But with different operating conditions, the maximum overall torque value is different. When the inlet velocity is v=1.5m/s, overall torque for NACA4415 is the largest, followed by NACA0015, NASANLF0416 and S1210 at last. And when v=2.0m/s, overall torque value for runner with airfoil NACA0015 is much larger than that of the other three airfoils. NACA4415 and NASANLF0416 are close to each other with the former slightly larger than the latter, but S1210 falls far behind from above three airfoils.

In general, under different operating conditions, the optimal airfoil with the largest torque value will also alter.

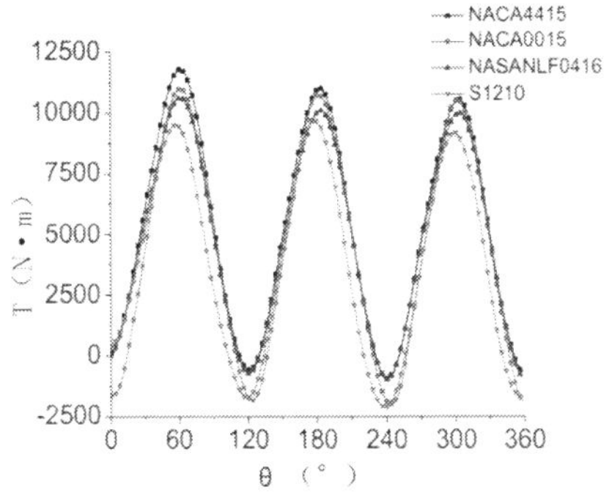

(a) v=1.5m/s

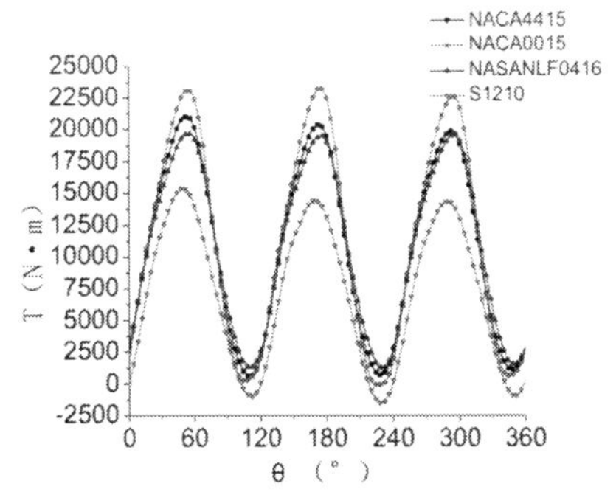

(b) v=2.0m/s

Figure 3. Comparison graph for turbine runner's torque under different airfoils

4.2 Influence from airfoil to runner's hydropower utilization coefficient

Through utilizing FLUENT Software, this search has applied simulation computation on the vertical axis tidal current energy turbine model. Given it is non-steady numerical simulation, when the iteration convergence turns steady, runner's torque value is manifested as a flat curve with regular fluctuations. And by integrating on the torque curve with certain period of time, average torque T is computed and then via computation formula (4):

$$C_p = \frac{P}{\frac{1}{2}\rho v^3 A} = \frac{T\omega}{\frac{1}{2}\rho v^3 A} \quad (4)$$

Wherein, P is the power actually obtained by the turbine's runner with the unit as W ; ρ is the fluid density with the unit as kg/m^3 ; v is the flowing speed with the unit as m/s ; A is the swept area of the runner with the unit as m^2 ; T is the torque with the unit as $N \cdot m$; ω is the rotating speed of the runner with the unit as $r/\min$.

The hydropower utilization coefficient, which is also known as the efficiency, can be derived from above formula. Efficiency output comparison is shown as Table 2.

Table 2. Efficiency output comparison

ω (r/min)	v (m/s)	λ	Cp/%			
			NACA 4415	NACA 0015	NASANLF 0416	S1210
13.38	1.00	2.45	-0 .01	-10.8	-3.72	8.30
13.38	1.35	1.81	28.46	20.61	26.30	21.68
13.38	1.50	1.63	31.21	26.44	28.89	21.25
13.38	2.00	1.23	26.74	28.31	26.77	17.79
13.38	2.50	0.98	21.24	23.49	21.74	13.65
13.38	3.00	0.82	15.55	17.53	15.59	10.05

According to the statistics from the table, efficiency change curve per blade tip speed is presented as below Fig. 4.

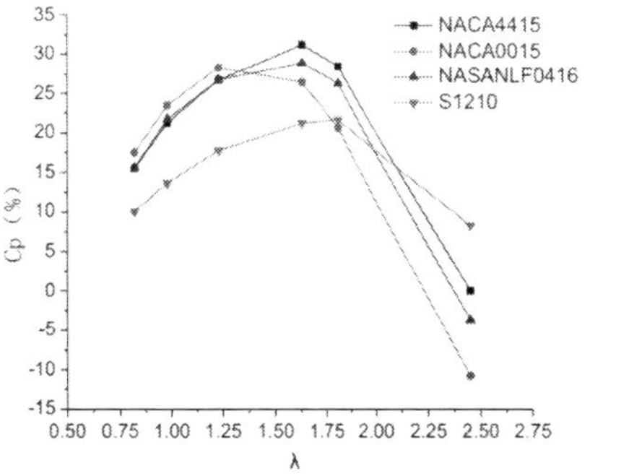

Figure 4. Efficiency curve comparison graph for turbines under different airfoils

Efficiency curve comparison graph for turbines under different airfoils is shown as Fig. 4. Through comparing the efficiency of the runners under the four airfoils, it's indicated: 1. With different airfoils, the highest efficiency of the vertical axis tidal current energy turbine is matched with different optimal operating conditions and different optimal blade tip speed ratio. To be specific, the optimal blade tip speed ratio for airfoil NACA0015 is 1.23, that for NACA4415 and NASANLF0416 is 1.64 and that for S1210 is 1.81. Generally speaking, blade tip ratio λ at the high efficiency area for all airfoils is within 1~2. 2. When the blade tip ratio λ is smaller than 1.3, the vertical axis turbine efficiency for airfoil NACA0015 is higher than that of the other three airfoils; efficiency of NACA4415and NASANLF0416 is close to each other and S1210 ranks last. When the blade tip ratio λ falls 1.3~1.8, efficiency of airfoil NACA4415 ranks the top, followed by NASANLF0416, NACA0015 and S1210. 3. When the blade tip ratio λ is larger than 2.1, turbine efficiency of all airfoils are relatively lower. And at this point, hydropower utilization coefficient of airfoil S1210 is higher than that of the other three airfoils, reaching as high as 31.21% and obtaining a better stability. At last, as indicated from Fig. 4, when the blade tip speed ratio λ is relatively larger, such as larger than 2.35, then the hydropower utilization coefficient of vertical axis tidal current energy turbine is very low with that of airfoil NACA4415, NACA0015 and NASANLF0416 even turning to negative value. Under such condition, the incoming fluid will turn around the runner's blade and generate larger resistance than elevating force. Hence, the tidal current energy turbine will not do any work and can't operate.

5 Conclusion

Through utilizing the three-dimensional numerical simulation method and analyzing the hydrodynamic performance of the straight blade H type of tidal energy current turbine under four different airfoils of runner blades, the conclusion is derived as below:

a. as far as the singe blade of tidal current energy turbine is concerned, torque value of airfoil NACA0015 is significantly higher than that of the other three airfoils; torque generated by NACA4415 and NASANLF0416 is extremely close to each other and the torque value of single blade under airfoil S1210 is relatively lower as a whole. Besides, when the phase degreeθ is within the range of 120°~270°, then the torque will first increase and then decrease and do positive work; whereas, for the other phase degree, it will mainly do a negative work. As for the overall runner, with different operating conditions, airfoil for the optimal runner which obtains the maximum torque value will also vary.

b. the optimal blade tip speed ratio for airfoil NACA0015 is 1.23 and that for NACA4415 and NASANLF0416 is both 1.63, while that for S1210 is 1.81. Besides, the blade tip ratio λ at the high efficiency area for all airfoils is within 1-2. And as a whole, hydropower utilization efficiency of asymmetric airfoil NACA4415 is better than that of the other three airfoils, reaching as high as 31.21% with better stability. This is followed by airfoil NACA0015 and NASANLF0416 with airfoil S1210 ranking the last.

Acknowledgments

The work is financially supported by the State Key Program of National Natural Science of China:51579080, the Fundamental Research Funds for the Central Universities (No. 2015B33814) and Jiangsu Provincial Graduate Student Innovation Project (Project No.: CXZZ13_0257).

References

1. T. J. Chu, *Development and utilization of ocean energy resources*, Beijing: Chemical Industry, (2015)
2. G. Calcagno, F. Salvatore, L. Greco, et al. An experimental investigation and a theoretical and computational methodology to study an innovative technology for marine current exploitation: the Kobold turbine, Bollettino della Comunità Scientifica in Australasia, (2006)
3. Verdant Power L L C. GCK Technology, Inc. Amesbury Tidal Energy Project: Integration of the Gorlov Helical Turbine into Optimized Hardware/ Software System Platform, MTC Final Report, (2005)
4. S. J. Wang, Study on hydrodynamic performances of a tidal current energy conversion device with flexible blade turbine, (2009)
5. L Zhang, Z. C. Li, X. W. Liang, et al, The status of research and application of vertical axis tidal turbine, Applied Energy Technology, 9 ,1 (2011)
6. Z. C. Li, K. Sun, F. M. Jing, et al, Study on the blade preset pitch of a fixed pitch vertical axis tidal turbine, Journal of Harbin Engineering University, 32, 7(2011)
7. Y. W. Li, *Marine energy exploitation*, Beijing: Ocean, (2008)
8. H. K. Versteeg, W. Malalasekera, *An Introduction to Computational Fluid Dynamics: The Finite Volume Method*.New York: Wiley, (1995)
9. Falconer R A. Ahmadian R, Assessment of array shape of tidal stream turbines on hydro-environmental impacts and power output, Renewable Energy, 44: 318-327, (2012)

Computational Design and Experimental Validation of SAR Mixers

Md Readul Mahmud, Vladimir Viktorov and Carmen Visconte

DIMEAS, Politecnico di Torino, Corso Duca degli Abruzzi, 24-10129, Torino, Italy

Abstract. In this study, two novel split and recombine (SAR) 'H-C' and 'Y-Y' micromixers along with a known SAR 'Chain' mixer are presented. The efficiency and pressure drop at Reynolds numbers (Re) up to 100 were investigated numerically as well as experimentally. Numerical and experimental values of mixing efficiency and pressure drop coincide considerably, which validate the numerical model. Results show that the mixing efficiency of the Chain mixer is good only at Re≥50 and the pressure drop is relatively high, whereas the H-C and the Y-Y mixers give a mixing efficiency higher than 90% over all the range of Reynolds numbers examined and their energy requirement is less than that of the Chain mixer. Furthermore, numerical residence time distribution (RTD) was explored, which successfully predicts the experimental results.

1 Introduction

Microfluidic devices have had a considerable impact in many fields such as biomedical diagnostics, drug development, and are widely used in the food and chemical industries [1]. Their function is to mix the reactants for initiation of the reaction process and they are an indispensable component in lab-on-chip (LOC) platforms for complex chemical reactions. Micromixers can also be integrated in a microfluidic system or work as stand-alone devices [2].

In design and evaluation of micromixers, the Reynolds number ($Re = \frac{\rho v d}{\mu}$) is of major concern, where ρ is the fluid density, v is the velocity of fluid, μ is the dynamic viscosity, and d is the characteristic length of the mixer. The characteristic dimension of a micromixer is in the sub-millimeter range and thus conventional methods to stir the fluids are not applicable [3]. Due to low Reynolds number, the flow in a microchannel is laminar and thus mixing is dominated by molecular diffusion rather than convection. Therefore, to get a good mixing performance, considerable time and long channel length are necessary, which yield high-pressure drop and high production cost [4]. Different kind of micromixers has been studied to overcome these problems. Micromixers can be classified into two categories, active and passive, depending on their mixing principle [5]. In active micromixers, active controls over the flow field are introduced to manipulate fluids to be mixed by using moving parts, acoustics waves or electro-magnetic fields or varying pressure gradient [6]. In the case of passive mixer, the mixing enhancement can be achieved by carefully designing the channel so as to create vortices and disturb the flow stability which, in turn, leads to the chaotic flow regime. As a consequence, the interfacial area among fluid segments is increased, reducing the diffusion length for heat and mass transport without any additional equipment or energy added to the system [7]. Passive micromixers are less expensive, easy to manufacture and maintain.

Good mixing performance and low pressure loss are two important design factors of any mixer. Low pressure drop is very crucial for industrial application. Good mixing efficiency is particularly suited for fast complex chemical reactions, where the reaction yields are strongly affected by mixing quality [7]. Degree of mixing can be achieved directly including experimentally [8] and numerically [9]; indirect method includes Poincare section analysis [10] and residence time distribution (RTD) [11]. Knowledge of the RTD enables the exact prediction of yield for first order reactions; the general reactor performance and mixing can be evaluated. Since micromixers are available in a great variety of geometries using different mixing principles, the selection of a suitable device for a given application becomes more challenging. Among various characteristics (like efficiency, energy loss) the knowledge of the RTD of a certain microfluidic device would be a useful help to identify the most appropriate one [12].

A number of experimental and numerical studies have been carried out on different split and recombine (SAR) mixers lately. SAR mixers show good mixing efficiency at low Reynolds number (Re<20) and high Reynolds number (Re>100), whereas mixing performance is relatively low at middle range [13]. Two novel SAR mixers, namely the Y-Y and the H-C, are developed and studied at Reynolds numbers ranging from 1 to 100. The proposed mixers show good efficiency (more than 90%) and low pressure drop. Their performance is also compared to those of a known SAR mixer, the Chain one.

2 Micromixer design

Two new passive SAR mixers, the Y-Y and the H-C, are designed and constructed. Fig. 1 shows the fluid volume of the Chain, the Y-Y and the H-C micromixers with some important dimensions. As shown, all mixers are made up of four identical elements connected after one another, denoted as 1 to 4. Micromixer geometries were micromilled in polycarbonate by a computer numerical control (CNC) Milling and Engraving Machine P20S, KUNLMANN (Germany) at I.T.D sas di Borgna Daniela e Borgna Eugenio & C. (Turin, Italy).

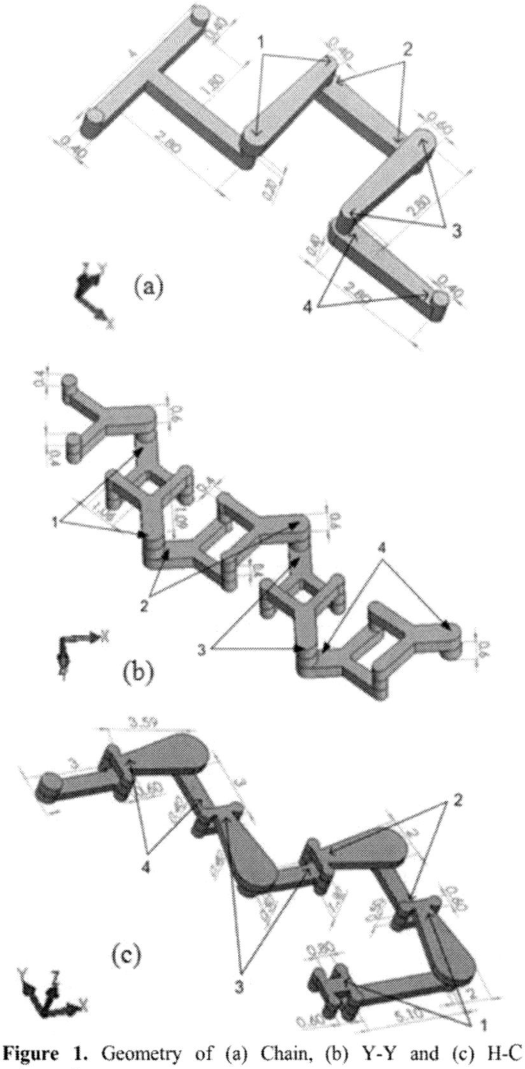

Figure 1. Geometry of (a) Chain, (b) Y-Y and (c) H-C micromixers.

3 Mixing effectiveness

To quantify the mixing performance the following equations were employed:

$$\sigma = \sqrt{\frac{1}{N}\sum_{i=1}^{N}(C_i - C_m)^2} \qquad (1)$$

$$\eta = 1 - \sqrt{\frac{\sigma^2}{\sigma_{max}{}^2}} \qquad (2)$$

where σ is the standard deviation of mass fraction; N is the number of cells at a target area; C_i is the mass fraction at a cell i; C_m is mean mass fraction at the target area; η is the mixing efficiency; σ_{max} is the maximum standard deviation, namely 0.5 for symmetrical boundary conditions, used for normalization. The experimental efficiency and pressure drop were computed following the steps described in [13].

4 Numerical methods

4.1 Numerical model of mixing

The commercial CFD software Ansys Fluent 15 was employed for the numerical analysis. The fluid flow was assumed to be steady, laminar, Newtonian, and incompressible; therefore, ignoring the body force and gravity, governing equations are the continuity equation, Navier-Stokes equation and species convection-diffusion equation, given respectively by:

$$\nabla V = 0 \qquad (3)$$

$$\rho V\, \nabla V = -\nabla \mathrm{P} + \mu \nabla^2 V \qquad (4)$$

$$V \nabla\, C = \mathrm{D}\nabla^2\, C \qquad (5)$$

where V is the fluid velocity vector, ρ is the fluid density, P is the pressure, μ is the fluid dynamic viscosity, C is the species mass concentration, and D is the diffusion coefficient of the species.

Water properties were applied to the mixing species: density and viscosity were set to constant values of 998 kg/m^3 and 1×10^{-3} kg/m·s, respectively. The diffusion coefficient was set at 10^{-9} m^2/s, which is the typical value for an aqueous solution at room temperature. Both entrances were set to "velocity inlets"; the exit port was set to "pressure outlet", with a gauge pressure of 0 Pa. At all walls no-slip boundary condition was applied.

Hexahedral cells were used to discretize the computational domain. Grid sensitivity test was performed for the mixing efficiency to determine the optimum number of grid points for all the micromixers. A grid system with approximately 3.53×10^5, 5.12×10^5 and 5.25×10^5 nodes was chosen for the Chain, the Y-Y and the H-C mixers, respectively.

4.2 Numerical model of RTD

The RTD is obtained commonly by injecting a tracer instantaneously (a pulse input) or at a constant rate (a step input) at the inlet of a flow system, and then measuring the tracer concentration at the outlet as a function of time. The RTD function, E(t), which describes quantitatively how much time different fluid elements have spent in a continuous flow system is defined mathematically so that E(t)dt is the fraction of fluid elements that have spent a time between t and t+dt in the flow system [11]. In the determination of RTD in this study, the pulse input method is used in order to avoid the numerical errors from data differentiation that is inherent in the step input method. The time steps for each simulation are appropriately adjusted so as to limit the necessary number of calculation steps needed by the time-consuming transient simulations. For pulse tracer input method, E(t) is given by equation (6). When the mixing performance of flow systems of different sizes or flow conditions is to be compared, a normalized RTD function, E (θ), is used instead of E (t). Both functions are related by equation 7 [14] where t_m and θ are the mean residence time and dimensionless time, respectively.

$$E(t) = \frac{C(t)}{\int_0^{\infty} C(t)dt} \tag{6}$$

$$E(\theta) = t_m E(t) \tag{7}$$

$$t_m(t) = \frac{\int_0^{\infty} tC(t)\,dt}{\int_0^{\infty} C(t)dt} \tag{8}$$

$$\theta = \frac{t}{t_m} \tag{9}$$

Carrying out the CFD simulations, the plots of E (θ) vs. θ are constructed for all the mixers. Some other important parameters, such as coefficient of variance (CoV), variance ($\sigma(t)^2$) and skewness (s), are defined by equations (10)-(12):

$$CoV = \frac{\sigma(t)}{t_m} \tag{10}$$

$$\sigma(t)^2 = \int_0^{\infty} (t_i - t_m)^2 E(t)dt \tag{11}$$

$$s = \frac{\int_0^{\infty} (t_i - t_m)^2 E(t)dt}{\sigma(t)^{3/2}} \tag{12}$$

Once the RTD function is obtained, residence time, coefficient of variance, variance and skewness were calculated to quantify the degree or extent of mixing in continuous flow systems. The RTD of a microchannel mixer will deviate from that of an ideal plug flow mixer depending on the hydrodynamics within the microchannel. In the context of the static microchannel mixers designed to improve radial/transverse mixing, a CoV of zero would imply complete plug-flow mixing while a non-zero CoV implies that there is axial dispersion or mixing caused by non-uniform or laminar velocity profile and molecular diffusion. In this case, the smaller the variance or the CoV, the narrower is the RTD, the closer is the distribution to the mean residence time, and the better the mixing quality [11], [15].

5 Results and discussion

The purpose of this study was to examine and compare the performance of three SAR micromixers.

Firstly, mixing efficiency of the Chain, the Y-Y and the H-C mixers was evaluated experimentally and numerically over a range of Re numbers from 1 to 100. Fig. 2 shows the concentration distribution of fluids along the length of mixers in both experimental tests and numerical simulations. As can be seen, a reasonable agreement between numerical and experimental results can be reported. However, there are still apparent differences in some sections. The main reason lies in the fact that simulation images can only show horizontal mid-plane. Therefore, it was not possible to take into consideration the multiple laminations of fluids layers. By contrast, the experimental images cover the whole depth of the channel viewable from below.

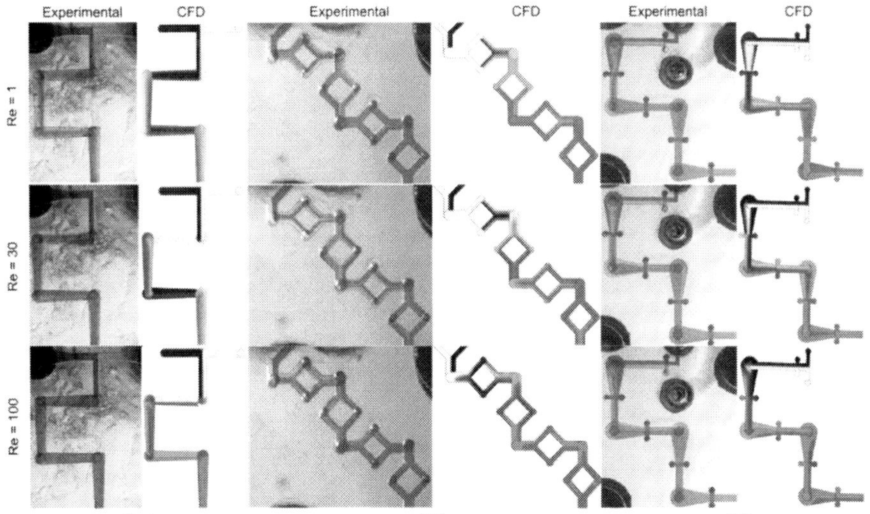

Figure 2. Comparison between the images taken from the numerical simulation and the experiments

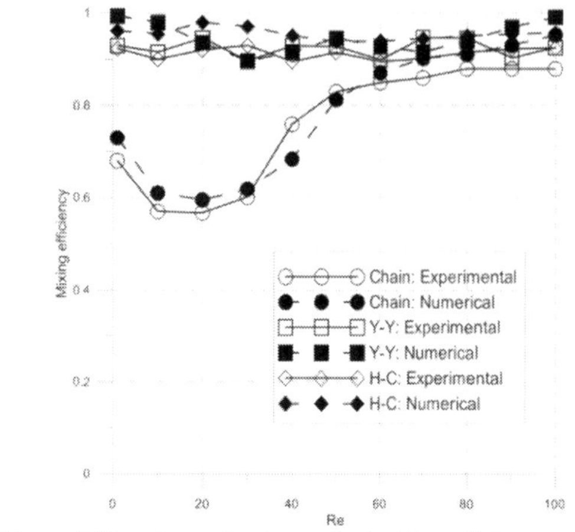

Figure 3. Experimental and numerical mixing efficiency

Figure 3 compares the mixing efficiency of the Chain, the Y-Y and the H-C mixers at the exit of the fourth element; as shown, experimental and numerical data have the same trend and reasonable agreement. The efficiency of the Y-Y and the H-C mixers does not depend on Reynolds numbers contrary to that of the Chain mixer. All three mixers show reasonable mixing efficiency at low Reynolds numbers due to the stratified flow. As the Reynolds numbers increase, there is a shorter residence time for the fluid particles to diffuse, and as a result worse mixing efficiency could be expected. Nevertheless, the Y-Y and the H-C mixers show good efficiency due to the three-dimensional change of direction of flow in the input of each element, which promotes a helical fluid motion; as a consequence, the fluid path is lengthened [13]. However, this swirling effect is less evident in the Chain mixer. Therefore, efficiency of the Chain mixer strongly depends on Reynolds numbers; in particular, at the low-middle range ($1 \leq Re \leq 50$) is less than 80%. At high Reynolds numbers, the engulfment flow regime is reached and species get enough time to diffuse; therefore all mixers show an efficiency of about 90%.

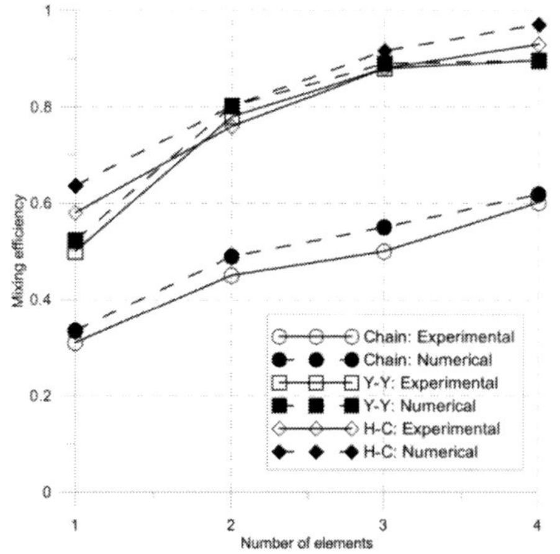

Figure 4. Mixing efficiency at Re = 30

Fig. 4 demonstrates the dependency of mixing efficiency on the number of elements at Re = 30. Experimental and numerical efficiency curves have the same trend: mixing efficiency increases with the number of elements. It is clear that the efficiency of the Chain mixer is always less than those of the Y-Y and the H-C mixers; after the first and the third element the difference in efficiency is 10% and 30%, respectively. Furthermore the Y-Y and H-C mixers show a mixing efficiency of almost 90% at the exit of the fourth element, whereas the Chain mixer shows an efficiency of only about 60%.

Thirdly, the pressure drop of the micromixers was computed experimentally and numerically as shown in Fig. 5. Experimental and numerical pressure drop increases with the increase of flow rate, as expected. In addition, experimental and numerical values show satisfactory agreement. It can be concluded that the pressure drop is linearly dependent on the flow rate up to 2×10^{-8} m^3/s due to the multi-laminated flow of fluid. As the flow rate increases, vortexes are introduced in the fluid stream and consequently the pressure drop curve moves away from the linearity. The H-C mixers show less pressure drop than other two mixers. Therefore, the Y-Y and the H-C mixers require less energy to operate at same flow condition.

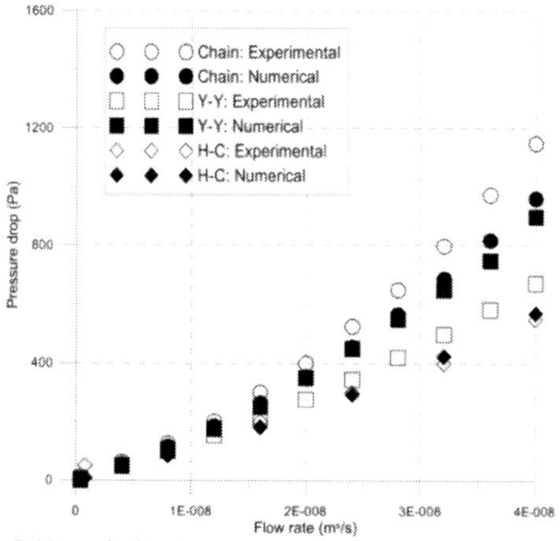

Figure 5. Numerical and experimental pressure drop

Table 1. Results of the numerical RTD analysis for at different Reynolds numbers

Re		Residence time (s)	CoV	Variance	Skewness
Chain	1	16.656	0.318	44.046	34.548
	10	1.770	0.388	0.849	2.592
	30	0.516	0.371	0.752	2.325
	100	0.137	0.225	0.002	0.029
H-C	1	41.897	0.259	204.329	151.876
	10	4.293	0.266	2.773	8.643
	30	1.331	0.192	0.128	0.707
	100	0.421	0.153	0.007	0.063
Y-Y	1	32.142	0.319	186.393	132.367
	10	3.252	0.327	1.952	4.064
	30	1.056	0.308	0.189	0.782
	100	0.297	0.196	0.006	0.051

Finally, using the results from the CFD simulations, the plots of E (θ) vs. θ were constructed at different Reynolds number as shown in Fig. 6. Generally, all micromixers show similar behavior. At low Reynolds numbers the distributions are broader and at higher Reynolds numbers the RTD curves become narrower and more symmetric. A curve with a sharp peak or narrower peak, which also gives low value of CoV, denotes better mixing efficiency. It is notable from Table 1 that the values of CoV and variance decrease with the increase of Reynolds numbers for both the Y-Y and the H-C mixers, which indicates the generation of secondary flow due to the helical fluid motion; as a result radial mixing have effect on total mixing performance. On the contrary in the case of the Chain mixer at Re = 30, the value of CoV is relatively high, which predicts a low mixing efficiency, as shown in Fig. 3. In addition, the residence time of the Y-Y and the H-C mixers is higher than that of the Chain mixer due to their longer channel length.

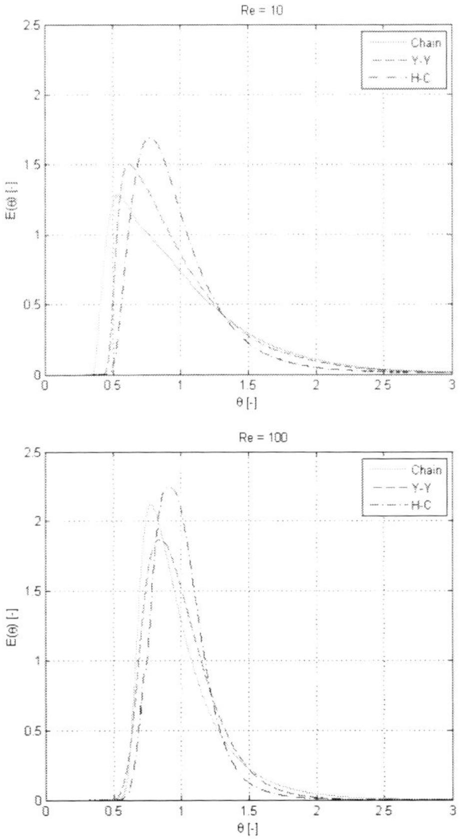

Figure 6. Normalized RTD of mixers at Re = 10 and 100

6 Conclusions

Three passive SAR micromixers, the Chain, the Y-Y and the H-C, were studied numerically at Reynolds numbers ranging from 1 to 100. Afterwards, experiments were done to validate the numerical work. All micromixers are made up of four identical elements connected one after another. The mixer structures are simple and no grooves are provided. An image analysis technique was used to evaluate mixture homogeneity at four target areas along the mixer. A good agreement is found between numerical and experimental results. The Y-Y and the H-C micromixers show an almost flat mixing characteristic over the examined range of Reynolds numbers, with a mixing efficiency higher than 90%; conversely, the Chain mixer shows a relevant decrease of efficiency at low-mid range. In addition, pressure drop within the Y-Y and the H-C micromixers is less than that of the Chain mixer. Furthermore, RTD is analyzed at low, middle and high Reynolds number, which also predicts the better performance of the Y-Y and the H-C mixers than the Chain one. Therefore, the new Y-Y and H-C micromixers have high potential for chemical and food applications.

References

1. C. Lee, W. Wang, L. Fu, Micromachines **6**, 1-24 (2015)
2. I. Sabotin, G. Tristo, M. Junkar, J. Valentincic, ChemE **91**, 778-788 (2013)
3. A. A. Mouza, C. M. Patsa, F. Schonfeld, ChemE **86**, 1128-1134 (2008)
4. A. Alam, A. Afzal, K. Kim, ChemE **92**, 423-434 (2014)
5. S. Hossain, K. Kim, Micromachines **5**, 913-928 (2014)
6. T. G. Kang, P. D. Anderson, Micromachines **5**, 1270-1286 (2014)
7. N. Solehati, J. Bae, A. P. Sasmito, Computers & Fluids **96**,10-19 (2014)
8. Y. Liu, Y. Deng, P. Zhang, Z. Liu, J. Micromech. Microeng. **23**, 1-10 (2003)
9. Y. Du, Z. Zhang, C. Yim, M. Lin, X. Cao, Micromachines **1**, 19-33 (2010)
10. D. J. Beebe, R. J. Adrian, G. O. Olsen, M. A. Stremler, H. Aref, B. Jo, Mechanics & Industries **2**, 343-348 (2001)
11. J. T. Adeosun, A. Lawal, Chemical Engineering Science **64**, 2422–2432 (2009)
12. D. Boskovic, S. Loebbecke, Chemical Engineering Journal **135S**, S138–S146 (2008)
13. V. Viktorov M. R. Mahmud, C. Visconte, Micromachines **6**, 1166-1179 (2015)
14. A. G. Kanaris, A. A. Mouza, Chemical Engineering Science **66**, 5366-5373 (2011)
15. J. T. Adeosun, A. Lawal, Chemical Engineering Science **65**, 1865-1874 (2010)

A Novel Approach in Designing PID Controller for Semi-active Quarter Car Model

Vedant Mehta, Yash Gandhi, Mayuri Patel, Bhargav Gadhvi, Anil Markana, Jay J. Vora

Pandit Deendayal Petroleum University, Raisan, Gandhinagar — 382007, India

Abstract. This paper implements Teaching-Learning based optimization (TLBO) to obtain optimized value of spring stiffness for better ride comfort. Further, this optimized value is then used in a semi-active quarter car setup to remove any discrepancies due to non-optimized spring. This paper also introduces a novel approach to control the Semi-active suspension parameter (damping coefficient) for a better performance. For controlling semi-active parameters, PID controller has been used. PID controller output is fed to the quarter car setup as a damping coefficient. Thus changing the damping coefficient dynamically as the disturbance occurs, and thus improving the ride comfort. The sprung mass acceleration and rattle space of semi-active quarter car has been compared with sprung mass acceleration and rattle space of passive quarter car model to show the difference in results and thereby, results and conclusions are drawn.

1 Introduction

Various researches have been made in automotive industry to improve the suspension system. Suspension system is one of the most vital parts of any vehicle. It makes the vehicle ride smoother over rough roads having bumps and potholes. Designing of a suspension system is a compromise between vehicle handling and ride comfort. Sports car has good vehicle handling but doesn't provide good ride qualities, whereas, a sedan or a luxury car has outstanding ride comfort but low road handling quality.

Suspension system is classified as: passive, active and semi active.

Passive suspension system is the traditional suspension consisting of springs and dampers. A passive suspension uses spring to store energy and to dissipate it, a damper is used. Generally, its parameters are fixed and they are chosen in such a way that there is a certain level of compromise between road handling, load carrying and ride comfort [1].

An active suspension system is externally controlled and may vary its parameters. It contains an actuator (usually hydraulic, pneumatic or electromechanical) which creates force to control the motion of sprung mass and relative velocity between sprung and unsprung mass. This actuator is controlled with the help of a controller. Different control strategies are developed for the same. Active suspension system is a closed loop control system whereas passive suspension system is an open loop control system [1, 2].

Semi-active suspension system consists of a varying damper. The force applied by semi-active suspension system is created by changing the fluid flow of damper. Magnetorheological (MR) fluid damper with many advantages such as high viscosity control range, comparatively low cost, low power consumption and quick response, small size etc. is an excellent choice for semi active damper fluid [3].

Due to its low energy consumption with similar vibration control performance to the active control methods Semi-active control has shown many advantages in vehicle suspension systems [4]. Its ability to control the damping force is also a major advantage. A compromise between the simplicity of passive systems, and the cost of higher- performance fully active suspension system can be offered by Semi-active suspension system [5].

Over years different optimization techniques have been used for optimization of various parameters [6, 7]. Bagheri-2011 [8] presented a paper in which a multi-objective genetic algorithm with a recently developed diversity preserving mechanism called as NSGA-II, has been used to optimally design vehicle vibration model.

Badran-2012 [9] worked on optimization of a four-degrees-of-freedom vehicle's human with seat suspension system using genetic algorithms (GA) to determine vehicle suspension parameters to achieve the best comfort of the human.

Qazi-2013[10] reached to result that fuzzy logic based particle swarm optimized semi-active suspension system is better than passive and active suspension model in terms of road handling and ride comfort. Particle swarm optimization was used to optimize scaling factors keeping normalized ranges of inputs and outputs of fuzzy logic controller.

Teacher learning based optimization (TLBO) was used to optimize parameters of PID integrated active suspension system. When optimized values of suspension stiffness and damping constants were compared with non-optimized, a 36% reduction in effort of actuator force and 41% reduction in suspension travel was observed [11].

In this paper, stiffness constant of sprung mass of passive suspension has been optimized using TLBO. For comparison purpose, a PID integrated semi-active suspension model has been used. Comparison between optimized and non-optimized values is done on the basis of rattle space and sprung mass acceleration.

2 Optimization of passive quarter car model

Passive quarter car model consists of sprung mass, unsprung mass, and suspension (consisting of spring and damper). Sprung mass consists of the upper body of the car including chassis weight. Unsprung mass consists of wheel assembly, and the parts below the suspension. Tire also has its own stiffness and damping properties. Stiffness of tire is very large and damping coefficient of tire is very negligible. Road input is given to tire and this input is transferred to unsprung mass and then to sprung mass via suspension. So sprung mass experiences less vibrations as compared to unsprung mass.
Figure shown below resembles the quarter car model.

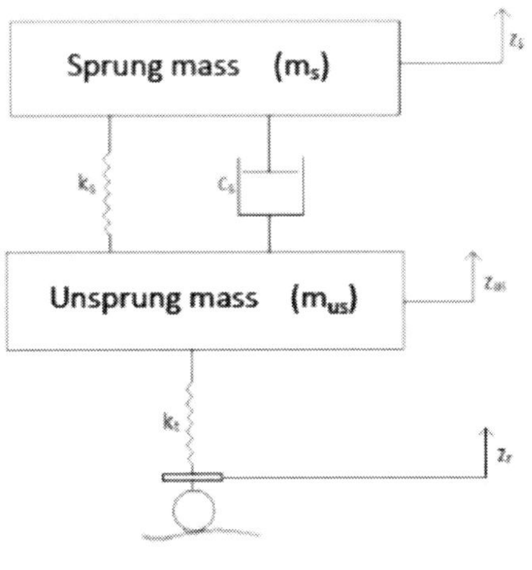

Figure 1: Quarter car model (Passive suspension system)

Optimizing the spring stiffness of the suspension helps in better ride comfort. Optimized value reduces the amplitude of vibrations felt by sprung mass and helps improve the damping of vibrations.

2.1 Teaching learning based optimization (TLBO)

A new optimization algorithm, called 'Teaching–Learning-Based Optimization (TLBO) [10] is used for the optimization of passive suspension system parameters. This method works on the basis of types of interaction between a teacher and learner. TLBO is a population-based method where it uses various solutions generated from each population to obtain a global solution.
This algorithm consists of two phases:

1. Teacher phase' which means interaction between teacher and a learner.
2. Teacher phase' which means interaction between teacher and a learner.

TLBO shows better performance as compared to other nature-inspired optimization algorithms such as Artificial Bee Colony (ABC), Particle Evolutionary Swarm Optimization (PESO), Cultural Differential Evolution (CDE), etc. for the constrained benchmark functions for different performance criteria, such as success rate, mean solution, average number of function evaluations required, convergence rate, etc. [12]. Different algorithms require their own algorithm-specific control parameters. The improper tuning of algorithm-specific parameters either increases the computational effort or yields the local optimal solution. Thus, TLBO proves to better as compared to other algorithms.

2.2 Optimization using TLBO

Since the design of suspension system is trade-off between ride comfort and road handling, this paper considers a weighted objective function consisting of ride comfort as well as road handling [13].
Objective function:

$$y = rms(\ddot{z}_s) + (1000 \times rms(z_{us} - z_r)$$

$\ddot{z}_s$ – represents sprung mass acceleration (ride comfort)
$(z_{us} - z_r)$ – represents tire deflection (road handling)
Quarter car parameters are shown in the table below

Table 1. Quarter car parameters

Symbol	Name	Value
ms	Sprung mass	327 kg
mus	Un-sprung mass	116 kg
ks	Suspension stiffness constant (optimized)	60186 N/m
kt	Tire stiffness constant	256740 N/m
cs	Suspension damping coefficient (non-optimized)	kg/s

3 Semi-active model and controller scheme

In this paper, semi-active suspension system has been taken into consideration. Semi-active suspension system has a variable damping system. Magneto-rheological fluid (MRF) is used as a damping fluid. This fluid has magnetic properties. The viscosity of the fluid can be controlled by subjecting it to magnetic field. This feature helps improve ride comfort. The damping coefficient varies within a defined range. Thus, semi-active is a kind of adaptive system which adapts to the varying road and thus improves the ride comfort.

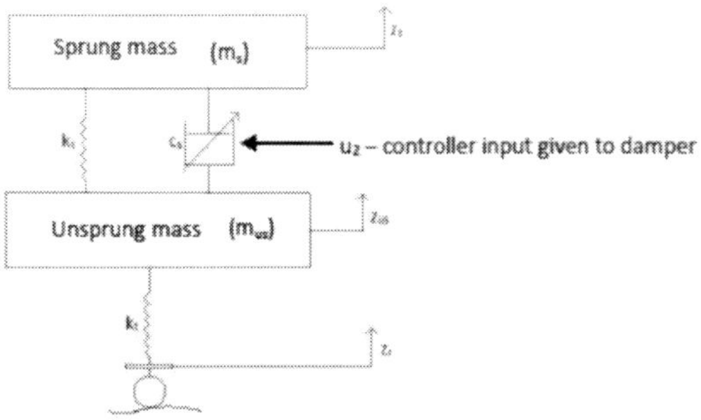

Figure 2: Quarter car model (Semi-active system)

In a conventional system, there is only a road input. But this paper incorporates damping coefficient as an input. So the quarter car model considered in this paper takes two inputs i.e., the road input and damping coefficient. This damping coefficient changes according to the PID controller used in the system. PID controller takes the input of sprung mass displacement and controller output is decided according to the criterion of minimizing sprung mass displacement (to improve ride comfort).

Given below are the state variables:

$$x_1 = z_s$$

$$x_2 = z_{us}$$

$$x_3 = \dot{z}_s$$

$$x_4 = \dot{z}_{us}$$

Given below are the state space equations:

$$\dot{x}_3 = \ddot{z}_s = \frac{k_s}{m_s}(x_2 - x_1) + \frac{u_2}{m_s}(x_4 - x_3)$$

$$\dot{x}_4 = \ddot{z}_{us} = \frac{k_t}{m_{us}}(z_r - x_2) - \frac{k_s}{m_{us}}(x_2 - x_1) - \frac{u_2}{m_{us}}(x_4 - x_3)$$

Where, z_s – represents sprung mass displacement
z_{us} – represents unsprung mass displacement
z_r – represents road input
u_2 – represents damping input taken from PID controller

4 Results and discussion

Simulink model of semi-active suspension system as well as passive suspension system was developed. The sprung mass acceleration of semi-active and passive suspension system was compared. The rattle space in the case of semi-active as well as passive suspension system was also compared.

As shown in **figure 3**, it depicts sprung mass acceleration of a Passive system, whereas **figure 4** shows the sprung mass acceleration of a Semi-active system. In figure 3, peak point is **15.5621 m/s2.**

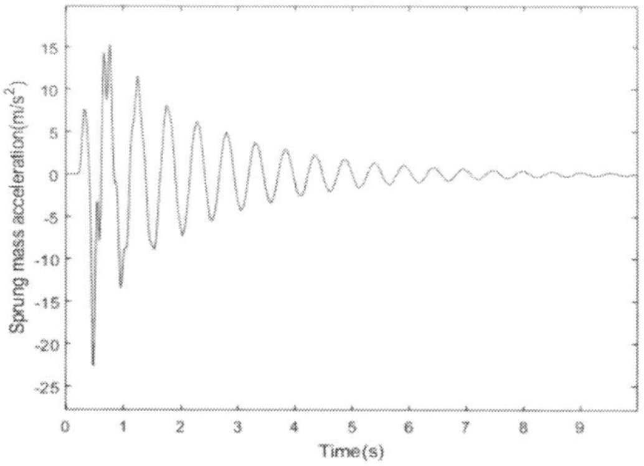

Figure 3: Sprung mass acceleration (Passive)

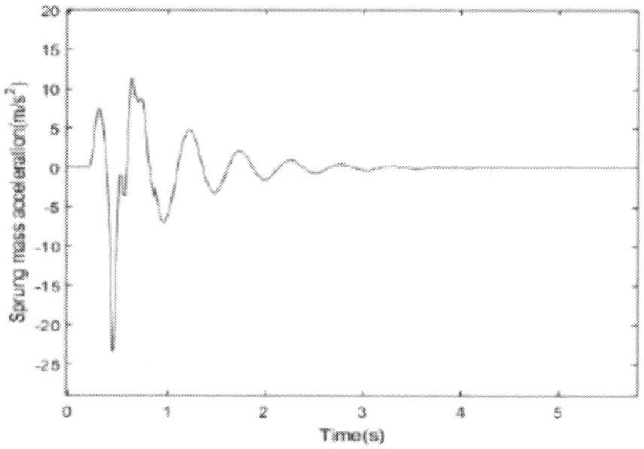

Figure 4: Sprung mass acceleration (Semi-active)

In figure 4, the peak point is **11.2767 m/s^2**. The given peak points show the highest vertical acceleration of sprung mass. So, semi-active system shows an improvement of **27.53%** over a passive system. Which means ride comfort is better in semi-active system. Also, sprung mass settles down in almost 4 seconds in semi-active system as compared to 10 seconds in passive system. Hence, semi-active suspension system provides a better ride comfort as compared to passive suspension system.

Rattle space is a relative displacement of sprung and unsprung mass. **Figure 5** shows rattle space in passive system, whereas **Figure 6** shows rattle space in semi-active system.

As shown below, in figure 5, highest displacement in Y direction is **0.12105m.**

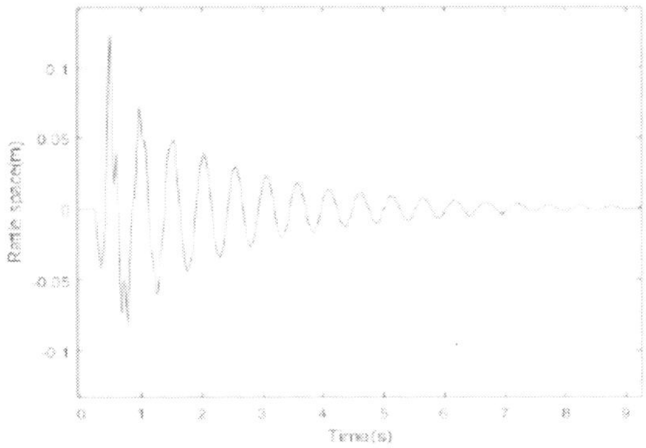

Figure 5: Rattle space (Passive)

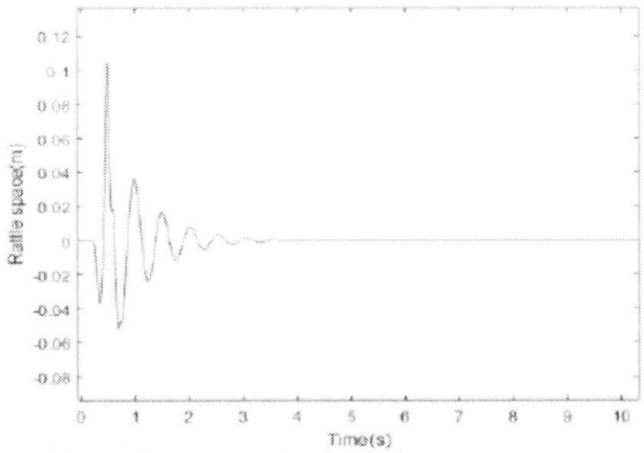

Figure 6: Rattle space (Semi-active)

In figure 6, highest displacement in Y direction is **0.10448m**. So, semi-active system has lesser displacement as compared to passive system. Which means rider discomfort is less in semi-active system as compared to passive system. Also, the vehicle using semi-active system settles down in 4 seconds as compared to 10 seconds in vehicle using passive system. Hence there is an improvement of **13.7%**.

5. Conclusion

The general approach in semi-actice suspension system is to control the damping constant using a controller. But it is considered as a constant in the mathematical model. However, this paper implements a novel approach to semi-active suspension system. The controller input (i.e., the damping coefficient) is considered variable and is given as input to the quarter car model. So there are two inputs in the system, first being the road input and second is the damping coefficient. Moreover, it has been found from the analysis that the semi-active suspension system provides a better ride comfort as compared to passive suspension system. And installing the controller helps achieve minimum acceleration quickly as compared to a normal semi-active suspension system.

The vibrations felt in Semi-active system are much less as compared to Passive system. And sprung mass acceleration can be directly correlated with ride comfort. So there has been a significant improvement in the ride comfort in a vehicle which uses Semi-active suspension system as compared to a vehicle which uses Passive suspension system.

Rattle space (relative displacement) is less in semi-active as compared to passive system, which can be concluded the simulation results. The improvement is significant. Since the settling time is relatively less in semi-active system, it provides much better ride comfort as compared to passive system.

References

1. A. Agharkakli, G. Shafiei Sabe, A. Barouz, International Journal of Engineering Trends and Technology Volume **3**, 5 (2012)
2. N. H. Amer, R. Ramli, Wan Nor Liza Mahadi, Mohd. Azman Zainul Abidin , *International Conference on Electrical, Control and Computer Engineering Pahang, Malaysia, June 21-22, 2011*
3. Mohammad Meftahul Ferdaus, M M Rashid,Moinul Bhuiyan , Advanced Materials Research Vol. **939** (2014) Page – 615-622
4. M. Biglarbegian, W. Melek, F. Golnaraghi, *Fuzzy Information Processing Society, 2006*
5. T. Ram Mohan Rao, G. Venkata Rao, K. Sreenivasa Rao & A. Purushottam, IJRRAS Vol. **5**, 1 (2010)
6. B. Gadhvi, V. Savsani, *Proceedings of the ASME 2014 International Mechanical Engineering Congress & Exposition, Quebec, Canada, November 14-20, 2014*
7. J. Patel, V. Savsani, R. Patel, *ASME 2015 9th International Conference on Energy Sustainability collocated with the ASME 2015 Power Conference, the ASME 2015 13th International Conference on Fuel Cell Science, Engineering and Technology, and the ASME 2015 Nuclear Forum, June 2015.*
8. A. Bagheri, M.J. Mahmoodabadi, H. Rostami, S. Kheybari, IJE TRANSACTIONS A: Basics Vol. **24**, 3 (2011) Page – 291-300
9. S. Badran, A. Salah, W. Abbas, O. Abouelatta, Research Bulletin of Jordan ACM, Vol. **2**, 8 (2012) Page – 42-51
10. A. Jamal Qazi, U. A. Farooqui, A. Khan, M. Tahir Khan, F. Mazhar, A. Fiaz, *AASRI Conference on Intelligent Systems and Control* (2013)
11. Y. Gandhi, V. Mehta, M. Patel, B. Gadhvi and A. Markana, *Proceedings of the 2015 International Conference on Operations Excellence and Service Engineering* (2015)
12. R.V. Rao, V.J. Savsani, D.P. Vakharia, *Computer-Aided Design*, Vol. **43**, 3 (2011)
13. T. Gillespie, *Fundamentals of Vehicle Dynamics*

Performance Analysis of EKF and UKF for Estimating Tire Longitudinal and Side Angle Slip

Shin Hoi Song[1], Myonghwan Ahn[2], Kwanghoon Kim[2] and Jin Bae Park[1]

[1]*Yonsei University, Department of Electrical and Electronic Engineering, Seoul, Korea*
[2]*LIG Nex1 Co., Ltd. 333, Pangyo-ro, Bundang-gu, Seongnam-City, Korea*

Abstract. Accurate tire slip estimation might be regarded as a small portion of the vehicle safety but is important criterion. Furthermore, as an autonomous vehicle system gets sophisticated, this type of technique will be more necessary. In this paper performance analysis for slip estimation in various situations is presented with several commonly known filter- extended Kalman filter (EKF) and unscented Kalman filter (UKF). Tire slip behaves differently depending on the surface type, the motion of the robot, and other environmental factors. Therefore, different kinds of situations and conditions are considered to estimate more accurate tire slip. Also as far as the tire slip is not based on actual data, it will be assumed to imitate the real tire slip behavior based on other study data. Finally, the performances of two filtering algorithms are compared to find more adequate algorithm with respect to the given condition for the future experimental results.

1 Introduction

Autonomous vehicle has been a popular research area for a long time though it is almost impossible task to realize in the past. However, many different types of techniques such as computer visions or localizations have improved and combined to accomplish the current level. For example, Google car or Mercedes Benz Bertha have proved that it is feasible to operate in the normal road whether the road is in civic or country area. The technical issue is no more considered as a blockage, instead some legal issues and cost problems are left to make perfect autonomous vehicle available in the market.

For the cost problem, a range sensor used in Google car, Velodyne LiDAR, is tremendously expensive to install to a regular vehicle. Likewise, directly measuring tire slips also requires expensive sensors such as speed over ground sensor as mentioned in [1], [2]. Therefore, many researches have been proposed to reduce the cost. In [3], [4], GPS (Global Positioning System) and INS (Inertial Navigation System) sensors are used to estimate vehicle tire slips. And a different algorithm or method have been suggested for more accurate estimation in [5]. However, most proposed researches are not proceeded under severe conditions such as icy or unpaved road, and have not directly compared the performance of estimation. Therefore, this paper focuses on the performance analysis of extended Kalman filter (EKF) and unscented Kalman filter (UKF) depending on different type of circumstances regarding that the robot replaces an actual vehicle. And the simulation condition is given similar to the circumstances that are expected to happen in real-world with larger-scaled magnitude of slip parameters. The first case is linear motion with frequent stop and go. This is the simplest motion that mobile robots can have and a good example to observe the longitudinal slip. The second case the robot is assumed to have is a gradual cornering. In most highway, they tend to avoid abrupt cornering due to safety reason, except some geographical issues. The last case that is made is a radical cornering. If the second case is easily seen in highway where vehicles move fast, the last case can be usually seen in civic area where vehicles move at low speed. With these three virtual but highly possible driving situations, the slips which have nonlinear characteristics are estimated by EKF and UKF algorithms. And then, the performance of each algorithm is evaluated.

The rest of the paper is organized as follows: the next section, Section 2, presents the robot kinematic model that is only considering horizontal plane and configuration of mobile robot with mounted position sensors. Section 3 presents the settings that are made to represent actual slip behaviors and this data is partially reflected by the actual test data from [1], [6], in which the sideslip angle is behaving approximately same as yaw rate. Section 4 discusses the system and measurement equations and states the noise for each system and measurement, and the slip parameters which are being estimated. Finally, the paper ends with conclusions.

2 Robot kinematic model

A kinematic model used in this paper is a skid-steering kinematic model which includes slip parameters-representing inner wheel slip, outer wheel slip, and tire sideslip angle, respectively. In this equation, an assumption is made that the ground surface is planar throughout the simulation time so that the other variables such as roll and bank angle are not subject to be exclusive to the system. Skid-steering is one of steering methods that the robot body changes its direction depending on the difference of the speed on each wheel, and this method can be seen in the mobile robot steering method in common. And a suitable model for the robot kinematic equations can be written as follows [7]:

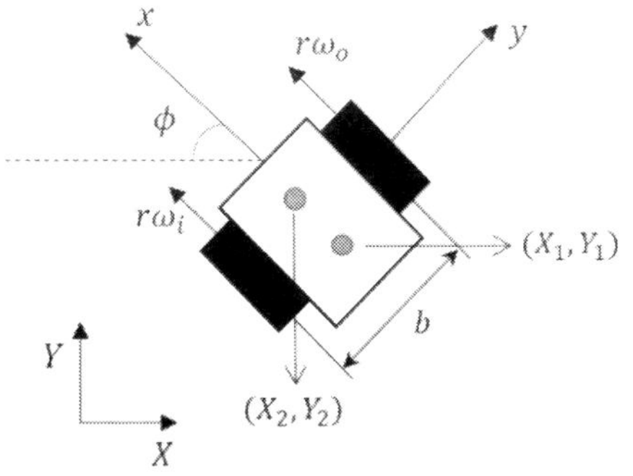

Figure 1. Robot & mounted sensor configuration.

$$\dot{X} = \frac{r}{2}[\omega_o(1 - i_o) + \omega_i(1 - i_i)](\cos\phi + \tan\alpha \sin\phi)$$

$$\dot{Y} = \frac{r}{2}[\omega_o(1 - i_o) + \omega_i(1 - i_i)](\sin\phi + \tan\alpha \cos\phi)$$

$$\dot{\phi} = \frac{r}{2b}[\omega_o(1 - i_o) + \omega_i(1 - i_i)] \quad (1)$$

where r is the effective radius of the robot's wheel, b is the distance between left and right wheels, and ω_o, ω_i are the angular speed for each wheel and considered as input through the paper.

Longitudinal slip is determined by the ratio of the longitudinal velocity of the robot and the longitudinal velocity of the robot exerted by the input. Also sideslip angle is determined by the angle made from the velocity vector of the robot in the body frame. And longitudinal slip and sideslip angle formulation is defined as:

$$i = 1 - \frac{V_x}{r\omega}, \quad \alpha = \arctan(\frac{V_y}{V_x}) \quad (2)$$

where x axis is forward axis of the robot, and y is the lateral axis of the robot.

Sensors are mounted on the vertical axis of the centre of gravity with the same distance. And they are assumed to be immensely accurate with a standard deviation of 0.03 cm. By calculating its geometric relations, the centre position of the robot in global coordinate and the heading angle are available as measurements.

$$X, Y = \left(\frac{X_2 - X_1}{2}, \frac{Y_2 - Y_1}{2}\right) \quad (3)$$

$$\phi = \arctan(\frac{y}{x}) \quad (4)$$

where the subscription 2 denotes the front sensor, the same manner follows for the rear sensor and capital x and y denotes the global frame, and vice versa the lower case the body frame.

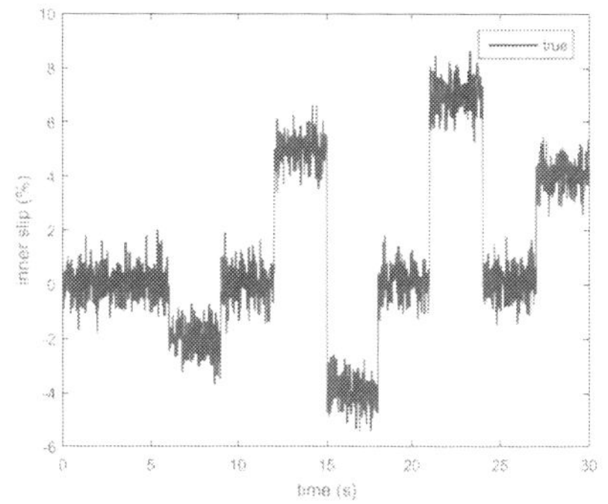

Figure 2. Generated inner slip for linear motion.

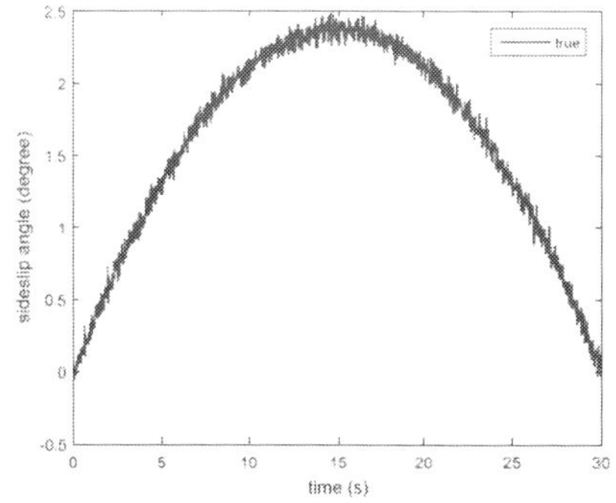

Figure 3. Generated sideslip angle for gradual motion.

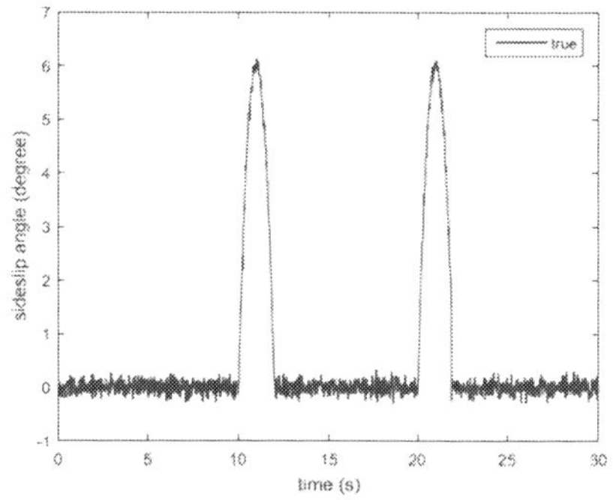

Figure 4. Generated sideslip angle for radical cornering.

And the physical formation of the sensors and the robot can be seen in Figure 1.

3 Slip allocation

To describe common vehicle motion and the actual vehicle characteristics, the simulation is run with 3 conditions as mentioned earlier. The first motion is linear motion, the second motion is gradually turning robot motion, and the third motion is radically turning robot motion.

3.1 Linear motion

The most fundamental and frequent motion of the robot is linear motion. In linear motion, sideslip angle is assumed not to be developed. Hence, sideslip angle is not subject to be estimated. On the other hand, longitudinal slip is assumed to be developed with respect to its acceleration and deceleration and it is generated as shown in Figure 2.

3.2 Gradual cornering

Gradual cornering can be detected when vehicles have cornering motions in high way. The characteristic of slip in this situation is as follows: low constant longitudinal slip, gradually increase and then decrease side angle slip (represented by a parabola [4]). For gradual cornering, longitudinal slip has also insignificant characteristic. Therefore, sideslip is subject to estimate and generated sideslip is shown in Figure 3.

3.3 Radical cornering

During radical cornering, sideslip angle spikes because the vehicle is changing direction in a short period. Therefore, in this case, sideslip angle estimation like in gradual cornering is only subject to be considered and the generated sideslip can be shown in Figure 4.

4 Performance evaluation

In this section system and measurement equations that are applied in the EKF and UKF algorithms are defined. Root mean square errors (RMSEs) are presented for comparing the performances of each filtering algorithm.

4.1 System identification

To cope with computer-based simulation, the continuous time system equation needs to be discretized. The following equations are discrete time system and measurement equations that are used to implement EKF and UKF.

$$X(k+1) = f\big(X(k),u(k)\big) + \omega(k) = \begin{bmatrix} X(k) + V_x[cos\phi(k) + \alpha(k)sin\phi(k)]\Delta T \\ Y(k) + V_x[sin\phi(k) + \alpha(k)cos\phi(k)]\Delta T \\ \phi(k) + \frac{2}{B}V_x\Delta T \\ i_i(k) \\ i_o(k) \\ \alpha(k) \end{bmatrix} \quad (5)$$

where $V_x = \frac{r}{2}\left[\omega_o(k)[1-i_o(k)] + \omega_i(k)[1-i_i(k)]\right]$ and denotes the longitudinal velocity of the robot in body fixed coordinate, input $u_k(k) = [\omega_i(k) \quad \omega_o(k)]^T$, the state $X(k) = [X(k) \quad Y(k) \quad \phi(k) \quad i_i(k) \quad i_o(k) \quad \alpha]$, and $\omega(k)$ is the system noise.

The measurement equation is the same as the system equation except the three slip parameters because the slip parameters are considered as unknown values. And it can be described as a nonlinear function below:

$$z(k) = h\big(X(k),u(k)\big) + v(k) \quad (6)$$

where $z(k) = [X(k) \quad Y(k) \quad \phi(k)]^T$

Note that the system noise $\omega(k)$ and measurement noise $v(k)$ are assumed to be zero mean, Gaussian and uncorrelated.

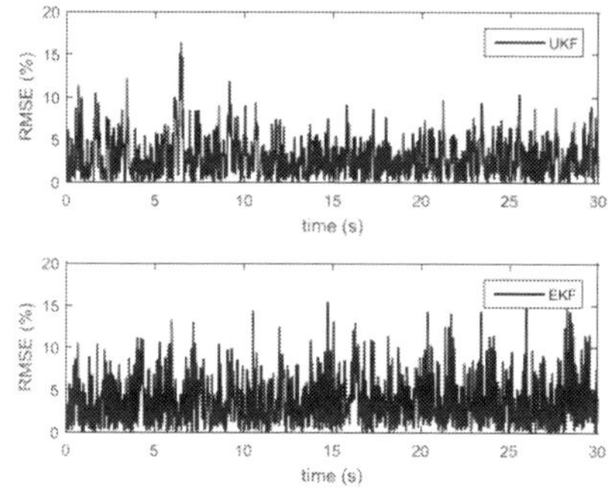

Figure 5. Longitudinal slip RMSE for linear motion.

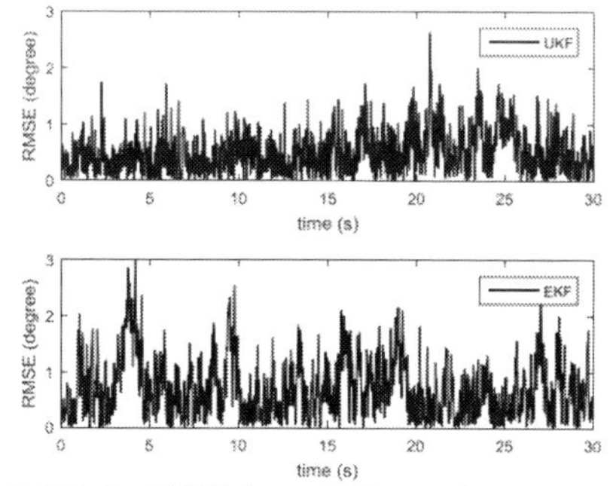

Figure 6. Sideslip RMSE for gradual cornering.

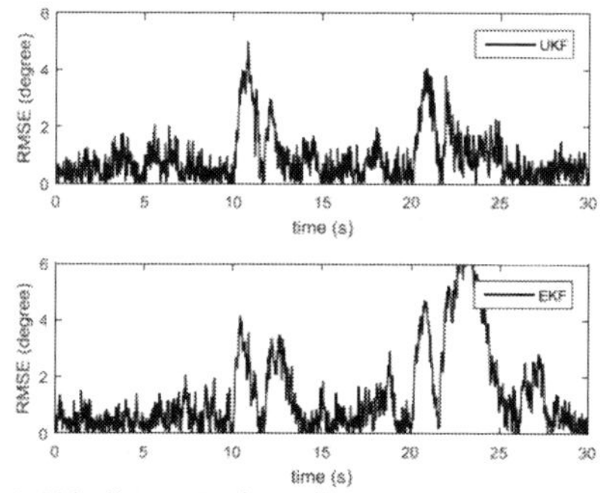

Figure 7. Sideslip RMSE for radical cornering.

4.2 Filter specification

Before estimating slip using EKF and UKF, some numbers such as system noise or initial state need to be justified. For EKF, the position and heading angle are assumed to be known and slip parameters unknown. Hence, initial error variance is positive only for slip states, inner outer and sideslip, and zero for the other. For UKF, there are three parameters- which adjust properties of the sigma points. However, since UKF also stems from the idea that the random variables (RVs) are distributed in Gaussian distribution and further they distributed in a little range, varying the parameters does not affect the filtering performance. Hence, acceptable values are arranged for the UKF parameters.

4.3 Comparison of the results

Slip parameters, longitudinal and side angle slip, are

estimated by using EKF and UKF with different situations and the performance are compared in Figure 5-7 with respect to given situation.

For linear motion, the performances of both filters are inferior due to radically changing longitudinal slips. It is believed that the convergence rate is not too fast to estimate time to time. But for the other cases, the performances are in acceptable range. Overall, UKF shows more consistent and close estimation to the true values than EKF.

5 Conclusions and future works

In this paper longitudinal and lateral tire slips are estimated based on robot kinematic equation by using most commonly used filtering algorithms, EKF and UKF. Normally, it requires either complex sensor fusion or high-end sensors to estimate vehicle longitudinal and lateral slips. However, in proposed algorithm the estimation of slips can be executed in simple and cheap way. Although the performance of the two filtering algorithms shows exceedingly accurate estimation, UKF is slightly accurate over EKF. For experimental session, the more suitable algorithm is UKF than EKF. In this paper, the slip parameters are generated by the author to mimic the real unknown parameters. Therefore, with the same situations, taking the true data is necessary to find more suitable algorithm. Also, other filtering algorithms such as sliding mode observer, particle filter or the variants of EKF, UKF need to be investigated in order to apply the superior algorithm for accuracy reason in estimating tire slips as EKF is compared with newly designed algorithm called NVSO in [8]. Additionally, it is assumed that the ground is planar so that no roll or bank angle is not considered as variables that affect the system, therefore the model equations which take account of roll and bank angles need to be introduced for the robustness in the experiment.

Acknowledgment

This work was supported by the LIG Nex1 Co., Ltd. under the contract LIGNEX1-2015-11-0038

References

1. J. Dakhlallah, S. Glaser, S. Mammar and Y. Sebsadji, "Tire-Road Forces Estimation Using Extended Kalman Filter and Sideslip Angle Evaluation," *American Control Conference* (2008)
2. D. Bevly, J. C. Gerdes, and C. Wilson, "The Use of GPS Based Velocity Measurements for Measurement of Sideslip and Wheel slip," *Vehicle System Dynamics*, **38**, 2, pp. 127-147 (2002)
3. D. Bevly, J. Ryu, and J. C. Gerdes, "Integrating INS sensors with GPS measurements for continuous estimation of vehicle sideslip, roll, and tire cornering stiffness," *IEEE Trans. Intelligent transportation systems*, **7**, 4, pp. 483-493 (2006)
4. M. A. Javed, W. Owen, M. Biglarbegian, and W. Melek, "A hybrid EKF/KF state estimator for a skid-steered ATV," *International Journal of Vehicle Mechanics and Mobility*, **52**, 1, pp. 85-110 (2014)
5. Z. Song, Y. H. Zweiri, and L. D. Seneviratne, Non-linear observer for slip estimation of skid-steering vehicles, *IEEE International Conference on Robotics and Automation* (2006)
6. M. Doumiati, A. Victorino, A. Charara, and D. Lechner, A method to estimate the lateral tire force and the sideslip angle of a vehicle : Experimental validation, *American Control Conference* (2010)
7. A. T. Le, D. C. Rye and H. F. Durrant-whyte, Estimation of Track-soil Interactions for Autonomous Tracked Vehicles, *IEEE International Conference on Robotics and Automation* (1997)
8. H. F. Grip, L. Imsland, T. A. Johansen, J. C. Kalkkuhl, and A. Suissa, Vehicle Sideslip Estimation, *IEEE Control Systems Magazine*, **29**, 5, pp. 36-52 (2009)

MATEC Web of Conferences 70, 04003 (2016) DOI: 10.1051/matecconf/20167004003
ICMIT 2016

Reducing Range Anxiety by Unifying Networks of Charging Stations

Khalil Salah and Nazri Kama

Advanced Informatics School, Universiti Teknologi Malaysia

Abstract. Availability of charging infrastructure is an important factor in penetration of electric vehicles into daily transportation system. Several factors in electric vehicle industry have caused range anxiety including insufficient charging stations, limited range of electric vehicles, long charging time, inaccurate estimation of available range, and energy consumption of auxiliary in-vehicle devices. However less attention has been paid to universality in charging station networks. This paper reviews the solutions to range anxiety. With regards to availability of charging station as one of the solutions, accessibility of charging stations by electric vehicle owners is also represented as another cause of range anxiety and a possible solution is provided.

1 Introduction

Charging Infrastructure (CI) is an important entity of power grid. Availability of this infrastructure is a key factor in general acceptance of Electric Vehicles (EV) [1]-[3]. The lack of CI as well as recharging time of an EV together with its limited range may discouraged many drivers from switching to EV [4], [5] and will lead to "Range Anxiety". Range anxiety is the fear of being left on the road with an empty battery where there is no charging station [6], [7] or lack of any CI in which the EVO has registered. Range anxiety can be considered as another impact on market acceptance of EVs [7], [8]. Therefore, availability of public CI is an important factor in penetration of EVs [1], [5], [6], [9].

In addition to the physical charging facilities, CI is a central communication interface among EV, Electric Vehicle Supply Equipment (EVSE), Power Grid, and energy suppliers [10], [11]. The data returning from EV through Charge Point (CP) can be used for monitoring, scheduling, energy distribution controlling during peak hours, and managing the energy consumption during off-peak hours. Therefore, a universal charging platform which is reliable and available in both public and private spots will accelerate user acceptance of EVs [1], [6], [9].

In this paper, causes and effects of range anxiety and solutions to ease this phenomenon is reviewed in section 2. Section 3 discusses on current situation of charging station networks and how to overcome range anxiety through unification of these networks. Finally, the paper is summarized by some conclusions in section 4.

2 Range anxiety

CI plays an important role in development and acceptance of EVs. Accuracy and safety of its components will bring satisfaction and reliability to the industry. New techniques and technologies in design and production of battery packs and lots of research and developments funded by governments together with different Battery Management Systems (BMS) with improved calculation algorithms are the signs of the EV's desire to stay in the market.

However there is still a big obstacle against general acceptance of EVs which is uncertainty of availability of CI upon EV Owner's (EVO) demand. Considering the limited driving range of an EV compared to the conventional internal combustion engines, lack of CI and the long charging time of an EV can cause anxiety and stress while driving or prior to driving an EV without any auxiliary power source. This situation, which is referred to as range anxiety will affect penetration of EVs in transport system [12]-[15].

Although range anxiety was first observed in General Motor's project in 1997 [16], lots of researches have been carried out on this issue since then. Some solutions are provided in order to reduce or overcome this issue, which are described in the following sections:

2.1 Availability of charging infrastructure

One of the major issues which is highlighted by researches on range anxiety is lack or insufficient number of public charging stations. Although the charging process mostly happens during the night at home or in parking lots during the day, availability of public charging stations is an important factor in penetration of EVs [1], [6], [9]. Therefore deployment of widely distributed networks of fast charging stations will provide flexibility to EVOs in planning long trips [4], [5], [12], [17]-[30]. However the long charging time in compared to Internal Combustion Engines (ICE) is still an issue.

Thus some researchers suggest to provide battery swapping stations or Charging While Driving (CWD) systems. While battery swapping might take as much time as it takes to fill the tank in ICE vehicles [27], CWD systems, in which the driver does not need to worry about charging station and spending long time to charge up the EV [31]-[35], will increase the driving range, and provide convenience and safety to EVOs. Even though the CWD system using electrified roadways and wireless power transfer methods may have a great impact on range anxiety of EVOs, until then development of current CI is required to encourage drivers to switch to EVs.

2.2 Routing and simulating energy consumption

Besides availability of CI, efficient route selection can also address range anxiety [36]. In [18] the analysis of the impact of a distributed CI on the EVs range shows that daily EV's range can be extended if suitable CI exists along the planned route therefore the range anxiety can be reduced. In [36] and [17], the authors provide smartphone applications to plan efficient routes. Topography of the route, wind speed, weight of EV, real-time road traffic, historical consumption of energy by EV, historical EVO's driving behavior, speed, temperature, operation of the air conditioner/heater, accurate State of Charge (SoC) of the battery, driving time, and the availability of CI are important factors while planning an efficient route and estimating consumption rate [17], [18], [20], [26], [36], [37].

2.3 SoC estimation

EV battery is an important element in an EV with direct effects on performance and rapid development of EVs [38]. Since batteries are sensitive to deep charge and discharge and in order to avoid it, a BMS capable of estimating accurate SoC is required [38]-[40].

Researchers approve that estimating the remaining power of the battery and displaying it to the user is important for effective EV use [12]-[14], [17], [20], [24]-[26], [30], [41]-[45]. Nevertheless, SoC and range estimation are only valid under certain conditions, therefore improvement in estimation algorithms is required to make it more accurate [20]. On the other hand in [13] the authors suggest that displaying error-prone information ambiguously might be advantageous rather than hiding them from users. In fact the knowledge of error-prone information can lead to improvements in driving experience, behavior and trust towards the EV and it might consequently decrease range anxiety.

Although displaying SoC information to EVO is critical, the user also needs to be informed about the appropriate methods of maximizing EV's range [29]. This can be interpreted as providing instructions to the EVO in order to adapt his/her driving style or to reduce using of auxiliary in-vehicle devices [44]. Therefore increasing actual range of an EV might be less important than providing the EVOs with reliable information [41] and educate them on how to use that information to control and manage the remaining range. For that reason in [46] and [47] the authors shows that current range of EVs might be enough for daily travels of EVOs if they understand the fact that "run-until-empty followed by a fill-up-to-full at a public gas station is not how EVs will be used". Therefore in their opinion, availability of public fast chargers might not be an issue.

2.4 In-Vehicle energy management

Although knowing the SoC and planning a suitable route with public fast charging stations might make the EVO convenient enough to drive an EV, researchers suggest to improve the efficiency of in-vehicle components and manage the energy consumption in order to maximize the global efficiency, minimize energy loss, extend the range, and reduce range anxiety [30], [43], [48]-[50]. While in-vehicle energy management might help, using of range extenders is also considered as another solution for reducing range anxiety [15], [51]-[54].

3 Discussion

Despite the rapid growth in EV industry, a few studies were conducted on optimization of charging networks to reduce cost of charging station selection [55]. In a network of charging station, the satisfaction level of the EVOs relies on the appropriate location of charging station, availability of the service when it is required, cost, and timing of getting the service [56], [57].

In the current charging station networks, before the EVOs start the charging process, they need to be identified and when the charging finished, billing and payment information will be sent to them. Therefore the charging station is only accessible by the users who have already subscribed to that network. This will cause isolation in network of charging stations and consequently bring up range anxiety.

To reduce range anxiety, improvements in all areas are required. Even though an accurate SoC estimation and displaying the real SoC might help the EVOs to adapt their driving style with the remaining power, and also choosing the most efficient route might extend the driving range, the EVOs have to have access to charging stations whenever they need one.

However, availability of charging stations is not a solution to range anxiety unless the stations are universally standardized. As mentioned earlier current charging networks only provide facilities to their own registered users. Therefore when the EVOs want to plan a trip there will be another question besides "Is there any charging station?" which is: "Can I use that charging station?"

Today most of the charging stations are using Open Charge Point Protocol (OCPP) as the standard communication protocol between their CPs and their central system. Moreover Standard Development Organizations (such as ISO, IEC, and SAE) provide standards addressing connectors, communication, charging topology, safety, and interoperability of EVSEs, and improvements are in progress to achieve a universal charging type for EVs. But more work is required in unification of networks of charging stations.

Networks of charging stations are suffering from isolation. This issue can be resolved with a unified communication protocol between service providers to allow cross-network charging. Thus by having a unified network of charging stations with universal charging types, EVOs can charge there EVs regardless of charging type or service provider, consequently it will reduce their range anxiety.

4 Conclusion

This paper reviewed the possible solutions to reduce or overcome range anxiety. Providing sufficient charging stations, displaying the SoC and improving the SoC estimation algorithms, routing, and providing real-time information about the remaining rage are among the solutions which are suggested by researchers for overcoming range anxiety. Although availability of networks of fast charging stations may reduce range anxiety, accessibility of these networks by EVO is also an important factor. Therefore providing cross-network charging capability to EVOs will bring user satisfaction.

References

1. C. Y. Chung, "Electric Vehicle Smart Charging Infrastructure," University of California, Los Angeles, Ann Arbor, (2014)
2. B. Römer, T. Schneiderbauer, and A. Picot, "How to Charge Electric Vehicles: A Comparison of Charging Infrastructure Concepts and Technologies," in *Driving the Economy through Innovation and Entrepreneurship*, C. Mukhopadhyay, K. B. Akhilesh, R. Srinivasan, A. Gurtoo, P. Ramachandran, P. P. Iyer, M. Mathirajan, and M. H. Bala Subrahmanya, Eds. Springer India, (2013), pp. 487–498
3. A. M. Foley, I. Winning, and B. P. Ó Gallachóir, "Electric vehicle: infrastructure regulatory requirements," *Proc. Inaug. Conf. Irish Transp. Res. Netw.*, (2010)
4. P. Dutta, "Coordinating rendezvous points for inductive power transfer between electric vehicles to increase effective driving distance," *International Conference on Connected Vehicles and Expo (ICCVE)*. pp. 649–653, (2013)
5. S. Bai and S. M. Lukic, "Unified Active Filter and Energy Storage System for an MW Electric Vehicle Charging Station," *IEEE Transactions on Power Electronics*, **28**, 12. pp. 5793–5803, (2013)
6. C. H. Dharmakeerthi, N. Mithulananthan, and T. K. Saha, "Planning of electric vehicle charging infrastructure," in *IEEE Power & Energy Society General Meeting*, (2013), pp. 1–5
7. J. Dong, C. Liu, and Z. Lin, "Charging infrastructure planning for promoting battery electric vehicles: An activity-based approach using multiday travel data," *Transp. Res. Part C*, **38**, pp. 44–55, (Jan. 2014)
8. J. Chynoweth, C.-Y. Chung, C. Qiu, P. Chu, and R. Gadh, "Smart Electric Vehicle Charging Infrastructure Overview," in *IEEE PES Innovative Smart Grid Technologies (ISGT) Conference*, (2014), pp. 1–5
9. M. Yilmaz and P. T. Krein, "Review of Battery Charger Topologies, Charging Power Levels, and Infrastructure for Plug-In Electric and Hybrid Vehicles," *IEEE Trans. Power Electron.*, **28**, 5, pp. 2151–2169, (May 2013)
10. A. M. A. Haidar, K. M. Muttaqi, and D. Sutanto, "Technical challenges for electric power industries due to grid-integrated electric vehicles in low voltage distributions: A review," *Energy Convers. Manag.*, **86**, pp. 689–700, (Oct. 2014)
11. T. Winkler, P. Komarnicki, G. Mueller, G. Heideck, M. Heuer, and Z. A. Styczynski, "Electric vehicle charging stations in Magdeburg," in *IEEE Vehicle Power and Propulsion Conference (VPPC)*, (2009), pp. 60–65
12. A. Lundström, C. Bogdan, F. Kis, I. Olsson, and L. Fahlén, "Enough Power to Move: Dimensions for Representing Energy Availability," in *Proceedings of the 14th International Conference on Human-computer Interaction with Mobile Devices and Services*, (2012), pp. 201–210
13. M. F. Jung, D. Sirkin, T. M. Gür, and M. Steinert, "Displayed Uncertainty Improves Driving Experience and Behavior: The Case of Range Anxiety in an Electric Car," in *Proceedings of the 33rd Annual ACM Conference on Human Factors in Computing Systems*, (2015), pp. 2201–2210
14. J. Ferreira, P. Pereira, P. Filipe, and J. Afonso, "Recommender system for drivers of electric vehicles," *2011 3rd International Conference on Electronics Computer Technology (ICECT)*, **5**. pp. 244–248, (2011)
15. M. Rodrigues, S. King, D. Scott, and D. Wang, "Advanced Energy Management Strategies for Range Extended Electric Vehicle." SAE International, (2015)
16. R. van Haaren, "Assessment of Electric Cars' Range Requirements and Usage Patterns based on Driving Behavior Recorded in the National Household Travel Survey of 2009," Solar Journey USA, New York, (2011)
17. Y. Du and G. de Veciana, "Mobile applications and algorithms to facilitate electric vehicle deployment," *IEEE Consumer Communications and Networking Conference (CCNC)*. pp. 130–136, (2013)
18. K. Davis, P. Rowley, and S. Carroll, "Assessing the viability of electric vehicle technologies for UK fleet operators," *48th International Universities' Power Engineering Conference (UPEC)*. pp. 1–6, (2013)
19. G. Michailidis, "Power allocation to a network of charging stations based on network tomography monitoring," *18th International Conference on Digital Signal Processing (DSP)*. pp. 1–6, (2013)
20. S. Heath, P. Sant, and B. Allen, "Do you feel lucky? Why current range estimation methods are holding back EV adoption," *IET Hybrid and Electric*

Vehicles Conference 2013 (HEVC 2013). pp. 1–6, (2013)
21. L. Gonzalez, H. Novella, E. Gutierrez, J. Ventura, and P. Mogas, "EVIC (Electric Vehicle Intelligent Charging)," *2013 World Electric Vehicle Symposium and Exhibition (EVS27)*. pp. 1–8, (2013)
22. M. Gharbaoui, B. Martini, R. Bruno, L. Valcarenghi, M. Conti, and P. Castoldi, "Designing and Evaluating Activity-Based Electric Vehicle Charging in Urban Areas," *2013 IEEE International Electric Vehicle Conference (IEVC)*. pp. 1–5, (2013)
23. J. Tan and W.-H. Lin, "A stochastic flow capturing location and allocation model for siting electric vehicle charging stations," *2014 IEEE 17th International Conference on Intelligent Transportation Systems (ITSC)*. pp. 2811–2816, (2014)
24. A. G. Bianchessi, G. Cugola, S. Formentin, A. C. Morzenti, C. Ongini, E. Panigati, M. Rossi, S. M. Savaresi, F. A. Schreiber, L. Tanca, and E. G. Vannutelli Depoli, "Green Move: A Platform for Highly Configurable, Heterogeneous Electric Vehicle Sharing," *IEEE Intelligent Transportation Systems Magazine*, **6**, 3. pp. 96–108, (2014)
25. W. He, N. Williard, C. Chen, and M. Pecht, "State of charge estimation for electric vehicle batteries under an adaptive filtering framework," *IEEE Conference on Prognostics and System Health Management (PHM)*. pp. 1–5, (2012)
26. J. C. Ferreira, V. Monteiro, and J. L. Afonso, "Dynamic range prediction for an electric vehicle," *2013 World Electric Vehicle Symposium and Exhibition (EVS27)*. pp. 1–11, (2013)
27. M. R. Sarker, H. Pandzic, and M. A. Ortega-Vazquez, "Electric vehicle battery swapping station: Business case and optimization model," *International Conference on Connected Vehicles and Expo (ICCVE)*. pp. 289–294, (2013)
28. M. Noro and S. Koizumi, "Present Application and Future Possibilities of Quick Charger for Electric Vehicle," in *1st International Electric Vehicle Technology Conference*, (2011)
29. C. Walsh, S. Carroll, and A. Eastlake, "UK ELECTRIC VEHICLE RANGE TESTING AND EFFICIENCY MAPS," in *1st International Electric Vehicle Technology Conference*, (2011)
30. P. Weissler, "Staying within range," *SAE International*, pp. 21–25, (Feb. 2011)
31. H. H. Wu, A. Gilchrist, K. Sealy, P. Israelsen, and J. Muhs, "A review on inductive charging for electric vehicles," *IEEE International Electric Machines & Drives Conference (IEMDC)*. pp. 143–147, (2011)
32. G. R. Nagendra, J. T. Boys, G. A. Covic, B. S. Riar, and A. Sondhi, "Design of a double coupled IPT EV highway," *39th Annual Conference of the IEEE Industrial Electronics Society, (IECON)*. pp. 4606–4611, (2013)
33. S. Lukic and Z. Pantic, "Cutting the Cord: Static and Dynamic Inductive Wireless Charging of Electric Vehicles," *IEEE Electrification Magazine*, **1**, 1. pp. 57–64, (2013)
34. O. C. Onar, J. M. Miller, S. L. Campbell, C. Coomer, C. P. White, and L. E. Seiber, "A novel wireless power transfer for in-motion EV/PHEV charging," *Twenty-Eighth Annual IEEE Applied Power Electronics Conference and Exposition (APEC)*. pp. 3073–3080, (2013)
35. F. Deflorio and L. Castello, "Traffic Modeling of a Cooperative Charge while Driving System in a Freight Transport Scenario," *Transp. Res. Procedia*, **6**, pp. 325–350, (2015)
36. R. Yaqub and Y. Cao, "Smartphone-based accurate range and energy efficient route selection for electric vehicle," *IEEE International Electric Vehicle Conference (IEVC)*. pp. 1–5, (2012)
37. M. Neaimeh, G. A. Hill, Y. Hübner, and P. T. Blythe, "Routing systems to extend the driving range of electric vehicles," *Intelligent Transport Systems (IET)*, vol. 7, no. 3. pp. 327–336, (2013)
38. M. Brandl, H. Gall, M. Wenger, V. Lorentz, M. Giegerich, F. Baronti, G. Fantechi, L. Fanucci, R. Roncella, R. Saletti, S. Saponara, A. Thaler, M. Cifrain, and W. Prochazka, "Batteries and battery management systems for electric vehicles," *Design, Automation & Test in Europe Conference & Exhibition (DATE), 2012*. pp. 971–976, (2012)
39. H. Rahimi-Eichi, U. Ojha, F. Baronti, and M. Chow, "Battery Management System: An Overview of Its Application in the Smart Grid and Electric Vehicles," *Industrial Electronics Magazine, IEEE*, **7**, 2. pp. 4–16, (2013)
40. G. Carli, A. Shafiei, and S. Williamson, "Advanced Electric Vehicles," in *Power Electronics for Renewable and Distributed Energy Systems*, S. Chakraborty, M. G. Simões, and W. E. Kramer, Eds. Springer London, (2013), pp. 525–566
41. R. C. McIlroy, N. A. Stanton, C. Harvey, and D. Robertson, "Sustainability, Transport and Design: Reviewing the Prospects for Safely Encouraging Eco-driving," in *Proceedings of the 5th International Conference on Automotive User Interfaces and Interactive Vehicular Applications*, (2013), pp. 278–284
42. J. B. Gläser and J. O. Berg, "Presentation of a General Purpose Simulation Approach for Enabling the Realization of Electromobility Concepts for the Transportation Sector," in *Proceedings of the 2014 Winter Simulation Conference*, (2014), pp. 2072–2081
43. K. Knoedler, J. Steinmann, S. Laversanne, S. Jones, A. Huss, E. Kural, D. Sanchez, O. Bringmann, and J. Zimmermann, "Optimal Energy Management and Recovery for FEV," in *Proceedings of the Conference on Design, Automation and Test in Europe*, (2012), pp. 683–684
44. C.-H. Lee and C.-H. Wu, "An Incremental Learning Technique for Detecting Driving Behaviors Using Collected EV Big Data," in *Proceedings of the ASE BigData & SocialInformatics 2015*, (2015), pp. 10:1–10:5
45. A. Lundström, C. Bogdan, F. Kis, I. Olsson, and L. Fahlén, "EVERT: Energy Representations for Probing Electric Vehicle Practice," in *CHI '12*

Extended Abstracts on Human Factors in Computing Systems, (2012), pp. 2141–2146
46. R. Bruninga, "Overlooking L1 charging at-work in the rush for public charging speed," *IEEE International Electric Vehicle Conference (IEVC)*. pp. 1–5, (2012)
47. R. Bruninga and J. A. T. Sorensen, "Charging EVs Efficiently NOW While Waiting for the Smart Grid," *IEEE Green Technologies Conference*. pp. 1–7, (2013)
48. N. M. A. Aziah, H. M. I. Nizam, S. Ahmad, R. Akmeliawati, S. F. Toha, B. S. K. K. Ibrahim, and M. K. Hassan, "Control strategy of segregation on HVAC energy efficiency as non propulsion electrical hotel load in EV," *IEEE International Conference on Power and Energy (PECon)*. pp. 611–615, (2012)
49. Y. Kondo, H. Kato, R. Ando, T. Suzuki, and Y. Karakama, "To what extent can speed management alleviate the range anxiety of EV?," *World Electric Vehicle Symposium and Exhibition (EVS27)*. pp. 1–8, (2013)
50. M. A. Jeffers, L. Chaney, and J. P. Rugh, "Climate Control Load Reduction Strategies for Electric Drive Vehicles in Warm Weather." SAE International, (2015)
51. P. Guttenberg, S. Chubbock, and T. Gilbert, "A New Philosophy for Hybrid Vehicle Efficiency Optimisation: Simultaneous Optimisation of Component Size and Control Strategy for Real-World Efficiency in Hybrid and Electric Vehicles," in *1st International Electric Vehicle Technology Conference*, (2011)
52. H. E. A. Abis, C. Schwab, R. Kirchberger, "An Innovative Two-Stroke Twin-Cylinder Engine Layout for Range Extending Application," in *JSAE/SAE 2013 Small Engine Technology Conference*, (2013)
53. J. Yu, Y. Zhang, N. Liu, and B. Wang, "Modeling and control strategy simulation of extended-range electric vehicle," *International Conference on Transportation, Mechanical, and Electrical Engineering (TMEE)*. pp. 829–832, (2011)
54. P. Sharer and A. Rousseau, "Benefits of fuel cell range extender for medium-duty vehicle applications," *2013 World Electric Vehicle Symposium and Exhibition (EVS27)*. pp. 1–12, (2013)
55. J. J. Escudero-Garzas and G. Seco-Granados, "Charging station selection optimization for plug-in electric vehicles: An oligopolistic game-theoretic framework," in *2012 IEEE PES Innovative Smart Grid Technologies (ISGT)*, (2012), pp. 1–8
56. I. S. Bayram, G. Michailidis, I. Papapanagiotou, and M. Devetsikiotis, "Decentralized control of electric vehicles in a network of fast charging stations," in *2013 IEEE Global Communications Conference (GLOBECOM)*, (2013), pp. 2785–2790
57. Wang Meng and Liu Kai, "Optimization of electric vehicle charging station location based on game theory," in *Proceedings 2011 International Conference on Transportation, Mechanical, and Electrical Engineering (TMEE)*, (2011), pp. 809–812

The Role of Higher Order Thinking Skills in Green Skill Development

Yee Mei Heong[1], Lai Chee Sern[1], Tee Tze Kiong[2] and Mimi Mohaffyza Binti Mohamad[1]

[1]Universiti Tun Hussein Onn Malaysia, Faculty of Technical and Vocational Education, Department of Engineering Education, 86400 Parit Raja, Batu Pahat, Johor, Malaysia
[2] Universiti Pendidikan Sultan Idris, Faculty of Technical and Vocational Education, 35900 Tanjong Malim, Perak.

Abstract. In the era of rapid development, industries sector will bring negative effect of environment. Therefore, green technologies should be exposed in education system for transforming economy model to greener economy modal. HOTS should be applied in green skill development in order to solve problems, generate new products and ideas. The application of HOTS can enhance someone in processing new information for various alternatives, ideas, actions, and design solutions to solve a problemIn this regard, a study on the role of Marzano HOTS in green skill development has been designed for the above purpose. This article suggests an initial idea of the study. The key elements identified in green skills development include green economy, green industry and green skills. Meanwhile, elements related to the Marzano HOTS is comparing, classifying, induction, deduction, error analysis, constructing support, abstracting and analyzing perspectives. As conclucion, HOTS requires someone to critically evaluate information, make inferences, and make generalizations. With HOTS, we enable to apply knowledge, skills and values to make reasoning and reflection to solve problems, make decisions, innovate and strive to create something in green skills development.

1 Introduction

The problems related to environment pollution will bring about devastating consequences such as rising sea level, extreme droughts or monsoons, and changing weather pattern [1]. These consequences will negatively affect the sustainability of social and economic activities across the world.

In order to deal with the environmental issues, governmental bodies as well as private sectors have to work hand in hand to ensure the sustainability of the social, economic, and environment development. For instance, the Malaysian government has announced that one of the primary components under the 11th Malaysian Plan 2016-2010 (11MP) is to pursue green growth for sustainability and resilience which aims at achieving a resilient, low-carbon, and resource-efficient economy model that leads to higher living quality, more conducive living environment, and sustainable living condition [2]. In addition, governmental agencies should come out with more efficient policies and the enforcement of rules and regulations related to environmental conservation must be strengthened. In this aspect, the European Union (EU) has developed the "EU 2020 strategies" which targets to elevate the consumption of renewable energy and reduce greenhouse gas emission within EU member countries [3].

Some developed countries, such as Japan and Germany, have taken pragmatic long-term approach to mitigate the environmental problems by transforming the existing economy model to a more environmental friendly and greener economy model [4]. The industries that are less environmental friendly and less energy efficient are now moving towards to those that are less polluting and more efficient. For example, the manufacturing processes produce lower carbon emissions and consume less unrenewable energy.

The transformation of economy model to a greener model is increasing the pace of change in manpower market. The green industry has created some entirely new green jobs (e.g., energy auditor) and "greening" the existing jobs in the industries. For instance, in automobile industry, engineers require green skills to work with fuel efficient technologies; and an architect needs green skills to integrate natural elements into building design.

The change of skill requirements is pervasive and it calls for a huge effort to revise the training programmes. The training institutions should embed "green" elements into the training curricula. Students should be exposed to green technologies during the training. In addition, higher order thinking skills (HOTS) should be taught to the students because the nature of the green jobs is dealing with problem solving and create new green technologies, products or ideas. For instance, how to reduce energy consumption, and how to reduce environmental pollution.

Before HOTS is further discussed, the concepts of green economy, green industry, and green skills will be presented in the following sections in order to provide some general insights to the readers.

2 Concepts of green economy, green industry, and green skill

2.1 Green economy

A green economy can be considered as an economy model for sustainable development as it promotes advancement and sustainability in terms of social, economy, and environment.

Many definitions have been put forward by a number of organisations based on different research works and different target audiences. For example, United Nation Environment Program (UNEP) defines green economy as an economy model that is socially inclusive, low carbon, and energy resource efficient that significantly cuts down environmental risks and substantially improves social well-being [4]. Additionally, the green economy which is supported public and private investment will bring growth in employment and income and at the same time reduce pollution, enhance energy and resource efficiency, and promote biodiversity and ecosystem conservation.

Apart from UNEP, United Nations Conference on Trade and Development (UNCTAD) defines green economy as follows [5]:

> *An economy that results in improved human well-being and reduced inequalities, while not exposing future generations to significant environmental risks and ecological scarcities. It seeks to bring long-term societal benefits to short-term activities aimed at mitigating environmental risks. A green economy is an enabling component of the overarching goal of sustainable development.*
>
> *(page IV)*

This definition puts the focus on sustainable development that emphasizes environmental conservation and protection for the sake of next generation.

In addition to the above definitions, Green Economy Coalition (GEC) also provides a brief concept of green economy. According to GEC, green economy can be defined as a resilient economy that provides better quality of life for all within the ecological limits of the planets [6].

Green economy has been defined differently by different agencies. However, two elements have been mentioned in all those definitions, namely, "nature conservation and environmental protection" and "human well-being improvement and social development". Taken together, green economy can be conceptualised as an economy model which emphasises nature conservation and environmental protection that leads to improved human well-being and sustainable social development.

2.2 Green industry

Green economy will bring sustainable development in social and economy without affecting the environment. Supports and cooperation from industries are required in order to achieve the sustainable development. Industries should opt for a more sustainable partway of growth by undertaking environmentally friendly investments.

According to Schwerzer [7], can be any industry that involves in production, distribution, and/or services which emphasises the aspects of environmental protection and nature conservation. In general, green industry is usually related to fields of renewable energy, green building, fuel efficient transportation, recycling, forestry, and ecotourism. For example, recycling companies is one of the green industries as its nature of business is promoting environmentally friendly living condition. Likewise, solar panel production company, power generation companies that utilise renewable energy, construction companies that apply eco-friendly concept in design and building construction can also be categorised as green industries.

Green industries create green jobs. Green jobs can be any jobs pertaining to business that produce goods or provide services that benefit the environment and conserve natural resources. Additionally, green job also can be regarded as jobs that aims at reducing natural resources consumption or making the production more environmentally friendly [8].

Hard and soft skills are no more sufficient for an individual to secure a job especially green job. Therefore, workers must equipped themselves with green skills in order to help them secure a green job.

2.3 Green skill

Green skill is referred to as skill for sustainable development. According to [9], green skills are regarded as skills for sustainability which are related to the technical skills, knowledge, values and attitudes needed in the workforce to develop and support sustainable social, economic and environmental outcomes in business, industry and the community. Likewise, CEDEFOP defines green skills as the knowledge, abilities, values and attitudes needed to live in, develop and support a society which reduces the impact of human activity on the environment [10]. Taken together, green skills is generally composed of three dimensions, namely, knowledge (cognitive dimension), skills/abilities (psychomotor dimension), and attitudes/values (affective dimension) needed by workers to promote sustainable development in social, economy, and environment. From the cognitive dimension, the knowledge concerning environmental protection can be regarded as an element of green skills. From psychomotor perspective, green skill refer to the ability to, for instance, minimise energy consumption, or reduce greenhouse gases. Green skills also refer to affective aspect, for example, motivation of an individual to conserve natural resources.

In most cases, the new jobs created by the green industries will require a mix of both traditional and new green skills. For example, construction companies that carry out building and housing projects will require workers with traditional construction skills and up-to-date training in energy efficiency [11]. In general, it can be said that for low to middle-skilled jobs, there will be a need for traditional skills complemented by green skills, most of which can be offered by on-the-job training programmes. Higher skilled occupations, such as those found in eco-consulting, will require a broader and more specific set of new green skills. Such new skills are best delivered by educational and in-depth training programmes which involve higher order thinking skills.

3 Higher Order Thinking Skills (HOTS)

In the 21st century, thinking skills especially creativity is crucial in order to support economic prosperity. Human creativity has become a major economic resource [12]. Industries of the 21st century will increasingly depend on the generation of knowledge through creativity and innovation [13]. Employers often suggest the importance of students to be equipped with skills to make decisions and think critically and creatively before entering the working world.

In addition, the increase in labor productivity also depends on the quality of each individual's talent, namely creativity [14]. Creativity and innovation are important keys to success in any field in this era of rapid development [15]. This is so because business management activities such as processes to increase productivity, solve problems, motivate employees, make decisions and rapid technological change are in dire need of creative ideas. Idea generation thus, is a crucial part in resolving a problem [16].

Reference [17] defines idea as a basic element of thought which can be visual, concrete, or abstract. Idea is all stages of the cycle of abstract thinking [18] and it also can be visualized in our mind. Reference [19] and [20] categorized the generation of ideas as a higher order thinking skills (HOTS) activities that require high level creative thinking and action.

However, not everyone is able to generate good ideas because ideas cannot be generated easily. Idea generation occurs in our brain through the cognitive, metacognitive, chemical and biological process [19]. Hence, complex thinking skills such as problems solving, creating, analyzing, evaluating and others are needed to process the collected information [21]-[22] for generating an idea.

HOTS is one component of the creative thinking skills and critical thinking. According to [23], creative thinking can develop individual to be more innovative, have good creativity, ideal and imaginative. When we know how to use both of these skills, it means that we have applied HOTS. Everyone is capable to think, but most of them need to be encouraged, taught and assisted to the higher order thinking processes.

These HOTS are teachable and learnable. Everyone has the right to learn and apply thinking skills, just like other disciplines of knowledge. Optimal use of thinking skills involve HOTS as the skills to analyze, synthesize, evaluate, and create them [24]. Higher order thinking is defined as the expanded use of the mind to meet new challenges [25]. It requires someone to apply new information or prior knowledge and manipulate the information to reach possible answer in new situation [26]-[28].

HOTS allows someone should be able to do a comparison, evaluation, justification and make inferences [29]. HOTS transformation process will happen when students combine facts and ideas through the ability to synthesize, generalize, explain, hypothesize, or produce conclusions and interpretations [30]. The process of manipulating information and ideas will enable students to solve problems, find understanding and gain new insights in learning [24].

A question to be answered or a problem to be solved cannot be done through routine application of previously acquired knowledge [31]. But it can be solved only when expanded use of mind occurred that a person must interpret, analyze or manipulate information. This is because higher order thinking is characterized as non algorithmic, complex, self regulative, meaningful, effortful and providing multiple solutions, nuanced judgments, multiple criteria and uncertainty [32]. HOTS is an imperatives aspect in green skill development.

4 Marzano higher order thinking skills

Marzano's research on thinking skills is important to students and educators in higher education institution primarily. Marzano identifies eight HOTS, namely *comparing, classifying, inducing, deducing, error analysis, constructing supporting, analyzing perspectives,* and *abstracting,* which work within the Dimensions of Learning (1992) framework. The Dimensions of Learning model assumes that the process of learning involves the interaction of the following five types of thinking:

(i) positive attitudes and perceptions about learning
(ii) thinking involved in acquiring and integrating knowledge
(iii) thinking involved in extending and refining knowledge
(iv) thinking involved in using knowledge meaningfully
(v) productive habits of mind

Marzano's Dimensions of Learning is a comprehensive model that uses what researchers and theorists know about learning to define the learning process. Dimensions of Learning offer a way of thinking about the extremely complex process of learning so that study can be attended to each aspect and gain insights into how they interact. The five types of thinking are premised as five dimensions of learning that are essential to successful learning [33]. The Dimensions framework will help us to

- Maintain a focus on learning;
- Study the learning process; and
- Plan curriculum, instruction, and assessment that takes into account the five critical aspects of learning.

These 8 higher order thinking skills are identified in Dimension 3 which will help students extend and refine knowledge. The most effective learning occurs when we use knowledge to perform meaningful task.

Table 1. The definition of five Marzano HOTS.

Marzano HOTS	Definition
comparing	Identifying and articulating similarities and differences among items.
classifying	Grouping things into definable categories on the basis of their attributes.
inductive reasoning	Inferring unknown generalization or principles from information or observation.
deductive reasoning	Using generalization and principles to infer unstated conclusion about specific information or situations.
analyzing errors	Identifying and articulating error in thinking.
constructing support	Building system of support for assertions.
analyzing perspectives	Identifying multiple perspectives on an issue and examining the reasons or logic behind each.
abstracting	Identifying and articulating the underlying theme or general pattern of information.

5 The role of Marzano HOTS

Marzano HOTS selected for this study because HOTS model has clear description of the process of the eight HOTS. Marzano HOTS has four major components that comprise the definitions, procedures, thinking tools and assessment (Fig. 1). All eight KBAT Marzano has four components. The first component describes the definition so that students clear on the function, characteristics and use of each HOTS. After knowing the definition, the step of using every HOTS are also described in the procedure with questioning technique.

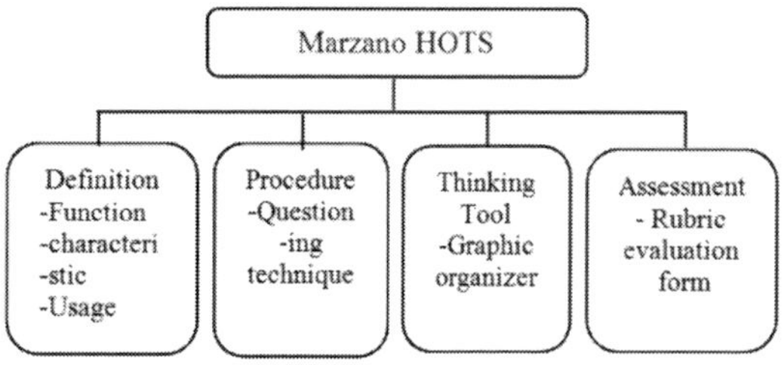

Figure 1. The main components of Marzano HOTS.

Questioning technique is used for probing questions to get information in every step of the process HOTS. Use of this technique allows students to observe the thinking process actively and to think reflectively (Othman & Rahman, 2011). In addition, this technique can also trigger and enhance students' thinking (Solomon et al., 2011) to analyze and reason in a more systematic and effective (Tee et al., 2010b).

This model also provides a thinking tool namely graphic organizer. Graphic organizer appropriate and relevant to the students recorded their thoughts. It's a form of illustration or sketch that was used for the purpose of planning, organizing, researching, analyzing or thinking of ideas to produce something more systematic, clear and effective (Solomon et al., 2011). By using graphic organizer, students can not only see the process of thinking through concrete and explicit, but also be able to think and process information actively (Othman & Rahman, 2011). So, applying HOTS will be more effective.

In the meantime, the thinking tool and thinking skills are essential and indispensable in solving problem and creating idea process so that students can make judgments and decisions wisely (Caviglioli, Harris & Tindall, 2002; Splitter & Sharp, 1995). The advantages of the combined use of questioning technique and graphic organizer in the long term allows students who have a tendency to accept information for concrete (Concrete perceivers) became more abstract (abstract perceivers) (McCarthy, 1987).

Finally, after adopting the HOTS, all thoughts were listed in graphic organizer and rated based on the rubric evaluation form. Rubric is an assessment tool that lists the criteria for a work to score in good rate (Andrade, 1997). Rubric serves to identify the level of student applying HOTS (Peirce, 2006) on problems solving and generating ideas. Each HOTS have their evaluation form. The evaluation criteria in the evaluation form is based on the steps of each process HOTS. With the four clear and systematic components, application of all HOTS become stronger and more organized.

6 Conclusion

Each student must take the initiative to improve the quality of thinking and to foster a positive attitude in developing green skill for any industries. If the industries know how to use the correct thinking skill such as how to take effective action and make the right decision for solving problem, so the industries must be able to obtain excellent solution in protecting environment. Through HOTS and problem solving, idea will become real when industries solve a problem (Mohd Radzi, 2010). To stimulate someone thinking, they must be given the opportunity to respond to the high-level questions. The use of more challenging high level questions will encourage someone to express opinions, make decisions, to find a solution to a problem, or to generate new ideas on matters relating to the content of education (Tee, 2012).

Acknowledgment

The authors would like to thank the Ministry of Higher Education, Malaysia for supporting this research under the Research Acculturation Grant Scheme (RAGS) vot no. RO51.

References

1. M. E. Mousavi, J. L. Irish, A. E. Frey, F. Olivera, & B. L. Edge, *Climatic Change*, 104(3-4), pp. 575-597 (2011)
2. A. Asohan, *Digital News Asia*. (2015, Mei 21). Retrieved from https://www.digitalnewsasia.com/digital-economy/11th-malaysia-plan-broadband-gets-some-love
3. CEDEFOP. *Skills for Green Jobs: European Synthesis Report*. (2010)
4. UNEP, *UNEP Year Book: emerging issues in our global environment*. 2011. Retrieved from www.unep.org/yearbook/2011
5. UNCTAD. *UNCTAD Annual Report*. 2011. Retrieved from http://unctad.org/en/PublicationsLibrary/dom2012d1_en.pdf
6. GEC. *Green economy: an analysis of the UNCSD submissions*. 2012 Retrieved from: http://www.greeneconomycoalition.org/document/green-economy-analysis-uncsd-submissions
7. J. Schwarzer. *Industrial Policy for a Green Economy*, (2013)
8. D. G., De Silva, T. P., Hubbard, R. P, McComb, & A. P. Schiller, *Entry, growth, and survival in the green industry*. 2011. Retrieved from: http://mpra.ub.uni-muenchen.de/34412/
9. G., McDonald, L. Condon, & M. Riordan, *The Australian Green Skills Agreement: Policy and Industry Context, Institutional Response and Green Skills Delivery*, (2012)
10. CEDEFOP. *Green skills and environmental awareness in vocational education and training: Synthesis Report*, (2012)
11. Apollo Alliance. *Green-Collar Jobs in America's Cities*. (2008)
12. R. Florida, *The Rise of the Creative Class. And How It's Transforming Work*. Leisure and Everyday Life. Basic Books, (2002)
13. The Third Outline Perspective Plan. Prime Minister April 2001.
14. B .Ario. *Managing Creativity in the Workplace: Using Creativity of Your Workers*. Associated Content, (2006)
15. K. J. Wheelihan. *Creativity for Success, The Creativity Institute*, 2011, Retrieved from: http://EzineArticles.com/
16. Sharp, A. (2008). Generating Ideas is an Important Part of Creative Problem Solving
17. B. Jonson. Design Ideation: The Conceptual Sketch in the Digital Age. *Design Studies*, 26 (6), 613–624, (2005)
18. D. Graham & T. Bachmann. *Ideation: The Birth and Death of Ideas*. John Wiley and Sons Inc. (2004)
19. M. A. Abd Hamid. *Pengenalan Pemikiran Kritis &Kreatif*. Skudai, Johor, Universiti Teknologi Malaysia, Cetakan pertama. (2001)
20. B. K. Beyer. *Practical Strategies for the Teaching of Thinking*. London: Allyn and Bacon, Inc. (1992)
21. A. R. Abd. Rashid. *Falsafah Budaya dalam Pendidikan*. Kuala Lumpur: Penerbit Universiti Malaya, (2003)
22. R. E. Meyer. Rotes Versus Meaningful Learning, *Theory into Practice*, 41 (4), 226-232, (2002)
23. S. H. Poh, "Kemahiran Berfikir Secara Kritis dan Kreatif," 2nd Ed, Kuala Lumpur: Kumpulan Budiman Sdn. Bhd, (2000)
24. L. W. Anderson, D. R. Krathwohl, P. W. Airasian, K. A. Cruikshank, R. E. Mayer, P. R. Pintrich, J. Raths, M. C. Wittrock. *A Taxonomy for Learning, Teaching, and Assessing: A Revision of Bloom's Taxonomy of Educational Objectives*. New York : Longman, (2001)
25. N. S. Rajendran, "Teaching and Acquiring Higher-Order Thinking Skills Theory & Practice," Tanjong Malim Perak: Penerbit Universiti Pendidikan Sultan Idris, (2008)
26. S. Z. Mohamed, "Kesan Pendekatan Penyebatian Kemahiran Berfikir Kreatif Dalam Pengajaran Karangan Deskriptif Dan Karangan Imaginatif Dalam Kalangan Pelajar Tingkatan IV," Universiti Sains Malaysia: Disertasi PhD, (2006)
27. J. Ea, A. Chang, and O. S. Tan, "Thinking about Thinking: What Educators Need to Know," Singapore: National Institute of Education, Nanyang Technological University, McGrawltill Education, pp. 72, (2005)
28. A. Lewis, and D. Smith, "Defining Higher Order Thinking. Theory Into Practice," 32, 131-137, 1993.
29. G. Sykes, R. Floden, & C. Wheeler. *Improving Teacher Learning in Thailand : Analysis and Options*. Bangkok : The National Educational Commission. (1997)
30. T. K. Tee. *Pengintegrasian Kemahiran Berfikir dan Peta Minda Buzan Bagi Penguasaan Kemahiran Berfikir Aras Tinggi*. Universiti Tun Hussein Onn Malaysia: Tesis Ph.D. (2012)
31. F. M. Newmann, "Higher Order Thinking In Teaching Social Studies: A Rationale for The Assessment of Classroom Thoughtfulness," Journal of Curriculum Studies, 22, 41-56, (1990)
32. L. B. Resnick, "Education and Learning to Think," Washington, DC: National Academy Press, (1987)
33. R. J. Marzano, D. J. Pickering, D. E. Arredondo, G. J. Blackburn, R. S Brandt, C. A. Moffett, D. E. Paynter, J. E. Pollock, and J. S. Whisler, "Dimension of Learning Trainer's Manual," 2nd Ed, Aurora, Colorado: McREL, (1997)

MATEC Web of Conferences 70, 05002 (2016) DOI: 10.1051/matecconf/20167005002
ICMIT 2016

Case Study on Lean Manufacturing System Implementation in Batch Printing Industry Malaysia

Tang Saihong[1], Ng Tanching[1, 2], Chong Weijian[2], Chen Kahpin[2]

[1]*Department of Engineering, Universiti Putra Malaysia, Malaysia*
[2]*Department of Mechanical and Material Engineering, Universiti Tunku Abdul Rahman, Malaysia*

Abstract. Lean Manufacturing is a popular tool to be implemented in printing industry fields for the purpose of achieving the successful production goals and it is a well-organized method used to eliminate the waste or non-value added activities. Lean Manufacturing consists of several tools like 5S, TPM, Kanban, Kaizen and others. In this paper, the researcher is focus on TPM (Total Productive Maintenance). TPM is a concept used for maintaining plants and equipment with the involvement of all employees in the company in order to increase the overall equipment effectiveness (OEE). In this paper, the researcher will identify the problem faced in company's production line and then implement lean tools in order to improve the productivity. Afterward, the researcher will carry out time study on company's production line. The purpose of time study is to getting the result for OEE and the time taken for producing one piece of product. The researcher is then designed and suggested one-piece flow system to the management with showing the result on simulation that getting improvement in productivity. The result was simulated by using Flexsim and shows that 7.59 seconds was saved in producing one piece of calendar. At the end, the result shows that OEE increase from 34.3% to 60% and the company total save around 6 hours per month by implement one-piece flow. At last, by verifying the distribution of collected data, the researcher will perform Anderson Darling Normality test to ensure the distribution of data are normal.

1 Introduction

The company that involved in this study is ABC Sdn Bhd. There are several problems faced by ABC SdnBhd which included low productivity, untidy working environment that affect the effectiveness of process flow, lack of labeling for tools, inventory that cause the space not enough, wastes such as defect, unnecessary motion and lastly is lack of multi-skill workers. Therefore, the researcher aims to overcome the problem faced by introducing the lean concept.

In this research, the company's production flow was identified and 5S was implemented in the production floor for effective flow. Besides that, the company's Overall Equipment Effectiveness (OEE) in production layout and redesign layout by using Flexsim software were analyzed. The scope of this research involves areas within the factory focusing on cleanliness and tidiness criteria. Also, this study will focus on the assembly line by combining the part of stand, coil and content of calendar together. Meanwhile, the distance between the stations is far apart from each other which extend the cycle time production as well as the overall transportation distance of work.

2 Literature review

Total Productive Maintenance (TPM). In order to implement TPM successfully, there are 5 important pillars to be followed for the purpose of getting improvement in performance or outcome. They are (i) training and Education, (ii) Autonomous Maintenance, (iii) Equipment Effectiveness Improve, (iv) Maintenance Efficiency and Effectiveness Improve and (v) Early Equipment Management and Maintenance.

For first aspect, all employees were ensured have the skills and knowledge which able to build up a TPM environment. Thus, training for employees is compulsory. Secondary, autonomous maintenance is one of the key factors which lead to the successful of TPM. Autonomous maintenance refer activities of worker to carry out daily check, lubrication, inspections, repair and so forth on equipment for the purpose of keeping equipment always in perfect condition. This aims the equipment which in design state to ensure all the equipment developed reliable, user friendly and easy to maintain or repair it [1]. Meanwhile, by improve on equipment effectiveness which means to ensure that the equipment performs itself to achieve its design specification. In addition, maintenance efficiency and effectiveness improve is important to reduce wastes in maintenance process. This improvement can lead to minimal impact on up time and unavailability of equipment. Lastly, total maintenance costs which equipment exactly needs can be reduced by early equipment management and maintenance prevention. The overall maintenance cost will increase at the end due to unnecessary maintenance without awareness on this aspect.

The benefit which can gain by organization after implement TPM is to ensure a better quality for products, reduce the equipment breakdown, reduce the total costs, able to create a working environment which full with motivate, and improve the working attitude of employees [2]. Most of the organizations able to achieve the result of getting improvement on reduce any breakdowns of equipment, minimize the idle time, increase the productivity, reduce numbers of accident case, and encourage the involvement of employee [3].

3 Types of wastes

The main goal of lean manufacturing is eliminate wastes byidentifying seven types of waste. They are overproduction, waiting times, unnecessary transportation, unnecessary processing, unnecessary motion, waiting inventory and defected products [4], [5]. There is no any doubt that elimination of wastes is important in today's world. All companies were fight for high quality product together with low production cost. Production cost can be reduced by elimination of wastes.

4 One-Piece flow

The objective of one piece flow is to build up the parts of product one at a time in a more organizing and sequencing processes in order to avoid the problem of long queue times or producing in batches. In addition, anotherobjectives of one-piece flow is used to avoid the non-value added movement which means the unnecessary movement [6].

5 Methodology

This research begins with choosing a suitable manufacturing company in Malaysia after sufficient amount of literature has been reviewed. ABC SdnBhd was chosen in this research. After that, the company chosen was pays a visit for the purpose of problem identification and investigate the whole layouts of the entire factory. During factory visitation, data needed was collected via time measurement for each working processes. This data was used further in calculating the Overall Equipment Effectiveness (OEE) before and after lean implementation. After investigation, one-piece-flow system was suggested with other lean tools implementation such as 5S, Kanban, TPM and kaizen. Flexsim, simulation software, was chosen for data analysis. For data analysis, mathematical operation such as OEE calculation and time taken for producing one piece of calendar in production line was applied. Afterward, comparison among current production line with the simulated one-piece flow designed was done. Further recommendations were proposed afterward.

6 Results and discussion

TPM focused on solving problem for wastes that related with production system. Overall Equipment Effectiveness (OEE) was employed to measure the performance of productive system. OEE is an improvement process that focused on productivity. It started from the management awareness of total productive manufactured and commitment from management to be more focus on the factory worker trainingto work as a team and cross functional equipment problem solving.

As result, OEE was increased gigantically from 34.3% to 60%. This can conclude that TPM produced significant result in raising the productivity and improve the performance of machine by lower down the breakdown time for the machine. The overall equipment effectiveness still can be improved by enhancing the TPM and to avoid this current situation return back to original situation.

Besides TPM implementation, one-piece flow layout that designed by researcher was simulated using Flexsim software. The purpose of implementing one-piece flow is to reduce the unnecessary transportation time and unnecessary process. In addition, waiting time, queue time and total distance moved by operators can be reduced with this layout. Originally, the punching process for the content and caser was separately. With the help of one-piece flow, these two processes can be combined together and carried out on the same time. As mentioned, those unnecessary transportation times can be reduced with the help of conveyor. As the result from simulation onFlexsim, the result obtained for processing one piece of calendar in 67.41seconds.

Total time save with implementing one-piece flow was calculated as shown as Table 2. As the result shown, time save for producing each piece of calendar is 7.59 second. Researcher has set 50 as targeted amount to be produced in this one-piece flow production line per hour. Thus, this can help in saving 6.325 minutes in producing targeted amount. As monthly working hour, one-piece flow save nearly 6 working hours compare with implementing batching production.

Lastly, Minitab was used to plot Anderson Darling Normality test based on the data collected. This test was used for data validation. Theoretically, when P-value > 0.05 means the null hypothesis is accepted and the distribution is normal or vice-versa. The test result was shows that the P-value for machine1, machine2, machine3, machine4 and machine5 is 0.808, 0.336, 0.573, 0.861 and 0.52. This represent the normal distribution of data collected and hypothesis is accepted.

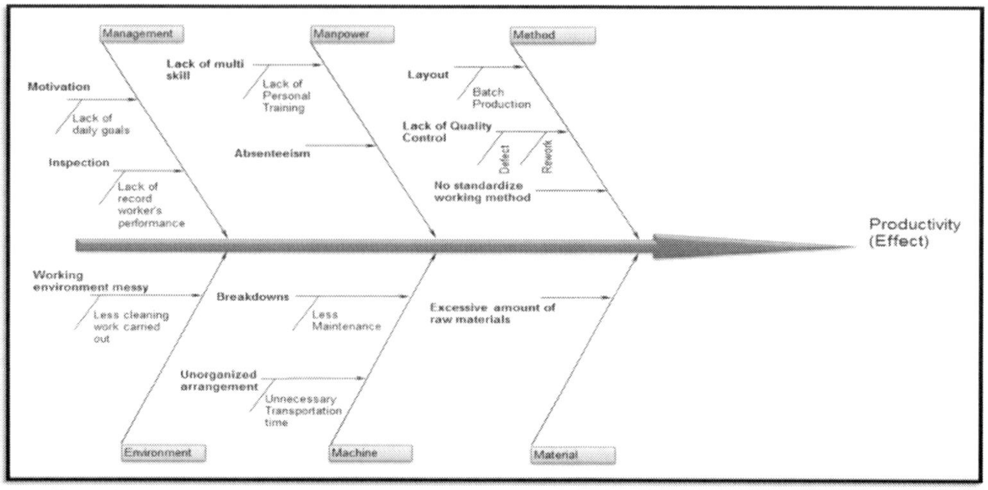

Figure 1. Fishbone diagram of problems faced by Company ABC's productivity.

Table 1. Calculation of TPM.

Elements	Before TPM Implementation	After TPM Implementation
Kolbus –DAS case maker breakdown	39 hours	22.5 hours
Paper Gluing Machine breakdown	13 hours	7.5 hours
Double Coil Binding Machine breakdown	19.5 hours	12.5 hours
Paper Hole Punching Machine breakdown	6.5 hours	2.5 hours
Set-up and others conditions	52 hours	52 hours
Total loss (sum of all breakdown)	130 hours	97 hours
Management loss	26 hours	26 hours
Start up loss	13 hours	13 hours
Total working hours	312 hours	312 hours
Net loss (total working hours – total loss)	312 – 78 = 182 hours	312 – 45 = 267 hours
Availability Rate (Net loss / total working hours) x 100	(182/312) x 100 = 58.33%	(267/312) x 100 = 70.83%
Percentage of Quality [(Total steam produced – defected steam) / total steam produced] x 100	[(3120-780) /3120] x 100 = 75%	[(3120-450) /3120] x 100 = 87.5%
Performance Rate [Net loss- (management loss + start up loss) / Net loss] x 100	[182 – (26+13)] / 182] x 100 = 78.57%	[267 – (26+13)] / 267] x 100 = 82.35%
OEE	58.33% x 75.00% x 78.57% = 34.3 %	85.58% x 85.58% x 82.35% = 60 %

Table 2. Total time save per month for targeted amount.

Time save for produce one piece calendar in seconds (s)	75 – 67.41 = 7.59 s
Researcher's target amount produced per hour	50
Time save per hour	(50 x 7.59s) / 60s = 6.325 minutes
Working days per month (30 days as standard)	22 days
Time save per month	(22x 6.325 minutes) / 24 hours = 5.80 hours

7 Conclusion

Due to the machine breakdown time has been reduce hugely, the availability rate for the machine after the implementation of TPM reached 70.83%. Moreover, the percentage of quality has been increase from 75% to 87.5% with the implement of TPM. At last, the performance rate was also shows the significant increase from 78.57% to 82.35% after implement TPM into the company. As conclusion, overall equipment effectiveness of ABC Sdn Bhd was improved from 34.3% to 60% through implementation of TPM with the help of reducing in the number of machine breakdowns. The entire working environment was change dramatically and mistakes reduced with 5S implementation. Lastly, a suggestion about changing current production layout was proposed. One-piece flow implementation was suggested to replace batching production line. According to simulation on Flexsim, one-piece flow implementation can reduce production time for one piece of calendar from 75 seconds to 67 seconds. The company can save total around 6 hours in one month.

After in-depth investigation in ABC SdnBhd, a serious problem in quality management was found by researcher. Due to the absent of quality control in production line, company get complaints from customers about receiving defect products. Thus, the researcher suggested that ABC SdnBhd have to set up a work station for quality control purpose in the company immediately before the company getting complaints or negative feedbacks from the customers.

References

1. Shirose, K., Equipment effectiveness, chronic losses, and other TPM improvement concepts in TPM development program: Implementing total productive maintenance. Productivity Press, Portland, OR, (1989)
2. Seth, D. and Tripathi, D., Relationship between TQM and TPM implementation factors and business performance of manufacturing industry in Indian context. *International Journal of Quality & Reliability Management*, **22**(3), pp. 256-277, (2005)
3. Suzuki, T., TPM in process industries. 1st ed. Portland, Or.: Productivity Press, (1994)
4. Womack, J. and Jones, D., Lean thinking. London: Simon & Schuster, (2003)
5. Jones, D. T., & Roos, D. Machine that Changed the World. Simon and Schuster, (1990)
6. Nyhuis, P., & Vogel, M. Adaptation of logistic operating curves to one-piece flow processes. *International Journal of Productivity and Performance Management*, **55**(3/4), 284-299, (2006).

The Three-dimensional Digital Factory for Shipbuilding Technology Research

Wei Xu, Mingyu Li and Wenjun Zhang

Wuhan Second Ship Design and Research Institute, Wuhan, Hubei, China

Abstract. The three-dimensional digital factory technology research is the hotspot in shipbuilding recently. The three-dimensional digital factory technology not only focus on design the components of the product, but also discuss on the simulation and analyses of the production process.Based on the three-dimensional model, the basic data layer, application control layer and the presentation layer of hierarchical structure are established in the three-dimensional digital factory of shipbuilding in this paper. And the key technologies of three-dimensional digital factory of shipbuilding are analysed. Finally, a case study is applied and the results show that the three-dimensional digital factory will play an important role in the future.

1 Introduction

Digital factory technology is an integrated virtual technology, which is focused on process planning and production planning. With the development of virtual reality and computer technology, the digital factory theory and technology have become matured gradually. The digital factory system has been designed and used by many shipbuilding companies and organizations in the world.

Based on the expert system, S. Akagi and K. Fujita achieved digital design for shipbuilding preliminary design [1]. Myung researched the Model-Based Definition (MBD) of shipbuilding design, and developed the software systems for design of shipbuilding [2]. Based on the CAD system, K.H Lee applied the design method to the layout for shipbuilding process [3]. Depended on the STEP data format, Kim researched database storage and data exchange for shipbuilding methods [4]. M.Y Li applied the discrete particle swarm optimization and evolutionary direction operation to realized a hybrid assembly sequence planning [5-6]. In addition, the digital technology of design industry abroad has been more mature and the powerful shipbuilding software system has been developed [7-9].

In terms of digital integration, the research is focused on the development of application software systems from the ship design to shipbuilding [10]. Along with the development of digital technology, the digital simulation of internal outfitting design, analysis and process can be realized[11]. The system of digital shipbuilding is developed by the Samsung Shipbuilding Industry, and the whole shipbuilding process can be simulated and in the virtual environment, the results are used in shipbuilding, directly.

At present, the digital factory technology is not yet fully throughout the whole life cycle of the shipbuilding. The realization of data exchange technology, information integration technology, shipbuilding cycle optimization and other functions should be researched. To realize the combination of the three-dimensional factory and shipbuilding, the hierarchy of the three-dimensional digital factory for shipbuilding is discussed, the key technology of the three-dimensional digital factory for shipbuilding is analyzed, and the shipbuilding process of the three-dimensional digital factory is established in this paper.

The rest of paper is organized as follows. Section 2 introduces the hierarchy of the three-dimensional digital factory for shipbuilding. Section 3 discusses the key technology of the three-dimensional digital factory for shipbuilding. A process flow of shipbuilding with three-dimensional digital is established in Section 4. In section 5, a verified case study is given. Finally, this paper is concluded in Section 6, and the prospect of future research is given.

2 The hierarchy of the three-dimensional digital factory

The hierarchy of the three-dimensional digital factory for shipbuilding can be divided as basic data layer, application control layer and presentation layer, as shown in Fig. 1.

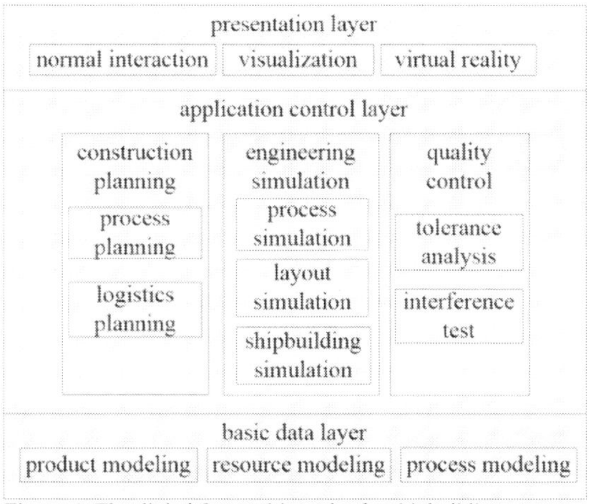

Figure 1. The digital factory hierarchy for shipbuilding

2.1 Basic data layer

The basic data layer is the foundation of the three-dimensional digital factory for shipbuilding, which is comprised of product modeling, resource modeling and process modeling. And the role of the basic data layer is to realize the interaction of basic data information.

Accompanied by the use of MBD modeling methods, the information of the traditional two-dimensional drawing is redefined by the three-dimensional digital. The shipbuilding information can be fully expressed in the three-dimensional model. And the three-dimensional model is used as the unique shipbuilding data source in shipbuilding design and process.

Resource modeling is comprised of workshop, equipment, tools and other resources, which are the necessary resources in the digital factory. The resource modeling is also comprised of virtual human resource, which is used in the virtual simulation process. The feasibility and authenticity of the digital factory can be determined by the integrity of the resource modeling.

Process modeling is the necessary crafts, which is comprised of processing craft, installation process and disassembly craft. The process modeling is the basis of the construction simulation.

2.2 Application control layer

The application control layer is the core layer of the three-dimensional digital factory technology for shipbuilding. The layer includes the construction planning, engineering simulation and quality control.

Construction planning is comprised of process planning and logistics planning. The goal of the process planning is to change product design information to manufacturing information, thus the manufacturing strategies can be determined. Logistics planning is divided into external logistics and production logistics. And the external logistics is also the factory logistics, which is comprised of material routes and bus routes. The production logistics is expressed as the process of production, products from the production step into the next, in accordance with the processing and storage process.

Engineering simulation is comprised of process simulation, layout simulation and shipbuilding simulation. The process simulation includes the changes of the forms of geometric and physical factors. Layout simulation includes the equipment layout and the location of tools layout. Shipbuilding simulation is the core of the simulation, whether the construction of products in the production line can be finished and verified by the simulation.

Quality control is comprised of tolerance analysis and interference test. The purpose of the quality control is to ensure the authenticity and validity of the simulation.

2.3 Presentation layer

The presentation layer is comprised of normal interaction, visualization and virtual reality. The purpose of interaction between human and computer can be realized. The interaction is comprised of desktop virtual construction, semi-immersive virtual construction and immersive virtual construction.

3 The key technology of the three-dimensional digital factory

The key technology of the three-dimensional digital factory is comprised of digital modeling technology, optimization simulation technology, virtual reality technology, and application production technology, as shown in Fig. 2.

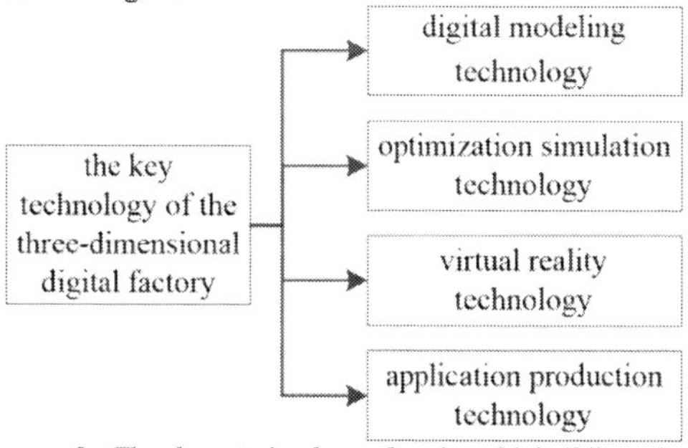

Figure 2. The key technology for the shipbuilding three-dimensional digital factory

3.1 Digital modeling technology

The digital modeling technique is the foundation for the three-dimensional digital factory. The models of products, manufacturing resources and tools are integrated with simulation system, and the underlying data of the digital factory can be established. The model of three-dimensional component is the actual component modeling in the three environments. The modeling of product function information, product structure information, product process information and other modeling methods are used in digital factory modeling technology. With the deepening of the research and application of the three-dimensional model, the two-dimensional drawings will be replaced gradually, and the paperless drawings will be realized.

3.2 Optimization simulation technology

The optimization simulation includes simulation and optimization. The simulation technology is comprised of application process simulation, welding process simulation, assembly process simulation, ergonomics simulation, logistics simulation and verification process simulation. The purpose of simulation is to analyse the original design, and find product design defect, finally, the process of preliminary shipbuilding is put forward.

Optimization is comprised of process methods optimization, assembly sequence optimization, process planning optimization, ergonomics optimization and so on. The optimizations of the simulation are often achieved by the methods of optimization algorithm or experience optimization. And the optimization methods are established on the basis of the simulation methods. Through the analysis of the process simulation, the logistics path optimization and the efficiency can be achieved.

3.3 Virtual reality technology

Virtual reality is based on human-computer interaction, whose purpose is to realize the information interaction between real environmental and virtual environment. And the depth of design and the degree of influence to the demand can be determined by the development level of virtual reality and application depth of virtual reality in the digital factory.

3.4 Application production technology

The purpose of application production technology is to accelerate the conversion between design and application of products manufacture. And to establish a process from parts design to product manufacturing.

4 The process flow of three-dimensional digital factory

The three-dimensional digital factory for shipbuilding can be divided into information modeling, sequence planning, path planning and process simulation.

4.1 Information modeling

The three-dimensional model is the basis of the information modeling for shipbuilding. On the basis of the completion of the three-dimensional model, the three-dimensional model of attribute information is edited, in order to finish building information modeling.

Information modeling including size information, datum information, attribute information, geometric information and environmental information. As shown in Fig. 3.

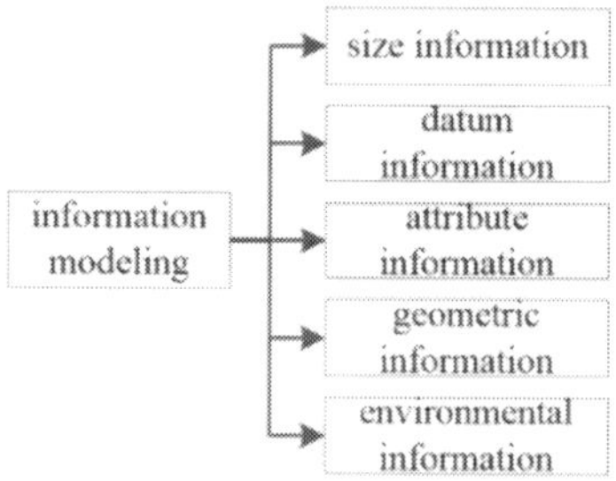

Figure 3. The constitution of information modeling for shipbuilding

4.2 Sequence planning

Sequence planning method is an important part of the digital factory. The planning method design and planning optimization are included in sequence planning of shipbuilding. And the best sequence planning of shipbuilding can be achieved by optimal simulation, finally.

Sequence planning consists of evaluation index system, sequence stability, sequence redirection number, optimization algorithms and the generation of the optimal sequence, as shown in Fig. 4.

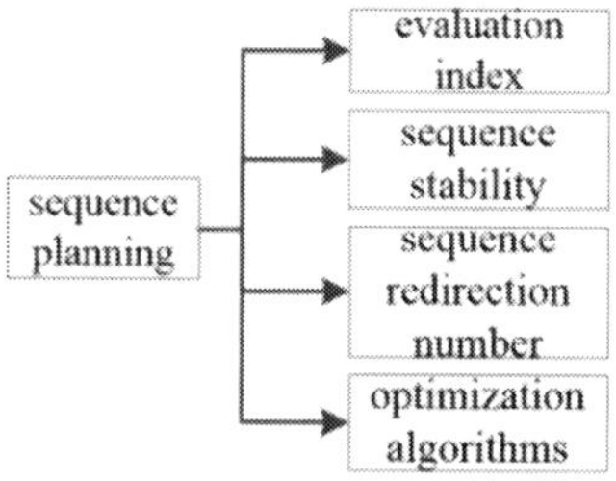

Figure 4. The constitution of sequence planning for shipbuilding

4.3 Path planning

The path planning is the equipment assembly path in the process of construction. Path planning consists of path resource planning, path optimization algorithm and the path evaluation method. The logistics planning level of digital factory shipbuilding process is determined by the quality of path planning.

4.4 Process simulation

Process simulation is used to verify and analyse the initial shipbuilding process in the construction of the virtual environment. Through the sequence optimization, path optimization and many other methods, the shipbuilding process optimization can be achieved.

Process simulation consists of assembly process interference inspection, ergonomics analysis, assembly reachability analysis and simulation environment. And the assembly reachability analysis is focused on the three-dimensional digital factory of shipbuilding. The process simulation is shown in Fig. 5.

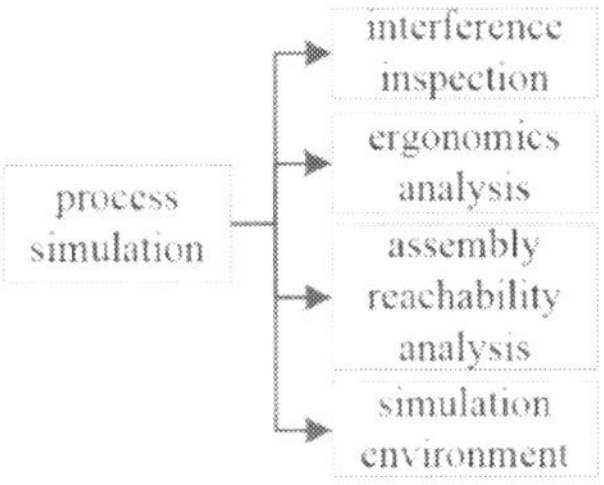

Figure 5. The combination of process simulation

The flow chart of three-dimensional digital factory for shipbuilding is shown in Fig. 6.

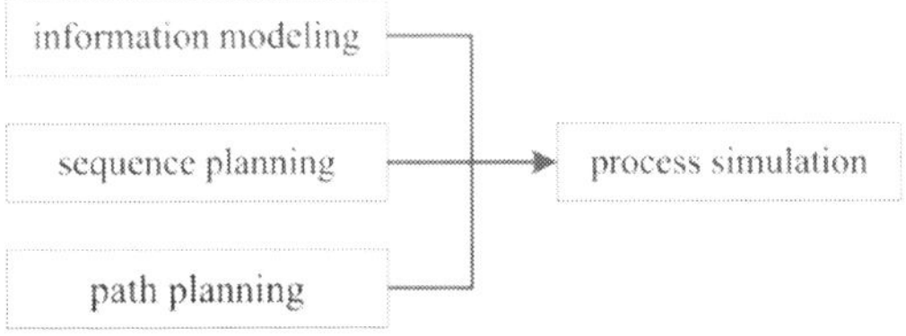

Figure 6. The flow chart of three-dimensional digital factory for shipbuilding

5 Case study

As shown in Fig. 7, a three-dimensional digital factory example was used to study and analyze, which was tested and analyzed in DELMIA software of Dassault Systemes. In this study, the information modeling was built, and the sequence planning was also optimized, the path planning was researched. Finally, the process simulation was verified completely.

The assembly process interference inspection, ergonomics analysis, assembly reachability analysis and simulation environment also are shown in Fig. 7.

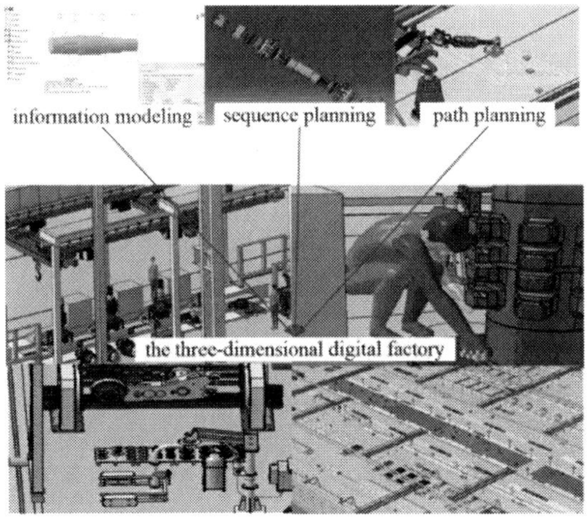

Figure 7. The three-dimensional digital factory

6 Conclusions and future work

The three-dimensional factory technology for shipbuilding is one of the hotspots in current shipbuilding. In this paper, a hierarchy of the three-dimensional digital factory for shipbuilding is researched, a three layer of structure for shipbuilding is established, the key technology of the three-dimensional digital factory for shipbuilding is discussed, and the process flow of shipbuilding is established. The case study shows that the three-dimensional digital factory for shipbuilding was suitable.

With the deep application of the three-dimensional factory technology, virtual shipbuilding, virtual maintenance and collaborative shipbuilding will be the focus of future research. As the future of shipbuilding advanced planning and mode of operation, the research of three-dimensional digital factory technology is still in the preliminary stage. The technical and systematic theoretical of three-dimensional digital factory should be discussed and studied in further.

References

1. S. Akagi, K. Fujita, Building an expert system for the preliminary design of ships, artificial intelligence for engineering design, analysis and manufacturing, **1**, 191-205 (1987)
2. M. Roh, K. Y. Lee, An initial hull structure modeling system for computer-aided process planning in shipbuilding, Advances in Engineering Software, **37**, 457-476 (2006)
3. K. H. Lee, J. K. Lee, N. S. Park, Intelligent approach to a CAD system for the layout design of a ship engine room, Computers & Industrial Engineering, **34**, 599-608 (1998)
4. J. Kim, S. Han, Encapsulation of geometric functions for ship structural CAD using a STEP database as native storage, Computer-Aided Design, **35**, 1161-1170 (2003)
5. M. Y. Li, B. Wu, Y. M. Hu, C. Jin, T. L. Shi, A hybrid assembly sequence planning approach based on discrete particle swarm optimization and evolutionary direction operation, International Journal of Advanced Manufacturing Technology, **68**, 617-630 (2013)
6. M. Y. Li, B. Wu, P. X. Yi, C. Jin, Y. M. Hu, T. L Shi, An improved discrete particle swarm optimization algorithm for high-speed trains assembly sequence planning, Assembly Automation, 33, 360-373 (2013)
7. D. Lee, K. H. Lee, An approach to case-based system for conceptual ship design assistant, Expert Systems with Application, **16,** 97-104 (1999)
8. J. H. Park, R. L. Storch, Overview of ship-design expert systems, Expert Systems, **19,** 136-141 (2002)
9. S. Helvacioglu, M. Insel, An expert system approach to container ship layout design, International Shipbuilding Process, **50,** 19-34 (2006)
10. S. H. Lee, K. Lee, Sumire system in sumiyomo oppama shipyard, Advances in Engineering Software, **4,** 356-365 (2001)
11. C. H. Lu, Y. Lin, Z. S. Ji, Virtual tanks division and capacity calculation based on NURBS ship form, Journal of Ship Mechanics, **11**, 435-443 (2007)

Energy Efficiency Tracking in Thai Manufacturing Sector by Decomposition Technique

Wongkot Wongsapai, Atchariya Phuangyod and Det Damrongsak

Energy Technology for Environment Research Unit, Faculty of Engineering, Chiang Mai University, Chiang Mai, 50200, Thailand

Abstract. This paper presents an analysis of energy saving and changes in energy intensities in Thai manufacturing sector by Logarithmic Mean Divisia Index decomposition technique. This method includes three effects consists of the energy intensity effect, the structural effect and the effect of the economic growth on the energy consumption in Thailand by using the 25-year annual data from 1990 to 2014, carried out in four phases; (i) before National Energy Conservation law, (ii) during the effect of the law, (iii) Transition period of the law from first to second version, and (iv) during the effect of the law (No.2). We found that the most effective intensity effect is in the third phase due to the effect of the implementation of new energy efficient equipment from the second phase by enforcement of the law, especially in non-metallic sector, while the first phase illustrates the lowest intensity effect due to the energy conservation law had not been occurred. However, due to the highest economic growth of the country and change from agricultural to industrial development direction, the first phase presents the most effective structural effect, then this effect continuously decreased by time. We also conclude that the energy conservation law have direct effect to energy efficiency of the country however, strictly individual regulation which have target to enforce to energy intensive industries is still required for sustainable energy efficiency improvement.

1 Introduction

The final energy consumption in Thailand increases continuously during the past 25 year. The economic growth rate in the period 1990-2014 averaged at 4.36 percent. In period 1990 - 1997 was 4.71 percent. In period 1997 - 2007 was 3.71 percent. In period 2007 - 2010 was 2.33 percent and the last period 2010 - 2014 economic growth increased average at 2.95 percent [1]. The growth of manufacturing sector has grown rapidly and continuously, so in energy planning essential to learn more effect and factors that affect to energy use in this sector.

This paper studies on the comparative analysis of factors influencing energy intensity changes. The analyse is also includes the energy saving in Thai manufacturing sector by using the decomposition techniques, which is the powerful tool for indicate the energy performance in manufacturing energy demand analysis [2],[3].

2 Methodology

The methodology consists of two topics, as follows;

2.1 Economic and energy aggregation.

During the last 25 years, Thailand's economic growth increase continuously while manufacturing sector has the highest share of the country. Share of the manufacturing sector GDP in total GDP increased from 25.41 % in the year 1990 to 28.49 % in the year 2014 and share of final energy consumption of the manufacturing sector in total final energy consumption increased from 27.94 % in the year 1990 to 36.76 % in the year 2014 , as presented in Fig. 1.

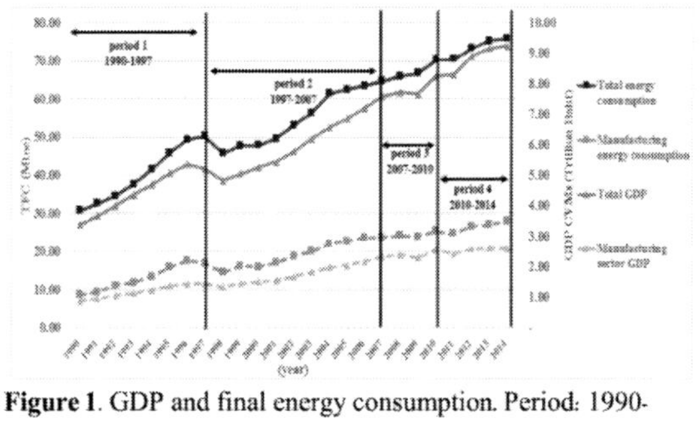

Figure 1. GDP and final energy consumption. Period: 1990-2014.

Fig. 1. The manufacturing sector and all economic sector of final energy consumption which has the same trend as the GDP growth of the country.

By considering the energy and economic significant impact, the twentyfive-year period of this study was divided into four period; i.e. (i) The first sub-period is from 1990 to 1997 which presents the high economic progress of the country before The National Energy Conservation law A.D.1992 and the time before the 1997 economic crises, (ii) The second, from 1997 to 2007, during the effect of the law and the period of economic slowdown after the crises. (iii) The third transition period of the law from 2007 to 2010 from first to second version and the global economic slowdown , and (iv) The last from 2010 to 2014, during the effect of the law (No.2) and the political crisis in the Thai economy.

2.2 Decomposition technique

To analyze technique, the amount of energy consumption data set is required to study the multiplicative decomposition analysis by Logarithmic Mean Divisia Index method I (LMDI I) [4], [5] for estimate an index value.

The main equation of this method shows in equation (1). Assume that all variables are in time series dataset and given as of time t applying the theory of instantaneous growth rate.

$$D_{tot} = D_{act} \times D_{str} \times D_{int} \quad (1)$$

where D_{tot} is the total effect, D_{act} is the activity effect, D_{str} is the structural effect and D_{int} is the intensity effect.

The activity effect considers changes in the scale of economic activity or in the level of output of the entire economy, under the assumption that an increase in the level of this output in volves an increase in energy consumption. At the aggregated and sectoral level usually measured by monetary indicators is Gross Domestic Product (GDP) or value-added[6].

The structural effect considers changes in the mix of activities. These changes have an impact on energy consumption due to each activity has different energy intensiveness : for example, energy consumption will grow if a high intensity industry such as oil extraction, increases its output share with in the economy [7].

The intensity effect considers the impact on energy consumption of changes in energy intensity. This is considered a good proxy of energy efficiency changes. Energy intensity reductions occur either because of more efficient technology or better processes, which can include substitution off actor inputs such as substitution of renewable energy for fossil fuels in the production processes [8].

The compute the method as shown in equations (2)-(6) [4], [9].

$$D_{act}=exp \sum_{i=1}^{n} \left[w_i^* \times In \left[\frac{A_{i,t}}{A_{i,0}} \right] \right] \quad (2)$$

$$D_{str}=exp \sum_{i=1}^{n} \left[w_i^* \times In \left[\frac{S_{i,t}}{S_{i,0}} \right] \right] \quad (3)$$

$$D_{int}=exp \sum_{i=1}^{n} \left[w_i^* \times In \left[\frac{I_{i,t}}{I_{i,0}} \right] \right] \quad (4)$$

$$w_i^* = \frac{L[w_{i,t}, w_{i,0}]}{\sum_i^n L[w_{i,t}, w_{i,0}]} \quad (5)$$

$$L[w_{i,t}, w_{i,0}] = \frac{w_{i,0}-w_{i,t}}{In\left[\frac{w_{i,0}}{w_{i,t}}\right]} = \frac{\frac{E_{i,0}}{E_0} - \frac{E_{i,t}}{E_t}}{In\left[\frac{\frac{E_{i,0}}{E_0}}{\frac{E_{i,t}}{E_t}}\right]} \quad (6)$$

where w_i^* is The weight function, A is value-added in economic sub-sector, S is share of activities in economic sub-sector, I is energy use per unit of activity and i is economic sub-sector within a given sector.

3 Results and discussion

3.1 Analysis of economic

This part presents the result of the economic. We analyze share of final energy consumption, value-added [1],[10] and energy intensity of all economic sectors and the manufacturing sectors, as presented in Table 1 to 2.

Table 1. Share of final energy consumption, share of value-added and energy intensity in Thai economic sectors, Period 1990-2014.

Economic sector	1990-1997	1997-2007	2007-2010	2010-2014
Final energy consumption (%)				
Agriculture	4.2	5.0	5.1	5.2
Mining	0.1	0.2	0.2	0.2
Manufacturing	32.2	35.0	36.2	36.0
Construction	0.6	0.3	0.2	0.2
Residential	18.3	14.8	15.2	15.1
Commercial	5.1	6.4	7.5	7.8
Transport	39.5	38.2	35.7	35.6
Total	100.0	100.0	100.0	100.0
Value-added (%)				
Agriculture	8.8	8.6	7.7	7.4
Mining	2.0	2.4	2.7	2.6
Manufacturing	26.2	29.1	30.6	29.3
Construction	6.9	3.2	2.7	2.6
Commercial	53.5	53.5	53.0	55.0
Total	100.0	100.0	100.0	100.0

Economic sector	1990-1997	1997-2007	2007-2010	2010-2014
Energy intensity (ktoe/billion Baht)				
Agriculture	4.3	5.3	5.7	5.8
Mining	0.5	0.7	0.6	0.6
Manufacturing	11.1	11.0	10.2	10.2
Construction	0.8	1.0	0.6	0.6
Residential	1.6	1.4	1.3	1.3
Commercial	0.9	1.1	1.2	1.2
Transport	3.6	3.5	3.1	3.0
Total	9.0	9.1	8.6	8.3

Source: Department of Alternative Energy Development and Efficiency [10] and Office of the National Economic and Social Development Board, 2014 [1]

Table 1. shows that during 1990 to 2014, all economic sectors that consumed the largest amount of energy were transport and manufacturing sectors [11] share average at 37.3% and 34.9% respectively.The growth of energy consumption of the manufacturing sector continued to increased 3.8% whereas the growth of energy consumption of the transport sector declined to 3.9% in last period. This implies that energy consumption trend in the manufacturing sector would higher than the transport sector for the next periods, as for the energy intensity of the manufacturing and transport sector declined to 0.9% and 0.6% respectively whereas the commercial sector increased to 0.3%. It shows that the efficiency of energy use increased in the transport and manufacturing sector but declined in the commercial sector.

Regarding in the growth rates of GDP, Chained Volume Measures (CVM) of agriculture and construction sector continued to declined 1.4% and 4.3% respectively whereas manufacturing and commercial sector continued to increased due to the export growth and increased direct foreign investment particularly in the fabricated metal sub-sector in the second period during the effect of the law (No.2) from 1997 to 2007.

Table 2. shows that during 1990 to 2014, the manufacturing sector which consumed the largest amount of energy ware food and beverages and non-metallic sub-sector share average at 32.8% and 28.8% respectively. In term of growth, energy consumption of labour-intensive production of textiles continued to decline to 4.9% whereas the food and beverages sub-sector declined to 4.1% in the first period due to the effect of economic slowdown in the after the crises but increased to 5.4% under of the law (No.2) A.D.2007.

The highest energy intensity were non-metallic sub-sector average 63.9 ktoe per billion Baht, followed by basic metal sub-sector average 25.8 ktoe per billion Baht and the lowest energy intensity were fabricated metal sub-sector average at 2.6 ktoe per billion Baht while the energy intensity of less energy-intensive industries were chemical and textiles sector, the high energy intensive industries were food and beverages, paper, non-metallic and basic metal sub-sector, showed a rapid rise in energy intensity throughout the period.

Regarding to the growth rates of GDP of textiles and food and beverages sub-sector had declined to 9.7% and 4.3% respectively whereas the fabricated metal and chemical sub-sector were as high as 10% and 3% respectively. The decline growth of the labour-intensive industries was the result of rising competition from the emergence of many low-wage countries such as China, India, and Vietnam, in the world market [9].

Moreover, the result show that the manufacturing sector use less energy has become more value-added was fabricated metal sub-sector, in the other hand high energy has become less value-added was non-metallic sub-sector. However, In the last period, energy efficiency of non-metallic still higher than fabricated metal sub-sector and effect to conducive to declining energy intensity of manufacturing sector.

3.2 Analysis of decomposition method.

This part presents the results of the multiplicative decomposition analysis by Logarithmic Mean Divisia Index method 1. We analyzed the change of energy consumption of all economic sector and manufacturing sector. The result are presented by graphical instruments, as presented in below.

Table 3 and Fig. 2 shows the increase in activity effect in all periods while the structural effect declined during 2007-2010 to 0.1 and increased to 1.45 during 2010-2014 due to the manufacturing sector changed. For intensity effect, it continue to decline to 0.09 by manufacturing sector change in the first and the second period during 1990-2007 and commercial sector change in the third and the last period during 2007-2014.

As presented in Table 3, the energy consumption has significantly changed during 2010-2014 which is during the law No. 2. In addition, the structural changes impact to energy consumption change more than intensity effect due to the manufacturing sector is also key part.

Table 2. Share of final energy consumption, share of value-added and energy intensity of manufacturing sector, Period 1990-2014.

Manufacturing	1990-1997	1997-2007	2007-2010	2010-2014
Final energy consumption (%)				
Food and beverages	34.1	30.0	31.5	35.4
Textiles	8.0	5.8	4.0	3.1
Wood and furmiture	0.8	0.8	0.9	1.2
Paper	4.0	4.2	6.6	6.0
Chemical	9.4	12.0	9.4	9.2
Non-Metallic	28.7	29.1	31.1	26.2
Basic metal	4.1	4.7	4.6	5.7
Fabricated metal	3.6	6.2	6.2	6.6
Others	7.4	7.2	5.8	6.7
Total	100.0	100.0	100.0	100.0
Value-added (%)				
Food and beverages	25.3	22.5	20.1	21.0

Manufacturing	1990-1997	1997-2007	2007-2010	2010-2014
Textiles	18.9	14.7	11.3	9.2
Wood and furnituer	2.1	1.1	1.0	1.1
Paper	3.0	3.3	2.8	2.6
Chemical	13.7	18.0	16.3	16.7
Non-Metallic	5.8	4.9	4.3	4.4
Basic metal	1.7	2.4	2.1	1.8
Fabricated metal	16.4	20.8	28.1	26.5
Others	13.1	12.5	14.1	16.6
Total	100.0	100.0	100.0	100.0
Energy intensity (ktoe/billion Baht)				
Food and beverages	14.8	14.6	15.9	17.4
Textiles	4.6	4.3	3.5	3.4
Wood and furnituer	4.1	8.1	9.0	10.9
Paper	14.8	14.1	23.7	23.5
Chemical	7.6	7.3	5.8	5.7
Non-Metallic	54.6	65.6	74.1	61.1
Basic metal	26.6	21.9	22.6	32.0
Fabricated metal	2.4	3.2	2.2	2.6
Others	6.2	6.3	4.1	4.1
Total	11.0	11.0	10.1	10.3

Source : Department of Alternative Energy Development and Efficiency [10] and Office of the National Economic and Social Development Board, 2014 [1]

Table 3. Change in energy consumption in Thailand (all economic sectors), Period 1990-2014.

Period	Total Effect	Structural Effect	Intensity Effect	Activity effect
1990-1997	1.40	1.07	0.99	1.33
1997-2007	1.91	1.07	0.97	1.83
2007-2010	2.12	0.97	0.91	2.41
2010-2014	3.46	1.45	0.90	2.67

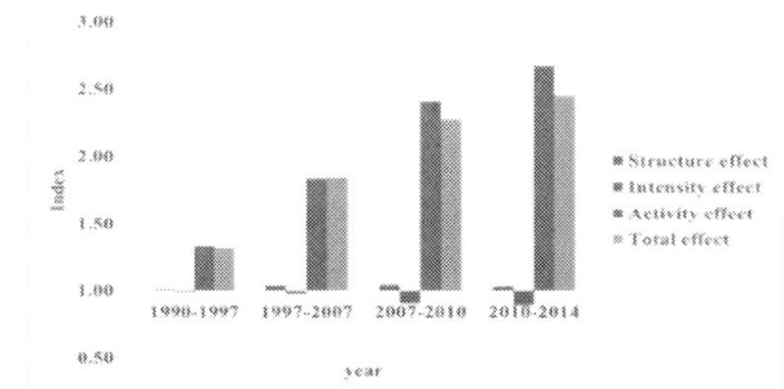

Figure 2. Results of decomposition in Thai economic sectors, Period: 1990-2014.

Table 4 and Fig. 3 show that six percent increase in the intensity effect in manufacturing sector due to the changes of energy intensity from non-metallic and basic metal sub-sector respectively. The effect of the increases of the energy intensity of individual sectors was greater than the structural change effect. The structural change did not lead to a large decline of the energy intensity in the manufacturing sector. The energy intensity increased mainly from energy intensive industries.

As a result, the energy comsumption had significantly change in manufacturing sector during the 2010-2014, which consisit of structural and intensity change by non-metallic sub-sector (cement products) and the most energy consumption due to activity increase from food and beverage sub-sector.

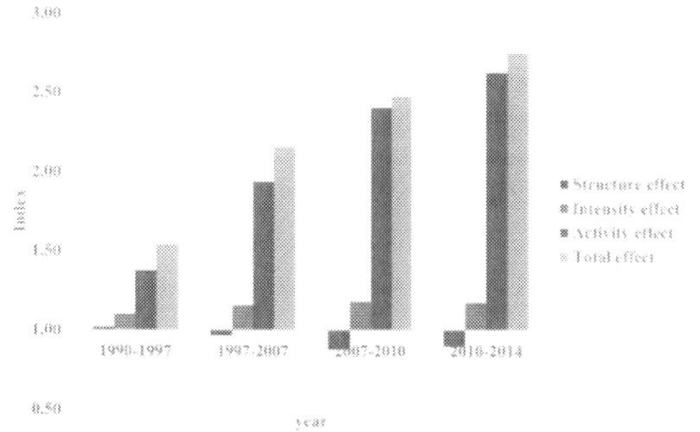

Figure 3. Results of decomposition in manufacturing sector , Period: 1990-2014.

Table 4. Analysis of change in energy consumption of manufacturing sector, Period 1990-2014.

Period	Total Effect	Structural Effect	Intensity Effect	Activitye ffect
1990-1997	1.54	1.02	1.10	1.38
1997-2007	2.16	0.96	1.16	1.94
2007-2010	2.48	0.87	1.18	2.41
2010-2014	2.75	0.89	1.17	2.63

4 Conclusions

The energy and economic significant impact, the twentyfive-year period of this study was divided into four period; i.e. (i) The first sub-period is from 1990 to 1997 which presents the high economic progress of the country before National Energy Conservation law and the time before the 1997 economic crises, (ii) The second, from 1997 to 2007, during the effect of the law and economic slowdown after the crises. (iii) The third, from 2007 to 2010, transition period of the law from first to second version and the global economic slowdown, and (iv) The last from 2010 to 2014, during the effect of the law (No.2) and the political crisis in the Thai economy. As a result in all economic sector, the energy use is the most change in the third phase, increasing the share of GDP by the transport and manufacturing sector, contribute greatly to the change of power. In addition, the structural changes impact to changing energy use more than intensity effect due to manufacturing sector is a key part. For manufacturing sector we found that the most effective intensity effect is the third phase due to the effect of the implementation of new energy efficient equipment from the second phase by enforcement of the law, while the first phase illustrates the lowest intensity effect due to the energy conservation law had not been occurred. However, due to the highest economic growth of the country and change from agricultural to industrial development direction, the first phase presents the most effective structural effect, then this effect continuously decreased by time. We also conclude that the energy conservation law have direct effect to energy efficiency of the country though the main industry more difficult to increase energy efficiency however, strictly individual regulation which have target to enforce to energy intensive industries is still required for sustainable energy efficiency improvement.

References

1. NESDB, QGDP-CVM. (2014)
2. W. Wang, X. Liu, M. Zhang, X. Song, LMDI Anal. CHN Energ. Con. Energy 67, 617–622 (2014)
3. K. Du, B. Lin, Comp. Dec. Energy **90**, 570-577 (2015)
4. IEA, Annual. Energ. DMI. IEA. (2013)
5. B.W. Ang, H. Wang, Dec. Anal. Energ Econ. **51**, 67-76 (2015)
6. E. Duran, C. Aravena, R. Aguilar, Anal. Dec. Energ. CL. Ind. Energ Policy. **86**, 552–561 (2015)
7. S. Rue du Can, A. Hasanbeigi, J. Sathaye, Anal. Dec. CA. Energ Policy. **46**, 234–245 (2012)
8. J.W. Sun, Dec. model. Energy Econ. **20**, 85–100 (1998)
9. J. Chontanawat, P. Wiboonchutikula, A. Buddhivanich, Dec. Anal. Energ. Int. Manuf Ind. THA. Energy **77**, 171–182 (2014)
10. DEDE, THA. Energ. Stat. (2014)
11. K. Punyong, J. Taweekun, P. Prasertsan, Eval. Energ. THA. Ind. 3-D DM. (SEE, 2006)

An Integrated Method for Sustainable Manufacturing Systems Design

Reda Nujoom, Qian Wang and Nick Bennett

School of Engineering, Faculty of Technology, University of Portsmouth, Portsmouth, UK

Abstract. In the past decade, there has been an increasing awareness in development of sustainable manufacturing systems as governments in many countries have been enforcing ever-stricter environmental policies and regulations in industry by promoting energy saving and low emissions manufacturing activities. Lean manufacturing can be helpful for achieving a sustainable manufacturing system as it can reduce production wastes and increase manufacturing efficiency. Nevertheless, this lean approach does not include a consideration in energy consumption and carbon dioxide (CO_2) emissions when designing a lean manufacturing system. This paper presents a methodology which can be useful for measuring energy consumption and CO_2 emissions for a typical manufacturing system design at an early stage. A case study was carried out for obtaining computational results using the developed methodology based on data collected from a real production line.

1 Introduction

To develop a sustainable manufacturing system, system designers need not merely to apply traditional methods to improve system efficiency and productivity but also to examine the environmental impact on the developed system. This is because governments in many countries have been enforcing ever-stricter environmental policies and regulations by promoting energy saving and low-emission manufacturing activities in industry. A manufacturing system can be defined as a process or system of transforming raw materials, components or parts into final products that meet customer's needs and specifications [1]. The traditional manufacturing system design is involved in determination and analysis of such as material-handling methods, system capacities, production methods, material flow, shop-floor layouts, system flexibilities and operations. Nevertheless, there is an environmental consideration that needs also to be addressed today; this leads to a new challenge for manufacturing system designers to develop an effective approach by incorporating environmental parameters or constraints [2].

The concept of manufacturing sustainability may be defined as the creation of manufactured products by minimizing negative environmental impacts on usage of energy or natural resources. The environmental issues in manufacturing activities focus on energy consumption and CO_2 emissions. Koc and Kaplan presented an investigation on energy consumption for a particular ring type yarn manufacturing system [3]. Lind *et al.* developed a production simulation tool that was used for making a trade-off decision based on considerations of ergonomic constraints, levels of automation and environmental impacts [1]. Wang *et al.* proposed the process integration (PI) method that was used for evaluating CO_2 emissions for a steel industry [4]. Gutowski *et al.* conducted a thermodynamic analysis by examining the resources used in manufacturing processes [5]. Branham *et al.* used the quantitative thermodynamic analysis for quantifying energy in different categories applied into manufacturing processes or systems [6].

Lean concepts are widely adopted by many manufacturing plants as a popular model for improving system efficiency and productivity without additional investments. Lean manufacturing is a systematic approach to eliminate non-value added wastes in various forms and it enables continuous improvement [7]. These wastes are overproduction, waiting for parts to arrive, unnecessary movement of materials, the waste in processing, unnecessary inventory, excess motion and the waste of rework [8]. However, traditional lean manufacturing does not consider environmental wastes which also need to be identified as these wastes add no values on manufactured products. This paper presents an integrated method by incorporating parameters of energy consumption and CO_2 emissions into a manufacturing system design at an early stage. A case study was used for obtaining computational results based on data collected from a real production line.

2 Energy consumption and CO_2 emissions of a typical manufacturing system

In a manufacturing system, energy is used for operating machines, air conditioning systems, illumination systems and other supportive equipment such as compressors which supply compressed air to some machines [3].

Energy and CO_2 emissions are generated by either combusting fossil fuels directly or using electricity which is generated indirectly by using fossil fuels or renewable resources. To describe amounts of energy consumption and CO_2 emissions, the following notations are used:

Notations:

m: number of processes in a manufacturing system

n_i: number of machines involved in process i, where $i \in \{1, 2,...., m\}$

E_i (kWh): energy consumption for a machine involved in process i

E_i^{cond} (kWh): energy consumption of an air conditioning system

E_i^{illum} (kWh): energy consumption of an illumination system

$E_i^{air\ comp}$ (kWh): energy consumption of a compressed air needed for a machine involved in process i

TE (kWh): total energy consumption of a manufacturing system

N_i (kw): installed power for a machine involved in process i

R_i (kg/h): manufacturing rate for a machine involved in process i

τ_i (hr): operating time for a machine involved in process i

μ_i (%): efficiency for a machine involved in process i

∂_i (kg): mass of materials transferred from a machine involved in process i

G_i(kg): mass production per month ¥$_i$ (%): total waste ratio for a machine involved in process i

$\underset{\cdot}{E}_i$ (kWh): energy consumption of air conditioning per month

$\check{E}_i$ (kWh): energy consumption of illumination per month

$\zeta_i^{air\ comp}$ (kWh/m^3): energy consumption per cubic meter of a compressor

υ_i (m^3/h): compressed air used for a machine involved in process i per hour

$\rho_i^{air\ comp}$ (m^3/h): capacity of compressed air in cubic meter per hour of a compressor

$N_i^{air\ comp}$ (kWh): installed power for a compressor

e_i (kg/kWh): amount of CO_2 emissions per kWh released from a machine involved in process i

Te_i (kg/kWh): amount of CO_2 emissions per KWh released from a machine, an air conditioning system and an illumination system, which involved in process i

ω : CO_2 emission factor using different energy sources

Te (kg/kWh): total amount of CO_2 emissions released from the manufacturing system

q_i (kg): mass of materials involved in process i

2.1 Energy consumption

The energy consumption E_i for a machine involved in process i is given by

$$E_i = \tau_i \times N_i \times n_i \qquad (1)$$

The operating time τ_i for a machine involved in process i is calculated by:

$$\tau_i = \frac{q_i}{R_i \times \mu_i} \qquad (2)$$

Mass of materials q_i transferred from a machine involved in process i is obtained by:

$$q_i = \partial_i \times (1 + ¥_i) \qquad (3)$$

The energy consumption of an air conditioning system E_i^{cond} in a manufacturing system is given by:

$$E_i^{cond} = \underset{\cdot}{E}_i \times \frac{\partial_i}{G_i} \qquad (4)$$

The energy consumption of an illumination system E_i^{illum} is calculated by:

$$E_i^{illum} = \check{E}_i \times \frac{\partial_i}{G_i} \qquad (5)$$

The energy consumption of a compressed air needed for a machine involved in process i $E_i^{air\ comp}$ is calculated by:

$$E_i^{air\ comp} = \tau_i \times \zeta_i^{air\ comp} \times \upsilon_i \times n_i \qquad (6)$$

where $\zeta_i^{air\ comp}$ can be determined by:

$$\zeta_i^{air\ comp} = \frac{N_i^{air\ comp}}{\rho_i^{air\ comp}} \qquad (7)$$

The total energy consumption TE for a manufacturing system is given below:

$$TE = \sum_{i=1}^{m} (E_i + E_i^{air\ comp} + E_i^{cond} + E_i^{illum}) \qquad (8)$$

where $i \in \{1,2,...., m\}$

Hence, equation (8) will be as follows:

$$TE = \sum_{i=1}^{m} [\frac{\partial_i \times (1 + ¥_i)}{R_i \times \mu_i} \times N_i \times n_i + \tau_i \times \zeta_i^{air\ comp} \times \upsilon_i \times n_i + \underset{\cdot}{E}_i \times \frac{\partial_i}{G_i} + \check{E}_i \times \frac{\partial_i}{G_i}]$$

2.2 CO_2 emissions

The amount of CO_2 emissions e_i released from a machine involved in process i is calculated by:

$$e_i = \omega \times E_i \qquad (9)$$

where, the CO_2 emission factor ω can be defined as shown in table 1 [9].

Table 1. Amount of CO_2 emission factor per kWh using deferent energy sources.

Energy source	Emission factor ω kg/kWh
Oil as direct energy source when oil is combusted to generate thermal energy	0.5
Oil as indirect energy source to generate electricity	0.6895
Solar as indirect energy source to generate electricity	0.05

The total amount of CO_2 emissions Te can be calculated as follows [10].

$$Te = \sum_{i=1}^{m}[e_i \times q_i + 0.6895 \times (E_i^{air\ comp} + E_i^{cond} + E_i^{illum})] \quad (10)$$

where $i \in \{1, 2,...., m\}$

Hence, equation (10) will be as follows:

$$Te = \sum_{i=1}^{m}\{\omega \times \tau_i \times N_i \times n_i \times \partial_i \times (1+¥_i) + 0.6895 \times [\tau_i \times \zeta_i^{air\ comp} \times \upsilon_i \times n_i + E_{\zeta_i} \times \frac{\partial_i}{G_i} + \breve{E}_i \times \frac{\partial_i}{G_i}]\}$$

3 A case study

Figure 1 illustrates a process flow chart or precedence diagram for producing plastic and woven sacks at a company. The study was carried out for analysing the energy consumption and CO_2 emissions for a manufacturing system consisting of machines for carrying out process tasks, an air conditioning system, an illumination system and a compressor system. Air conditioning is necessitated to maintain reasonable temperatures required for operators, effective machining operations and quality of products. The production line comprises 10 different processes tasks, each process task involves a number of machines and each machine has energy, mass inputs and different specifications. Table 2 shows the symbols representing the manufacturing processes used by the factory.

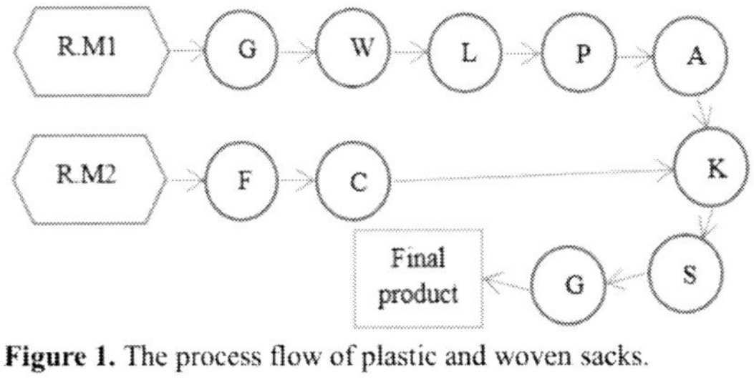

Figure 1. The process flow of plastic and woven sacks.

Table 2. Manufacturing processes tasks for producing plastic and woven sacks.

Tasks	Description	Predecessors
R.M1	Raw material (Polypropylene)	None
R.M2	Raw material (Polyethylene)	None
G	Extruding the Polypropylene to make stands	R.M1
W	Weaving the strands into rolls of sacks	G
L	Laminating the rolls	W
P	Printing and branding	L
A	Cutting the rolls into bags	P
F	blowing the polyethylene into Inner-film bags	R.M2
C	Cutting blown inner-film in to bags	F
K	Liner stick, inserts and smoothes out blown film	F,C
S	Film sewn into bag	K
B	End product compressed using baling machines	S

Table 3 shows collected data from the company. These include the mass of materials involved in process *i* per month, number of machines involved in process *i*, waste ratio, manufacturing rate and installed power.

Table 3. Data collected from a plastic and woven sacks company.

Tasks	∂_i (kg/month)	n_i	$¥_i$ (%)	R_i (kg/h)	N_i (kW)
G	254800	4	0.02	408	200
W	244608	40	0.04	392	14
L	300309	3	0.015	481	70
P	297306	5	0.01	476	20
A	291360	13	0.02	467	7
F	11050	3	0.002	18	40
C	11028	2	0.002	18	7
K	301536	13	0.003	483	0
S	298521	54	0.01	478	0.8
B	298521	4	0	478	4

4 Comparative results

Table 4 shows computational results in energy consumption and CO_2 emissions of machines involved in a process task and Figure 2 illustrates these results. Energy consumption of the machines involved in a process task of a manufacturing system depends on the installed power *Ni* and the operating time τ_i. As shown in Figure 2, the machines for completing process task *G* have the largest energy consumption and the machines for completing process task *K* have the least energy consumption. This is because the machines involved in process task *G* have the highest installed power *Ni* (200 kw) and the highest operating time τ_i (646 h), while the machines involved in process task *K* have the lowest installed power *Ni* (zero kw) as this process is a manual task.

CO_2 emissions released from the machines involved in a process task of a manufacturing system depends on the

DOI: 10.1051/matecconf/20167005005

energy consumption used by the machines, the emission factor and the energy source. Table 4 alos shows the amount of CO_2 emissions which are subject to the CO_2 emission factor ω using different energy sources and Figure 3 illustrates these results. As shown in Figure 3, the machines in completing process task *G* have the highest amount of CO_2 emissions Te_i (91244910178 kg/kWh using oil as indirect energy to generate electricity, 66167449497 kg/kWh using oil as direct thermal energy and 6616751966 kg/kWh using solar as indirect energy to generate electricity), respectively.This is because the machines involved in process task *G* have the highest amount of energy consumption. By contrast, the machines involved in process task *K* generate zero CO_2 emissions as this process is a manual task. The amount of CO_2 emissions that released from this task is released from the air conditioning and illumination system only which is 4431 kg/kWh. The results in Table 4 also show that using the solar source of energy has the lowest total of CO_2 emissions *Te* of 15679203081 kg/kWh, followed by oil as a direct energy source to generate thermal energy of 1.6×10^{11} kg/kWh and oil as indirect energy source to generate electricity of 2.2×10^{11} kg/kWh, because solar energy has the least emission factor ω (0.05 kg/kWh) followed by oil as direct energy source (0.5 kg/kWh) and oil as indirect energy source (0.6895 kg/kWh). The amount of total CO_2 emissions *Te* using oil as indirect energy source is higher than using oil as direct energy source because electricity is already generated through oil and the emission factor ω for using oil as indirect energy source is higher than the emission factor for using oil as direct energy source.

Table 4. Calculated results for energy consumption and CO_2 emissions

			Source of energy		
			Oil		Solar
			Indirect energy to generate electricity	Direct energy when oil is combusted to generate thermal energy	Indirect energy to generate electricity
Tasks	E_i (kWh)	τ_i (h)	Amount of CO_2 emissions (kg/kWh)		
			Te_i	Te_i	Te_i
G	509184	646	91244910178	66167449497	6616751308
W	363417	640	63744725119	46225327628	4622536751
L	133005	636	27953731584	20271018351	2027108300
P	63024	633	13048677512	9462422354	946249157
A	57919	625	11868414917	8606547911	860689587
F	75029	630	572799149	415374486	41542833
C	8753	625	66700011	48370533	4844068
K	0	624	4431	0	0
S	27226	623	5660055586	4104465022	410450490
B	9984	620	2055013982	1490221654	149026153
$E_i^{air\ comp}$	69309				
E_i^{cond}	41184				
E_i^{illum}	23088				
(TE)	1381125				
		(Te)	2.2×1011	1.6×1011	15679203081

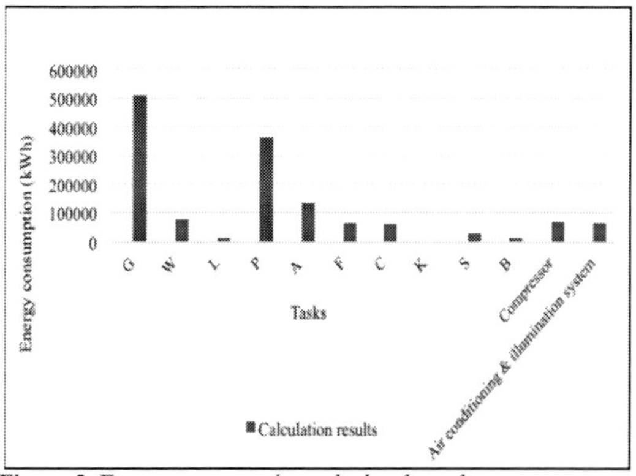

Figure 2. Energy consumption calculated results.

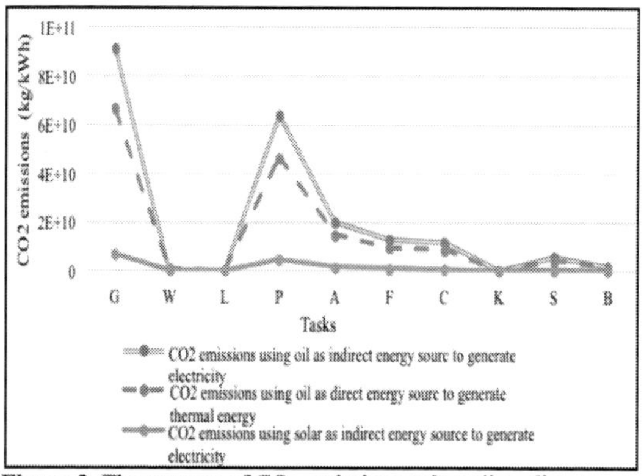

Figure 3. The amount of CO_2 emissions using oil as direct energy source and indirect energy source and solar as indirect energy source to generate electricity.

5 Conclusions and discussions

When designing a manufacturing system, engineers used to focus on key performance indicators in terms of system productivity and capacity; environmental considerations are often overlooked. This paper presents an integrated method which addresses environmental sustainability relating to manufacturing activities. The developed method was aimed at helping decision-makers to design a manufacturing system incorporating a number of environmental parameters in terms of energy consumption and CO_2 emissions. The computational results were validated based on data collected from a real case in industrial case. In addition, the developed method allows a joint analysis of the system performance using environmental constraints.

References

1. S. Lind, B. Krassi, B. Johansson, J. Viitaniemi, J. Heilala, S. Stahre, S. Vatanen, A. Fasth, C. Berlin, SIMTER: A Production Simulation Tool for Joint Assessment of Ergonomics, Level of Automation and Environmental Impacts, *In the 18th FAIM*, Stockholm, 5-49 (2008)
2. M. Paju, J. Heilala, M. Hentula, A. Heikkila, B. Johansson, S. Leong, K. Lyons, Framework and indicators for a sustainable manufacturing mapping methodology, *In the WSC, IEEE*, Baltimore, MD, 3411-3422 (2010)
3. E. Koç, E. Kaplan, An investigation on energy consumption in yarn production with special reference to ring spinning, J. Fibr & Texti in Eas. Eur, **4**, 18-24 (2007)
4. C. Wang, M. Larsson, C. Ryman, C. E. Grip, J. O. Wikström, A. Johnsson, J. Engdahl, A model on CO_2 emission reduction in integrated steelmaking by optimization methods, IJER, **32**, 1092-1106 (2008)
5. T. G. Gutowski, M. S. Branham, J. B. Dahmus, A. J. Jones, A. Thiriez, D. P. Sekulic, Thermodynamic analysis of resources used in manufacturing processes, J. Env. sci. tech **43**, 1584-1590 (2009)
6. M. Branham, T. G. Gutowski, A. Jones, D. P. Sekulic, A thermodynamic framework for analysing and improving manufacturing processes, *In ISEE. IEEE*, San Francisco, CA, 1-6 (2008)
7. A. Susilawati, J. Tan, D. Bell, M. Sarwar, Fuzzy logic based method to measure degree of lean activity in manufacturing industry, JMSY, **34**, 1-11 (2015)
8. Q. Wang, S. Lassalle, A. R. Mileham, G. W. Owen, Analysis of a linear walking worker line using a combination of computer simulation and mathematical modeling approaches, JMSY, **28**, 64-70 (2009)
9. EPA, The Lean and Environment Toolkit. U.S. Environmental Protection Agency, *http://www.epa.gov/lean/toolkit/index.htm* accessed June 26, (2008)
10. K. P. Nurjanni, M. S. Carvalho, da L. A. A. F. Costa, Green Supply Chain Design with Multi-Objective Optimization, *In the IEOM*, Bali, 7-9 (2014)

MATEC Web of Conferences 70, 05006 (2016) DOI: 10.1051/matecconf/20167005006
ICMIT 2016

The Ideal Criteria of Supplier Selection for SMEs Food Processing Industry

Rohaizan Ramlan[1], Engku Muhammad Nazri Engku Abu Bakar[2], Fatimah Mahmud[3] and Hooi Keng Ng[1]

[1]*Department of Production and Operations, Universiti Tun Hussein Onn Malaysia, Malaysia*
[2]*Department of Decision Sciences, Universiti Utara Malaysia, Malaysia*
[3]*Faculty of Industrial Management, Universiti Malaysia Pahang, Malaysia*

Abstract. Selection of good supplier is important to determine the performance and profitability of SMEs food processing industry. The lack of managerial capability on supplier selection in SMEs food processing industry affects the competitiveness of SMEs food processing industry. This research aims to determine the ideal criteria of supplier for food processing industry using Analytical Hierarchy Process (AHP). The research was carried out in a quantitative method by distributing questionnaires to 50 SMEs food processing industries. The collected data analysed using Expert Choice software to rank the supplier selection criteria. The result shows that criteria for supplier selection are ranked by cost, quality, service, delivery and management and organisation while purchase cost, audit result, defect analysis, transportation cost and fast responsiveness are the first five sub-criteria. The result of this research intends to improve managerial capabilities of SMEs food processing industry in supplier selection.

1 Introduction

Food processing industry is identified as a top priority for industrial development and it will be expanded and diversified towards making Malaysia a regional food production and distribution hub during the (IMP3) period between 2006 until 2020 [1]. The food processing industry is dominated by small and medium scale companies where sectors are the fish processing and canning, processed meats, confectionary, canned fruits and vegetables, dairy products, noodles, bread and other bakery products [2].

The increasing trend of Malaysian standard of living, lifestyle and purchasing power have resulted to increase the demand for convenience food which in turn has led to more establishments of food-processed based SMEs in the Malaysia [3]; [4]. Therefore, availability of raw materials is essential to produce food processing products [5]. Unfortunately, this is the main problem faced by SMEs food processing industry in Malaysia which is raw materials are of substandard grades because of inconsistency in supply and availability of suppliers [6].

Supplier plays an important role in the food processing industry that provides the machineries, equipment, after-sales maintenance services and especially the raw materials [4]. Ting and Cho [7] stated that the selecting ideal supplier made by purchasing department of food processing industry has a direct effect on cost reduction, flexibility and profitability of company.

The objective of supplier selection is to identify suppliers with the highest potential for meeting a firm's needs consistently [8]. Over the last decade the research community has extensively studied the problem of evaluating and selecting suppliers by adopting numerous multi-criteria decision support tools such as Analytic Hierarchy Process (AHP) model [9]; [10]. According to Ahmad and Pirzada [11], the application of AHP model is widely adopted decision supported technique. Therefore, this research aims using Analytical Hierarchy Process (AHP) model to determine ideal criteria of supplier selection in SMEs food processing industry.

Selecting the ideal supplier is a difficult mission for purchasing department of SMEs food processing industry due to suppliers have varied strengths and weaknesses that require careful assessment before making decision [12]. Ren and Song [13] stated that the SMEs food processing industry are always facing supplier selection problem.

However, SMEs food processing industries in Malaysia are paying too much attention on price offered by supplier lead to incorrect supplier selection [12]. Moreover, SMEs food processing industry in Malaysia are facing plenty of obstacles to stay as competitive industries in market such as lack of managerial capability [14] and a slower adopter of management practices at various levels for quality and operational improvement [15] that lead to incorrect supplier selection. Therefore this research aims to identify ideal supplier criteria for SMEs food processing industry using AHP model.

2 Literature review

2.1 Malaysia SMEs food processing industry

Food processing industry is one of the important sub-industries in the development of the manufacturing sector in the Malaysia and it has been identified as one of twelve prioritized industries in the Second Industrial Master Plan 2 (IMP2) and Third Industrial Master Plan 3 (IMP3) [1]. In the Industrial Malaysian Plan 2006-2020 (IMP3) period, the food processing industry's investment target have been set at RM24.6 billion (Ministry of International Trade and Industry, 2012, as cited in Ayupp and Tudin [16]). The following sub-sectors in Malaysia are the most important in the overall food processing industry, which are fish processing and canning; processed meats; confectionary; canned fruits and vegetables; dairy products; noodles, bread and other bakery products and processed meat [5].

Malaysian food industry is dominated by small and medium scale firms [17] that represent more than 80 per cent of the total number of establishments in the processed food segment [18]. According to Alam et al. [3], the numbers of SMEs food processing industry in Malaysia increased due to processed food products have become the choice of many Malaysians.

2.2 Supplier selection in SMEs food processing industry

Supplier plays a vital role in the food processing industry that provides the machineries, equipment, after-sales maintenance services and especially the raw materials [4]. Supplier selection process requires a correct guideline or adequate dimension of supplier's criteria to determine a suitable supplier [19].

The previous studies done by researchers conclude that five criteria with sub-criteria are used to evaluate food supplier. These criteria and sub-criteria are service (fast responsiveness, supply capacity and follow up), management and organisation (good reputation among industry, geographical location and financial status), cost criteria (purchase cost, transportation cost and operational cost), delivery (delivery reliability, delivery lead time and fill rate) and quality (audit result, defects analysis and documentation) ([20-23]; [10]; [24]). Thus, these five criteria and sub-criteria (Figure 1) are selected in this research to determine the ideal supplier of SMEs food processing industry.

2.3 AHP in supplier selection

The AHP was introduced by Saaty [25]. The AHP is one of the most well-known multi-criteria decision making (MCDM) methods to assist in the complex task of reaching the best decision from a set of possible criteria [26].

According to Kahraman et al., [8], selection is a broad comparison of suppliers using a common set of criteria and measures. These criteria include such as service (fast responsiveness, follow up and supply capacity), management and organization (good reputation among industry, geographical location and financial status), cost criteria (purchase price, transportation cost and operational cost), delivery (delivery reliability, fill rate and delivery lead time) and quality (audit result, documentation and defect analysis) ([20]; [21]; [22]; [23]). There are numerous studies done which summarizes the application of Analytic Hierarchy Process (AHP) model in decision making problems of supplier selection by Xia & Wu [20]; Vijayvagy [21]; Shen, Liu, & Tzeng [22]; Anyaeche & Abegunde [23]; Reza, Essmaeel, Hossein, & Mohammad [24]; Ramlan & Lee [10].

3 Methodology

This study adopts a quantitative research approach, with questionnaires distributed to 50 respondents of SMEs food processing companies in Batu Pahat, Johor, Malaysia that has authority to make decision on supplier selection. The data is analysed using Expert Choice Software. In this study, the AHP implementation step was simplified by using the Expert Choice software. The following are steps for AHP method.

Step 1: Develop the matrix of criteria.

The problem is decomposed into a hierarchy of goal, criteria, sub-criteria and alternatives as shown in figure 1.

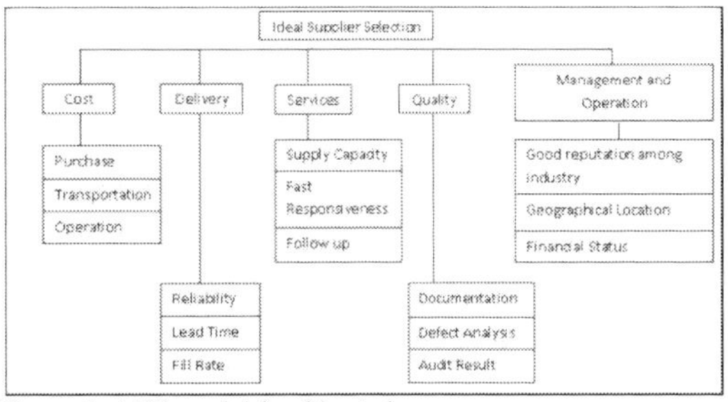

Figure 1. AHP model for this study.

Step 2: Make a pairwise comparison of alternatives on a qualitative scale.

Table 1. AHP pairwise scale [27].

Intensity	Definition	Explanation
1	Equal importance of criteria	Two criteria contribute equally to the property
2	Weak	
3	Moderate importance of one criterion over another	Experience and judgement slightly favor one criterion over another
4	Moderate plus	
5	Essential or strong importance of one criterion over another	Experience and judgement strongly favor one criterion over another
6	Strong plus	
7	Very strong or demonstrated importance of one criterion over another	A criterion is very strongly favored, and its dominance is demonstrated in practice

8	Very, very strong	The evidence favoring one criterion over another is of the highest possible order of affirmation
9	Extreme importance of one criterion over another	
Reciprocals	If activity A has one of the preceding numbers assigned to it when compared with activity B, then B has the reciprocal value when compared with A	

Step 3: The pairwise comparisons of various criteria generated are organized into a square matrix.

Let C = $\{Cj \mid j = 1, 2, \cdots, n\}$ be the set of criteria. Equation (1) is the pairwise comparison shown by a square and reciprocal matrix.

$$A = a_{ij} = \begin{pmatrix} a_{11} & \cdots & a_{1n} \\ \vdots & \ddots & \vdots \\ a_{n1} & \cdots & a_{nn} \end{pmatrix} \tag{1}$$

Step 4: The principal eigenvalue and the corresponding normalized right eigenvector of the comparing matrix give the relative importance of the various criteria being compared. The elements of the normalized eigenvector are termed weights with respect to the criteria or sub-criteria and ratings with respect to the alternatives. Equation (2) showed formula of each matrix that needs to be normalized.

$$A_w = \lambda_{\max} \cdot W \tag{2}$$

Saaty [28] demonstrated that $\lambda_{\max} = n$ is a necessary and sufficient condition for consistency. Inconsistency may arise when $\lambda_{\max}$ deviates from n due to varying responses in the pairwise comparisons. Therefore, the matrix A should be tested for consistency by using the equation of (3) and (4).

$$CI = \frac{(\lambda_{\max} - n)}{(n-1)} \tag{3}$$

$$CR = \frac{CI}{RI} \tag{4}$$

4 Results and discussion

Total population of 50 SMEs Food Processing Industry consists of 30 Small enterprises and 20 Medium enterprises. All the questionnaires distributed by e-mail, phone calls, and by hand and all 50 questionnaires returned. However, there are three respondents are not met consistency ratio less than 0.1. Therefore, only 47 questionnaires are reliable, trustworthy and proceed to analysed.

The ranking of main criteria and sub-criteria shown in Table 2 and Table 3. From the perspective of SMEs respondents, the most important main criteria are cost (0.383), followed by quality (0.326), service (0.153), delivery (0.095) and management and organisation, (0.043) while purchase cost, audit result, defect analysis, transportation cost and fast responsiveness were the first five major sub-criteria with a global weight of 0.201, 0.145, 0.139, 0.128 and 0.082 respectively.

Table 2. The ranking of main criteria.

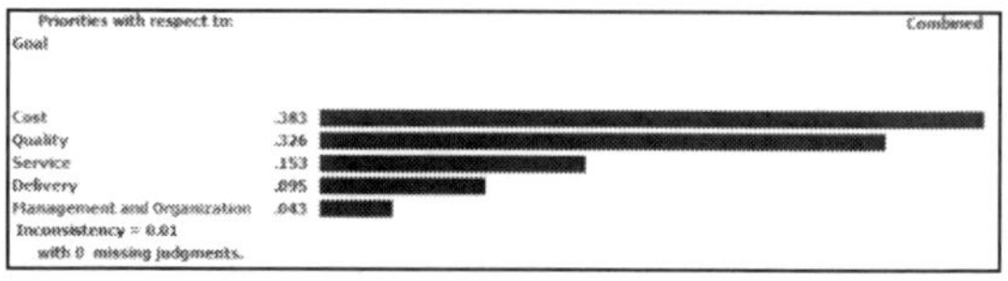

Table 3. The ranking of sub-criteria.

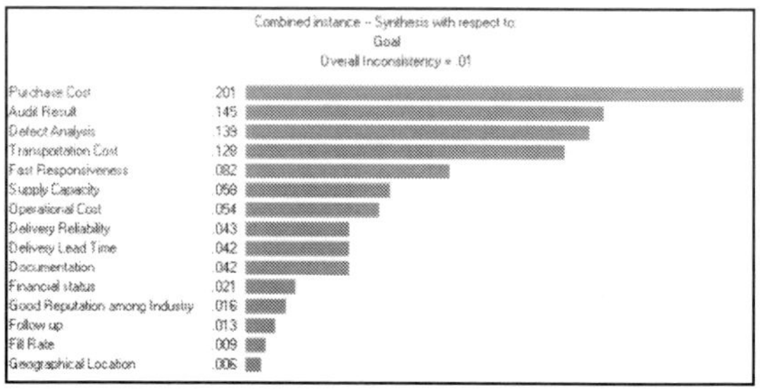

Based on criteria ranking in Table 4, obviously shows that SMEs strongly agreed that cost was the most priority criteria. This finding was similar with the research done by Choi & Hartley [28]; Chan & Niraj [29]; Magdalena [30] indicated that cost criteria is the most emphasis criteria that purchasing team will consider when they selecting supplier. Next, quality and service were ranked as second and third priority criteria. SMEs strongly agreed that quality of raw material supplied is an essential to make sure there are no any defect raw materials item that will be interrupt during the food production. Chan & Niraj [29]; Ware et al. [31], indicated that quality provided by supplier has directly effect on production process and as well as end of product, therefore provided quality must be assessed carefully.

According to the SMEs perspective, service was the third priority criteria that they will consider when selecting supplier even they claimed that most of the time service provided by suppliers almost similar and they did not concern it too much. Sim et al. [32], Sagar & Singh [19] summarized that company emphasizes service provided even most of the service provided by suppliers is similar with each other but supplier who able provides better customer services and maintain a good relationship in long term is more preferred. However, delivery and management and organization were fourth and fifth priority criteria. SMEs agreed that delivery and management and organization should be the last two considerations for supplier selection as these two criteria were would not be much emphasized by them.

The first five sub-criteria were purchase cost (0.201), audit result (0.145), defect analysis (0.139), transportation cost (0.128) and fast responsiveness (0.082). SMEs agreed that purchase cost is the most priority sub-criteria when selecting supplier as they have to choose a lower cost given by supplier in order increased quantity of food product manufactured. Transportation cost was ranked at fourth of cost criteria by SMEs as they have consider to minimize the expenses involved in moving ordered raw materials from location of supplier to company.

Moreover, SMEs agreed that audit result and defect analysis were important as they must review audit results of raw material supplied to determine the appropriate food safety practices at food processing industry and using defect analysis to determine the potential safety risks in food products and seek to reduce or eliminate them. Lastly, fast responsiveness was the fifth important sub-criteria that ranked by SMEs even there was different point of view from medium enterprises respondents which they ranked operational cost more important.

Table 4. Ranking value of supplier selection.

Criteria (Rank)	Sub Criteria	Weight	Global Weight (Rank)
Cost, 0.383 (1)	Purchase Cost	0.524	0.201 (1)
	Transportation Cost	0.335	0.128 (4)
	Operational Cost	0.142	0.054 (7)
Quality, 0.326 (2)	Audit Result	0.445	0.145 (2)
	Defect Analysis	0.427	0.139 (3)
	Documentation	0.128	0.041 (10)
Service, 0.153 (3)	Supply Capacity	0.376	0.058 (6)
	Fast Responsiveness	0.537	0.082 (5)
	Follow Up	0.087	0.013 (13)
Delivery, 0.095 (4)	Delivery Reliability	0.457	0.0043 (8)
	Delivery Lead Time	0.448	0.042 (9)
	Fill Rate	0.096	0.009 (14)
Management and Organization, 0.043 (5)	Financial Status	0.489	0.021 (11)
	Good Reputation among Industry	0.375	0.016 (12)
	Geographical Location	0.136	0.006 (15)

5 Conclusions

This research provided useful information and reference for the person who involved in process of selecting supplier or has the authority in supplier selection. The objective of this research is to determine the ideal criteria of supplier selection in SME food processing industry using AHP. A total of 47 collected questionnaires have been analysed by using Expert Choice Software. The results shows SMEs agreed that cost (0.383), quality (0.326), service (0.153), delivery (0.095) and management and organisation (0.043) are important. Meanwhile, purchase cost (0.201), audit result (0.145), defect analysis (0.139), transportation cost (0.128) and fast responsiveness (0.082) are the first five sub-criteria ranking

The study cannot be generalised to the entire SMEs food processing companies in Malaysia since a survey was conducted among respondents in Batu Pahat, Johor areas only. Future study should be conducted in a representative sample of the SMEs food processing companies in Malaysia for a better understanding of ideal supplier selection.

Acknowledgments

This research is supported by Ministry of Higher Education of Malaysia (MOHE), under Exploratory Research Grant Scheme (ERGS) to the Universiti Tun Hussein Onn Malaysia. Grant number 045/phase I 2013.

References

1. C. M. Z. Che Omar, & T. Anas, European Journal of Business and Social Science, **3**(5), 76–87 (2014)
2. Global Agricultural Information Network. USDA Foreign Agricultural Service (2014). Retrieved May 24, 2015, from http://agriexchange.apeda.gov.in/marketreport/Reports/Food_Processing_Ingredients_Kuala_Lumpur_Malaysia_11-10-2014.pdf
3. S. S. Alam, M. F. M. Jani, Z. C. Senik, & A. K. A. Domil, International Business Research, **4**(1), 252 (2011)
4. N. R. N. M. Masdek, & M. F. Othman, Economic and Technology Management Review, **9**, 19-28 (2014)
5. M. K. Ghani, *APO Multi-Country Study Mission*,1-145 (2004)
6. Mulyaningrum, B. Edna, & J. Jabar, *REACH UTEM conference*, 2007 1-7 (2007)
7. S. C. Ting, & D. I. Cho, Logistics Information Management, **16**, 395–400 (2003)
8. C. Kahraman, U. Cebeci, & Z. Ulukan, Logistics Information Management, **16**(6), 382-394 (2003).
9. O. Larichev and D. Olson, Springer Science & Business Media (2001)
10. R. Ramlan, & W. K. Lee, *3rd International Conference on Global Optimization and Its Application*. pp. 150-160, (2014)
11. Y. Ahmad, & D. S. Pirzada, SAGE Open, **4**(4) (2014)
12. F. Tahriri, M. R. Osman, A. Ali, , R. Mohd Yusuff, & A. Esfandiary, Journal of Industrial Engineering and Management, **1**(2), 54–76 (2008)
13. W. Ren, & N. Song, *Second Pacific-Asia Conference on Web Mining and Web-based Application*, 249-252 (2009)
14. M. Khalique, A. H. Md Isa, J. A. N. Shaari, & A. Ageel,. International Journal of Current Research, **3**(6), 398–401 (2011)
15. S. S. Omar, R. Ramlan, K. Ahmad, & S. W. Chan, Advanced Science Letters, **21**(5), 1089-1093 (2015).
16. K. Ayupp, & R. Tudin,. International Journal of Business and Social Science, **4**(16) (2013)
17. Malaysian Investment Development Authority. *Industries in Malaysia* (2013). Retrieved on March 23, 2013 from http://www.mida.gov.my/home/food-technology-and-sustainable-resources/posts/
18. Malaysian-German Chamber of Commerce & Industry, Market Watch, *Market Watch Malaysia 2010 Food Industry* (2010). Retrieved April 5, 2015, from http://malaysia.ahk.de/fileadmin/ahk_malaysia/Dokumente/Sektorreports/Market_Watch_2010/Food_2010__ENG_.pdf

19. M. K. Sagar, & D. Singh, International Journal of Engineering Research and Development, **4**, 34–39 (2012)
20. W., Xia, & Z., Wu, Omega. **35**(5), 494-504 (2007)
21. L. Vijayvagy, International Journal of Management & Innovation, **4**(2) (2012)
22. J. L. Shen, Y. M. Liu, & Y. L. Tzeng, Journal of Advanced Computational Intelligence and Intelligent Informatics, **16**(5), 567–575 (2012)
23. C. O. Anyaeche, & A. S. Abegunde, IOSR Journal of Mechanical and Civil Engineering, **9**(1), 10–15 (2013)
24. A. Reza, R. Essmaeel, V. Hossein, & H. P. Mohammad, International Journal of Current Life Sciences, **4**(5), 1984-1988 (2014)
25. T. L. Saaty, Management science, 26(7), 641-658 (1980)
26. T. L. Saaty, *RWS Publication 2 nd ed* (pp. 73) (2006)
27. T. L. Saaty, Mathematics and computers in simulation, 20(3), 147-158 (1978)
28. T. Y. Choi, & J. L. Hartley, Journal of Operations Management, **14**(4), 333–343 (1996)
29. F.T.S. Chan, & K. Niraj, Omega, **35**, 417-431 (2007).
30. R. Magdalena, Asian J Technol Manage, **5**(1), 13-22(2012)
31. N. Ware, S. Sing, & D. Banwet, Management Science Letters, **2**(5), 1465-1490 (2012)
32. H. K. Sim, M. K. Omar, W. C. Chee & N. T., *11th* Asia Pacific Industrial Engineering and Management Systems Conference (2010)

Modifying Ways of Modern Local Furniture Design to The Ming-style Circle Chairs

Xue Ruizhu[1], Zhao Hongsheng[2]

[1] Taiyuan University of technology, Taiyuan 030024, China
[2] Taiyuan University of technology, Taiyuan 030024, China

Abstract. In recent years, the scales and works for local Chinese original furniture designs have been rapidly increasing. This article elaborated the inner requirement for modern families to local furniture designs. And by comparing circle chairs of "Moreless", "Fnji", "pusu","Lost&Found"and "LEXTO" with those made in Ming Dynasty , deeply analysis the modifying ways of modern local furniture design to the circle chairs with Ming style, such as tenon structure, matierals and so on. It aimed at increasing peoples' understanding for local furniture designs, improving peoples' attention for local original furniture design. Finally, this article supplied reference for the development of the local furniture design.

1 Introduction

Now more and more Chinese furniture designers escape the dominant ideology of Western design, using traditional heritage combined with the trend of Western ways instead, creating boutique furniture design. Ming-style furniture with its concise shape, well-structured, fine workmanship, decorated properly, simplified affordable, hard wood, texture and elegant temperament characteristics and vulgarity known to the world [1]. In the history of all kinds of furniture in the world, Only in China beautiful round-backed armchair can be found [2], it presents a unique oriental feeling, full of traditional cultural qualities of the Chinese nation. Foreign furniture design industry is projected to look for reference Ming-style chair and further development, such as Hans • Wagner's "Chinese chair" series, the Ming-style furniture, simplified strips, light and beautiful shape is formed [3]. Domestic independent design has made considerable progress in recent years, "Moreless", "Fnji", "pusu", "Lost&Found"and "LEXTO" and other original brand in China to the Ming-style chair inspired design variety of modern chair, this paper analyze the shape, structure, pattern, materials, we hope these native design ideas generates inspiration for the future of furniture design.

1.1 Contemporary local furniture design

Contemporary furniture design is under local traditional aesthetic norms, applying modern materials and technology, showing a new interpretation of the classic essence of traditional Chinese culture, making furniture not only has an elegant, dignified Chinese flavor, but also has obvious modern features [4-5]. Contemporary designers through the transformation of traditional furniture, exhibits contemporary furniture design unique characteristics of the times, in which the expression of the Chinese-style, elegant and subtle, dignity, showing the Eastern spiritual pursuit, advocating a Chinese contemporary furniture design way "make the past serve the present".

1.2 The evolution of Ming furniture

Ming Dynasty furniture, especially from the Jiajing, Wanli Ming Dynasty to the Qing Dynasty Emperor Kangxi, Yongzheng (1522--1735 years), the products both from the quantitative point of view, or from the artistic value point of view is perfect, this period of two hundred years called traditional furniture Golden Age is well-deserved[6]. Ming-style have concise modeling, ergonomic structure of science, tenon joint structure, and on the aspect of materials selection and matching, attending to the wood itself natural texture and color, which bring the Ming-style furniture into a quiet and elegant literati aesthetic connotation. after a long history, Ming-style furniture increase more charm.

Modern Transformation of the Ming-style furniture to meet the people's pursuit to peaceful and elegant mood and simple lifestyle. Soothing lines, smooth clear wood texture, artistic effect, meet the modern sense of belonging, a sense of belonging to find psychological symbols, to meet the people's need for furniture humanistic value. Local furniture and Scandinavian furniture, Japanese furniture in their own development process, resulting in a lot of thinking and the pursuit in common: respect for nature and the nice feeling from natural materials, pursue the harmony between man and nature; easy to use, avoiding affectation decoration, balance and moderation; humility, leisurely and elegant life attitude, so that the modernization of Ming furniture can better integrate modern home style, matching with modern furniture to create a favorable living environment.

1.3 Contemporary domestic furniture design for the transformation of the Ming-style circle chair

Ming-style circle chair have simple elements in general, on the seat pan is round wood, under the seat pan is square in cylindrical, simple rods in three direction, the back plate with a relief, a small angle between the rods and the goose neck, this can be regarded as a fundamental form of circle chair (refer with: Figure 1 Ming-style chair). Because of the elegant shape and unique, Ming-style circle chair has become a major reference for contemporary design.

Although the design of contemporary Native develop their own characteristics, but all vaguely embodies the spirit of Chinese ancient furniture. The following contrasts the "Moreless", "Fnji", "pusu", "Lost&Found" and "LEXTO" with Ming-style circle chair, deeply analyze the transformation method of contemporary indigenous design on the Ming-style circle chair.

Ming-style chair

more or less-autumn chairs

more or less- rare words chair

more or less-together chair

Pu Su-Yunyao chair

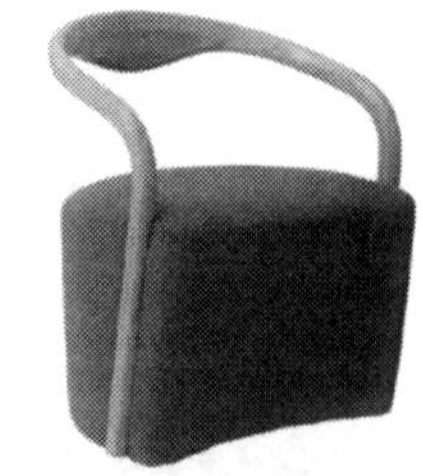

LEXTO -wood armchair

fnji-back chair

Lost&Found –qiji chair

Figure 1. modern local circle chairs

1.4 Simplified Shapes

Different forms of modern circle chair, brought changes in use-pattern and mental function, meanwhile, as a map of the level of social development and people's social life. Modern circle chair no longer need to reflect this symbolic hierarchy, which makes the Ming-style chair from the solemn image increasingly developing in the direction of ease and rich flavor of life [7]. Contemporary local design reinforces the "concise shape "features of Ming Dynasty chair, so simplified design is the main method of modernize Ming-style circle chair(refer with Fig.1).

Most designs retain the coherent curve between brain and armrest, but transform it to more simple and smooth lines, or a little quadrate curve. Such as "more or less rare words chair " and " more or less-together chair ", using straight lines connected with arcs a dignified and generous character instead of the prime arc; and "LEXTO-wood armchair" use smooth rounded curves, reflecting the graceful and poetic.

There are three types to remould the brain and the back plate: the first type reserved ergonomic spine curves back plate, the second type is using the odd cylindrical root with the same curvature replaced the back plate, and the third type is simply remove backrest board, through widening the brain to support the body back instead;

All of designs removed the goose neck which is not actually supporting role, some also removed upside leg; on the transformation of seat pan, some reserve flat or transform to slightly curved surfaces, and others increase comfortable sit sense by adding cushions or rattan class material; on the transformation of foot doorpost and rods differentiated into two general renovation: remove the foot doorpost reserved rods or removed rods to increase the crossbar, while maintaining stable to make the leg more slender and light. especially " more or less-together chair " the lower part of the upper portion is almost symmetric, "

LEXTO-wood armchair " transplant sofa form in the lower part , "Pu Su-Yunyao chair" increase in the bentwood chair bottom to become a rocking chair. The legs of "Fnji" and "Lost&Found " are tapered, making it more lightweight.

Without exception, the transformation of formal ties with function together, based on the understanding of the scientific principles and the taste of potential users, to ensure the functional premise simplified the form, it retains unique characteristic curve of the brain and handrails, with simple flowing lines reflect the modern, light and overall slim, elegant and modern, carrying the culture, way of thinking and relationship with nature. It is worth mentioning that the progress and development of molding process combined with the designer's creativity, so chairs show a different shape and style." LEXTO-wood armchair" take wood armchair armrest and the back combined to form a complete structure, not only comfortable, but also has good toughness, smooth formal beauty.

On the handling of decorative patterns, decorative relief modeling traditional elements, although beautiful, but is not suitable for the needs of modern aesthetic taste, remove the decorative patterns can make furniture have much modern taste, and can save material, also avoid difficulties in cleaning.

1.5 Simplifies the Tenon and Modern Technology Combine

In order to improve production efficiency, adapt to mechanized production methods, appropriate to simplify the structure is imperative, which requires the relevant process improved. Tenon is major coupling member of traditional Ming-style circle chair, wherein the protruding part is called "sun", concave part is called "Mao" [8-9]. Traditional Ming Dynasty chair chairs circle combined with a dovetail, the back plate and the brain and seat pan connect in slot tenon, goose neck, together with the stick and armrests connected to the seat pan with a circular sleeve joints, legs connect seat pan with the length tonon.

Tenon joint structure in ancient times is a simple utility connections, but in modern it always as a decorative element to reflect the literati temperament and the traditional skills of furniture designers. More modern techniques can simplify the structure and reduce the difficulty of mechanization. Figure 1 is still a few cases of local modern armchair to analyze:

the tenon was made by machine instead of hand-made mechanical tenon. Dovetail, straight mortise and tenon and other simple internal structure can be mechanically forming. For example, the circle of "more or less- Autumn chair" use the three-mile circle straight tenon in connection, instead of the traditional wedge tenon, handrails and goose neck tenon connect by Yandai tenon, cylindrical thin teeth connect chair legs with a simple plug tenon; the circle of "Fnj- back chair " is square, so connect the brain and handrails by dovetail stitching, handrails and use dug Yandai goose neck tenon, sticks connect both sides insert with straight tenon, seat pan do not use the traditional grid corner tenon, using the three plate slot tenon splicing;

Ming-style chair modern multi-tenon joint structure combining hardware technology is. In ancient times make manufacture nails is difficult, but modern alloys such as copper, chrome and zinc alloy are more rigid and does not rust. In addition, non-hardwoods use tenon joint structure prone to cracking or loose change. For example, the seat pan and legs of " more or less-autumn chairs" connected each other by hardware, but is well hidden up;

useing bent wood technology, such as " LEXTO-wood armchair" use cantilevered armrests and backrest integrated design, modeling simple, use a wood bending technology, not with tenon.

1.6 Combination of Wood and a Variety of Materials.

Most of Ming and early Qing Period furniture was made of precious hard wood. Most of them have solid texture, elegant color, vivid and beautiful patterns [10]. Such as rosewood, Xichi wood, iron force, ebony and mahogany and so on. But these woods are not suitable for modern production.

Modern chair still using wood as the main material, the wood itself contains irreplaceable natural and simple emotions,is not only quite environmentally friendly materials if cut properly , but also play the role of moisture balance in the living room. Contemporary chairs of the local brand shown in Fig.1, where "More or Less", "Lost&Found" transformation chair used black walnut, black walnut is a tough medium-density hardwood and diameter relatively uniform , can be used for large cross-section of the plate, the heartwood is light brown to dark brown, often with a crack purple and dark stripes, string section as large parabolic pattern, wood drying , low rigidity, good steam bending performance, ease of hand tools and machining, good fixing with nails, screws and glue, and strong corrosion resistance heartwood. "Pu Su-Yunyao chair" made of beech, yellow Burmese rosewood or birch. Where beech sturdy, even texture, good steam bending, texture clear and smooth, pastel colors, but in a dry environment easily deformed. Burmese rosewood wood densityis high , grain crisp and clear, with incense and hold nails performance is good, but the scarcity of expensive and difficult to dry and easy to moth-eaten. Birch is relatively heavy and hard, light brown to reddish-brown wood, easy processing, cut smooth, ring obviously, paints and glue good performance, but its moisture absorption, it is decayable , not wearable , and fiber shear force is poor .

In the use of materials can also be mixed and matched on a rattan palm, sponges and cloth and so on. Cloth makes people feel soft, warm, Figure 1 "More or Less –rare words Chair" and "together Chair" embedded in the surface of a piece of cloth chair cushion, not only enhance the comfort but also enhanced contemporary; " LEXTO -wood armchair " is use of whole cloth seat, a simple structure, to make the shape simple and elegant. The wood and rattan palm as a complex of two natural elements makes a sense of leisurely and affinity, such as the soft seat of "Pu Su-Yunyao chair" , is all handmade rattan palm stretched.

2 Summary

The local design has became an important part of Chinese culture , and will greatly promote the development and prosperity of China's manufacturing industry in the future . By summarizing the contemporary of local furniture design of the Ming-style circle chair transformation method to provide more local furniture design ideas. As China's domestic original brand, they have a natural Chinese culture gene, transformation of the shape of the chair reflects the commonality of simplification and slim, but performed in different ways; tenon traditional Chinese art, which had reservations, but reflect some thoughts and changes under the background of mechanization; in terms of materials, they have showed the insistence of wood materials and the absorb of modern furniture materials .

References

1. HUANG Wei, Zhang Meng, Wu Jian-feng Influences Of The Ming Literati'S Design On The Ming-Style Furniture Style Formation. Packaging Engineering, 2015, **10**: 80-84.
2. CAO Yong Ming-style chair in perspective the essence of traditional Chinese culture. Forward Position, 2012, **22**: 178-180.
3. LI Wei-wei Innovative Design of Ming Style Furniture from the Perspetive of "Chinese Chair" by Hans Wegner . Packaging Engineering, 2015, **02**: 95-98.
4. HU Tian-jun, WU Ya-wei Characteristics of Contemporary Furniture Design. Design Art (Shandong College of Arts and Crafts), 2010, **05**: 11-12.
5. WANG Pei, Application Of Traditional Furniture Element In The Modern Furniture Design . Packaging Engineering, 2014, **24**: 87-90.
6. LIAO Zhi-min Scholars and collectors Ming dynasty furniture. art market, 2013, **07**: 54-58.
7. HUANG Mei-wei, XU Xue-min Study on The Evolution of The Modern Wenzhou Round-Backed Armchair By Analyzing Practical Case . Art and Design, 2013, **06**: 74-75.
8. LI Min, WU Zhi-hui, MAO Yi-chao The Types of Mortise and Tenon In Modern Mahogany Furniture . Packaging Engineering, 2015, 0**4**: 88-92.
9. WANG Di. Chinese Traditional Furniture Wood Structure and Technology Analysis Round-hacked Armchair . Northwest Forestry University, 2014, **05**: 207-213.
10. LIU Shu-zhen, SUN Jian-ping Analysis on the Material Selection and Design of the Back of Ming-style Chairs. Forest Engineering, 2014, 02.

Multi-objective Optimization of a Transportation Network of a HMSC

Ahmed Mohammed and Qian Wang

School of Engineering, Faculty of Technology, University of Portsmouth, Portsmouth, UK

Abstract. In this paper, we investigated a transportation network of a tri-layer Halal meat supply chain (HMSC) in which Halal meat transportation process was monitored by a Radio Frequency Identification (RFID) communication system to ensure safety and integrity of Halal meats. This monitoring system is subject to an extra cost in investment that needs to be taken into account. Thus, a multi-objective linear programming model (MOLPM) was developed aiming to minimize the total cost in transportation and number of transportation vehicles and maximize the service level in product quantity as requested by abattoirs and retailers. The facility location-allocation problem in farms, abattoirs and retailers needs also to be addressed in relevance to the quantity flow of products from farms to abattoirs and from abattoirs to retailers. The utility function method was employed to obtain Pareto-optimal solutions and the global criterion method was used for searching the most suitable Pareto solution by minimizing the distance to its ideal objective value. The research work shows that the developed model can be useful for supply chains design through a case study based on numerical results.

1. Introduction

A food supply chain network includes food production, distribution and consumption. Design of a supply chain network involves strategic decisions in allocating facilities that should be established and tactical decisions in terms of product flows throughout the supply chain network. To maintain a competitive share in the market, it requires an effective and efficient supply chain network by minimizing costs and maximizing profits. For Muslim people, Halal food accounts for over sixteen percent of the global food trade. Halal is an Arabic word which means lawful or permitted according to the Islamic law. Integrity of Halal meat products refers to a Halal meat production and distribution which must comply with the Islamic Shari'ah (laws). This includes processes of livestock feeding, slaughtering, packing, storing and transporting in a Halal meat supply chain (HMSC) network.

Based on the reviewed literature, few research works applied multi-objective optimization models in the context of food supply chain networks. Vanek and Sun [1] suggested an energy minimizing model that considers both the shipment and the perishable produce production. These Authors proposed to ship an additional produce quantity to guarantee the consumer demand satisfaction, due to the produce deterioration. Bosona and Gebresenbet [2] investigated the existing flow of food products from producers to consumers. To this aim, an integrated logistics network embracing producers, retailers, a collection center and a distribution center was proposed. Rong et al. [3] developed a mixed integer linear programming model for a fresh food supply chain aimed at guaranteeing the produce quality for consumers. Sahar et al. [4] proposed a multi-objective optimization model for a two-layer dairy supply chain. The model aimed at minimizing CO_2 emissions from transportation and total costs in the distribution chain.

Furthermore, different researches exist in the literature regarding the application of multi-objective optimization models into different supply chain network design and distribution problems. Revelle and Laporte [5] addressed supply chain design problems, these problems may be solved by a number of compromising solutions known as Pareto solutions. Sabri and Beamon [6] developed a multi-objective programming model to obtain an optimum supply chain network performance by considering two conflicting objectives including minimization of the total cost and maximization of volume flexibility of plants. Chan et al. [7] presented a hybrid-genetic algorithm to solve the distribution problem of a supply chain network. Three objectives were optimized including total cost, completion lead time and capacity utilization. Chen and Lee [8] developed a multi-objective model for a multi-echelon supply chain network seeking a compromise among conflicting objectives including fair profit distribution, safe inventory levels, customer service levels, and uncertain demands of products. Altiparmak et al. [9] proposed a genetic algorithm focusing on minimization of inbound and outbound distribution costs and maximization of customer services in terms of delivery time and capacity for distribution centers. Schütz et al. [10] formulated a decision support system using a two-stage stochastic program with respect to minimizing costs of investment and operations of a supply chain. Shankar et al. [11] investigated a two-echelon supply chain for locating distribution centers with a minimal cost using the genetic algorithm, which was compared with using the Lagrangian heuristic approach.

This paper presents a development of a multi-objective linear programming model for a RFID-based HMSC proposed by Mohammed and Wang [12]. The model aimed at solving the facility location-allocation problem and the quantity flow of products considering three objective functions by minimizing the total transportation cost, number of transportation vehicles and maximizing the service level in satisfying the quantity of Halal meat products requested by abattoirs and retailers.

2. The HMSC model

Figure 1 illustrates a tri-layer Halal meat supply chain (HMSC), which consists of farms, abattoirs and retailers. The objectives of this model include: (1) minimization of the total transportation cost (N_1); (2) minimization of the number of required vehicles for transportation (N_2); and (3) maximization of service level (N_3). The model is also aimed at helping design the HMSC network in numbers and locations of farms and abattoirs in response to quantity of products flow from farms to abattoirs and from abattoirs to retailers.

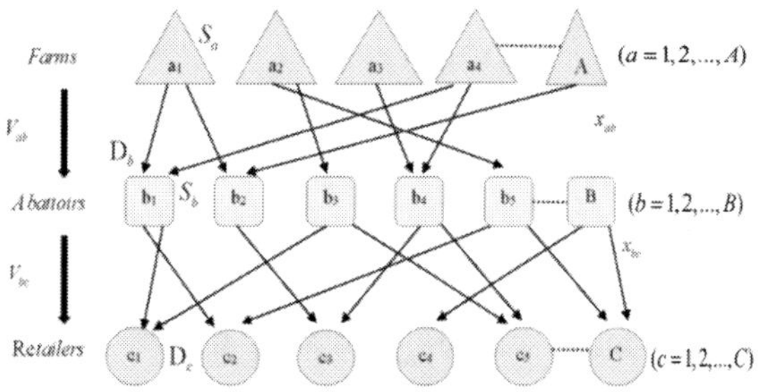

Figure 1. The tri-layer HMSC network.

The mathematical notations, parameters and decision variables are as follows:

Indices

A index of farms $(a=1,2,...,A)$

B index of abattoirs $(b=1,2,...,B)$

C index of retailers $(c=1,2,...,C)$

Parameters

R_{ab} unit transportation cost (GBP) per mile from farm a to abattoir b

R_{bc} unit transportation cost (GBP) per mile from abattoir b to retailer c

T_{ab} RFID tag cost (GBP) per item transported from farm a to abattoir b

T_{bc} RFID tag cost (GBP) per item transported from abattoir b to retailer c

SM_{ab} RFID system cost (GBP) required per lorry travelling from farm a to abattoir b

SM_{bc} RFID system cost (GBP) required per lorry travelling from abattoir b to retailer c

C transportation capacity (units) per lorry

S_a maximum supply capacity (units) of farm a

S_b maximum supply capacity (units) of abattoir b

D_b minimum demand (in units) of abattoir b

D_c minimum demand (in units) of retailer c

Variables

x_{ab} quantity of livestock transported from farm a to abattoir c

x_{bc} quantity of processed meats transported from abattoir b to retailer c

V_{ab} number of expected required vehicles to transport livestock from farm a to abattoir b

V_{bc} number of expected required vehicles to transport processed meats from abattoir b to retailer c

$$y_a = \begin{cases} 1: \text{if farm } a \text{ is opened} \\ 0: \text{otherwise} \end{cases}$$

$$y_b = \begin{cases} 1: \text{if abattoir } b \text{ is opened} \\ 0: \text{otherwise} \end{cases}$$

Thus, by minimizing the total transportation cost (N_1), number of transportation vehicle (N_2) and maximizing service level (N_3), the model formulation is given as follows:

$$Min\ N_1 = \sum_a \sum_b T_{ab} y_a + \sum_b \sum_c T_{bc} y_b + \sum_a \sum_b SM_{ab} V_{ab} + \sum_b \sum_c SM_{bc} V_{bc} \sum_a \sum_b R_{ab} x_{ab} + \sum_b \sum_c T_{bc} x_{bc} \tag{1}$$

$$Min\ N_2 = \sum_a \sum_b R_{ab} V_{ab} + \sum_b \sum_c R_{bc} V_{bc} \tag{2}$$

$$\max N_3 = \frac{\left[\left(\frac{\sum_a \sum_b x_{ab}}{\sum_b D_b} \right) y_a + \left(\frac{\sum_b \sum_c x_{bc}}{\sum_c D_c} \right) y_b \right]}{2} \tag{3}$$

s.t.

$$\sum_a x_{ab} \le S_a y_a \quad \forall b \in B \tag{4}$$

$$\sum_b x_{bc} \le S_b y_b \quad \forall c \in C \tag{5}$$

$$\sum_a x_{ab} \ge D_b \quad \forall b \in B \tag{6}$$

$$\sum_b x_{bc} \ge D_c \quad \forall c \in C \tag{7}$$

$$D_b \ge \sum_c x_{bc} \quad \forall b \in B \tag{8}$$

$$\sum_e V_{ab} = x_{ab} / C_l \quad \forall b \in B \tag{9}$$

$$\sum_{b} V_{bc} = x_{bc} / C_l \quad \forall c \in C \tag{10}$$

$$V_{ab}, V_{bc} \ integer \tag{11}$$

$$x_{ab}, x_{bc} \geq 0 \quad \forall a,b \tag{12}$$

$$y_a, y_b \in \{1,0\}, \quad \forall a,b \tag{13}$$

Equations 4 and 5 are constraints in capacity at farms and abattoirs, respectively. Equations 6, 7 and 8 determine the demands of abattoirs and retailers. Constraints 9 and 10 provide the estimated number of vehicles indicated in the objective function 2. Constraints 11, 12 and 13 prohibit non-negativity and the non-binary restrictions on decision variables.

2.1 Solution methodology

To obtain Pareto solutions, the utility function method was used and the most suitable Pareto solution was selected using the global criterion method. The utility of each Pareto solution is determined by a sum of the scaled objective functions. The scalar value λ for each objective is determined by decision makers according to the importance of each objective [13]. In this work, the objective function (or utility function) U is expressed as follows:

$$U(N_1, N_2, N_3) = \left\{ \sum_{i=1}^{3} \lambda_i N_i \,|\, \lambda_i > 1, \sum_{i=1}^{3} \lambda_i = 1 \right\} \tag{14}$$

The next step after obtaining the Pareto-optimal solutions is to find the best trade-off solution. From the decision maker's view-point, choosing a solution of Pareto-optimal solutions is called a posteriori method [14]. There are several methods for selecting the most suitable solution in a multi-objective problem. In this case, the global criterion method was used in this study for determining the best solution by minimizing the distance to the ideal objective value (N_i^*) [15], it can be given by:

$$\min \text{N} = \left(\sum_{i=1}^{3} \left| N_i - N_i^* \right|^{\rho} \right)^{1/\rho} ; \quad 1 \leq \rho \leq \infty \tag{15}$$

Figure 2 shows the block diagram of the MOLPM development.

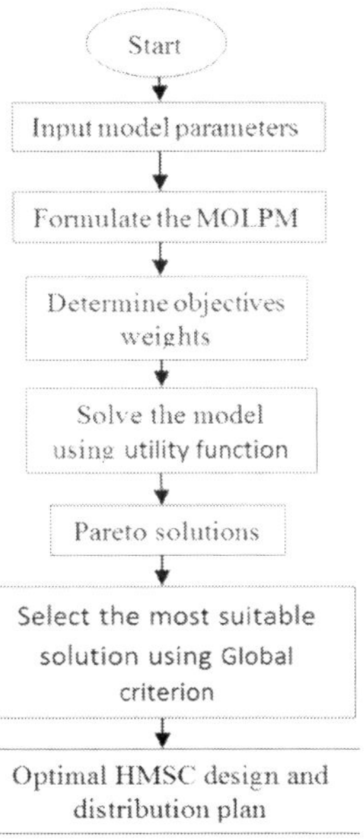

Figure 2. Block diagram of the MOLPM development.

Table 1. Collected HMC data.

A = 4	R_{bc} = 15-20	D_B = 600-1.5K
B = 5	S_a = 1.2K-2.5K	D_C = 100-200
C = 11	S_b = 1K-1.8K	SM_{bc} = 800
R_{ab} = 15-20	C_l = 20-30	T_{ab} = 0.15
T_{ab} = 0.15	SM_{ab} = 800	

3. Implementation and evaluation

A case study was used for examining the validity of the developed mathematical model. Table 1 shows data collected from the Halal Meat Committee in the UK [16]. Figure 3 illustrates the locations of four farms, five abattoirs and eleven retailers in a region. The computational solutions and their optimization were conducted using LINGO. First, each objective function (OF) was individually optimized to obtain Table 2.

The optimal boldface value for each OF was used as an ideal value for the decision making method (described in sub-section 2.1). In order to optimize the three objective functions and obtain Pareto solutions, ten different scalar values (λ_i) were assigned. Table 3 shows the obtained seven solutions of the Pareto frontier. It is noticed that the utility function method has a disadvantage as it only obtains seven rather than ten results

Table 2. Computational results with solutions provided based on each objective function to be minimized (N_1, N_2) and maximized (N_3) individually.

Objective functions	min N_1	min N_2	max N_3
OF_1	**55455**	56384	187673
OF_2	26	**25**	73
OF_3	0.899	0.895	**0.999**

when the scalar value of the first objective function was set to be less than 0.4 (solutions 8, 9 and 10). This may limit decision makers for measuring the importance of the three objectives. Table 3 also shows that the objective functions are conflicted as it is impossible to achieve optimal values for all the three objectives at a time. This conflict among the three objective functions is further illustrated in Figure 4. It is noteworthy in Table 3 that a number of facilities were eliminated. For instance, solution 3 was obtained with the weight (0.8, 0.1, 0.1), the results only gave one farm (located in Yorkshire). This farm can supply the required livestock to three abattoirs located in York, Bradford and Sheffield with respect to the demands of the eleven retailers. When designing the transportation network of a HMSC, decision makers need to determine the best solution based on a number of alternative possibilities. In this study, the global criterion method was used to select the most suitable solution. Accordingly, if ρ is set to 1 for equation 15, solution 4 is the most suitable solution as it gives the minimum distance to the ideal objective value. This solution has a transportation cost of 84,569 GBP with 48 vehicles and a service level of 93.3%. This solution requires an establishment of three farms to supply livestock to three abattoirs. Table 4 shows the quantity of products flow between farms (1, 2, 3) and abattoirs (1, 2, 4) and eleven retailers. For instance, u2, 1 implies that farm 2 is demanded to supply 800 livestock to abattoir 1, and v1, 1 implies that abattoir 1 is demanded to supply 850 packages of processed meats to retailer 1.

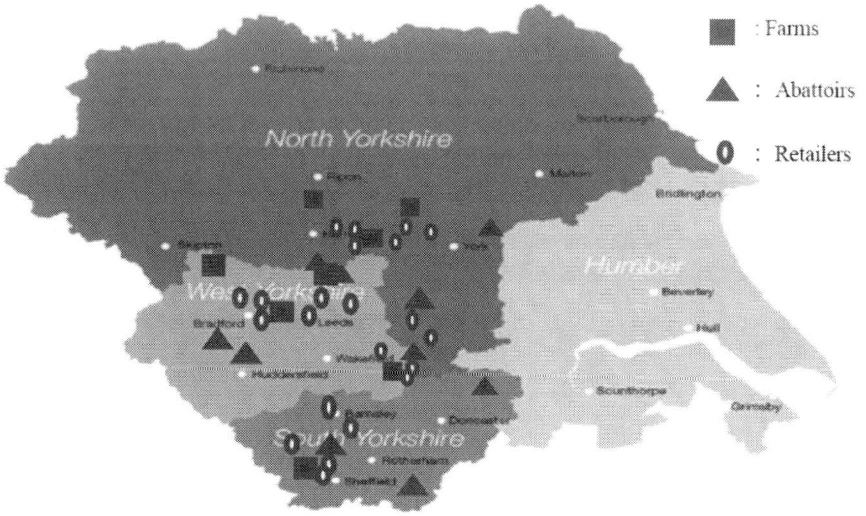

Figure 3. The HMSC network.

Table 4. The quantity of products flow of the HMSC.

	Quantity	Facilities	Quantity
$u_{1,4}$	1200	$v_{1,1}$	850
$u_{2,1}$	800	$v_{1,3}$	210
$u_{2,4}$	1200	$v_{1,6}$	690
$u_{3,1}$	1000	$v_{2,5}$	290
$u_{3,2}$	290	$v_{2,10}$	100
$v_{2,11}$	700	$v_{2,8}$	160
$v_{4,6}$	850	$v_{2,9}$	110
$v_{4,7}$	450	$v_{2,2}$	350
$v_{2,4}$	220		

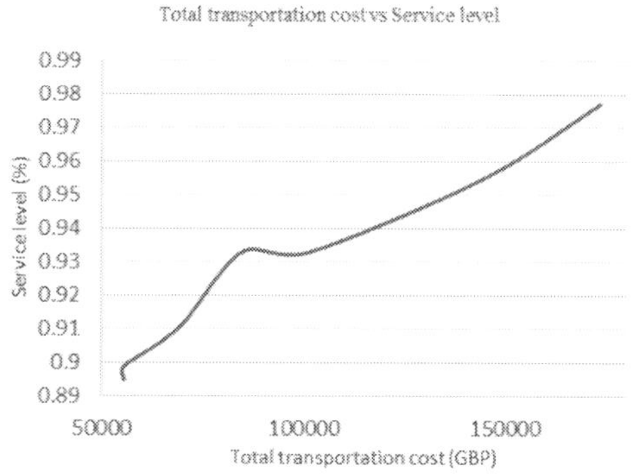

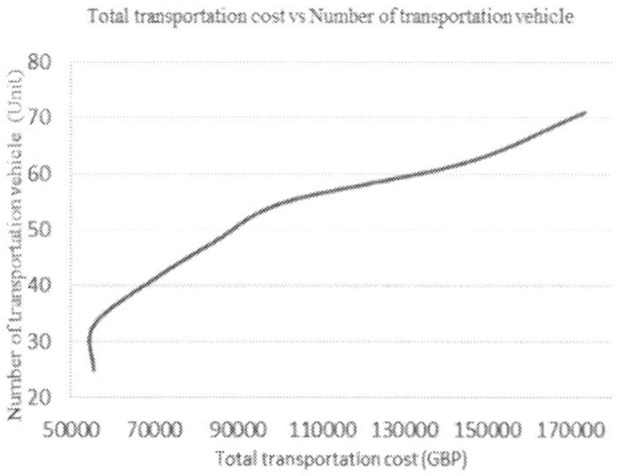

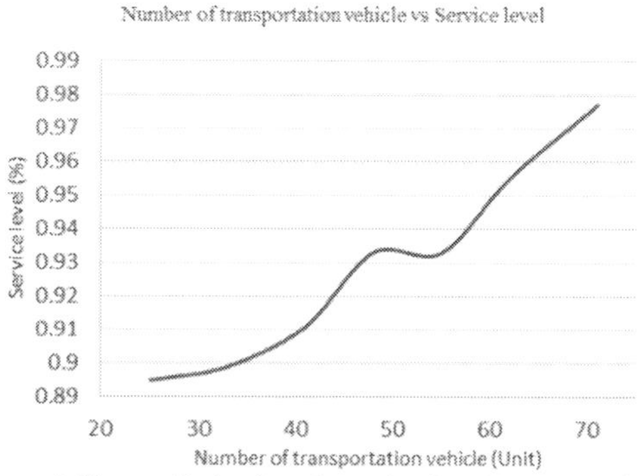

Figure 4. The confliction in values related to OF_1, OF_2 and OF_3.

Table 3. Pareto solutions for the HMSC aiming to three objective functions.

	Objectives weights	Objectives values			Open facilities	
Solution	(λ_1, λ_2, λ_3)	Min OF_1 (GBP)	Min OF_2 (Units)	Max OF_3 (%)	Farms	Abattoirs
1	1,0,0	55400	25	0.895	1, 2	4, 5
2	0.9,0.05,0.05	55430	33	0.899	1, 3	1
3	0.8,0.1,0.1	69343	41	0.911	1	3, 4, 5
4	0.7,0.15,0.15	84569	48	0.933	1, 2, 3	1, 2, 4
5	0.6,0.2,0.2	101234	55	0.933	1, 3, 4	1, 2, 3, 4, 5
6	0.5,0.25,0.25	144653	62	0.955	1, 2, 3	1, 2, 4, 5
7	0.4,0.3,0.3	173450	71	0.977	2, 3, 4	1, 3, 4, 5
8	0.3,0.35,0.35	-	-	-	-	-
9	0.2,0.4,0.4	-	-	-	-	-
10	0.1,0.45,0.45	-	-	-	-	-

4. Conclusions

This paper presented a development of a multi-objective mathematical model used for obtaining a compromise solution by optimizing three objectives simultaneously, which include (1) minimization of the total transportation cost, (2) minimization of the number of vehicles required for transportation and (3) maximization of the service level by satisfying all demands in quantity of meat products as requested by abattoirs and retailers. The model solves the facility location-allocation as well as the optimal quantity of product flows. The computed results showed that there were no feasible solutions when the scalar value for the first objective was set less than 0.4. A case study was used for verifying the developed mathematical model and the performance of the proposed solution methodology. Research findings indicate that the developed model is applicable and can be used as a reference for food supply chain designers. The model can be developed further as a fuzzy multi-objective model incorporating the uncertainty of input data such as varying demands and costs.

Acknowledgment

The authors would like to express their gratitude to the Higher Committee for Education Development (HCED) in Iraq for the financial support in this study.

References

1. F. Vanek, Y. Sun, Transport versus perishability in life cycle energy consumption: a case study of the temperature-controlled food product supply chain, Transp. Res. Part D Transp. Environ., **13**, 383-391 (2009)
2. T. Bosona, G. Gebresenbet, I. Nordmark, D. Ljungberg, Integrated logistics Network for the supply chain of locally produced food, part I: location and route optimization analyses, J. Serv. Sci. Manag, **4**, 174–183 (2011)
3. A. Rong, R. Akkerman, M. Grunow, An optimization approach for managing fresh food quality throughout the supply chain, Int. J. Prod. Econ., **131**, 421-429 (2011)
4. V. Sahar, B. Arijit, P.J. Byrne, A case analysis of a sustainable food supply chain distribution system—A multi-objective approach, Int. J. of Pro. Eco., **152**, 71-87 (2014)
5. C.S. Revelle, G. Laporte, The plant location problem: New models and research prospects, Ope. Res., **44**, 864–874 (1996)
6. B.H. Sabri, A. Beamon, A multi-objective approach to simultaneous strategic and operational planning in supply chain design, Omega, **28**, 581–598 (2011)
7. F.T.S. Chan, S.H. Chung, S. Wadhwa, A hybrid genetic algorithm for production and distribution, Omega, **33**, 345–355 (2005)
8. C.L. Chen, W.C. Lee, Multi-objective optimization of multi-echelon supply chain networks, with uncertain product demands and prices, Com. & Chem. Eng., **28**, 1131–1144 (2004)
9. F. Altiparmak, M. Gen, L. Lin, T. Paksoy, A genetic algorithm approach for multiobjective optimization of supply chain Networks, Com. & Ind. Eng., **51**, 197-216 (2006)
10. P. Schütz, A. Tomasgard, S. Ahmed, Supply chain design under uncertainty using sample average approximation and dual decomposition, Eur. J. of Ope.Res., **199**, 409–419 (2006)
11. B.L. Shankar, S. Basavarajappa, C.H. Jason, S. Chen, Location and allocation decisions for multi-echelon supply chain network – A multi-objective evolutionary approach, Exp. Sys. with App., **40**, 551–562 (2013)
12. H. W. Stoll, *Product Design Methods and Practices*, (New York-Basel: Marcel dekker, Int, 1999)
13. A. Mohammed, Q. Wang, Integrity of an RFID-enabled HMSC network, Proceedings of the 3rd ICDEIS, China, 79-86 (2015)
14. A. Ahmadia, H. Moghimib, A.E. Nezhadc, G. Vassilios, M. Sharafd, Multi-objective economic emission dispatch considering combined heat and power by normal boundary intersection method, Ele. Pow. Sys. Res., **129**, 32–43 (2015)
15. R.G. Pandu, Multi-Objective Optimization: Techniques And Applications, In Chem. Eng. (Adv. in Pro. Sys. Eng.), Singapore: WSP (Dec 2008)
16. HMC, UK. Available: http://www.halalhmc.org/test-demo.htm, (accessed October 26, 2014)

Fuzzy Goal Programming Approach for Integrating Production and Distribution Problem in Milk Supply Chain

Achraf Touil[a], Abdelwahed Echchatbi and Abdelkebir Charkaoui

Laboratory of Engineering, of Industrial Management and Innovation, FST, BP: 577, Settat, Morocco

Abstract. In this paper, a bi-objective mixed integer programming model is proposed to deal with the production-distribution problem found in a dairy company in Morocco. The supply chain containing three echelons: multi-sites, multi-distribution centers and multi-customers. The model seeks to integrate two conflicting simultaneous objectives: maximizing benefit by considering the shelf life of products and the total cost (quantitative objective), including production, storage, and distribution, as well as maximizing the service level (qualitative objective), which relates to providing satisfactory services to customers. This is subject to several technological constraints that typically arise in the dairy industry, such as sequence-dependent changeover time, machine speed and storage capacity. Due to imprecise aspiration levels of goals, an interactive approach is proposed based on fuzzy goal additive variants to find an efficient compromise solution. Numerical results are reported to demonstrate the efficiency and applicability of the proposed model.

1 Introduction

The food industry is growing in Morocco in recent years, mainly the dairy industry, which represents an important activity — one that contributes to the development of the economy by an agricultural GDP of 30%, and a coverage rate of approximately 90% of the national demand for milk and dairy products. However, the sector is characterized by several challenges, namely high market competition, demand remaining below requirements recommended by nutritional standards, the specific characteristics of the raw material (perishability), and diversification of ranges of products. Due to of these factors, the enterprises investigate the possibility to turn these challenges to a competitive advantage while continuously improving their operation. One of the key activities to achieve this goal is by an efficient integrate of production and distribution planning.

In supply chain management, the production and distribution operations can be decoupled if there is a sufficient buffer between them. Related to the dairy SC, poor integration within a network may consist of excessive inventories at the manufacturers' warehouses. Regarding the perishability characterized, the finished products should be kept refrigerated at a controlled temperature that results in a significant cost of storage, which may influence efficiencies. Therefore, we consider a model to integrate the production and distribution problem.

In academic literature, a considerable number of studies have examined the importance of the production-distribution problem in supply chain management. A comprehensive review on integrating production-distribution problems has been presented by [1-3].

Related to the dairy supply chain, a comprehensive study is proposed to deal with the production-distribution problem. However, milk production is a significant activity in the dairy industry, and the literature regarding this subject is very scarce. Nonetheless, some research has been conducted considering other dairy products, such as yogurt and ice cream. The studies related to milk supply chain management can be classified according to the problem characteristics and the proposed solution methodologies. The problem characteristic refers to decision levels, supply chain processes, product type, production process, perishability, working time and changeovers. The proposed solution methods vary from MILP and CP to stochastic programming (SP) and simulation (SM). A comprehensive review of quantitative models for the dairy industry was discussed by [4] and [5].

In this paper, we study a multi-product, multi-echelon milk supply chain containing three echelons: multi-sites, multi-distribution centers and multi-customers, which is inspired by the formulation previously considered in [4]. The problem lies in assigning the products to the processing units and determining the inventory level of each product at production sites and distributions centers, subject to sets of constraints that must be satisfied, such as capacity constraints, production, and storage. The model seeks to simultaneously incorporate two objectives: the first objective refers to maximizing the benefit by considering the shelf life of products [6] and the total costs, including production, storage, distribution and back

[a] Corresponding author: ac.touil@uhp.ac.ma

order, while the second objective is relevant to maximizing the service level. In addition, the model takes into account the transport lead time between different echelons of the supply chain. Because the imprecise aspiration level of objectives, a fuzzy bi-objective mixed integer model is proposed to deal with this issue. The main contributions of this paper can be summarized as follows: First, we propose a MILP formulation for the production-distribution problem encountered by a milk company. Second, the maximization and minimization goals for each objective are given and are converted to fuzzy goals using fuzzy membership values. Each fuzzy goal is assigned with a weight and all fuzzy goals are combined together to frame the Weighted Additive Fuzzy Goal model. Taking the importance of each individual objective into consideration, the weight for each objective is given.

The remainder of this paper is organized as follows. The proposed bi-objective mixed integer programming model is elaborated in Section 2. The proposed solution method is given in Section 3. Computational experiments are reported in Section 4 and finally, the concluding remarks and some directions for future research are given in Section 5.

2 Bi-objective mathematical model

2.1. Problem statement

The following considerations further define and delimit the problem:

• The dairy supply chain consists of several production sites and several DCs that deliver the final products to various customers.

• Each production site comprises several not necessarily identical parallel processing units.

• Due to restrictions on hygiene and security in the dairy industry, a setup operation between products is required, which results in setup time and cost.

• The scheduling horizon is supposed to be five days. The time horizon of short-term planning for dairy manufacturing is usually one week [7].

• The time for the quality control process is considered in the model. It is not allowed that the products are delivered before they complete the required time for the quality control process. A quarantine time is defined as a precondition for achievement of the desired sensory qualities [6].

• Freshness, which plays a significant part in the competitiveness, is taken into account for profitability. The objective function aims at maximizing the contribution margin given a shelf life-dependent pricing component, as stated in the model developed by [6].

• The available working time for the processing units is defined according to the shifts. Overtime production is considered on the third shift of every working day and under an environment of heavy demand, and if necessary, the production day can proceed in weekdays that are defined as overtime. The cleaning and sterilization operations are ensured on the third shift of each production day.

• Customer demand for each product type is daily collected from DCs.

• The processing unit capacity should not exceed a maximum capacity.

• The capacity of storage at sites and DCs is limited. In this model, we considered the safety stock level and the storage capacity related to each type of product to each site and each DC. Also, the total storage capacity of both sites and centers.

• The transport parameters, such as transport costs and transport lead times, are taken into consideration.

The key decisions of the model are:

• The production amount that is produced by each processing unit and in each period.

• The assignment of products to each unit, and in each period.

• The dispatched amount between (Sites-DCs) and (DCs-Customers) in each demand period.

• The inventory level of each product at each site and center at the end of each period.

• Sales and demand backlog of each product in each demand period.

• Service level desired by each customer.

2.2. Model formulation

Indices | Sets

$t \in T$: Planning periods (days).

$n \in N$: Demand periods (days).

$p, k, l, e \in P$: Finished products (products).

$a \in A$: Distribution centers (DCs).

$s \in S$: Production sites (sites).

$j \in J$: Processing units (units).

$c \in C$: Customers.

Subsets

J_p : Units j that can processed product p.

J_s : Units j that are installed on site s .

S_a : Sites s that can supply center a .

A_s : DCs a that are supplied by site s

A_c : DCs a that can supply customer c

C_a : Customers c that are provided by DC a.

Parameters

$D_{c,p,n}$: Demand of product p required by customer c on demand day d.

$Benefit_p$: Maximum benefit for meeting the highest shelf life of product p,

Cr_p : Critical rate for the shelf life of product p required by customers.

shl_p : Shelf life of product p, (day).

qqt_p : Quarantine time needed to control product quality, (day).

$rate_{p,s,j}$: Processing rate of product p on unit j at site s.

$\left[\underline{Lot_p}, \overline{Lot_p}\right]$: Minimum and maximum lot size of product p.

$VPC_{p,s,j,t}$: Variable production cost of one unit of product p in unit j of the site s in period t (including labor and utilities cost).

$IHCS_{p,s,t}$: Inventory holding cost for one unit of product p at site s in period t.

$SSS_{p,s,t}$: Safety stock of product p at site s in period t.

$SCS_{p,s}$: Storage capacity associated with product p at site s.

$TSCS_s$: Total storage capacity of site s.

$\phi_{p,k,s,j,t}$: Changeover time between product p and product k on unit j of site s in period t.

$\chi_{p,k,s,j,t}$: Changeover cost between product p and product k on unit j of site s in period t.

$OTC_{s,j,t}$: Overtime cost associated with unit j of site s in period t.

$IHCD_{p,s,t}$: Inventory holding cost for one unit of product p at center in period t.

$SSA_{p,a,t}$: Safety stock of product p at center in period t.

$SCA_{p,a}$: Storage capacity associated with product p at center a.

$TSCA_a$: Total storage capacity of center a.

$TCD_{s,a}$: Transport cost from site s and center a.

$TCC_{a,c}$: Transport cost from center a and customer c.

$TLT1_{s,a}$: Transport lead-time between site s and center a.

$TLT2_{a,c}$: Transport lead-time between center a and customer c.

$BC_{p,c,t}$: Backorder cost per unit of product p from customer c at the end of period t.

$Cap_{s,j}^{\max}$: Maximum production capacity of unit j at site s.

$Rt_{s,j,t}$: Regular in use shift of unit j at site s in period t.

M: Extremely large number.

Variables

Binary variables

$Y_{p,s,j,t}$ = 1 if product p is assigned to a unit $j \in J_p$ at site s in period t.

$Z_{p,k,s,j,t}$ = 1 if product p is processed before product k when both are assigned to the same unit $j \in J_p$ at site s in period t.

Continuous variables

$QPS_{p,s,j,t,n}$: Amount of product p produced by unit j at site s in period t for demand period n.

$QTD_{s,a,p,t,n}$: Amount of product p sent from site s to center a in period t for demand period n.

$QTC_{a,c,p,t,n}$: Amount of product p sent from center a to customer c in period t for demand period n.

$IS_{s,p,t}$: Inventory level of product p at site s at end of period t.

$ID_{a,p,t}$: Inventory level of product p at center a at end of period t.

$BP_{c,p,n}$: Amount of product p backordered by customer c on demand period n.

$SL_{c,p,n}$: Desired service level at customer c related to product p in period t for demand period n.

$OT_{s,j,t}$: Overtime on unit j at site s in period t.

$PT_{p,s,j,t}$: Processing time of product p on unit j at site s in period t.

$ST_{p,s,j,t}$: Starting time of product p on unit j at site s in period t.

$CT_{p,s,j,t}$: Finishing time of product p on unit j at site s in period t.

$C_{s,j,t}$: Finishing time of the last product on unit j at site s in period t.

Objective function

Minimization of total cost included (Production-Storage-Distribution)

$$B - \sum_s \sum_p \sum_{j \in J_p \cap J_k} \sum_t \sum_d B_p * \frac{(1-Cr_p) * Shl_p - (d-t)}{(1-Cr_p) * Shl_p} * QPS_{p,s,j,t,n}$$

$$\begin{aligned}
TC = & \sum_s \sum_p \sum_{j \in J_s \cap J_p} \sum_t \sum_d VPC_{p,s,j,t} * QPS_{p,s,j,t,n} \\
& + \sum_s \sum_j \sum_t OTC_{s,j,t} * OT_{s,j,t} \\
& + \sum_s \sum_p \sum_{k \neq p} \sum_{j \in J_s \cap J_p \cap J_k} \sum_t \phi_{p,k,s,j,t} * \chi_{p,k,s,j,t} \\
& + \sum_s \sum_p \sum_t IS_{s,p,t} * IHCS_{s,p,t} \\
& + \sum_s \sum_{a \in A_s} \sum_t \sum_n QTD_{s,a,p,t,n} * TCD_{s,a} \\
& + \sum_a \sum_p \sum_t ID_{c,p,t} * IHCD_{c,p,t} \\
& + \sum_a \sum_{c \in C_a} \sum_t \sum_n QTC_{a,c,p,t,n} * TCC_{a,c} \\
& + \sum_c \sum_p \sum_n BP_{c,p,n} * BC_{c,p,n}
\end{aligned}$$

$$Max\ Z_{\text{Profit}} = B - TC \qquad (1)$$

The first component of the objective function (1) seeks to maximize benefit by considering the shelf life of products. It is assumed that the manufacturer yields a financial benefit if the products have a longer residual shelf life when being delivered [6]. The other component describes the setup, production, storage, distribution and backorder costs, respectively.

Maximization of service level

The last objective function aims to maximize the service level related to each product by the customer:

$$SL_{c,p,n} = 1 - \frac{\sum_n BP_{c,p,n}}{\sum_n D_{c,p,n}} \quad (2)$$

Constraints

Constraint (3) ensures the lower and upper bounds on lot product sizes.

$$\underline{Lot}_p * Y_{p,s,j,t} \leq \sum_{n=t} QPS_{p,s,j,t,n} \leq \overline{Lot}_p * Y_{p,s,j,t} \quad (3)$$

$$\forall s, j, p, t$$

The processing time of product p on unit j that equals the produced amount divided by the rate of the same unit is ensured by Constraint (4).

$$PT_{p,s,j,t} = \frac{\sum_n QPS_{p,s,j,t,n}}{rate_{p,s,j}} \quad (4)$$

$$\forall s, p, j \in J_p \cap J_s, t$$

The finishing time of product p on unit j at site s in period t is equal to the sum of its corresponding starting time and processing time on the same unit and is described by Constraint (5).

$$CT_{p,s,j,t} = ST_{p,s,j,t} + PT_{p,s,j,t} \quad (5)$$

$$\forall s, p, j \in J_p \cap J_s, t$$

Constraint (6) ensures that the starting time of the product that directly follows another product k on a processing unit j at production site s in period t (i.e., $Z_{p,k,s,j,t} = 1$) should be greater than the completion time of product k ($CT_{p,s,j,t}$), plus the necessary changeover time $\phi_{p,k,s,j,t}$ between those products. Similarly, Constraint (7) describes the opposite case.

$$CT_{k,s,j,t} \geq CT_{p,s,j,t} + \phi_{p,k,s,j,t} - M * Z_{p,k,s,j,t} - M * (1 - Y_{p,s,j,t} - Y_{k,s,j,t})$$

$$\forall s, p, j \in J_p \cap J_k \cap J_s, t \quad (6)$$

$$CT_{p,s,j,t} \geq CT_{k,s,j,t} + \phi_{k,p,s,j,t} - M * (1 - Z_{p,k,s,j,t}) - M * (1 - Y_{p,s,j,t} - Y_{k,s,j,t})$$

$$\forall s, p, j \in J_p \cap J_k \cap J_s, t \quad (7)$$

Constraints (8-10) describe the product sequencing and timing. Constraints (8) and (9) ensure that if a product p is assigned to processing unit j at production site s in period t (i.e., $Y_{p,s,j,t} = 1$) then at most one product k is processed before and after product p, respectively. According to Constraint (10), the total number of active sequencing binary variables $Z_{p,k,s,j,t}$ should be less than the total number of active allocation binary variables $Y_{p,s,j,t}$ minus 1 in a processing unit j at production site s in period t.

$$\sum_{k \in P, k \neq p} Z_{k,p,s,j,t} \leq Y_{p,s,j,t} \quad (8)$$

$$\forall s, p, j \in J_p \cap J_k \cap J_s, t$$

$$\sum_{k \in P, k \neq p} Z_{p,k,s,j,t} \leq Y_{p,s,j,t} \quad (9)$$

$$\forall s, p, j \in J_p \cap J_k \cap J_s, t$$

$$\sum_{k \in P, k \neq p} Z_{p,k,s,j,t} \leq Y_{p,s,j,t} \quad (10)$$

$$\forall s, p, j \in J_p \cap J_k \cap J_s, t$$

Constraint (11) ensures the total processing time of the last product on unit j at site s in period t.

$$C_{s,j,t} = CT_{P,s,j,t}, \ \forall s, p, j \in J_s, t \quad (11)$$

The constraint (12) ensures that the total processing time will not exceed the maximum available time for the same unit in period t.

$$C_{s,j,t} \leq Cap_{s,j}^{\max} \ \forall s, p, j \in J_s, t \quad (12)$$

The overtime on unit j at site s in period t is computed according to Constraint (13).

$$C_{s,j,t} = OT_{s,j,t} + Rt_{s,j,t}, \ \forall s, p, j \in J_s, t \quad (13)$$

Constraint (14) ensures that to meet the demand for demand day d, the products should be produced within a window of time according to quarantine time and shelf life. The quarantine time is required for the achievement of the desired level of quality. The product cannot be delivered before a quarantine time has reached $n - t \geq qqt_p$. The shelf life lost (d-t), may not exceed a threshold provided by the customer $(n-t) \leq (1 - Cr_p) * Shl_p$. [6]. Otherwise, it should be noticed that $Q_T = 0$.

$$Q_T = \sum_t \sum_n \sum_{j \in J_s} QPS_{p,s,j,t,n}, \ \forall s, p, t \quad (14)$$

where $n = 1 : (n-t) \geq qqt_p \wedge (n-t) \leq (1 - cr_p) * shl_p$

The inventory of product p in production site s at the end of period t is equal to the inventory of that product in the previous period, plus the total production quantity of product p at site s in period t, minus the amount of product p transported from site s to center $a \in A_s$ in period t, is ensured by Constraint (15)

$$IS_{s,p,t} = IS_{s,p,t-1} + Q_T - \sum_{a \in A_s} QTD_{p,s,a,t-TLT1,n} \quad (15)$$

$\forall s, p, t$

Constraints (16-21) ensure the safety stocks, product-dependent storage limitation, and the total capacity for product p at site s and distribution center, respectively.

$$IS_{s,p,t} \geq SSS_{p,s,t}\ ,\ \forall s,p,t \quad (16)$$

$$IS_{s,p,t} \leq SCS_{p,s}\ ,\ \forall s,p,t \quad (17)$$

$$\sum_p IS_{s,p,t} \leq TSCS_s\ ,\ \forall s,p,t \quad (18)$$

$$ID_{a,p,t} \geq SSA_{p,a,t}\ ,\ \forall a,p,t \quad (19)$$

$$ID_{a,p,t} \leq SCA_{p,a}\ ,\ \forall a,p,t \quad (20)$$

$$\sum_p ID_{c,p,t} \leq TSCA_a\ ,\ \forall a,p,t \quad (21)$$

Constraint (22) describes the quantity of product p transported from center a to customer a in period t to adequately meet the demand period d.

$$\sum_{a \in A_c} QTC_{p,a,c,t,n} = D_{c,p,n}\ \forall c,p,t,n \quad (22)$$

Constraint (23) guarantees that the inventory level of product p at the end of period t must be equal to the amount entering from site $s \in S_a$, minus the amount dispatched to customers $a \in A_c$

$$ID_{a,p,t} = ID_{a,p,t-1} + \sum_{s \in S_c} QTD_{p,s,a,t,n} - \sum_{c \in C_a} QTC_{p,a,c,t-TLT2,n} \quad \forall p,t,n \quad (23)$$

The backlog level of product p incurred by client c for demand period d equals the backlog level of that product in the previous period, plus the total demand of the product by client during that period, minus the total amount provided by the distribution center $a \in A_c$, is shown by Constraint (24).

$$BP_{c,p,t} = BP_{c,p,t-1} + D_{c,p,t} - \sum_{a \in A_c} QTC_{a,c,p,t,n} \quad \forall p,t,n \quad (24)$$

3 Solution approach

Fuzzy linear programming was proposed by [8], and has been typically used to solve fuzzy multi-objective programming problems by converting such problems into the single objective formulation. A modified fuzzy linear programming version called WAFGP was proposed by [9]. In this method, each objective function (Z_k) of a multi-objective optimization problem is resolved to obtain its minimum ($Z_k^{\min}$) and maximum ($Z_k^{\max}$) goals, individually satisfying all the constraints. All individual objectives are converted into fuzzy goals using their minimum and maximum goals, respectively [10].

$$\mu_{Z_k} = \begin{cases} 1 \ ; \ Z_k(x) \prec Z_k^{\min} \\ \dfrac{Z_k^{\max} - Z_k(x)}{Z_k^{\max} - Z_k^{\min}} \ ; \ Z_k^{\min} \prec Z_k(x) \prec Z_k^{\max} \\ 0 \ ; \ Z_k(x) \succ Z_k^{\max} \end{cases} \quad (Eq.1)$$

$$k = 1,2,\ldots,p$$

$$\mu_{Z_l} = \begin{cases} 0 \ ; \ Z_l(x) \prec Z_l^{\min} \\ \dfrac{Z_l(x) - Z_l^{\min}}{Z_l^{\max} - Z_l^{\min}} \ ; \ Z_l^{\min} \prec Z_l(x) \prec Z_l^{\max} \\ 1 \ ; \ Z_l(x) \succ Z_l^{\max} \end{cases} \quad (Eq.2)$$

$$l = p+1, p+2, \ldots q$$

Fuzzy goals (FG) are combined, and the WAFGP model is formulated as shown in Eq.1 subjected to constraints as in Eq.2 and solved to obtain optimal solutions. The optimal solutions depend on the weights (ω_k) assigned to each fuzzy goal:

$$\text{Max } \sum_k \omega_k * \theta_k$$

s.t

$$\theta_k \leq \mu_{Z_k}$$

$$\theta_k, \mu_{Z_k}, \omega_k \in [0,1]$$

$$\sum_k \omega_k = 1$$

where θ_k denotes the satisfaction degree of the kth objective (individual satisfaction degree of each objective).

4 Numerical results

In this paper, we analyzed an integrated production-distribution of a multi-echelon dairy supply chain in Morocco. The manufacturer is a big company that has four geographically-dispersed production sites. Two different products are considered in this paper (pasteurized and UHT milk). Due to confidentiality issues, the real data has been modified and is randomly generated from distributions given in Table 1. U (a, b) denotes uniform distribution between a and b. We define each membership function by defined upper and lower bounds of each objective function. The upper and lower bounds obtained by maximizing and minimizing each objective function separately are presented in Table 2. The proposed MILP was solved on a 2.5 GHz Intel Core i5 PC with 4 GB RAM using IBM ILOG CPLEX 12.4. The computational results are given in Table 3. The solution results is given in Table 4. It is clear that the value of weights influence on the relations between the different objectives functions. In Table 4, ω_{SL} and ω_{TC} denotes the weights of service level and total costs objectives for each instance and the degree of satisfaction of the objective functions for the proposed method.

Table 1. Random generation functions of parameters

$D_{c,pasteurized-milk,n}$	1000*U(120,150)
$D_{c,UHT-milk,n}$	1000*U(100,120)
Critical rate	066
$qqt_{pasteurized-milk}$	2
$qqt_{UHT-milk}$	30
$Benefit_{pasteurized-milk}$	200
$Benefit_{UHT-milk}$	200
$VPC_{p,s,j,t}$	U(10,15)
$Cap_{s,j}^{\max}$	U(18,20)
$Rt_{s,j,t}$	U(10,16)
$OTC_{s,j,t}$	U(150,200)
$IHCS_{p,s,t}$	U(10,20)
$IHCD_{p,s,t}$	U(10,20)
$SSS_{p,s,t}$	U(600,650)
$SCS_{p,s,t}$	U(70,75)
$TSCS_{p,s,t}$	U(350000,400000)
$SSA_{p,a,t}$	U(600,650)
$SCA_{p,a,t}$	U(70,75)
$\mathrm{TSCA}_{p,a,t}$	U(15000,20000)
$TCD_{s,a}$ - $TCC_{a,c}$	U(2,4)
$TLT1_{s,a}$ - $TLT2_{a,c}$	U(0.1,0.2)
$BC_{c,pasteurized-milk,n}$	U(20,30)
$BC_{c,UHT-milk,n}$	U(15,20)

Table 2. Lower and upper bounds of the objectives

Objectives	Lower bound	Upper bound
Total costs	2135890	2850070
Service level	95%	99%

Table 3. Computational results

No of variables	44303
No of integer variables	18340
No of constraints	56109
CPU (s)	840

Table 4. Solution results

Instance	ωSL	ωTC	μ_{SL}	μ_{TC}
1	0.65	0.35	0.9175	0.8655
2	0.20	0.80	0.8919	0.7760
3	0.50	0.50	0.988	0.7655
4	0.10	0.90	0.8851	0.8919
5	0.90	0.10	0.9245	0.9036

5 Conclusion

In this paper, we proposed a bi-objective mixed integer model to deal with the production-distribution problem found in the dairy supply chain in Morocco. The model aims to incorporate two objectives: the first objective function aims to maximize benefit by considering the shelf life-dependent pricing component and various cost components, such as setup costs, storage, distribution, and backorders. The second objective function aims to maximize the service level related to each product for customers. To deal with the imprecise preferences of the DM regarding the aspiration level of objectives. An approach based on Weighted Additive Fuzzy Goal Programming is used. The numerical experiments indicate that the proposed model is practical and tractable.

References

1. Erengüç, Ş Selçuk, Simpson, Natalie C, & Vakharia, Asoo J. Integrated production/distribution planning in supply chains: An invited review. European Journal of Operational Research, **115**(2), 219-236, (1999)
2. Chen, Zhi-Long. Integrated Production and Distribution Operations Handbook of quantitative supply chain analysis, 711-745, (2004).
3. Bilgen, Bilge, & Ozkarahan, Irem. Strategic tactical and operational production-distribution models: a review. International Journal of Technology Management, **28**(2), 151-171 (2004)
4. Bilgen, Bilge, & Çelebi, Yelda. Integrated production scheduling and distribution planning in dairy supply chain by hybrid modelling. Annals of Operations Research, **211**(1), 55-82, (2013)
5. Sel, C, Bilgen, B, Bloemhof-Ruwaard, JM, & van der Vorst, JGAJ. Multi-bucket optimization for integrated planning and scheduling in the perishable dairy supply chain. Computers & chemical engineering, **77**, 59-73, (2015)
6. Lütke Entrup, Matthias, Günther, H-O, Van Beek, Paul, Grunow, Martin, & Seiler, Thorben, Mixed-Integer Linear Programming approaches to shelf-life-integrated planning and scheduling in yoghurt production. International journal of production research, **43**(23), 5071-5100, (2005)
7. Nakhla, M. Production control in the food processing industry: the need for flexibility in operations scheduling. International Journal of Operations & Production Management, **15**(8), 73-88, (1995)
8. Zadeh, L. A. Outline of a new approach to the analysis of complex systems and decision processes. Systems, Man and Cybernetics, IEEE Transactions on, (1), 28-44, (1973).
9. Tiwari, R. N., Dharmar, S., & Rao, J. R. Fuzzy goal programming—an additive model. Fuzzy sets and systems, **24**(1), 27-34, (1987)
10. Zimmermann, H. J. Fuzzy, set theory—and its applications. Springer Science & Business Media, (2001)

An Investment Evaluation of a RFID-Enabled Meat Supply Chain: A Multi-Criteria Approach

Ahmed Mohammed and Qian Wang

School of Engineering, Faculty of Technology, University of Portsmouth, Portsmouth, UK PO1 RDJ, 3HF

Abstract. In recent years, consumers have become increasingly concerned about the safety and quality of meat they purchase from supermarkets. A study by Mohammed [1] proposed a RFID (Radio Frequency Identification)-enabled monitoring system for meat supply chains to improve the traceability of meat products throughout their entire supply chain with the aim of maintaining product safety. This paper extends that work to examine the economic feasibility for the proposed RFID-enabled monitoring system. To this aim, a multi-criteria optimization model was developed. The considered criteria were minimizing the total cost, maximizing consumer satisfaction, maximizing product freshness and maximizing profits. In order to obtain Pareto solutions from the developed model, a new solution approach was developed and its results were compared to two traditional solution approaches. A case study was applied conducive to an examination for the applicability of the developed model and the performance of the proposed solution approaches. Results have proved the feasibility of the proposed RFID-enabled monitoring system in terms of economic costs in addition to the capability of the developed optimization model in obtaining a trade-off among the considered criteria.

1. Introduction

Meat supply chains are a methodical connotation that constitute generally from four different echelons including farms, abattoirs, retailers and consumers. They are acquainted as a network of facilities that supply livestock which is transported into intermediate and final meat to be purchased by consumers. In the last few decades, increased safety and quality concerning meat purchased by consumers has been demanded in the UK [2, 3]. This refers to particular rules that should be maintained throughout the meats' entire supply chain including feeding and health of livestock, and method of slaughtering process at abattoirs [4]; Different commitments lead to a guide to decision making when purchasing types of meat. A study by Peattie [5] mentioned that consumers spend considerable time and effort seeking out fresh food and reading food labels in order to ensure they are purchasing good quality. Implementation of the RFID technology in meat supply chains was proposed to improve product safety [1]. Such a system leads to an increase in consumer satisfaction; on the contrary, it is also subjected to additional costs that should be considered.

Multi-criteria optimization is an optimization approach that encompasses multiple criteria and rather than a single optimal solution it has a set of solutions called "Pareto optimal solutions". Pareto optimal solutions represent the compromise among multiple conflicting criteria. During the last few decades, there has been increasing use of the multi-criteria optimization (or multi-objective optimization) approach to tackle a variety of problems in supply chain network design and its operations management [6-9]. These problems can be tactical such as the facility location-allocation problem or strategic such as product quantity flows in addition to considering different criteria such as costs or profit.

To the best of our knowledge, little or no research has been presented so far investigating the RFID-enabled supply chains seeking a compromise between the benefits of the RFID implementation in supply chains (e.g. consumer satisfaction and service level) and its need for additional costs along with the supply chain network design. This paper addresses this gap in the literature. It considers the proposed RFID-enabled three-echelon meat supply chain seeking a compromise among four criteria including total cost, consumer satisfaction in terms of percentage of satisfying consumers' demand, product freshness in terms of the number of fresh meat products as a result for the proposed RFID-enabled monitoring system and profits as a fourth criterion. To this aim, a multi-criteria mixed integer linear programming model was developed to simultaneously optimize the considered criteria. To reveal the solutions from the developed optimization model, three multi-criteria solution approaches were investigated including a new developed approach, compromise programming, and Weighted Tchebycheff. The developed model and its solution methodology can be used as a reference for decision makers to evaluate the performance and the economic feasibility of their food supply chains in implementing the RFID technology.

2. Related studies

Barbarosoglu [10] developed a mixed integer linear programming model to minimize the total fixed and variable costs for a two-echelon system. The Lagrangian relaxation method was used to solve the developed model. The Same method was also used by Sadjady [11] to tackle a two-level supply chain network design problem comprising both tactical and strategic concerns. The problem was formulated as a mono-criteria optimization model to minimize total costs including transportation, lead-times, and inventory holding costs for products, as well as opening and operating costs for facilities. Selim [12] presented a multi-criteria optimization model to cope with a production-distribution planning problem in a supply chain. Fuzzy goal programming was considered to incorporate decision maker's imprecise aspiration levels. Ferrio [13] formulated a mixed integer linear programming model for configuring and optimizing the design of a multi-product chemical supply network consisting production sites, an arbitrary number of distribution centers, and customers. Tuzkaya [14] studied a three-level supply chain including supplier, warehouses, and manufacturers seeking the best products distribution plan. The authors developed a multi-objective model aimed at minimizing the total cost of inventory, warehouse, manufacturer, and penalty costs for supplier, manufacturers, and warehouses.

3. Modeling the HMSC

The meat supply chain under consideration is comprised of three echelons including farms, abattoirs and retailers. In this chain, livestock is supplied from farms to abattoirs to be slaughtered then transported to retailers as a packed meat. The RFID technology was proposed to trace product production at farms and abattoirs, and during the transportation from abattoirs to retailers for maintaining the safety and freshness of products throughout its entire supply chain [1]. Fig. 1 depicts the schematic illustration of the investigated supply chain.

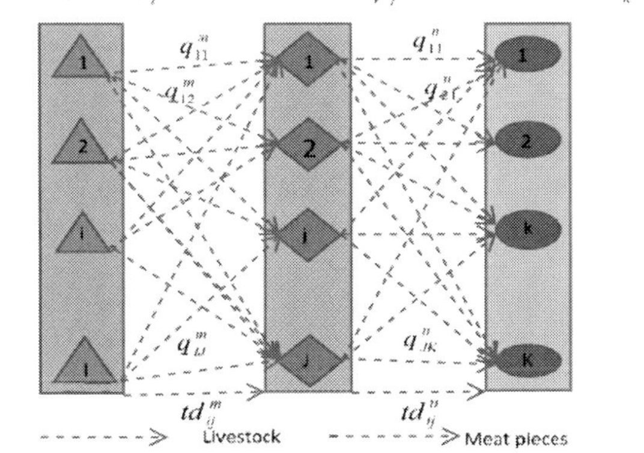

Figure 1. The schema of the investigated supply chain network.

3.1 Notations

To formulate the multi-criteria model, the following indices, parameters and decision variables are presented:

Indices

I index used for a potential farm i, $1 \le i \le I$

J index used for a potential abattoir j, $1 \le j \le J$

K index used for a fixed retailer k, $1 \le k \le K$

Cost parameters

C_i^{α} equipment and implementation cost of the RFID technology required for farm i

C_j^{β} equipment and implementation cost of the RFID technology required for abattoir j

TC_{ij} unit transportation cost per mile from farm i to abattoir j

TC_{jk} unit transportation cost per mile from abattoir j to retailer k

LC_i^{α} unit labor cost per hour at farm i

LC_j^{β} unit labor cost per hour at abattoir j

Parameters of capacity, demand and transportation distance:

C_i^{α} maximum supply capacity (units) of farm i

C_j^{β} maximum supply capacity (units) of abattoir j

W_v transportation capacity (units) per vehicle v

D_j^{β} minimum demand (in units) of abattoir j

D_k^{δ} minimum demand (in units) of retailer k

d_{ij} travel distance (mile) through the first transportation link m from farm i to abattoir j

d_{jk} travel distance (mile) through the first transportation link n from abattoir j to retailer k

Labor parameters

$R_i^{l\alpha}$ working rate (items) per laborer (l) at farm i

$R_j^{l\beta}$ working rate (items) per laborer (l) at abattoir j

$h_i^{l\alpha}$ minimum required number of working hours (h) for laborer l at farm i

$h_j^{l\beta}$ minimum required number of working hours (h) for laborer l at abattoir j

Other parameters

Q_{ij} quality percentage of products transported from farm i to abattoir j

F_{jk} freshness percentage of products transported from abattoir j to retailer j

P_i^{α} profits (GBP) per item at farm i

P_j^{β} profits (GBP) per item at abattoirs j

Decision variables

q_{ij} quantity of units transported from farm i to abattoir j

q_{jk} quantity of units transported abattoir j to retailer k

x_i^{α} minimum required number of laborers at farm i

x_j^{β} minimum required number of laborers at abattoir j

Non-negative and binary decision variables

$$y_i^{\alpha} = \begin{cases} 1: \text{if farm } i \text{ is located} \\ 0: \text{otherwise} \end{cases}$$

$$y_j^{\beta} = \begin{cases} 1: \text{if abattoir } j \text{ is located} \\ 0: \text{otherwise} \end{cases}$$

3.2 Optimization criteria

The criteria functions are formulated as follows:

Minimum the total cost (F_1) = costs of equipment and implementation for the RFID technology + transportations costs – labor costs saved after the RFID implementation as several manual operation activities were eliminated

$$MinF_1 = \sum_{i\in I} C_i^{\alpha} y_i^{\alpha} + \sum_{j\in J} C_j^{\beta} y_j^{\beta} + \sum_{i\in I}\sum_{j\in J} TC_{ij}\left\lceil q_{ij} / W_v \right\rceil d_{ij}$$
$$+\sum_{j\in J}\sum_{k\in K} TC_{jk}\left\lceil q_{jk} / W_v \right\rceil d_{jk} - \sum_{i\in I} LC_i^{\alpha} x_i^{\alpha} h_i^{l\alpha} - \sum_{j\in I} LC_j^{\beta} x_i^{\beta} h_i^{l\beta} \quad (1)$$

Maximum consumer satisfaction (F_2) = the fulfillment of consumers 'demand

$$MaxF_2 = \sum_{k=1}^{K}\left(\sum_{j=1}^{J} q_{ij} \Big/ D_k \right) \quad (2)$$

Maximum product freshness (F_3) = product quality of livestock shipped from farms to abattoirs + product freshness of meat pieces shipped from abattoirs to retailers

$$MaxF_3 = \sum_{i=1}^{I} Q_{ij} y_i^{\alpha} + \sum_{j=1}^{J} F_{jk} y_j^{\beta} \quad (3)$$

Maximum profits (F_4) = return of investment for farms + return of investment for abattoirs

$$MaxF_4 = \sum_{i\in I}\sum_{i\in J} R_i^{\alpha} x_{ij}^{u} + \sum_{j\in J}\sum_{k\in K} R_j^{\beta} x_{jk}^{v} \quad (4)$$

3.3 Defining the constraints

Several constraints were defined after formulating the criteria functions. These constraints are grouped in different categories:

Capacity constraints: ensure the flow balance of products from farms to abattoirs and from abattoirs to retailers.

$$\sum_{i\in I} q_{ij} \le C_i^{\alpha}\ x_i^{\alpha} \quad \forall\, j \in J \quad (5)$$

$$\sum_{j\in J} q_{jk} \le C_j^{\beta}\ y_j^{\beta}? \quad \forall k \in K \quad (6)$$

Demand constraints: ensure that all demands of abattoirs and retailers are fulfilled.

$$\sum_{i\in I} q_{ij} \ge D_j^{\beta} \quad \forall\, j \in J \quad (7)$$

$$\sum_{j\in J} q_{jk} \ge D_k^{\gamma} \quad \forall\, k \in K \quad (8)$$

$$D_j^{\beta} \ge \sum_{k\in K} q_{jk} \quad \forall j \in J \quad (9)$$

Working rate constraints: indicate the required number of laborers at farms and abattoirs.

$$\sum_{j\in J} q_{ij} \le x_i^{\alpha}\ R_i^{l\alpha} \quad \forall\, i \in I \quad (10)$$

$$\sum_{k\in K} q_{jk} \le x_j^{\beta}\ R_j^{l\beta} \quad \forall j \in J \quad (11)$$

Restriction constraints: restrict the decision variables to binary and non-negative.

$$q_{ij}, q_{jk} \ge 0,\ \forall i, j, k; s \quad (12)$$

$$y_i^{\alpha}, y_j^{\beta} \in \{0,1\},\ \forall i, j; \quad (13)$$

Finally, $0.75 \le Q_{ij} \le 1$ and $0.75 \le F_{jk} \le 1$ constraints which limit the quality percentage (Q) and the freshness percentage (F) to be between 0.75 and 1 (based on decision makers' preferences).

4. Solution approach

In this paper, three different solution approaches were examined to obtain three sets of Pareto solutions aiming to select the best one in terms of a solution value. Concisely, descriptions of the investigated solution approaches are presented in the next sub-section.

1) The developed approach: This approach transforms the multi-criteria model into a single-criterion model (F_s) which is formulated by considering each criterion individually. This single-criteria model aims to minimize the scalarized differences between each criterion and its optimal value. Undesired deviations (F_d) are proposed to be subtracted from F_s with the aim to achieve more accurate criteria values. These values are close enough to Pareto optimal solutions which lead to a clear insight of a compromise solution between conflicting criteria for decision makers. The solution approach function (F) can be formulated as follows:

$$Min\ F = F_s - F_d \quad (14)$$

$$\text{s.t.}\begin{cases}\mu_n=\left[\dfrac{F_n-F_n^*}{F_n^*}\right]\\ 0\geq w_n\geq 1;\ \sum_{n=1}^{4}w_n=1;\ \ n=(1,2,3,4)\end{cases}\tag{15}$$

Set $w_n^* = \dfrac{w_n F_n^{\bullet}}{F_n^{\bullet}-F_n}$, then

$$\begin{aligned}F_{sl}&=w_1^*F_1+w_2^*F_2+w_3^*F_3+w_4^*F_4\\&=\frac{w_1F_1^{\bullet}}{F_1^{\bullet}-F_1}F_1+\frac{w_2F_2^{\bullet}}{F_2^{\bullet}-F_2}F_2+\frac{w_3F_3^{\bullet}}{F_3^{\bullet}-F_3}F_3+\frac{w_4F_4^{\bullet}}{F_4^{\bullet}-F_4}F_4\end{aligned}\tag{16}$$

Finally, based on the aforementioned procedures the developed approach's solution function can be written as follows

$$\begin{aligned}Min\ F&=\left(w_1\mu_1-w_2\mu_2-w_3\mu_3-w_4\mu_4\right)\\&-\left(\frac{w_1F_1^{\bullet}}{F_1^{\bullet}-F_1}F_1+\frac{w_2F_2^{\bullet}}{F_2^{\bullet}-F_2}F_2+\frac{w_3F_3^{\bullet}}{F_3^{\bullet}-F_3}F_3+\frac{w_4F_4^{\bullet}}{F_4^{\bullet}-F_4}F_4\right)\end{aligned}\tag{17}$$

The constraints contain equations (5)-(13) and (15).

2) Compromise programming approach: The compromise programming approach is its ability to achieve efficient points in a non-convex Pareto curve [15]. This method based on optimizing one criterion function and shifting the other to the constraint set to restricted to an assigned value ($\hbar$). The equivalent solution formula (F) is presented as follow.

$$Min\ F_1\tag{18}$$

Additional constraints:

$$F_2\geq\hbar_1\tag{19}$$

$$(F_2)^{min}\leq\hbar_1\leq(F_2)^{max}\tag{20}$$

$$F_3\geq\hbar_2\tag{21}$$

$$(F_3)^{min}\leq\hbar_2\leq(F_3)^{max}\tag{22}$$

$$F_4\geq\hbar_3\tag{23}$$

$$(F_4)^{min}\leq\hbar_3\leq(F_4)^{max}\tag{24}$$

In this paper, criterion function one is selected to be optimized (Eq.18) and shifting criterion functions two, three and four to be constraints; where F_2, F_3 and F_4 are greater or equal to $\hbar_1, \hbar_2$ and $\hbar_3$ respectively (Eq.19,21 and 23). A dramatic increasing to the $\hbar$ values (Eq.20, 22 and 24) yields more Pareto solutions.

3) Weighted Tchebycheff approach: This approach transforms the multi-criteria model into a single-criterion model (F). This single-criteria model aims to minimize the distance between the ideal objective vector (F^*) and the feasible objective surface [16]. The solution approach function (F) can be formulated as follows:

$$Min\ F=\left(\sum_{i=1}^{4}l_i\left|F_i-F_i^{\bullet}\right|^p\right)^{\frac{1}{p}}\tag{25}$$

Subject to constraints defined in sub-section 3.3.

5. Application and evaluation

In order to demonstrate the application of the developed multi-criteria model and compare the performance of the proposed solution approaches, a case study is presented in this section. In the case study, the South East of London area encompasses 4 farms (I), 7 retailers (K) and 4 abattoir (J) suppliers. The given parameters were chosen in a defined range (based on assumption): equipment and implementation costs at farm i (C_i^{α}) = 4400K-8800K (GBP), equipment and implementation costs at abattoir j (C_j^{β}) = 1100K-8.7K (GBP), transportation costs from farm i to abattoir j and from abattoir j to retailer k (TC_{ij}^{m}, TC_{jk}^{n}) = 20 (GBP), supply capacity of farm i (C_i^{α}) = 2.5K-4.4K, supply capacity of abattoir j (C_j^{β}) = 1.2K-1.8K, demand of abattoir j (D_j^{β}) = 800-1.3K, demand of retailer k (D_j^{β}) = 100-800K, travel distance from farm i to abattoir j (d_{ij}^{m}) = 23-400, travel distance from abattoir j to retailer k (d_{jk}^{n}) = 110-162, vehicle capacity (W_V) = 100, quality percentage of livestock transported from farm i to abattoir j (Q_{ij}^{m}) = 0.75-1, freshness percentage of meat pieces transported from abattoir j to retailer k (F_{jk}^{n}) = 0.75-1, minimum required number of working hours h per laborer at farm i and abattoir j ($h_i^{l\alpha}, h_j^{l\beta}$) = 1,800 for each facility, labor cost per hour at farm i and abattoir j ($LC_i^{\alpha}, LC_j^{\beta}$) = 6.5 (GBP), working rate per labor l at farm i and abattoir j ($R_i^{l\alpha}, R_j^{l\beta}$) = 50 (item), profits per item at farm i (P_i^{α}) = 30 (GBP) and profits per item at abattoirs j (P_j^{β}) = 25 (GBP). It should be noted that the transportation distances between supply chains facilities were estimated using Google-Maps; also, the demand reported is the total demand over a one year period.

5.1 Computational Results and comparison

Using the above numerical data, the mixed-integer linear programming model described in Section 3.2 was solved using three approaches on a computer with corei5-CPU 2.60 GHz, RAM 4.00 GB, using the LINGO software.

Table 1 elucidates the four criteria solutions when solved individually. The total cost would be minimized to 194,180 GBP. If the criterion function one was only considered, while in this solution the criterion function two, three and four worsen to 75 %, 8,885 items and 363,080 GBP respectively. On the antithesis, if the second criterion function (F_2) was only considered, consumer satisfaction would increase to 100 %. However, the total cost was increased to 491,000 GBP in this solution. Considering the third criterion (F_3) individually, product freshness would increase to 13,099 items with an increase in the total cost (481,390 GBP), customer satisfaction (99 %) and profits (728,000 GBP). In this situation, the contradictory is manifested between these four criteria functions. However, moving to an improvement in consumer satisfaction and product freshness in the supply chain requires significantly higher cost investment. In Table 1, it can be noticed that no solution is optimal i.e. it is impossible to obtain an optimal solution for the four criteria when solving them individually.

Table 1. The values for each criterion function when considered individually.

Criterion function	Min F_1	Max F_2	Max F_3	Max F_4
F_1	194,180	0.75	8,885	363,080
F_2	491,000	1	13,099	728,000
F_3	481,390	0.99	13,099	728,000
F_4	491,000	0.99	13,099	728,000

Table 2. Pareto solutions obtained by using three different approaches.

Solution approach	#	Min (F_1) (GBP)	Max (F_2) (%)	Max (F_3) (items)	Max (F_4) (GBP)	Open farms	Open abattoirs
The developed approach	1	194180	0.75	8885	363080	1 0 0 1	0 1 0 1
	2	194180	0.75	8885	363080	1 0 0 1	0 1 0 1
	3	194180	0.77	9411	507260	0 0 1 1	0 1 0 1
	4	226150	0.815	10162	567860	1 0 1 1	1 0 1 1
	5	253475	0.85	10876	596000	1 1 1 1	1 1 0 1
	6	355336	0.9	11444	635240	1 1 1 1	1 0 1 1
	7	392720	0.94	12131	666800	1 1 1 1	0 1 1 0
	8	475660	0.99	13092	723620	1 1 1 1	1 1 0 1
Compromise programming	1	194180	0.75	8885	363080	1 0 0 1	0 1 0 1
	2	223257	0.776	9411	384534	1 0 1 1	0 1 0 1
	3	248214	0.8	9937	406025	1 0 1 1	0 1 0 1
	4	273171	0.826	10473	467927	0 0 1 1	0 1 0 1
	5	300475	0.85	10989	529011	1 0 1 1	1 0 1 1
	6	345228	0.91	11515	590501	1 1 1 1	1 1 0 1
	7	382940	0.95	12041	651995	1 1 1 1	1 0 1 1
	8	468475	1	13089	710610	1 1 1 1	0 1 1 0
Weighted Tchebycheff	1	194180	0.75	8885	363080	1 0 0 1	0 1 0 1
	2	194180	0.75	8885	363080	1 0 0 1	0 1 0 1
	3	249231	0.78	8920	395060	1 0 1 1	1 1 1 1
	4	288557	0.8	9808	4560100	1 1 1 1	1 1 1 1
	5	338858	0.85	10414	470000	1 1 1 1	1 1 1 1
	6	422451	0.91	11094	586880	1 1 1 1	1 1 0 1
	7	539128	0.96	12376	620120	1 1 1 1	1 1 1 1
	8	580471	0.99	13029	713480	1 0 0 1	0 1 0 1

To this aim, three solution approaches were employed seeking the Pareto sets derived from co-optimizing the four contradicting criteria functions being considered (simultaneously) as minimizing the total cost (F_1) in addition to maximizing consumer satisfaction (F_2), product freshness (F_3) and profits (F_4). In aiming to obtain Pareto-optimal solutions using the three approaches: (i) for the developed solution approach, the values of the four criteria functions illustrated in Table 1 were given as optimal values ($F^*_1, F^*_2, F^*_3, F^*_4$) along with an assignment of different weight values (w_1, w_2, w_3, w_4) for the solution function F (Eq.17); (ii) for the compromise programming approach, eight ђ values were increasingly assigned from 8,885 to 13,099 with step 526 for criterion two (Eq.19), from 0.75 to 1 with step 0.25 for criterion three (Eq.21) and from 360,000 to 728,000 with step 46,000 for criterion four (Eq.23); and (iii) for the Weighted Tchebycheff approach, the values of the four criteria functions illustrated in Table I were given as optimal values (F^*_1, F^*_2, F^*_3, F^*_4) along with an assignment of different values (l_1, l_2, l_3, l_4) for the solution function F (Eq.25). The obtained three sets of Pareto-optimal solutions using the three solution approaches are shown is Table 2. The third column of Table 2 represents the obtained values of the first criterion function (F_1) in terms of GBP, obtained values of the second, third and fourth criterion functions (F_2, F_3 and F_4) are presented in columns four, five and six respectively. The last two columns (right-end) correspond to the number of farms and abattoirs that should be established. For instance, solution#4 for the first solution approach obtained by an assignment of $w_1 = 0.6, w_2 = 0.2, w_3 = 0.1$ and $w_4 = 0.1$; accordingly, minimum total cost equals 226,150 GBP while maximum consumer satisfaction equals 81.5 %, maximum product freshness equals 10,162 items and maximum profits equals 567,860 GBP. This solution consists an establishment of farms number one, three and four (1 0 1 1) and abattoirs number one, three and four (1 0 1 1). As can also be observed in Table II, the Pareto optimal cannot get better in one criterion function except worsening its performance of the other three criterion functions. Fig. 2 illustrates a further comparison among the Pareto fronts obtained from the investigated solution approaches.

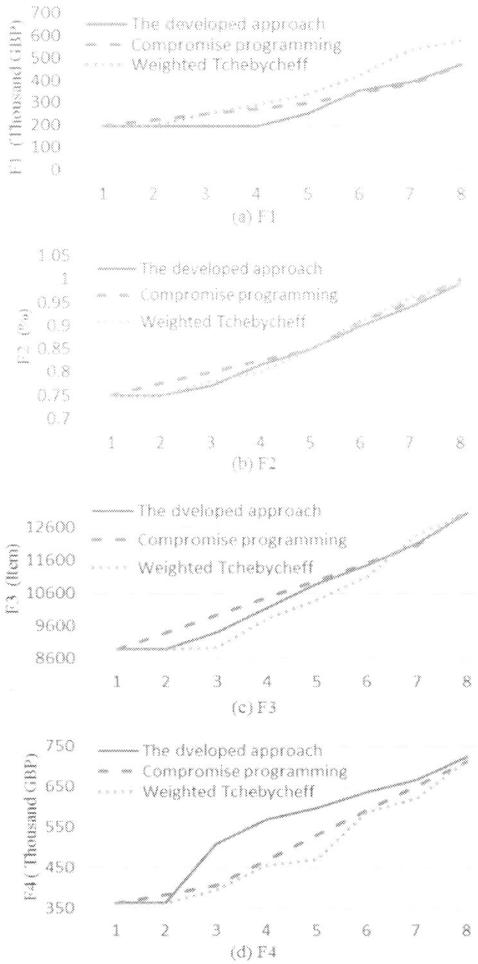

Figure 2. Pareto fronts using three solution approaches (a) F1, (b) F2, (c) F3, (d) F4.

It is noteworthy that in Table 2, the value of maximum profits for the eight solutions is more than the total costs which proves the feasibility in terms of economic costs of the proposed RFID-enabled meat supply chain after one year period for the RFID implementation.

6. Conclusions

This paper presents a study in investigating the economic feasibility of a proposed RFID-enabled monitoring system using a developed multi-criteria optimization model. The first criterion was the total cost; the second criterion was customer satisfaction which includes the fulfillment of consumers' demand; product freshness and profits were considered as third and fourth criteria respectively. The work findings show that high profits can still be achieved which proves that the RFID implementation in such a food supply chain is feasible in terms of economic costs. The direction for future research is to consider the effects of uncertain demand and costs regarding the investment evaluation.

Acknowledgment

The authors would like to express their gratitude to the Higher Committee for Education Development (HCED) in Iraq for the financial support in this study.

References

1. A. Mohammed, Q. Wang, Integrity of an RFID-enabled HMSC network, *Pro. the 3rd ICDEIS*, China, 79-86 (2015)
2. A. Ahmed, Marketing of halal meat in the United Kingdom, Bri. F. J., **10**, 655–670 (2008)
3. R.L. Meier, M.R. Williams, R.B. Singley, Supply Chain Management: Strategic Factors From The Buyers, Perspective, *J. Ind. Tec.*, **20**, 1-8 (2012)
4. W. Verbeke, P. Rutsaert. K. Bonne, L. Vermeir, Credence quality coordination and consumers' willingness-to-pay for certified halal labelled meat. Meat Sci., **95**, 790–797 (2013)
5. S. Peattie, K. Peattie, A. Jamal, Influences on child nutrition: Bri. Muslims. Pub. H. Nutri., **9/7a**, 181-182 (2006)
6. M. Gen, R. Cheng, Genetic *Algorithms and Engineering Design*, (New York: John Wiley & Sons 1997)
7. K. Deb, *Multi-Objective Optimization using Evolutionary Algorithms*, (New York, John Wiley & Sons, 2001)
8. Z.J. Max Shen, M.S. Daskin, Trade-offs between customer service and cost in integrated supply chain design. Man. & Ser. Op. Man., **7**, 188–207 (2005)
9. R.G. Pandu, Multi-Objective Optimization: Techniques And Applications In Chemical Engineering (Advances in Process Systems Engineering). Singapore: Wor. Sci. Pub., (2009)
10. G. Barbarosoglu, d. Ozgur, Hierarchical design no fan integrated production and 2 echelon distribution system, Eur. J. Op. Res., **118**, 464–84 (1999)
11. H. Sadjady, H. Davoudpour, Two-echelon, multi-commodity supply chain network design with mode selection, lead-times and inventory costs, Com. Op. Res., **39**, 1345–54 (2012)
12. H. Selim, C. Araz, I. Ozkarahan, Collaborative production–distribution planning in supply chain: a fuzzy goal programming approach, Tra. Res. Part E: Log. Tra. Rev., **44**, 396–419 (2008)
13. J. Ferrio, J. Wassick, Chemical supply chain network optimization, Com. Che. Eng., **32**, 2481–504 (2008)
14. U. Tuzkaya, S. Önüt, A holonic approach based integration methodology for transportation and warehousing functions of the supply network, Com. Ind. Eng., **56**, 708–23 (2009)
15. V. Chankong, Y. Haimes, *Multi-objective decision making theory and methodology* (New York: Elsevier Science, 1983)
16. K. Miettinen, *Nonlinear Multiobjective Optimization*, (Ed. 1st . Springer US: Springer Science+Business Media New York, 1998)

Influence Study of the Viscoelastic Fluids Features in Drag Reduction in Laminar Regime Flow in Pipeline

Guillermo Vilalta[1], Mário Silva[2] and Alejandro Blanco [3]

[1,2] *Federal University of São João del Rei, Thermal Sciences and Fluids Department, 36307-352 São João del Rei - MG, Brazil.*
[3] *Neurology and Neurosurgery Institute, Department of Engineering, Havana, Cuba.*

Abstract. The drag reduction by polymer addition is wide interest in several areas. It has been shown that the polymer addition cushions the dissipative effects in turbulent flows. The main objective of this work is to establish a methodology for the numerical simulation of viscoelastic fluid through internal subroutines implemented in the Fluent code, via UDF. The validation of this methodology is made for the laminar flow regime case in pipeline. To describe the viscoelastic effect, it was used the Finitely Extensible Nonlinear Elastic model closing with Peterlin model. To taking in account the viscous effects $50 \leq Re \leq 2000$ values were used. In addition, for the polymer concentration analysis it was used values which depend on the polymers molecular weight and the solution concentration that ranged from $0 \leq Cw \leq 20$. The molecular elasticity and extensibility were maintained at constant values. The results showed that the addition of polymers regardless of their molecular weight in laminar flow regime causes no change in power dissipation. This result, which is consistent with the literature, is a significant advance in defining a credible and appropriate methodology to viscoelastic fluid flow study by UDF implementation of constituent models that characterize these fluids.

1 Introduction

An ideal elastic material, under pressure, responds with reversible deformation. When the tension ceases, the body returns to its original shape and volume. A viscous material when subjected to a shear stress, responds irreversibly to deformations, dissipating energy. The viscoelastic fluid undergoes both effects, dissipates some energy and stores the rest. These properties of viscoelastic fluids, has caused a significant increase of its practical applications, mainly those related with the fluid flow in internal flows, particularly in long distance transport fluid. It has been shown that the addition of a small amount of polymer, surfactant or other substances can cause relevant reductions on the resistance force, which leads to a considerable reduction of the power consumption in pumping systems. [1-3].

The first evidence of drag reduction due to certain additives, in particular polymers and/or fibers, was reported by Toms [1]. His results showed that the addition of 10 ppm (by weight) of polymethyl methacrylate in a monochlorobenzene flow with high Reynolds number reduced substantially (30-40%) the pressure drop compared to the pure solvent at the same flow rate. Other later studies confirmed this phenomenon, which has become known in the literature as Toms phenomenon. Despite the development of new techniques of visualization and mapping flows and measurement instruments, even today, there is no consensus on what is the real influence of the addition of polymers on the flow structure.

The drag reduction depends on the type of polymer and its concentration. Experimental and numerical studies have demonstrated that the drag reduction increases with increasing concentration of the polymeric solutions [2] and [3], but it can not exceed an upper limit, which is called the "Maximum Drag Reduction Asymptote" (MDRA). This curve is independent of the polymer concentration or its characteristic [4].

Furthermore, the drag reduction by polymer addition depends not only the concentration but also the kind of polymer, its molecular weight and the flow regime.

The drag reduction becomes increasingly effective at high molecular weight.

In addition, there is a consensus in the researches conducted to date that it is possible to check that the effect of drag reduction, by polymers injection in pipes, is highly effective in turbulent regime, but does not seem to have an effect on laminar flow regime [3], [5], [6].

Zhang et al [7] have demonstrated by numerical simulations that the polymer solutions flow using FENE-P model in a sudden expansion geometry of 1:3 ratio and a rectangular cross section, in a laminar regime does not lead to a drag reduction. Oliveira [8] had also simulated for the same geometry and conditions of Zheng et al [7], but using DNS (Direct Numerical Simulation) and the FENE-CR viscoelastic model. His results from the point of view of the drag reduction coincide with the Zheng et al [7].

The major problem appears when the simulation of viscoelastic fluids because most commercial software does not have models for such fluids. This research has the aim of developing a methodology for simulation of viscoelastic fluids for consider a conformational model that allows modeling the polymer accurately and efficiently in confined flows. In order to do that it is intended to define an internal subroutine to be implemented in *Fluent* code, via UDF, in a way to incorporate the scalar equations and governing equations source terms for the calculations. For the first verification test of this methodology it was selected the simplest case that consist in a laminar flow in pipe of viscoelastic fluids at different concentrations defined by the FENE-P conformation model.

2 Numerical simulation

2.1 Problem description

It was simulated the flow inside a pipe with a commercial diameter of 2" (50.8 mm) and a unit of length. It is not necessary a greater length of pipe because it was applied the fully developed velocity profile at the entrance according to the Reynolds number, programmed by the UDF [9].

The mesh used in the simulations can be seen in Figure 1. It can be noted that near to the wall there is a further refinement of the grid, because this is a region of utmost importance in drag reduction question. More information at Silva [9].

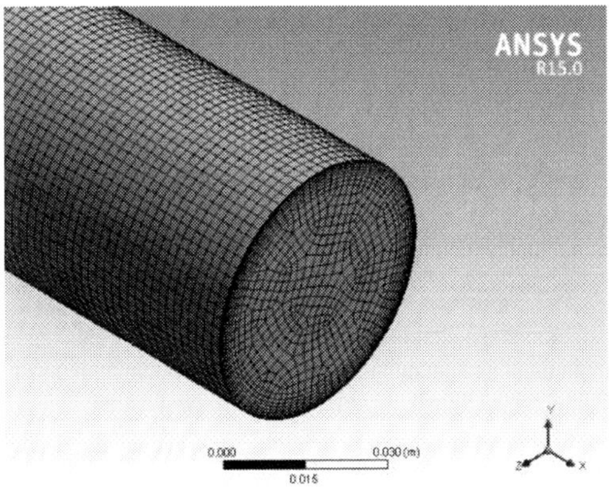

Figure 1. Mesh used in this study.

2.2 Government equations

When compared with the Newtonian fluids, it is remarkable the need for a more sophisticated constitutive relation to evaluate the stress tensor for viscoelastic fluids. FENE-P was the model used in this study because it is one of the polymers models used with some success in fluid dynamics simulation (CFD) since it eliminates the necessity of determining the statistical mean at each grid point in any moment in time. This model demonstrates to be able to capture some of the most important behaviors of polymeric flows such as drag reduction due to polymer addition in pipes [10]. In addition, FENE-P it is a non-linear model, which represent more accurately the features of viscoelastic fluids.

For FENE-P model, the continuity equation is unchanged compared to a Newtonian flow:

$$\frac{\partial \rho}{\partial t} + \frac{\partial}{\partial x_i}(\rho u_i) = 0 \tag{1}$$

Where u_i is the local instantaneous velocity component, ρ is the fluid density, x_i is the Cartesian coordinate component and t is the time. In the other hand, the Navier-Stokes equations change, because of the stress caused by the polymer in the flow, emerging a new term related to the solute [7]:

$$\rho \frac{\partial u_i}{\partial t} + \rho u_j \frac{\partial u_i}{\partial x_j} = -\frac{\partial p}{\partial x_i} + \frac{\partial \tau_{ij}{}^{s}}{\partial x_j} + \frac{\partial \tau_{ij}{}^{p}}{\partial x_j} \tag{2}$$

Where p is the pressure, $\tau_{ij}{}^{s}$ is the stress caused by the solvent being exactly the same Newtonian fluid flow stress (Equation (3)), and $\tau_{ij}{}^{p}$ is the elastic stress caused by the polymer (Equation (4)):

$$\tau_{ij}{}^{s} = \mu_s \left(\frac{\partial u_i}{\partial x_j} + \frac{\partial u_j}{\partial x_i}\right) \tag{3}$$

$$\tau_{ij}{}^{p} = N(x,t) \frac{\mu_p}{\lambda} \left[f(r) C_{ij} - \delta_{ij}\right] \tag{4}$$

C_{ij} is the tensor conformation component of the polymer molecule, μ_s is the dynamics solvent viscosity, $N(x,t)$ is the unit concentration and it is equal to 1 when the polymer or surfactant concentration is homogeneous along the flow (which is the case under consideration), μ_p is the solution viscosity, λ is the polymer time relaxation, δ_{ij} is the Kronecker delta and $f(r)$ is the Peterlin function expressed by:

$$f(r) = \frac{L^2 - 3}{L^2 - r_{mol}{}^2} \tag{5}$$

In the equation above, L is the molecular (or micellar) extensibility polymer in relation to its equilibrium length. r_{mol} is the molecular (or micellar) length at steady state, and it is equal to the square root of the conformation tensor thread:

$$r_{mol}{}^2 = \sum C_{ii} \tag{6}$$

C_{ij} can be formed from the inner product vector distance **r** from end to end molecule. Furthermore, according to Zhang et al [7] for this model $C_{ij} = C_{ji}$. In the "coiled" state (not extended or spiral) the polymers are usually spherical and C_{ij} can be normalized: $C_{ij}{}^{eq} = \delta_{ij}$. The tensor conformation transport equation C_{ij}, also according to Zheng et al [7], is:

$$\frac{\partial C_{ij}}{\partial t} + u_k \frac{\partial C_{ij}}{\partial x_k} = C_{ik} \frac{\partial u_j}{\partial x_k} + C_{kj} \frac{\partial u_i}{\partial x_k} - \frac{1}{\lambda}\left[f(r) \delta_{ik} C_{kj} - \delta_{ij}\right] + \Upsilon \frac{\partial}{\partial x_k}\left(\frac{\partial C_{ij}}{\partial x_k}\right) \tag{7}$$

The last term of this transport equation is the artificial viscosity term (which is attached to diffusivity) which is added to increase the stability and convergence of the calculation. In the present study, a convergent solution can be obtained with $\Upsilon = 0$ adjusting the sub-relaxation factors in the numerical simulations.

The Weissenberg number, *We*, expresses the relationship between the fluid relaxation time and a specific temporal process:

$$We = \dot{\gamma}\lambda = \lambda\frac{U}{D} \tag{8}$$

Where D is the pipe diameter, U is the mean velocity and $\dot{\gamma}$ is the strain rate. For a polymer solution, the Reynolds number becomes [7]:

$$Re = \frac{\rho U D}{\mu_p + \mu_s} \tag{9}$$

Finally, another important parameter also in the viscoelastic model is the relationship between the viscosities of the pure solvent and polymer solution, as shown in Equation (10).

$$\beta = \frac{\mu_s}{\mu_s + \mu_p} \tag{10}$$

2.3 Definition of parameters for simulations

To evaluate the influence of the material type in the viscoelastic fluids flow in the simulations it was used two types of polymer, one with a high molecular weight (commercially called *Iqapol*) and another one with a low molecular weight (commercially called *Iqapac*) varying their concentrations in weight, C_w. As the solvent, it was used water, which is a highly fluid used in the industry.

According to Vilalta [3], the polymeric solution density formed by the polymers here used does not vary substantially. So it is necessary only the viscosity coefficient of the solutions determined by the polymer concentrations. Thus the viscosity values used herein are according to the values found by Vilalta [3] as a function of shear rate how is shown in Table 1. And for the density it was used the constant value $\rho = 998{,}2\ kg/m^3$, which corresponds to the water density at ambient conditions of 20^0C temperature and 1 atm (101.325 Pa) pressure.

To take into account the viscous effects, the Reynolds number ranged between 50 and 2000 were used, so it can characterize the laminar regime and allowing understanding the viscoelastic fluid behavior near the transition zone.

The molecular extensibility was kept constant, $L^2 = 100$, which is a typical value for polymer solutions, since it allows considerable molecular extension before the fully extended state be achieved. Different values of 100 there is a very small influence on the flow [8].

The Weissenberg number was also kept constant and equal to 2, because according to Silva et al [9] by increasing the value of We it increases concomitantly the difficulties of convergence and the numerical method instability.

Thus, the values for the parameters discussed above were adopted according to Table 1.

Table 1. Parameters used in the studied.

Iqapol (High molecular weight)			
C_w (*ppm*)	$\mu_p(mPa\ s)$	β	Re
5	1.309	0.434	50
10	1.45	0.410	100
15	1.645	0.378	500
20	1.78	0.360	2000
Iqapac (Low molecular weight)			
C_w (*ppm*)	$\mu_p(mPa\ s)$	β	Re
15	1.21	0.453	50
18	1.30	0.435	100
20	1.37	0.423	500
			2000

The head loss (h_f) in a flow of Newtonian fluid inside pipes can be calculated according to Equation (11) [11].

$$h_f = \Delta z + \frac{\Delta p}{\rho g} + \frac{U_1{}^2 - U_2{}^2}{2g} = f\frac{L}{D}\frac{U^2}{2g} \tag{11}$$

Where Δz is the topological difference between the inlet and outlet pipe, Δp is the pressure difference also between the input and output, U_1 and U_2 are the mean velocities at the inlet and outlet pipe respectively, L is the pipe length, g is the local gravity acceleration and f is the friction coefficient, which is a dimensionless parameter used to calculate the pressure drop in pipeline.

As the behavior of the flow is associate to the value of f, the variation of this parameter was selected to comparing the Newtonian and viscoelastic fluids to analyze the drag reduction. For the particular case of the laminar flow regime, the friction coefficient of a Newtonian fluid is analytically given by Equation (12) [11].

$$f = \frac{64}{Re} \tag{12}$$

2.4 Numerical method and boundary conditions

At the pipe inlet, a fully developed velocity profile was programmed and implemented, via UFD, as inlet boundary condition, according to each value of the Reynolds number analyzed. Using the Neumann boundary condition at the inlet and using the derivative value equal to zero, the velocity profile for the viscoelastic fluid flow by using FENE-P method is the same as the Newtonian fluid [9]:

$$u = 2U\left[1 - \left(\frac{r}{R}\right)^2\right] \tag{13}$$

Where R is the pipe radius and r is the radial coordinate. At the outlet, it was selected the boundary condition linked to the pressure value, which assigns 0 Pa relative value. In the other words, it receives the absolute value of the reference pressure defined in the software such as atmospheric pressure with 101.325 Pa. Finally, in the wall region was defined the wall condition which imposes a non-slip condition.

The algorithm provided by *Fluent* software is based on the finite volume method. Double precision was used in the calculations. The convergence criteria (residuals) adopted for all variables was 10^{-5} and different *underrelaxation* factors were adopted for different conditions.

For spatial discretization of the momentum, continuity and transport equations for the tensor conformation was used the second order *Upwind* method. For the gradients numerical calculation, it was used the *Least Squares Method Cell-Based Gradient*.

For time discretization, it was used a second order implicit method known as *"Backward Differentiation"*, described in Ferziger & Peric [12].

The *Pressure-Implicit Method with Splitting of Operators (PISO)* that is more appropriate transient regime was adopted to solve the pressure-velocity coupling, and the standard scheme was used to the pressure equation discretization.

The tensor conformation starts considering its equilibrium state, when no force acts in the flow. And when in a state of non-equilibrium, for the tensor conformation for all surfaces (inlet, outlet and wall) it was adopted the Neumann boundary condition with derivatives equal to zero.

3 Results

3.1 Polymer concentration effect

Polymeric solutions concentration is certainly one of the most important parameters in respect to the drag reduction phenomenon by polymer addition. As above-mentioned, the drag reduction tends to increase with concentration increasing, as was found in Lumley [2] and Vilalta [3] studies. However, both studies are related to the turbulent flow regime.

The analysis of the polymer concentration effects on friction coefficient and in head loss was carried out by mean of two representative cases. Case 1 represents a polymeric solution formed with a high molecular weight polymer, varying the polymer concentration according to Table 1 and keeping constant the Reynolds number (*Re=50*), the elasticity parameter (*We=2*) and molecular extensibility (*L=10*). Case 2 represents a polymeric solution formed with a low molecular weight polymer, varying the polymer concentration according to Table 1 and keeping constant the Reynolds number (*Re=500*), the elasticity parameter (*We=2*) and molecular extensibility (*L=10*). The Newtonian cases were simulated using the same Reynolds number for each case above defined.

Figure 2(a) shows the relation between the friction coefficient and the polymeric solutions concentration. It is possible to observe that the friction coefficient remains essentially unchanged with the concentration variation and no significant difference exist when compared with Newtonian fluid. This result is in agree with the literature. In Figure 2(b) is shown the relation between the head loss and the concentration of the polymeric solutions. When the concentration increases (β decreasing), an increasing in head loss is verified.

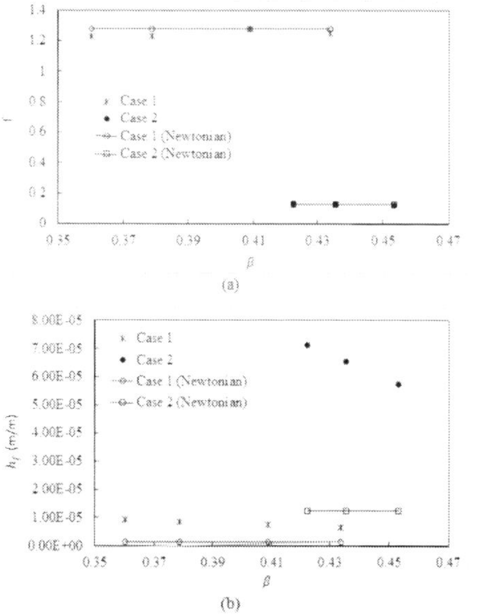

Figure 2. (a) Polymer concentration effect on the friction coefficient, *f*, for Cases 1 and 2; (b) Polymer concentration effect on the head loss, h_f, for Cases 1 and 2.

When compared with Newtonian fluid, the drag reducing effect is more significant in viscoelastic fluid. However, this is not directly associated to the effect of viscoelastic characteristics of the fluid. As already pointed out, no changes in friction coefficient in laminar flow of viscoelastic fluid.

The variation of head loss with the concentration of polymeric solution can be explained by the fact that the viscosity of viscoelastic fluid will always be higher than the Newtonian fluid, and tends to increase according to concentration increases. So, to keep constant the Reynolds number, the viscoelastic fluid flow velocity should increase proportionally with the viscosity. In the simulation here carried out, it is considered that the pipeline is horizontal ($\Delta z=0$) and has transversal section constant. Therefore, the head loss is equals to pressure loss, as expressed by Equation (11). This means that the increase in velocity, with friction factor, diameter and length being constant, the head loss increases.

3.2 Viscous effect

For analysis of the viscosity effect, the Reynolds number was varied, taking into consideration that this dimensionless number is associated to the viscous forces acting in the flow. This analysis was carried out by means of Case 3 and Case 4.

Case 3 represents a polymeric solution formed with a high molecular weight polymer varying the *Re* according to Table 1 and keeping the polymer concentration constant, $C_w = 20\,ppm$ ($\beta = 0{,}360$). In the Case 4, which corresponds to low molecular weight polymer, it was varied the *Re* according to Table 1 and keeping the polymer concentration constant, $C_w = 15\,ppm$ ($\beta = 0{,}453$). The Newtonian cases were simulated using the same *Re* for each case defined above for pure water.

As might be expected, in all viscoelastic cases the curves follow the Newtonian case, so there is no drag reduction due to the polymer interference. These results are showed in Figure 3(a). It can also be observed from Figure 3(b) that the viscoelastic fluids have a loss of pressure greater than in Newtonian fluid in all cases examined. This is explained by the velocity increases that occurs in these fluids while keeping Reynolds number constant and not due to the friction factor *f*, as it was explained above in the analysis of the polymer concentration effect.

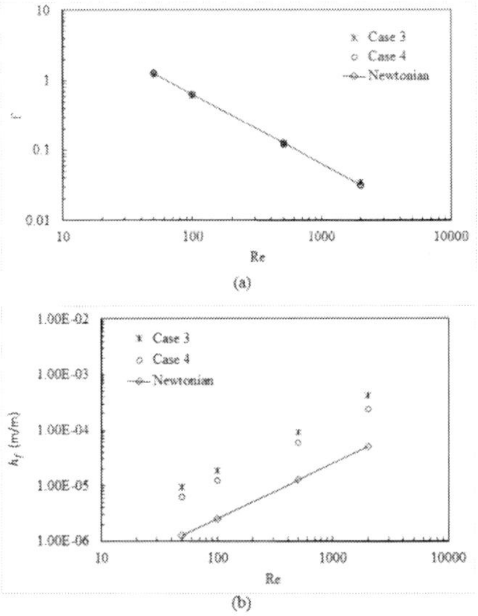

Figure 3. (a) Viscous effect in the friction coefficient, *f*, for Cases 3 and 4; (b) Viscous effect on the pressure loss, h_f, for Cases 3 and 4.

These results are in agreement with those found experimentally by Virk (apud Diamond [5] and Wójs [6]), even in their case the latter study has modeled the viscoelastic fluid by the power law. No drag reduction effects were also verified by Zheng et al [7] and Oliveira [8] for a rectangular cross section in laminar regime.

4 Conclusion

This work represents a numerical investigation about the influence of the main fluid viscoelastic features on drag reduction phenomenon in laminar flow. The viscoelastic fluid was modeled by the FENE-P model.

The results obtained in this work show that in laminar flow the concentration and the kind of polymer do not interfere on drag reduction phenomenon. From drag reduction point of view, no changes in friction coefficient behavior were verified for all cases studied.

An interesting result was obtained when the influence of viscous effects on drag reduction was analyzed. When the pressure loss between Newtonian and viscoelastic fluid are compared, for the same values of Reynolds number, a higher loss pressure in viscoelastic fluids is observed. This result may seem contradictory because what is expected is a reduction in the pressure loss. As already indicated, to keep the Reynolds number constant, the mean velocity of the flow must increase. If no chang in the friction coefficient is verified, then an increase in the pressure loss should occur.

Establishing a numerical methodology, via UDF, as here presented, has important numerical challenges that are necessary solve, due to the inclusion of the conformation tensor transport equation. Such challenge concerns about the simulations convergence and numerical instabilities. All the challenges faced in this study were solved by using different strategies mainly related to appropriate selection of the time step and of the spatial discretization method. Different time steps were used for each of the cases studied. For numerical instabilities, second order *Upwind* model was used instead of one of third order. The latest one is more accurate than the second order *Upwind*, but has serious instability problems and is computationally more complex, especially when it comes to three-dimensional mesh.

The methodology here presented constitutes a significant advance in defining a credible and appropriate methodology for the study of viscoelastic fluid flow, by implementing the constitutive model. Further advances in this field will allow to extend the study to other flow regimes to get better insight of the inherent physic of the drag reduction phenomenon, aiming to extend their application in other industrial applications.

References

1. B. A. Toms, Anais Processing of The International Congress on Rheology, 135-141 (1949)
2. J. L. Lumley, Symp. Mathematica, **9**, 315-334 (1972)
3. G. A. Vilalta, Ph.D. thesis, USP/Brazil (2000)
4. P. S. Virk, Aiche J. **21**, 625–656 (1975)
5. P. Diamond, et al. *Drag Reduction by Polymer Additives* (The MITRE Corporation, Virginia, 1992)
6. K. Wójs, J. of Non-Newt. Fluid Mec. **48**, 337-355 (1993)
7. Z. Zheng, F. Li, J. Yang, Hindawi Publishing Corporation, **2013**, 1-13 (2013)
8. P. J. Oliveira, G. N. Rocha, R. J. Poole, J. of Non-Newt. Fluid Mec. **114**, 33-63 (2003)
9. M. L. F. Silva, Master dissertation, UFSJ/Brazil (2015)
10. A. P. Mósca, et al, IVth National Conference on Fluid Mechanics, Thermodynamics and Energy, 169-178 (2012)
11. F. M. White, *Mecânica dos Fluidos* (Mc Graw Hill, Porto Alegre, 2011)
12. J. H. Ferziger, M. Peric, *Computational methods for fluid dynamics* (Rosewood, New York, 2002.

Effect of Fins on the Transition to Oscillating Laminar Natural Convection in an Enclosure

Emel Evren Selamet and Ahmet Selamet

Ohio State University, Mechanical and Aerospace Engineering, USA

Abstract. Flow characteristics and evolution of heat transfer are investigated numerically within a differentially heated square cavity in the presence of thin fins which are either insulated or highly conductive and at the base wall temperature. The transition from steady-state to oscillatory and chaotic convection is observed at certain Rayleigh numbers with varying lengths of fins positioned at different locations.

1 Introduction

Modification of heat transfer in cavities with fins attached to the wall(s) has been subject of several investigations. Depending on the application, either the suppression or the enhancement of heat transfer such as solar collectors and electronic cooling by natural convection may be desirable. This goal guides the design through choosing the relevant parameters appropriately. Winters [1] and Sathiyamoorthy and Chamkha [2], for example, considered a vertical divider in the middle of bottom wall; Faces [3] studied two fins on the hot, one on the cold wall in a tall cavity; Nag, *et al.* [4], Shi and Khodadadi [5], and Bilgen [6] investigated a thin fin on the hot wall; and Nada [7] examined multiple fins on the hot wall of a narrow enclosure both horizontal and vertical.

The present work is a numerical investigation of heat transfer and fluid flow characteristics in a square enclosure which incorporates two fins of varying lengths and locations on hot and cold vertical walls. Transition from steady-state to oscillatory and chaotic convection occurring at certain Rayleigh numbers is examined as such geometric parameters are varied.

The definition and formulation of the problem is next, followed by the validation of the code, discussion of results from selected configurations, and concluding remarks.

2 Problem definition

Similar to other natural convection flow models, the present study considers a square cavity. The horizontal walls are thermally insulated and the vertical walls are kept at temperatures of T_h (hot left wall) and T_c (cold right wall), respectively. In terms of the following dimensionless variables $x=x^*/L$, $y=y^*/L$; $u=u^*L/\alpha$, $v=v^*L/\alpha$; $t=t^*\alpha/L^2$; $p=p^*L^2/(\rho\alpha^2)$; $\theta=(T-T_c)/(T_h-T_c)$ where x^*, y^* are the coordinates; u^*, v^* the velocities; t^* the time; p^* the pressure; T the temperature; L the width of the enclosure, the governing equations for mass, x- and y-momentum, and energy may be written for buoyancy-driven incompressible fluids as

$$\frac{\partial u}{\partial x}+\frac{\partial v}{\partial y}=0,$$

$$\frac{\partial u}{\partial t}+\frac{\partial u^2}{\partial x}+\frac{\partial uv}{\partial y}=-\frac{\partial p}{\partial x}+\Pr\nabla^2 u,$$

$$\frac{\partial v}{\partial t}+\frac{\partial uv}{\partial x}+\frac{\partial v^2}{\partial y}=-\frac{\partial p}{\partial y}+\Pr\nabla^2 v+Ra\,Pr\,\theta,$$

$$\frac{\partial \theta}{\partial t}+\frac{\partial u\theta}{\partial x}+\frac{\partial v\theta}{\partial y}=\nabla^2\theta,$$

where ∇^2 the two-dimensional Laplacian; $\Pr=\nu/\alpha$, the Prandtl number; $Ra=g\beta\Delta TL^3/\nu\alpha$, the Rayleigh number which accounts for the coupling between energy and momentum balances, g being the gravitational acceleration, β the coefficient of thermal expansion of the fluid, ΔT the temperature difference between left and right boundaries; ν the kinematic viscosity; and α the thermal diffusivity. Boussinesq approximation, which treats density as a constant in all equations except the gravitational force term in the momentum equation, is used. Even though the computations are performed by using primitive variables, flow field results are presented in terms of the stream function Ψ ($u=\partial\Psi/\partial y$, $v=-\partial\Psi/\partial x$). No slip boundary conditions are imposed on all walls and fins. The governing equations with primitive variables are solved on a staggered grid by applying the Godunov scheme to convective terms and centered finite difference to diffusive terms. Details of the method are explained in the earlier work of the lead author [8].

The fluid is assumed to be air with Pr= 0.71. In the present configuration, two horizontal fins are attached to both vertical (hot and cold) walls. Two cases are then considered in the study:

1. Fins are thermally insulated,
2. Fins are highly conductive, thus isothermal at the base wall temperature.

2.1 Validation for an enclosure with single fin

The results from the present predictions are first compared with those from the earlier work of Shi and Khodadadi [5] who studied natural convection in a differentially heated square cavity with a single thin fin mounted on the hot wall. They considered the fin at the base wall temperature (highly conductive fin) and obtained the solution with steady-state governing equations. They also indicated that the solution didn't converge due to not retaining the unsteady terms in some cases at Ra=10^6 and 10^7. Although a steady solution exists at those Ra in the absence of fin, it becomes unsteady with the introduction of fin depending on the fin's length and location. A representative case is considered here with a fin located in the middle of vertical wall. Fig. 1 compares at Ra=10^5 the results from the present study with [5] in terms of absolute maximum value of stream function field Ψ_{max} – a measure of intensity of natural convection – for a fin length of l_f=0.5 and as a function of fin location s. Here, both l_f and s are nondimensionalized with respect to *L*, with s representing the distance from top of the enclosure. The agreement between two studies is excellent. For this case, the solution is steady regardless of the fin location. For higher Ra, however, a similar comparison cannot be made since Ψ_{max} oscillates, and so does the Nusselt number (Nu) – the ratio of heat transferred by convection to that of conduction, or, alternatively the dimensionless temperature gradient. The resulting unsteady nature of the thermal field is examined in the present study in terms of the evolution of Nu at

Ra=10^7. Fig. 2 depicts Nu on the left side of cavity, without fin and with fins of different lengths located at s=0.5. Nu variation as a function of time shown in Fig. 2a is calculated only at the hot wall vs. hot wall plus fin in Fig. 2b which yields the total heat transfer. Bilgen [6] studied this problem numerically with different fin conductivities after introducing a ratio of conductivity of fin to fluid and varying this ratio between 0 (case 1 in the present work) and 60. Steady solutions were obtained from unsteady-state equations by using commercial code SIMPLER, though no referral was made to unsteadiness of solutions. She calculated Nu at the wall only, excluded heat transfer from the fin, and indicated that Nu decreased as a function of fin length as observed (on the mean) in Fig. 2a. Fig. 2 shows regular sinusoidal oscillations for l_f=0.1, 0.2, and 0.35; non-sinusoidal but regular for l_f=0.5; and non-sinusoidal and irregular behavior for l_f=0.7. In Fig. 2b the frequency of oscillations decreases, while the amplitude increases with increasing fin length for $0.1 \le l_f \le 0.35$. Nu is smaller than that of without fin for $l_f \le 0.1$ at this fin location. At lower Ra (=10^6), this is true for $l_f \le 0.2$ with the same parameters of fin and the solution is steady for $l_f < 0.35$ and decaying oscillations are observed for l_f=0.35 and 0.5.

Flow and temperature fields are visualized in Fig. 3 for Ra=10^7 in terms of streamlines and isotherms at three fin lengths chosen from Fig. 2: l_f=0.1(a,d), l_f=0.35(b,e), and l_f=0.5(c,f). One or two counter clockwise rotating vortices are observed above the fin in the separation region with their strength and size depending on the fin length. It is interesting to examine the flow field at different times of the cycle for oscillating flows. For example, Fig. 4 shows the streamlines at four different times each separated by an interval of 0.0008 which is approximately a quarter of the period of irregular oscillations for l_f=0.7 at Ra=10^7. The evolution in this sequence suggests that the unsteadiness is essentially caused by the separation region. Two small vortices are observed in that region at t=.1208, then one of them gets stronger affecting the flow field on top of it, and making the weaker one disappear near the end of the cycle

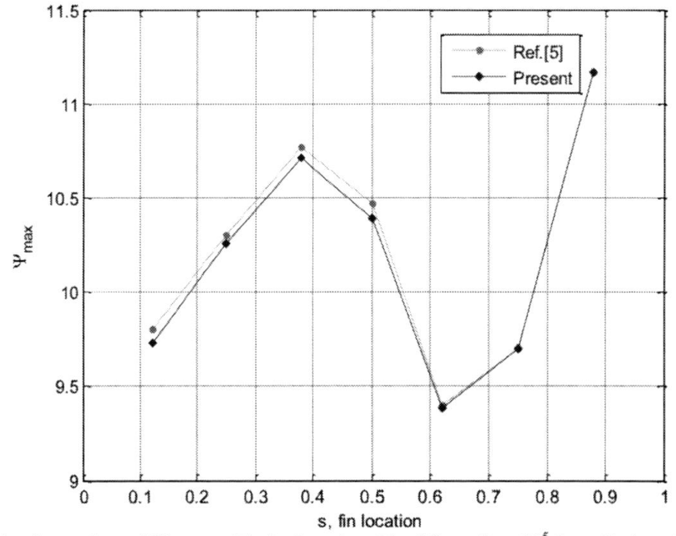

Figure 1. Variation of absolute value of Ψ max with the location (s) of fin at Ra=10^5 for a fin length of l_f=0.5.

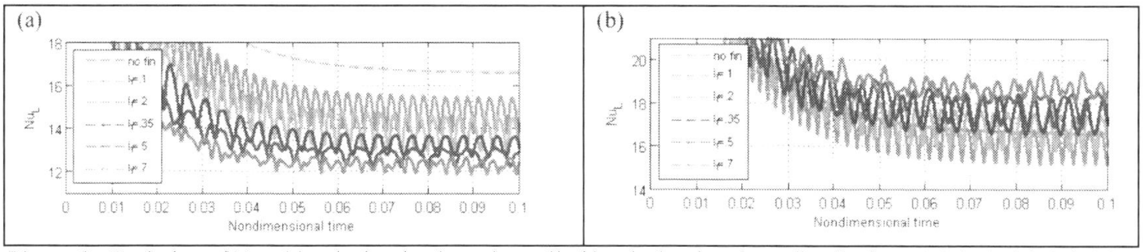

Figure 2. Evolution of Nu_L, (a) calculated only at the wall, (b) calculated at the wall + fin for s=0.5, l_f=0, 0.1, 0.2, 0.35, 0.5 and 0.7 for Ra=10^7.

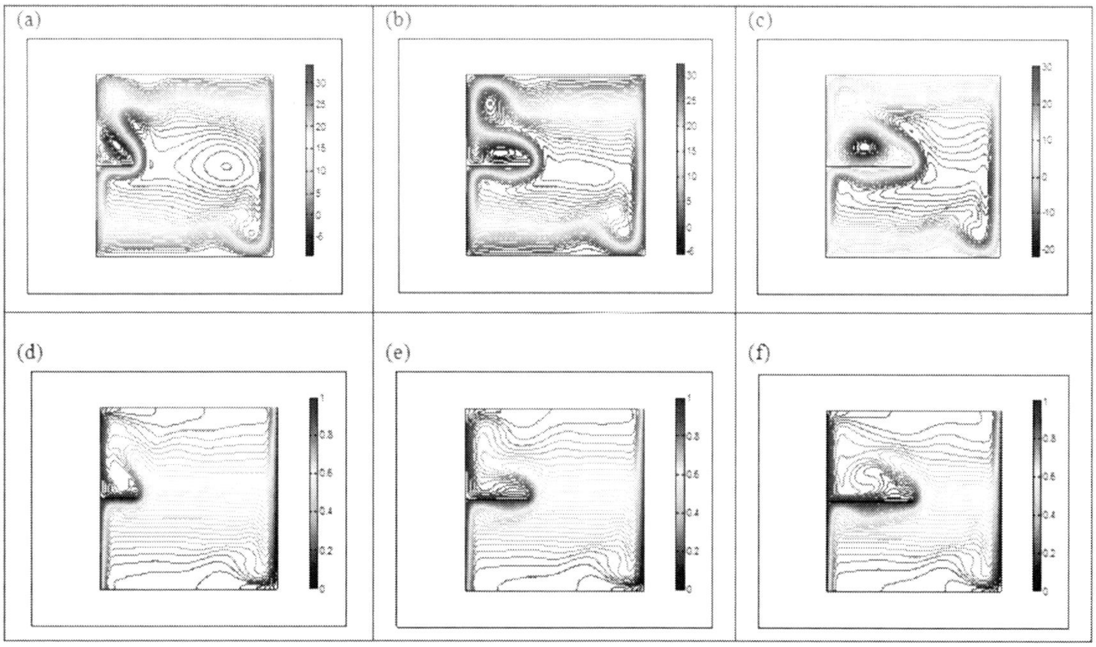

Figure 3. Streamlines and isotherms for l_f=0.1 (a,d), l_f=0.35(b,e) and l_f=0.5 (c,f,), respectively for Ra=10^7 at t=0.12.

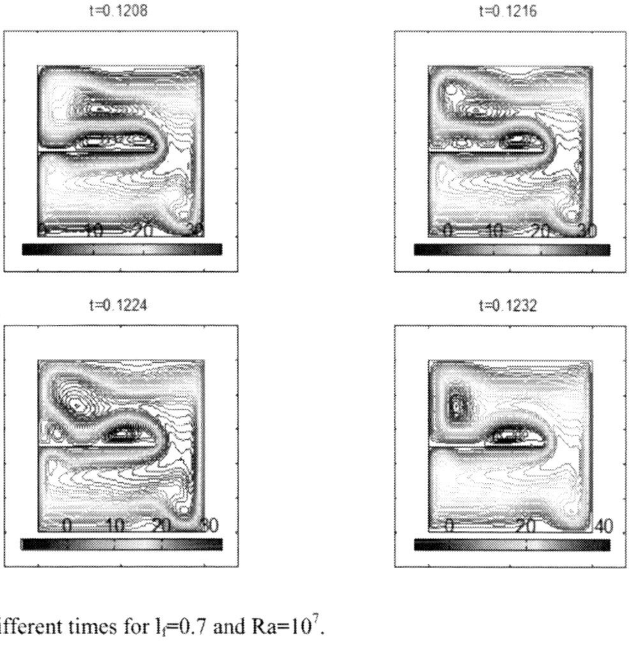

Figure 4. Streamlines at different times for l_f=0.7 and Ra=10^7.

3 Results for an enclosure with four fins

The schematic of the system studied here is shown in Fig. 5. The top and bottom walls are insulated, whereas the left and right walls are maintained at temperatures of T_h and T_c ($T_h > T_c$), respectively. Two fins are placed both at the left and right walls.

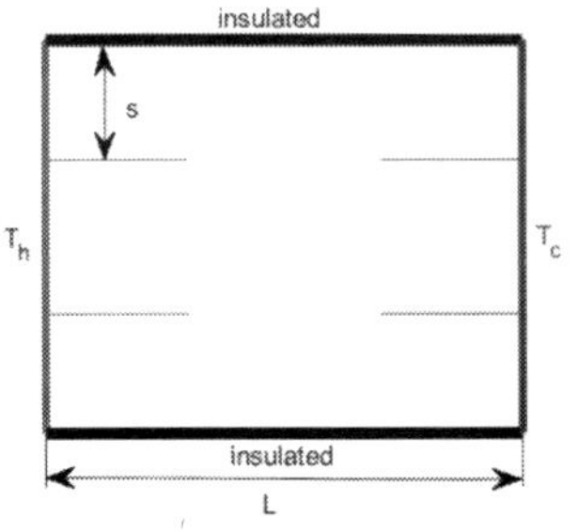

Figure 5. Schematic configuration.

The effects of fin length and spacing are examined for adiabatic (case 1) and highly conductive (thus isothermal at base wall temperature) fins (case 2) similar to [9] who considered a spherical container. Steady or pseudo-steady solution is obtained for $Ra<10^6$. Since the primary interest here is to study unsteady or oscillatory convection, the results only for high $Ra=7x10^6$ are presented and discussed. Figs. 6a, b depict Nu variation (with heat transfer from fin included) as a function of time for different l_f for both cases, at fin locations of s=[0.23, 0.77]

which represents a fin placement closer to adiabatic horizontal walls. With insulated fins, heat transfer decreases with fin length, whereas it is enhanced with highly conductive fins for all fin lengths except l_f=0.1. In comparison with case 2, case 1 yields a more steady behavior with Nu and a considerably less spread with respect to the fin length. Oscillating Nu is obtained only for l_f=0.25 in case 1, whereas larger oscillations are present for l_f=0.25 and 0.35 for case 2. Figs. 6c,d show Nu variation for fins placed evenly with s=[0.33, 0.67], exhibiting persistent sinusoidal oscillations particularly for l_f=0.35 in both cases. Nu oscillations are significant in Figs. 6b,d for l_f=0.25 and 0.35, indicating that highly conductive fins cause unsteady convection unless they are short. The comparison of Figs. 6b (with s=[0.23, 0.77]) and 6d (with s=[0.33, 0.67]) reveals higher frequency of oscillations in the former, with the oscillations for l_f=0.25 being regular in the former vs. irregular in the latter. Case 1 is visualized in Fig. 7 in terms of streamlines for three fin lengths (0.1, 0.2, and 0.35) and two prescribed fin locations at t=0.12. Each configuration has its own specific flow field also manifesting one or more vortices rotating counter-clock wise in the separation regions after the fluid passes the fins. Corresponding temperature field is given in Fig. 8. Case 2 is similarly visualized in Figs. 9 and 10. Note, particularly for l_f=0.35, how different the flow fields are between the two cases in both fin location

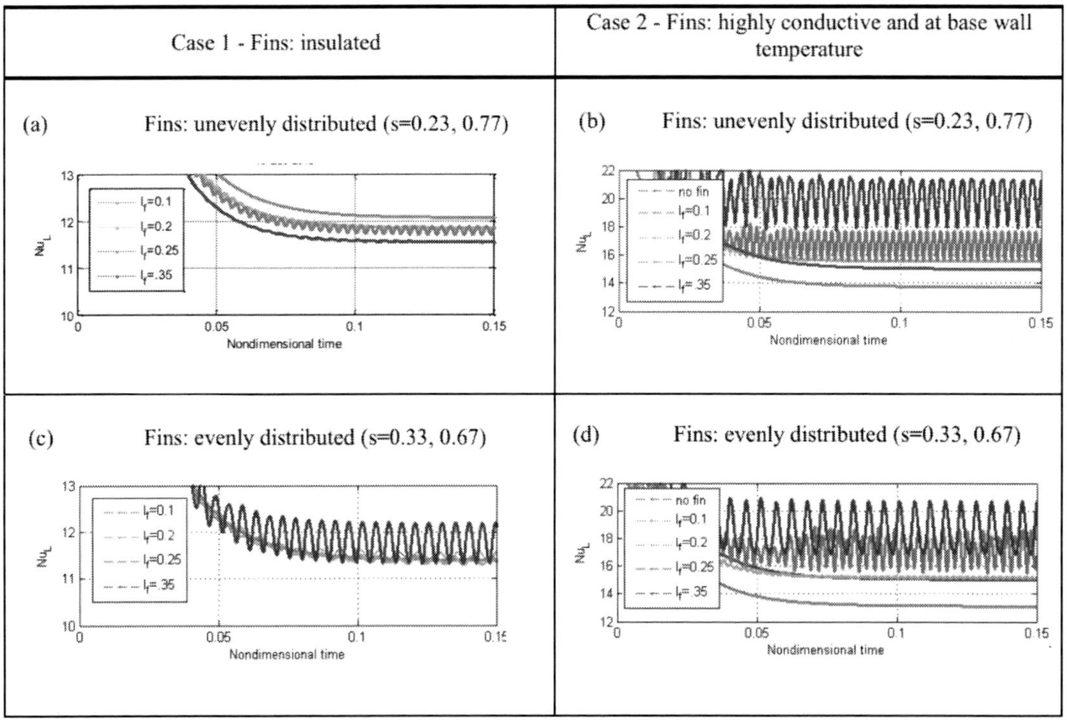

Figure 6. Evolution of Nu_L for l_f=0.1, 0.2, 0.25, and 0.35 for $Ra=7x10^6$, left column: fins are insulated, right column: fins are highly conductive and at the base wall temperature (Nu=15.07 for no fin).

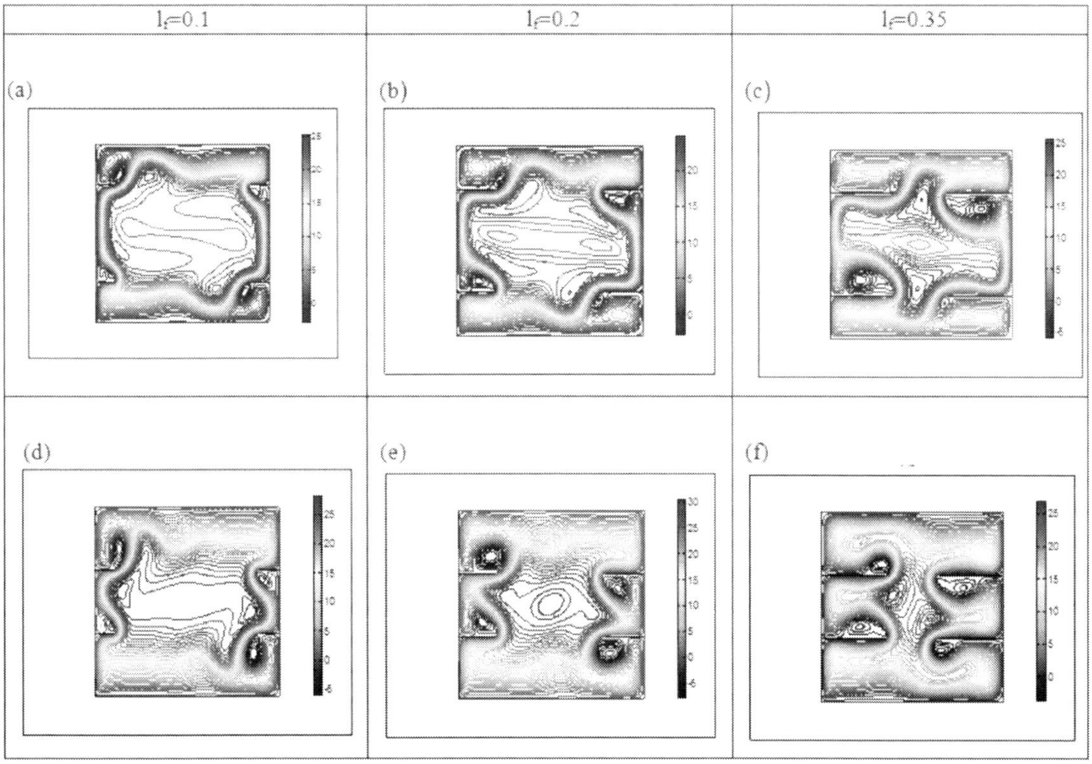

Figure 7. Streamlines for Case 1 (fins are insulated) with Ra=7x10^6 and fin lengths of 0.1(a,d), 0.2(b,e), and 0.35(c,f); s=[0.23, 0.77] (a,b,c) and s=[0.33, 0.67] (d,e,f).

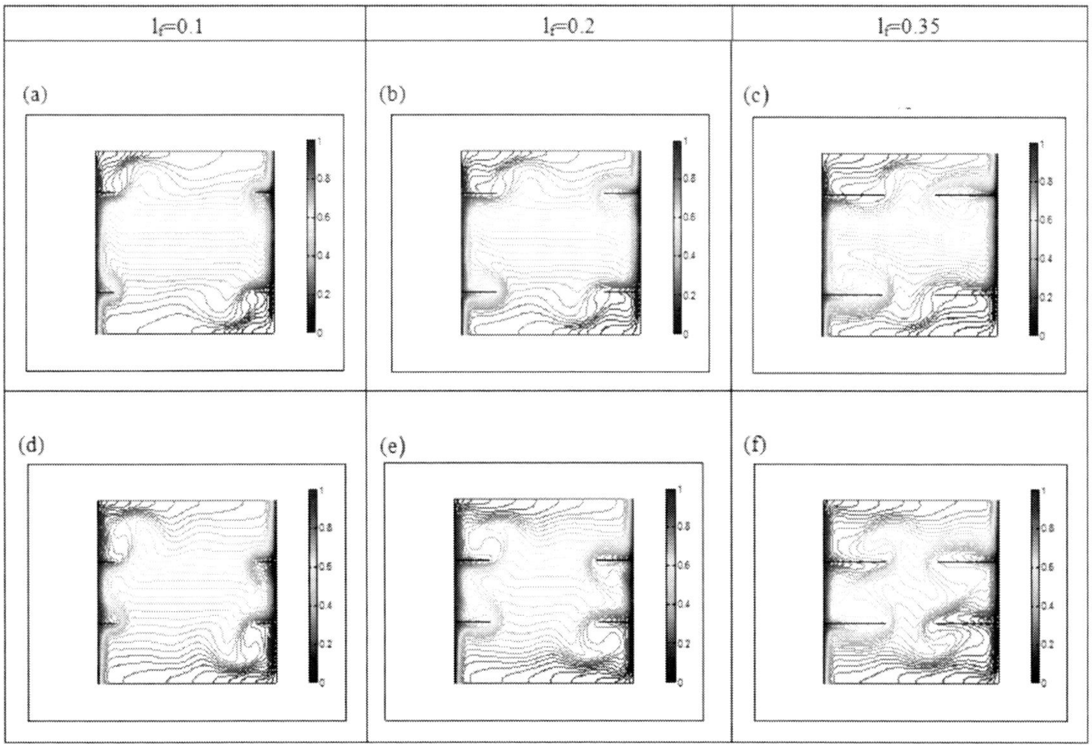

Figure 8. Isotherms for Case 1 (fins are insulated) with Ra=7x10^6 and fin lengths of 0.1(a,d), 0.2(b,e), and 0.35(c,f); s=[0.23, 0.77] (a,b,c) and s=[0.33, 0.67] (d,e,f).

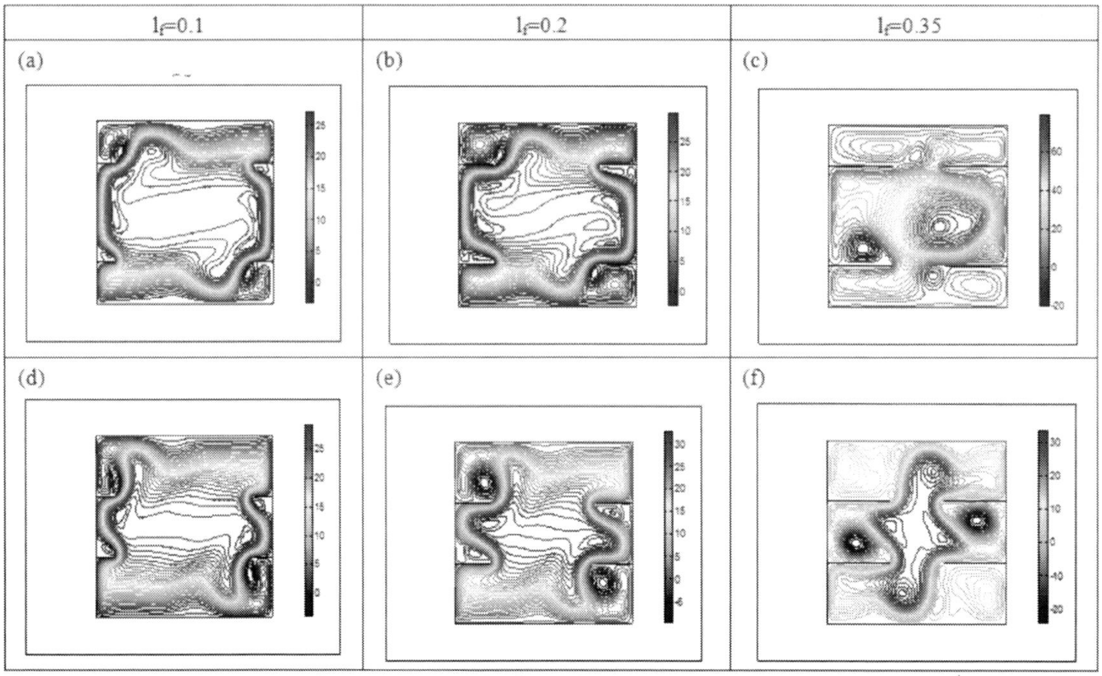

Figure 9. Streamlines for Case 2 (fins are highly conductive and at base wall temperature) with Ra=7x10^6 and fin lengths of 0.1(a,d), 0.2(b,e) and 0.35(c,f); s=[0.23, 0.77] (a,b,c) and s=[0.33, 0.67] (d,e,f).

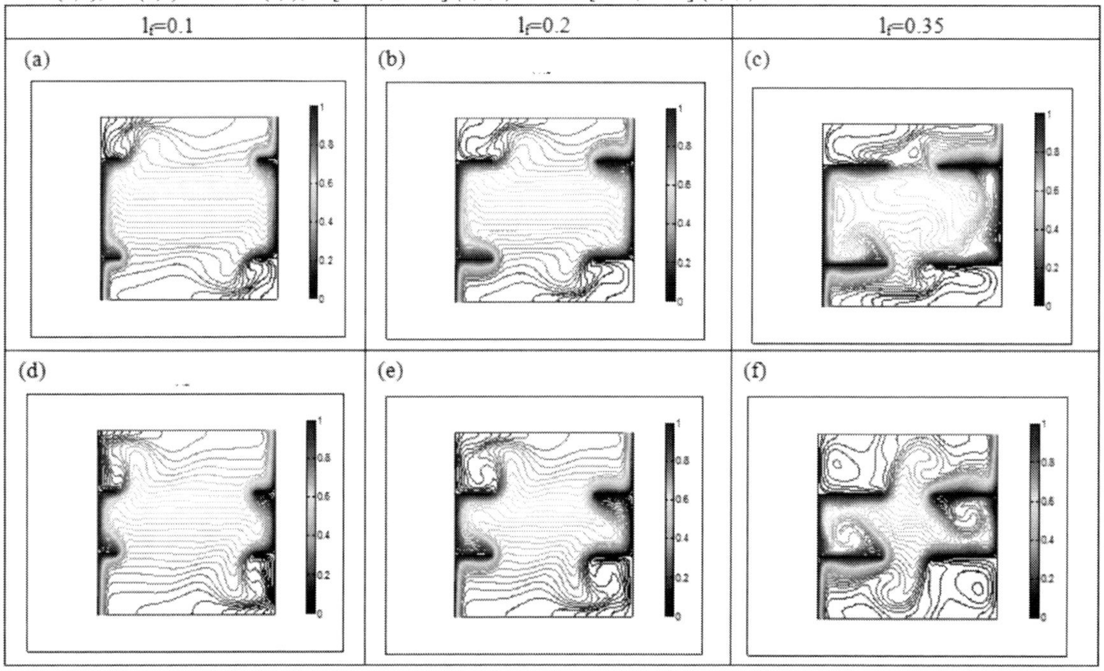

Figure 10. Isotherms for Case 2 (fins are highly conductive and at base wall temperature) with Ra=7x10^6 and fin lengths of 0.1(a,d), 0.2(b,e) and 0.35(c,f); s=[0.23, 0.77] (a,b,c) and s=[0.33, 0.67] (d,e,f).

4 Concluding Remarks

A computational study of laminar natural convection with Boussinesq approximation within a differentially heated square cavity in the presence of thin fins is presented. The suppression and enhancement of heat transfer are examined by considering adiabatic and highly conductive fins at varying lengths and locations. The transitions from steady-state to regular oscillating and from oscillating to chaotic flow occur depending on the location and length of fins and Ra. It is important to examine the fin lengths and locations for optimum performance of fins when heat transfer modification is desirable since the dynamic behavior of fluid is greatly influenced by the fin length and spacing.

References

1. K. H. Winters, Laminar natural convection in a partially divided rectangular cavity at high Rayleigh number, *International J. for Numerical Methods in Fluids* **8**, 247-281 (1988).
2. M. Sathiyamoorthy and Ali J. Chamkha, Analysis of natural convection in a square cavity with a thin partition for linearly heated side walls, *International J. Numerical Methods for Heat and Fluid Flow* **24**, 5, 1067-1072 (2014).
3. G. N. Faces, Natural convection in a cavity with fins attached to both vertical walls, *J. Thermophysics and Heat Transfer* **7**, 4, 555-560 (1993).
4. A. Nag, A. Sarkar and V. M. K. Sastri, Natural Convection in a differentially heated square cavity with a horizontal partition plate on the hot wall, *Computer Methods in Applied Mechanics and Engineering* **110**, 143-156 (1993).
5. X. Shi and J. M. Khodadadi, Laminar natural convection heat transfer in a differentially heated square cavity due to a thin fin on the hot wall, *Transactions of ASME* **125**, 624-634 (2003).
6. E. Bilgen, Natural convection in cavities with a thin fin on the hot wall, International *J. Heat and Mass Transfer* **48**, 3493-3505 (2005).
7. S. A. Nada, Natural convection heat transfer in horizontal and vertical closed narrow enclosures with heated rectangular finned base plate, *International J. Heat and Mass Transfer* **50**, 667-679 (2007).
8. E. Evren-Selamet, V.S. Arpaci, and C. Borgnakke: Simulation of laminar buoyancy-driven flows in an enclosure, *Numerical Heat Transfer*, Part A, **22**, 401-420 (1992).
9. Y. Duan, S. F. Hosseinizadeh, J. M. Khodadadi, Effects of insulated and isothermal baffles on pseudosteady-state natural convection inside spherical containers, *J. Heat Transfer* **132**, 6, 1-10 (2010).

Sliding Mode Control of Diesel Engine Air-path System With Dual-loop EGR and VGT Based on the Reduced-order Model

Sooyoung Kim [a], Seibum Choi

School of Mechanical, Aerospace & System Engineering, KAIST, Daejeon 305-701, Korea

Abstract. This paper presents the design of a model-based controller for the diesel engine air-path system. The controller is implemented based on a reduced order model consisting of only pressure and power dynamics with practical concerns. To deal with the model uncertainties effectively, a sliding mode controller, which is robust to model uncertainties, is proposed for the air-path system. The control performance of the proposed control scheme is verified through simulation with the valid plant model of a 6,000cc heavy duty diesel engine.

1 Introduction

As environmental regulations for diesel engines have become stricter, the worldwide automotive industries need to develop more environmental-friendly engine systems continuously. In these days, most diesel engines are commonly equipped with exhaust gas recirculation (EGR) system and variable geometry turbine (VGT). With the combination of EGR and VGT, the amount of emissions can be reduced significantly because it can adjust fundamental properties of combustion in the diesel engine. Recently, dual-loop EGR systems, consisting of two EGR systems: high pressure (HP) EGR and low pressure (LP) EGR, are often installed in production engines for the effective utilization of EGR functions.

As the hardware configurations of diesel engines have been developed, research on precise control of diesel engines has also been required. One of the main issues in the air-path control is to treat coupling properties of HP EGR and VGT effectively, both of which are driven by exhaust gases. The strong coupling effects between HP EGR system and VGT was investigated in [8]. To handle control problems of this highly coupled system, various model-based control methods of VGT and EGR systems have been investigated. Many studies have been conducted on the control of single-loop EGR systems [1]-[9] and a few studies on the control of dual-loop EGR systems, which is a relatively new research field.

Design of a controller for the air-path system is starting from selection of set-points which indicate the required engine performances such as desired engine torque or emissions. In specific, because the diesel air-path control is focusing on the simultaneous reduction of hazardous emitted materials of particulate matters (PM) and nitrogen oxide (NOx) while keeping generating desired torque, the set-points are concerned with desired values of air to fuel ratio (AFR), EGR rate, and engine torque. The problem is that those combustion variables such as EGR rate and AFR cannot be directly regulated by an engine controller. Hence, we need to select proper control states to be controlled for the combustion variables to satisfy the set-point requirements. In this paper, intake manifold pressure, exhaust manifold pressure, and fresh air flow rate are chosen as control states considering their physical importance in the operations of the air-path and availability of sensors [9, 10]. This paper investigates the design of a model-based controller for the states, which is based on the 5th-order reduced model of the air-path. The reduced order model is composed of only pressure and power dynamics considering availability of sensors. To treat the model uncertainties well, the method of a sliding mode control is adopted for the controller implementation.

2 Air-path Modeling

The structure of the air-path system with dual-loop EGR and VGT is described in Fig. 1. The main model equations for the system validated in many papers (e.g. [2], [4]), are introduced in this section. In the next section, a model-based control will be developed under the assumption that the model equations describe behavior of the air-path system well enough to control the system accurately.

The model equations consist of variables such as pressure (p), temperature (T), volume (V), and mass flow rate (W). The subscripts, i, x, c, t, uc, and dt stand for intake manifold, exhaust manifold, compressor, turbine, upstream of compressor, and downstream of turbine respectively.

[a] Corresponding author: westkllr@kaist.ac.kr

DOI: 10.1051/matecconf/20167008001

2.1 Pressure dynamics

Pressure dynamics are described as (1)-(4) based on the ideal gas law with mass flow rate. The equations are derived under the assumption that the temperature of which the dynamics is relatively slow is not changed in same operating points and volume is constant for all operating conditions.

$$\dot{p}_i = \frac{R\gamma}{V_i}\left(T_{ic}W_c + T_x W_{HPegr} - T_i W_{ie}\right), \quad (1)$$

$$\dot{p}_x = \frac{R\gamma}{V_x}\left(T_e W_x - T_x W_{HPegr} - T_x W_t\right), \quad (2)$$

$$\dot{p}_{uc} = \frac{R\gamma}{V_{uc}}\left(T_a W_{air} + T_{dt} W_{LPegr} - T_c W_c\right), \quad (3)$$

$$\dot{p}_{dt} = \frac{R\gamma}{V_{dt}}\left(T_x W_t - T_{dt} W_{LPegr} - T_{out} W_{out}\right), \quad (4)$$

where γ and R are the specific ratio and ideal gas constant. Note that a diesel particulate filter (DPF) is not included in the after-treatment systems in Fig. 1 that makes pressure drop through after-treatment systems in the LP-EGR loop is negligible. Hence, it is assumed that $p_{dt} \approx p_{DOC}$ throughout this study.

2.2 Turbocharger dynamics

Generally, a turbocharger in a diesel engine has the configuration that turbine is connected with compressor by a turbo-shaft. Turbocharger inertia and turbine speed affects the turbocharger operations, and this behavior can be described as turbocharger speed dynamics (5)

$$\frac{d}{dt}\left(\frac{1}{2}J_{tc}N_{tc}^2\right) = P_t - P_c, \quad (5)$$

where J_{tc} and N_{tc} are the turbocharger inertia and speed, respectively.

Power dynamics of a turbocharger is mathematically modelled as (2.10) based on the principle of power balance of compressor and turbine.

$$\dot{P}_c = \frac{1}{\tau_{tc}}\left(-P_c + \eta_m P_t\right), \quad (6)$$

where P_c , P_t , τ_{tc} η_m are the power generated by the compressor, the power delivered by the turbine, a turbocharger time constant, and mechanical efficiency respectively.

The turbine power can be calculated by

$$P_t = \eta_t c_p T_x \left(1 - \left(\frac{p_{dt}}{p_x}\right)^{\mu}\right) W_t, \quad (7)$$

$$\text{where } \mu = \frac{\gamma - 1}{\gamma},$$

where η_t is the turbine efficiency and c_p the specific heat at constant pressure.

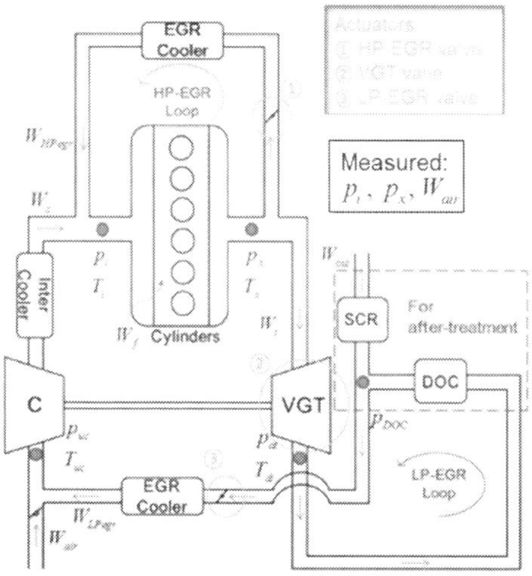

Figure 1. The air-path system with a dual-loop EGR and VGT

2.3 Flow equations

The mass flow rate through the compressor can be described by (8), which is determined mainly by pressure ratio across the compressor and compressor power.

$$W_c = \frac{\eta_c}{T_{uc}c_p}\frac{P_c}{\left(\frac{p_i}{p_{uc}}\right)^{\frac{\gamma-1}{\gamma}} - 1}, \quad (8)$$

where η_c is the compressor efficiency.

HP EGR flow rate is modeled as (9) using an orifice equation.

$$W_{HPegr} = \sqrt{\frac{2\gamma A_{HP}^2 p_x^2}{RT_x(\gamma-1)}\left((PR_{HP})^{\frac{2}{\gamma}} - (PR_{HP})^{\frac{\gamma+1}{\gamma}}\right)}, \quad (9)$$

$$\text{where } PR_{HP} = \max\left(\frac{p_i}{p_x}, \left(\frac{2}{\gamma+1}\right)^{\frac{\gamma}{\gamma-1}}\right).$$

In (9), PR_{HP} is the pressure ratio across HP EGR flow and A_{HP} the effective area of HP EGR flow determined by the valve position.

In the same manner, the mass flow rate of LP EGR is described as (10).

$$W_{LPegr} = \sqrt{\frac{2\gamma A_{LP}^2 p_{dt}^2}{RT_{dt}(\gamma-1)}\left((PR_{LP})^{\frac{2}{\gamma}} - (PR_{LP})^{\frac{\gamma+1}{\gamma}}\right)}, \quad (10)$$

$$\text{where } PR_{LP} = \max\left(\frac{p_{uc}}{p_{dt}}, \left(\frac{2}{\gamma+1}\right)^{\frac{\gamma}{\gamma-1}}\right).$$

A model for the fresh air flow rate in dual-EGR systems is designed as (11) to describe the mass flow rate of pure fresh air flow in the dual-loop EGR system using the orifice equation [10].

$$W_{air} = A_{air} p_a \sqrt{\frac{2\gamma}{RT_a(\gamma-1)}\left[\left(\frac{p_{uc}}{p_a}\right)^{\frac{2}{\gamma}} - \left(\frac{p_{uc}}{p_a}\right)^{\frac{\gamma+1}{\gamma}}\right]}. \quad (11)$$

Note that we assume that the intake throttle is fully opened throughout this study. Since the throttle position is fixed, the effective area A_{air} is a constant in (11).

3 Controller Design

3.1 The reduced order model

The 5th-order control-oriented model of the air-system with dual-loop EGR and VGT is designed as (12)-(16) for the controller design. The model consisting of (12), (13), and (16) is a well-known 3rd order control-oriented model for the diesel engine with the single-loop EGR and VGT system validated in references [2, 4, 5]. The 5th-order model is designed by adding two dynamics (14) and (15) representing the properties of the LP EGR loop, to the 3rd-order model [9, 10]. The 5th-order model is implemented under the following assumptions:

1. Fraction states of oxygen or burnt gas, which are hard to measure and not directly coupled with other dynamics, are omitted [7].
2. Temperature states whose dynamics are relatively slow are ignored [2, 5].
3. Variations of the temperature states are negligible.

$$\dot{p}_i = k_1\left(W_c + W_{HPegr} - k_e p_i\right), \quad (12)$$

$$\dot{p}_x = k_2\left(W_x - W_{HPegr} - W_t\right), \quad (13)$$

$$\dot{p}_{uc} = k_3\left(W_{air} + W_{LPegr} - W_c\right), \quad (14)$$

$$\dot{p}_{dt} = k_4\left(W_t - W_{LPegr} - W_{out}\right), \quad (15)$$

$$\dot{P}_c = \frac{1}{\tau_{tc}}\left(-P_c + \eta_m \eta_t c_p T_x\left(1-\left(\frac{p_{dt}}{p_x}\right)^{\mu}\right)W_t\right),$$
$$\text{where } k_1 = \frac{RT_i}{V_i},\ k_2 = \frac{RT_x}{V_i},\ k_3 = \frac{RT_{uc}}{V_{uc}},\quad k_4 = \frac{RT_{dt}}{V_{dt}},\ k_e = \frac{\eta_v N V_d}{120 RT_i}. \quad (16)$$

In (12), the cylinder-in flow rate W_{ie} is modelled by the speed-density equation and denoted as $k_e p_i$ where η_v, V_d, and N are the volumetric efficiency, displacement volume, and engine RPM, respectively.

In general, the higher the order the model has, the more information it can have in it. The reduced order model may not be accurate enough to describe the behavior of the real system. However, implementation of the model-based controller in an engine control unit (ECU) would be very difficult when the order of the model is too high. Thus, it is important to develop a practical controller using only limited information of the reduced order model. Uncertainties of the reduced model should be addressed by designing a robust control algorithm. The model is composed of four pressure states and one power dynamics. The order of the model is reduced more than 5 by omitting fraction and temperature states, and easy to be implemented for the controller design.

3.2 Desired trajectories

The objective of control of the air-path system is for the diesel engine in operation to satisfy the emission regulations without sacrificing engine performance. Those desired engine performances are specified by a set-point. The combustion variables such as EGR rate and AFR should be controlled to be in acceptable ranges to meet the emission regulations. However, the combustion performance variables are usually not measured by sensors, so they cannot be directly controlled. Therefore, it is necessary to select proper control states to be controlled for satisfying the target set-point. In addition, control algorithms should take into account practical aspects including sensor availability.

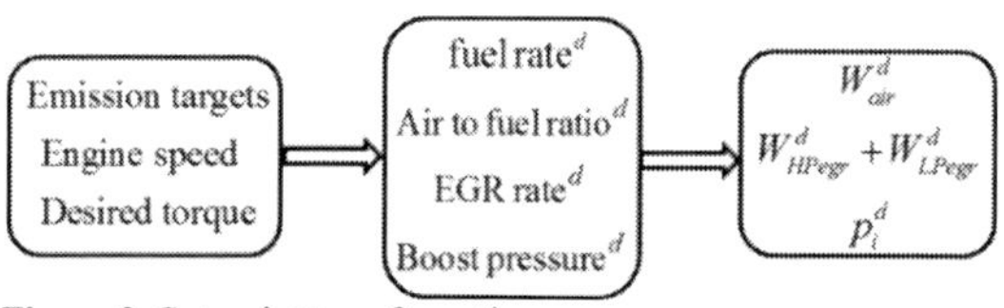

Figure 2. Set-point transformation

Fig. 3-1 shows the set-point transformation to get the desired values of control states. The desired values of the combustion variables including fuel rate, AFR, EGR rate are determined from the target performance criteria. Then, the values of these combustion variables can be transformed into those of the control variables such as EGR flow, fresh air flow, and intake manifold pressure, which can be dealt with in the control system while the fuel rate values are fixed in accordance with desired torque. In this way, the problem of emission regulations is interpreted as a reference tracking problem [6]. When reference values of fresh air flow rate, total EGR flow rate, and intake pressure are given, the control algorithm should make the system follow the values without deteriorating the system balance.

3.3 Controller design

There are three control actuators in the air-path system this paper deals with: HP EGR valve, LP EGR valve, and VGT rack. Besides, there exist three states to be controlled, p_i, p_x, and W_{air} for satisfying the emission requirements [9]. Proper model-based control algorithm is needed for the control states to track their desired trajectories well. Since the controller is developed based on the reduced order model, it is expected that the uncertainties of the model will degrade the control performance. Hence, a sliding mode controller is proposed to cope with the uncertainties and enhance the tracking performance. The design procedure of the model-based controller is presented as follows.

We define a new output state as (17).

$$y_1 = p_i + \frac{k_1}{k_2} p_x. \tag{17}$$

Differentiating (17) with regard to time and combining with (12) and (13), (18) is derived.

$$\begin{aligned} \dot{y}_1 &= \dot{p}_i + \frac{k_1}{k_2}\dot{p}_x, \\ &= k_1(W_c + W_f - u_1), \end{aligned} \tag{18}$$

where u_1 is the control input for VGT.

Equation (18) prescribes the direct relation of turbine operations with intake and exhaust pressures. Physically, the VGT operations directly affect both the intake and exhaust pressures. In (18), W_{HPegr}, the HP EGR input which was explicitly seen in the pressure dynamics, is not shown any more that enables up to decouple HP EGR operations from pressure dynamics.

Let the sliding surface be defined as:

$$S_1 = y_1 - y_1^d. \tag{19}$$

The time derivative of (19) is described as (20).

$$\dot{S}_1 = k_1(W_c + W_f - u_1) - \dot{y}_1^d. \tag{20}$$

To ensure $S_1 \to 0$, the following control law is derived :

$$\dot{S}_1 \triangleq -\lambda_1 \operatorname{sgn}(S_1), \tag{21}$$

$$u_1 \triangleq W_c + W_f + k_1^{-1}\lambda_1 \operatorname{sgn}(S_1) - k_1^{-1}\dot{y}_1^d, \tag{22}$$

where λ_1 is the control gain for VGT.

The control law u_1 makes $S_1 \to 0$, but does not guarantee that each tracking error of intake pressure and exhaust pressure converges to zero. To ensure both pressure states converge to their reference values, the second sliding surface is defined as:

$$S_2 = p_x - p_x^d. \tag{23}$$

The time derivative of (23) is described as

$$\dot{S}_2 = k_2(W_{ex} - u_2 - u_1) - \dot{p}_x^d, \tag{24}$$

where u_2 is the HP EGR input.

The corresponding u_2 satisfying $\dot{S}_2 = -\lambda_2 \operatorname{sgn}(S_2)$ is derived as (25).

$$u_2 \triangleq W_{ex} - u_1 + k_2^{-1}\lambda_2 \operatorname{sgn}(S_2) - k_2^{-1}\dot{p}_x^d, \tag{25}$$

where λ_2 is the control gain for HP EGR.

The state of fresh air flow rate can be described as (26) by differentiating (11) with respect to time [9-10].

$$\begin{aligned} \dot{W}_{air} &= \frac{\alpha}{p_a} k_3 (W_{air} + W_{LPegr} - W_c)\cdots \\ &\times \left(\frac{\gamma-1}{-2\gamma}\beta^{-0.5} + \frac{1}{\gamma}\left(\frac{p_{uc}}{p_a}\right)^{\frac{1-\gamma}{\gamma}}\beta^{0.5} \right) \end{aligned} \tag{26}$$

where $k_3 = \frac{RT_{uc}}{V_{uc}}$, $\alpha = A_{air} p_a \frac{1}{\sqrt{RT_a}}\sqrt{\frac{2\gamma}{\gamma-1}}$,

$$\beta = 1 - \left(\frac{p_{uc}}{p_a}\right)^{\frac{\gamma-1}{\gamma}}$$

The third sliding surface and its time derivative for the fresh air flow rate are defined as (27)-(28).

$$S_3 = W_{air} - W_{air}^d. \tag{27}$$

$$\begin{aligned} \dot{S}_3 &= \frac{\alpha}{p_a} k_3 (W_{air} + u_3 - W_c)\cdots \\ &\times \left(\frac{\gamma-1}{-2\gamma}\beta^{-0.5} + \frac{1}{\gamma}\left(\frac{p_{uc}}{p_a}\right)^{\frac{1-\gamma}{\gamma}}\beta^{0.5} \right) - \dot{W}_{air}^d, \end{aligned} \tag{28}$$

where u_3 is the LP EGR input.

The corresponding u_3 satisfying $\dot{S}_3 = -\lambda_3 \operatorname{sgn}(S_3)$ is derived as (29) if we define λ_3 as the control gain for LP EGR.

$$u_3 = W_{LPegr}^{in} = -\frac{\lambda_3 p_a}{k_3 \alpha c}\operatorname{sgn}(S_3) - W_{air} + W_c + k_3^{-1}\dot{W}_{air}^d, \tag{29}$$

where $c = \left(\frac{\gamma-1}{-2\gamma}\beta^{-0.5} + \frac{1}{\gamma}\left(\frac{p_{uc}}{p_a}\right)^{\frac{1-\gamma}{\gamma}}\beta^{0.5} \right)$.

u_3 is the control input for LP EGR valve to control mass flow rate of LP EGR while maintaining desired flow rate of fresh air into the compressor inlet [10].

4 Control verification

The control verification is carried out with the proven 10th-order model of a 6,000cc heavy duty diesel engine. The analysis on the validation of the 10th-order and the reduced order models comparing with the experimental data is conducted in [9]. The reference trajectories for the feedback controller are acquired experimentally from the actual engine test bench. The verification results are shown in Fig. 3-5.

The results show that though the controller is developed on the reduced order model, it shows high tracking performance, which means the controller is robust to the model uncertainties. Particularly, the intake pressure and fresh air flow rate follow the reference trajectories without large overshoot or oscillatory behaviors. There exist some steady state errors for the exhaust pressure control, but the errors are in an acceptable range because the exhaust pressure does not determine certain engine performances. The maximum error in steady states is 1.64% for intake pressure, 1.63% for fresh air rate, and 4.98% for exhaust pressure. The tracking errors are mainly caused by poor accuracy of the reduced order model at the corresponding operating condition.

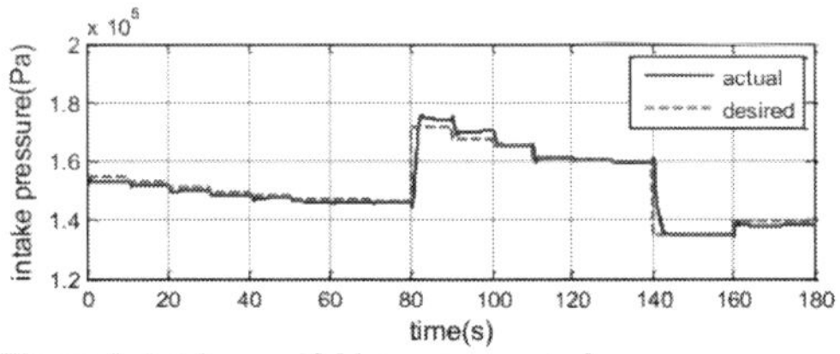

Figure 3. Intake manifold pressure control

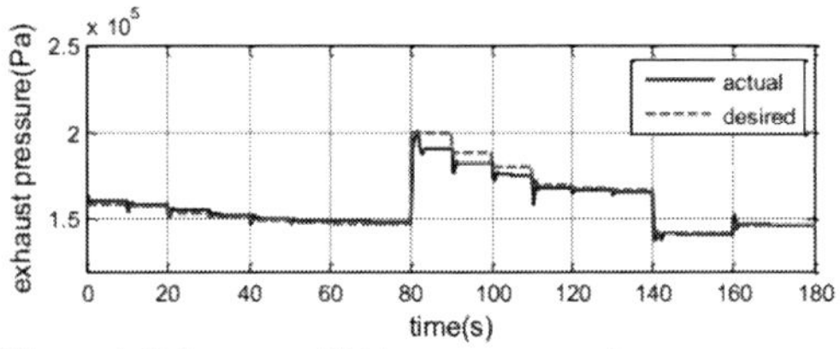

Figure 4. Exhaust manifold pressure control

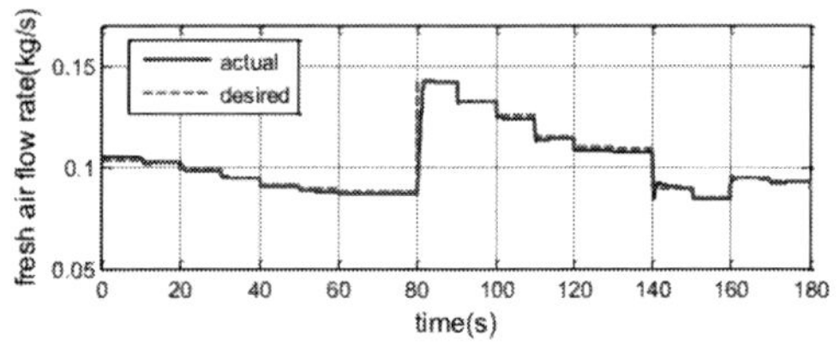

Figure 5. Fresh air flow rate control

Acknowledgement

This research was supported by the MSIP(Ministry of Science,
ICT & Future Planning), Korea, under the C-ITRC(Convergence Information Technology Research Center) (IITP-2015-H8601-15-1005) supervised by the IITP(Institute for Information & communications Technology Promotion) and the National Research Foundation of Korea (NRF) grant funded by the Korea government (**MSIP**) (No. 2010-0028680).

References

1. M. Ammann, N. P. Fekete, and L. Guzzella, Model Based Control of the VGT and EGR in a Turbocharged Common-rail Diesel Engines: Theory and Passenger Car Implementation, *SAE Technical Paper*, (2003)
2. D. Upadhyay, Modeling and Model based Control Design of the VGT-EGR system for Intake Flow Regulation in Diesel Engines, *Ph.D, dissertation, The Ohio University*, (2001)
3. M. J. van Nieuwstadt, Cooridinated EGR-VGT Control for Diesel Engines: an Experimental Comparison, *SAE 2000 World Congress*, (2000)
4. Y. Yoon, A Study of Turbocharged Diesel Engine Modeling and Robust Model Based Sliding Mode Controller Design, Master's Thesis, KAIST, (2011)
5. H. Jin, S. Choi and H. Jung, Simplified Multiple Sliding Mode Transient Control with VGT and EGR Diesel Engine, *SAE International*, (2013)
6. J. Sarlashkar, Sliding Mode Control For Diesel Engines with Airflow Dominant Fueling, *JSAE Annual Congress*, (2010)
7. J. Wang, Hybrid Robust Air-path Control for Diesel Engine Operating Conventional and Low temperature Combustion Modes, IEEE Trans. on Cont. Sys. Tech., **Vol. 16, No. 6**, 1138-1151, (2008)
8. I. Kolmanovsky, Issues in Modelling and Control of Intake Flow In Variable Geometry Turbocharged Engines, *18th IFIP Conference on System Modelling and Optimization*, (1997)
9. S. Kim, H. Jin, and S. Choi, Pressure and Flow Based Control of a Turbocharged Diesel Engine Air-path System Equipped with Dual-Loop EGR Systems and VGT, *American Control Conference*, 1493-1498, (2014)
10. S. Kim, H. Jin, and S. Choi, Pressure Model Based Coordinated Control of VGT and Dual-loop EGR in a Diesel Engine Air-path System, Int. J. of Automotive Tech. (to be published)

Designation and Investigate of a Full-Wave Controller Rectifier (FWCR) for Effect Source Inductance for Full Wave Rectifier

Kamarudin Kh[a], Ramli M. S., A. W. SitiSufiah, Razali N. F. and S. A. Nordin

Faculty of Electrical Engineering, Universiti Teknologi MARA, 81750 Masai, Johor, Malaysia

Abstract. The full wave rectifier with RL load consists of all the SCRs as controlled devices. This full wave converter can operate in quadrants and output voltage can be negative in case of inductive loads. The power factor is poorer than a half converter, as the full bridge converter is primarily used to control the DC motor speed [1].The purpose of this paper is to solve higher ripper power supply in a full wave controlled rectifier using an RL load of 240rms voltage, 50Hz with an attached source inductance, Ls of 10mH value. The average output voltage can be varied from 100-200VDC. The performance of a full wave rectifier with source inductance is to create conduction of four thyristor simultaneously with overlapping angle. The result obtained is with an efficiency of 50.18% at output RL Load.

1 Introduction

Power electronic is one of the important branches of electronic and electrical engineering. It deals with conversion and control of electrical energy. A power electronic system performs conversion of electrical energy. It also controls the amount of electrical energy to be given to the output. Power electronics combines power, electronics and control. Control deals with the steady-state and dynamic characteristics of a closed-loop system. Power deals with the static and rotating power equipment for generation, transmission and distribution of electric energy. Based on that, impedance source in power electronics provides an efficient power converter to a load with a wide range of electrical power conversion applications (dc-dc,dc-ac,ac-ac,ac-dc) [1]. The impedance-source network is broadly classified into two categories based on magnetics; 1) non-transformer based; and 2) coupled or transformer. Each topology has distinct features and advantages [2]. The single-phase full converter, the period of α to π, the input voltage Vs and the input current are positive which results in power flow from the power supply to the load. With this condition, the converter is in rectification mode. In the period π to $\pi + \alpha$, the input voltage is negative and the input current is positive which reverses power flow from the load to the supply. Hence, the converter is in inverse mode [3].

2 FWCR without source inductance (Ls)

Figure 1 shows a Full-Wave Rectifier without Source Inductance. This figure is basic on a full wave rectifier by controlling four SCRs with a firing angle controller. To obtain the output voltage of the circuit, derivations must be following as in equation (1).

$$\text{Vdc} = \frac{2}{2\pi}\int_{\alpha}^{\pi+\alpha} Vm \sin wtd(wt) = \frac{2\text{m}}{\pi}\cos\alpha \quad (1)$$

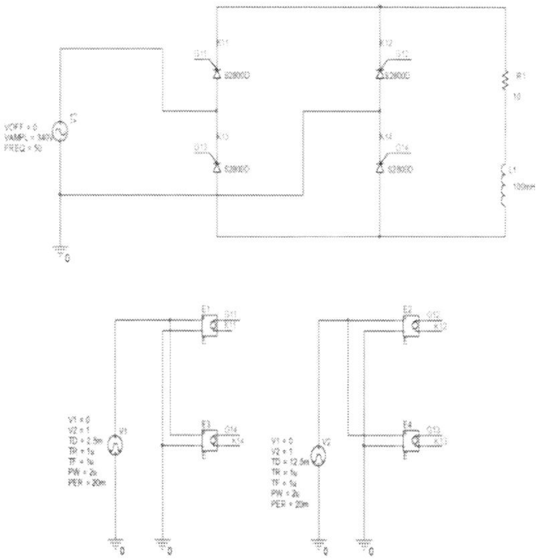

Figure 1. Full-Wave rectifier without source inductance.

The values of the firing angle must be calculated by a given target of an output voltage which is 160 Volts. The calculations involved are stated below:

$$\text{Vodc} = \frac{2(340)}{\pi}\cos\alpha$$

[a] Corresponding author: khairul@johor.uitm.edu.my

$$Cos\alpha = \frac{160}{216.45} = 0.7391$$

$$\alpha = 42.34^{\circ}$$

$$T = \frac{1}{f} = \frac{1}{50} = 20\text{ms}$$

$$42.34^{\circ} \rightarrow \frac{42.34^{\circ} \times 20ms}{360^{\circ}} = 2.35\text{ms}$$

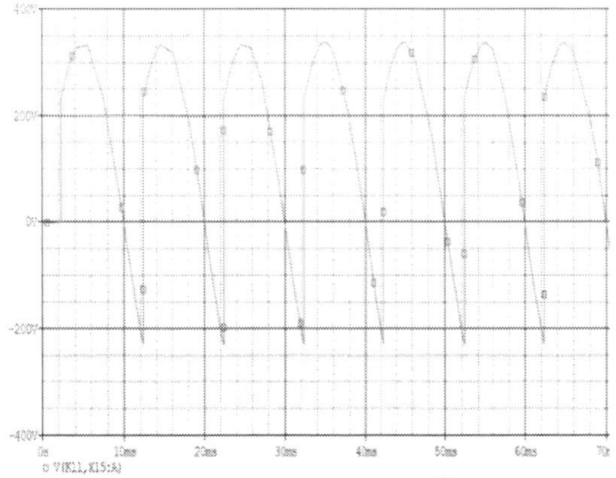

Figure 2. The output voltage without source inductance.

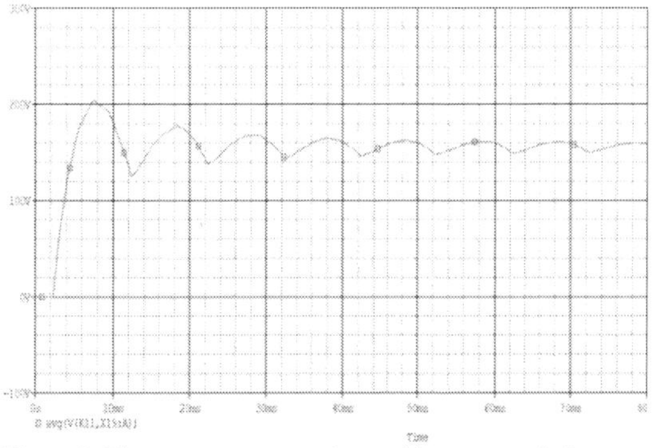

Figure 3. The average output voltage without source inductance.

By referring Figure 3, the value obtained from the simulation is 157.7 Volts. The measurement value is near the calculated value which is 160 Volts. Calculations done after this will have an average output of 160 Volts as a reference target when adding source inductance in section 3.

3 FWCR with the source inductance (Ls) is 10mH

Based on Figure 5, the simulation for α is 12.53ms and β is 12.905ms. These values can be calculated to find the values of an angle shifting source with the values of a total 360° cycle. Based on the calculation shown, the value of degrees in β is 52.31°. The function of β is to determine the effect of source impedance to the value of the average output voltage and rms output voltage. Figure 4 shows a Full-Wave Rectifier with Source Inductance.

To analyze the output voltage of the circuit, derivations are following as in equation (2).

$$Vodc = \frac{\text{Vm}}{\pi} \int_{\beta}^{\pi+\alpha} Vm\text{Sin wtd(wt)} \qquad (2)$$

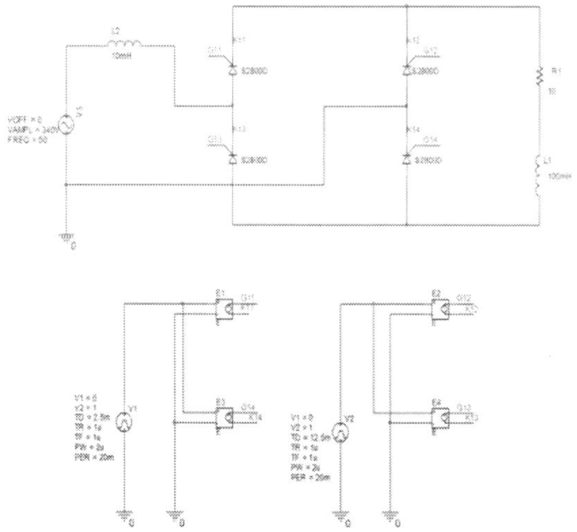

Figure 4. Circuit full rectifier with source inductance.

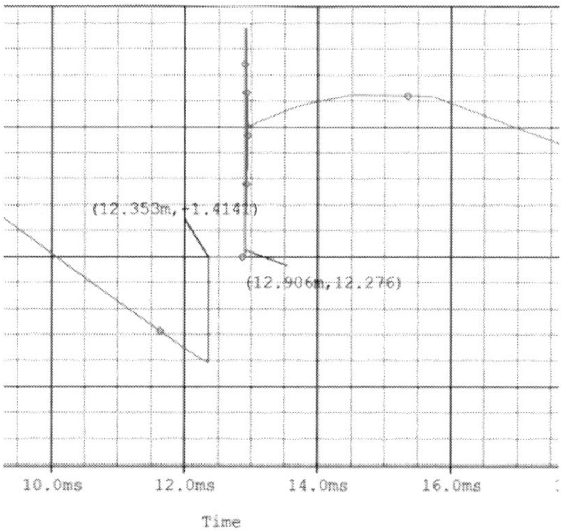

Figure 5. The values of β.

The value of average voltage which involves the shifting of β is 146.16 Volt. This value is less from the average output voltage from the result of not using the source impedance. By using equation (3), we get 13.84Volts.

$$\text{Vlosses} = \int_{\alpha}^{\beta} \text{Vm sin wt d(wt)} \qquad (3)$$

The result of calculation of average current and rms output current have different result, this happen because due to the effect of using with inductance at load. This results in voltage losses from the α to β period by the effect of an impendence source. The equation (4) is used for rms voltage output. The final calculation is 233.11 Volts.

$$Vorms^2 = \frac{1}{\pi}\int_{\beta}^{\pi+\alpha} Vm^2 Sin^2 wt \, \mathrm{d(wt)} \qquad (4)$$

$$Iodc = \frac{1}{\pi}\int_{\beta}^{\pi+\alpha}\left[10.312\sin(wt-1.2626)+20.64e^{-0.3183wt}\right]d(wt)$$

$$= \frac{10.312}{\pi}\begin{bmatrix} -0.3033\cos wt - 0.9529\sin wt \\ -6.2897e^{-0.3138wt} \end{bmatrix}_{0.913}^{\pi+0.739}$$

$$Iodc = 15.363\mathrm{A}$$

The calculation of rms output current is 19.12 Ampere. From the calculation of output current, the performance of efficiency can reach up to 50.14%.

$$Iorms^2 = \frac{10.312^2}{\pi}\int_{\beta}^{\pi+\alpha}\begin{bmatrix} \sin^{2(}(wt-1.2626) \\ +2.002^2 e^{-0.1013wt} \end{bmatrix}d(wt)$$

$$= \frac{53.17}{\pi}\int_{\beta}^{\pi+\alpha}\begin{bmatrix} 1-\cos(2wt-2.5252) \\ +8.016e^{-0.1013wt} \end{bmatrix}d(wt)$$

$$Iorms = 19.21A$$

Performance of rectifier with source inductance

$$\eta\% = \frac{\mathrm{VodcIodc}}{\mathrm{VormsIorms}}$$

$$= \frac{(146.16)(15.363)}{(233.11)(19.21)}$$

$$= 0.5014$$

$$\eta\% = 50.14\%$$

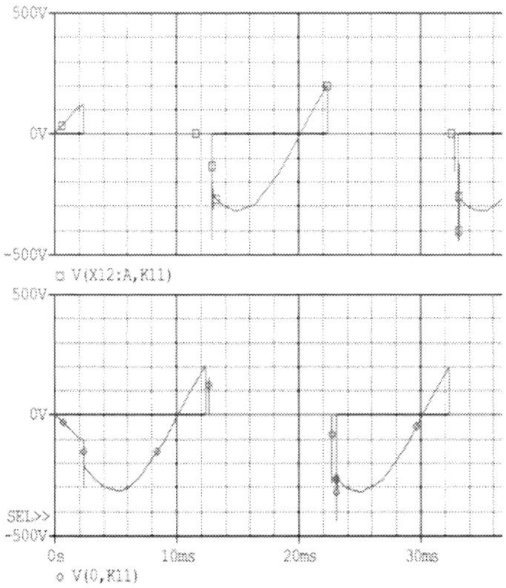

Figure 6. The voltage anode to cathode on thyristor.

Based on the simulation result, the thyristor is turned on by applying a positive voltage between its gate and cathode as in Figure 6. Once a thyristor is turned on by a gating signal and its anode current is greater than the holding current, the device continues to conduct due to positive feedback even if the gating signal is removed.

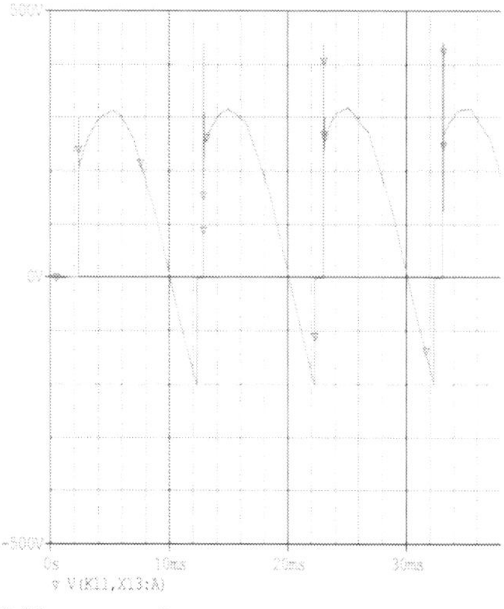

Figure 7. The output voltage.

Figure 7 and Figure 8 show the simulation has dropped V_{odc} as in the previous calculation. The simulation gives a value of 140.6Volts. This value is the nearest to the previous calculation.

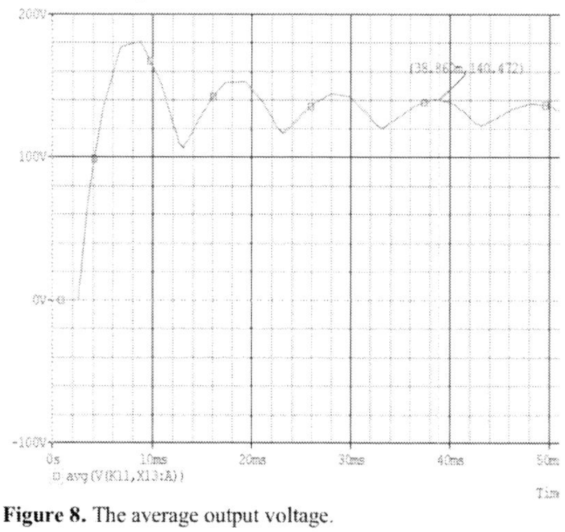

Figure 8. The average output voltage.

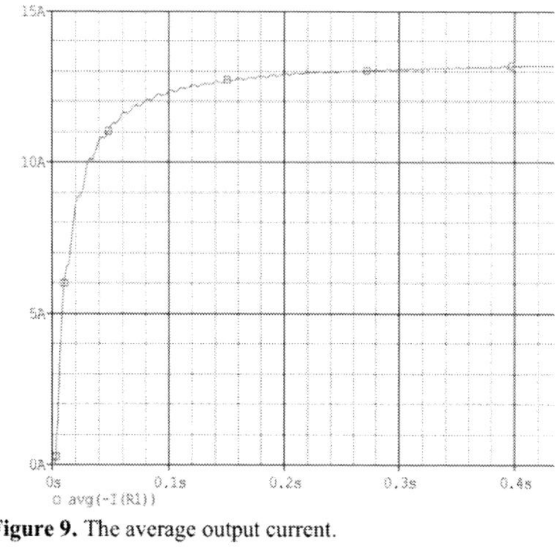

Figure 9. The average output current.

4 Conclusion

Based on the assumption of inductance is infinite, normally the converters are fed from the supply transformer. The inductance of the supply transformer becomes an input to the converter. This inductance is called source inductance, and it affects the performance of the converter. Common source inductance is the loop inductance shared by the gate driver and the main power transfer path [4]. Potential application of the full-wave rectifier circuit is implanted in medical instruments, inductive power transfer to weapons, power transfer using space reflectors and power generation in space [5]. The impedance source is also suitable in the implementation of the capital cost of a small wind turbine driven in a electrical power generation system, which is very competitive compared to that of other known non-polluting and renewable energy source [6].

When the load inductance is infinite, it is acceptable to assume that the load current is continuous and steady without ripple. Let the firing angle be 'a' and the commutation overlap period last from wt = a till wt = b. During the period, $\alpha <$ wt $<\beta$, the output voltage is zero because all the SCRs are in conduction. If the SCRs are ideal, the drop across an SCR in conduction is zero and hence the output voltage is zero. During $\beta <$ wt $< (\pi + \alpha)$, the output voltage equals V_m*Sin (wt) and then the average output voltage can be obtained like in Figure 16.The period from α to β is call overlap period. The angle β is called overlap angle. This overlap angle is to create simultaneous conduction between SCR 1 & SCR 2 and SCR3 &SCR4.

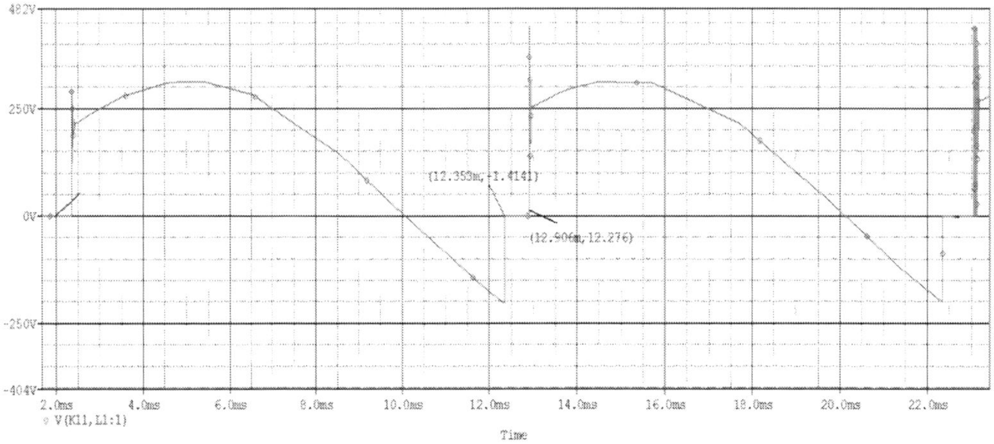

Figure 10. The output voltage analysis.

References

1. F. Z. Peng, "Z-source inverter," in Proc. Ind Appl. Conf, **2**, pp 775-781 (Oct 13-18, 2002)
2. Yam P. Siwakoti, Fang Zheng Peng, Frede Blaabjerg, Poh Chiang Loh and Graham E. Town, "Impedance-Source Networks for Electric Power Conversion Part I: A Topologies Review," IEE Transation Power Electronc, **30**, 2 (2015)
3. Muhammad H.Rashid, "PowerElectronic :circuit device and applications, 2nd edition," Prentice-Hall international, Inc., (1993), pp. 434-435
4. Bo Yang, Jason Zhang, "Effectandutilization of common source inductance in synchronous rectification," International rectifier (2005), IEEE, pp 1407-1400.
5. Richard J. Arthur, "Maximum Power Transfer for Full Wave rectifier circuit", member IEEE Noval surface weapons center, pp. 454-461
6. R. Bharanikumar, R. Senthikumar, and Nirmal Kumar, "Impedance Source for Wind Turbine Driven Permananet Magnet Generator," Power System Technology and IEEE Power India Conference Powercon, (2008)

MATEC Web of Conferences 70, 08003 (2016) DOI: 10.1051/matecconf/20167008003
ICMIT 2016

Power-level regulation and simulation of nonlinear pressurized water reactor core with xenon oscillation using H-infinity loop shaping control

Gang Li[1, 2], Bin Liang[1, 2, a], Xueqian Wang[1,2], Xiu Li[1, 2] and Kang Wang[1, 2]

[1]*Tsinghua University, National Laboratory for Information Science and Technology, 100084 Beijing, China*
[2]*Shenzhen Key Lab of Space Robotic Technology and Telescience, 518055 Shenzhen, China*

Abstract. This investigation is to solve the power-level control issue of a nonlinear pressurized water reactor core with xenon oscillations. A nonlinear pressurized water reactor core is modeled using the lumped parameter method, and a linear model of the core is then obtained through the small perturbation linearization way. The H∞loop shapingcontrolis utilized to design a robust controller of the linearized core model.The calculated H∞loop shaping controller is applied to the nonlinear core model. The nonlinear core model and the H∞ loop shaping controller build the nonlinear core power-level H∞loop shaping control system.Finally, the nonlinear core power-level H∞loop shaping control system is simulatedconsidering two typical load processes that are a step load maneuver and a ramp load maneuver, and simulation results show that the nonlinear control system is effective.

1 Introduction

Nuclear energy as a clean energy can be sustainable and renewable. Hence, developing technologies of nuclear power plants (NPPs) for electricity generation is inevitable. Of these technologies, it is necessary to investigate the power control technology of nonlinear nuclear reactor cores with xenon oscillations.

Several researchers have made great efforts for the improvement of reactor core power control systems.The state-feedback basedcontrol with classical control part was proposed by Edwards et al. [1-2] to devise power control systems of pressurized water reactor (PWR) cores. The linear quadratic Gaussian with loop transfer recovery (LQG/LTR) control wasadopted to contrivePWR core controllers for power regulations [3-4]. However, the controllers in references [1-4] are designed based on a linear core model without xenon oscillations, and not always optimal or even ineffective for nonlinear cores.

Based on this consideration, the H∞loop shaping control [5] with stronger robustness than that of the state-feedback or LQG/LTR control is adopted to design a nonlinear power control system ofa PWR core with xenon oscillations. Finally, the nonlinear PWR core power H∞loop shaping control system is simulated and conclusions are drawn.

2 Modeling for PWR core

The According to the point reactor core modeling [6-8], the nonlinear PWR core is modeled adopting the point kinetics equations with six groups of delayed neutrons and reactivity feedbacks due to control rod movement and variations in fuel temperature, coolant temperature and xenon concentration. Main model parameters are given in Table 1.The nonlinear core model is showed as Eq.(1).

$$\begin{cases} \dfrac{dP_r}{dt} = \dfrac{\rho-\beta}{\Lambda}P_r + \sum_{i=1}^{g}\dfrac{\beta_i c_{ri}}{\Lambda} \\ \dfrac{dc_{ri}}{dt} = \lambda_i P_r - \lambda_i c_{ri}, i = 1,2,\ldots,g \\ \dfrac{dI_r}{dt} = \dfrac{\gamma_I P_{f0}}{E_{eff} V I_{f0}}P_r - \lambda_I I_r \\ \dfrac{dX_r}{dt} = \dfrac{\gamma_X P_{f0}}{E_{eff} V X_{f0}}P_r + \dfrac{\lambda_I I_{f0}}{X_{f0}}I_r - (\lambda_X + \dfrac{\sigma_X P_{f0}}{E_{eff}\Sigma_f V}P_r)X_r \\ \dfrac{dT_f}{dt} = \dfrac{f_f P_0}{\mu_f}P_r - \dfrac{\Omega}{\mu_f}T_f + \dfrac{\Omega}{2\mu_f}T_i + \dfrac{\Omega}{2\mu_f}T_e \\ \dfrac{dT_e}{dt} = \dfrac{(1-f_f)P_0}{\mu_c}P_r + \dfrac{\Omega}{\mu_c}T_f + \dfrac{2M-\Omega}{2\mu_c}T_i - \dfrac{2M+\Omega}{2\mu_c}T_e \\ \rho = \rho_{rod} + \alpha_f(T_f - T_{f0}) + \dfrac{\alpha_c}{2}(T_i - T_{i0}) + \dfrac{\alpha_c}{2}(T_e - T_{e0}) \\ \quad + \alpha_x X_{f0}(X_r - X_{r0}) \\ \delta\rho_{rod} = \alpha_{rod}\delta rod \end{cases} \tag{1}$$

[a] Corresponding author: wychug@163.com

Table 1. Main model parameters.

Parameter	Name
P_r	core power level
c_{ri}	ith group normalized precursor concentration
c_r	normalized precursor concentration
I_r	normalized iodine concentration
X_r	normalized xenon concentration
T_f	fuel average temperature
T_i	coolant inlet temperature
T_e	coolant outlet temperature
δrod	position variation of the control rod (fraction of core length)
δ	deviation of a parameter from initial steady-state value

One group delayed neutron model is utilized and the coolant inlet temperature is treated as a constant [2-4].The small perturbation linearization methodology is utilized to linearize the nonlinear core model (1). The transfer function and the state equation of the core are separately calculated and expressed by

$$G = \frac{\delta P_r}{\delta rod} = \frac{\sum_{i=0}^{5} a_i s^i}{\sum_{i=0}^{6} b_i s^i} \tag{2}$$

$$\begin{cases} \dot{\mathbf{x}} = A\mathbf{x} + Bu \\ y = C\mathbf{x} + Du \end{cases} \tag{3}$$

where u=δrod-the input; y=δP_r-the output; a_i (i=0,1,2,5)-numerator coefficients; b_i (i=0,1,2,3,6)-denominator coefficients; s-laplace operator; $\mathbf{x}=[x_1, x_2, x_3, x_4, x_5, x_6]^T=[\delta P_r, \delta c_r, \delta I_r, \delta X_f, \delta T_f, \delta T_e]^T$-the state matrix; A-the $R^{6\times6}$ system matrix; B-the $R^{6\times1}$ input matrix; C-the $R^{1\times6}$ output matrix; D-the zero matrix.

The transfer function G is calculated by using parameters from Ref. [9], in which the total primary heat output is 2200 MW, the primary coolant inlet temperature is 285°C, the primary coolant outlet temperature is 317°C, the primary coolantaverage pressure is 15.5 MPa, and the primary coolant mass flow is 12861.1 kg/s.

3 H∞ loop shaping control

The H∞ loop shaping design approach was proposed by McFarlane et al. [5] to implement tradeoffs between performance and robust stability of a stable H∞ loop shaping control system for a plant, where a normalized coprime factor H∞ robust stabilization issue needs to be solved to guarantee closed-loop stability and a certain level of robust stability of the system at all frequencies.

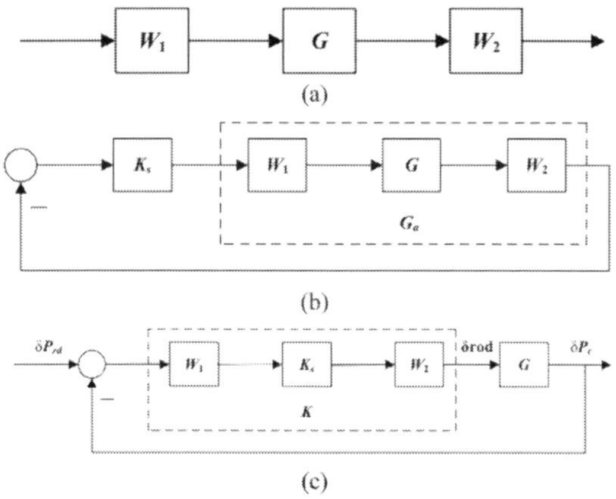

Figure 1. The loop shaping design procedure for the core.

This H∞ loop shaping design approach consists of two key techniques, namely loop shape and robust stabilization. The loop shaping design procedure is shown in Fig.1, where W_1 and W_2 are weight matrices or compensators; G_a is a shaped plant based on G; K_s is the H∞ controller of G_a; K is the controller to be devised; δP_{rd} denotes the reference input, namely the desired change of power level.

(1) Loop shaping

Singular values of a controlled plant G are regulated to obtain a desired open loop shape adopting an additional pre-compensator W_1 and/or an additional post-compensator W_2. The plant G is shaped to be G_a introducing W_1 and/or W_2, $G_a=W_2GW_1$. Suppose W_1 and W_2 are such that G_a contains no hidden modes with unstability.

(2) Robust stabilization

Step one: Calculation of $\varepsilon_{\max}$

$$\varepsilon_{\max} = \left[\inf_{K\,stabilizing} \left\| \begin{bmatrix} I \\ K \end{bmatrix} (I - G_a K)^{-1} M_s^{-1} \right\|_\infty \right]^{-1} \tag{4}$$

where M_s is a normalized coprime factor of G_a subjecting to $G_a= M_s^{-1}N_s$.

Step two: Select $\varepsilon \le \varepsilon_{\max}$, design a stabilizing controller K_s satisfying

$$\left\| \begin{bmatrix} I \\ K_s \end{bmatrix} (I - G_a K_s)^{-1} M_s^{-1} \right\|_\infty \le \varepsilon^{-1} \tag{5}$$

(3) The eventual feedback controller: $K=W_1K_sW_2$.

The stability margin ε is both a measurement for robustness of a loop shape control system and the indicator of the success of the loop shaping. Generally speaking, K_s is solved so that $\varepsilon \le \varepsilon_{\max}$ and the controller can guarantee that the control system possess satisfied robustness. A small value of $\varepsilon_{\max}$ indicates incompatibility among a desired loop shape, a controlled plant and closed-loop robust stability. In the presence of this case, W_1 and W_2 need to be regulated again to obtain a decent loop shape.

4 Design of nonlinear PWR core power control system

According to Section 3, solving K is the following.

(1) Loop shaping

G_a is exploited and defined as a shaped plant of G. Of selected weight matrices for G, W_1 is $I_{1\times 1}$; W_2 is expressed by Eq.(6), and parameters of W_2 are shown in Table 2, where $e^i=10^i$.

$$W_2 = \frac{\sum_{i=0}^{6} c_i s^i}{\sum_{i=0}^{6} d_i s^i} \qquad (6)$$

Table 2. Parameters of W_2.

Weight matrix	[c6 c5 c4 c3 c2 c1 c0] [d6 d5 d4 d3 d2 d1 d0]
W_2	[10 4014 6921 3898 432.1 0.03186 1.42e-6] [5088 7032 1868 141.2 0.2974 4.531e-5 8.151e-10]

(2) Solve controllers

The controller K_s is solved, and then the final controller K of G is calculated by $K=W_1K_sW_2$. K is expressed by Eq.(7), and parameters of K of G are shown in Table 3. The solved stability margin ε of K is shown in Table 3.

$$K = \frac{\sum_{j=0}^{11} e_j s^j}{\sum_{j=0}^{12} f_j s^j} \qquad (7)$$

Table 3. Parameters of K.

Controller	[e11 e10 e9 e8 e7 e6 e5 e4 e3 e2 e1 e0] [f12 f11 f10 f9 f8 f7 f6 f5 f4 f3 f2 f1 f0]
K	[8.06 3.625e4 1.326e7 2.288e7 1.291e7 1.452e6 2960 2.525 0.001146 2.914e-7 3.925e-112.184e-15] [1 8213 1.687e7 2.334e7 6.233e6 4.79e5 1839 2.546 0.001786 7.093e-7 1.62e-10 1.987e-14 1.015e-18]

(3) Effect of local controllers

When the reference input δP_{rd} is taken as a 0.1 step, responses of the output δP_r in G with K are shown in Fig.2, where the curve is step response of δP_r in G with K. According to the curve, the output δP_r in G with K in response to a step possesses good performance, such as the zero overshoot, the short adjustment time and the zero static error. Hence, the controller Kis desirable.

The solved controller K is applied to the nonlinear core model. The combination of the nonlinear core model and K is the nonlinear core power H_∞ loop shapingcontrol system. The schematic of the nonlinear core power H_∞ loop shaping control system is shown in Fig.3.

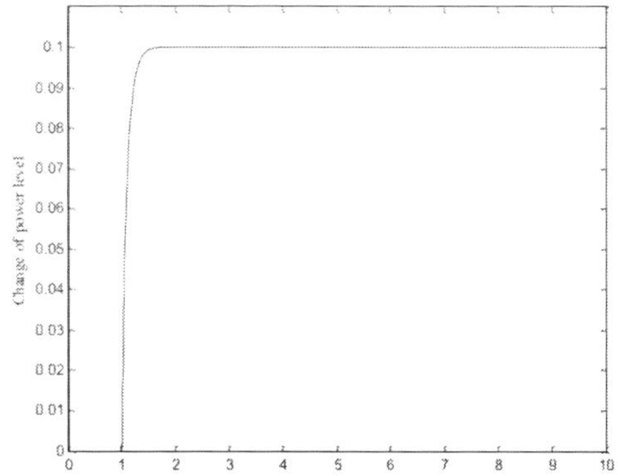

Figure 2. Step response of the output δP_r from G_i with K_i (i=1,...,7).

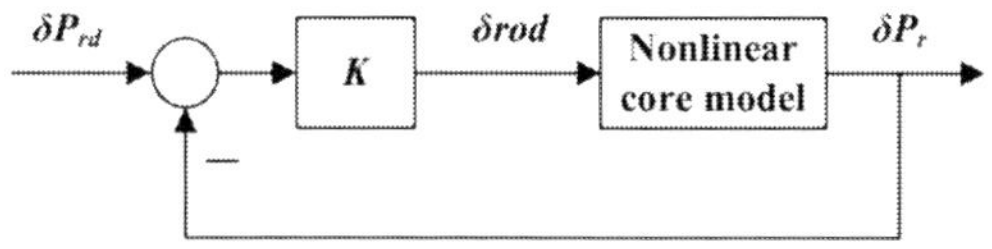

Figure 3. Schematic of the nonlinear core power H_∞ control system.

5 Simulations

The nonlinear core power H_∞ loop shaping control systemis simulated considering two typical load variations that are the -10% step process and the ramp process with ±0.0556%·min^{-1} variations. The direction from the bottom to the top of core is treated as the positive direction of y-axis for simulation work, the core bottom is the zero point of y-axis and the height of core is 366cm. When the core power level is at 100%, the initial rod position is at the top of the core, namely at 366cm.

When the reference input δP_{rd} is taken as a -10% step load in 24 hours, change trajectories of core parameters during the core following the step load (100%→90%) are shown in Fig.4, where S1 is the desired load change trajectory for the start of reactorfull power level operation, the middle of reactor shut-down operation with a -10% step variation, and the end of 90% power level operation; S2 is a change trajectory of the core power level; S3 is a change trajectory of the control rod position in the core.

When the reference input δP_{rd} is taken as a ramp load with ±0.0556%·min^{-1} variations, change trajectories of core parameters during the core following the ramp load (100%→90%→100%) are shown in Fig.5, where R1 is the desired load change trajectory for the start of reactor full power level operation, the middle ofreactor shut-down operation with a -6%·min^{-1} ramp variation and 90% power level operation, and the end of reactor start-up operation with a 6%·min^{-1} ramp variation; R2 is a change trajectory of the core power level; R3 is a change trajectory of the control rod position in the core.

Consequently, SL2 (RL2) can follow SL1 (RL1) in real time; meanwhile, the processes of SL3 and RL3 are suitable and acceptable.

Hence, S2 can follow S1, and R2 can follow R1 in real time. Besides, the dynamic processes of S3 and R3 are acceptable.

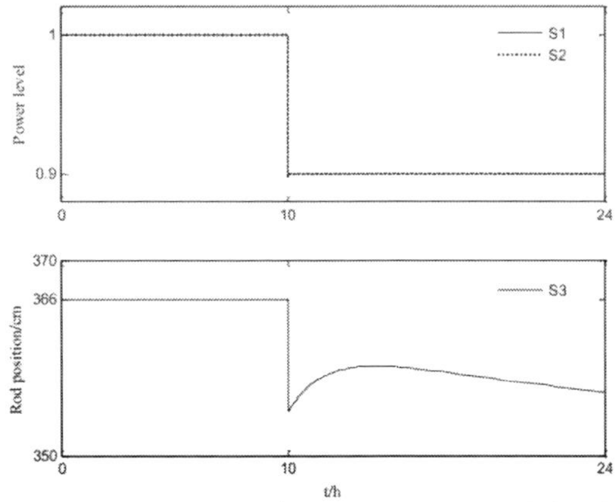

Figure 4. Change trajectories of core parameters during the core following a step load (100%→90%).

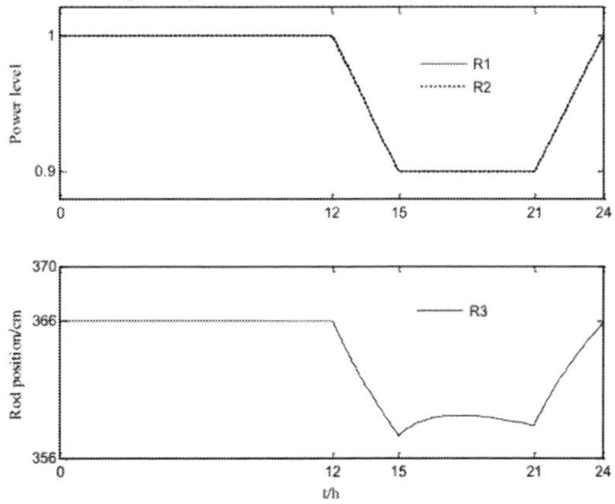

Figure 5. Change trajectories of core parameters during the core following a ramp load (100%→90%→100%).

6 Conclusions

The power-level control for PWR cores with xenon ocsillationsis difficult to be implemented, which is also a challenge reaesrch direction. In this paper, the power-level control problem for a PWR core is handled applying the H_∞loop shapingcontrol strategy to this nonlinear core.

The nonlinear PWR core and its linearized model are separately modeled using the lumped parameter method and the small perturbation linerization method that are capable of be reference modeling approaches to study other nonlinear plants.

In order to implement applications of the controller designed on a linearizedcore model to a nonlinearcore model, the solved H_∞loop shaping controller has strong robustness of accommodating model errors between the nonlinear model and the linear model.

The devised nonlinear PWR core power H_∞loop shaping control systemcan control the core power-levelsatisfactorily in response to different load changes such as a step or ramp. Hence, the nonlinear PWR core power H_∞loop shaping control system is effective.

Acknowledgement

The authors would like to thank anonymous reviewers for their valuable comments. The work is funded by National High Technology Research and Development Program of China (863 Program) (No.2015AAXX46611) and China Postdoctoral Science Foundation (No.20159200078).

References

1. R. M. Edwards, *Robust optimal control of nuclear reactors* (Pennsylvania State University, 1991)
2. R. M. Edwards, K. Y. Lee, M. A. Schultz, Nuclear Technology, **92**, 167-185 (1990).
3. A. Ben-Abdennour, R. M. Edwards, K. Y. Lee, IEEE Trans. on Nuclear Science, **39**, 2286-2294 (1992).
4. H. Arab-Alibeik, S. Setayeshi, IEEE Trans. on Nuclear Science, **50**, 211-218 (2003).
5. D. McFarlane, K. Glover, IEEE Trans. on Autom. Control, **37**, 759-769 (1992).
6. M. A. Schultz, *Control of nuclear reactors and power plants* (McGraw-Hill, 1961).
7. T. W. Kerlin, E. M. Katz, J. G. Thakkar, J.E. Strange, Nuclear Technology, **30**, 299-316 (1976).
8. M. N. Khajavi, M. B. Menhaj, A. A. Suratgar, Annals of Nuclear Energy, **29**, 751-760 (2002).
9. IAEA, *Directory of Nuclear Reactors Vol.IX*, (International Atomic Energy Agency, 1971).

Design of Feed-forward Controller with Stick-slip Friction Modeling in Electro-mechanical Brake System

Giseo Park, Sanghun Jung and Seibum Choi

Department of Mechanical Engineering, Korea Advanced Institute of Science and Technology, Daejeon, Korea

Abstract. Electro-Mechanical Brake (*EMB*) is expected to be one of the future brake system. Feedback controller with sensor measuring is commonly used for control of EMB. However, this controller has some issues like delayed response and extra cost about sensor installation. In this paper, Feed-forward controller in EMB is proposed for solving these problems of feedback control. Also, it is very necessary to describe dynamical phenomenon of friction in actual EMB system. The actual EMB system shows stick-slip friction of mechanical parts which is difficult to model and apply to design of controller. This research is focused on exquisitely describing this stick-slip friction. In order to do this, the experiment about EMB is proceeded in the open loop system with the motor current command and data from the experiment is used for identification of model parameters during stiction. Then, parameters during slip is estimated in the closed loop system. Finally, developed friction model of EMB is proposed and it is utilized for design of feed-forward controller. Matlab Simulink is used for design of EMB simulation model and EMB test bench is utilized for experiment. Performance of proposed control system is compared with that of feedback control system.

1 Introduction

Brake-By-Wire (BBW) system is expected to be one of the future brake system [4]. Because it has many advantages like faster response, shorter braking distance, good space efficiency and connectivity to vehicle chassis control systems like Anti-lock Brake System (ABS) and Electronic Stability Control (ESC) [7]. This system has no direct connection between brake pedal and wheel. Then, hydraulic line in hydraulic brake system is replaced by pedal sensor, brake pedal simulator, control unit and back-up system. Electro-Mechanical Brake (EMB) is one of the BBW system and has been studied in many companies in automotive industry [5].

Most of EMB is controlled by feedback controller embedded in ECU [3]. The control system has cascaded architectures, which are consisted of force controller, velocity controller, and current controller in series [1]. The cascaded architecture is shown in Fig. 1. However, this controller has some issues related to control performance and usefulness. Because controller based feedback, which is measured value by a sensor, cannot help having delayed response. This delayed response has a bad effect for control performance.

In this paper, new EMB controller design using feed-forward control is proposed for solving these issues of existing feedback controller. It is anticipated to have no delayed response.

Also, it is very necessary to utilize clamping force and friction torque estimator for generating feed-forward term. EMB simulation model is developed and used for accuracy and usefulness of this control system are shown with comparison of the simulation results in feedback control system.

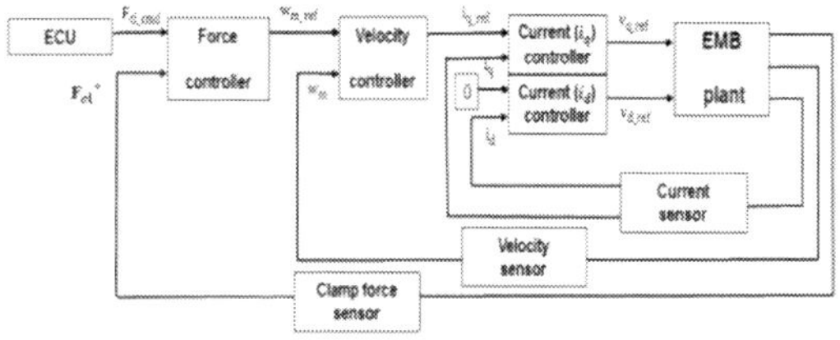

Figure 1. Cascaded architecture of EMB feedback controller

2 EMB dynamics

2.1 Simplified EMB model

The simplified model of EMB plant is necessary for controller design based on characteristics of EMB plant. Fig. 2 shows diagram of this model. Torque balance equation of EMB model is described as follows.

$$T_m = T_L + J_{tot}\frac{d^2}{dt^2}\theta_m + T_f \tag{1}$$

T_m means motor torque, which is generated by Permanent Magnet Synchronous Motor (PMSM) or Brushless DC motor. These motors are commonly used for motor of EMB. Also It can be expressed like that motor current is multiplied by motor-torque constant K_m [6].

$$T_m = K_m i_q \quad (2)$$

T_L is load torque, which makes clamping force on brake disk. Rotational motion of motor is converted into linear motion by the ball screw. p is pitch of ball screw and η is efficiency of ball screw. Also, GR is total gear ratio, which can be calculated from the number of gear teeth of all reduction gears in EMB. Load torque is represented as follows [1] :

$$T_L = \frac{1}{GR}\frac{p}{2\pi\eta}F_{cl} = k_{cl}F_{cl} \quad (3)$$

T_f, friction torque will be explained in chapter 3. Friction torque modeling.

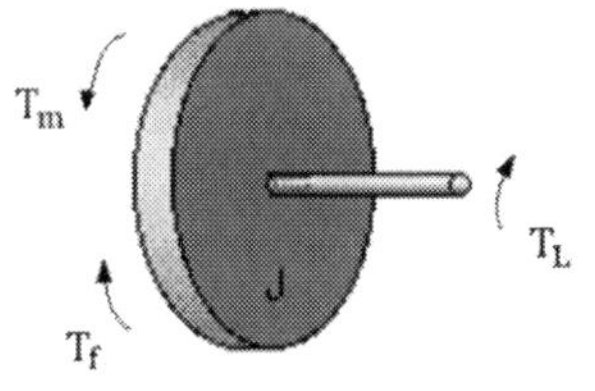

Figure 2. Diagram of Simplified EMB model

2.2. Clamping force estimation

It is recommended that a clamping force sensor is replaced with simple and precise algorithm due to high price of the sensor. Also, the sensor has low accuracy in high temperature and structural issue. Clamping force algorithm has been studied in a lot of researches [1, 2, 4]. In previous papers, clamping force is expressed as a function of motor angle or displacement of brake pad [2].

In this paper, clamping force versus motor angle is estimated with only measuring motor current and motor angle. In order to estimate clamping force, motor current versus motor angle is expressed by fitted curve, which is second-order polynomial function.

The curve of motor current is distinguished between clamping and releasing situations. In each situation, rotation of motor angle has forward and backward directions. If same motor angle in these two situations, each friction torque has same absolute value and different sign. Because friction toque in EMB has features of Coulomb friction and contain no viscous friction. In spite of the opposite direction, if the motor angle is same, the clamping force is equal. Clamping force in each situation is represented by the following equations.

$$\hat{F}_{cl}(\theta_m) = \frac{K_m i_{q-cl} - J_{tot}\frac{d^2}{dt^2}\theta_{m-cl} - T_f}{k_{cl}} \quad (4)$$

$$\hat{F}_{cl}(\theta_m) = \frac{K_m i_{q-rel} - J_{tot}\frac{d^2}{dt^2}\theta_{m-rel} + T_f}{k_{cl}} \quad (5)$$

Calculating average between (4) and (5), clamping force of any motor angle θ_m can be derived. Also, there is no consideration of friction torque T_f, which is estimated in chapter 3 Friction torque modeling.

$$\hat{F}_{cl}(\theta_m) = \frac{K_m(i_{q-cl}+i_{q-rel}) - J_{tot}\left(\frac{d^2}{dt^2}\theta_{m-cl}+\frac{d^2}{dt^2}\theta_{m-rel}\right)}{2k_{cl}} \quad (6)$$

Fig. 3 shows this process and estimated clamping force $\hat{F}$ versus motor angle. Characteristic curve is estimated with kissing point, which means zero value of the curve and calculated by measurement when pad contacts brake disk. In order to verificate this algorithm, experiment data is necessary.

Thus, EMB test bench is designed and composed of EMB hardwrare, ECU, Micro autobox and control desk in Fig. 4.

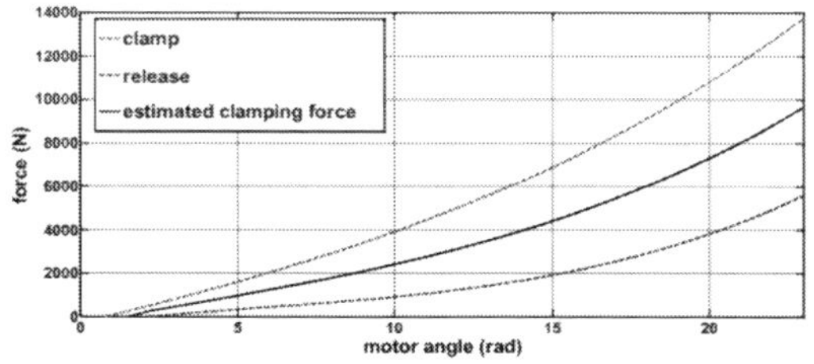

Figure 3. EMB characteristic curve

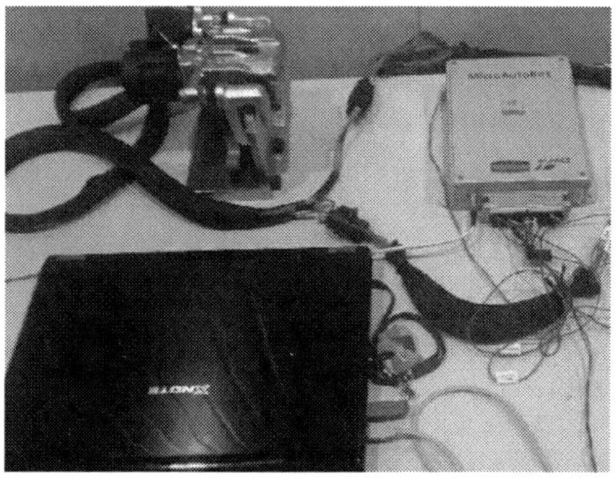

Figure 4. EMB test bench

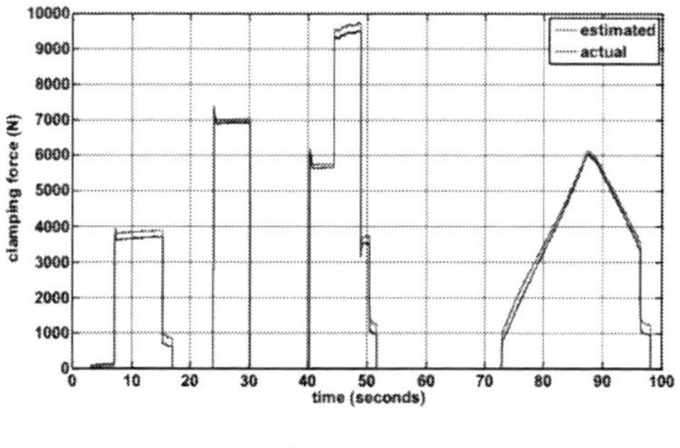

Figure 5. Estimated clamping force results

Fig. 5 represents real time esimation results of clamping force. In Fig. 5, compared to actual clamping force measured by sensor, estimated clamping force has contentable root mean square error, 0.3kN. clamping force esimation algorithm is utilized for design of feed-forward controller.

3 Friction torque modeling

Friction toque occuring in motor, gear and ball screw of EMB is complicated to define as mathmatical expression. However, estimation of friction torque is necessary for elaborate design of EMB feed-forward control system. Chris Line proposed friction toque model of EMB [3]. According to motor angular velocity, friction torque is divided to static and kinetic friction torque. Also, the friction torque is described as a function of clamping force and angular velocity.

In this paper, modificated friction torque model is proposed for exquisitely describing real dynamics of EMB. Especially, accuracy of modeling of kinetic friction torque is improved through clamping force estimation. Following expression shows this modeling.

$$T_{f,k}(\theta_m)=\begin{cases}\mathrm{sign}(w_m)\{\,T_{f0}+\alpha(\theta_m-\hat{\theta}_0)^2+\beta(\theta_m-\hat{\theta}_0)\} \\ T_{f0}\,\mathrm{sign}(w_m)\end{cases} \tag{7}$$

Upper expression in (7) is occured when motor angle is over kissing point. Lower expression in (7) is occured when motor angle is under kissing point. Because friction torque is proportional to clamping force, friction torque can be expressed by second order polynomial function of motor angle. Also, before generating clamping force such as under kissing point, friction torque is almost constant.

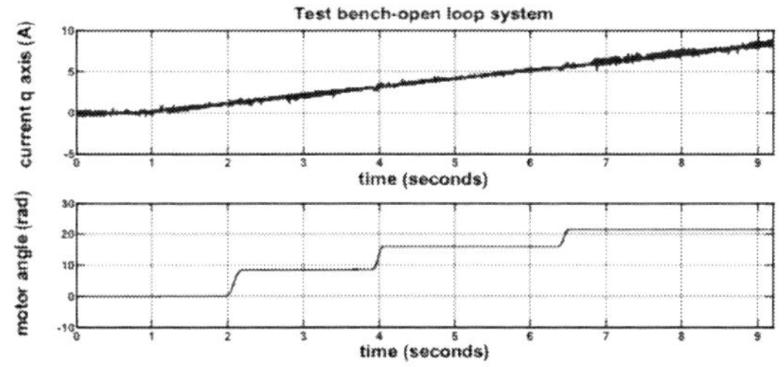

Figure 6. Stick-slip phenomenon of actual EMB

Fig. 6 shows stick slip phenomenon, which means spotaneous jerking motion. The rotation part of EMB slip over the surface when external torque is larger than friction torque. This external torque is motor torque minus load torque and increased can be increased by motor current command such as Fig. 6. The rotation stops when the external torque is smaller than friction torque. Because friction torque is nonlinearly increased according to motor angle, more larger torque is determined alternately between external torque and friction torque. Therefore, stick and slip motion of rotation part is occured alternately and repeatedly.

If a designed simulation model contains frcition torque model, which can express stick slip phenomenon, the simulation model can be identified as very exquisite model similar to actual EMB plant. In order to make this model, unknown parameters in (7), T_{f0}, α, β are estimated using actual experiment data like Fig. 6. Least square method is utilized for estimation of these parameters. Fig. 7 shows block diagram of friction torque in EMB simulation model.

4 Design of feed-forward controller

Using clamping force estimator and identified friction torque model, EMB feed-forward controller is designed. Clamping force estimator is transformed to inverse function, which yields the motor angle value according to desired clamping force. Feed-forward term equal additional motor current command. It is designed like follow equation.

$$i_{q-ff}=\frac{Jw+\gamma F_{cl}+T_f}{k_t} \tag{8}$$

Compared with feedback control system in Fig. 1, feed-forward term is added and clamping force sensor is replaced to characteristic curve. Diagram of feed-forward control system is shown in Fig. 8.

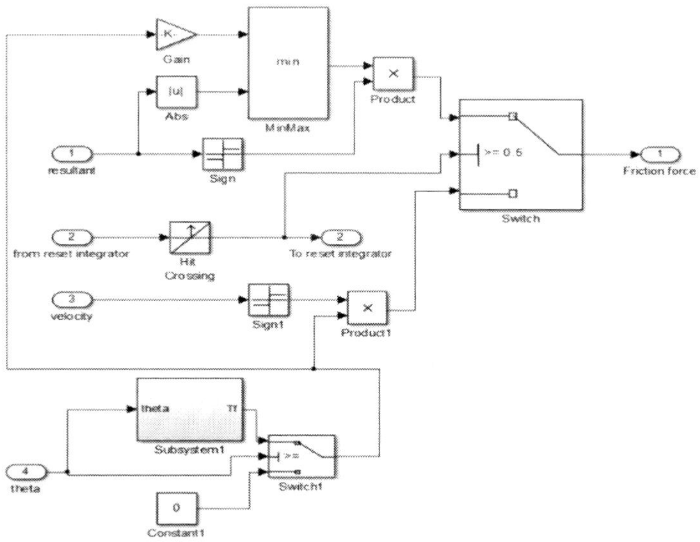

Figure 7. Simulation model of EMB friction torque

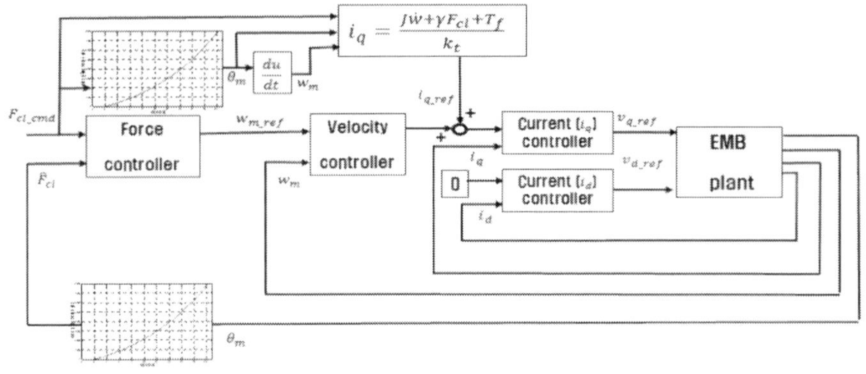

Figure 8. Diagram of feed-forward control system

5 Simulation results

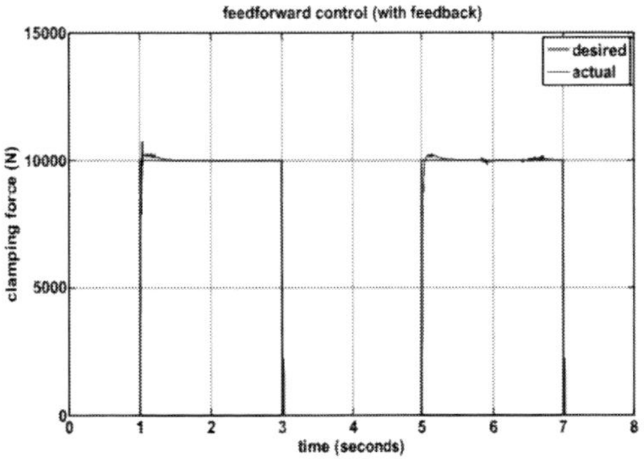

Figure 9. Diagram of feed-forward control system

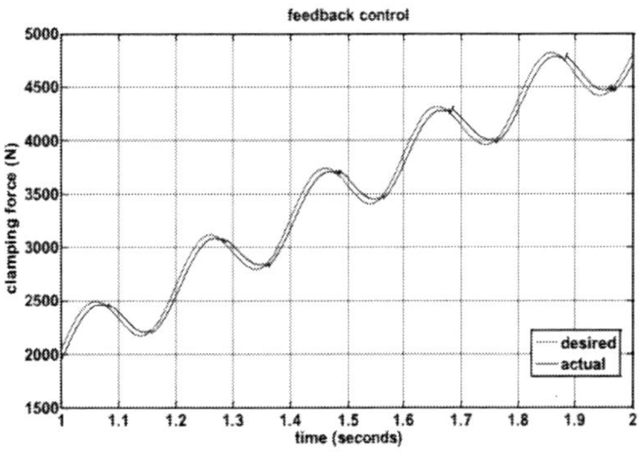

Figure 10. Control results in feedback control system

The control system is implemented in Matlab Simulink and performance of controller is evaluated. In Fig. 9, maximum step input is commanded and it is shown that actual clamping force are close to desired clamping force. As shown in Fig. 10 and Fig. 11, performance of feed-forward control system is better than that of feedback control system in terms of same sinusoidal input. Rms errors in each systems are 0. 15kN and 0.5kN.

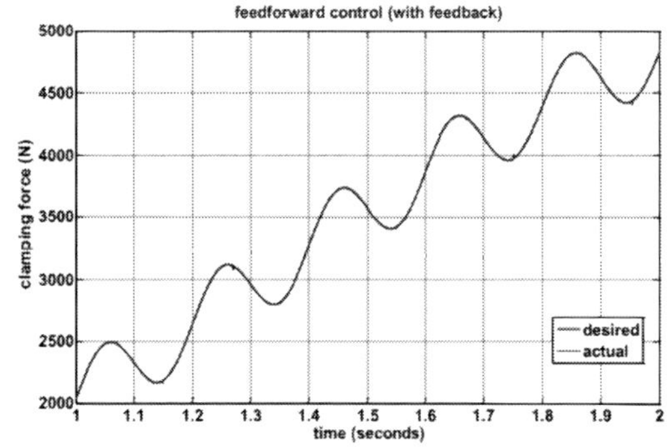

Figure 11. Control results in feed forward control system

6 Conclusion

In this paper, EMB feed-forward control system using clamping force estimator and friction torque model is proposed. Especially, friction torque model exquisitely describes actual EMB plant using experimental data in stick slip phenomenon. Feed-forward controller is effective to solve some problems of existing EMB feedback control system like control performance. Feed-forward term related to clamping force, inertia torque and friction torque is very effective to follow desired input with no delay. It is anticipated that these algorithm can be used for improvement of EMB system.

References

1. R. Schwarz, R. Isermann, J. Bohm, J. Nell, P. Rieth, SAE 980600 (1998)
2. Y. H. Ki, K. J. Lee, J. S. Cheon, H. S. Ahn, Int. J. Automotive. Tech. **14**, 739 (2013)
3. C. Line, C. Manzie, M. C. Good, IEEE Trans. Control Sys. Tech. **16**, 446 (2008)
4. C. Jo, S. Hwang, H. Kim, IEEE Trans. Vehicular Tech. **59**, 3205 (2010)
5. J. Choi, I. Kim, K. Huh, M. Kim, J. Kim. K. Kim, KSAE 653 (2009)

6. P. Pillay, R. Krishnan, IEEE Trans. Industry Applications, **25,** 265 (1989)
7. S. Saric, A. Bab-Hadiashar, R. Hoseinnezhad, IEEE Trans. Vehicular Tech. **57**, 778 (2008)
8. R. Hoseinnezhad, A. Bab-Hadiashar, T. Rocco, IEEE Trans. Vehicular Tech. **57**, 770 (2008)
9. R. Hoseinnezhad, A. Bab-Hadiashar, IEEE Trans. Industry Elec. **54**, 1052(2007)

A New Algorithm of Online Stator Faults Diagnosis of Three-Phase Induction Motors Using Duty Ratios of Half-Period Frequencies According to Phase Angle Changes

YoungJin Go[1], Myoung-Hyun Song[2], Jun-Young Kim[2], Buhm Lee[3], Wangrim Choi[4] and Kyoung-Min Kim[3]

[1]*Dept. of Electrical Automation, Suncheon JEIL College, Deogwol-dong, Suncheon-city, Jeollanam-do, Republic of Korea*
[2]*Dept. of Electric Control Eng., Sunchon National Univ., Maegok-dong, Suncheon-city, Jeollanam-do, Republic of Korea*
[3]*Dept. of Electrical and Semiconductor Eng., Chonnam National Univ., Dundeok-dong, Yeosu-city, Jeollanam-do, Republic of Korea*
[4]*Dept. of Biomedical and Electronic Eng., Chonnam National Univ., Dundeok-dong, Yeosu-city, Jeollanam-do, Republic of Korea*

Abstract. The causes of faults of induction motors are largely categorized into bearing fault, which causes a mechanical fault, and stator fault and rotor fault, which cause an electrical fault. A stator fault among these faults, which causes an electrical fault, occurs due to the breakdown of insulation, meaning the stator is directly connected with the power supply, and the direct connection is a direct cause of a major accident. For this reason, many studies are being performed to detect the faults. This paper explained the effects of a negative sequence on phase angle change by analyzing the effects of the existing negative sequence on the d-q transform of Park's vector approach. This paper suggested a new algorithm that identifies the causes of stator faults with the use of the change in the duty ratio of the half-period frequency of the frequency when a phase angle change occurs at that moment.

1 Introduction

Most of the faults of induction motors are caused by bearing fault (44%); wrong winding or stator fault due to a stator winding short circuit fault or a stator winding open circuit fault (26%); and fault of the bar of the rotor (8%). The causes of induction motors are largely categorized into mechanical faults and electrical faults - bearing faults causing mechanical faults; and stator faults and rotor faults causing electrical ones.

Among the causes, a stator fault occurs due to the breakdown of insulation of the stator, a deterioration in insulation of the stator winding is caused by various reasons, such as coil movement, thermal stress, overload, and mechanical vibration, and the deterioration in insulation of the winding results in a turn to turn short[1].

The occurrence of a turn to turn short causes a large circulating current to occur in the faulty loop, and the heat in the winding, which is proportional to the square of the stator current, plays a role in accelerating the occurrence of a more serious fault, such as phase-to-phase short fault and a phase-earth fault, which cause a serious damage to the winding and core of the stator.

Hence this is the time when early diagnosis is desperately needed, as detecting a turn to turn short in the early stage can prevent a sequential damage to the motor and considerably reduce economic loss by minimizing the outage time of the motor, the number of additional workers, and the cost of repair.

Among methods of diagnosis of ITSC fault of induction motors, studies on MESA (motor electrical signature analysis) using only electrical signatures are being performed actively. [2] was researched through Park's vector (PV) approach, which is a representative one; as for [3], faults were identified by setting the circular distortion ratios at the threshold values with the use of DR; [4,5] were researched through AI technique; [6] was researched through wevelet analysis. As for [7,8], negative sequence currents were utilized; as for [9], negative sequence impedance was utilized. [10] was researched through a matrix of impedances modeling cross-coupling between positive and negative sequence components; and recently as for [11], studies were performed considering asymmetric elements in the machines in which a negative sequence current component is bound to occur.

But other papers, analysis of the stator fault, it is difficult to analyze online.

Typically, the line fault diagnosis method has been used in the PV.

This paper suggested a new fault diagnosis method in online using the duty ratio of half-period frequency according to phase angle change, with the consideration of the effects of negative sequence components on PV's d-q transform.

2 Background theory

A three-phase induction motor can simplify MCSA by using Park's Vector Approach. The marks for a phasor space are located on the cross-section of an induction motor and are converted into arbitrary instantaneous values on the three-phase on the complex plane. The phasor space on this complex plane is rotated at the same angular rate as the angular frequency of the three-phase supply system.

Hence if the three-phase input current of an induction motor is applied to Park's Vector Approach, it is converted as follows (see the Eq. 1 and Eq. 2 below :

$$I_d = \left(\frac{\sqrt{6}}{2}\right) i_M \sin(\omega t) \quad (1)$$

$$I_q = \left(\frac{\sqrt{6}}{2}\right) i_M \sin\left(\omega t - \frac{\pi}{2}\right) \quad (2)$$

here i_M represents supply current's maximum value [A], ω represents supply angular frequency (rad/s), and t represents time parameter [s].

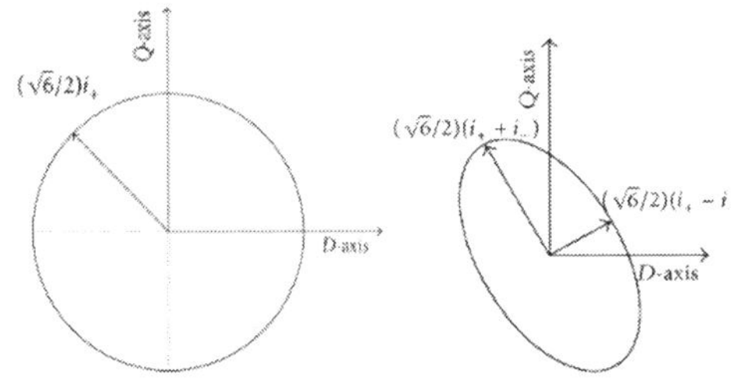

Figure 1. Pattern of Park's Vector

Fig. 1 shows the trace of a circle with its center at the origin on a coordinate plane. Under normal start conditions, the left pattern in Fig. 1 is shown. If a stator winding turn fault occurs, the right pattern (oval shape) in Fig. 1 is shown. More turn faults mean more asymmetric components of supply current, and more asymmetric components of supply current mean a higher degree of distortion of the trace of the pattern of Park's Vector[12].

The degree of distortion of the pattern has recently been defined as *DR* (Distortion Ratio), which has been numerically calculated using the Eq. 3 and Eq. 4[3].

$$r = \sqrt{i_d^{\,2} + i_q^{\,2}} \quad (3)$$

$$DR = \frac{r_{max}}{r_{min}} \quad (4)$$

here *r* represents the scalar of Park's Vector, r_{max} represents the broadest amplitude of it, and r_{min} represents the narrowest amplitude of it.

If the maximum value r_{max} is the same as the minimum value r_{min} , the induction motor is in the most ideal state (*DR*: 1). As a larger difference between them means a higher *DR*, the phase current when the difference is large is under highly unbalanced conditions due to winding turn fault.

2.1 Phase changes

$r = \sqrt{I_d^2 + I_q^2}$ represents the maximum value of current, actually determines the shape of a circle in Park's Vector Approach, and is an important factor representing the amplitude of current. Here the components cosθ and sinθ of R_e and $I_m$determine the position of I_s.

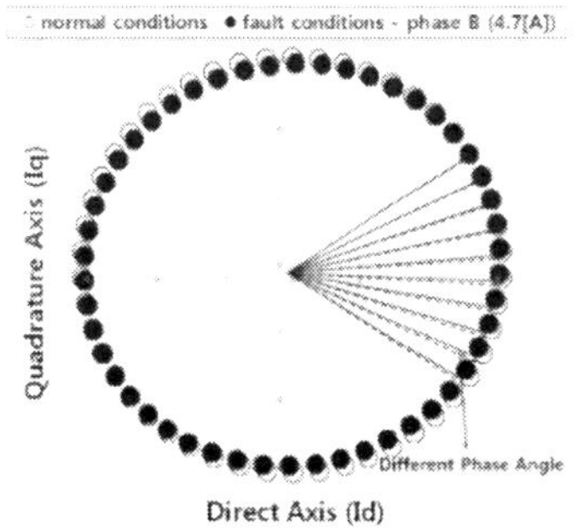

Figure 2. Different Phase Angle

When a fault occurred, a particular pattern was created by the actual phase angle θ although an analysis was performed only using the maximum value of r. This can be seen in Fig. 2.

As for Fig. 2 a simulation was performed based on a normal condition current of 5[A]. A current fault signal was generated in phase B among three phases (the phase current: 4.7[A]). If the difference in fault current is large, the change of the phase angle θ is also large.

To use the characteristics of the pattern according to the change of the phase angle *θ*, the following Eq. 5 can be found using Euler's theorem[13].

$$e^{j\theta} = \frac{I_d + jI_q}{r} = \frac{I_d + jI_q}{\sqrt{I_d^2 + I_q^2}} \quad (5)$$

2.2 Negative sequence

In steady state, the relationship between the vector operat or 'a' and the phase angle ө can be interpreted through a vector operator: This can be expressed as a theorem for th e operator 'a' as follows:

$$a = 1\measuredangle 120° = cos120° + jsin120° = -\frac{1}{2} + j\frac{\sqrt{3}}{2} \quad (6)$$

$$a^2 = 1\measuredangle 240° = cos240° + jsin240° = -\frac{1}{2} - j\frac{\sqrt{3}}{2} \quad (7)$$

$$a^3 = 1\measuredangle 360° = 1 \quad (8)$$

$$a + a^2 = -\frac{1}{2} + j\frac{\sqrt{3}}{2} - \frac{1}{2} - j\frac{\sqrt{3}}{2} = -1 \quad (9)$$

$$1 + a + a^2 = 0 \quad (10)$$

That is, in steady state, the total power of 3-phases must be zero.

Here a zero sequence component is not shown on a revolution indicator, and the component can be ignored. Therefore, when the total voltage of the positive sequence component of the stator is represented as $\tilde{V}_{1s}$ and that of the negative sequence component is represented as $\tilde{V}_{2s}$, it is possible to write the stator voltage of $\tilde{V}_s$ using the relationship $V_d + jV_q = \tilde{V}$ as follows:

$$\tilde{V}_s = \tilde{V}_{1s} + \tilde{V}_{2s} \quad (11)$$

When seeing Eq. 11 based on d-q transform, it can be expressed as follows:

$$v_{dqs} = \bar{V}^{e}_{1dqs}\, e^{-j(\theta - w_s t)} + \bar{V}^{-e}_{2dqs}\, e^{-j(\theta + w_s t)} \quad (12)$$

here the formula $v_{dqs} = v_{ds} + jv_{qs}$ can be devised; it represents the positive sequence component of $\bar{V}^{e}_{1dqs} = \tilde{V}_{1s}$ and the negative sequence component of $\bar{V}^{-e}_{2dqs} = \tilde{V}^{*}_{2s}$; and t represents time and θ represents an angle of rotation.

In the Eq.12, the domain of e is created by the positive sequence component that rotates together with the magnetic field and is expressed as such when only the positive sequence component in the 3-phase system exists; and v_{dqs} equals V_{1s} and is expressed as V^{e}_{1dqs}.

The domain of as $-e$ is created by the negative sequence component that rotates together with the magnetic field and is expressed as such when only the negative sequence component in the 3-phase system exists; and v_{dqs} equals V^{*}_{2s} and is expressed as V^{-e}_{2dqs}.

These two domains rotate in the opposite direction to which the two pairs of frequencies of d-q transform rotate at the same speed.

This can be expressed as follows:

$$V^{e}_{1dqs} = v^{e}_{1ds} + jv^{e}_{1qs} = \tilde{V}_{1s} \quad (13)$$

$$V^{-e}_{2dqs} = v^{-e}_{2ds} + jv^{-e}_{2qs} = \tilde{V}_{2s} \quad (14)$$

$$V^{e}_{dqs} = v^{e}_{ds} + jv^{e}_{qs} = V^{e}_{1dqs} + V^{-e}_{2dqs} e^{-j2w_s t} \quad (15)$$

In this way, it is also possible to infer currents and magnetic fluxes[14].

That is, it is possible to approach the interpretation of imbalance through the negative sequence component.

2.3 Imbalance between frequencies by interpreptation of imbalance

When supposing that the positive sequence component, the negative sequence component, and the zero sequence component of 3-phases exist as in these Figures,

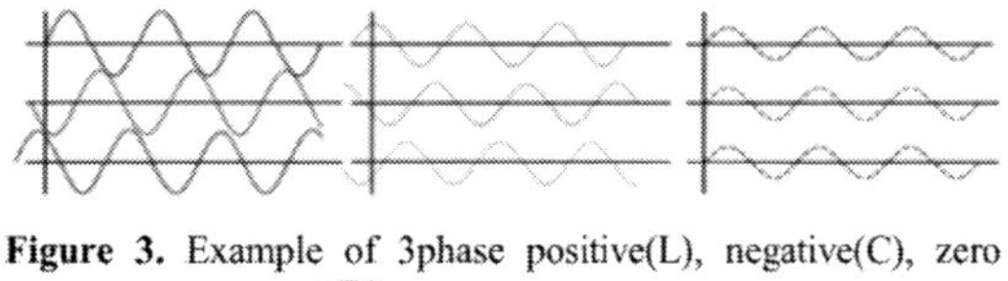

Figure 3. Example of 3phase positive(L), negative(C), zero sequence component(R)

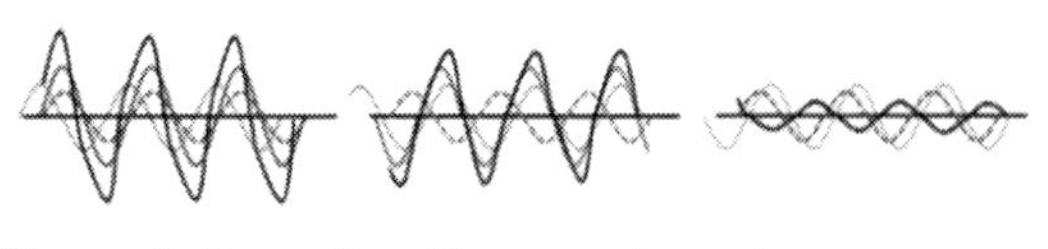

Figure 4. Sum of positive, negative and zero sequence component

As you can see in Fig. 4, an imbalance between frequencies occurs due to the negative sequence component, which causes $e^{-j2w_s t}$ to affect the total phase angle θ in the element $-e$, the domain of the negative sequence component in the Eq.15.

2.4 An analysis of duty ratio of frequency

This paper suggests a simplified algorithm of PWM (pulse width modulation) by detecting the rising times and falling times of half-period frequencies and analyzing distortions of the frequencies - the use of duty ratios of the frequencies for the analysis.

The amount of time for the half-period frequency at 60 Hz is approximately 0.0083s, and in steady state, the falling time must be the same as the rising time: approximately 0.04167s. But as you can see in Fig. 5, 0.0167s (the total frequency: 60Hz) remains the same because of the negative sequence component; but there is a time difference between the rising time and falling time in the duty ratio of the half-period frequency.

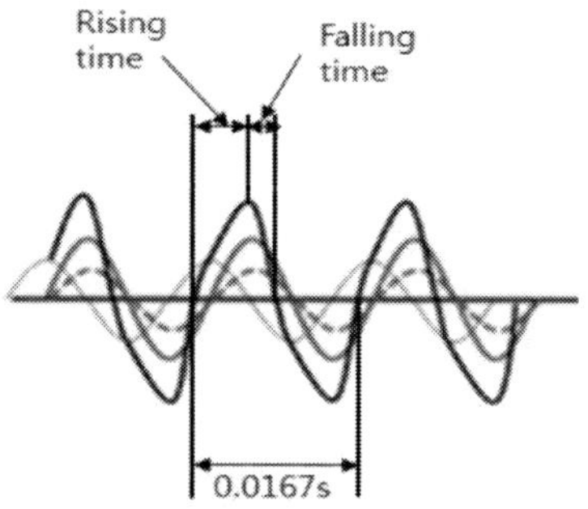

Figure 5. Half-period frequency duty ratio analysis algorithm

In addition, the value of the duty ratio represented a change in the falling time compared to the amount of time for the half-period frequency.

$$\text{Duty Ratio}[\%] = \frac{Falling\ time}{Rising\ time + Falling\ time} * 100\% \quad (16)$$

The measured frequency was expressed using Euler's theorem Eq. 5 and the duty ratio of the frequency in the formula above Eq. 16 was analyzed.

3 Experiment

3.1 Experiment condition

In this experiment, 1[HP], 220[V], 4 poles, 60Hz, a three-phase induction was used.

The numbers of rotor slot and stator are 44, 36.

In addition, the experiment was performed under the following conditions to consider changes in the load conditions.

- Operation of an induction motor by using inverter
- Adjustment of the motor operation speed by using dynamometer
- The initial speed - the rated speed 1690[rpm]

- The frequency of samplings, sampling rate and the No. of samplings: 1[s], 10,000[S/s], 10,000[s]
- Full load current and torque: 3.8[A]/2.2[A] and 0.43[kg m]

As an induction motor is operated by an inverter unlike a simulation, noise components exist as in the actual data in Fig. 6. For removal of noises, a filtering (IIR butterworth lowpass filter) was performed.

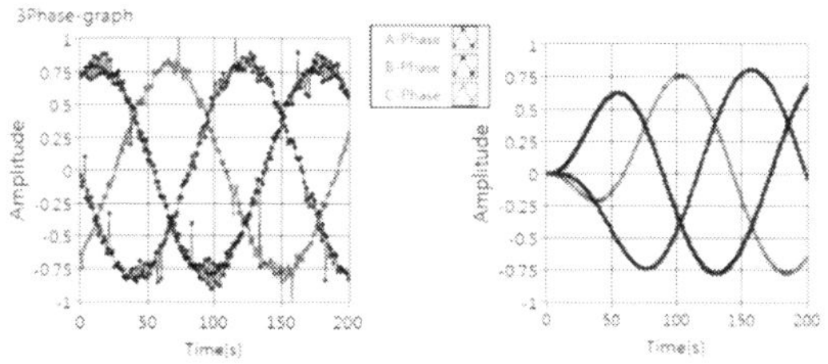

Figure 6. IIR filtering: before(L) and after(R)

3.2 Measurement error range

At 60 Hz (power frequency), the amount of time for a period is approximately 0.0166-0.0167 [s] and that for a half period is 0.00833 [s]. The ideal rising time or falling time should be measured as 0.00416-0.00417 [s], half of the amount of time for the half period.

To measure this, a sampling rate and synchronization must be achieved; for this paper, data collection was performed: 10,000 at the speed of 10,000 [S/s]. With the sampling rate used for the paper, the amount of time for one period was measured as 166-167 [$10^{-4}s$]. The frequency for the power should be accurately expressed with decimal points - analogue data - but the errors that occur due to the measured sampling rate should be considered as follows:

Table 1. Measurement error

	One period (166)	One period(167)
Half period	83	83 or 84
Rising time	41 or 42	42
Falling time	42 or 41	42

The maximum error range for a half-period is 83 [$10^{-4}s$]; and if the falling time is 41 [$10^{-4}s$], as high as approximately 49.4% can be set as the error range for the normal duty ratio when there is no particular threshold setting method of the fault diagnosis.

3.3 Frequency distortion phenomenon in case of stator fault and duty ratio

An experiment was performed at 1690 rpm, the full-load state, to check the actual distortions of frequencies in steady state Fig. 7(L) and when an 8-turn winding fault has occurred Fig. 7(R); and its result is as follows:

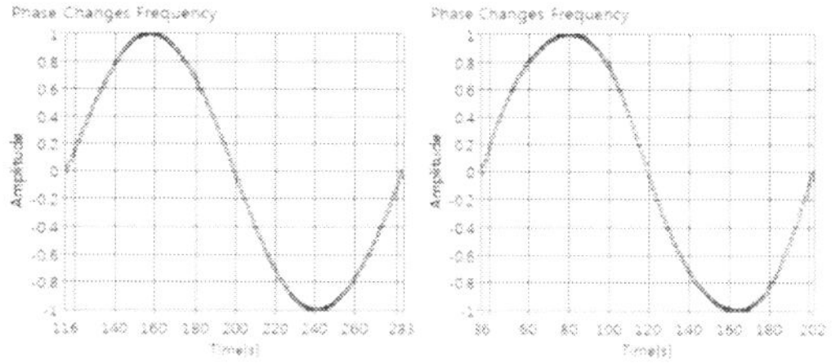

Figure 7. Frequency of 1690rpm by steady state(L), 8-turn windings(R)

The measurements were performed at the sampling rate of 10,000 [S/s], at frequency, the actual frequency of 60 Hz times 10,000. In Fig. 7, the x-axis represents time and [$10^{-4}s$] represents the units. As for the sine wave in Fig. 7, it can be seen that the amount of time for one period, approximately 167 [$10^{-4}s$] (from 116 to 283 [$10^{-4}s$]), is the same as that of one period at 60 Hz, 0.0167[s].

As for these two frequencies, the result of a comparison of the rising times and falling times in the sections showing positive numbers is shown in Table 2 below:

Table 2. Average duty ratio at full-Load

	rising time	falling time	Duty ratio[%]
Steady state	41.7736	41.6038	49.8982
2-turn winding	41.9231	41.4615	49.7232
4-turn winding	42.6	40.88	48.9698
6-turn winding	43.2857	40.0816	48.0783
8-turn winding	44.3696	39.2391	46.9318

The data in Table 2 are the result of the measurements performed at 1690rpm, the measurements being performed for only one second per each state, from steady state to 8-turn winding. The Table shows duty ratios and the data are the average figures for the falling times and falling times in the frequencies - the sections of positive numbers.

It can be seen that an increase in seriousness of turn-to-turn short fault means an increase in rising time and a decrease in falling time.

Table 3. Average duty ratio at No-Load

	rising time	falling time	Duty ratio[%]
Steady state	41.7736	41.6226	49.9095
2-turn winding	42.4118	40.9804	49.1418
4-turn winding	43.8333	39.5208	47.4131
6-turn winding	45.4091	38	45.5568
8-turn winding	47.7179	35.7436	42.8265

Table 3 shows the result of a comparison of the rising times and falling times and duty ratios for each turn to turn short fault at 1790 rpm in no-load state.

As the fault becomes more serious, the amount of rising time started to increase while that of falling time decreased, and it can be seen that there is to be a change in duty ratio.

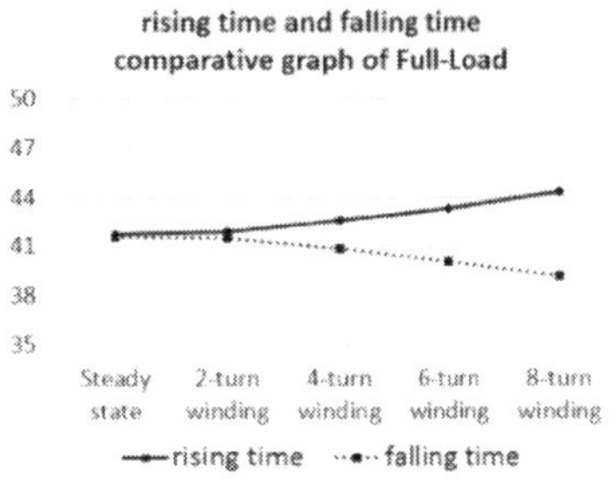

Figure 8. Rising time and falling time comparative graph of Full-Load

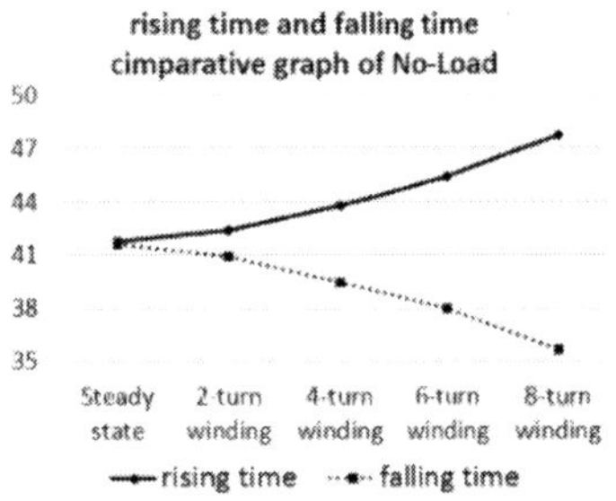

Figure 9. Rising time and falling time comparative graph of No-Load

Fig. 8 is a graph showing the result of a comparison of the rising times and falling times at full-load. Judging by each slope, it can be seen that there is no change in the overall frequency.

Fig. 9 is a graph showing the result of a comparison of the rising times and falling times at no-load. A pattern similar to that in Fig. 8 can be seen. Therefore, it can be seen that the duty ratio more dramatically changes as the fault becomes more serious at no-load, rather than full-load.

Fig. 10 is a graph showing the result of a comparison of duty ratios of average frequencies at full-load and no-load.

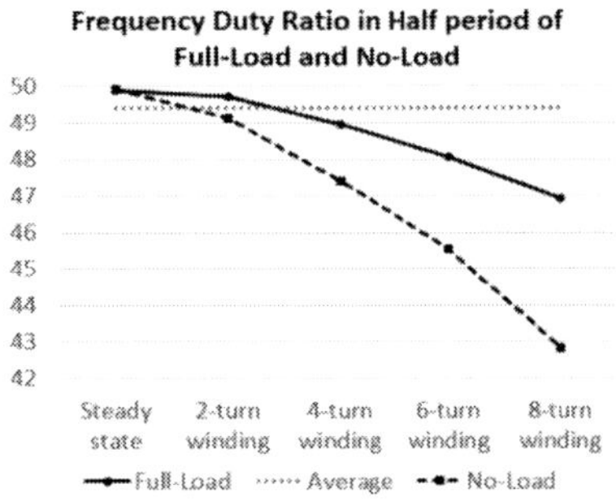

Figure 10. Frequency duty ratio of Full-Load and No-Load

When considering measurement error range, it is possible to judge accurately whether a 4-turn winding fault or a more serious turn-to-turn short fault has occurred although the ideal duty ratio 49.4% is considered as a reference threshold value (a 2-turn winding fault cannot be detected accurately in this case); and a 2-turn winding fault diagnosis can also be performed through the suggested method when being operated at no-load.

Fig. 11 below compares all of the data on stator fault.

The result of the preceding experiment was expressed as the sum of successive average figures for frequencies measured for one second. For the measurements for Fig. 11, 10 rpm was increased every 30 seconds, from 1690 rpm (full-load) to 1790 rpm, and the Fig. 11 shows the successive average figures for the duty ratios of the half-period frequencies measured every 30 seconds.

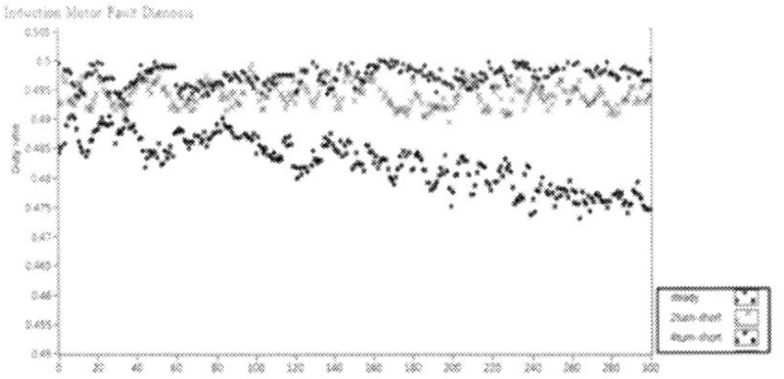

Figure 11. Result of induction motor fault diagnosis in electrical fault

In Fig. 11, the black circle represents steady state; the 'X' mark 2-turn short fault; the quadrangle 4-turn short fault.

As for stator fault, it can be seen that linear separation is impossible at the ideal threshold value that the paper suggested when a 2-turn short fault has occurred; but linear separation from steady state is possible at the value when a 4-turn short fault has occurred.

4 Conclusion

The paper suggested a new technique that online stator faults with the use of the duty ratios of the half-period frequencies created when a phase angle change occurs, induction motor faults being caused by an electrical fault.

According to the findings of the paper, when a stator fault has occurred, the rising times of specific frequencies of the phase angle are longer. With these characteristics, which appear because of a phase angle change, now it is possible to judge swiftly whether a fault has occurred.

In addition, as for stator fault, it could be seen that linear separation from steady state is perfectly possible in case of the occurrence of a 4-turn short fault, complete fault state.

It is judged that analysis studies on the characteristics through long-term operation will need to be performed in the future, to find out a method of identifying 2-turn short faults, the situation where the fault has not become more serious yet.

Acknowledgements

This study was supported by the Ministry of Trade, Industry and Energy(MOTIE) through the Regional Innovation Centre Programme.

References

1. S. Nandi and H. A. Toliyat., *Record of the IEEE Industry Applications Conference*, 197-204, (1999)
2. A. Ferrero and G. Superiti-Furga., IEEE Trans. Instrum. Meas **40**, 10 (1991)
3. M. H. Song, K. N. Park, D. G. Han, and C. O. Yang., Trans. KIEE (in Korean) **14**, 8 (2006)
4. F. Filippetti, G. Frauceschini, C. Tassoni and P. Vas, IEEE Trans. On Energy Conversion **21**, 8 (2006)
5. R. M. Tallam, T. G. Habetler, R. G. Harley, O. J. Gritter, and B. H. Burton., *Proceeding of IEEE 1ST Annual Meeting*, **5** (2000)
6. M. A. Awadallah, M. M. Marcos, S. Gopalakrishman and T. W. Nehl., IEEE Trans. on Energy Conversion **21**, 9 (2006)
7. G. B. Kliman, W. I. Premerlani, R. A. Koegl and D. Hoeweler., Electric Machines and Power Systems **28**, 13 (2000)
8. R. M. Tallam, T. G. Habetler, and R. G. Harley., IEEE Tran. On Industry Applications **38**, 6 (2002)
9. J. L. Kohler, J. Sottile, and F. C. Trutt, IEEE Trans. On Industry Applications **38**, 6 (2002)
10. S. B. Lee, R. M. Tallam, and T. G. Habetler, IEEE Trans. On Power Electronics, **18**, 8 (2003)
11. S. Bakhri, N. Ertugrul, and W. L. Soong., IECON 2012-38th Annual Conference on IEEE Industrial Electronics Society, **6** (2012)
12. A. J. M. Cardoso, S. M. A. Cruz, and D. S. B. Fonseca., *IEEE Trans. on Energy Conversion*, **14**, 4 (1999)
13. Y. J. Go, B. Lee, M. H. Song, K. M. Kim., Institute of Control, Robotics and Systems(in Korean) **20**, 6 (2014)
14. R. Reginatoo and R. Bainy., *IEEE PES General Meeting*, 5 (2014)

Coordinated Low Voltage Ride through strategies for Permanent Magnet Direct Drive Synchronous Generators

Zhang Ge[a], Yang Yiyun, Xiao Jing, Gao Like and Xiao Yuanyuan

Guangxi Power Grid Company Electric Power Research Institute, Nanning, 530000

Abstract. By analyzing the mechanism of the low voltage ride through on the permanent magnet direct drive synchronous wind power generating units, this paper proposes a coordinated control strategy for permanent magnet synchronous generator. In order to avoid over speed operation of the generation units, over voltage on DC capacitor and over current on convert, the improved pitch angle control and inverter control are used. When the grid voltage drops, the captured wind power is cut down by the variable pitch system, which limits the speed of the generator, the generator side converter keeps the DC capacitor voltage stabile; and the grid side converter provides reactive power to the grid to help the grid voltage recover. The control strategy does not require any additional hardware equipment, with existing control means, the unit will be able to realize low voltage ride through. Finally, based on Matlab/Simulink to build permanent magnet direct drive wind power generation system, the simulation results verify the correctness and effectiveness of the control strategy.

1 Introduction

As an important new energy power generation, wind power generation rapidly develop in China. In 2014, the whole country (except Taiwan) new capacity is 23196MW as an increase of 44.2% [1]. As the proportion of wind power in the power grid upgrade, due to grid failure or grid disturbances caused by voltage drop cause the wind turbines taking off network. Therefore, the study of wind turbines LVRT capability has important practical significance wanted [2].

Compared with double-fed unit, direct drive Permanent Magnet Synchronous Generator (PMSG) do not need rose -speed gearbox so has a small mechanical loss; no slip rings and brushes, to improve the operational reliability of the system [3], [4]. After a lot of PMSG connected to the grid, if you do not LVRT capability, when the grid fails, the unit will be a large area off the network, causing a tremendous impact on the stability of the grid [5], [6]. At present, domestic and foreign research workers have done a lot of research on the low voltage ride through of PMSG. Literature [7]-[9] by adding unloading resistance or energy storage element to ensure the unit PMSG in low voltage ride through, but additional hardware increases system cost and maintenance difficulty [10], [11]. Literature [12], [13] by limiting the grid failure PMSG power to limit electromagnetic converter from the input side to the DC side capacitance and grid-side power converter, the program improved the traditional converter control strategy to avoid an increase in redundant hardware protection device, but it does not restrict the ability of fans during a grid failure to capture wind energy, and therefore can't effectively reduce the burden LVRT unit.

In this paper, the performance characteristics of PMSG are analyzed in detail from the viewpoint of energy balance. Based on the theoretical analysis, combined with the existing low voltage ride through method, a new permanent magnet direct drive wind turbine low- voltage ride through coordinated control strategy has been put forward. The strategy take avoiding generator over speed and DC side capacitor voltage for the purpose, using the improvement of the pitch angle control and inverter control. Using the pitch control system, reduce the fan captured during grid fault wind energy, in order to avoid generator over speed; Monitoring DC bus voltage, once the capacitor voltage reaches the threshold, the controller responds quickly and avoids DC capacitor; according to the degree of grid voltage drop, providing dynamic reactive power to grid converter side through the network, to help restore the grid voltage.

2 Operation characteristic analysis of power network fault

Figure 1 is permanent magnet direct drive wind turbine system which consists of wind turbine, permanent magnet generator, generator side converter, DC capacitor, grid side converter, transformer and control system, etc.

The wind turbine is used to capture wind energy, which determines the power output of the whole wind

[a] Corresponding author.

power system. The power of the wind turbine P_w can be expressed as:

$$P_w = 0.5\pi\rho R^2 v^3 C_p \quad (1)$$

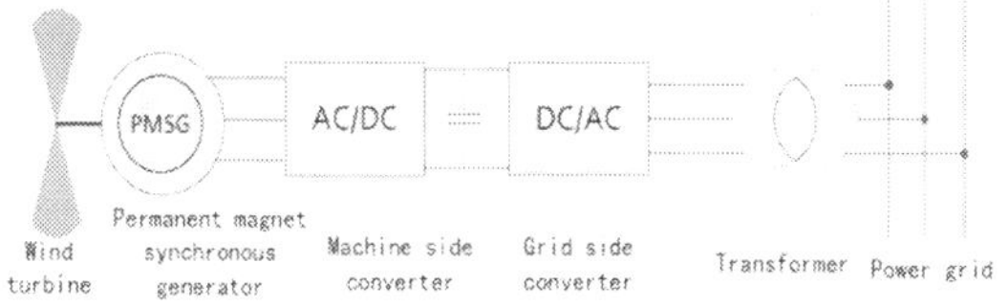

Figure 1. Structure diagram of wind turbines.

In the formula, ρ is air density, unit is kg/m3; R is radius of wind wheel, unit is m; v is wind speed, unit is m/s; C_p is wind energy utilization coefficient [14].

Analysis of transient characteristics of permanent magnet direct drive wind power system when the grid voltage drops from the viewpoint of energy conservation, Figure 2 is Internal power transfer diagram of wind turbines. i_s, i_g, i_{dc} is the input current of the machine side, the network side and the DC capacitor; U_{dc} is capacitor terminal voltage; P_s, P_g is the input power of the machine side and the grid side converter.

$$i_s = i_g + i_{dc} = i_g + C\frac{d}{dt}U_{dc} \quad (2)$$

$$P_s = i_s U_{dc} \quad (3)$$

$$P_g = i_g u_{dc} \quad (4)$$

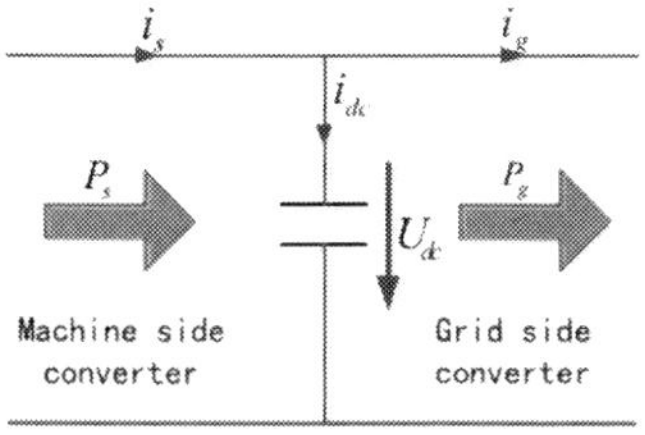

Figure 2. Internal power transfer diagram of wind turbines.

Under normal circumstances, the wind energy and the power of the wind turbine are equal to the electric energy of the transmission, that is i_s equals i_g. At this point, the capacitor charge and discharge current i_{dc} is 0, rate of change of capacitance is 0, capacitor terminal voltage is a fixed value at this time.

When power grid faults occur, the power of the converter to the power grid transmission power is greatly reduced, but due to the independence of the machine side converter, the input power of the machine side converter will remain relatively fixed, so that the excess input energy of the machine side converter will remain in the unit and the DC capacitor charging, leading to a surge in the capacitance of the capacitor, the mathematical formula is shown as follows:

$$U_{dc}(t) = \sqrt{U_{dc}^2(t_0) + \frac{2}{C}\int_{t_0}^{t_r}\left(P_s(\tau) - P_g(\tau)\right)d_\tau} \quad (5)$$

In the formula, t_0 is time of low voltage occur; t_r is time of grid voltage recovery.

Therefore, it is necessary to use the appropriate control strategy to reduce the power grid fault in the unbalanced energy, in order to improve the low voltage ride through PMSG.

3 Coordinated control strategy

3.1 Pitch control strategy

The pitch angle control system determines the ability of fan wind energy capture. Wind speed at rated value and pitch angle control system does not work, pitch angle is 0.When the wind speed is higher than the rated value, control system by increasing the pitch angle to reduce wind energy capture, in order to avoid generator overspeed. Figure 3 shows the pitch angle control system structure diagram [15].

Under normal circumstances, the rotor speed ω_m is the input signal of pitch angle control system, when the rotor speed ω_m exceeds the given maximum speed value, the pitch angle control system works in order to reduce the wind turbine output. When power grid fault happens, pitch angle reference value replaces by given grid tolerable lower limits value U_{gref}, when the power grid voltage drop is detected, the pitch angle control system work immediately and pitch angle increases rapidly. The servo link can simulate the dynamic response process of pitch system, limits changing speed and changing range of the pitch angle.

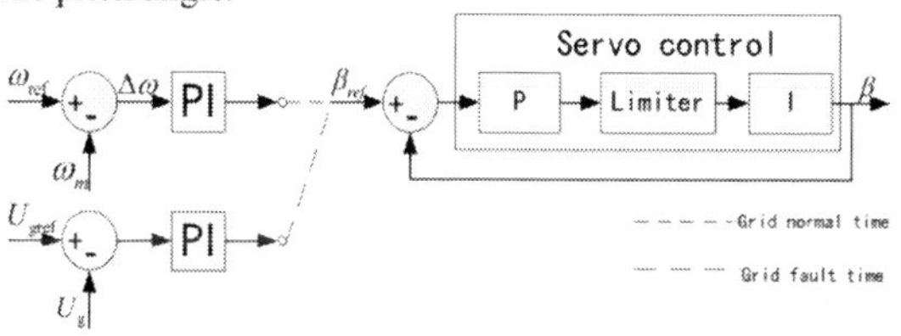

Figure 3. Pitch control system structure diagram.

3.2 Generator-Side converter control strategy

The motor side converter can control the speed of the generator running at optimal tip speed ratio, so as to achieve the purpose of maximum wind energy capture. Stator voltage equation and torque equation of permanent magnet synchronous motor [16]:

$$\begin{cases} u_{sd} = R_s i_{sd} + L_d \frac{d}{dt} i_{sd} - \omega_s L_q i_{sq} \\ u_{sq} = R_s i_{sq} + L_q \frac{d}{dt} i_{sq} + \omega_s L_d i_{sd} + \omega_s \psi_f \\ T_e = 1.5 n_p i_{sq}\left[\left(L_q - L_d\right) i_{sd} + \psi_f\right] \end{cases} \quad (6)$$

In the formula: u_{sd}, u_{sq} and i_{sd}, i_{sq} are respectively indicate d and q axis components of stator voltage and stator current, R_s indicates the stator resistance, L_d, L_q are respectively indicate d and q axis components of stator inductance. ω_s is the generator power angular velocity; n_p is the generator pole number; ψ_f is the permanent magnet flux.

For the non-salient pole permanent magnet direct drive generator, the excitation flux and the pole number are constant, L_d equals L_q, and the electromagnetic torque equation of the generator can be rewritten:

$$T_e = 1.5 n_p i_{sq} \psi_f \tag{7}$$

The generator side converter adopts double closed loop control based on stator flux orientation, the reference value of generator d-axis current i_{sdref} is zero, q-axis current controls the output active power of generator, and the control structure of generator side converter is shown in Figure 4.

Under normal condition, the speed reference value of the generator rotor ω_{ref} is obtained by the maximum wind power tracking algorithm. The active power current reference value i_{sdref} is controlled by the error between ω_{ref} and ω_m through PI controller. During the grid fault, the DC capacitor voltage is controlled by the generator side converter control. When the DC voltage U_{dc} is higher than the rated value U_{dcref}, the controller works to avoid DC capacitor over voltage.

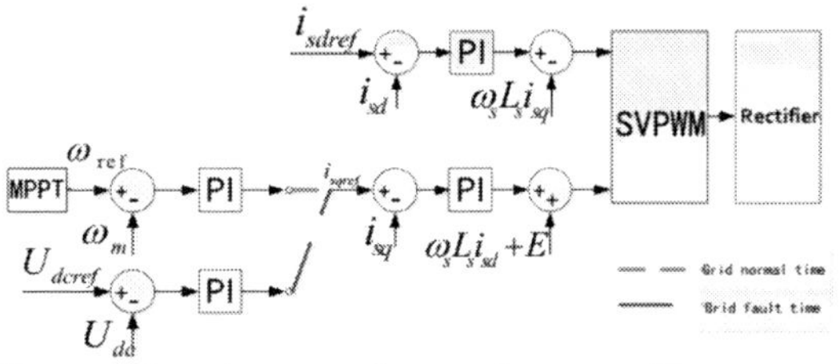

Figure 4. Control structure diagram of generator side converter.

3.3 Control strategy of grid side converter

On the one hand, the grid side converter makes the DC capacitor voltage remain stable and the active power injected into the power grid. On the other hand, according to the needs of the power grid, outputs a certain reactive power.

In the synchronous rotating d-q coordinate system, the voltage equation of the grid side converter is:

$$\begin{cases} u_d = u_{gd} - Ri_d - L\dfrac{d}{dt}i_d + \omega L i_q \\ u_q = u_{gq} - Ri_q - L\dfrac{d}{dt}i_q - \omega L i_d \end{cases} \tag{8}$$

In the formula: u_{gd}, u_{gq} are respectively indicate d and q axis components of grid voltage, i_d, i_q are respectively indicate d and q axis components of grid current. R and L are the resistance and inductance of the reactor.

Grid side converter using vector control based on voltage oriented. The d-axis in the synchronous rotating d-q coordinate system is selected as grid side voltage space vector direction, and the q-axis is 90 degrees advanced q-axis, then the decoupling control of active power and reactive power is realized. The active power of the power grid is only related to the d-axis current, the reactive power is related to the q-axis current, active power P and reactive power Q of the grid side converter are as follows:

$$\begin{cases} P_g = 1.5 u_{gd} i_d \\ Q = 1.5 u_{gq} i_q \end{cases} \tag{9}$$

The control structure of the grid side converter is shown in Figure 5. Under normal circumstances, the grid side converter uses the voltage outer loop and current inner loop control structure. The voltage outer loop can maintain the stability of the DC capacitor voltage and the inner loop can realize the stable operation and fast dynamic response of the control system. When the grid fault happens, the network side converter will output a certain reactive power to help the recovery of the grid voltage based on the voltage sag degree of the power grid. First, according to the need of compensate reactive power to calculate reactive current reference value i_{gqref}, and then calculate the active current reference value i_{gdref} in accordance with the requirements of the current limiter.

$$i_{gdref} = \sqrt{I_N^2 - I_{gqref}^2} \tag{10}$$

In the formula: I_N is the rated current of the power grid.

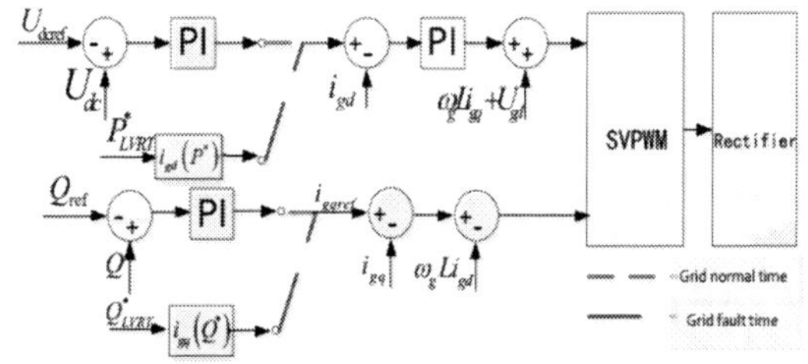

Figure 5. Control structure diagram of grid side converter.

4 Simulation study

4.1 Data preparation

To test correctness and validity of the proposed low voltage ride through control strategy on permanent magnet direct drive wind turbine, this paper respectively emulate two examples of different grid voltage dropping degree based on MATLAB/Simulink. Before the grid fault occurs, the permanent magnet direct drive wind turbine operates under the rated wind speed, and the system is running in the unit power factor condition. The simulation parameters are shown in Table 1.

Table 1. System simulation parameters.

Device name	Technical parameters
Wind turbine	Air density 1.225 kg/m3, Blade radius 38.8m
permanent magnet direct drive	Rated power 1MW, rated voltage 690V, pole number 40
Grid side converter	Reactor resistance 0.002, inductance 0.3mH, grid voltage 690V
Capacitor	DC capacitor 100mF, set up voltage 1500V

4.2 Case analysis

Case 1: The light drop of grid voltage simulation

At 0.4s, the grid voltage occurs the three-phase symmetrical drop fault, the drop depth is 30%, the duration time is 200ms, and at 0.6s grid voltage recovers, as is shown in Figure 6(a). After the grid fault occurs, pitch angle control system begins to work, the pitch angle increases to decrease the wind energy captured by wind turbine, and the rotating speed of the generator increases as is shown in Figure 6(b) and Figure 6(c). By Figure 6(d), the capacity of the grid side converter to transfer power to the grid is reduced to 0.4MW. The DC capacitor terminal voltage will have a transient overshoot, but under the machine side converter control system, the terminal voltage slowly drops until the normal value, as is shown in Figure 6(e). Grid connected current increases, however, under the grid side converter control system, it is limited to the normal range, as is shown in Figure 6(f). At the same time, the grid side converter will issue 0.4MVAR reactive power to help recovery the grid voltage under grid fault, as is shown in Figure 6(g). When the grid occurs light drop fault, the generator speed, DC capacitor voltage and grid side current are all in the safe range, and PMSG achieves the low voltage ride through.

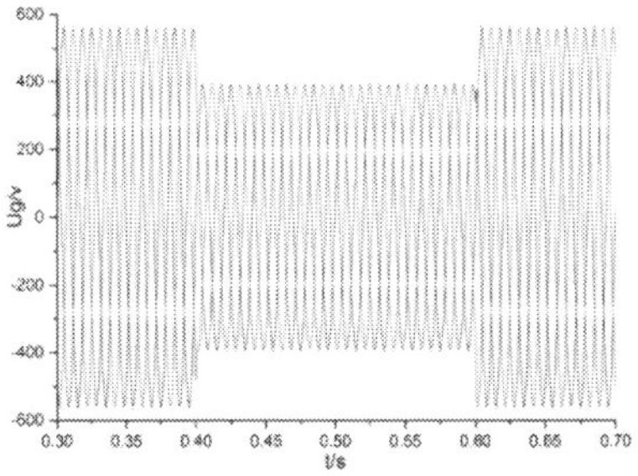

(a) Grid voltage

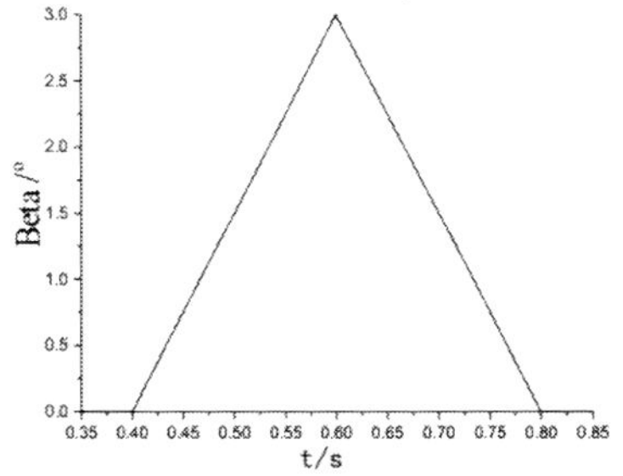

(b) Wind turbine pitch angle

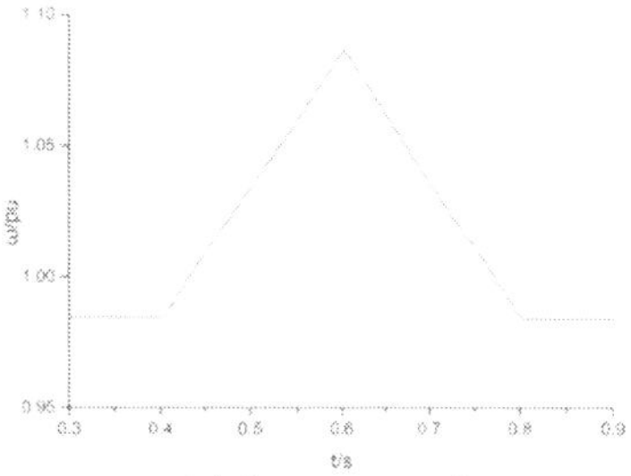

(c) Generator speed

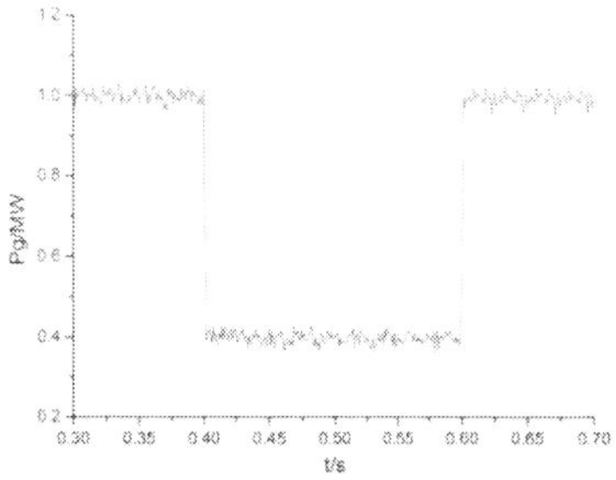

(d) Output active power of grid side converter

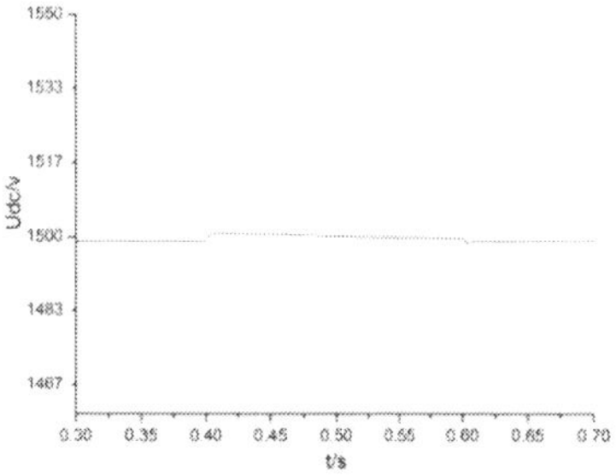

(e) DC side capacitor voltage

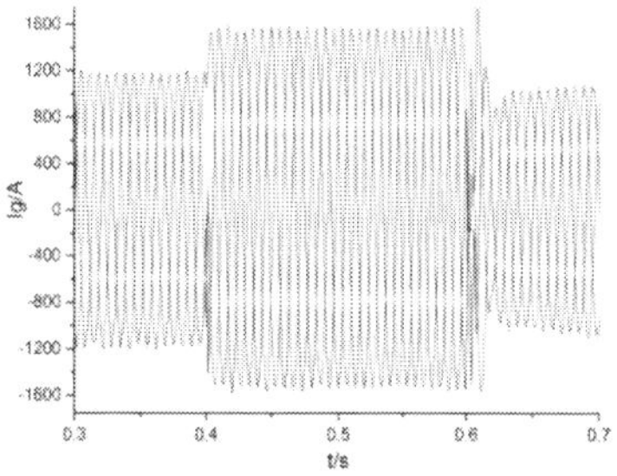

(f) Grid current

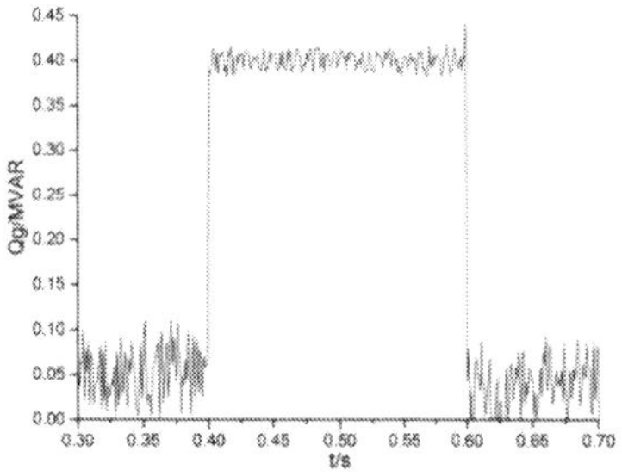

(g) Output reactive power of grid side converter

Figure 6. Response waveforms of the grid voltage light drop.

Case 2: The deep drop of grid voltage simulation

At 0.4s, the grid voltage occurs the three-phase symmetrical drop fault, the drop depth is 60%, the duration time is 200ms, and at 0.6s grid voltage recovers, as is shown in Figure 7(a). After the grid fault occurs, pitch angle control system begins to work, the pitch angle increases to decrease the wind energy captured by wind turbine, and the rotating speed of the generator increases as is shown in Figure 7(b) and Figure 7(c). By Figure 7(d), the capacity of the grid side converter to transfer power to the grid is reduced to 0.1MW. The DC capacitor terminal voltage will have a transient overshoot, but under the machine side converter control system, the terminal voltage slowly drops until the normal value, as is shown in Figure 7(e). Grid connected current increases, however, under the grid side converter control system, it is limited to the normal range, as is shown in Figure 7(f). At the same time, the grid side converter will issue 0.63MVAR reactive power to help recovery the grid voltage under grid fault, as is shown in Figure 7(g). When the grid occurs light drop fault, the generator speed, DC capacitor voltage and grid side current are all in the safe range, and PMSG achieves the low voltage ride through.

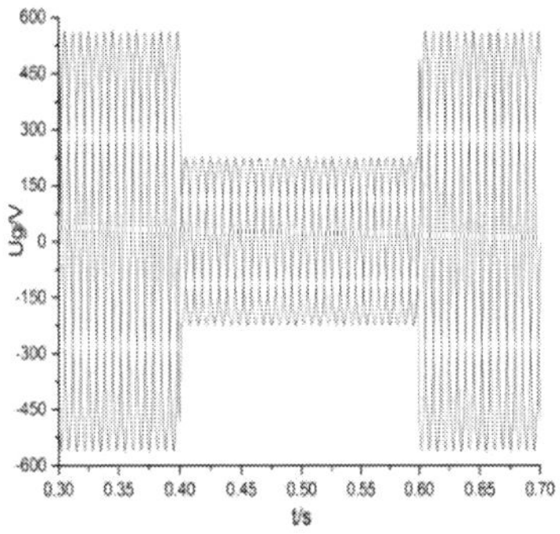

(a) Grid voltage

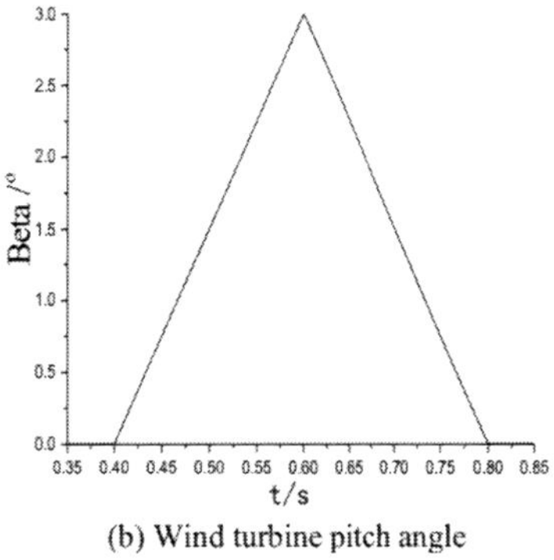

(b) Wind turbine pitch angle

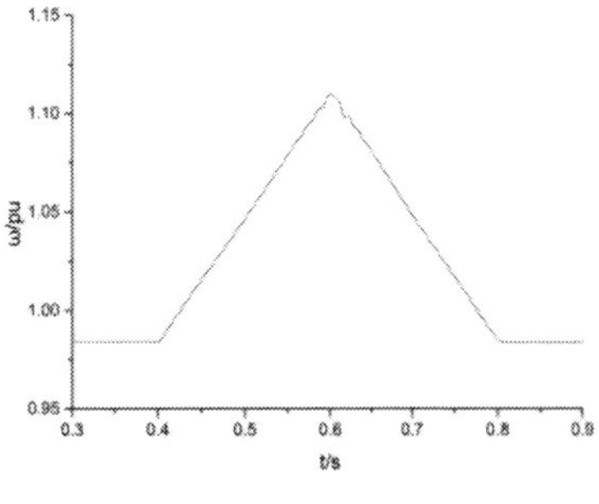

(c) Generator speed

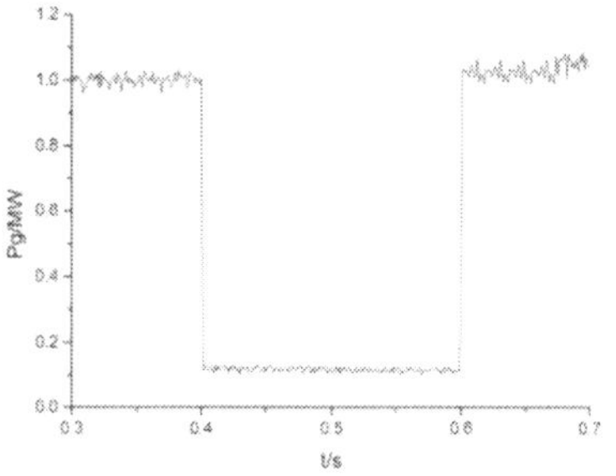

(d) Output active power of grid side converter

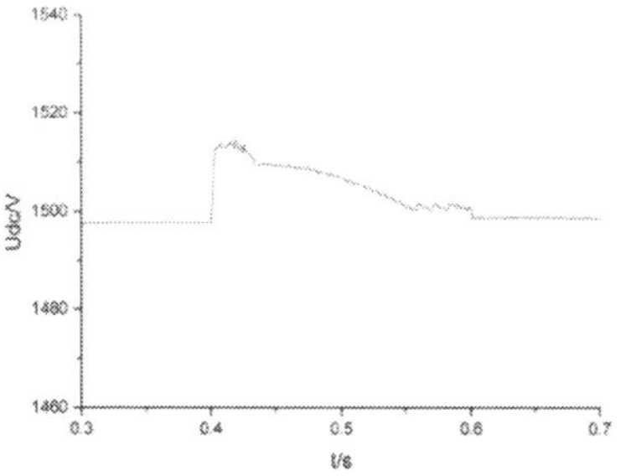

(e) DC side capacitor voltage

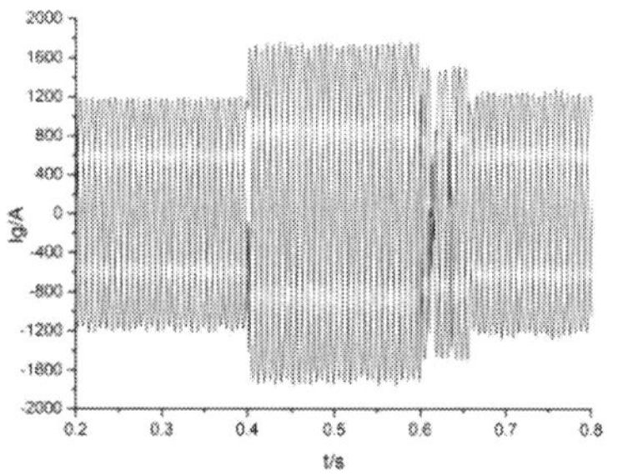

(f) Grid current

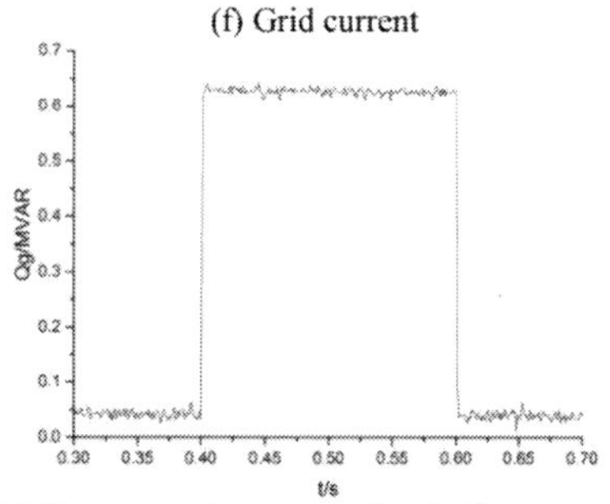

(g) Output reactive power of grid side converter

Figure 7. Response waveforms of the grid voltage deep drop.

5 Conclusion

In this paper, a model of direct drive permanent magnet wind turbine under grid voltage drops is analyzed. Aiming at the energy imbalance between machine side convert and grid side convert, a coordinated control strategy including pitch angle control and inverter control is proposed and can avoid excessive generator over speed and the DC capacitor over voltage. Based on Matlab/Simulink, light drop and deep drop of grid voltage are simulated respectively. The simulation results show that the speed of generator and the DC capacitor voltage are limited to the safe range under grid voltage drops, and the permanent magnet direct drive wind turbine realizes low voltage ride through successfully.

Acknowledgement

This work was supported by Guangxi Power Grid limited liability company Scientific Project (GXKJ00000006).

References

1. XU, China's wind power installed capacity statistics of 2014, (2015)
2. W. P. Wang, A new LVRT method applied on the integration of distributed wind power generation: Electricity Distribution (CICED), 2012 China International Conference on, Shanghai, (2012)
3. Y. Liao, J. B. He, J. Yao, et al., Power Smoothing Control Strategy of direct-driven Permanent Magnet Synchronous Generator for Wind Turbine with Pitch Angle Control and Torque Dynamic Control, Proceedings of the CSEE, **18** (2009)
4. W. L. Wang, Grid-Connected Operation Control for Energy Storage based Direct-drive Permanent Magnet Wind Power system, Beijing Jiaotong University, (2010)
5. X. Zhang, L. Y. Zhang, S. Y. Yang, et al., Low Voltage Ride-through Technologiesin Wind Turbine Generation, Proceedings of the CSU-EPSA, **02** (2008)
6. Z. Chen, J. He, D. Wang, et al., A simplified wind farm simulation method for LVRT capability assessment: Power System Technology (POWERCON), 2014 International Conference on, Chengdu, (2014)
7. T. H. Nguyen, D. Lee, Advanced Fault Ride-Through Technique for PMSG Wind Turbine Systems Using Line-Side Converter as STATCOM, Industrial Electronics, IEEE Transactions on, **60**(7):2842-2850, (2012)
8. J. Yao, X. M. Chen, X. F. Xia, et al., A Low Voltage Ride-through Control Strategy for Direct-driven Permanent Magnet wind Power Generation System with Flywheel Energy Storage Unit, Automation of Electric Power Systems, **13** (2012)
9. S. Y. Hou, Y. Fang, J. X. Zeng, et al., Application of supercapacitors to improve wind power system's low voltage ride through capability, Electric Machines and Control, **05** (2010)
10. Low Voltage Ride-Through Technique for PMSG Wind Turbine Systems Using Interval Type-2 Fuzzy Logic Control
11. S. Alepuz, A. Calle, Busquets-Monge S, et al., Use of Stored Energy in PMSG Rotor Inertia for Low-Voltage Ride-Through in Back-to-Back NPC Converter-Based Wind Power Systems, IEEE Transactions on Industrial Electronics, **60**(5):1787-1796, (2013)
12. J. Yao, Y. Liao, K. Zhuang, A Low Voltage Ride-through Control Strategy of Permanent Magnet Direct-driven Wind Turbine Under Grid Faults, Automation of Electric Power Systems, (12):91-96, (2009)
13. S. Dong, Y. Wang, H. Li, Coordinated control for active and reactive power of PMSG-based wind turbine to enhance the LVRT capability: Electrical Machines and Systems (ICEMS), 2012 15th International Conference on, Sapporo, (2012)
14. H. Chen, X. B. Hu, G. G. Yan, et al., A Coordinated Control Strategy for Low Voltage Ride-Though of Direct-Drive Permanent Magnet Wind Power Generating Units, Power System Technology, (05):1464-1470, (2013)
15. Z. Y. Liu, C. R. Liu, G. Y. Li, Coordinated Power Control Method for Improving Low Voltage Ride Through Capability of Wind Turbines with Permanent Magnet Synchronous Generators, Automation of Electric Power Systems, (03):23-29, (2015)
16. X. Yang, X. Gong, W. Qiao, Mechanical sensorless maximum power tracking control for direct-drive PMSG wind turbines: Energy Conversion Congress and Exposition (ECCE), 2010 IEEE, Atlanta, GA, (2010)

An Approach to Determine the Weibull Parameters and Wind Power Analysis of Saint Martin's Island, Bangladesh

Khandaker Dahirul Islam, Dr. Natthawud Dussadee and Dr. Tanate Chaichana

School of Renewable Energy, Maejo University, Sansai, Chiang Mai 50290, Thailand

Abstract. This paper explores wind speed distribution using Weibull probability distribution and Rayleigh distribution methods that are proven to provide accurate and efficient estimation of energy output in terms of wind energy conversion systems. Two parameters of Weibull (shape and scale parameters *k* and *c* respectively) and scale parameter of Rayleigh distribution have been determined based on hourly time-series wind speed data recorded from October 2014 to October 2015 at Saint Martin's island, Bangladesh. This research has been carried out to examine three numerical methods namely Graphical Method (GM), Empirical Method (EM), Energy Pattern Factor method (EPF) to estimate Weibull parameters. Also, Rayleigh distribution method has been analyzed throughout the study. The results in the research revealed that the Graphical method followed by Empirical method and Energy Pattern Factor method were the most accurate and efficient way for determining the value of *k* and *c* to approximate wind speed distribution in terms of estimating power error. Rayleigh distribution gives the most power error in the research. Potential for wind energy development in Saint Martin's island, Bangladesh as found from the data analysis has been explained in this paper.

1 Introduction

Wind data analysis and accurate wind energy potential assessment is critical for proper and efficient development of wind power application and is highly site-dependent [1]. Bangladesh possesses nearly 580km of coast in the southern part of the country along with a couple of offshore areas as well. Saint Martin's is one of those small islands situated about 15km apart from Teknaf, the southern-most tip of Bangladesh mainland. Energy situation of the island depends fully on off-line power sources. Currently annual electric energy demand of the island is about 359MWh [2].

Due to the prominence of tourism, Saint Martin's island has been grown with commercial shops, restaurants and residential hotels which have their own or shared electricity mean of diesel generator. This stands as a mean to pollute the green environment of the island. A balanced, diversified, and environmental friendly energy system is the target for the whole world to curb problems associated with fossil fuel energy systems, which is the main cause of the worrying climate change [3]. Besides this, recent increases in the cost of fossil fuels have revived the interest in modern exploitation of energy from wind power [4].

But in order to get the maximum benefits from wind, detailed knowledge of wind characteristics and distribution are crucial parameters to select optimum wind energy conversion system to optimize energy output and minimization of electricity generation cost [5]. Wind resources are seldom consistent and vary with time of the day, season of the year, height above the ground, type of terrain, and from year to year, hence should be investigated carefully and completely [6]. Since Bangladesh lacks in having an accurate and reliable wind atlas, necessary attempts like pre-selecting wind sites, wind survey especially in coastal and offshore regions etc. could not be undertaken as it were needed to diagnose the wind potential of the country. It is expected that the importance of this economical issue and the environmental pollution problem associated with the use of oil, will boost over time the development of renewable energy resources, which have gained huge magnitude due to their sustainability, inexhaustibility and ecological awareness [7]. Renewable energy for the current research is estimated for Saint Martin's island wind data for a complete year. Weibull Probability Distribution Function (PDF) has been adopted in this research, for this method is usually considered as the most qualified function due to its simplicity and high accuracy [8]. Again Rayleigh distribution function has been also analysed in the research for better understanding of the wind. The Weibull PDF has been employed almost unanimously by researchers involved in wind speed analysis and it has also extensively been used in wind power analysis for many decades [9].

2 Data source and methods

The wind speed data recorded in every minute in a time-series format over a one year (October 2014 - October

2015) have been collected and statistically analyzed. Saint Martin's wind speed data were recorded at a height of 10 meter from the nearest local meteorological station situated in Teknaf, about thirteen kilometres apart from the island, and also the southern-most tip of Bangladesh mainland. Following is the table (Table 1) that provides the geographical information of the island.

Table 1. Geographical coordinates of Saint Martin's Island, Bangladesh

Variables	Value
Latitude	20°34' - 20°39' N
Longitude	92°18' - 92°21' E
Anemometer height	10 M
Maximum elevation from sea level	3 M

Bangladesh counts six seasons in a year thought the country observes prominently three seasons namely - summer, monsoon and winter. This paper analyzes the data in these three seasonal manners.

3 Measurements of mean wind speed and standard deviation

For every season comprised of four months in a year, the wind speed data recorded every minute are taken to sort out as hourly basis. Mean wind speed and standard deviation have been estimates using the following equations:

$$v_m = \frac{1}{N}\left(\sum_{i=1}^{N} v_i\right) \quad (1)$$

$$\sigma = \left[\frac{1}{N-1}\left(\sum_{i=1}^{N}(v_i - v_m)^2\right)\right]^{\frac{1}{2}} \quad (2)$$

where, v_m is the mean wind speed, m/s, σ is standard deviation of the data (m/s), v_i means hourly wind speed (m/s) and N is the number of measured hourly wind speed data.

4 Wind data analysis

The wind speed at 10 metres from the ground was converted to the speed at 50 metres up from the ground using power law. The optimistic height from the ground is 40-70m, where the wind speed is high which is more applicable for wind energy generation [10]. The equation for power law is given here:

$$\frac{v_z}{v_{z_{ref}}} = \left(\frac{z}{z_{ref}}\right)^{\alpha} \quad (3)$$

and

$$\alpha = \frac{0.37 - 0.088\ln(v_{ref})}{1 - 0.088\ln\left(\frac{z_{ref}}{10}\right)} \quad (4)$$

where, V_z = wind speed at height z, V_{Zref} = wind speed at reference height (Z_{ref}), Z = height above ground level, Z_{ref} = reference height, α = power law exponent. After calculating the wind speed at 50 metres from the above equation, monthly mean wind speed of the following table (Table. 2) can be identified using the following formula:

$$V' = \frac{i}{N}\sum_{i=1}^{N} m_i f_i \quad (5)$$

The same formula can be applied to measure the mean wind speed for other two seasons. Now, the whole three seasons mean wind speeds are being displayed in Table 2.

Table 2. Mean seasonal wind speed for Saint Martin's Island.

Seasons	Mean wind speed (10M)	Mean wind speed (50M)
	m/s	m/s
Summer	2.74	4.23
Monsoon	2.91	4.45
Winter	2.39	3.76
Whole year	2.67	4.23

Wind resources are seldom consistent and vary with time of the day, season of the year, height above the ground, type of terrain, and from year to year, hence should be investigated carefully and completely [11]. So it needs more to investigate how wind behaves in different times in a year. That's why, after finding the mean speeds of different time range (daily, hourly, monthly etc.), some of figures in the following sections can be shown in order to understand the wind characteristics. Study of wind velocity regime over a period of time in a locality can really help to optimize the design of the wind energy conversion system by ensuring less energy generating costs [12]. It can be shown that how wind interact in everyday of a year. For 365 days of a year to show in a figure is difficult to understand, that's why hourly and monthly figure should also be analysed for more better understanding.

Wind speed recorded over minutely or hourly basis over 24 hours in a day, and thus is continued to record the data for a couple of months, or even a number of years. Average wind speed for a complete year with respect to 24 hours is so far the best way to represent wind characteristics. It means that, the first hour of every day of the year is recorded and averaged. In this way, the other hours average wind speed is calculated. The complete frequency distribution of the year or the month is expected only by annual or monthly average wind [13]. In order to determine frequency distribution of the wind speed, it is a must first to divide the wind speed domain into a number of intervals, mostly of equal width of 1 m/s [14].

5 Determination of probability distribution

Weibull distribution can be characterized by its probability density function *f(v)* and cumulative distribution function *F(v)* as follows [15]:

$$f(v) = \left(\frac{k}{c}\right).\left(\frac{v}{c}\right)^{k-1}.\exp\left(-\left(\frac{v}{c}\right)^{k}\right) \quad (6)$$

and

$$F(v) = 1 - \exp\left(-\left(\frac{v}{c}\right)^k\right) \quad (7)$$

where; $f(v)$ is the probability of observing wind velocity v_i, F(v) is the cumulative distribution function, c is the Weibull scale parameter (m/s) and k is the dimensionless Weibull shape parameter. There are many methods to estimate k and c. Three methods upon which the structure of the research is grown are being analysed in the following sections.

5.1 Graphical method

In this method, the wind speed data are interpolated by a straight line, using the concept of least squares regression [16], [17]. Equation (7) is converted to a logarithm form like the following:

$$\ln(-\ln(1 - F(v)) = k.\ln(v) - k.\ln(c) \quad (8)$$

Now, the Weibull shape and scale parameters are estimated by plotting $\ln(v)$ against $\ln(-\ln(1 - F(v))$ in which a straight line is derived. The shape parameter k is determined by the slope of the line, and scale parameter from the following relation: *- k. ln(c) = y-intercept.*

5.2 Empirical method

Shape and scale parameter k and c are measured in this method by the following equations. Shape and scale parameters in this method are estimated from the mean wind speed (v_m) and standard deviation (σ) of wind data.

$$k = \left(\frac{\sigma}{v_m}\right)^{-1.086} \quad (9)$$

Once getting k from equation 9, v_m and k help finding c from the following equation:

$$c = \frac{v_m}{\Gamma\left(1 + \frac{1}{k}\right)} \quad (10)$$

where Γ is the standard Gamma function expressed as:

$$\Gamma = \int_0^\infty t^{x-1} e^{-t} dt \quad (11)$$

Gamma function was used by Manwell et al. [18], and can be expressed as:

$$\Gamma(x) = \sqrt{2\pi x}.x^{x-1}.e^{-x}.\left(1 + \frac{1}{12}x + \frac{1}{288}x^2 - \frac{139}{58140}x^3 + ...\right) \quad (12)$$

5.3 Energy pattern factor method

This method to estimate k and c is related to the averaged data of wind speed. It has simpler formulation, easier implementation and also requires less computation. The method is defined by the following equations:

$$E_{pf} = \frac{(v^3)_m}{(v_m)^3} \quad (13)$$

where $(v^3)_m$ is denoted as the mean of the cubed wind speed, and can be mathematically written as:

$$(v^3)_m = \frac{1}{N}\sum_{i=1}^{N} v_i^3 \quad (14)$$

And $(v_m)^3$ is the cube of mean wind speed. This can be shown as equation (15):

$$(v_m)^3 = \left(\frac{1}{N}\sum_{i=1}^{N} v_i\right)^3 \quad (15)$$

here E_{pf} is called energy pattern factor. Once E_{pf} is calculated from equation 13, Weibull shape parameter k can be estimated as:

$$k = 1 + \frac{3.69}{(E_{pf})^2} \quad (16)$$

Once k is found, from the equation, scale parameter c can be measured from the following, as equation (17) shows:

$$v_m = c\Gamma\left(1 + \frac{1}{k}\right) \quad (17)$$

where Γ is known as Gamma function as expressed in equation (11) and (12).

5.4 Rayleigh probability distribution function

If c is obtained from equation (8) by putting k=2, then probability distribution functions for Rayleigh distribution can be found like the following equation:

$$v_m = c\Gamma\left(1 + \frac{1}{2}\right) \quad (18)$$

Now from the above methods explained, the shape and scale parameters k and c are found like the following table (Table 3):

Table 3. *k* and *c* for Weibull and Rayleigh distribution

Methods	Summer		monsoon		winter	
	K	*C* m/s	*k*	*c* m/s	*k*	*c* m/s
Graphical Method	1.46	4.18	1.43	4.52	1.53	3.94
Empirical Method	1.57	4.71	1.76	4.99	1.93	4.22
Energy Pattern Factor	1.66	4.78	1.91	5.11	2.05	4.28
Rayleigh Distribution	2	4.77	2	5.02	2	4.23

6 Result

6.1 Weibull probability measurement

The following figure shows for each season the Weibull Frequency plotted with the measured wind speed. These curves illustrate the Weibull methods that fit best to the measured wind speed data.

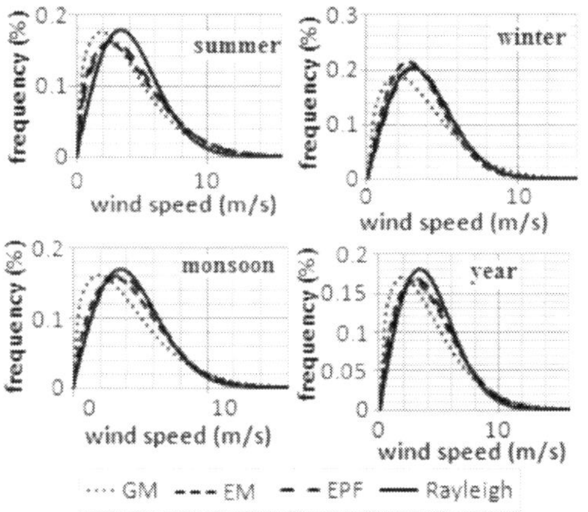

Figure 1. Weibull frequency against wind speed frequency distribution

6.2 Wind power density calculations

The available power of the wind that crosses the rotor of a wind turbine generator is [19]:

$$p_w(v) = \frac{1}{2}\rho A v^3 \tag{19}$$

where $p_w(v)$ is the power associated with wind speed v; A is the rotor area; and ρ is the air density (typically 1225 kg/m3 at sea level [20]). But when evaluating the available energy at a wind site, the following function which is known as wind power density distribution is used:

$$e_w(v) = \frac{1}{A} f(v) p_w(v) \tag{20}$$

This function represents wind energy distribution at different wind speed per unit of time and rotor area in the unit of W/m^2, where f(v) is known as Webull Probability Distribution Function (PDF).

Wind Power Density (WPD) based on the Weibull PDF (with the help of k and c) can be calculated using expression given as [21]:

$$p(v) = \frac{1}{2}\rho c^3 \Gamma\left(1 + \frac{3}{k}\right) \tag{21}$$

where, $p(v)$ is the wind power density in W/m^2, ρ is the air density at the site, (kg/m^3), k and c are shape and scale parameters for Weibull distribution, Γ is Gamma function as described in equation (12). Equation (21) determines the average power density for different seasons that is shown in the following figure.

From Figure 2 it is clear that, monsoon and winter have the highest and lower average power respectively. Figure 2 also identifies that empirical and energy pattern factor method give good power output for summer, monsoon and yearly gain.

For monsoon and winter season as shown in Figure 2, graphical method tops in giving power. However, Rayleigh method lags in all the seasons with an exception of winter. Rayleigh places third in the winter.

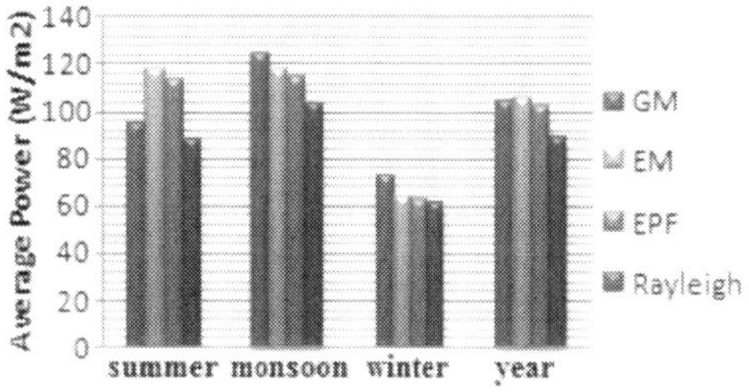

Figure 2. Average wind power of all seasons and year vs. distribution methods

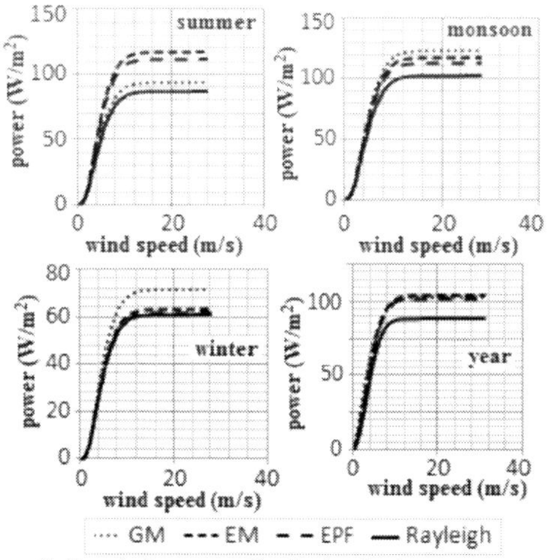

Figure 3. Cumulative power distribution for three seasons

Now, Figure 3 above shows the cumulative power distribution for three seasons. Yearly gain of the wind power as the data analysis generates is also shown in the figure.

6.3 Relative error of the available power density

In order to determine if a particular Probability Distribution Function is suitable to describe the wind speed data, indicators can be used. For each wind site and fitting method, this paper considers power density errors to understand the wind power for different seasons for different distribution methods. The errors in calculating the power densities using the distribution models (Weibull and Rayleigh) in comparison to the values of the probability density distributions derived from Field data values can be found using the following formula [22]:

$$error_{PDF} = \frac{P_{PDF} - P_m}{P_m} X100\% \tag{22}$$

where $error_{PDF}$ is the error in calculating power densities with probability distribution function models, P_{PDF} is denoted as the wind power density for the probability distribution functions, and P_m is the power derived from the data directly.

With equation (22), deriving the error for different distribution methods for three seasons signifies the usability of the methods to choose for the current data.

Power of summer season for graphical and Rayleigh methods give biggest error. For monsoon and winter season, it is seen that, in terms of giving smaller error, the ranks are GM, EM, EPF and Rayleigh. For the whole year data analysis, almost same scenario generates with an exception of GM and EM. For whole year power error analysis EM is followed by GM.

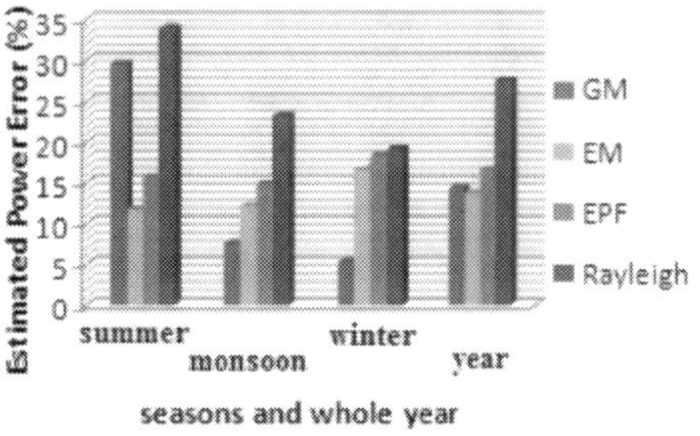

Figure 4. Relative error of power density

7 Conclusions

The hourly wind speed data in time-series format from October 2014 to October 2015 for Saint Martin's island, Bangladesh have been statistically analyzed based on Weibull and Rayleigh PDF. The aim of the study was to investigate the wind power of the island with different shape and scale parameters within Weibull PDF domain. The research as the result has been produced, suggests that suitability of the methods used here may vary with the sample data size, sample data distribution, sample data format etc. is an efficient method for determining the k and c parameters to fit Weibull distribution for the wind speed data at Saint Martin's island. The most important outcomes of the study can be written as follows:

1. k and c for Weibull and Rayleigh PDF have been measured.
2. Maximum wind power was derived in the season of monsoon.
3. Rayleigh distribution estimates less power for all seasons.
4. Graphical method gives less error in terms of power calculations with an exception in summer season.
5. MM and EPF method exhibit consistency in giving power error for all seasons (Figure 4).
6. The wind of Saint Martin's island gives power densities between (88.36 – 118.07) W/m^2, (103.00 – 125.05) W/m^2 and (61.62 – 72.28) W/m^2 for summer, monsoon and winter seasons respectively with different distribution methods at 50 meters height. This has been shown in Figure 3.
7. The power from the wind is satisfactory for the island.

The most widely used methods were selected for Weibull analysis here. Power errors are sought to be significantly less in this research.

References

1. S. Darbandi, M. T. Aalami, H. Asadi, Comparison of Four Distributions for Frequency Analysis of Wind Speed, Environment and Natural Resources Research, **2**(1):96-105, (2012)
2. Electricity demand survey report at the St. Martin's Island, SRE, LGED, Dhaka, Bangladesh, (May 2004)
3. M. Grein, B. Nordell, A. Al Mathnani, Energy consumption and future potential of renewable energy in North Africa, Revue des Energies Renouvelables ICRESD-07 Tlemcen, (2007) 249-254.
4. T. Chaichana, S. Chaitep, Wind power potential and characteristic analysis of Chiang Mai, Thailand, Journal of Mechanical Science and Technology, **24**(7):1475-1479, (2010)
5. R. Pallabazzer, Parametric analysis of wind siting efficiency, J. Wind Eng. Indus. Aerod., **91**:1329-52, (2003)
6. S. Rehman, Long-Term Wind Speed Analysis and Detection of its Trends Using Mann– Kendall Test and Linear Regression Method, Arabian Journal for Science and Engineering, **38**(2):421-437, (2013)
7. D. K. Kidmo, R. Danwe, S. Y. Doka, N. Djongyang, Statistical analysis of wind speed distribution based on six Weibull Methods for wind power evaluation in Garoua, Cameroon, Revue des Energies Renouvelables, **18**(1):105-125, (2015)
8. A. Mostafaeipour, M. Jadidi, K. Mohammadi, A. Sedaghat, An Analysis of Wind Energy Potential and Economic Evaluation in Zahedan, Iran, Renewable and Sustainable Energy Reviews, **30**:641-650, (2014)
9. A. N. Celik, Energy Output Estimation for Small-Scale Wind Power Generators Using Weibull-Representative Wind Data, Journal of Wind Engineering and Industrial Aerodynamics, **91**(5): 693-707, (2003)
10. A. Roy, Reliable estimation of Density Distribution in Potential Wind Power Sites of Bangladesh, International Journal of Renewable Energy Research, **2**(2), (2012)
11. S. Rehman, Long-Term Wind Speed Analysis and Detection of its Trends Using Mann– Kendall Test and Linear Regression Method, Arabian Journal for Science and Engineering, **38**(2):421-437, (2013)
12. K. Ulgen, A. Hepbasli, Determination of Weibull parameters for wind energy analysis of Izmir, Turkey. Int. J. Energy Res., **26**(6):495-506, (2002)
13. A. K. Azad, M. M. Alam, A Statistical Tools for Clear Energy: Weibull's Distribution for Potentiality Analysis of Wind Energy, International Journal of Advanced Renewable Energy Research, **1**:240-247, (2012)
14. E. H. Lysen, Introduction to Wind Energy, Consultancy Services Wind Energy Developing Countries, CWO 82, Second Edition, (1983)
15. A. N. Celik, A Statistical Analysis of Wind Power Density Based on the Weibull and Rayleigh Models at the Southern Region of Turkey, Renewable Energy, **29**(4):593-604, (2004)
16. P. A. C. Rocha, R. C.o de Sousa, C. F. de Andrade, M. V. da Silva, Comparison of Seven Numerical Methods for Determining Weibull Parameters for Wind Energy Generation in the Northeast Region of Brazil, Applied Energy, **89**(1):395-400, (2012)

17. C. G. Justus, W. R. Hargraves, A. Mikhail, D. Grabber, Methods for Estimating Wind Speed Frequency Distributions, Journal of Applied Leteorology, **17**(3):350-353, (1978)
18. J. F. Manwell, J. G. McGowan, A. L. Rogers, Wind Energy Explained: Theory, Design and Application, John Wiley and Sons Ltd, (2002)
19. T. Burton, D. Sharpe, N. Jenkins, E. Bossanyi, Wind Energy Handbook, Chichester, UK: John Wiley & Sons, Ltd., (2001)
20. Wind Turbines—Part. 12-1: Power Performance Measurements of Electricity Producing Wind Turbines; IEC 61400-12-1 Edition 1.0; International Electrotechnical Commission (IEC): Geneva, Switzerland, (2005)
21. H. S. Bagiorgas, G. Mihalakakou, S. Rehman, L. M. Al-Hadhrami, Wind Power Potential Assessment for Seven Buoys Data Collection Stations in Aegean Sea Using Weibull Distribution Function, Journal of Renewable and Sustainable Energy, **4**(1):013119-1 - 013119-16, (2012)
22. N. C. Ali, A Statistical Analysis of Wind power Density Based on The Weibull and Rayleigh Models at the Southern Region of Turkey, Renewable Energy, **29**:593-604, (2003)

Research on Short-Term Wind Power Prediction Based on Combined Forecasting Models

Chi Zhang[1], Jie Zeng[1], Ning Xie[1], Ping Yang[2,3], Yujia Zhang[2,3] and Zhen Zhang[4]

[1]Electric Power Research Institute, Guangdong Power Grid Corporation, 510640, Guangzhou, China
[2]School of Electric Power, South China University of Technology, 510640, Guangzhou, China
[3]Guangdong Key Laboratory of Clean Energy Technology, South China University of Technology, 511458, Guangzhou, China
[4]Guangdong Intework Energy Technology Co., Ltd, 511458, Guangzhou, China

Abstract. Short-Term wind power forecasting is crucial for power grid since the generated energy of wind farm fluctuates frequently. In this paper, a physical forecasting model based on NWP and a statistical forecasting model with optimized initial value in the method of BP neural network are presented. In order to make full use of the advantages of the models presented and overcome the limitation of the disadvantage, the equal weight model and the minimum variance model are established for wind power prediction. Simulation results show that the combination forecasting model is more precise than single forecasting model and the minimum variance combination model can dynamically adjust weight of each single method, restraining the forecasting error further.

1 Introduction

Over the past 10 years, the wind power has been developing rapidly. By the end of 2013, the world's wind-power capacity exceeded 290GW [1], [2]. However, the generated energy of wind farm is fluctuating, which brings great difficulties to grid-connection. In order to guarantee the reliability of power grid, an accurate method for wind power prediction must be found.

At present, there are two major methods for the wind-power prediction: physical forecasting model and statistical forecasting model. Physical forecasting model established mainly based on NWP (Numerical Weather Prediction) [3], transferring the wind speed and the wind direction of NWP into those of Turbine hub height, helps forecasting the wind power of each wind turbine [4], [5]. The main methods of statistical prediction commonly contains several methods, such as time series method, improved time series method [6], Feedforward Neural Network (FNN) [7], Regression Neural Network, Back propagation Neural Network (BP) [8], Support Vector Machine (SVM) [9], grey theory, genetic algorithm and so on.

Both physical and statistical methods have their own advantages and disadvantages. Combination model put different prediction model together through the appropriate combination method, in order to utilize a single model's merit and overcome the limitations of the disadvantage, so as to improve the accuracy of prediction. The most critical problem in combination forecasting model is how to set the weight of each single model for making the accuracy of prediction higher [10]. In this paper, we present a wind power combination forecasting method, combining physical model and statistical model. This combination forecasting model uses minimum variance method to determine the weight of each single model to obtain the wind power forecasting value [11]. This method combines the advantages of physical model and statistical model, which leads to higher wind power forecasting accuracy.

2 Short-Term wind power forecasting method

2.1 Physical wind power forecasting model

The physical wind power forecasting model tries to obtain the wind speed of each turbine hub from NWP data. Based on physical model, the wind power prediction method includes wind-speed transformation model, wake model and wind speed-power curve.

In this paper, the format of the NWP is WRF, whose spatial resolution is 27 kilometers and the temporal resolution is 15 minutes. WRF, researched and developed by the USA, is the world's advanced meteorological numerical model at present.

The wind-speed transformation model uses Inverse Distance Weighted function to transform wind speed of NWP into that of wind turbine. The Inverse Distance Weighted function is shown as follows.

$$v^{*} = \sum \frac{1/l_i}{\sum 1/l_i} v_i \quad (1)$$

where l_i is the distance between the position of NWP data and that of wind turbine, v^* is wind speed of the ith wind turbine, v_i is wind speed of correlative NWP data.

Considering the wind shear, the power function is used to calculate wind speed at the height of hub. The power function in this paper is shown as follows.

$$v_2 = v_1 \frac{\ln h_2 - \ln h_0}{\ln h_1 - \ln h_0} \tag{2}$$

In the actual wind farm, the position of each turbine is different. It is obviously that the downwind wind turbine will receive less wind resource than upwind wind turbine. This phenomenon is called wake effects.

When the wind is absorbed by the wind turbine, the wind speed of the wind turbine will be lower than the wind speed of the wind turbine. By improving the Jensen model, we can get the wind speed of the actual wind turbine hub height. According to improved Jensen model, we calculate the actual wind speed reaching the hub of each wind turbine.

At last, the wind power of each wind turbine can be calculated through wind speed-power curve.

The complete flowchart of physical wind power forecasting model based on NWP is shown as follows.

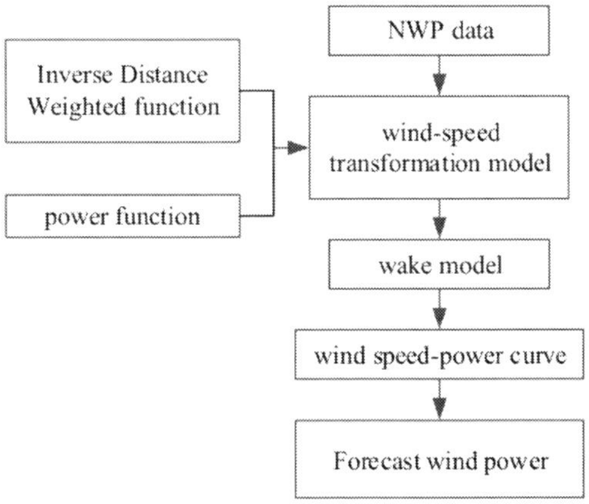

Figure 1. Flowchart of physical wind power forecasting model.

2.2 Statistical wind power forecasting model

In this paper, the MIV-BP method is used to forecast wind power.

First, BP neural network is used to calculate the Mean Impact Value (MIV) of each input variable. Then, the variables will be sorted in order to find the key input items. At last, the BP the neural network is trained with the selected key items as the input variables and the actual power as the output variable.

The process of calculating Impact Value Mean (MIV) is as follows.

First, we use the training set to get the training model, increasing and decreasing each training input variable values by 10% to constitute two new sample set. Taking the two new sample sets as simulation samples, the trained model is utilized to calculate the output value. The Subtraction of two output results of the two data sets is the output value of the Impact Value (IV), the mean value of the IV of each variable is Mean Impact Value (MIV). Finally, in order to find the relative influence of each variable, the MIV value of each variable is seriously ordered. According to the relative influence, we can pick the input variables of neural network out.

The complete flowchart based on MIV-BP is shown as follows.

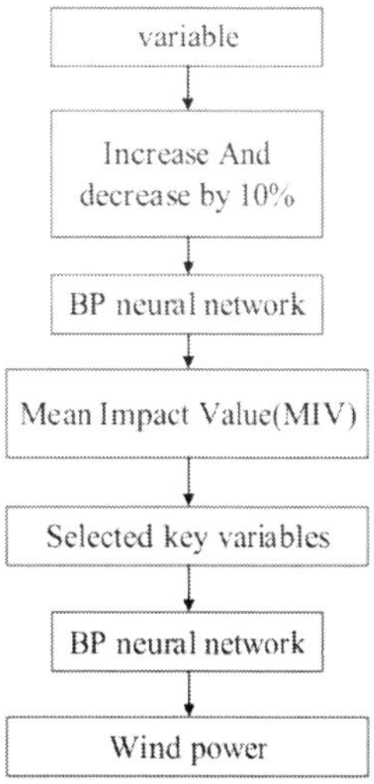

Figure 2. Flowchart of MIV-BP model.

3 Combination wind power forecasting method

3.1 Equal weight combination model

The most common and simple combination forecasting algorithm is equal weight method. Assuming there exist a single forecasting model whose number is m. The wind power forecasted of the i model at the time of J is $\hat{y}_{ji}$ and the actual wind power at the time of J is y_j. Then the forecasting value of equal weight method at the time of J is shown as follows.

$$\hat{y}_j = \frac{1}{m}\sum_{i=1}^{m} \hat{y}_{ji} \tag{3}$$

The relative error of the equal-weight combined forecasting method is shown as follows.

$$\delta_{mean} = \frac{\hat{y}_j - y_j}{y_j} \tag{4}$$

According to the upper formula, equal-weight combined forecasting method is trying to reduce the prediction error of the final power by averaging the forecasting value of single forecasting methods. Equal weight which is simple, combined with forecasting method can reduce the random error of the forecasting model to some extent. The problem of equal weight model is that the weight of each single model cannot be adjusted dynamically according to the difference between the predicted value and the actual value. So the improved space of this algorithm is not good as other algorithms.

Due to the characteristics of the equal weight combination forecasting model, the weight of combination model is used as a benchmark to evaluation the performance of the other algorithms.

3.2 Minimum variance combination model

The minimum variance combination forecasting model can solve the problem of equal weight method which cannot be adjusted dynamically. The minimum variance method can get higher accuracy than equal weight method.

Assuming that there are m different forecasting models, the forecasting wind power of the i model at the time of J is $\hat{y}_{ji}$ and the actual wind power at the time of J is y_j. The weight of the i model is w_i. Then the following constraints need to be satisfied.

$$\begin{cases} \sum_{i=1}^{m} w_i = 1 \\ w_i \geq 0, i = 1, \ldots, m \end{cases} \tag{5}$$

The forecasting value at the time of J is shown as follows.

$$\hat{y}_j = \sum_{i=1}^{m} w_i * \hat{y}_{ji} \tag{6}$$

The error of forecast at the time of J is shown as follows.

$$\delta_j = \hat{y}_j - y_j \tag{7}$$

In order to determine the weight coefficient w_i, the forecasting value and actual value are used to fitting the weight coefficient w_i. This paper uses the minimum variance as the objective function. The expression of objective function and constraints is shown as follows.

$$\begin{cases} \min Z = \sum_{j=1}^{n} \delta_j^{2} \\ st: \sum_{i=1}^{m} w_i = 1 \\ w_i \geq 0, i = 1, \ldots, m \end{cases} \tag{8}$$

where n is the number of actual value and m represent the number of single forecasting model.

The formula shown above is a quadratic programming problem, which can be written as the following form.

$$\begin{cases} \min f(x) = \frac{1}{2} x^T G x + r^T x (x \in R^n) \\ st: A_i^T x - b_i = 0, (i = 1, \ldots, m) \\ A_i^T x - b_i \geq 0, (i = m+1, \ldots, m+l) \end{cases} \tag{9}$$

where G is n-orders symmetric matrix, r, $A_i (i=1, 2, \ldots, m+l)$ is n-dimensional column vector and $b_i (i=1, 2, \ldots, m+l)$ is real number.

The objective function can be expanded as follows.

$$\begin{aligned} \min Z &= \sum_{j=1}^{n} (\hat{y}_j - y_j)^2 = \sum_{j=1}^{n} \left(\sum_{i=1}^{m} w_i * (\hat{y}_{ij} - y_{ij}) \right)^2 = \sum_{j=1}^{n} \left(\sum_{i=1}^{m} w_i * \delta_{ij} \right)^2 \\ &= \sum_{i=1}^{m} \left[w_i^2 \left(\sum_{j=1}^{n} \delta_{ij}^{2} \right) \right] + 2 \sum_{1 \leq k, j \leq m, k \neq j} \left[w_i * w_k * \sum_{j=1}^{n} \left(\delta_{ij} * \delta_{kj} \right) \right] \end{aligned} \tag{10}$$

Simplifying the formula above, we need to set:

$$\begin{cases} g_{jk} = g_{kj} = \sum_{j=1}^{n} \left(\delta_{ij} * \delta_{ki} \right), (j, k = 1, 2, \ldots, m) \\ G = \begin{bmatrix} g_{11} & \cdots & g_{1m} \\ \cdots & \cdots & \cdots \\ g_{m1} & \cdots & g_{mm} \end{bmatrix} \\ W = \left(w_1, w_2, \ldots, w_m \right)^T \\ e = \left(1, 1, \ldots, 1 \right)^T \end{cases} \tag{11}$$

Then, the expression of objective function and constraints is shown as follows.

$$\begin{cases} \min Z = W^T G W \\ e^T W - 1 = 0 \\ W^T \geq 0 \end{cases} \tag{12}$$

The extreme value of the above formula can be calculated through the Lagrange function. The N actual power and N forecasting power is constantly updated as time goes by. Moreover, w_i can also be constantly updated. The wind power sequence is a time series with very strong temporal correlation. Thus, by constantly updating the last N data and determine the weight w_i at different times, we will effectively improve the forecasting accuracy.

4 Simulation verifications

4.1 Initial data

In this case, the format of NWP data is WRF. The actual output data of wind farm is provided by the Power Grid Corp. The selected data is active power of high voltage side in the No. 1 main transformer, whose time interval is 15 minutes. The observed data vary from 00:00:00 2013-01-01 to 23:45:00 2013-07-31. The latitude and longitude of NWP points is shown as follows.

Table 1. Latitude and longitude of NWP points.

longitude	108.42	108.39	108.37
Latitude	18.79	19.02	19.25

The spatial interval of WRF data is 27 kilometers, and the temporal resolution of WRF data is 15 minutes. In this wind farm, there are 24 wind turbines, whose rated capacity are all 1500kW and rated wind speed 12m/s and the height of wind turbines are 70 meters.

Combination forecasting model combining physical forecasting model and BPNN model is used to forecast wind power and compare the accuracy with that of single physical forecasting model and BPNN model in this case.

4.2 Result analysis of single forecast method

In this case, we analyzed the forecast result of two single models shown above. The results of physical forecasting methods based on WRF and statistical forecasting methods with BP neural network are shown as follows.

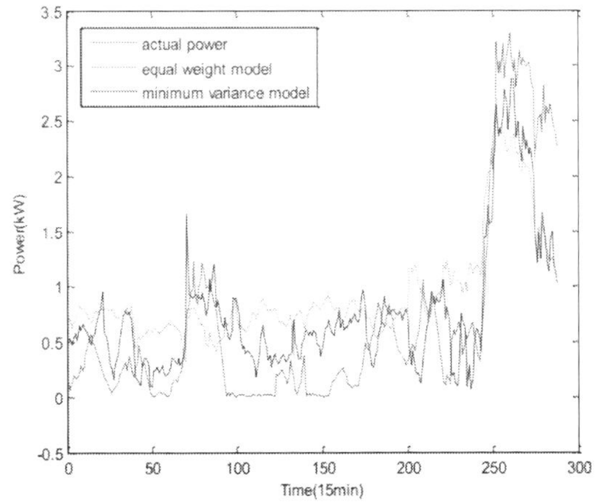

Figure 3. Forecasting results of single model.

The forecasting error curves of physical forecasting methods based on WRF and statistical forecasting methods with BP neural network are shown as follows.

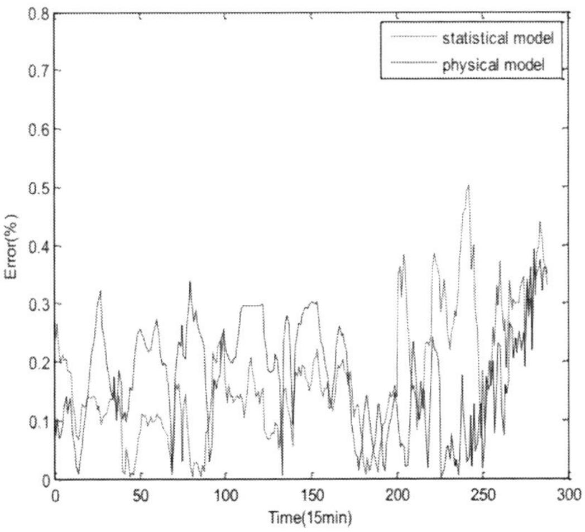

Figure 4. Forecasting error curves of single model.

4.3 Result analysis of combination forecast method

In this case, the combined forecasting method is used to predict wind power by combining the physical forecasting methods based on WRF and statistical forecasting methods with BP neural network.

In equal weight combination model, the weights of two methods are both 0.5. In minimum variance combination model, the weights of two methods are changing dynamically.

The forecasting result of combination method is shown as follows.

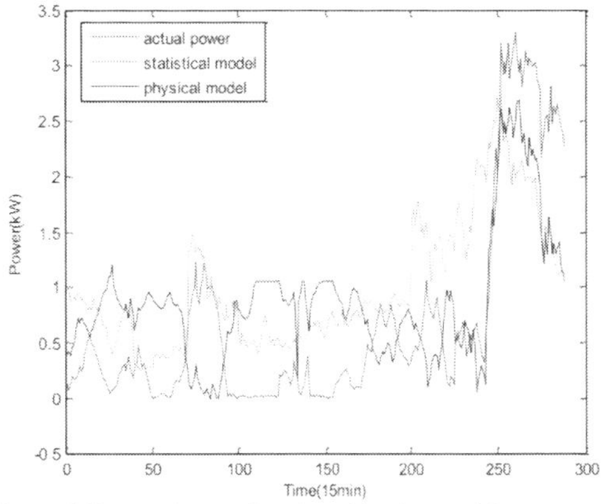

Figure 5. Forecasting results of combination model.

The forecasting error curves of equal weight combination method and minimum variance combination method are shown as follows.

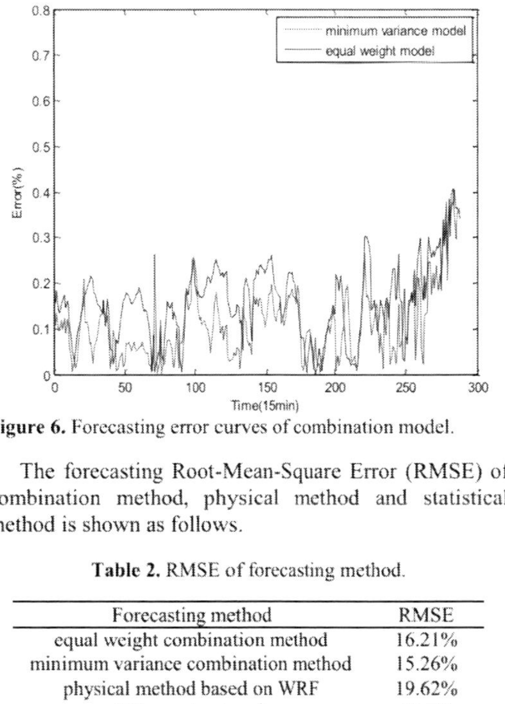

Figure 6. Forecasting error curves of combination model.

The forecasting Root-Mean-Square Error (RMSE) of combination method, physical method and statistical method is shown as follows.

Table 2. RMSE of forecasting method.

Forecasting method	RMSE
equal weight combination method	16.21%
minimum variance combination method	15.26%
physical method based on WRF	19.62%
BP neural network	18.6%

It can be seen from the above that equal weight combination method cannot adjust its weight according to the error of the previous forecast namely it cannot improve forecast accuracy significantly. However, the equal weight combination method also behaves better than single forecasting method though it is the most simple combination method, which shows that the combination forecasting method is a direction of improving accuracy.

It concludes that the minimum variance combination forecasting method can make full use of the wind farm output temporal correlation information and adjust the weight of each simple method, which reduces forecasting error greatly, simultaneously, the minimum variance method also has its own limitation. According to the picture above, we consider that the minimum variance results can't follow the actual power when the actual power varies widely for the adjustment of minimum variance method lags behind the changes of actual power, which causes that the minimum variance method can't effectively reduce the forecasting error in the case of widely changes of output power.

5 Conclusions

Due to the nonlinear and unstable character of wind power, wind power forecasting is a tough and complicated work in the application of wind. In this paper, according to various wind power prediction mathematical model, we build two kinds of single prediction model based on NWP physical forecasting model and the model with optimized initial value in the method of BP neural network. For integrating the advantages of these models, we formulate the combination forecasting algorithm of equal weight model and minimum variance model. Simulations show as follows:

1) The physical forecasting model based on NWP keeps away from the limitation of a large amount of historical data while the prediction accuracy is relatively low. The statistical model based on BP neural network needs to accumulate a large amount of historical data to maintain high prediction accuracy.

2) The combined forecasting model can improve the forecasting accuracy for reducing the accidental error caused by single forecasting model.

3) The minimum variance combination forecasting model can adjust the weight of each simple model according to the accumulated error, which can achieve higher accuracy combined with equal weight combination forecasting model.

Acknowledgments

This work was supported by Guangdong Strategic Emerging Industry Core Technology Research Projects (2012A032300013); National Science and Technology Support Program (2015BAA06B02).

References

1. M. A. Islam, et al., Global renewable energy-based electricity generation and smart grid system for energy security. Scientific World Journal, **2014**: 197136, (2014)
2. World Wind Energy Association, Half-Year Report, (2013)
3. E. Pelikán, et al., Wind power forecasting by an empirical model using NWP outputs, in Environment and Electrical Engineering (EEEIC), 2010 9th International Conference on, 2010, IEEE: Prague, Czech Republic, p. 45
4. S. Al-Yahyai, A. Gastli, Y. Charabi, Probabilistic wind speed forecast for wind power prediction using pseudo ensemble approach, in Power and Energy (PECon), 2012 IEEE International Conference on. 2012, IEEE: Kota Kinabalu, p. 127
5. S. Buhan, I. Cadirci, Multi-Stage Wind-Electric Power Forecast by Using a Combination of Advanced Statistical Methods. Industrial Informatics, IEEE Transactions on, (99):1, (2015)
6. J. Taylor, P. McSharry, Short-Term load forecasting methods: An evaluation based on European data, IEEE Trans. Power Syst., **22**(4):2213-2219, (Nov. 2007)
7. H. Quan, D. Srinivasan, A. Khosravi, Short-Term load and wind power forecasting using neural network-based prediction intervals, IEEE Trans Neural Netw Learn Syst,. **25**(2):303-15, (2014)
8. H. Quan, D. Srinivasan, A. Khosravi, Incorporating Wind Power Forecast Uncertainties Into Stochastic Unit Commitment Using Neural Network-Based Prediction Intervals, IEEE Trans. Neural. Netw. Learn. Syst., (2014)
9. J. Wen, et al., Short-Term wind power forecasting based on lifting wavelet transform and SVM, in Power Engineering and Automation Conference (PEAM), 2012 IEEE, IEEE: Wuhan, p. 1, (2012)
10. M. B. Ozkan, P. Karagoz, A Novel Wind Power Forecast Model: Statistical Hybrid Wind Power Forecast Technique (SHWIP), Industrial Informatics, IEEE Transactions on, **11**(2):375, (2015)
11. L. Landberg, Short-Term prediction of local wind conditions, No.RisØ-R-702(EN), RisØ National Laboratory, RisØ, Denmark, (1994)

Wind Farm Dynamic Equivalence Based on the Wind Turbine Output Active Power Sequence Clustering

Zhang Ge, Yang Yiyun and Xiao Jing

Guangxi Power Grid Company Electric Power Research Institute, Nanning, 530000

Abstract. In order to reduce the complexity of simulation model containing wind farms in the context of keeping the accuracy static, this paper put forward a kind of Dynamic Equivalence method aiming at making output characteristic of the connecting point of wind farm consistent. Based on the output power sequence of wind turbines, geometric template matching algorithm is used to obtain the characteristic of that power sequence and then Attribute Threshold Clustering Algorithm is used to classify wind turbine. In each cluster, the parameter of wind turbine is made equal according to the principle of constant power output character and then be distinguished according to AMPSO. At last, this paper takes a practical wind farm as an example and respectively simulates the conditions of fault of system side and variation of wind speed, which is used in comparing the output characteristic of detailed model and Equivalent model. Results show that the output characteristic of the connecting point of wind farm keeps consistent after equivalent and that the Clustering Algorithm can reflect the operating characteristics of the wind turbine in the whole moment of any time period. It can also be saw that Equivalent method is reasonable and effective, which has certain value in engineering application.

1 Introduction

In the simulation of large wind farm connecting into power grid, it is a heavy work if we modelling every wind turbine, which will also affect the speed, accuracy and convergence of algorithm and thus it is necessary to make Dynamic Equivalent model of wind farm. Currently, Dynamic Equivalent method of wind farm mainly includes single-generator equivalent method, semi equivalent method and multi-generator equivalent method [1], [2]. Among the three methods, single-generator equivalent method is simple but has the lowest accuracy and multi-generator equivalent method has the highest accuracy but is complicated while semi equivalent method behaves in between of two methods.

Multi-machine equivalent firstly need to divide wind turbines into a plurality of clusters and then make equivalence of each clusters, among which it is essential to solve how to realize the equivalence of each clusters [3]-[6]. Reference [7] classifies the wind turbine with the standard of transient track of terminal voltage but its adaptability is not wide for there is less condition of fault in actual operation. Moreover, reference [8], [9] put forward a feasible classification method of time data from wind turbine that classifying time sequence of measured active and reactive power by spectral clustering algorithm based on diffusion mapping theory.

This paper takes the output active power of wind farm as time sequence and put forward a new classification method based on geometric template matching and attribute threshold clustering algorithm, and then distinguish the parameter of wind turbine by AMPSO. At last, based on DIgSILENT/ Power Factory, the accuracy and adaptability of the equivalent model is verified.

2 The wind turbine clustering algorithm

2.1 Similarity algorithm for geometric template matching

The geometric sample matching algorithm can take into account the dynamic change process of the system, and the calculation efficiency is higher [10], [11]. Firstly, the time delay embedding method is used to analyze the phase space reconstruction of active power sequence, and then to analyze the dynamic characteristics of the wind turbine, and then evaluate the difference of each reconstruction vector with the data window and similarity concept. Finally, the similarity is transformed into the distance matrix of the sample, and it provides the basis for the subsequent clustering algorithm.

By meaning time delay, the phase space reconstruction can be reconstructed into a dimensional space vector at different times and the chaotic behavior of the system is studied. Assuming a single variable of time series is $x(t)(t=1,2,\dots,n)$. After the time delay, the new m dimension vector X (t) can be obtained:

$$X(t)=\{x(t),x(t+\tau),...,x(t+(m-1)\tau)\}$$
$$t=1,2,...,M,M=n-(m-1)\tau \quad (1)$$

In the formula: τ is the time delay; m is the embedding dimension; M is the phase point number.

The embedding dimension m and time delay τ is given as follows: given embedding dimension m, the time delay is increased, and the complex auto correlation function R (τ) is calculated as:

$$R(\tau)=\frac{1}{M}\sum_{i=1}^{M}\sum_{j=1}^{m-1}\left\{\left[f(x(i))-f\left(\bar{x}\right)\right].\left[f(x(i+j\tau))-f\left(\bar{x}\right)\right]\right\} \quad (2)$$

$$f(x)=1+x+x^2+...+x^{m-1} \quad (3)$$

In the formula: $\bar{x}=\frac{1}{n}\sum_{t=1}^{n}x(t)$;

Geometric sample matching algorithm can be used to compare the similarity of two time series. For time series $x_1(t)$ and $x_2(t)$ (t=1, 2… n) , the same time delay τ and embedding dimension m are used to reconstruct the phase space. Two new m dimensional vector $X_1(t)$ and $X_2(t)$ (t=1,2,…,M) can be obtained. Nearest neighbor elements of K in vector $X_1(t)$, which is the nearest to the distance vector $X_2(i)$, $X_1(n_{i,1})$, … $X_1(n_{i,k})$, The expression of $n_{i,k}$ is:

$$\begin{cases} n_{i,1}=\arg\min_{j} \| X_2(i)-X_1(j) \| \\ \cdots \\ n_{i,k}=\arg\min_{j\neq n_{i,1},...,n_{i,1}} \| X_2(i)-X_1(j) \| \end{cases} \quad (4)$$

In the formula, function $\arg\min_{x} f(x)$ is to find the x so that the value of F(x) is the smallest.

The similarity of vector $X_1(t)$ and vector $X_2(t)$ can be expressed as:

$$S(X_1,X_2)=\frac{1}{M-1}\times\sum_{i=1}^{M-1}\frac{\left[\tilde{X}_1(i+1)-\tilde{X}_1(i)\right].\left[X_2(i+1)-X_2(i)\right]}{\left\{\max\left[|\tilde{X}_1(i+1)-\tilde{X}_1(i)|,|X_2(i+1)-X_2(i)|\right]\right\}^2} \quad (5)$$

In order to use the distance to replace the similarity, definition:

$$d(X_1,X_2)=\exp(-S(X_1,X_2)) \quad (6)$$

2.2 Characteristic threshold clustering algorithm

Characteristic threshold clustering algorithm does not need to specify the number of K in advance [12], and for the same sample, it is the same every time the running results. In fact, the steps are as follows:

1) Calculate the distance between the two samples.

2) Given the classification of clustering radius d.

3) Set up all samples candidate classes, and make candidate class contain samples within radius d.

4) If the samples in a candidate class are maximum, then it is the first class, and will be removed.

5) Repeat steps 3 and 4 until the remainder of the sample is classified into a certain class.

Based on the cluster method described above, wind farm cluster can be obtained, as shown in Figure 1.

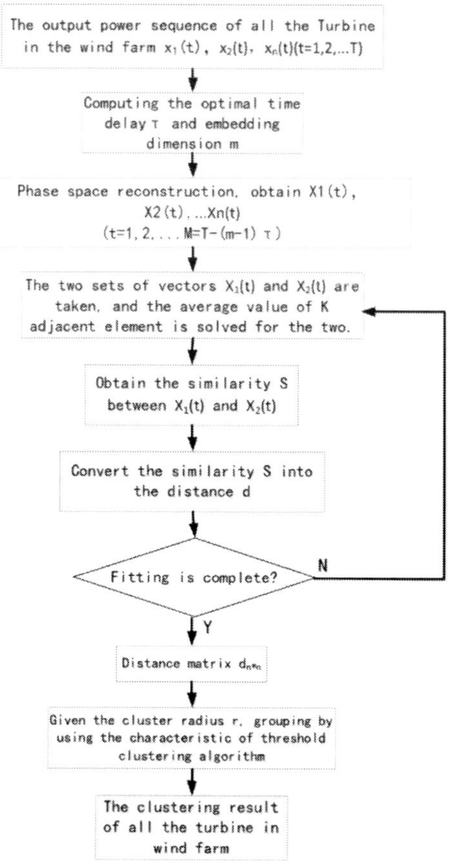

Figure 1. The classification flow chart of wind farm.

3 PMSG parameter identification based on AMPSO

In view of the close correlation of the output active power and its dynamic characteristics of PMSG, this paper chooses the square sum of the output active power of the wind farm before and after the equivalent function to find the optimal parameters of the PMSG. The specific implementation steps of PMSG parameter identification based on AMPSO are as follows:

1) Get system input data including u_d, u_q, i_d, i_q, ω_m, these five variables are measured by the actual motor, so as to ensure that the algorithm is feasible in practical application. In order to improve the accuracy of parameter identification, it is necessary to sample enough samples; this paper intends to sample 1800 sets of data.

2) The position and velocity of the particles are initialized, the size of the particle population, the number of iterations, the time step, the convergence speed, the random attenuation factor, and the upper and lower limits of the parameters L_d, L_q, R_s, ψ_f are set up.

3) To calculate the fitness value of particles, compare the fitness value of individual particles and individual extreme value, update the existing individual extreme value, and compare the fitness value of individual particle and global extreme value, and update the existing global extreme value.

4) To update the speed and position of particles, and to carry out mutation operation of 2% particles, thus forming a new particle swarm. Among them, the first iteration, the particle velocity and position update rules are as follows:

$$v_{ik}(n+1) = wv_{ik}(n) + c_1 \times rand1(0,1) \times (p_{ik}(n) - x_{ik}(n)) + c_2 \times rand2(0,1) \times (p_{gk}(n) - x_{ik}(n)) \quad (7)$$

$$x_{ik}(t+1) = x_{ik}(t) + v_{ik}(t+1) \quad (8)$$

ω is the inertia weight coefficient; c_1, c_2 are the learning factors; the rand1 (0,1) and rand2 (0,1) are two random numbers of (0,1).

5) Repeat steps 3 and 4 to calculate the fitness value of each particle, if the fitness value of a certain iteration is less than the threshold set in advance, then stop searching, the solution vector is the optimal parameter of PMSG.

4 Numerical simulation

4.1 Division of wind power plants

Select the active power sequence of 24 typhoon machine output in Figure 3 $x_1(t)$, $x_2(t)$, ... $x_{24}(t)$ (t=1, 2, ..., 1501). The phase space reconstruction of 24 active power series is carried out, so we got the reconstruction vector $X_1(t)$, $X_2(t)$, ... $X_{24}(t)$ (t=1, 2, ..., 1501), optimal embedding dimension=4, time delay τ =11. To reconstruct the vector as the input, solve adjacent element number k=5 between any two sets of vector, a matrix $n_{M\times k}$ was obtained between every two groups; the similarity $S_{24\times 24}$ between two sets of vectors is obtained; converted the similarity S to the distance d, so got the distance matrix $d_{24\times 24}$; Finally, clustering algorithm is used to classify the clusters based on attribute threshold clustering algorithm, set clustering radius $r_0 = 0.5$, the clustering result is shown in Figure 2.

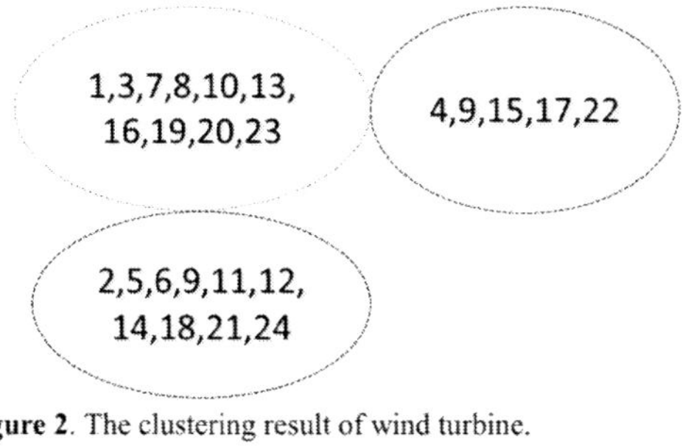

Figure 2. The clustering result of wind turbine.

4.2 Equivalent model parameter identification

Set the maximum number of iterations for the 200 time, adaptive mutation probability is 0.002%, the fitness curves of AMPSO are shown in Figure 3.

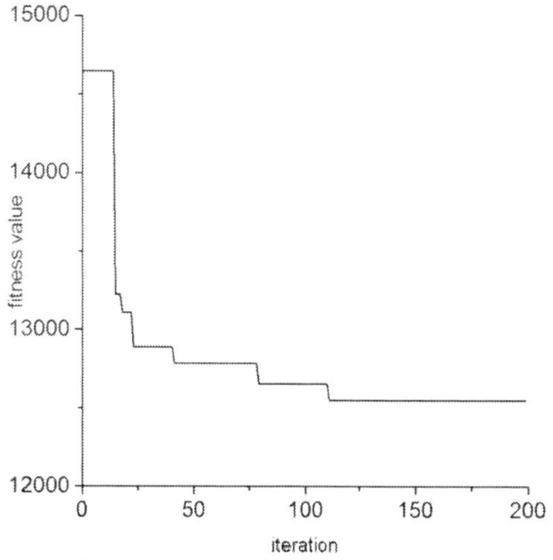

Figure 3. The AMPSO fitness curve.

By using AMPSO identify PMSG, the optimal parameters of the wind turbine can be obtained: $L_d = L_q = 1.432\,\text{mH}$, $R_s = 0.00082\Omega$, $\psi_f = 6.774Wb$.

4.3 Dynamic simulation

In order to verify the correctness of the proposed method in this paper, we use DIgSILENT/Power Factory to simulate, the active power and reactive power are shown in Figure 4-6.

1) Gust, start at 2s, stop at 8s, maximum wind speed is 3m/s.

2) Wind speed increase, start at 2s, stop at 8s, maximum wind speed is 4m/s.

3) Three-phase short-circuit fault occurred at grid side, the failure occurs at 8.2s, disappeared at 8.2s.

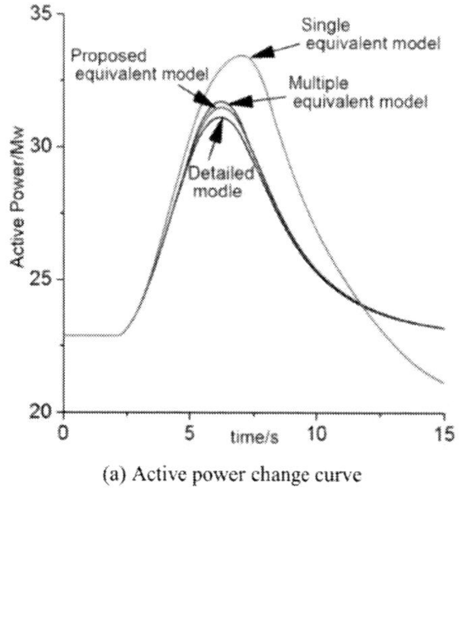

(a) Active power change curve

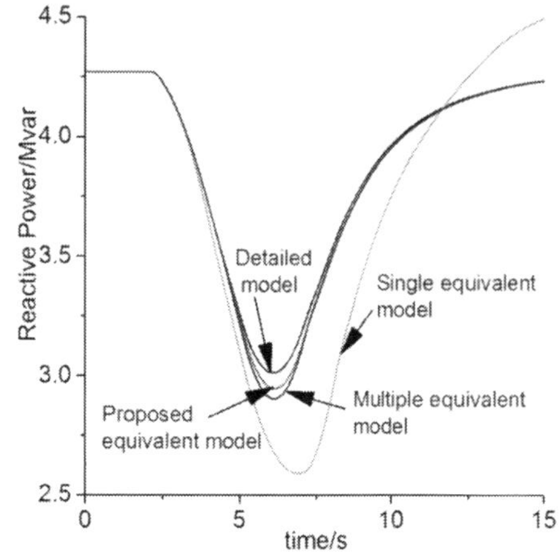

(b) Reactive power change curve

Figure 4. The Output curve of wind power plants in gust.

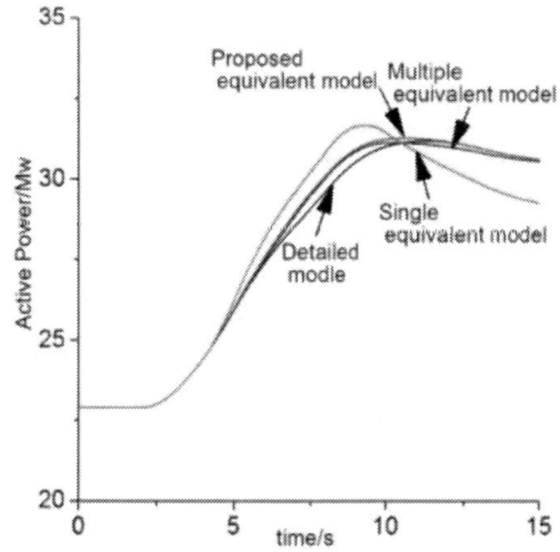

(a) Active power change curve

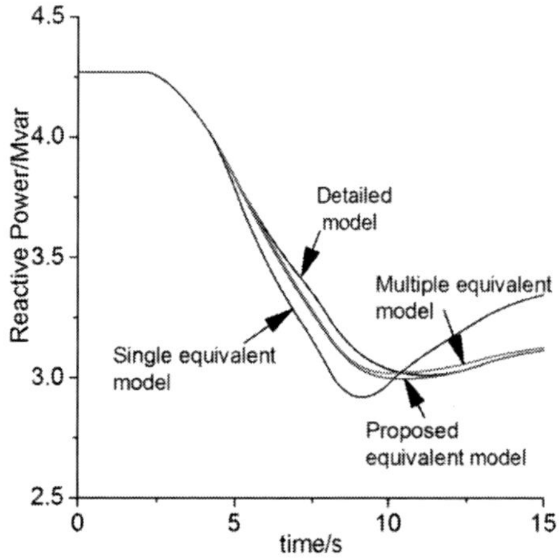

(b) Reactive power change curve

Figure 5. The Output curve when wind speed increasing.

It can be seen in Figure 4, Figure 5 and Figure 6 the equivalent effect on the output characteristics of the equivalent model and the detailed wind power plants model is of high degree coincidence; compared with the equivalent model of single well, the multi machine equivalent model is better, which also explained the necessity of wind power generators. From the view of adaptability, under the gradient wind and gust disturbance, the equivalent model can response and maintain the consistency of the detailed model with, this also proves that all the time in the equivalent method of clustering result can be applied to the period of time.

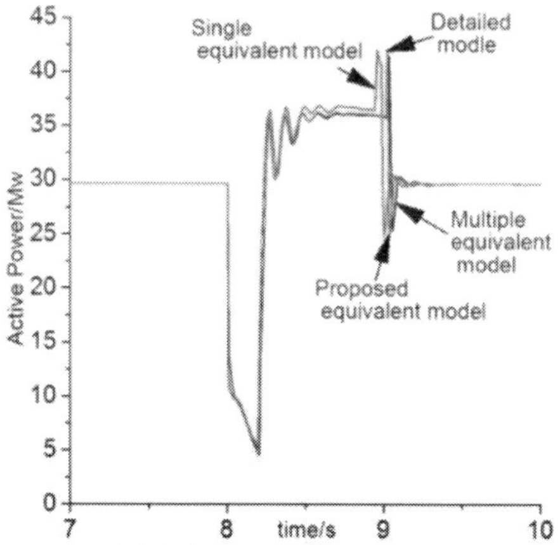

(a) Active power change curve

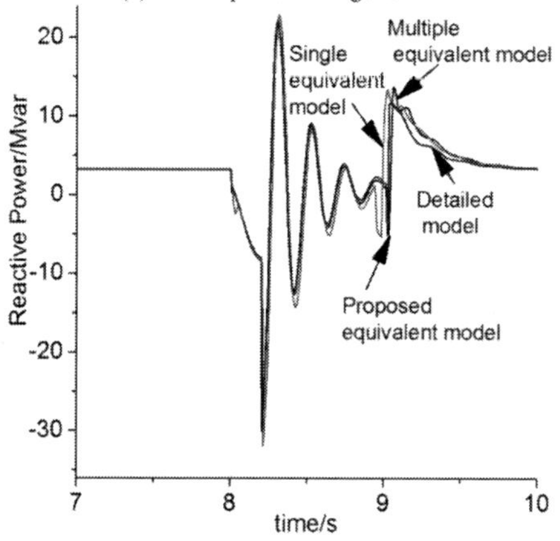

(b) Reactive power change curve

Figure 6. The wind power plants output curve of the system side fault.

5 Conclusions

The wind power plants dynamic equivalence method is proposed in this paper to the time series data of the power output of the wind turbine for clustering the sample data, using the geometric template matching and attribute threshold clustering algorithm for clustering, the clustering method can reflect the operating characteristics of wind turbine, and the clustering result is suitable for all the time during the period of time, can avoid wind turbine multiple groups.

The output characteristics of the wind power plants in three different conditions are compared with the simulation results in DIgSILENT/ Power Factory. The results show that the dynamic equivalent method of the wind power plants proposed in this paper can reflect the dynamic characteristics of the wind power plants.

References

1. M. A. Chowdhury, W. X. Shen, N. Hosseinzadeh, et al., A novel aggregated DFIG wind farm model using mechanical torque compensating factor, Energy Conversion and Management, (67):265-274, (2013)
2. H. Z. Liu, Z. Chen, Aggregated modelling for wind farms for power system transient stability studies, 2012 Asia-Pacific Power and Energy Engineering Conference (APPEEC), Shanghai, China: IEEE, (2012):1-6
3. V. Akhmatov, H. Knudsen, An aggregate model of a grid-connected, large-scale, offshore wind farm for power stability investigations importance of windmill mechanical system, Electrical Power Energy Systems, **24**(9):709-717, (2002)
4. Z. Q. Mi, X. W. Su, Y. Yang, et al., Study on dynamic equivalent model of wind farms with wind turbine driven doubly fed induction generator, Automation of Electric Power Systems (in Chinese), **34**(17):72-77, (2010)
5. S. Y. Chen, C. Wang, H. Shen, et al., Dynamic equivalence for wind farms based on clustering algorithm, Proceedings of the CSEE (in Chinese), **32**(4):11-19, (2012)
6. Y. Q. XU, N. Wang, Study on dynamic equivalence of wind farms with DFIG based on clustering analysis, Journal of North China Electric Power University (in Chinese), **40**(3):1-5(2013)
7. Z. Q. Mi, X. W. Su, Q. X. Yang, et al., Muti-machine representation method for dynamic equivalent model of wind farms, Transaction of China Electrotechnical Society (in Chinese), **25**(5):162-169, (2010)
8. Y. Ma, T. Runolfsson, J. N. Jiang, Cluster analysis of wind turbines of large wind farm with diffusion distance method, IET Renewable Power Generation, **5**(2):109-116, (2011)
9. L. Li, Y. Chen, Wind turbine grouping with spectral clustering algorithm based on diffusion mapping theory, Electric Power Automation Equipment (in Chinese), **33**(6):113-118, (2013)
10. F. Jordan, M. Shie, P. Joelle, et al., Time series analysis using geometric template matching, IEEE Transactions on Pattern Analysis and Machine Intelligence, **35**(3):740-754, (2013)
11. Xu rock, the estimation of the parameters of the phase space reconstruction of chaotic time series, Chongqing: Chongqing School, (2013)
12. J. H. Laurie, K. Semyon, Y. Shibu, Exploring expression data: Identification and analysis of coexpressed genes, Genome Research, **9**:1106-1115, (1999)

Comparison of Parameter Identification Techniques

Rafael Eder[1], Christian Zehetner[1] and Wolfgang Kunze[2]

[1]*Linz Center of Mechatronics GmbH, 4040 Linz, Austria*
[2]*Salvagnini Maschinenbau GmbH, 4482 Ennsdorf, Austria*

Abstract. Model-based control of mechatronic systems requires excellent knowledge about the physical behavior of each component. For several types of components of a system, e.g. mechanical or electrical ones, the dynamic behavior can be described by means of a mathematic model consisting of a set of differential equations, difference equations and/or algebraic constraint equations. The knowledge of a realistic mathematic model and its parameter values is essential to represent the behaviour of a mechatronic system. Frequently it is hard or impossible to obtain all required values of the model parameters from the producer, so an appropriate parameter estimation technique is required to compute missing parameters. A manifold of parameter identification techniques can be found in the literature, but their suitability depends on the mathematic model. Previous work dealt with the automatic assembly of mathematical models of serial and parallel robots with drives and controllers within the dynamic multibody simulation code HOTINT as fully-fledged mechatronic simulation. Several parameters of such robot models were identified successfully by our embedded algorithm. The present work proposes an improved version of the identification algorithm with higher performance. The quality of the identified parameter values and the computation effort are compared with another standard technique.

1 Introduction

The trend to highly accurate products – often with arbitrary lot size and low production time – requires a deep understanding of the production process with highly complex mechatronic machines e.g. automatic panel benders as provided by Salvagnini (www.salvagninigroup.com). In order to satisfy this demand, it is necessary to obtain detailed knowledge of the mechatronic behavior of each machine component. After representing this knowledge as a mathematical model, the methods of model-based control lead to improvements in the performance and accuracy of the production process. Thus, it is important to develop highly accurate mathematical models with accurate parameter values. Usually some important parameters of the mathematic model of a mechatronic machine part are unknown, so it is important to compute those missing parameters with an appropriate parameter identification technique.

A manifold of parameter identification techniques can be found in the literature. Main focus of the present work is to compare several techniques, to find criterions for choosing the appropriate strategy for parameter identification for mathematic models of specific mechatronic components. Those models may also contain nonlinearities and discontinuities, so it is necessary to use a zero-order method, which does not need gradients information. Furthermore, it is important, that those techniques should also be suitable to automatically assembled models, e.g. [1]. In [2, 3], such models have been obtained by multibody simulation code HOTINT. In a previous work, a novel zero-order parameter identification algorithm has been introduced and applied successfully to parameter identifications of test examples with well-known parameters as well as to real mechatronic systems, e.g. the identification of drive parameters from an industrial robot with parallel kinematics [4]. In a further work and among investigations with this identification algorithm, the performance assessment and the comparison against another heuristic technique, i.e. the Particle Swarm Optimization (PSO) [5, 6], showed higher accuracy and precision in most of the tested cases as well as lower computation times and a lower number of optimization steps [7]. Our experience has shown, that Newton Search is not very suitable for such problems, especially when the initial parameter values are not close to the optimal parameters [8].

Various Genetic Algorithms (GA's) were applied successfully to highly complex mechatronic optimization problems [9, 10, 11]. Increasing computation power as well as the capability to parallelize GA's has lead generally to an upward trend of this way to solve optimization problems, compare Fig. 1.

The present work is focused on the performance and quality of the identification results, especially when the local minima of the cost functions increase due to the number of unknown parameters. In contrast to previous works, an improved parameter identification algorithm is proposed, in the following labeled as Special Genetic Algorithm (SGA), i.e. a special type of the GA's and described in section 2.1. The results of the parameter identification are compared to a standard GA.

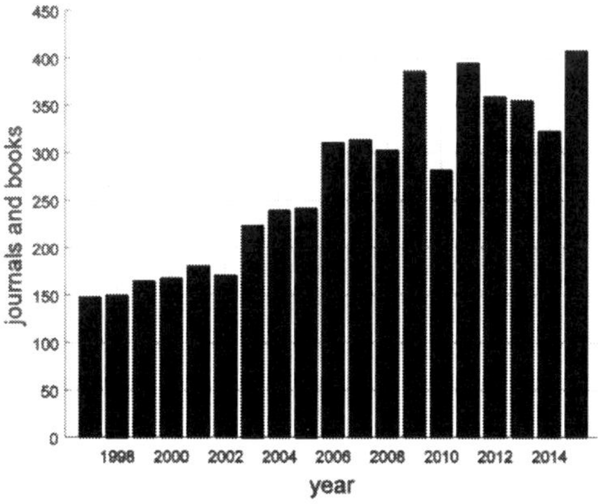

Figure 1. Trend of GA – publications (source: ScienseDirect).

2 Identification techniques

The proposed SGA is described in section 2.1. Standard parameter identification techniques for comparisons are shortly described in section 2.2 and 2.3.

2.1 SGA

The algorithmic efficiency of the SGA has been increased in order to compete with other heuristic techniques. The improved SGA is described in this section – the previous version of the algorithm has been described in [4].

The SGA is an evolutionary algorithm. Therefore, the fitness of a population of parameter vectors dedicated to a mathematic model is optimized. Each fitness value is computed by an evalutation of the cost function. Note, that in parameter identification problems, the cost function is a measure of the accordance of simulated and measured quantities. The best parameter vector is dedicated to the global minimum. In order to find it, following major steps are processed:

1. Compute generation g of parameter vectors (Initialisation/Mutation) and compute fitness values
2. Choose fittest parameter vectors (Selection)
3. Check stop criterion
4. If not stopped: $g \rightarrow g+1$, then repeat 1.

2.1.1 Step 1: Initialization or mutation

In step 1 of the SGA, the components of the parameter vectors are computed. Note, that each component may be dedicated to an unknown parameter value of a mathematic model.

Due to the fact, that the first generation of parameter vectors has no predecessor, the first generation (g=1), is uniformly distributed between the user defined limits of each parameter.

For all further generations, $g \in \{2, 3, \ldots\}$, the surviving parameter vectors (also denoted as parent parameter vectors) are well-known after the selection process, as described in section 2.1.2. Thus, the components of those mutated parameter vectors of the further generations depend on the selected surviving parameters. The main idea is that the mutation of the parent parameter vectors leads to a user defined number of children parameter vectors, which are located near the surviving parameter vectors. Thus, each surviving parameter is surrounded by its normally distributed children parameters after the mutation process is finished. Remark, that the children parameters are placed in a normally distributed parameter space, which shrinks each generation, so the SGA is able to estimate the optimal parameter vector very close to the real optimum.

After the computation of a generation of children parameter vectors due to the latter described initialization or mutation, their cost functions are evaluated.

2.1.2 Step 2: Selection

In the selection step, the fitness function of each parameter vector is well-known. Our experience with the SGA showed, that the algorithm for the selection process in the SGA is crucial for the computation speed. Thus, during the selection step, only a user defined number of the fittest children parameter vectors survive, i.e. chosen from the union of the set of surviving parameter vectors of the previous generation and the set of children parameter vectors of the actual generation. Note, that it is important to choose an adequate number of surviving parameter vectors for searching in the whole parameter space in a sufficient way.

2.1.3 Step 3 – Stop criteria

The extended stop criteria of the SGA define the end of the computation:

1. Stop, if a user defined, maximal number of generations g is reached
2. Stop, if the best cost function value is below an user-defined accuracy limit

After stopping the algorithm, the fittest parameter vector with the lowest cost function is returned.

2.1.4 Step 4 – Compute next generation

If the latter described stop criteria are not fulfilled, the number of generations g is increased. Then the algorithm proceeds with the mutation process, as described in section 2.1.1.

2.2 Genetic Algorithm (GA)

An overview over GA's can be found in [12]. In contrast to the latter described SGA, a common GA uses also the principle of crossover of parent parameters to obtain children parameters. The latest released genetic optimization code, the Octave package "ga" was chosen for further investigations. In section 4, the results of the comparison of the GA with the SGA can be found.

2.3 Nelder-mead simplex method

A highly popular direct search method is the Nelder-Mead Simplex algorithm (NMS). The convergency properties for low dimensions are investigated in [13]. Experience with this technique has shown, that the choice of the start value of the parameter search should be chosen near the global minimum, otherwise the optimization result might be only a global minimum. If the cost function has multiple minima, e.g. in cost function shown in Fig. 2,

$$e(\theta) = [\cos(\theta - 20) + 0.1 (\theta - 20) + 1]^2 + 0.1 \mid \theta - 20 \mid \quad (1)$$

In contrast to the GA and SGA, the NMS only can find a local minimum near the starting parameter value, i.e. 50, see Fig. 2. The computation time of NMS is nearly equal to the GA (section 2.2). The SGA, see section 2.1, finds the minimum of eq. 1 nearly four times faster than the NMS.

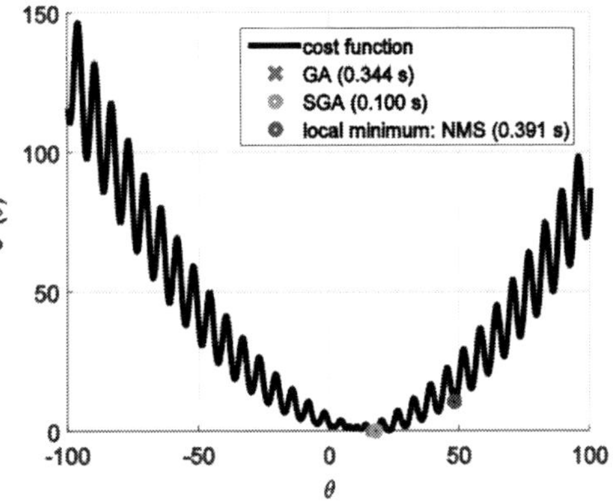

Figure 2. Local minimum found by NMS; GA and SGA find better solutions.

Due to this convergence to a local minimum, the further investigations are more focussed on the comparison of the SGA and GA. Although the GA needs 3.44 times higher computation time for minimizing eq. 1, only the SGA finds the global minimum, see Fig. 3.

These results motivate to perform further comparison of the performance and quality of the SGA and standard heuristic techniques for parameter identification in more detail. Therefore, a set of cost functions has been defined in section 3.

3 Benchmark cost functions

In the following benchmark, the identification algorithms search the optimal N-dimensional parameter vector,

$$\boldsymbol{\theta}_{\mathbf{opt}} = [\theta_{1,opt}, \theta_{2,opt}, \dots \theta_{N,opt}], \quad (2)$$

i.e. minimize the N-dimensional cost functions.

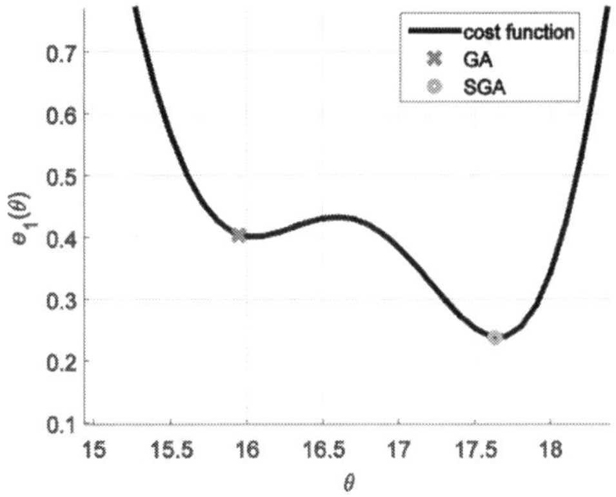

Figure 3. SGA find global minimum of function e, GA only a local minimum.

For the comparison of the GA and the SGA, the cost functions in Table 1 are defined, which are well-established for testing optimization methods.

Table 1. Cost functions for benchmark.

Rosenbrock	$\sum_{k=1}^{N-1} 100(\theta_{k+1} - \theta_k^2)^2 + (1-\theta_k)^2$
Rastrigrin	$\sum_{k=1}^{N} \theta_k^2 - 10\cos(2\pi\theta_k)$
Griewank	$\sum_{k=1}^{N} \frac{\theta_k^2}{4000} + \prod_{k=1}^{N} \cos\left(\frac{\theta_k}{\sqrt{k}}\right) + 1$
Sphere	$\sum_{k=1}^{N} \theta_k^2$

The optimal parameter value of each component of the optimal, N-dimensional parameter vector is searched in the very large interval [-25, 100]. This leads to a high number of local minima, especially in the Rosenbrock, Rastrigrin and Griewank function.

4 Results

In the following, the parameter identification algorithms are tested with the cost functions in section 3. The settings of the algorithms are described in section 4.1 and the identification residuals and the computation times are shown in section 4.2.

4.1 Settings of Algorithms

In order to compute comparable results with the GA and the SGA, the auxiliary variable

$$p = \min(5^N, 5^4) \quad (3)$$

is defined. It is a tradeoff between the increasing number of local minima, which strongly depend on the dimension N and the required computation times. Remark, that the algorithms might find the optimum also with a lower number p, but the present work is interested in the comparison of different comparable parameter identification algorithms.

Table 2. Settings of parameter identification algorithms.

Option	SGA	GA
Surviving population size	p	No GA option
Number of children	p	No GA option
Initial population size	p	No GA option
Number of maximal generations	15	p
Accuracy limit	10^{-4}	10^{-4}
Population size	No SGA option	p^2

The options of SGA and GA are summarized in Table 2. Note, that with the default options of the SGA, a number of 15 generations is sufficient, due to the reduction parameter range around the surviving parameter vector, see section 2.1.1.

The SGA and the GA have slightly different options, see Table 2, but the population size is equivalent to the surviving parameter size multiplied with the number of children in the SGA.

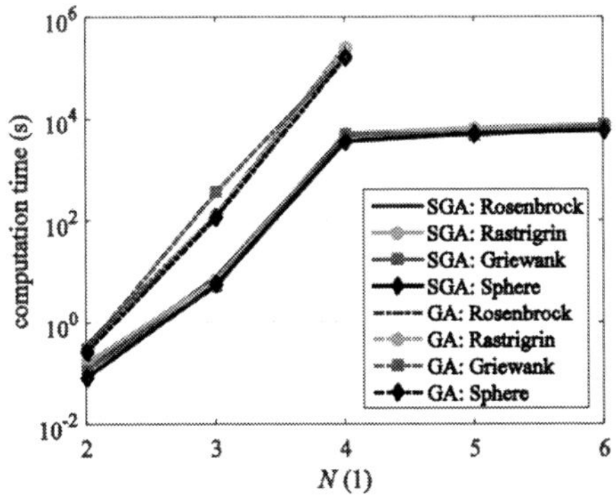

Figure 4. Computation time (s): SGA is faster than GA.

4.2 Computation results of GA and SGA

contrast to previous work, the SGA has been computed against a standard GA for the first time in a successful

The following results are obtained by applying the GA and SGA (section 2), to minimize the benchmark cost functions, as defined in section 3 with rising dimension of the parameter space N. Note, that it was not possible to compute all results with the GA within reasonable computation times.

Table 3. Cost function residuals.

	GA N={2,3,4}	SGA N={2,3,...,6}
Rosenbrock	[4.55×10^{3}, 1.97×10^{-5}, 4.96×10^{-5}]	9.51×10^{-6} 1.32×10^{-5} 9.50×10^{-5} 3.06×10^{-5} 3.72×10^{-4}
Rastrigin	[4.66×10^{-1}, 5.32×10^{-5}, 5.19×10^{-5}]	1.48×10^{-5} 6.03×10^{-5} 9.67×10^{-6} 4.20×10^{-5} 1.92×10^{-4}
Griewank	7.40×10^{-3}, 7.40×10^{-3}, 7.98×10^{-5}]	8.42×10^{-6} 2.00×10^{-5} 7.40×10^{-3} 7.66×10^{-6} 9.86×10^{-3}
Sphere	[5.59×10^{-5}, 3.91×10^{-5}, 4.76×10^{-5}]	5.33×10^{-5} 2.48×10^{-5} 3.62×10^{-5} 5.85×10^{-5} 3.39×10^{-5}

5 Conclusion

In the present work, the comparison of quality and computation time of a common GA with a proposed parameter identification technique SGA leads to the conclusion, that the SGA solved the minimization problem of the benchmark cost functions faster than the GA, see Figure 4. Furthermore, the cost function residuals were solved more accurately than the standard GA in the majority of the tested cases (Table 3). Even, in cases where no reasonable computation time has been possible with the standard GA, the SGA solved also problems in a six-dimensional parameter space and a high number of multiple local minima. The ability to deal also with a higher number of unknown parameters is needed for parameter identification of mechatronic systems. In way. The proposed method needs no differentiation of the cost function and algebraic manipulations of the mathematic model of a mechatronic system and is immediately usable to arbitrary identification problems. Further work will focus on possible improvements of the SGA in higher dimensions of the parameter vector.

Acknowledgement

This work has been supported by the Linz Center of Mechatronics (LCM) in the framework of the Austrian COMET-K2 programme.

References

1. R. Ludwig, J. Gerstmayr, C. Augdopler, C. Mittermayr, Flexible Robot with Clearance, Proc. of 4ECSC **2**, (2008)
2. J. Gerstmayr, M. Stangl, High-order implicit Runge-Kutta methods for discontinuous mechatronical systems, Proc. XXXII Summer Sch. AP, St. Petersburg, (2004)
3. J. Gerstmayr et. al., HOTINT-a script language based framework for the simulation of multibody dynamics systems, Proc. of ASME 2013 IDETC&CIE Conf., Oregon, (2013)
4. R. Ludwig, J. Gerstmayr, Automatic Parameter Identification for Generic Robot Models, Proc. of ECCOMAS Conf., Brussels, Belgium, (2011)
5. J. Kennedy, R. Eberhart, Particle Swarm Optimization, Proc. IEEE Int. Conf. on Neural Networks **4** (1995)
6. J. Kennedy, R. C. Eberhard, *Swarm Intelligence* (Morgan Kaufmann Publishers, San Francisco, California, 2001)
7. R. Eder, J. Gerstmayr, Special Genetic Identification Algorithm with Smoothing in the Frequency Domain, Adv. Eng. Softw. **70C** (2014)
8. R. Ludwig, J. Gerstmayr, Identification of vibration-relevant parameters in robotic systems, Proc. of Austrian Robotics Workshop, Linz, (2014)
9. P. Boyraz, M. Gündüz, Dynamic modeling of a horizontal washing machine and optimization of vibration characteristics using Genetic Algorithms, Int. J. Mechatron. **23**, 6 (2013)
10. C. Chan, G. Liu, Hysteresis identification and compensation using a genetic algorithm with adaptive search space, Int. J. Mechatron. **17**, 7 (2007)
11. Z. Affi, B. El-Kribi, L. Romdhane, Advanced mechatronic design using a multi-objective genetic algorithm optimization of a motor-driven four-bar system, Int. J. Mechatron. **17**, 9 (2007)
12. I. Rechenberg, *Evolutionsstrategie* **1** (Friedrich Frommann Verlag, Stuttgart, 1994)
13. J. C. Lagarias, J. A. Reeds, M. H. Wright, P. E. Wright, Convergence Properties of the Nelder-Mead Simplex Method in Low Dimensions, SIAM J. of Optim. **9**, 1 (1998)

Hydrodynamic Analysis for a Jacket-Type Support Structure of the Offshore wind Turbine

Ming-Hong Chen[1], Chun-Wei Lin[2], Tso-Ren Wu[2] and Chin-Cheng Huang[1]

[1]*Mechanical and System Engineering Program, Institute of Nuclear Energy Research, Atomic Energy Council, Longtan, Taoyuan, Taiwan (R.O.C.)*
[2]*Graduate Institute of Hydrological and Oceanic Sciences, National Central University, Taoyuan, Taiwan (R.O.C.)*

Abstract. In the present study, the hydrodynamic load due to extreme external condition, such as typhoon, on the offshore wind turbine (OWT) has been investigated by numerical simulations. The hydrodynamic loads including the sea current, regular and irregular wave on the structure are taken into account. The sea current is considered with the constant flow speed, whereas the wave is considered as regular and irregular based on the observed data at considered locations. Using Splash3D, a three-dimensional numerical model is developed for performing the force evaluation of the OWT. Multi-phase flow is evaluated by the volume of fraction (VOF) model, and the turbulent flow is described by the large-eddy-simulation (LES) model. It can be deduced that using the regular wave as the hydrodynamic load for the evaluation of the OWT system will vanish some instantaneous extreme wave. In the case of using the irregular wave, the instantaneous extreme wave is also taken into account. With such more practical hydrodynamic load, the average force for the case of irregular wave is 64% larger than that of the regular wave.

1 Introduction

Comparing to the onshore wind farm, the offshore wind farm has higher quality of wind power source, including higher average wind speed and lower turbulence. However, harsh environmental impacts, such as typhoon, wave, sea current, and earthquake, must be taken into the consideration for the development of a reliable offshore wind turbine (OWT) system. In order to provide safety and reliability for the construction of the OWT, site specific loads induced by the extreme wind, wave, current, and earthquake, should be employed to the analysis and the evaluation for the verification of the designed wind turbine and its support structure.

In the present study, a numerical model based on the reference 5MW OWT from the report of NREL is developed. The objective of this work primarily focuses on the evaluation of the hydrodynamic load. Specifically, the effects of the site-specific sea current and wave on the support structure of the OWT are investigated. By including the effect of the hydrodynamic loads on the support structure into the design procedure, the reliable and robust power production of the OWT system could be achieved.

The computational fluid dynamics (CFD) model is capable to predict the interaction between the wind turbine and hydrodynamic loads. The complicated effects of turbulence, wave, sea current, and structure should be considered comprehensively in the design analysis of the OWT. In recent studies, overly simplified assumptions were employed in the simulation of the ocean flow. For instance, it was commonly assumed that the flow is inviscid and irrotational to solve the potential equation. The semi-empirical Morison's formula was employed to describe the wave load. Those assumptions made the proposed model unable to evaluate the unsteady and highly turbulent flow [1]. Solving the Navier-Stokes equation with the method of fluid-structure interaction (FSI) is an effective way to evaluate the hydrodynamic load on the OWT. Another critical issue for the effective simulation is that the investigated environment is a multi-scale problem. The scale of the oceanic effect is usually measured by kilometer, while it is meter for the OWT. This scale difference makes the simulation more difficult. The method of the direct numerical simulation (DNS) results in the most correct solution. However, the flow field with large Reynolds number will slow down the calculation, making it an impractical option.

A reliable wave model should be able to predict the effect of the nonlinear and breaking wave to simulate the interaction among large waves, strong current, and structure. Wienke and Oumeraci [2] coupled the breaking wave theory and parameter model to evaluate the wave force of wind turbine due to the plunging and breaking waves. Yang and Shen solved the equations of the potential flow to calculate the behaviors of the wave [3]. Calderer et al. accounted for the effects of wave by the simplified linear theory [4]. Marino et al. [5], [6] coupled the aero-elastic model with non-linear wave theory to solve the dynamic force on the structure. It was found by the studies of Marino and Lugni [5]-[7] that the wave force and local pressure increase in a short period of time when wave breaking near the support structure of the wind turbine. Marino et al. [8] employed the nonlinear wave theory and statistical method to analyze the force on the wind turbine for a long time.

Since the traditional wave theories, whether linear or non-linear, are unable to appropriately describe the combined effects of typhoon induced wave and current, the study implements the internal wave maker module to produce the target wave. The behavior of the surface breaking wave is described by the volume of fluid (VOF) method. For the turbulent effect, the large-eddy simulation (LES) method is employed to describe its behaviors. By the LES model, the small scale turbulence, which is unable to be resolved by the computational grid, can be thus predicted with reasonable accuracy. The LES equation is solved by the projection method. The model of the wind turbine structure is developed by the partial-cell treatment.

2 Numerical model

Using Splash3D, a three-dimensional numerical model is developed for perfomring the force evaluation of the OWT. The LES model is incorporated into the simulation to describe the turbulent flow. The Slash3D model was modified from the Truchas developed by Los Alamos National Laboratory (LANL)[9, 10]. With the access of the source code, one can modify or add necessary functions. The developed model is capable to solve the three-dimensional, incompressible N-S equations, and track the position of the fluid surface by the VOF model. Parallel computation is available by using the Message Passing Interface with Chameleon portability system (MPICH) to increase the grid resolution and reduce the required computation time. The developed code is able to predict the breaking wave and the turbulence around the solid structure. Governing equations and important numerical models are described in the following sections.

2.1 VOF model

The VOF equation and its distribution during the calculation are discretized by the Finite Volume Method (FVM) as:

$$\frac{\partial f_m}{\partial t} + \nabla \cdot (f_m u) = 0 \tag{1}$$

$$f_m^{n+1} = f_m^n - \frac{1}{V_i} \sum_f \Delta t \left[A \cdot u^n \right]_f f_{m,f}^n \tag{2}$$

The free surface in each grid is re-built during the transient calculation to avoid the numerical diffusion at the interface. The Piecewise Linear Interface Calculation (PLIC) [11, 12] method is employed to calculate the position of the free surface in grids, and the interface between fluids are re-built accordingly. The PLIC method evaluates the fluid interface as a tiltable, horizontal surface. The normal vector of the free surface is calculated by the volume fraction of fluid in the cell.

2.2 LES model

For a simulation with the effect of breaking wave, a proper model to describe the turbulent flow should be considered. The LES method is capable to describe the turbulent flow [13]. Without evaluating the motion of all vortexes, the LES method employed the Spatial Filtered method to classify the physical properties into the grid-scale and the subgrid scale catalog. Effects of geometry and flow conditions on grid-scale properties are significant, while these are minor on the subgrid scale properties. Thus, vortexes of grid-scale are solved by the N-S equations, while those of subgrid scale are evaluated by the Turbulence Closure Model (TCM). If the size of a grid is smaller than that of the involved minimum vortex, results by the LES model would be consistent with that by the DNS method. If not (due to limited computational resources), vortexes with difference scale are described by different scheme. As a result, the trade-off is reached on the precision and practicality by the LES method. In this work, the LES method is employed to describe the complicated turbulent flow as the wave and current go through the support structure of the OWT.

The scale of the vortex is classified by the Spatial Filtered method. The physical property is filtered by the filter function as:

$$\bar{\phi}(x_i,t) = \frac{1}{V} \int_{C.V.} \phi(x_i,t) G(x_i) dV \tag{3}$$

The over-bar denotes that it is spatial filtered, and V is the volume of the computational cell. The filter function $G(x_i)$ is described as:

$$G(x_i) = \begin{cases} \frac{1}{V} & x \in V \\ 0 & otherwise \end{cases} \tag{4}$$

With the spatial filter, the momentum equation (N-S) can be re-written as:

$$\frac{\partial \overline{u_i}}{\partial x_i} = 0 \tag{5}$$

$$\frac{\partial \rho \overline{u_i}}{\partial t} + \frac{\partial \rho \overline{u_i u_j}}{\partial x_i} = -\frac{\partial \overline{P}}{\partial x_i} + \rho g + \frac{\partial}{\partial x_j} \left[\overline{\mu_{eff}} \left(\frac{\partial \overline{u_i}}{\partial x_j} + \frac{\partial \overline{u_j}}{\partial x_i} \right) \right] \tag{6}$$

where the subscripts i, j, and k represent the direction of x, y, and z. The effective viscosity is defined as:

$$\overline{\mu_{eff}} = \mu + \mu_{SGS} \tag{7}$$

The concept of the subgrid scale (SGS) is usually employed with the Smagorinsky model [14], and the viscosity of the SGS is described as:

$$\mu_{SGS} = \rho(C_s\Delta)^2\sqrt{2\overline{S_{ij}}\,\overline{S_{ji}}} \qquad (8)$$

$$\overline{S_{ij}} = \left(\frac{\partial \overline{u_i}}{\partial u_j} + \frac{\partial \overline{u_j}}{\partial u_i}\right) \qquad (9)$$

where S_{ij} is the shear strain rate, C_s is the coefficient of SGS model. Depending on the considered environment, it ranges from 0.1 to 0.2. It is 0.15 for the present study. The term of Δ is the minimum resolvable scale of turbulence, and it depends on the grid size as:

$$\Delta = (\Delta x \Delta y \Delta z)^{1/3} \qquad (10)$$

where Δx, Δy, and Δz, are the length of grid on each direction, respectively.

2.3 Wave model and input data

In the present simulation, the considered wave is generated based on the observed data from the Central Weather Bureau of Taiwan during 2013 and 2014 at considered locations. Samples were taken every 10 minutes, and data were collected every hour. Measurement was made within 0.03 to 0.4 Hz with 41 frequencies and resolution of 0.01 Hz. In order to select the representative event, the case with relatively large significant wave height and single peak spectrum is preferred. Selecting the event with higher significant wave height is based on the engineering consideration. The single crest spectrum is preferred since there is only one crest for the employed Joint North Sea Wave Project (JONSWAP) spectrum. Furthermore, event with two or more crests spectrum is usually not representative and it is observed with special conditions. Based on the two selection criteria, the event observed on July 13 2013 is employed as the input data to generate the required wave for the simulation. After rearrangement and statistical methods, the wave parameters for the 3D simulation are obtained. For the observed data, the significant wave height is 5.78 m. The significant wave period is 9.8 seconds, and the peak wave period is 10.4493 seconds. The angular frequency of the transformed peak wave is 0.601301 rad/s. The average wave period is 7.5 seconds, and the corresponding probability density is about 7 m^2-s/rad.

Based on the dispersion relationship, the modified internal wave maker module is implemented to generate required wave as a site-specific load for the OWT. In addition to the generated wave, the effect of current is also considered as a coupled load. Numerical simulation is conducted with this coupled load for the analyzed referenced 5MW OWT.

3 Results and discussion

3.1 Force analysis for the jacket-type support structure under the regular wave with different flow direction

By the developed numerical model, the effect of the regular wave on the force of the structure is investigated with different flow directions. The investigated inflow direction includes the angle of 0° and 45° as illustrated in Fig. 1. Other parameters for the investigated case are listed in Table 1.

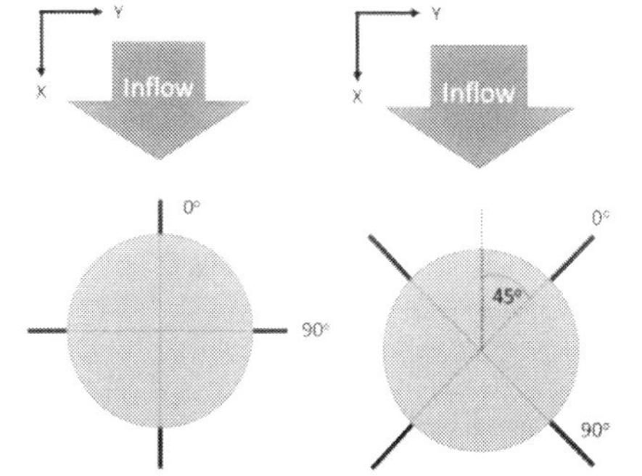

Figure 1. Schematic diagram for the flow direction.

Table 1. Parameters for the investigated case

Bottom Friction Effect	Not considered
Flow Velocity(m/s)	1.41
Finest Grid Resolution (m)	1.0
Residual of Convergence Condition	1.0e-6
Mean Water Depth (m)	25.0
Courant Number	0.85

The computational domain and layout is shown in Fig 2. The wave generating module is installed near the inlet region. Sponge zone is applied downstream to alleviate the numerical reflection from outlet. By a preliminary test without structure, the desired wave is obtained at the position of 350 m. Therefore, the structure is installed at the position of 350 m to investigate the wave-driven force as sown in Fig. 2.

For the computational cell, the grid refinement is applied near the region of the structure, while it is uniform grid for the region away from the structure. The ratio for the grid refinement in x-direction is 1.2, and it is 1.1876 in y-direction. In z-direction, the domain is divided with uniform grid size. The finest grid resolution is 1.0 m applied in the structure region.

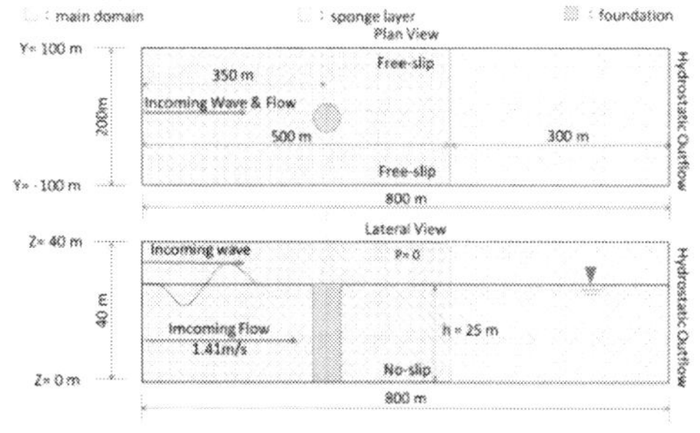

Figure 2. Geometry of the computational domain.

 DOI: 10.1051/matecconf/20167009008

Results are taken at the position of the installed structure (x=350 m). The variation of the elevation and force on the structure for the case of regular wave and 0° flow direction is shown in Fig. 3. Force is positive if its direction is consistent with the coming flow, otherwise it is negative. Periodic variation of the elevation and force can be observed with the synchronized frequency. As the crest of the regular wave reaches the structure, the maximum force on the structure is up to +1.3 MN. Thereafter, the maximum force is -2.0 MN as the trough of the regular wave reaches the structure.

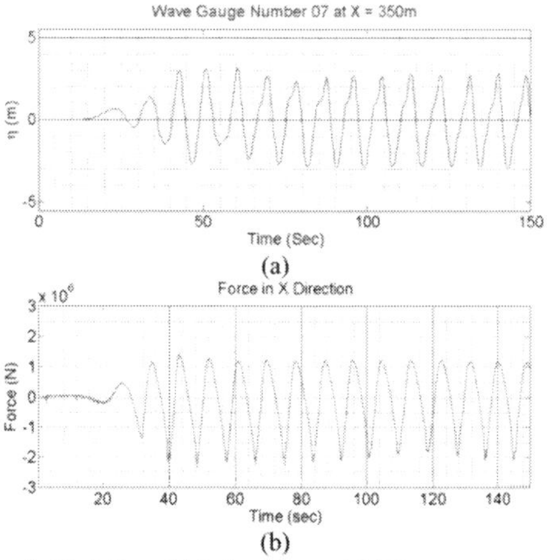

Figure 3. Variation of (a) elevation and (b) force on structure with time under regular wave and flow direction of 0°.

The variation of elevation and force on the structure for the case of regular wave and 45° flow direction is shown in Fig. 4. Periodic variations of the elevation and force are also observed with the synchronized frequency. As the crest of the regular wave reaches the structure, the maximum force on the structure is up to +1.9 MN. Thereafter, the maximum force is -1.7 MN as the trough of the regular wave reaches the structure.

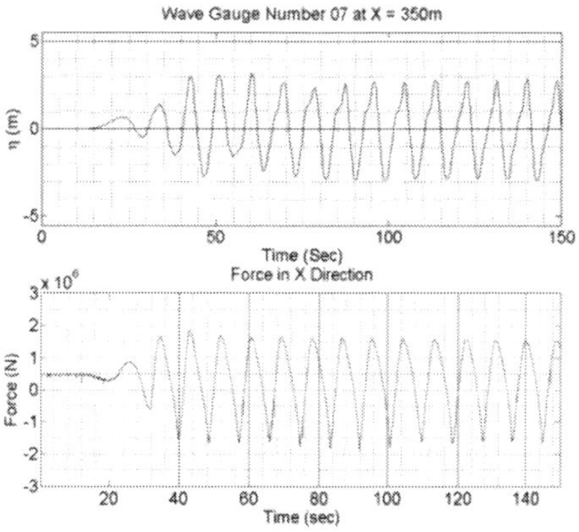

Figure 4. Variation of (a) elevation and (b) force on structure with time under regular wave and flow direction of 45°.

3.2 Force analysis for the jacket-type support structure under the irregular wave with different flow direction

In this section, the elevations of free surface and force on the structure due to the irregular wave and current are investigated. The inflow direction of 0° and 45° are implemented as well for the 5 MW OWT with jacket-type support structure. Similar trends on the variation of the elevation and force are observed for the case using the irregular wave as shown in Fig. 5 and Fig. 6.

In the case of 0° inflow direction, the maximum force is +4.5 MN as the crest of the irregular wave reaches the structure, and it is -1.5 MN at the moment of the trough reaching the structure. In the case of 45° flow direction, the maximum force is +3.3 MN as the crest of the irregular wave reaches the structure, and it is -2.0 MN at the moment of the trough reaching the structure.

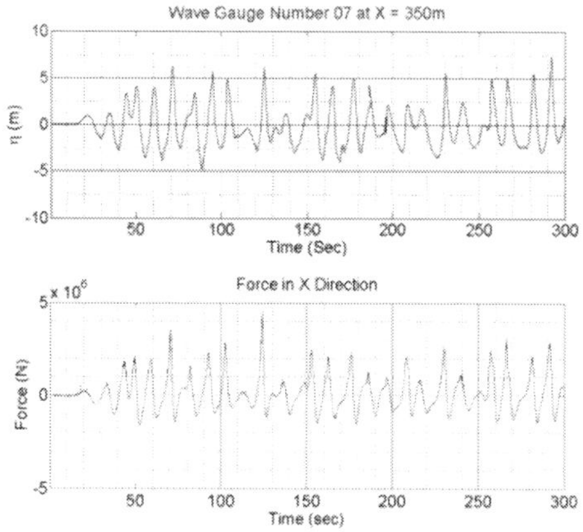

Figure 5. Variation of (a) elevation and (b) force on structure with time under irregular wave and flow direction of 0°.

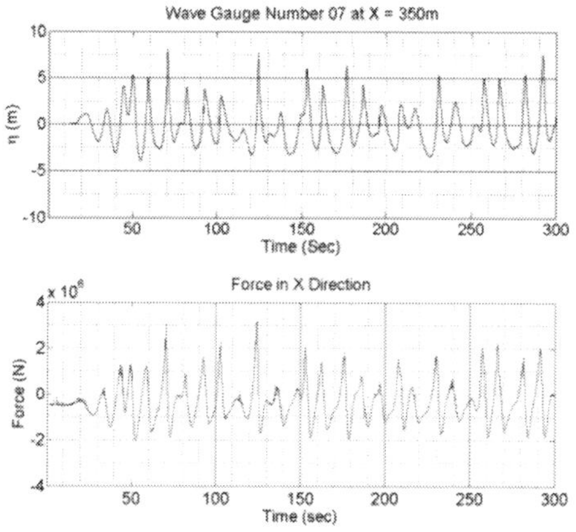

Figure 6. Variation of (a) elevation and (b) force on structure with time under irregular wave and flow direction of 45°.

A comprehensive comparison for the maximum force among the investigated cases is made in Fig. 7. In the case using the regular wave, the trough leads to the larger maximum force. The difference comparing to the crest is also larger (0.7 MN) at 0° of the inflow direction. In the case of 45° of the inflow direction, however, the larger maximum force is observed by the crest. The difference compared to the trough is also smaller (0.2 MN).

For the case of considering the irregular wave, the larger maximum forces are observed by crest for two considered inflow directions. The crest-generated force is significantly larger than the trough. Using the irregular wave also results in the much larger maximum force compared to the regular wave.

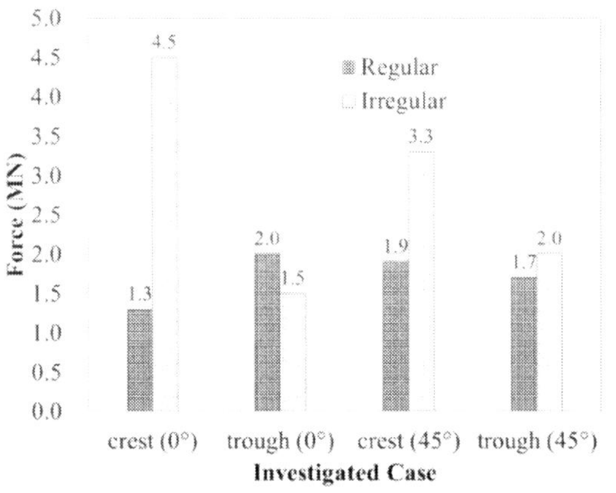

Figure 7. Comparison of the maximum force among investigated cases.

It can be deduced that using the regular wave as the hydrodynamic load for the evaluation of the OWT system will vanish some instantaneous extreme wave. The average force by the regular wave is about 1.7 MN.

In the case of using the irregular wave, the instantaneous extreme wave is also considered. With such more practical hydrodynamic load, the average force for the case of irregular wave is about 2.8 MN, and it is 64% larger than that of the regular wave.

4 Conclusion

In the present study, the hydrodynamic load due to extreme external condition, such as typhoon, on the OWT has been investigated by the numerical simulations. The hydrodynamic loads including the current, regular and irregular wave on the structure are taken into account. The sea current is considered with the constant flow speed, whereas the wave is considered as regular and irregular based on the observed data at considered locations. Using Splash3D, a three-dimensional numerical model is developed for performing the force evaluation of the OWT. Multi-phase flow is evaluated by the VOF model, and the turbulent flow is described by the LES model.

Besides the constant speed current, the regular and irregular waves are considered with different inflow directions (0 and 45 degree). For the computational cell, grid refinement is applied near the region of the structure, while it is uniform grid for the region away from the structure. The finest grid resolution is 1.0 m applied in the structure region.

Comprehensive comparison for the maximum force among the investigated cases is made. The trough leads to the larger maximum force in the case of using the regular wave. The difference comparing to the crest is also larger at 0° of the inflow direction. In the case of 45° of the inflow direction, the larger maximum force is observed by the crest. The difference comparing to the trough is also smaller. For the case considering the irregular wave, the larger maximum forces are observed by crest for two considered inflow directions. The crest-generated force is significantly larger than the trough. Using the irregular wave also results in the much larger maximum force comparing to the regular wave.

It can be deduced that using the regular wave as the hydrodynamic load for the evaluation of the OWT system will vanish some instantaneous extreme wave. In the case of using the irregular wave, the instantaneous extreme wave is also considered. With such more practical hydrodynamic load, the average force for the case using the irregular wave is 64% larger than that of the regular wave.

References

1. Sebastian, T. and M.A. Lackner, Wind Energy, (2012)
2. Wienke, J. and H. Oumeraci, Coastal Engineering, 435-462, **52**, (2005)
3. Yang, D. and L. Shen, Journal of Computational Physics, 5488-5509, **230**, (2011)
4. Calderer, A., S. Kang, and F. Sotiropoulos, Journal of Computational Physics, 201-227, **277**, (2014)
5. Marino, E., C. Borri, and U. Peil, Journal of Wind Engineering and Industrial Aerodynamics, 483-490, **99**, (2011)
6. Marino, E., An Integrated Nonlinear Wind-Waves Model for Offshore Wind Turbines, Ph.D. dissertation, (2010)
7. Lugni, C., M. Brocchini, and O.M. Faltinsen, Physics of Fluids, 122101, **18**, (2006)
8. Marino, E., C. Lugni, and C. Borria, Computer Methods in Applied Mechanics and Engineering, 275-288, **255**, (2013)
9. U, P.L.-F.L.I., T.-R.W. U, F. Raichlen, C.E. Synolakis, and J.C. Borrero, Journal of Fluid Mechanics, 107-144, **536**, (2005)
10. Wu, T.R., A Numerical Study of Three-Dimensional Breaking Waves and Turbulence Effects, PhD dissertation, (2004)
11. Kothe, D.B., R.W. J., S.J. Mosso, J.S. Brock, and J.I. Hochstein, AIAA Meeting Papers, 96-0859, (1996)
12. Rider, W.J. and D.B. Kothe, Journal of Computational Physics, 112-152, **10**, (1998)
13. Deardorff, J.W., Journal of Fluid Mechanics, 453-480, **41**, (1970)
14. Smagorinsky, J., Monthly Weather Review, **91**, (1963)

Study on Modelling Standardization of Double-fed Wind Turbine and Its Application

Xiang Li, Yongning Chi, Haiyan Tang and Linjun Wei

China Electric Power Research Institute, Beijing, China

Abstract. Based on the standardized modelling of the International Modelling Team, study on double-fed induction generator (DFIG) wind turbine is processed in this paper, aiming at capability of universally and reasonably reflecting key performance related to large scale system analysis. The standardized model proposed is of high degree of structural modularity, easy functional extension and universalization of control strategy and signal. Moreover, it is applicable for wind turbines produced by different manufacturers through model parameter adjustment. The complexity of the model can meet both needs of grid-connected characteristic simulation of wind turbine and large scale power system simulation.

1 Introduction

The planning and operation of large interconnected electric power system need analysis and computer simulation every day. Power system planners and operators evaluate the potential impact of credible emergency plans and also the capability of power system in bearing these incidents and maintaining its stability and interconnection, by using simulation study. To ensure reliable and correct evaluation and minimize investment cost, power system model shall be able to reasonably present the actual performance of the equipment in its simulation. In terms of safe and stable operation of power system, an accurate electromechanical transient model of wind turbine is needed for analysing various operation modes of the power grid. In the past, most wind turbine models were developed by wind turbine manufacturers and software suppliers, to be as the special models for users to program on them. Generally speaking, these models, as object files, can only be read by software, while users cannot read the source code.

But, to integrate the user programming model in-to large interconnected power system network mod-el may bring multiple problems. Firstly, when adopting the models developed by individual users and companies, it is difficult to investigate whether the development of these models can accurately reflect the characteristics of wind turbines in various working conditions; secondly, since these user-specific models are usually in the mode of "black box", users cannot check it and moreover, special models highlight differently on time scale and exterior features. Thus, it is difficult for users to understand the features of the models profoundly, which would be disadvantageous for system study and brings more stringent requirements on model maintenance and upgrade; thirdly and most importantly, the models established by wind turbine manufacturers are often based on different software versions. When a large wind farm consists of different types of wind turbines of different manufacturers, study on power system stability would be extremely complicated and difficult.

In order to gain a transient model of wind turbine used for power system stability study, which can accurately and reasonably reflect the characteristics of grid-connected wind turbine in real cases and provide powerful support for power system stability analysis, working groups have been set up internationally for developing a universal model of wind turbine. Based on the standardized modelling of the International Modelling Working Group, this paper processed study on double-fed wind turbine, aiming at "capability of universally and reasonably reflecting key performance related to large scale system analysis". The standardized model proposed in this paper is of high degree of structural modularity, easy functional extension and universalization of control strategy and signal. Moreover, it is applicable for wind turbines produced by different manufacturers through model parameter adjustment. The complexity of the model can meet both needs of simulation about the characteristics of grid-connected wind turbine and large power system simulation.

2 Standardized Modeling of International Working Group

2.1 WECC wind turbine modelling

In June 2005, North American Utility Wind Integration Group (UWIG), Institutes of Electrical and Electronics

Engineers (IEEE), Sandia National Laboratories (SNL) and National Renewable Energy Laboratory (NREL) of the U.S. Department of Energy (DOE) have jointly studied and set up the WECC Wind Turbine Modeling Team. Most technical activities are accomplished collaboratively by Western Electricity Coordinating Council (WECC), WGMG and wind power dynamic characteristics working group of IEEE. This team has completed the study of Phase I in 2010 and issued the study report, WECC WIND GENERATOR DEVELOPMENT. In the report, four kinds of wind turbines were defined based on wind turbine technology, the basic model structure were proposed based on each kind of wind turbine and the model of sub-module of wind turbine was provided. Furthermore, the WECC Team has also put forward relevant proposals on basic parameters of wind turbine model, validation of wind turbine model, modelling and equivalent methods of wind farms in this report. But according to the analysis to the report, the model of wind turbine module proposed by WECC Team is mainly based on the technologies of specific wind turbine manufacturers. Thus, its universality needs to be further improved [1, 2].

2.2 IEC wind turbine modelling

The International Electrotechnical Commission (IEC) established a WG27 Team in October 2009 for wind power simulation model, to solve the issues involved in universal model development of wind turbines. This Team aims at seeking for model specification and validation procedure. Their expectation is to set up a standardized modeling method and model for wind turbines and wind farms (similar to regular generator model).

This Team divides their work into two parts: on one hand, stipulate the topological structure / concept / configuration of universal wind turbine in current market and also the method for modifying the model after introducing the topological structure / concept, aiming at developing and bringing forward a universal model for wind turbine and avoiding the constraint of any special software platforms; on the other, stipulate the topological structure / configuration of wind farms and bring forward the universal dynamic simulation model of wind farms, including wind farm control and auxiliary facilities.

IEC WG27 Team has accomplished IEC 61400-27-1 CD1: "Standard for Electrical Simulation Model and Model Validation of Wind Turbines" by early 2012 and submitted it for TC88's discussion. At present, IEC WG27 Team is focusing on standardized model of wind farms [3].

Some members of IEC Team are also the members of WECC, IEEE Teams and also NERC Special Team. Hence, IEC WG27 Team carries on some study outcomes of WECC Team in the standardized model of wind turbines and on this basis, improves and develops it. The three teams collaborate with each other with clear work division, in order to ensure maximum benefit for the global industry and share the acquired knowledge to the greatest extent, under the joint efforts of WECC and IEEE.

2.3 Colour illustrations

Currently running wind turbines mainly consist of the following four types: constant speed wind turbine, variable slippage wind turbine, double-fed variable speed wind turbine and full-power conversion wind turbine. Each type of wind turbine differs both in terms of topological structure and performance [4-10].

As one of the most predominant wind turbines in current market, type C wind turbine (Fig. 1) has integrated all the advantages of the above types of wind turbines and achieved some improvements in terms of power electronics technologies. The rotor of the wound-rotor induction generators are connected with the power grid through an insulated gate bipolar transistor (IGBT) based power converter by back to back connection mode. The power converter can control both the amplitude and frequency of rotor current. About 40% of the power output is injected into power grid through this converter and the rest are directly through stator side. This design plan can achieve variable speed within ±40% of the rated speed and then maximize wind energy capture. Such converter achieves decoupling control of active power and reactive power, can realize flexible voltage control without the need of installing a reactive power compensation device and also achieve rapid voltage recovery and abnormal voltage cross. Type C wind turbines are also equipped with a pitch control system [11-13].

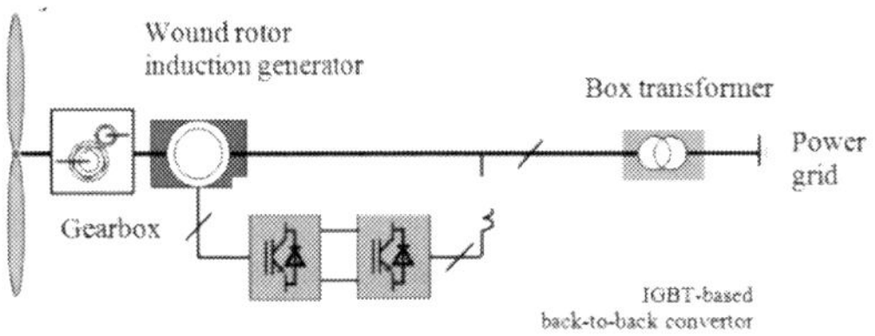

Figure 1. Topological structure of double-fed variable speed wind turbine.

3 Study on standardized modelling for doubly-fed wind turbine model

In the four model types of wind turbines, since the model of doubly-fed wind turbine contains both a motor model and a converter control model, its modeling is the most complicated. Usually, the sub-module of double-fed wind turbine model con-sists of aerodynamic module, maximum power tracking (MPT) module, pitch control, driving chain model, motor model, converter control and wind turbine protective module. The double-fed wind turbine model can be described with the help of the above modules, but these basic modules may neither be sufficient nor be necessary for the wind turbines of specific manufacturers. Besides, the processing strategies of sub-module models differ. Therefore, the standardized model of doubly-fed wind turbine shall be considered with accurate simulation of in-terconnection characteristics, procurability of pa-rameters and interface with power grid model.

As for the universal model studied and estab-lished currently which contains each basic sub-module of the double-fed wind turbine, its mod-el structure is consistent with that of the double-fed wind turbine proposed by IEC Team. The sub-module consists of aerodynamic module, shaft system double-mass model, pitch control, speed control and MTP module, motor model, converter control model and Crowbar protective module. Based on the latest study outcome of WECC and IEC, at this stage, we further study and modify rele-vant sub-module of the universal model of dou-ble-fed wind turbine, mainly including converter control and protection, shaft system and damping module, motor model and interface with power grid and aerodynamic model.

3.1 Converter control and protection

Converter control is a core part of wind turbine control. Normally, some major electrical functions and grid connected characteristics of wind turbines are achieved by converter control. Although converter control technologies of different wind turbine manufacturers differ and are regarded as extremely confidential technical information, as the development of wind turbine control technology and the adaptation to interconnection rule, the converter controls of different wind turbine manufacturers have more similar functional modules or electrical characteristics progressively. Thus, in the study on standardized modeling of wind turbine, converter model of wind turbine is an important part.

IEC WG27 Team will focus from the angle of the system on the converter control logic which influences exterior characteristics of wind turbines and provide a reference model. The converter control model proposed by IEC WG27 Team is given the following control logics, in addition to achieving active power and reactive power control under normal operation conditions:

(1) Voltage and reactive power control mode;

(2) Capability of injection of reactive power during a fault;

(3) Flow capacity limit of the converter during a fault; reactive power prioritized / active power prioritized;

(4) Amplitude limit and gradient limit of active power recovery after a fault resolved;

(5) Selection of torque reference during fault recovery;

(6) Whether reactive power compensating capacity possessed during fault recovery.

In the preliminary study, control models were established respectively for rotor-side converter and network-side converter of wind turbines. Rotor-side inverter is a double closed-loop structure of power outer loop/current inner loop. It achieves control over active power and reactive power of wind turbines in normal operation. The network-side converter is used to control DC link voltage and also the reactive power released by network-side inverter. Furthermore, in converter's control logic, the simulations of injection capacity of reactive current during a fault, recovery characteristics of active power after a fault (amplitude limit and gradient limit) and also flow capacity limit of a converter (reactive power prioritized) are given considerations to.

Based on the study outcomes of IEC WG27 Team and also on survey of the exterior characteristics of mainstreamed wind turbines domestically, our study added the simulation of reactive power compensating capacity during fault recovery and modified the simulation of active power recovery characteristics. The modified active power recovery characteristics can simulate the four types shown in Fig. 2. Thus, the universality of simulating the characteristics of different wind turbines is improved [16-18].

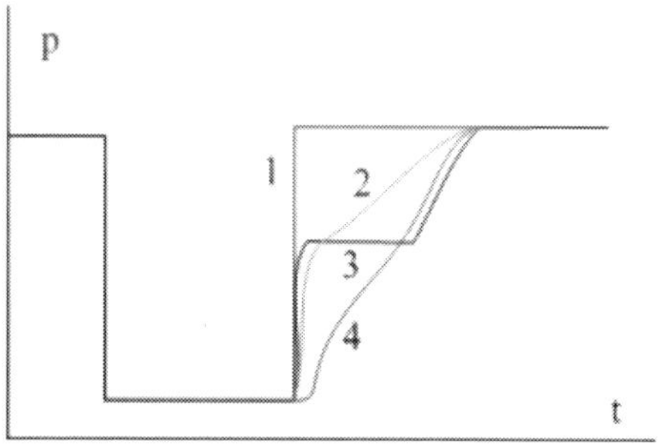

Figure 2. Different recovery characteristics of active power.

Converter control status of a wind turbine can usually be grouped to normal operation control mode, low-voltage ride-through control mode and also uncontrollable mode under protection. In the working condition of normal operation, the control characteristics of active power and reactive power are subject to normal operation control mode; in case of machine-end voltage drop caused by power grid faults, wind turbines will enter into low-voltage ride-through control mode, with its operation characteristics subject mainly to relevant control logic of the converter; when a system fault leads to machine-end voltage drop and thus triggers protective actions, the rotor current of a wind turbine will enter into uncontrollable mode, with its operation characteristics subject to the protective actions.

3.2 Shaft system double-mass model and its damping control

During low-voltage ride-through of a wind turbine, the energy stored in a driving chain will be released to make oscillating component exist in the rotating speed of a generator. To better simulate the moment of force and rotating speed change of a wind turbine during voltage drop in power grid, while considering the order number of wind turbine model to be reduced, double-mass dynamic model is usually employed in electromechanical transient study to describe the driving chain.

WECC Modeling Team hold their view in the simulation study that, when a wind farm has multiple wind turbines, based on the analysis to electrical characteristics observed at grid connection points, the scattered distribution of wind turbines will have smoothing filtering effect on random perturbation of wind speed, but influences little the dominant shaft system oscillating components. When the frequency of the dominant shaft system oscillating components of MW-wind turbines ranges in 2~5Hz, it is necessary to employ a double-mass driving chain model to evaluate its influence to the system.

In addition, to reduce damage to shaft system of wind turbines, some wind turbine manufacturers process rapid control by using power electronic technologies and design a damping link to restrict shaft system fluctuating component. The damping control technologies designed by different wind turbine manufacturers differ, are often regarded as patented technological secrets and will not be publicized. But since the damping control link influences distinctly the electrical characteristics of dominant shaft system oscillating components, WECC Modeling Team and IEC WG27 Modeling Team suggested simulation.

Meanwhile, WECC Team holds the view that, in stability study, wind turbine shaft system shall be employed with a double-mass mathematical model and the damping characteristics of the shaft system shall also be simulated. However, since wind turbine manufacturers regard it as technologically confidential when they employ control to improve shaft system damping, we can adjust system shaft damping coefficient for approximate fitting, instead of simulating active damping control technology.

3.3 Aerodynamic model

IEC WG27 Team proposed a universal modeling method for aerodynamic models. They aimed to avoid asking wind turbine manufacturers to provide their confidential parameter C_p curve and also to achieve procurability of model parameters. Thus, they can simulate the mechanical power characteristics of a wind turbine, like linear model of traditional turbogenerators.

C_p is the function of λ and β. λ can be expressed by the equation below:

$$\lambda = \frac{\omega_{tur} R}{V_w} \tag{1}$$

Thus, when wind speed changes, to track wind speed change ω_{tur} can achieve operation at the optimal working point λ.

For double-fed wind turbines:

$T_{tur} \cdot \omega_{tur} = T_{gen} \cdot \omega_{gen} = P$, and $\omega_{gen} = G \cdot \omega_{tur}$, where G indicates the gear ratio of the gear box.

Therefore, double-fed wind turbines can track and control generator speed $\Delta\omega_{gen}$ by MPT modules to achieve maximum power tracking.

In the stability study of POWER SYSTEM, we often assume wind speed as constant; besides, since the inertia time constant of MW-level wind turbine rotor is often at second-level, generator speed change $\Delta\omega_{gen}$ is usually relatively small in electromechanical transient stability study process. Hence, $\Delta\lambda$ changes little in electromechanical transient stability study process. This means we can model the mechanical power P near the working points by linear method.

3.4 Standardized model of a wind turbine

Based on the work of experts in IEC and WECC Teams, the structure of double-fed wind turbine model should include :

- Generator converter system module includes generator and converter models.
- The control system module shall contain the main control system model of wind turbines.
- Electrical equipment module includes grid-connected switch.
- For wind turbines achieved low-voltage ride-through functions by installing a supporting device, we shall establish a simulation model of the installed device.

An interface of a wind turbine model shall at least include the following variables and parameters:

(1) Input variable:

1) Machine-end voltage of a wind turbine;

2) Operating speed of wind turbine; for types C and D wind turbines, wind speed input can be replaced by active power;

3) Reactive power or power factor order of wind turbine operation;

4) Controller order received by a wind turbine from a wind farm, e.g., active power and reactive power orders.

(2) Output variable:

1) Machine-end current of a wind turbine.

2) Active power of a wind turbine.

3) Reactive power of a wind turbine.

Based on the actual structure of wind turbine models, the output variables may also include: parameters of generators and mechanical driving systems, e.g., rotational inertia, generator speed or wind turbine speed, pitch and control parameter.

(3) Parameters that can be set:

1) Operation mode of wind turbines, e.g., voltage control mode, power factor control mode and reactive power control mode.

2) Model parameters to be adjusted in model simulation, e.g., operation mode parameter and controller parameter.

3) Rated model parameters;

4) Initial model parameters.

4 Model comparisons

We simulated the same system working conditions, set the same system faults and adjusted some parameters of the model; then, compared and analyzed the testing data and simulation results of some actual double-fed wind turbines under the working conditions of small load / full load; validated and compared the simulation curve and actual curve of machine-end voltage, active power and reactive power in the standardized model of a double-fed wind turbine proposed.

Fig. 3 illustrates the fitting result of the actual curve and simulation curve of a wind turbine in case of full load / three-phase short circuit fault / 20% voltage drop. The diagrams show the per-unit values of machine-end voltage, active power and reactive power.

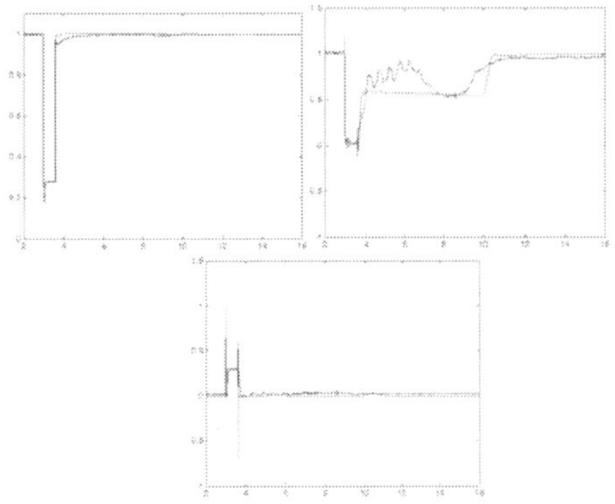

Figure 3. Comparison of actual and simulated curves of wind turbines with full load.

Fig. 4 illustrates the fitting result of the actual curve and simulation curve of a wind turbine in case of small load / three-phase short circuit fault / 20% voltage drop. The diagrams show the per-unit values of machine-end voltage, active power and reactive power.

The fitting results of actual curve and simulation curve show that, the fitting is good under small load, but there is distinct error in the fitting of active power characteristics under full load. Thus, local modules need to be modified based on a standardized model of a wind turbine, on the premise that detailed control strategies of wind turbine manufacturers are obtained.

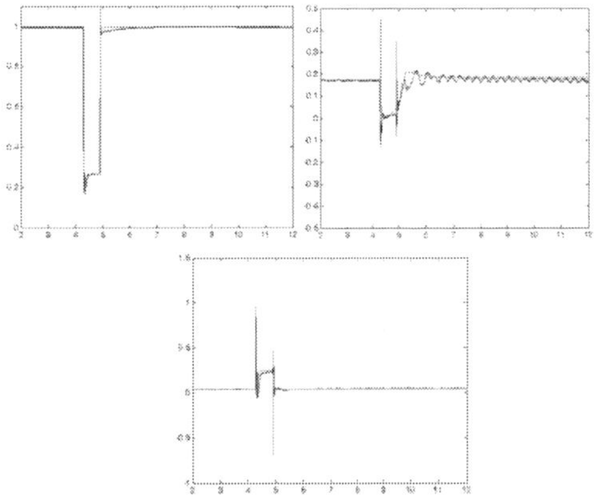

Figure 4. Comparison of actual and simulated curves of double-fed wind turbines with small load.

5 Conclusion

As constant increase of wind power scale in China, wind power has become the third largest power source after thermal power and hydropower. Thus, we need to process model standardization study to meet the simulation demand of great power grid and big perturbation. Moreover, we need to refer to the advanced experience overseas, promote our technology by standard and provide technical support for promoting our wind turbines in the global market, so as to make simulation models more accurately reflect the overall exterior characteristics of a wind turbine in various working conditions.

Acknowledgements

This work is sponsored by National Basic Research Program of China (973 program) under Grant 2012CB215105 and 2015 State Grid Corporation S&T Project named Security Analysis and Research on Basic Technology of Protection under New Grid Development Situation.

References

1. WECC wind generator modeling group. No.NFT-1-11342-01 with NREL, proposed changes to the WECC WT3 generic model for type 3 wind turbine gen-erators model. USA: Electric Power Research Institute, (2013)
2. WECC wind generator modeling group No.NFT-1-11342-01 with NREL, proposed changes to the WECC WT4 generic model for type 4 wind turbine gen-erators model[S]. USA: Electric Power Research Institute, (2013)
3. IEC National Committees. 61400-27-1, Electrical simulation models for wind power generation[S]. International Electrotechnical Commission, (2013)
4. Liu Tao, Ye Xiaohui, Wu Guoyang, et al. An active power control model of wind power generating unit suitable for medium- and long-term dynamic simulation of power grid [J]. Power System Technology, **38(5)**: 1210-1215, (2014)
5. Boukhezzar B, Siguerdidjane H. Nonlinear control of a variable-speed wind turbine using a two-mass model [J]. IEEE Transactions on Energy Conversition, **26(1)**: 149-162, (2011)
6. Santoso S K, Thu L H. Fundamental time-domain wind turbine models for wind power studies [J]. Renewable Energy, **32(14)**: 2436-2452, (2007)
7. Lang Binbin, Mu Gang, Yan Gangui, et al. Research on wind speed vs output power characteristic curve of wind power generator interconnected with power grid [J]. Power System Technology, **32(12)**: 70-74, (2008)
8. Wang Songyan, Li Bijun, Yu Jilai, et al. Analysis on time-varying characteristics of probability error in fore-cast of wind speed and wind power [J]. Power System Technology, **37(4)**: 967-973, (2013)
9. Yang Lihui, Ma Xikui. Impact of doubly fed induction generator wind turbine on power system low-frequency oscillation characteristic, Proceedings of the CSEE, **31(10)**: 19-25, (2011)
10. Zhao Haoran, Wu Qiuwei, Ioanis Margaris, et al. Imple-mentation of IEC generic models of type 1 wind turbine generator in DIgSILENT/PowerFactory, Automation of Electric Power Systems, **33(8)**: 26-33, (2013)
11. Zhang Yanxia, Tong Rui, Zhao Jie, et al. Transient char-acteristics analysis and low voltage ride-through scheme of double-fed wind turbine generators. Automation of Electric Power Systems, **37(6)**: 7-11., (2013)

A Photovoltaic Power Station Equivalent Method Based on Real-Time Digital Simulator

Ping Yang[1,2], Qunru Zheng[1], Sibo Song[1], Ting He[1], Xu Yin[3] and Jinyong Lei[4]

[1]School of Electric Power, South China University of Technology, Guangzhou 510640, China
[2]National-Local Joint Engineering Laboratory for Wind Power Control and Integration Technology, South China University of Technology, Guangzhou 510640, China
[3]Guangdong Intework Energy Technology Co. Ltd, Guangzhou, 511458, China
[4]Electric Power Research Institute, CSG, Guangzhou 510080, China

Abstract. As a clean energy power generation technology, photovoltaic generation has the characteristics of low carbon, green and environmental protection. More and more countries pay close attention to it, more than one megawatt demonstration projects of grid photovoltaic power station has been commenced or built. In order to study interconnection characteristics of large-scale photovoltaic power station better, reduce the effects of photovoltaic power station to the grid, we need to set up photovoltaic power station simulation model. However, the model of photovoltaic power station is complex, the speed of system simulation become slow after establish. In this paper, a real-time digital simulator based on the photovoltaic power station equivalent method will be given, through the proposed hierarchical equivalence method can simplify the process of equivalent of photovoltaic power station, shortening the period of equivalent model, and can be carried out in the process of each layer of the equivalent model accuracy validation, improve the accuracy of the equivalent model.

1 Introduction

Due to the rapid consumption of traditional fossil fuels and worsening environmental pollution, there is growing concern over the development of new energy all over the world. PV generation, as a kind of clean energy generation technology, attract more and more countries' attention, Many governments have successively introduced favorable policies to support PV industry research, and established a number of large-scale PV power stations. With the rapid development of PV generation, large-scale PV generation system will become the future trend.

Since the 1970s, many developed countries have begun to be concerned about PV generation. Germany, Japan and the United States have established a lot of PV power stations, and invested a series of pilot projects. After 2000, the total installed capacity of PV power stations increased rapidly, especially in Japan in 2006 the cumulative PV installed capacity reaching 1760MW to become the world leader in the photovoltaic industry [1]-[3]. PV power station mainly consists of photovoltaic array, controller and inverter. Photovoltaic array is composed by the multi-block photovoltaic modules in series and parallel structure, converting solar energy into electricity. The inverter is a power conversion equipment of PV power station, converting DC to single-phase or three-phase alternating current. Because PV array is made of semiconductor, and controller and inverter are electronic devices, PV power station has no rotating parts, reliable operation, and low failure rate [4]-[6].

With the increase of the capacity and scale of the photovoltaic power station, the influence of the photovoltaic power station on the access point grid is gradually emerging. Because the PVPS is different from conventional energy generation, the access point is a great challenge to the security, stability and reliability of the power access point [7]. At present, most of large-scale grid-connected photovoltaic power station in China is still in the experimental stage and there is no united design specification and access standard, so the security and stability of power system can't be guaranteed.

In order to study the grid connection characteristic of large-scale PV power station, and reduce the impact of PV power station on the grid, the simulation model of photovoltaic power station is needed. A detailed PVPS model will greatly increase the complexity of the simulation, resulting in lowering the speed of system simulation, costing a lot of computing time and reducing the efficiency of the system simulation, so it is important to establish the equivalent model of large-scale PV power station.

The Real-Time Digital Simulator (RTDS) has two advantages: (1) RTDS can simulate in real time. (2) RTDS can be connected with the external device through the interface. Compared to the physical experiment, the real-time digital simulation is more convenient and has better expansibility and compatibility, and it is more close to the actual situation, providing more accurate simulation results [8]-[11]. This paper provides an equivalent modeling method for the photovoltaic power stations, to shorten the equivalent modeling time and simplify the process of the equivalent modeling. This method can be verified in each equivalence procedure, which improves the accuracy of the equivalent model.

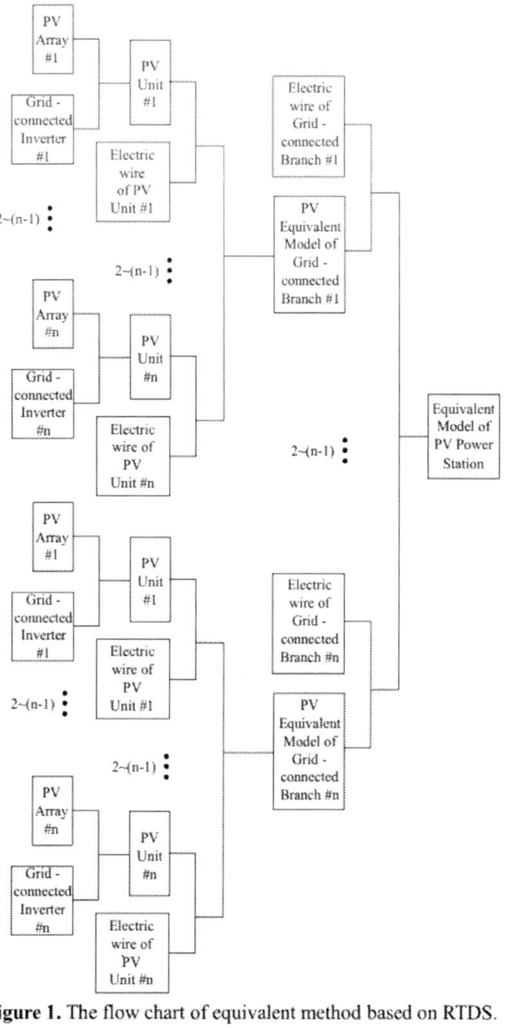

Figure 1. The flow chart of equivalent method based on RTDS.

2 Equivalent modeling of PVPS

2.1 Process of equivalent modeling

As shown in Figure 1, the equivalent method of PV power station based on real-time digital simulator is carried out as follows:

(1) Divide the PV power station into three layers: bottom layer, middle layer and top layer.

(2) Establish the generation model of PV array as bottom layer, including PV array model, grid-connected inverter model, and electric wire model. Obtain the equivalent model of bottom layer and verify its accuracy, including the parameters of above models.

(3) Establish the middle layer model made of PV arrays, including the generation model of PV array, and electric wire model between generation units. Obtain the equivalent model of middle layer and verify its accuracy.

(4) Establish the top layer model made of grid-connected branches, including the equivalent model of all branches and electric wire. To obtain the equivalent model of top layer, complete the equivalent model of entire PV power station and verify its accuracy are needed.

2.2. The parameter of equivalent modeling

Equivalence modeling of the PV power station is to aggregate the whole detail model of PV station as an integration. It is required that the Equivalence Model (EM) and the Detail Model (DM) show the same external characteristic in the utility connection points of the grid. As a result of that, the parameter should be set as follows:

(1) The voltage of EM and DM are equal at utility connection point of grid.

(2) The rated capacity of EM is equal to the sum capacity of each PV power generation units, as formula (1) shows:

$$S_{eq} = \sum_{i=1}^{n} S_i \tag{1}$$

In the formula, n is the number of PV power generators in the PV power generation station; S_{eq} is the rated capacity of EM; S_i is the rated capacity of generator i in the PV power station.

(3) The grid-connected active power of EM is equal to the sum grid-connected active power of each PV power generators.

$$P_{eq} = \sum_{i=1}^{n} P_i \tag{2}$$

In the formula, P_{eq} is the grid-connected active power of EM; P_i is the grid-connected active power of PV power generators i.

(4) The grid-connected reactive power of EM is equal to the sum grid-connected reactive power of each PV power generators.

$$Q_{eq} = \sum_{i=1}^{n} Q_i \tag{3}$$

In the formula, Q_{eq} is the grid-connected reactive power of EM; Q_i is the grid-connected reactive power of PV power generators i.

Generally speaking, the equivalence modeling of PV station is to integrate each models of PV generators in the PV station as a whole in order to make the two model show same external characteristic at the utility connection point of grid. In the EM, for guaranteeing the active and reactive output power of EM and DM are equal, the circuit parameter of EM should be integrated based on each DM. If the PV station is made up of n PV generators, the integrating formula of EM could be as follow:

➢ **The parameter of converter**

The parameter of converter mainly includes DC side capacitance and AC side filter inductance. Their calculation formula is:

$$\begin{cases} C_{dc\_eq} = C_{dc} \times n \\ L_{eq} = \dfrac{L}{n} \end{cases} \quad (4)$$

In the formula, C_{dc_eq} is DC side capacitance of EM; C_{dc} is DC side capacitance of each generators; L_{dc} is AC side filter inductance of EM; L is AC side filter inductance of each generator.

In the simulation, three phase π type equivalent circuit was used as model of electric wire, as it is shown in Figure 2. When stimulating, the parameter of electric wire is set under the premise that its active and reactive power loss are equal after doing equivalent transform.

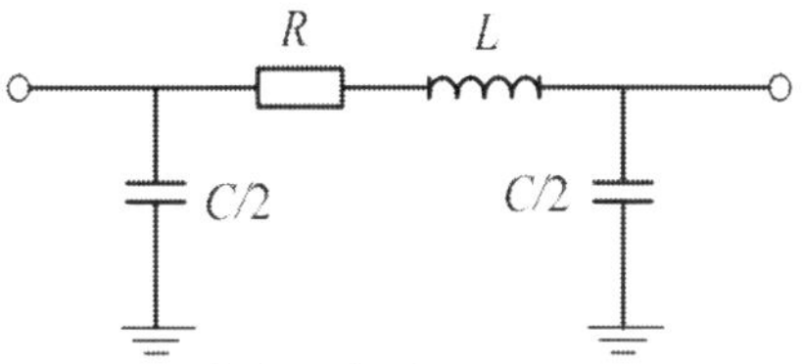

Figure 2. π type equivalent circuit.

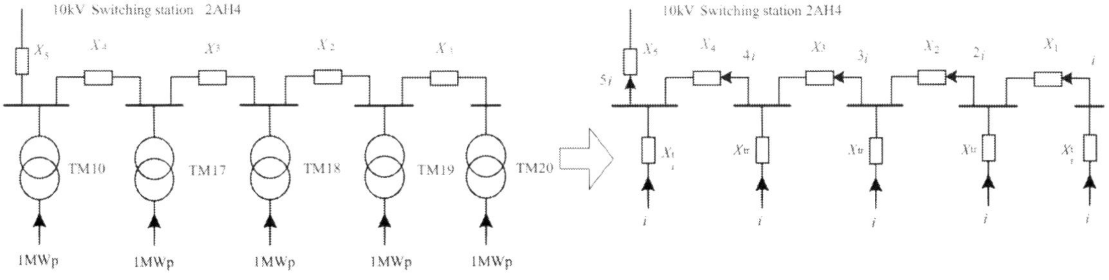

Figure 3. 4#line structure.

➢ **The parameter of transformer and electric wire**

The diagram of 4# electric wire is shown as Figure 3. As the structure shown in Figure 3, the power loss on electric wire and transformer are:

$$\begin{cases} Q_{tr} = 5i^2 X_{tr} \\ Q_{line} = i^2 X_1 + (2i)^2 X_2 + (3i)^2 X_3 \\ \qquad + (4i)^2 X_4 + (5i)^2 X_5 \end{cases} \quad (5)$$

Xtr is the impedance of transformer, X1, X2, X3, X4, X5 is the impedance of each wire. After taking equivalent measure, the line structure of Figure 3 turns into Figure 4, the power loss on electric wire and transformer become:

$$\begin{cases} Q_{tr\_eq} = (5i)^2 X_{tr\_eq} \\ Q_{line\_eq} = (5i)^2 X_{eq} \end{cases} \quad (6)$$

X_{tr_eq} is equivalent impedance of transformer, X_{eq} is equivalent impedance of wire.

According to formula (5) and (6):

$$\begin{cases} X_{tr\_eq} = \dfrac{X_{tr}}{5} \\ X_{eq} = \dfrac{X_1 + 2^2 X_2 + 3^2 X_3 + 4^2 X_4 + 5^2 X_5}{5^2} \end{cases} \quad (7)$$

As a result, when n PV generator connected as the structure shown in Figure 3, the parameter of transformer and electric wire are shown as listed:

➢ **Parameter of transformer**

$$\begin{cases} S_{T\_eq} = S_T \times n \\ X_{tr\_eq} = \dfrac{X_{tr}}{n} \end{cases} \quad (8)$$

S_{T_eq} is the rated capacity of transformer in EM, S_T is the rated capacity of transformer, X_{tr_eq} is the impedance of transformer in EM, X_{tr} is the impedance of transformer.

➢ **Parameter of electric wire**

$$X_{eq} = \frac{X_1 + 2^2 X_2 + 3^2 X_3 + \cdots + n^2 X_n}{n^2} \quad (9)$$

According to the formula of capacitance reactive power: $Q_c = \omega CU^2$, when the voltage remain constant, the reactive power is direct ratio to capacitance. As for electric wire, the voltage on wire is approximately constant, so the capacitance of wire is:

$$C_{eq} = C_1 + C_2 + \cdots + C_n \quad (10)$$

C_{eq} is capacitance of wire, C_1, C_2, C_n is capacitance of each wire.

3 RTDS modeling of photovoltaic power station

The Photovoltaic Power Station (PVPS), whose capacity is 20MW, is composed of 20 photovoltaic array generating units with a capacity of 1MW, each photovoltaic array generating unit consists of 2 subunits of 500kWp rated capacity, and a total of 40 grid-connected inverters with a capacity of 500kW are included. In the design of the PVPS, the photovoltaic arrays is integrated into power grid through 4 electric wires, each electric wire contains 5 photovoltaic array generating units with a capacity of 1MW, which is shown as Figure 4(a).

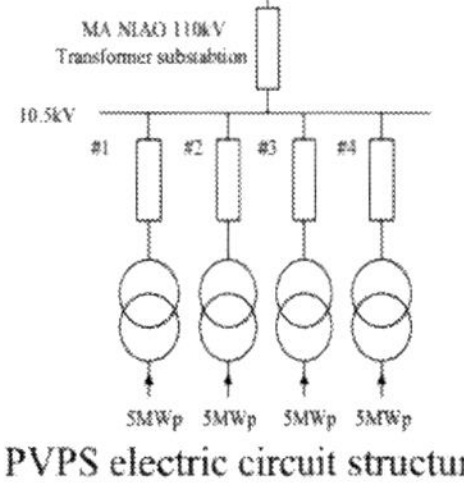

(a) PVPS electric circuit structure

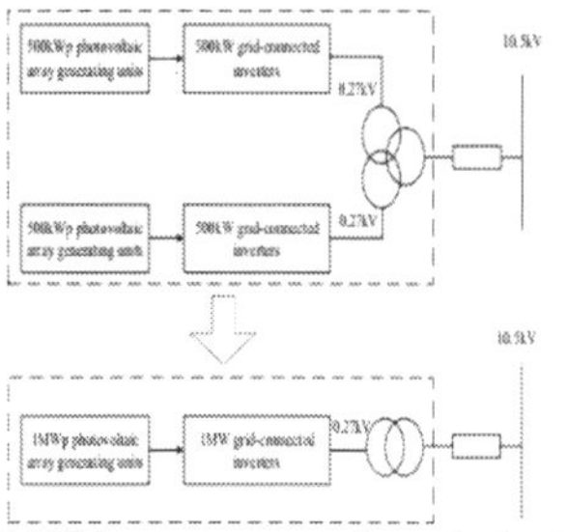

(b) PV array power units equivalent

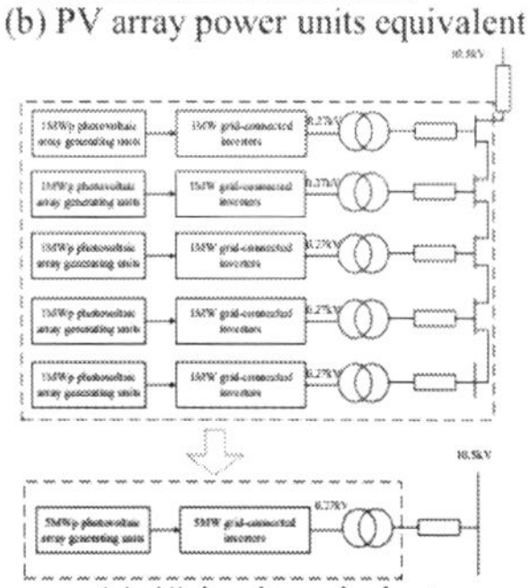

(c) 4#circuit equivalent

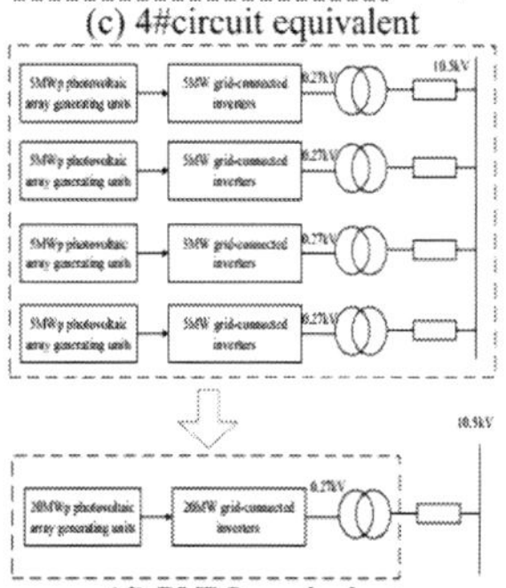

(d) PVPS equivalent

Figure 4. Equivalent model of PV.

The RTDS modeling process can be listed as followed, first of all, two grid-connected inverters with a capacity of 500kW are aggregated into a 1MW grid-connected inverter. Secondly, 5 photovoltaic array generating units of 1MWp in each electric wire are aggregated into a 5MWp photovoltaic array generating unit according to the electrical structure of PVPS. Finally the whole PVPS is aggregated into a 20MWp photovoltaic grid-connected power generation unit. The whole equivalent process are shown as Figure 4(b), (c), and (d).

After the equivalent process above, the system parameters can be aggregated according the parameter aggregation formula. When the circuit parameters is obtained and put them into the equivalent model, the RTDS modeling of PVPS is completed.

4 RTDS simulation of the photovoltaic power station

4.1 Simulation and modeling

According to the equivalent modeling method of the photovoltaic power station, the detailed model and the equivalent model of the photovoltaic power station are built in RTDS. In order to avoid the repetitive narration, only the detailed model and equivalent model shown in Figure 4(d) are discussed. The RTDS model corresponding to Figure 4(d) is as shown in Figure 5.

A detailed model is compared with the equivalent model to verify the correctness of the RTDS equivalent model. The following chapter will verify the accuracy of the equivalent model, based on the consistencies of the responses to external characteristics between the DM and EM when the light intensity changes and the three-phase short circuit happens in the external grid.

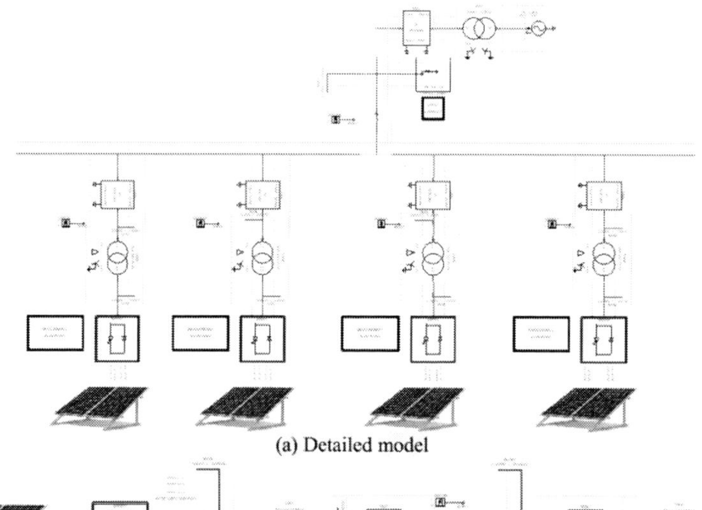

(a) Detailed model

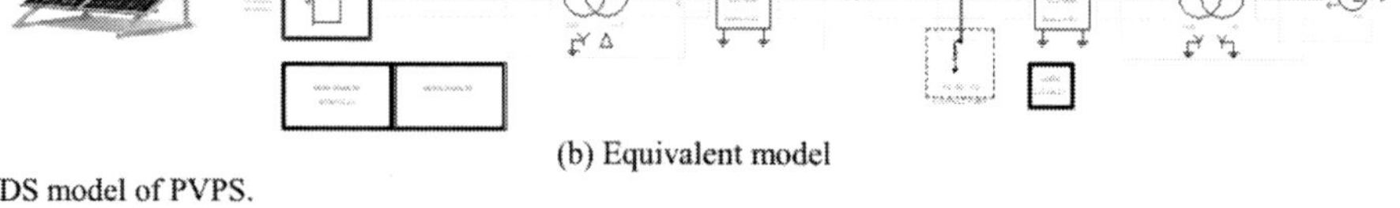

(b) Equivalent model

Figure 5. RTDS model of PVPS.

4.2 Simulation and comparison under varying light intensity

Figure 6 shows the simulation results under the varying light intensity. The curve of the light intensity is shown in the Figure 6(a). The initial illumination intensity is 1000W/m^2, and declines in a short time, and finally mutates to 700 W/m^2. The Figure 6(b) shows the curve of the grid-connected active power. The output active power of the photovoltaic power station changes with the changing illumination intensity, and the trend of active power is the same as that of the illumination intensity. Figure 6(c) shows the curve of the consumed reactive power. The consumption of the reactive power is caused by the reactive power loss of the lines and the transformers. Figure 6(d) shows the curve of the high voltage side current effective value. The voltage per-unit value of PCC is shown in Figure 6(e).

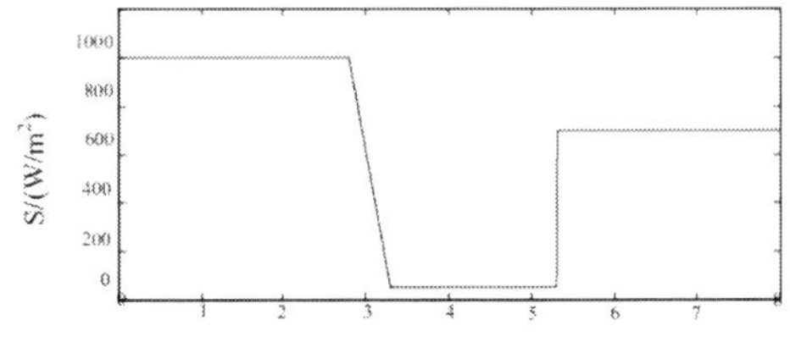

(a) Change of illumination intensity

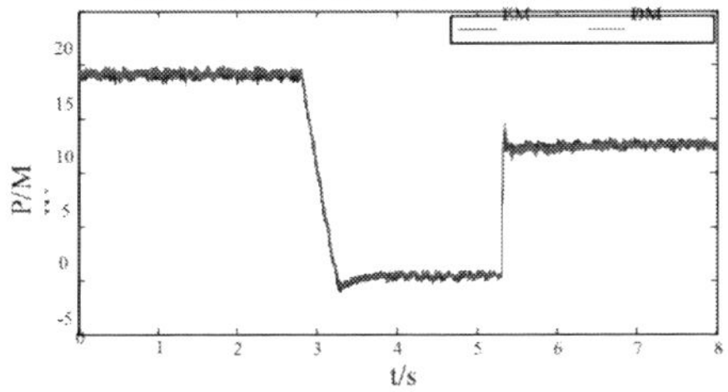

(b) Active power

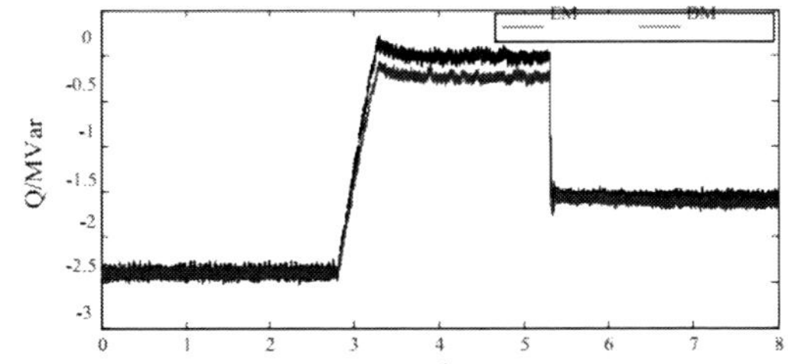

(c) Reactive power

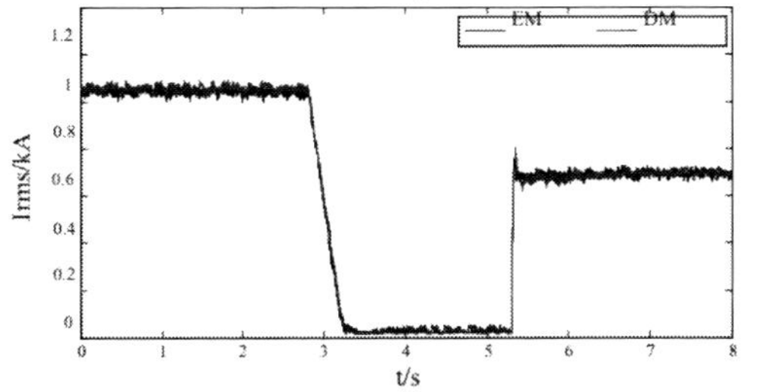

(d) Grid current valid values of PCC point

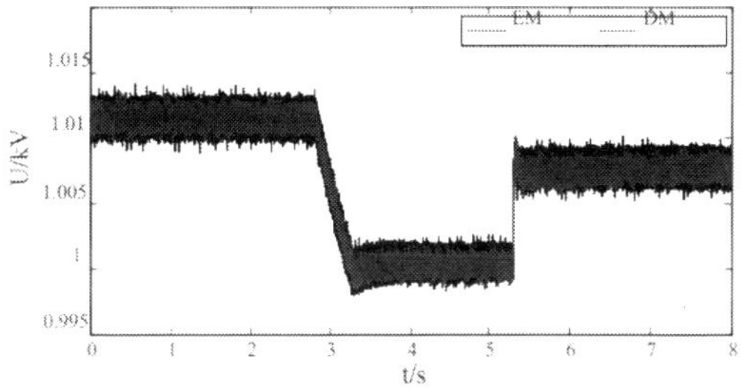

(e) Voltage per-unit value of PCC point

Figure 6. Simulation result of PVPS

In Figure 6, the trends of the voltage, current, active power, and reactive power are the same that of the light intensity. It means that the equivalent model and the detailed model have the similar external characteristics, which verifies the validity of the equivalent modeling of photovoltaic power station.

4.3 Simulation and comparison in three-phase line-to-ground fault

The simulation result of PVPS with the three-phase line-to-ground fault in the radiation of 1000W/m^2 is shown in Figure 7. Firstly, PVPS worked steadily while the three-phase line-to-ground fault happened in 0.1s with 0.1Ω grounding resistance and continued 0.1s. When the three-phase line-to-ground fault happened, the voltage of PCC decreased rapidly and the output power of PVPS decreased. The unbalance between the DC input power and the AC output power of grid-connected inverter, increased the DC voltage quickly. Power grid returned to be normal after 0.1s and the DC input power and the AC output power of grid-connected inverter became still balanced through the short adjustment of PVPS. During the process of short circuit, the output active power of PVPS was shown in Figure 7(a). The PV grid-connected inverter worked in the unity power factor and didn't have the ability of reactive power compensation. As shown in Figure 7(b), the reactive power consumption of system was mainly caused by the transformer and line loss and changed little.

From the Figure 7(C), the grid-connected current about 1.6 times during the three-phase line-to-ground fault for the inherent I-V characteristics of photovoltaic. The change of the PCC voltage during the three-phase line-to-ground fault can be seen in Figure 7(d).

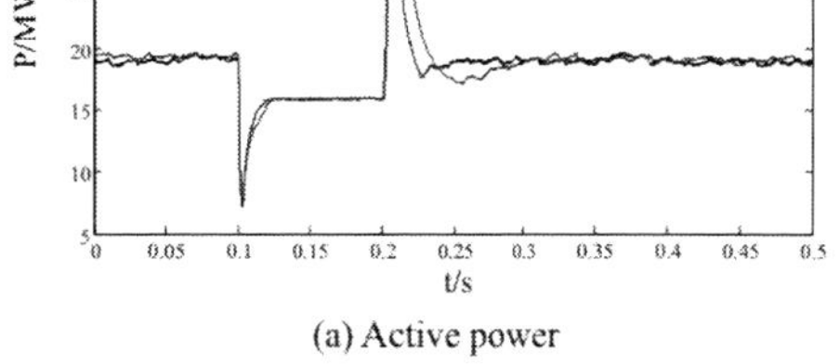

(a) Active power

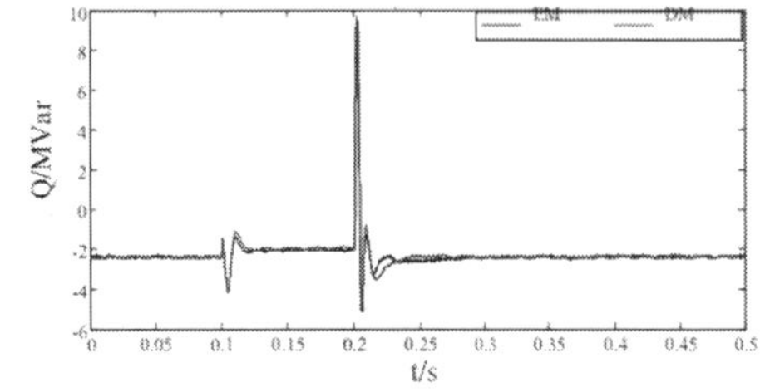

(b) Reactive power

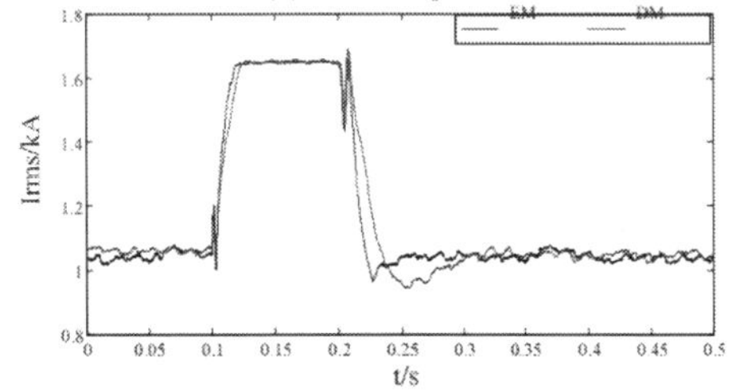

(c) Grid current valid values of PCC point

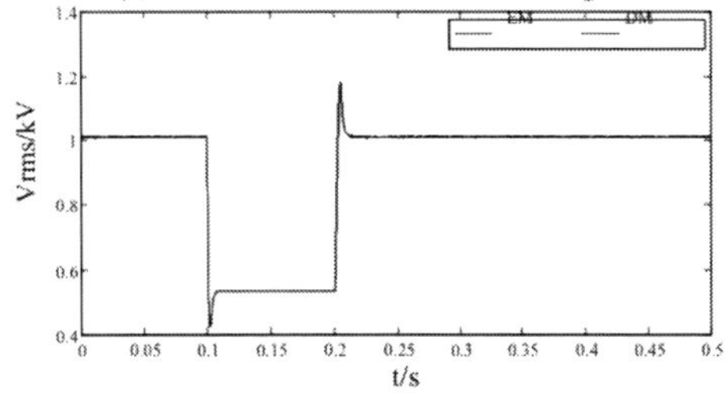

(d) Voltage per-unit value of PCC point

Figure 7. Simulation result of PVPS when three-phase line-to-ground fault

During the three-phase line-to-ground fault, the external characteristics response of the equivalent model and the detailed model are consistent. In other words, the response of the equivalent model in active power, reactive power, current and voltage are consistent with the detailed model, which proved the correctness of the equivalent model.

5 Conclusion

In this paper, the basic concepts of PVPS and RTDS have been presented at first and the mathematical model of PV array and grid-connected inverter have been established. Combined with the actual PVPS, the RTDS model of PV grid-connected inverter, the detailed model and the equivalent model of PVPS have also been established. At last, the correctness of the equivalent model of the PVPS is verified by comparing the detailed model and the equivalent model in different conditions.

Acknowledgment

This work was supported by the national high technology research and development program (863 Program, 2014AA052001); Guangdong Province science and technology plan project (2012B040303005); Electric Power Research Institute, CSG, science and technology project (SEPRI-K143003); China Southern Power Grid, CSG, science and technology project (K-KY2014-009).

References

1. The development of photovoltaic industry and enlightenment in Japan. http://www.docin.com/p-871467.html
2. Y. W. Zhao, D. C. Wu, S. C. Wang, et al., Report on China PV industry development, Solar Energy (in Chinese), (8):6-13, (2008)
3. C. Rodriguez, G. A. J. Amaratunga, Dynamic stability of a grid connected photovoltaic systems, IEEE PES General Meeting, Denver: IEEE, (2004) 2193-2199
4. Q. Q. Zhang, Technology and features study when PVPS connect to grid, Zhejiang: Zhejiang University, (2011)
5. H. F. Feng, D. L. Ma, L. J. Xu, Simulation analysis of single-stage grid-connected photovoltaic power generation system, Computer Simulation, **25**(3):245-250, (2008)
6. G. Sui, H. Gao, Analysis of key problem in grid-connected photovoltaic power generation system, Shandong Electric Power, **03**:21-30, (2009)
7. Y. Lei, Z. G. Zhao, The summarize of large capacity photovoltaic generation key technology, Power Electronics, **03**:16-23, (2010)
8. M. Park, I. K. Yu, A novel real-time simulation technique of photovoltaic generation systems using RTDS, IEEE Trans on Energy Conversion, **19**(1):164-170, (2004)
9. Y. Zhou, L. M. Liu, H. Li, et al., Real Time Digital Simulation (RTDS) of a novel battery- integrated PV system for high penetration application, IEEE International Symposium on Power Electronics for Distributed Generation Systems, **10**(2):786-791, (2010)
10. W. Zhou, P. C. Zhang, X. X. Yang, A real-time closed-loop digital simulation system for computer protections based on RTDS, Power System Protection and Control, **38**(16):127-131, (2010)
11. N. Xie, A. Luo, Y. D. Chen, et al., Dynamic modeling and characteristic analysis on harmonics of photovoltaic power stations, Proceedings of the CSEE (in Chinese), **33**(36):10-17, (2013)

A Hybrid Model for Short-Term Wind Power Forecasting Based on MIV, Tversky Model and GA-BP Neural Network

Jie Zeng[1], Chi Zhang[1], Ning Xie[1], Ping Yang[2,3], Chenyu Xu[2,3] and Zhen Zhang[4]

[1]Electric Power Research Institute, Guangdong Power Grid Corporation, 510640, Guangzhou, China
[2]School of Electric Power, South China University of Technology, 510640, Guangzhou, China
[3]Guangdong Key Laboratory of Clean Energy Technology, South China University of Technology, 511458, Guangzhou, China
[4]Guangdong Intework Energy Technology Co., Ltd, 511458, Guangzhou, China

Abstract. Wind power forecasting, which is necessary for wind farm, is significant to the dispatch of power grid since the characteristics of wind change intermittently. In this paper, a hybrid model for short-term wind power forecasting based on MIV, Tversky model and GA-BP neural network is formulated. The Mean Impact Value (MIV) method is used to monitor the input variable of BP neural network which will simplify the neural network model and reduce the training time. And the Tversky model is used for cluster analysis, which keeps watch over the similar training set of BP neural network. In addition, the Genetic Algorithm (GA) is used to optimize the initial weights and thresholds of BP neural network to achieve the global optimization. Simulation results show that the method combined with MIV, Tversky and GA-BP can improve the accuracy of short-term wind power forecasting.

1 Introduction

The output power of the wind farm fluctuate frequently, bringing great difference to grid-connection [1]. In order to facilitate the power grid dispatching, it is critical to forecast short-term wind power accurately. At present, there are two main methods to forecast short-term wind power, physical method and statistical method. There are two ways to achieve the physical method [2], [3]. One way is to fit the parameters of the wind speed and the direction from the numerical weather forecast and transfer it to wind speed in each wind turbine in order to get the wind power of each wind turbine and then calculate wind power of wind farm [4]. Another way is to establish a high-accuracy three-dimensional atmospheric physical model and calculate the wind speed and wind direction of each wind turbine based on the boundary conditions of the wind farm, which can finally be transferred to wind power of wind farm [5]. The main methods of statistical prediction are time series method [6], neural network [7], Support Vector Machine (SVM) [8], etc. The time series method establishes a time series model based on a large number of historical data and forecasts wind power with the time series model. This method is simple, but it is only suitable for the ultra-short-term wind power forecasting because the forecasting accuracy is lower when ahead time is too high. Neural network method is based on the human brain information mechanism. It has the advantages of fast calculating, good robustness and good learning ability. But there are also disadvantages of neural network. The training time of neural network is long. Moreover, neural network has high demand for data and easily fall into local optimum [9]. The support vector machine method transfers the non-linearly unseparable samples in low dimensional space into linearly separable one in high dimensional space via kernel function, which can greatly avoid local optimum. But this method don't make enough analysis of input variables so the accuracy is not high. In order to improve the accuracy of the forecasting of wind power, it is important to make pretreatment so that the data can better meet the needs of model.

In this paper, we argue a short-term wind power forecasting method based on MIV, Tversky model and GA-BP neural network. First, we calculate the Mean Impact Value (MIV) of each input variable based on BP neural network [10], [11]. Then, the variables will be sorted in order to find the key input items. The Tversky model is used to cluster analysis. Finally, the Genetic Algorithm (GA) optimizes the initial weights and thresholds of Back Propagation (BP) neural network to achieve the global optimization [12]. This method can optimize the input variables to improve the accuracy of short-term wind power forecasting.

2 Optimized feature vector based on MIV

In wind power forecasting, the input variables include three layers of WRF generally. In each layer, there are 7 input variables include wind speed of longitude direction, wind speed of latitude direction, vertical wind speed, air pressure, geopotential height, temperature and humidity. The height of the general wind turbine is close to second layer, so we use the variables in second layer to calculate. The process of calculating Mean Impact Value (MIV) is as follows.

Firstly, we use the training set to get the training model, increasing and reducing each training input variable values by 10% and respectively constitute two new sample set. Secondly, we take the two new sample set as the simulation sample set and use the trained model to calculate the output value of the samples and the subtraction, which is the Impact Value (IV) of influence of the changing argument, and the mean value of the IV of each variable is Mean Impact Value (MIV). Finally, the MIV value of each variable should be arranged to find the relative influence of each variable, from which we can select the input variables of neural network out.

3 Samper cluster analysis based on Tversky model

In this paper, a similar sample clustering method is used to pretreat the data samples. Write forecasting samples and the training samples into vector and then filter out the training data which is similar to forecasting samples by calculating vector similarity. Then use the pretreated data as train sample of neural network and establish model for forecasting.

The method to measure vector similarity can be divided into 2 types, Geometric Model and Feature Contrast Model. In geometric model, each point represents a sample, and the distance between the points similarity represents the similarity of the vector. The distance of each dimension in the geometric model is symmetrical, however, that in wind power forecasting is not. So it's not suitable to use geometric model in short-term wind power forecasting. In feature contrast model, it is common to use Tversky model. Using Tversky model means assuming that the similarity of vector satisfies three characteristics, which is the independence, monotonicity and matching. Based on the assumption, the Tversky model gives the following theorem:

Assuming that s is a similarity function, which satisfies the three characteristics of the assumption. Then we can find a similarity function *Sim*, a non-negative function f and constant θ, α, β. For all objects a, b, c, d, we have:

$$\begin{aligned} &Sim(a,b) \geq Sim(c,d) \Leftrightarrow s(a,b) \geq s(c,d) \quad and \\ &Sim(a,b) = \theta f(A \cap B) - \alpha f(A-B) - \beta f(B-A) \end{aligned} \tag{1}$$

where A and B are sets of vector a and b, α and β describe the inconsistent feature vector.

In this paper, we use the cosine function to compare the similarity between two vectors. Assuming that S,S' are feature vector in the sample set. The method to calculate similarity using cosine function can be expressed as follows:

$$Sim_{cf}(S,S') = \frac{\sum_{i=1}^{n} S_i * S_i'}{\sqrt{\sum_{i=1}^{n} S_i^2} * \sqrt{\sum_{i=1}^{n} S_i'^2}} \tag{2}$$

By comparing the similarity of vector with cosine function, we can pick out the samples which are similar to forecasting set. The process of selecting samples is shown as follows:

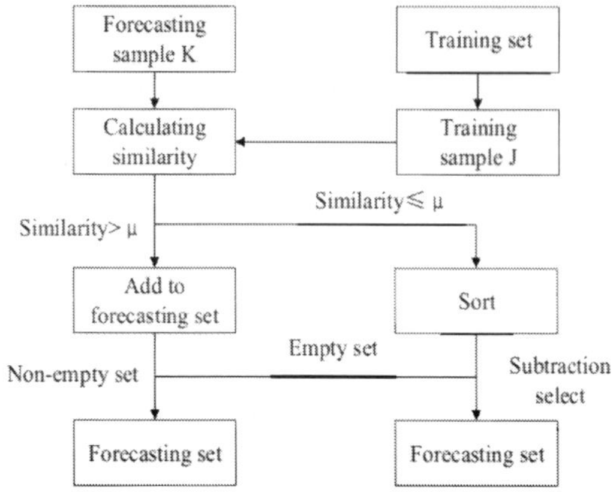

Figure 1. The process of selecting samples.

4 Genetic Algorithm (GA) optimizes initial features of neural network

4.1 Steps of genetic algorithm

In this paper, Genetic Algorithms (GA) is used to optimize initial weights and thresholds of Back Propagation (BP) neural network. Commonly BP neural network has random initial values. With the help of GA, BP neural network can be forecasted better.

Genetic algorithm is an optimization algorithm, which can be realized by simulating the natural genetic law. According to the natural genetic law, GA takes each solution as an individual which is possibly the optimal solution. Similar to the theory of evolution, GA set up a fitness function to distinguish the individual fitting well. After selection, crossover and mutation, we will save the individuals who fit well and eliminate those who fit bad in order to ensure that the new generation fit better than previous generation. Finally we can obtain the optimal solution.

Genetic algorithm includes the following steps.

4.1.1 Initialize population

Each individual should be a possible solution. In this paper, the individual should include the weights and thresholds of the BP neural network. In the case of BP neural network structure determined, each individual can determine the weights and thresholds of the neural network. So we can use BP neural network to judge whether this solution is the optimal solution.

4.1.2 Calculating fitness function

Calculate the fitness of each individual by fitness function. According to each individual we can determine the weights and thresholds of neural network. By using

BP neural network, we can obtain the predicted wind power output. It is obvious that the smaller difference between the actual output and the predicted output, the higher fitness of the individual. Therefore, we can define the individual's fitness function as the error between the predicted output and the actual output, which can be expressed by the following formula.

$$f = \frac{1}{\sum_{i=1}^{n} abs(y_i - o_i)} \tag{3}$$

where n is the nodes number of output layer of neural network, y_i is predicted output of node i and o_i is actual output of node i.

As can be seen from the above, the smaller error of the individual's fitness is, the larger fitness function is. The Individual with great fitness pass information to the next generation ensuring the generation better than the former.

4.1.3 Operation of selection, crossover and mutation

In order to enrich the kinds of operator, we need to use three operations such as selection, crossover and mutation.

The operator selected is based on fitness strategy which contains the methods of roulette and championship commonly. We chose the method of roulette in this paper. The probability of selecting an individual is expressed as follows:

$$p_i = \frac{f(i)}{\sum_{i=1}^{N} f(i)} \tag{4}$$

where N represents the number of operators. The probability of each operator is proportional to the fitness. So the greater fitness of the operator is, the higher probability to be chosen.

Since each operator is showed by encoding, the method of crossover can be used to deal with the cross operation of each operator. To explain the process of cross algorithm, assume that there are two chromosomes called m and n, the crossover operation between the two chromosomes at position K is expressed as follows:

$$\begin{cases} a_{mk} = a_{mk}(1+b) - a_{nk}b \\ a_{nk} = a_{nk}(1-b) + a_{mk}b \end{cases} \tag{5}$$

where b is a random number in the interval [0,1].

The purpose of mutation operation is to increase the diversity of the operator. By mutating the operator of a certain position, the type of operator will increase. Hypothesizing changing the gene a_{ij} of j in the individual I, the operation method is shown as follows:

$$a_{ij} = \begin{cases} a_{ij} + (a_{ij} - a_{\max}) * f(g), r \geq 0.5 \\ a_{ij} - (a_{\min} - a_{ij}) * f(g), r < 0.5 \end{cases} \tag{6}$$

$$f(g) = r_2(1 - \frac{g}{G_{\max}}) \tag{7}$$

where a_{max} and a_{min} represent the maximum boundary and minimum boundary of gene a_{ij}, r_2 is a random number in the interval [0, 1], g is the present number of iteration, G_{max} is the maximum number of evolution.

4.2 The process of GA-BP neutral network

The process of GA-BP neutral network is shown as follows.

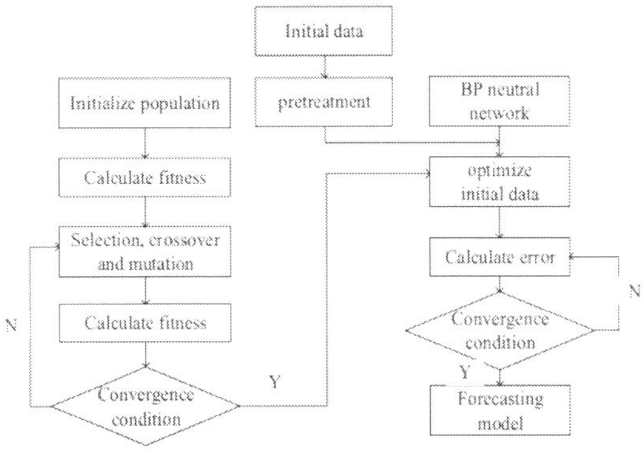

Figure 2. The process of GA-BP neutral network.

From the graph shown above, we can see that the genetic algorithm optimize the weights and thresholds of BP neural network, which is used to establish forecasting model of wind farm. Genetic algorithm can improve the accuracy of forecasting model by optimizing the initial weights and thresholds of BP neutral network.

5 Simulation verification

5.1 Initial data

In this case, the mode of numerical weather forecast data is WRF mode and the actual output data of wind farm is provided by the Power Grid Corp. The selected data is active power of high voltage side in the No. 1 main transformer, the time interval of data is 15 minutes. The observed data vary from 00:00:00, 2013-01-01 to 23:45:00, 2013-07-31. The latitude and longitude of NWP points is shown as follows,

Table 1. Latitude and longitude of NWP points.

longitude	108.42	108.39	108.37
Latitude	18.79	19.02	19.25

The time interval of NWP data is 15 minutes, the distance between two adjacent points is 27km. In this wind farm, there are 24 wind turbines, whose rated capacity are all 1500kW and rated wind speed 12m/s. The height of wind turbines is 70m.

5.2 Results analysis

First, we forecast wind power output without cluster analysis and directly use sample set for training. In this way, we obtain the forecasting wind power output curve. Then we use the method this paper explains to forecast wind power output and obtain the forecasting wind power curve. Finally, comparing with two output curves above and actual output curve we can obtain the result shown as follows.

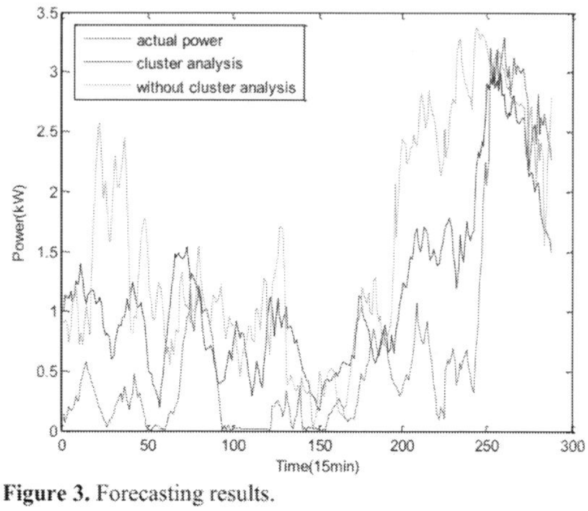

Figure 3. Forecasting results.

In the figure above, the red curve is actual wind power output curve of the wind farm, the blue curve is forecasting wind power output curve with the paper's explaining method and the green one is forecasting wind power output curve without cluster analysis. From the figure above, it is obvious that the forecasting method this paper explains can better forecast wind power output, for it can choosing the sample with higher similarity as the input parameter of BP neural network. Thus the neural network can be better trained and get more accurate results.

The forecasting error curves of forecasting method with cluster analysis and the method without cluster analysis are shown as follows.

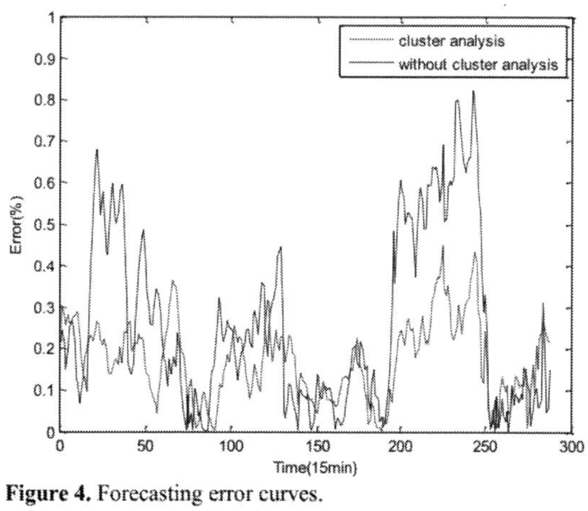

Figure 4. Forecasting error curves.

The forecasting root-mean-square errors (RMSE) of the two methods above are shown as follows.

Table 2. RMSE of the two methods.

Method of this paper	18.6%
Method without cluster analysis	32.4%

The statistical forecasting model is suitable for the wind farm with a large amount of historical data. According to the forecasting result, the forecasting accuracy is obviously improved by Optimized the initial data.

6 Conclusions

In this paper, a BP neural network algorithm with MIV and Tversky model monitoring input variable and Genetic Algorithms optimizing initial the weights is presented. The BP neural network is used to calculate the Mean Impact Value (MIV) of each input variable so that the most useful data can be selected for training according to MIV. Then, for filtering out the training data which is similar to forecasting samples, the way of cluster analysis is put forward. At last, to optimize the initial weights and thresholds of BP neural network, GA-BP neutral network is used to establish forecasting model and obtain the forecasting wind power output value. From the case shown above, it is concluded that the algorithm formulated can greatly enhance the accuracy of short-term wind power forecasting with shorter time.

Acknowledgment

This work was supported by Guangdong Strategic Emerging Industry Core Technology Research Projects (2012A032300013); National Science and Technology Support Program (2015BAA06B02).

References

1. M. A. Islam, et al., Global renewable energy-based electricity generation and smart grid system for energy security, Scientific World Journal, **2014**:197136, (2014)
2. E. Pelikán, et al., Wind power forecasting by an empirical model using NWP outputs, in Environment and Electrical Engineering (EEEIC), 2010 9th International Conference on, IEEE: Prague, Czech Republic, (2010) 45
3. S. Al-Yahyai, A. Gastli, Y. Charabi, Probabilistic wind speed forecast for wind power prediction using pseudo ensemble approach, in Power and Energy (PECon), 2012 IEEE International Conference on, IEEE: Kota Kinabalu, (2012) 127
4. S. Buhan, I. Cadirci, Multi-Stage Wind-Electric Power Forecast by Using a Combination of Advanced Statistical Methods. Industrial Informatics, IEEE Transactions on. (99):1, (2015)
5. N. O. Jensen. A note on wind generator interaction, (1983)
6. J. Taylor, P. McSharry, Short-term load forecasting methods: Anevaluation based on European data, IEEE Trans. Power Syst., **22**(4):2213-2219, (Nov. 2007)
7. H. Quan, D. Srinivasan, A. Khosravi, Short-term load and wind power forecasting using neural network-based prediction intervals. IEEE Trans. Neural. Netw. Learn. Syst., **25**(2):303-15, (2014)
8. J. Wen, et al., Short-term wind power forecasting based on lifting wavelet transform and SVM, in Power Engineering and Automation Conference (PEAM), 2012 IEEE, IEEE: Wuhan, (2012) 1
9. H. Quan, D. Srinivasan, A. Khosravi, Incorporating Wind Power Forecast Uncertainties Into Stochastic

Unit Commitment Using Neural Network-Based Prediction Intervals, IEEE Trans. Neural. Netw. Learn. Syst., (2014)
10. Y. Bashon, D. Neagu, M. J. Ridley, Fuzzy set-theoretical approach for comparing objects with fuzzy attributes, in Intelligent Systems Design and Applications (ISDA), 2011 11th International Conference on, IEEE: Cordoba, (2011) 754
11. A. Tversky, Features of Similarity, Psychological Review, **84**(4):327-352, (1977)
12. Y. Wang, S. Wang, N. Zhang, A novel wind speed forecasting method based on ensemble empirical mode decomposition and GA-BP neural network, in Power and Energy Society General Meeting (PES), 2013 IEEE, IEEE: Vancouver, BC, (2013) 1

Evaluation Index of Equipment Utilization Efficiency in Substation Considering Distributed Generation

Xiaojian Chen[1], Xiang Li[1], Fenglu He[2] and Ziheng Zhang[3]

[1]*China Southern Power Grid Co.Ltd, Shaoguan Power Supply Bureau, No. 66, Gongyexi Road, Shaoguan, China*
[2]*Guangzhou Power Electrical Technology Co. Ltd, System Research Division, Room 440, Building C, Wushan Technology Square, Wushan Road Tianhe, Guangzhou, China*
[3]*South China University of Technology, School of Electric Power, No. 381, Wushan Road Tianhe, Guangzhou, China*

Abstract. With the development of Distributed Generation (DG), traditional evaluation index of equipment utilization efficiency is not objective in some situations. To solve this problem, a comprehensive evaluation method of equipment utilization efficiency in substation is proposed in this paper. Based on the characteristics of distribution network including DG, two indexes, the burden rate and the load rate, are considered. Then, the indexes are further analyzed that they are specified into the injected indexes and the bidirectional indexes. From the operation data of transformers, the specified indexes are calculated representatively and it is indicated that bidirectional indexes could reflect the equipment utilization efficiency in substation more objectively. The presented method provides a new perspective of evaluating the equipment utilization efficiency in substation considering DG.

1 Introduction

In the rural areas with low load density, 35kV and 10kV distribution systems play an important role in power distribution, and they are significant connection between Distributed Generation (DG) and the power Grid in rural areas. Evaluating transformer equipment utilization in 35kV substation scientifically is the base of operation economy and power system planning for power supply enterprise.

At present, power supply reliability, voltage profile and line loss are the most concerning indexes in evaluation of distribution network for power supply enterprise. Most of the equipment utilization efficiency research is carried on by other industries [1]-[3], and research in power grid is still relatively fewer. Indexes like maximum and average burden rate of the main transformer are chosen to reflect the equipment utilization efficiency in some papers. Reference [4] studies on how to improve the equipment utilization efficiency in medium-voltage distribution network from the perspective of optimal burden rate of equipment. Based on the burden rate, the evaluation index system of equipment utilization efficiency in distribution network is established in [5]. Reference [6] proposes that the burden rate and the load rate should be considered simultaneously in evaluation of equipment utilization efficiency in distribution network. So the burden rate and the load rate could be used to analyze the transformer utilization efficiency in substation, but in distribution network including DG, like small hydropower, the traditional evaluation index is controversial. As electricity seller, the power supply enterprises generally focus on the injected load. But in rural distributed power grid, if only evaluate the transformer equipment utilization efficiency by injected burden rate and load rate, the index can't reflect the real contribution of the transformers when DG in operation.

Therefore, it is necessary to specify the indexes of burden rate and load rate to reflect the transformer utilization situation objectively when DG transmits power reversely.

2 Definition of burden rate and load rate

Burden rate is defined as the ratio of practical load to transformer capacity. Since there is evaluation period of equipment utilization efficiency in substation, the index of burden rate is specifically defined as the ratio of maximum active load to transformer capacity. The transformer loading capability is represented by burden rate, which shows the utilization level of equipment. However, burden rate only reflects the best utilized situation at a certain time point, and the utilized situation in a long period is unknown. Therefore, using the burn rate as the sole evaluation index would lead to imperfect and unilateral problem. So the evaluation indexes including load rate are needed.

Load rate is defined as the ratio of average load in a regulation period divided by maximum load, which reflects load variation in the regulation period and transformer utilization level. Under the consideration of economic operation, the transformer utilized situation would be better when load rate is closer to 1, which means the load is smoother.

Therefore, burden rate and load rate are complementary indexes in the evaluation of transformer utilization efficiency, in which burden rate represents the ultimate transformer utilized situation at a time point, and load rate represents the transformer utilized situation in a long period.

3 Analysis of equipment utilization efficiency in substation including DG

Traditionally, it is net load that power supply enterprises focus on. In substation, net load is given by subtracting DG output from overall load, which is injected load from substation and the equation is shown in (1).

$$P_{eq} = \sum_{i \in S_L} P_{Li} - \sum_{i \in S_{DG}} P_{DGi} \quad (1)$$

where P_{eq} is net load in substation, P_{Li} is load in power grid, P_{DG} is the output of DG, S_L is sets of loads in power grid and S_{DG} is sets of DGs.

When DG inversely transmitting power to main network, the uncertainty of net load is increase because of the DG fluctuation, which requires more reserve capacity of transformer and reduces the utilization efficiency.

3.1. Specify the burden rate

With DG in operation, it may result in inverted power transmission and the effect on burden rate is analyzed in Figure 1.

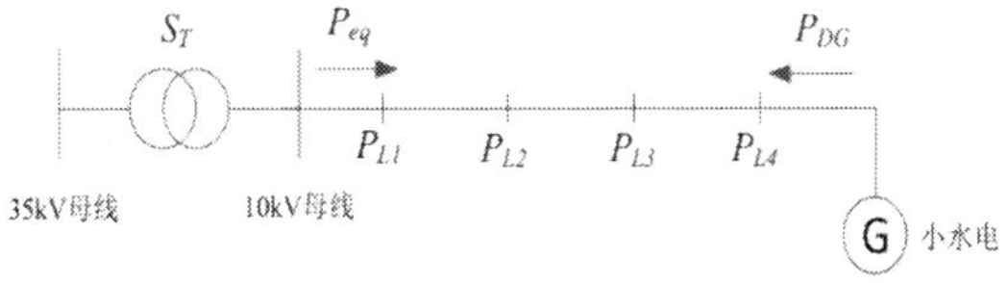

Figure 1. The inverse power flow with DG in operation.

Neglecting the line loss, it is obtained that $P_{eq} = \sum_{i=1}^{4} P_{Li} - P_{DG}$.

Assuming in evaluation period, the burden rate of transformer is referred to the ratio of maximum $|P_{eq}|$ to transformer capacity S_T in traditional way. But when the DG output reaches a relatively large value like small hydropower in wet season, it may occur in $\max(|P_{eq}|) > \max(P_{eq})$. So it is not objective to evaluate the equipment utilization efficiency in substation by traditional burden rate.

However, from the perspective of power supply enterprises, considering the burden rate of transmission power in forward direction is reasonable in some way because only injected power could bring economic profit. So the burden rate is specified into the injected burden rate α_1 representing the economic profit and the bidirectional burden rate α_2 representing the equipment utilization efficiency in substation.

The formula of the injected burden rate is shown in (2).

$$\alpha_1 = \frac{P_{ud-\max}}{S_T} \quad (2)$$

where $P_{ud\text{-}max}$ is the maximum value of injected load and S_T is transformer capacity.

The formula of the bidirectional burden rate is shown in (3).

$$\alpha_2 = \frac{P_{bd-\max}}{S_T} \quad (3)$$

where $P_{bd\text{-}max}$ is the maximum load when considering the bidirectional load.

The specific algorithm of the burden rate index is shown in Table 1.

Table 1. Algorithm of injected and bidirectional burden rate.

Index	Calculation method	Physical meaning
The injected burden rate α_1	Neglect the negative value of load data, and the max positive value is $P_{ud\text{-}max}$	Burden rate index that only considers the economic profit
The bidirectional burden rate α_2	Choose absolute value of the load data, and the max value is $P_{bd\text{-}max}$	Burden rate index that consider utilization efficiency objectively

3.2 Specify the load rate

From the equation $P_{eq} = \sum_{i=1}^{4} P_{Li} - P_{DG}$ drawn by Figure 1, it is pointed out that the fluctuation of P_{eq} would be reduced if the characteristic of operating DG is complementary with the load. However, the DG output is influenced greatly by season and weather. Moreover, the characteristic of DGs is usually inverse with the strategy of peak shaving. So the fluctuation of P_{eq} would be larger with DG in operation. And the load rate would be smaller according to the definition, which means the index of transformer utilization efficiency would be smaller. Therefore, it is necessary to specify the load rate into injected load rate β_1 which represents the economic profit and the bidirectional load rate β_2 which represents the equipment utilization efficiency in substation.

The formula of the injected load rate is shown in (4).

$$\beta_1 = \frac{P_{ud-avg}}{P_{ud-\max}} \quad (4)$$

where $P_{ud\text{-}avg}$ is the average injected load.

The formula of the bidirectional load rate is shown in (5).

$$\beta_2 = \frac{P_{bd-avg}}{P_{bd-\max}} \quad (5)$$

where $P_{bd\text{-}avg}$ is the average bidirectional load.

The specific algorithm of the load rate index is shown in Table 2.

Table 2. Algorithm of injected and bidirectional load rate.

Index	Calculation method	Physical meaning
The injected load rate β_1	Neglecting the negative value of load data, the average positive value is $P_{ud\text{-}avg}$, and the max value is $P_{ud\text{-}max}$	Load rate index that only considers the economic profit
The bidirectional load rate β_2	Choose absolute value of the load data, and the average value is $P_{bd\text{-}avg}$ while the max value is $P_{bd\text{-}max}$	Load rate index that consider utilization efficiency objectively

From Figure 2 and Table 3, in interval of 0~30%, the proportion calculating by bidirectional burden rate is reduced to 5%, which is smaller than the proportion calculating by injected burden rate 14%. In interval over 70%, the proportion calculating by bidirectional burden rate is increased to 59%, which is larger than the proportion calculating by injected burden rate 51%. And in other intervals the outcome is the same between them. The results indicate that the injected burden rate which only considers the economic profit would cover up the objective utilization efficiency of transformer in some extent.

4 Application analysis

Simulation cases are chosen from Shaoguan, a city located in south of China. Shaoguan is a typical city including distributed generation in China. In 2012, the average burden rate of Shaoguan is 36% while it is 35% in China, and the average load rate of Shaoguan is 79% while it is 80% nationwide. The management of power grid operation and electric equipment in Shaoguan is in average level of China. Therefore, the equipment utilization efficiency in Shaoguan is representative.

4.1 Comparison of burden rate indexes

According to the operation data of 37 transformers with 35kV class in Shaoguan in 2013, the calculating results of injected and bidirectional burden rate are given by means of Table 1.

The comparison result is shown in Figure 2 and Table 3.

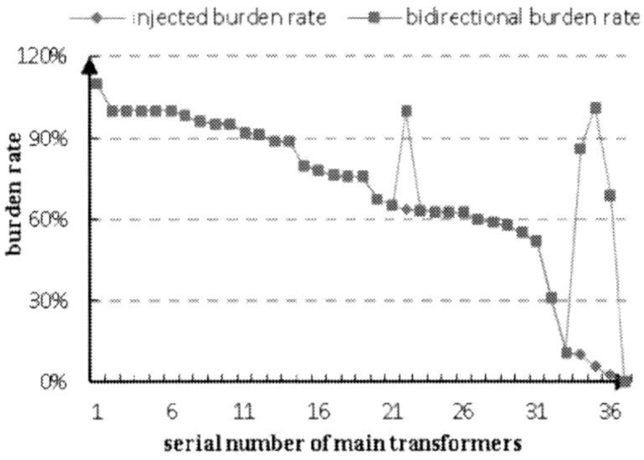

Figure 2. Comparison between injected burden rate α_1 and bidirectional burden rate α_2.

Table 3. Comparison between the proportion of two burden rates when in different interval.

Interval of burden rate	0~30%	30%~70%	>70%
Proportion when calculating by injected burden rate	14%	35%	51%
Proportion when calculating by bidirectional burden rate	5%	35%	59%

4.2 Comparison of load rate indexes

According to the operation data of 25 transformers with 35kV class in Shaoguan in 2013, the calculating result of injected and bidirectional load rate is given by means of Table 2.

The comparison result is shown in Figure 3 and Table 4.

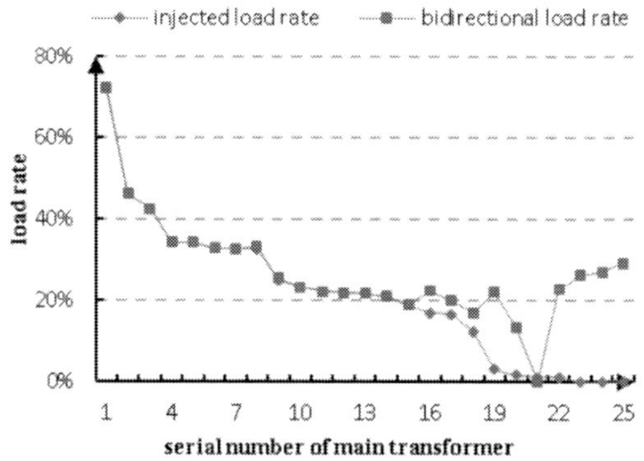

Figure 3. Comparison between injected load rate β_1 and bidirectional burden rate β_2.

Table 4. Comparison between the proportion of two load rates when in different interval.

Interval of load rate	0~15%	15%~30%	>30%
Proportion when calculating by injected load rate	32%	36%	32%
Proportion when calculating by bidirectional load rate	8%	60%	32%

From Figure 3 and Table 4, in interval of 0~15%, the proportion calculating by bidirectional load rate is reduced to 8%, which is smaller than the proportion calculating by injected load rate 32%. In interval of 15%~30%, the proportion calculating by bidirectional load rate is increased to 60%, which is larger than the proportion calculating by injected burden rate 36%. And in other intervals, the outcome is the same between them. The results indicate that the injected load rate which only considers the economic profit would cover up the objective utilization efficiency of transformer in some extent.

5 Conclusion

The equipment utilization efficiency is a significant index in economic evaluation of distribution network construction. However, the importance of it didn't take seriously before and the power supply enterprises only consider the contribution of transformer from the perspective of injected load, which is not objective to the substation including DGs.

This paper concentrates on the transformer utilization efficiency in substation. The burden rate and load rate are selected as the evaluation indexes, and specify them under the consideration of the particularity of distribution network including DGs. Then, compare the traditional injected burden rate α_1 and the bidirectional burden rate α_2, as well as the traditional injected load rate β_1 and the bidirectional load rate β_2. Compared with traditional evaluation index α_1 and β_1, the application results show that the equipment utilization efficiency in substation is reflected more objectively by bidirectional calculation. Based on the proposed method, the bidirectional calculation and specific indexes should be considered in evaluation of equipment utilization efficiency in distributed network including DGs, which represent the contribution of equipment when the power transmission of DG is inverse.

References

1. Z. Q. Jiang, Z. Q. Guo, Z. Wei, Research on equipment utilization in computer composition and maintenance experiment curriculum, *ICFCC*, **3**, 561-564 (2010)
2. M. Ogawa, H. Endo, H. Fukuda, H. Kodama, T. Sugimoto, H. Soneda, M. Kondo, Cooling control restraining effects due to ICT equipment utilization of disturbance based on model predictive control for modular data center, *CCA*, 183-190 (2014)
3. E. Merrill, J. Bostwick, C. Gilhoi, K. Mikkelsen, Yield and equipment utilization improvements achieved through fab conversion to carbon fiber/PEEK wafer carriers and carbon fiber/polypropylene storage boxes, *SEMI*, 41-45 (2000)
4. S. Ge, L. Cao, H. Liu, Study on the best utilization rate of equipment in medium-voltage distribution network, *ISGT Asia*, 1-4 (2012)
5. W. Yang, H. Liu, Z. Wang, W. Cui, L. Cao, Study on evaluation index system of equipment utilization on distribution network, *ISGT Asia*, 1-5 (2012)
6. Z. Hu, Y. Zhang, C. Li, Utilization efficiency of electrical equipment within life cycle assessment: Indexes, analysis and a case, Energy, **88**:885-896, (2015)

Research on Harmonic Characteristics Influence for Distribution Network with Renewable Energy

Wei Huang, Yuyao Yang and Xiangmin Huang

School of Electric Power, South China University of Technology, No.381, Wushan Road, Tianhe, Guangzhou, China

Abstract. In order to study the influence on harmonic characteristics of distribution network with renewable energy, a simplified model of distribution network is established to analyse theory while considering the influence of cable on the system capacitive current. Establishing an actual distribution network model by Digsilent to research the harmonic characteristics impact when photovoltaic power with PV incorporated the distribution network in a variety of different access , while using constant current source model as harmonic source model which often been used in engineering practices. The simulation results show that: optimizing the access location and dispersion of PV can decrease the waveform distortion levels in distribution network and a certain number of high-order harmonic will magnify by a specific grid structure, deteriorating the distribution network power quality.

1 Introduction

Fossil fuels, mainly including oil, coal and natural gas-based, is gradually coming to exhaustion, and in response to the energy crisis, governments began to put renewable energies, such as wind power and solar power onto the strategic agenda [1]-[3]. The proportion of non-fossil energy sources in total energy will increase to about 20%, according to the <Energy Development Strategy Action Plan (2014-2020)> proposed by Chinese government.

Photovoltaic (referred to as PV) roof and small household wind power, which are both classified as renewable energy, will change the power structure of the distribution network, resulting in complex effects on the network [4]. Most of the renewable energies have to be transformed into industrial frequency current using power electronic devices before grid-connected, causing inevitable harmonics injection [5], which will not only increase extra losses and accelerate the aging of transformers, capacitors, cables and other power equipment, but also brings higher production error to enterprise customers who have significant waveform quality sensitivity, leading to unnecessary losses.

Distributed PV is one of the most potential renewable energy technologies, but which produces significant harmonics, so a study on how renewable energy would influence the harmonic characteristics of distribution network , where the three-phase PV inverters [6] act as study object, has been made in this paper. However, the complexity of distribution network [7], as well as diversity of how and where the PV generation will be connected to the public grid, brings the harmonic analysis of distribution network great challenges. Models of power electronic devices and distribution network have been built in [8], where harmonic characteristics of distribution network integrated with DG are analysed in detail, but only the effects of high-order harmonics under ideal conditions are taken into major consideration. The linear and nonlinear models of DP in ideal distribution network system have been respectively studied in [9], but it distribution network' take the impact of cable earth capacitance in actual distribution network into account.

In the actual generated by PV inverter harmonic component is very complex, harmonic frequency covers from 2th to 25th and even higher. And there are a lot of branch in practical distribution network, capacitive parameters of branch will make some harmonic amplification, different grids will form different amplification characteristics. In addition to that, two or more PV system may be accessed in the same distribution network, making voltage distribution become more complicated. The actual parameters must be selected so that the simulation result will be close to actual situation.

In this paper an actual model of inverter is chosen to simulate the PV systems, and a model of PV harmonic source whose fundamental information came from factory test data has been built. Effects of grid-connection of renewable energy generation on harmonic characteristics in distribution network, was exhaustively studied by changing the factors including how and where the PV is grid-connected, PV capacity, etc. A simulation of actual distribution network has been made using Digsilent Power Factory, indicating that, the level of waveform distortion in distribution network would be lowered by optimizing location and manner of dispersion of PV grid-connection, and some particular grid structures could amplify one or several degrees' harmonics, which causes severe harmonic distortion that the system can't bear damaging the system.

2 Harmonic analysis

PV inverters, which would generate a mass of harmonic currents, must be installed between the solar power plant and public grid, causing voltage distortion on buses. The amount of distortion is related to location of PV grid-connection and PV capacity, and in order to simplify the analysis, a simple distribution network with single radiation feeder has been designed, as shown in Figure 1, where:

- M, L: Distance from equivalent bus to PV and from load to equivalent bus, in km
- δ_G, V_g: Voltage angle and amplitude in V at PV connection node
- P_G, Q_G: Active power and reactive power exported by PV, in kW and kvar
- P_D, Q_D: Active power and reactive power consumed by load, in W and var
- φ_G, φ_D: Power factor angle of PV and load
- V_D: Voltage amplitude of load node, in kV

And the unit impedance of the transmission line is set to $r + jx$, in Ω/km, while b in S/km for susceptance.

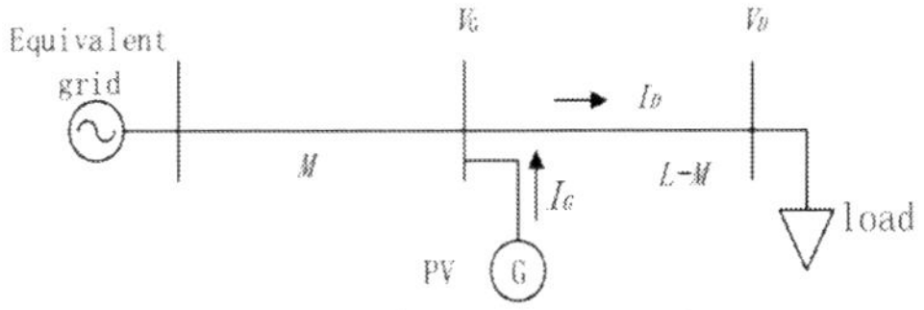

Figure 1. PV connected to distribution network.

3 Influence of PV access on harmonics

Amplitude of fundamental current IG flowing out from PV is:

$$I_G = \left| P_G / (V_G COS\varphi_G) \right| \tag{1}$$

Similar to Hth-degree harmonic of load current, that of PV injection current is expressed by ratio of its amplitude to that of fundamental current, referred to as I_G^h, then amplitude of Hth harmonic injection current is:

$$I_{Gh} = I_G I_G^h = \left| P_G / (V_G COS\varphi_G) \right| I_G^h \tag{2}$$

And its angle is:

$$\theta_{Gh} = \theta_G^h + h(\delta_G - \varphi_G) \tag{3}$$

Assuming that θ_G^h, angle of Hth harmonic current of PV, is 0, and the power factor of PV is close to 1, the φ_G has a very small value near 0. And because a line in distribution network is short, with a small value of r/x, resulting in δ_G close to 0, then the θ_{Gh} is 0. Similarly, angle of Hth harmonic injection current of load is also 0. Amplitude of Hth harmonic voltage at terminal node can be represented as:

$$V_{Dh} = \sqrt{\frac{(hx)^2 + r^2}{(2hbrM)^2 + \left(2h^2bxM - \frac{1}{M}\right)^2}} I_{Gh} + \sqrt{\frac{(hx)^2 + r^2}{(2hbrL)^2 + \left(2h^2bxL - \frac{1}{L}\right)^2}} I_{Dh} \tag{4}$$

The system equivalent model after PV grid-connected is shown in Figure 2.

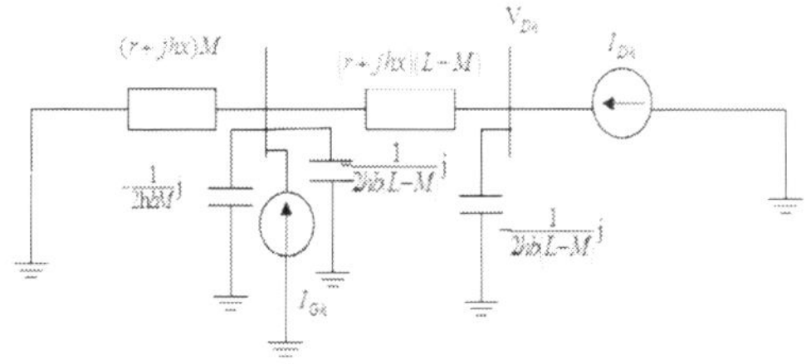

Figure 2. PV connected to distribution network model.

Change of the amplitude of Hth harmonic voltage at terminal node after PV grid-connected:

$$\Delta V_{Dh} = \sqrt{\frac{(hx)^2 + r^2}{(2hbrM)^2 + \left(2h^2bxM - \frac{1}{M}\right)^2}} \left| \frac{P_G}{V_G} \right| I_G^h \tag{5}$$

As it can be found out from Equation (5), after PV grid-connected, voltage distortion at nodes along with lines is related to connected capacity P_G, access location M, and the distortion of harmonic source itself, which depends on manufacturing level of electronic devices. Besides, the higher the degree of distortion of harmonic source itself or the connected capacity raising the increment of harmonic voltage amplitude, the severer the system voltage distortion. For the impact of the access location, it should be discussed in two cases below:

1) Lines behind connection point

Assume Z, a position ahead of the connection point, where Z<M, then amplitude of Hth harmonic voltage at that point could be expressed as:

$$V_{Zh} = \sqrt{\frac{(hx)^2 + r^2}{(2hbrZ)^2 + \left(2h^2bxZ - \frac{1}{Z}\right)^2}} (I_{Gh} + I_{Dh}) \tag{6}$$

It is called constant power model if PV is generally not involved in voltage regulation, and it's indicated by voltage quality analysis that, a greater access location causes a greater V_G, then a smaller I_{Gh}. Similarly, voltage fall decreases and then terminal voltage goes up with the access location M increasing. The I_{Dh} would decrease with the terminal voltage rising, so that the V_{Zh} in Equation (6) would be reduced, so is the voltage distortion.

2) Lines behind connection point

Assume Z, a position behind the connection point, where Z>G, then amplitude of Hth harmonic voltage at that point can be expressed as:

$$V_{Zh}=\sqrt{\frac{(hx)^2+r^2}{(2hbrM)^2+\left(2h^2bxM-\frac{l}{M}\right)^2}}I_{Gh}+\sqrt{\frac{(hx)^2+r^2}{(2hbrZ)^2+\left(2h^2bxZ-\frac{l}{Z}\right)^2}}I_{Dh} \tag{7}$$

As indicated in Equation (7), although b value of a cable line is relatively great, but still at a small magnitude. If h is small, the greater M, the smaller the overall value of $(2hbrM)^2+(2h^2bxM-1/M)^2$; When h has a greater value, and the harmonic current injection I_{Gh} is small, the total voltage distortion would be less affected by corresponding degrees' harmonics voltage increment. Therefore, voltage distortion on the lines behind the connection point goes greater as the connection position M does.

4 Analysis on influence of system response on harmonics

To improve the voltage quality at terminals and power supply reliability, power supply enterprises gradually enhance the reactive compensation devices in 10kV distribution networks in the process of renovation of urban and rural power grids. Inductive reactance of a long branch and capacitive reactance of a compensation capacitor exactly form a series circuit, besides, to facilitate the analysis of the corresponding harmonic characteristics, the resistance is ignored here, as shown in Figure 3.

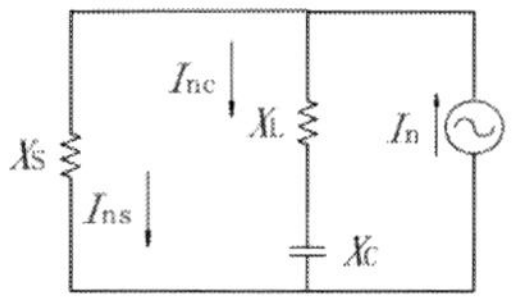

Figure 3. Equivalent network model.

I_n, I_{nc}, I_{ns}	Total Nth-degree harmonic current, and that flows into branch side and system side
X_C, X_L, X_S	Capacitive reactance of compensator, inductive reactance of branch and system

And harmonic impedance of an inductor is proportional to frequency, while that of a capacitor is inversely proportional to frequency. Thus parallel resonance occurs as harmonic degree increases to n_x, and n_x and around n_x degrees' harmonic currents will be amplified by the resonant circulation, corresponding harmonic voltage distortion will be amplified too. In addition, a greater branch capacitance results in a smaller n_x.

5 Simulation and analysis

5.1 Example introduction

The distribution network model used in this paper comes from a chain feeder along the direction of the switching station in 10kV distribution network in a certain city in Guangdong Province, which is a lone line to stretch from the switching station to the load nearby. Its grid structure and specified parameters are shown in Figure 4 and Table 1.

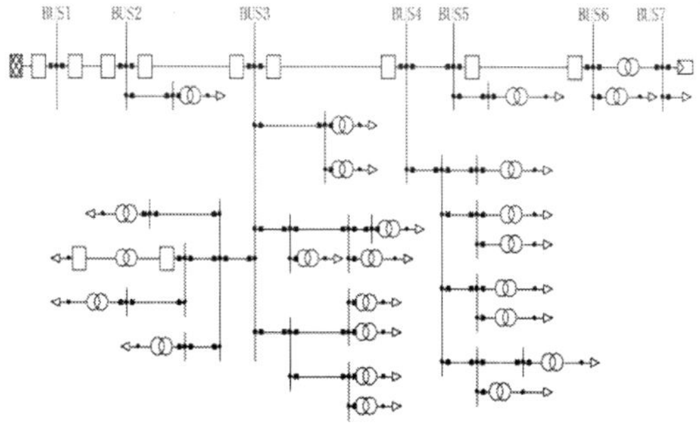

Figure 4. Structure of China feeder.

Bus 1 is the 10kV bus of substation, while other buses are switching stations along line. Loads mainly consist of schools, malls and residential area, whose changes has a certain amount of time characteristics.

Table 1. Equivalent parameters of a feeder.

No.	Line	Line model	Length/ km	Active power/ MW	Reactive power/ Mvar
1	0-1	YJV-300	1.95	0.460	0.044
2	1-2	YJV-300	1.28	1.905	0.342
3	2-3	YJV-3×300	1.80	1.383	0.252
4	3-4	YJLV-240	0.25	0.725	0.104
5	4-5	YJLV-240	0.10	0.440	0.116
6	5-6	YJV-150	0.47	0.219	0.057

5.2 Description of simulated cases

Access location, capacity, methods and different load conditions of distributed photovoltaic power will cause different effects on the distribution network of power quality. Several typical access cases are chosen to make a simulation analysis. Details are shown in Table 2.

Table 2. Simulation cases illustrate.

-	Access location	Load	Power factor	Output
1	Bus7	30%	-0.98	100%
2	Variety	80%	-0.98	100%
3	Dispersed	80%	-0.98	100%

Note: Percentage in the above table is percentage of maximum load capacity.

5.3 Influence of change of access location on harmonics

Figure 5, a diagram of the total voltage distortion rate of each node in case 2, shows that without the harmonic source model, access method and access capacity changing, the closer to the system bus the distributed PV position is, the lower the level of harmonic distortion of the line is as a whole. On the contrary, the closer access point is to the end, the greater impact harmonic voltage has on the system. Through altering fundamental voltage, access points make influence on the harmonic injection level. When access capacity does not exceed the total load capacity, voltage continues to drop along the line so that voltage connected to the head of line is higher than that at the end. Meanwhile, for the PV power with constant *PQ* model, fundamental current is smaller with the incensement of fundamental voltage. Hence, the smaller overall harmonic current injected into the PCC is caused by the closer position of photovoltaic power to the system bus. When access location is closer to the end of the line, conversely, harmonic current injected into the PCC is greater.

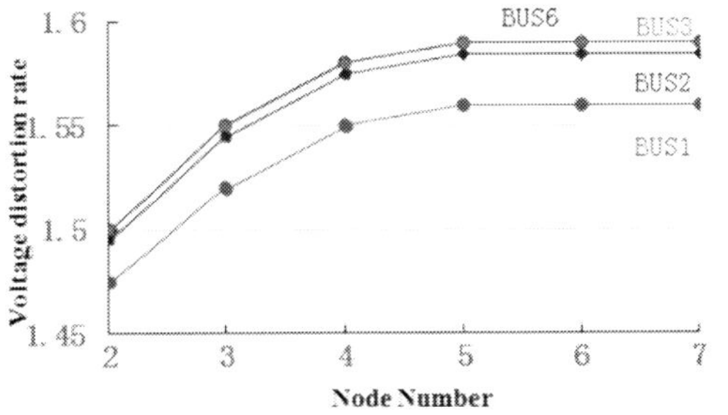

Figure 5. Total distortion rate under different access location.

5.4 Influence of dispersed mode on harmonics

Taking the actual needs of users and resource distribution issues into account, in the actual planning, it is extremely difficult to centralize distributed photovoltaic and other renewable energy sources in one load point. When energy sources own larger access capacity, they are usually distributed to each node. Case 3 was simulated under the condition that its decentralized mode was respectively set to be evenly, randomly (computer-generated random number), and in proportion of loading (the ratio of access capacity of the point to the total access capacity equal to the ratio of access load of the point to the total capacity of the load), Results of simulation are shown in Figure 6.

As can be seen from Figure 6, the overall level of harmonics in evenly distributed mode and in proportion of load is slightly higher than that in randomly distributed mode. There are two reasons for this phenomenon. For one thing, from the analysis by the access position that access location close to the system bus can drop the level of injected harmonics, the capacity of distributed power access to the head line is larger in random mode. For another reason, each PV inverter can be seen as a small harmonic source, and there is a superposition of harmonic relationship among the harmonic sources. When two harmonic sources are relatively close to each other, the phase angles may be more close to linear superposition. When the number of grid-connected points increases, it will appear the case that the phase angle is positive and negative offset, and harmonics is smaller than it is in the more decentralized access.

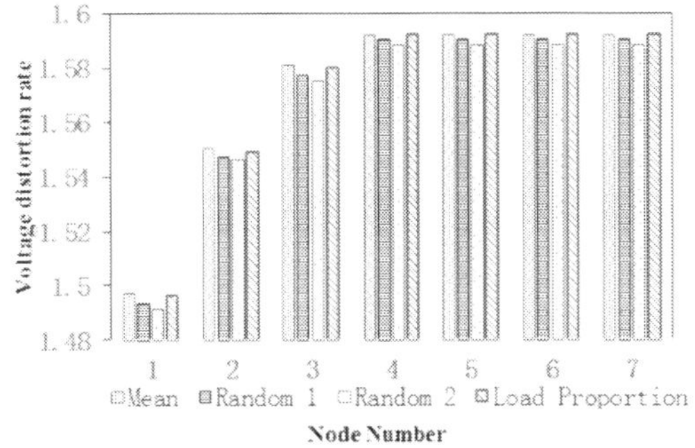

Figure 6. Total distortion rate under different dispersity.

5.5 Influence of cable capacitance on the system response characteristics

Simulate case 1 with harmonic source at the end of the feeder. Earth capacitance parameters of all cable lines are set to 0 firstly. Then observe the injection current, current in middle line and current inflow to the external network of each harmonic. The values are shown in Figure 7.

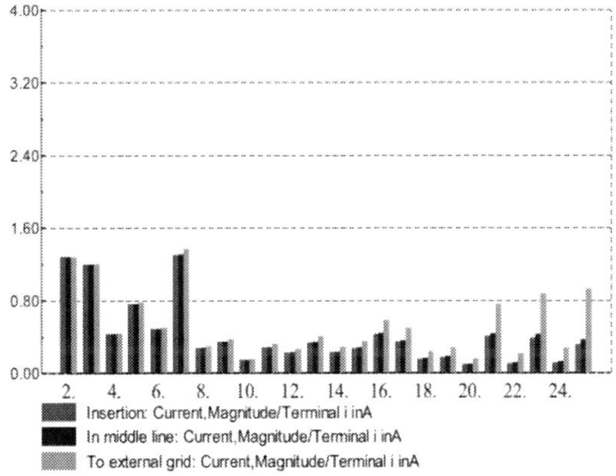

Figure 7. Harmonics in system without cable capacitance.

The capacitance parameter of cable lines is set to the actual value. The injection current, cut-line current and current in flow to the external network of each harmonic are shown in Figure 8.

Something can be drawn from Figure 8. On condition that the cable capacitance is out of consideration, parallel resonance can be realized when times of harmonic waves gets to over 25. And photovoltaic power harmonic times are mainly concentrated on 2-25 times, with 11 below and 25 above of that ignored especially. Therefore, resonance makes no significant effect on the overall harmonic amplification when the harmonic order of that is in 25 outside. As is shown in Figure 8, cable capacitance counted, the resonant point is migrated forward significantly, which extremely amplify the nearby 14th harmonics and harmonic current of resonance point wave is amplified to a certain value, rather than theoretical infinity, since the resistance value is not small enough to be ignored actually. Usually, filter channel is installed in frequency with larger harmonics, but the higher harmonics amplified after system response will also cause harm to the system. Therefore, you should also consider system characteristic and adjust series reactance ratio when setting the filter channel, so that the resonance point is within the minor harmonic bandwidth in [10].

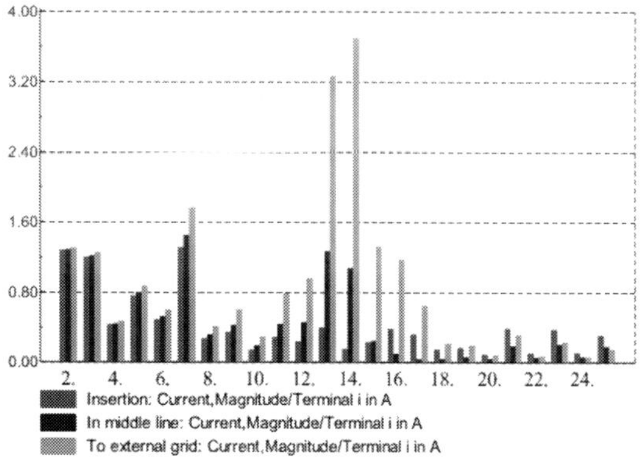

Figure 8. Harmonics in system without cable capacitance.

6 Conclusion

1) Cable line with specified length will result in amplification of harmonic waves around. To limit singular and high-order harmonic generation, filtering device is a need to install.

2) Keeping the capacity of the access invariant, the closer PV access location to the end of the line will also cause higher harmonic distortion level of distribution network. In order to reduce the harmonic distortion, the renewable energy can be considered being connected to the middle of node with heavy load, instead of the end of the lines.

3) After meeting the user's installation requirements and checking capacity of lines the position PV grid-connected should be controlled to be similarity, because counteraction between harmonic sources can assist in reducing the diffusion of harmonics, the verification of power capacity of line.

4) PV access makes value of 10-25 times of harmonics within the distribution network not be ignored. In case that some of the higher harmonics is amplified to damage grid security, not only injection characteristic of harmonics but the response characteristics of the system is to be considered when filter channel is to be set.

Currently, the paper only considered the inverter access. And with the development of renewable energy, there will be more types of power connected into distribution network, which will produce interaction, impact of which on distribution network harmonics remains to be further studied.

References

1. A. R. Metke, R. L. Ekl, Security Technology for Smart Grid Networks, IEEE Transactions on Smart Grid, **1**, 99-107, (2010)
2. S. Munir, Y. W. Li, Residential Distribution System Harmonic Compensation Using PV Interfacing Inverter, IEEE Transactions on Smart Grid, **4**, 816-827, (2013)
3. S. A. Kaabi, H. Zeineldin, V. Khadkikar, Planning Active Distribution Networks Considering Multi-DG Configurations, IEEE Transactions on Power Systems, **29**, 785-793, (2014)
4. L. R. Camargo, R. Zink, W. Dorner, Spatio-temporal modeling of roof-top photovoltaic panels for improved technical potential assessment and electricity peak load offsetting at the municipal scale, Computers Environment & Urban Systems, **52**, 58-69, (2015)
5. J. F. Z, S. M. D, Y. L. H, Research on distributed generation source placement, Sustainable Power Generation and Supply, 2009. SUPERGEN '09. International Conference on. IEEE, (2009)
6. Y. Ye, M. Kazerani, V. H. Quintana, Modeling "Control and implementation of three-phase PWM converter, IEEE Trans. on Power Electronics, **18**, 57-864, (2003)
7. Modeling and simulation of large-scale grid-connected photovoltaic system, Proceedings of the Power System Technology of IEEE, Aviva, (2010)
8. X. Bingxin, W. Zong, F. Shi-in, Influence on harmonic characteristics of distributed network with distributed generations, Power System Protection and Control, **40**, (2012)
9. Z. Junfang, Y. Qiang, H. Yinli, Research on effect of distributed generation source on harmonic of distributed network, Electrotechnical Application, **30**, (2011)
10. GB/T 14549-1993Quality of electric energy supply. (1993)

Optimal Sizing of Decentralized Photovoltaic Generation and Energy Storage Units for Malaysia Residential Household Using Iterative Method

Hasimah Abdul Rahman and Nur Dalilah Nordin

Centre of Electrical Energy Systems, Institute of Future Energy, Universiti Teknologi Malaysia, Johor, Malaysia
Faculty of Electrical Engineering, Universiti Teknologi Malaysia, Johor, Malaysia

Abstract. World's fuel sources are decreasing, and global warming phenomena cause the necessity of urgent search for alternative energy sources. Photovoltaic generating system has a high potential, since it is clean, environmental friendly and secure energy sources. This paper presents an optimal sizing of decentralized photovoltaic system and electrical energy storage for a residential household using iterative method. The cost of energy, payback period, degree of autonomy and degree of own-consumption are defined as optimization parameters. A case study is conducted by employing Kuala Lumpur meteorological data, typical load profile from rural area in Malaysia, decentralized photovoltaic generation unit and electrical storage and it is analyzed in hourly basis. An iterative method is used with photovoltaic array variable from 0.1kW to 4.0kW and storage system variable from 50Ah to 400Ah was performed to determine the optimal design for the proposed system.

1 Introduction

World's electricity production is severely depending on coal, oil and natural gas as energy supply. However, fuel sources are decreasing, and global warming phenomena cause the necessity of urgent search for alternative energy sources. The use of renewable energy (RE) reduces the dependency on fossil fuels, and it is proven that RE has great potential and can be utilized to fulfill world energy demand [1].

Among other alternative energy available, photovoltaic (PV) system is the most promising RE in Malaysia. In the future, it is expected more installation of decentralized PV generation units combined with electrical storage in Malaysia's distribution network.

Grid tied decentralized PV generation system has numbers of advantages; modularity system, reduce dependency on conventional energy supply, environmental friendly [2] and the excess energy can be sold to utility [3], which entitled for bill reduction. Even though its potentials are proven, the coincidence of local load requirement with PV generation needs to be studied.

This paper introduces the new power management for interaction between electrical household demand, decentralized PV generation and electrical storage unit. This paper identifies the optimum configuration for PV array and storage capacity required to supply a specific load demand over a year. The optimization of system's parameters were performed under criteria of favoured cost of energy (COE), payback period (PP), degree of autonomy (DA) and degree of own consumption (DO).

2 System configurations

Figure 1 shows a schematic diagram for the proposed system. Referring to the figure, PV generator is used as a decentralized power source operating in grid parallel operation mode with public network. Electrical storage unit acts as the main backup power supply. Excess energy from PV plant can be sold to utility, and utility also act as backup energy when electrical storage reaches maximum depth of discharge (DOD)

The residential loads considered in the case study are only resistive loads. Figure 2 shows a typical hourly load profile for a residential house in Malaysia [4], [5] with the total daily load demand of 2.215 kWh. It is assumed that the energy requirement remains the same in each day for a year period.

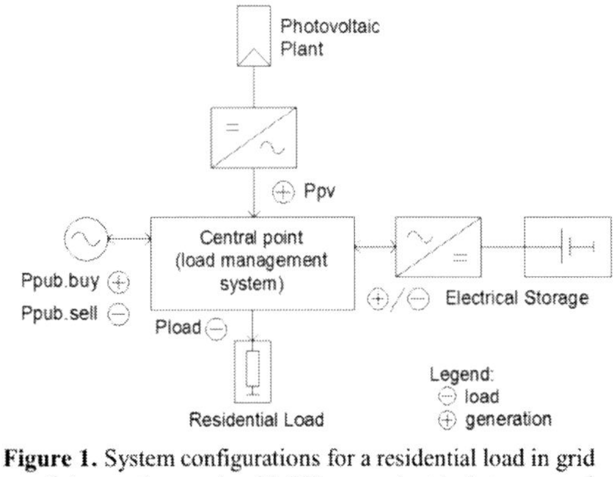

Figure 1. System configurations for a residential load in grid parallel operation mode with PV array, electrical storage and utility.

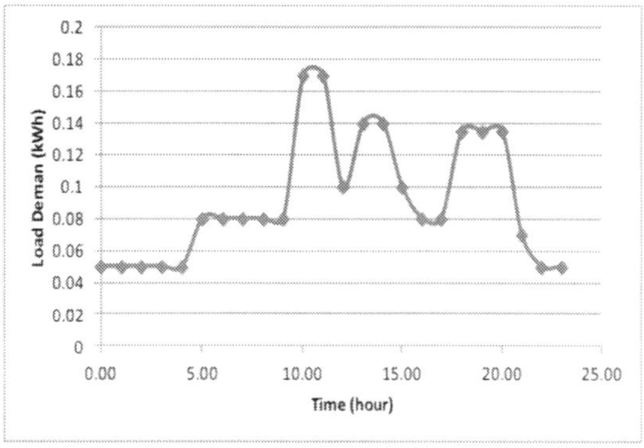

Figure 2. Hourly load profile [4], [5].

3 System design

In this paper, the proposed power management strategies algorithm is summarized in Figure 3. The simulation is performed by iterating the PV array variable (0.1kW - 4.0kW) and also the storage system variable (50Ah - 400Ah). The priority of the management strategies is to optimize all PV energy output to supply demand. When PV output is higher than load demand, it is used to charge the battery and sell the surplus power to public network. However, when PV energy generation is not enough to supply the household load, battery will be discharge. If battery reaches lowest DOD, the additional energy for the load is supply or bought from utility.

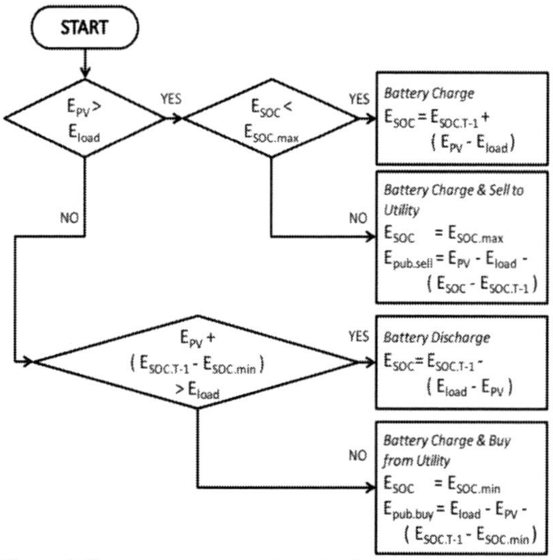

Figure 3. Power management strategies algorithm.

3.1 PV array modeling

PV output size must able to fulfill load demand with extra energy to cover system losses. In the analysis, Kuala Lumpur's hourly climate data is used. The hourly energy generated by PV array, E_{pv} is calculated using Equation (1) [3], [6]. *PSH* is peak sun hour over the period of interest (one hour), P_{mp_stc} is PV's maximum power output in standard test condition, *STC*, N_{pv} is number of PV module in the array, η_{temp} is temperature de-rating factor, η_{wire} is cable efficiency, η_{inv} is inverter efficiency, η_{mm} is module mismatch, and η_{dirt} is dirt derating factor.

$$E_{pv}(t) = PSH(t) * P_{mp\_stc} * N_{pv} * \eta_{temp} * \eta_{wire} * \eta_{inv} * \eta_{mm} * \eta_{dirt} \quad (1)$$

3.2 Battery modeling

Estimation of battery state of charge (SOC), E_{bat} analysis is shown as Equation (2) [4], [7] (charging/discharging and self discharge efficiency is ignored). I_{bat} is battery capacity in Ah, and V_{bat} is battery's nominal voltage. Minimum SOC, $E_{SOC.min}$ is 50% of its full capacity.

$$E_{bat}(t) = Q_{bat}(t) * V_{bat} \quad (2)$$

3.3 Inverter size selection

Inverter is expected to deliver maximum AC load in household. Hence, inverter's power rating for battery system, P_{inv_bat} is selected using Equation (3), where P_{ACload} is total power from AC load demand and 1.25 is set as oversized factor [2].

$$P_{inv\_bat} = P_{ACload} * 1.25 \quad (3)$$

Inverter's power rating for PV array, P_{inv_pv} is selected using Equation 4:

$$P_{inv\_pv} = SF_{inv\_pv} * P_{mp\_stc} * N_{pv} \quad (4)$$

where SF_{inv_pv} is inverter-to-PV sizing factor (0.80) [2], and N_{pv} is quantity of PV modules in the system.

4 System optimization criteria

In order to select an optimum combination, four analyses were used to determine the ideal parameters for PV plant and electrical storage unit. To find optimum configuration with the lowest investment possible, it is inevitable to use economic analysis. Therefore, Cost of Energy (COE) and Payback period (PP) is used. Meanwhile, Degree of Autonomy (DA) and Degree of own-much energy generated from PV plant are used in residential household.

4.1. Cost of energy

COE (RM/kWh) is defined as the average cost per kWh of electrical energy produced by the system when a lifetime, investment cost, replacing, operation & maintenance, and capital cost is considered [8]. Based on technical datasheet, the batteries need to be replaced for every 5 years and inverter need to be replaced every 10 years [5]. Lower COE is desirable and it is calculated by dividing producing electricity annualized life cycle cost, LCC_{1year} with E_{PV} [9]. The market price for system components is summarized as in Table 1 [9], [10].

$$LCC = C_{pv} + C_{controller} + C_{bat} + C_{inv} + C_{install} + C_{batrep} + C_{invrep} + C_{O\&M\_25years} - C_{salvage} \quad (5)$$

$$LCC_{1year} = \frac{LCC}{\left[\frac{(1+i)^N}{i(1+i)^N}\right]} \quad (6)$$

$$COE = \frac{LCC_{1year}}{E_{PV}} \quad (7)$$

4.2. Payback period

PP is used to determine the time it would take to recover or break even on initial investment cost and ongoing O&M cost [11]. E_{sell} is energy fed back to utility during surplus energy from PV and E_{buy} is energy bought from utility during lack of PV output. Meanwhile, feed in tariff, *FIT* is different for each country, and for Malaysia, it is RM 1.35/kWh in 2015 [10]. Each energy unit bought from utility is provided by Tenaga Nasional Berhad, *TNB* [12]. Shorter payback period are more desirable.

$$PP = \frac{LCC}{\left(\sum_{365}^{T} E_{sell} * FIT\right) - \left(\sum_{365}^{T} E_{buy} * TNB\right)} \quad (8)$$

Table 1. Components' pricing.

Component		Unit	Price
Module		RM/Wp	5.000000
Battery		RM/Output Watt Hour	0.041667
Inverter		RM/Continuous Watt	1.000000
Support Structure & Installation cost		RM/Wp	4.000000
FIT	Basic FiT rates up to and including 4kW	RM/Wh	0.000917
	Use as installation in buildings or building structures	RM/Wh	0.000172
	Use as building materials	RM/Wh	0.000166
	Use of locally manufactured or assembled solar PV modules	RM/Wh	0.000050
	Use of locally manufactured or assembled solar inverters	RM/Wh	0.000050
TNB Bill pricing tariff		RM/Wh	0.000218

4.3. Degree of autonomy

DA shows how much energy provided from decentralized PV generator and directly used in residential household compared to the whole energy demand. It is calculated using Equation (9) [13]:

$$DA = \frac{\sum_{365}^{T} E_{pv} - \sum_{365}^{T} E_{sell}}{\sum_{365}^{T} E_{pv} - \sum_{365}^{T} E_{sell} + \sum_{365}^{T} E_{buy} - E_{bat\_start} + E_{bat\_end}} \quad (9)$$

where E_{bat_start} is initial battery's energy and E_{bat_end} is battery's energy at the end of the period. Higher DA is desirable for optimum configuration.

4.4. Degree of own-consumption

Meanwhile, DO shows how much energy is provided from decentralized PV plant that is directly used by the residential household compared to the provided power by PV generator. It is calculated as Equation 10 [13]. However, contrast with DA, optimum design tends to have lower DO.

$$DO = \frac{\sum_{365}^{T} E_{pv} - \sum_{365}^{T} E_{sell}}{\sum_{365}^{T} E_{pv} - E_{bat\_start} + E_{bat\_end}} \quad (10)$$

5 Results and discussion

Figure 4, 5, 6 and 7 show the analyses results based on the optimization criteria as stated in equations 7, 8, 9 and 10. Referring to Figure 4, it shows the results from COE analysis as the optimization criteria. It can be seen that when battery capacity ranged from 50Ah to 100Ah is paired with PV capacity from 3kW to 4kW, the system obtained the lowest COE of RM 0.57/kWh.

Figure 5 is obtained after payback period analysis was performed; it is observed that for all combination of battery capacity at the PV ranged from 0.3kW to 0.9kW gives the payback period more than 20 years. Hence, it can be concluded that any configuration within the specified range of PV capacity is not feasible. This is due to insufficient PV output, which causes the total benefit obtained from selling energy to utility is inadequate to break-even on LCC system. However, for all pairs of battery capacity at the PV ranged from 3.45kW to 4kW gives PP lower than 8 years.

Hence, the lowest PP and DO values obtained are 7.333 years and 0.1436p.u respectively at the combination of 4kW PV array and 50Ah battery capacity. Meanwhile, the highest DA obtained using the configuration of 4kW PV capacity and 100Ah is 0.9998p.u.

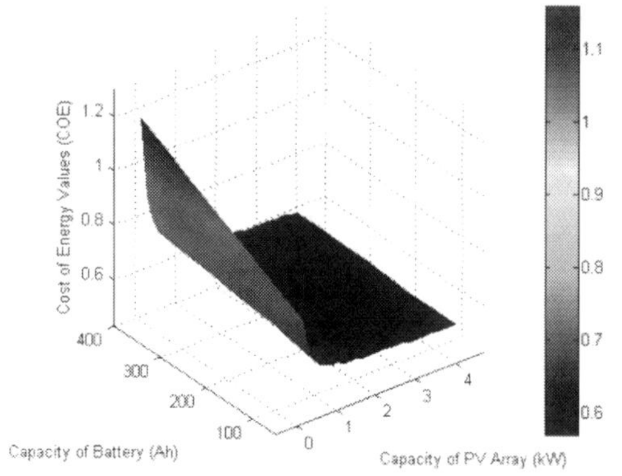

Figure 4. Results of COE analysis.

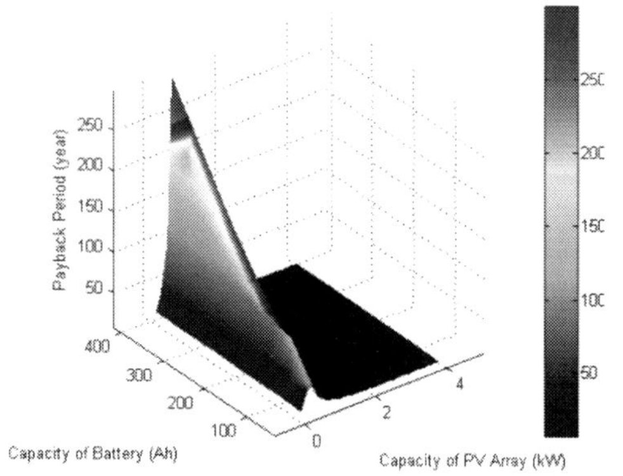

Figure 5. Results of PP analysis.

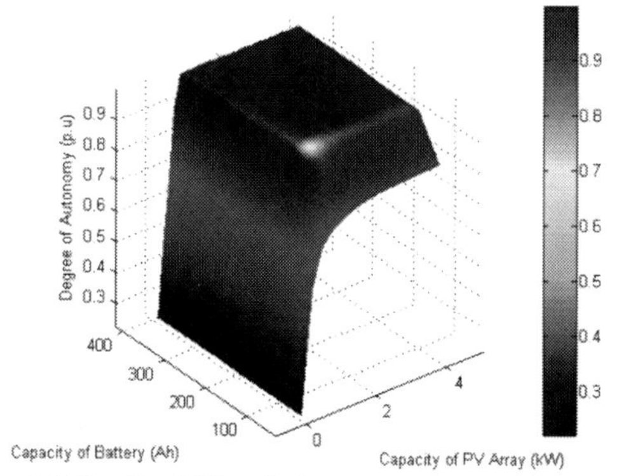

Figure 6. Results of DA analysis.

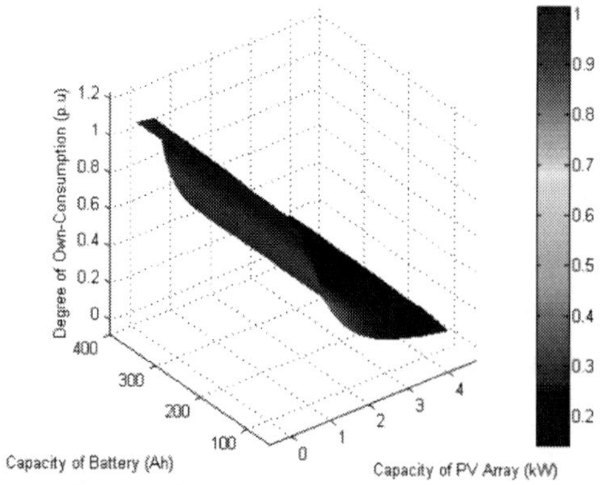

Figure 7. Results of DO analysis.

6 Conclusions

The optimal sizing of decentralized photovoltaic generation and energy storage units for Malaysia residential household has been presented. From the results obtained, the most favorable design is either a system of PV capacity 4kW and 50Ah or 100Ah battery capacity, which is obtained from the lowest DO or the highest DA value. However, based on the observation, it is wiser to select the best combination parameters using DA analysis. It is because DA indicates which design has high degree of energy independence from local utility with lowest COE and low PP. The selected parameter pair is able to use most energy from PV generator directly in residential household, with low energy need to be bought from utility and has small changes between initial and ending battery's capacity.

This has proven that it is possible to effectively use RE sources and storage that allows the system to be independent from utility, and gain profit from solar FIT.

Acknowledgement

Authors wish to acknowledge the supports for this paper received from Malaysian Ministry of Higher Education (MOHE) and Universiti Teknologi Malaysia through Research University Grant (RUG; Vote no: Q.J13000.2509.07H54). This research is also partially supported by MOSTI (Vote no: RJ130000.7923.4S068).

References

1. A. Mellit, and S. A. Kalogirou, Artificial intelligence techniques for photovoltaic applications: A review. *Progress in Energy and Combustion Science*, **34**(5): pp. 574-632 (2008)
2. S. I. Sulaiman, et al., Sizing grid-connected photovoltaic system using genetic algorithm. *Industrial Electronics and Applications (ISIEA), 2011 IEEE Symposium* on IEEE (2011)
3. S. I. Sulaiman, et al., An intelligent method for sizing optimization in grid-connected photovoltaic system. *Solar Energy,* **86**(7): pp. 2067-2082 (2012)
4. W. Shen, Optimally sizing of solar array and battery in a standalone photovoltaic system in Malaysia. *Renewable Energy,* **34**(1): pp. 348-352 (2009)
5. N.D. Nordin, and H.A. Rahman. An optimization method for designing stand alone photovoltaic system using iterative method. in *Smart Energy Grid Engineering (SEGE), 2015 IEEE International Conference* (2015)
6. R.Dufo-López, J.M. Lujano-Rojas, and J.L. Bernal-Agustín, Comparison of different lead–acid battery lifetime prediction models for use in simulation of stand-alone photovoltaic systems. *Applied Energy,* **115**: pp. 242-253 (2014)
7. J. Li, W. Wei, and J. Xiang, A Simple Sizing Algorithm for Stand-Alone PV/Wind/Battery Hybrid Microgrids. *Energies,* **5**(12): pp. 5307-5323 (2012)

8. S. Kamel, and C. Dahl, The economics of hybrid power systems for sustainable desert agriculture in *Egypt. Energy*, **30**(8): pp. 1271-1281 (2005)
9. J. Abdulateef, et al., Economic analysis of a stand-alone PV system to electrify a residential home in Malaysia. *10th IASME/WSEAS International Conference on Heat Transfer, Thermal Engineering and Environment (HTE'12)* (2012)
10. FIT Malaysia. Available from: http://seda.gov.my/.
11. Simple Photovoltaic Economic Calculations. Available from: http://www1.eere.energy.gov/solar/pdfs/simple_pv_economic_calculations_tn.pdf
12. TNB tariff. Available from: https://www.tnb.com.my/residential/pricing-tariffs/
13. T. Wieland, et al., Optimal sizing of electric and thermal energy storage units for residential households with decentralized generation units in the low voltage grid. *Electric Power Quality and Supply Reliability Conference (PQ), IEEE* (2014)

Thermal Rating of Overhead Insulation-Covered Conductors in the Steady-State Regime

Stanislav Girshin, Vladimir Goryunov, Evgenii Kuznetsov, Elena Petrova, Anton Bubenchikov, Dmitrii Batulko

Omsk State Technical University, Energy Department, 644050 pr. Mira 11, Omsk, Russian Federation

Abstract. One can offer based on the solution of the heat equation and the heat balance equation a mathematical model of steady thermal regime of the conductor, which allows to determine the temperature of bare and insulation-covered conductors of overhead power lines, considering weather conditions, as well as to perform the calculation of electricity losses with conductors temperature. The expressions are for the gradient of temperature distribution in the current-carrying conductor, as well as conductor insulation with and without dielectric losses. The accuracy of the created model is checked when compared with the methods of CIGRE, IEEE and the Finite Element Method. High precision of matching results is achieved.

1 Introduction

Currently, there is a need to increase the capacity of transmission lines due to the accelerated growth of consumption power. One way to increase the ampacity through the transmission line is a reliable determination of the temperature of the wires in real time [1-6]. The calculation of the conductors temperature is possible with integrating operating and weather factors [7-9]. Along with the increase in load current the task is to reduce the losses of energy in electric networks. The loss of electricity is reduced with introducing measures to reduce losses. When selecting actions one must assume the network elements temperature. For example, when installing capacitor banks the energy loss is not only due to the actual reduction of the transmitted reactive power, but also by reducing the resistance. It reduces losses also to the transfer of active power, and not just reactive. Calculation excluding temperature will not consider these additional factors.

Paper considers the mathematical model of steady-state mode of insulation-covered conductor on the basis of the heat balance equation. The model allows to determine the temperature and the active power losses in the conductors of overhead lines. For creation of the model have been defined temperature gradient of conductor, the temperature changes in the conductor insulation and influence of dialectical losses on thermal mode. A positive feature of the mathematical model is a form of analytical equations obtained. It simplifies the implementation of the computer model and analysis results. The accuracy of equations is confirmed with good matching with the results of determining the temperature methods of IEEE [8], CIGRE [9] and the Finite Element Method (FEM) [10]. The practical value of the model is the ability to analyze both bare and insulation-covered conductors of overhead lines.

2 Mathematical model of steady-state thermal rating

Calculation of the temperature on the surface of the conductor Θsur is performed on the basis of the heat balance equation. At steady-state the heat balance of insulated and bare overhead line conductor is given by equation (1)

$$\Delta p'_0(1 + \alpha\Theta_{sur}) = d_c[\pi\alpha_c(\Theta_{sur} - \Theta_{amb}) + \pi\varepsilon C_0(T^4_{sur} - T^4_a) - A_s q_s] \quad (1)$$

where α is a temperature coefficient of resistance; d_c is outer diameter of the conductor; α_c is a coefficient of heat transfer with convection; Θ_{sur} is temperature of the outer surface of the conductor (°C); Θ_{amb} is ambient temperature (°C); T_{amb} is ambient temperature(K); T_{sur} is the temperature of the outer surface of the insulation (K); ε is the emissivity of the surface of the conductor; A_s is the absorption capacity of the surface for solar radiation; C_0 is radiation coefficient of blackbody; q_s is flux density of solar radiation on the conductor; $\Delta p'_0$ is active power losses in the conductor per length unit at $\Theta_{sur} = 0$ °C.

The values of q_s, $\Delta p'_0$, Θ_{sur} will be in equations (2-4):

$$q_s = k_{sh}q_{s.dir}\sin\varphi_s + \pi q_{s.dif} \quad (2)$$

$$\Delta p'_0 = \frac{I^2 r_0}{1 - \alpha I^2 r_0 S_{ins}} \quad (3)$$

$$\Theta_{sur} = -\Delta p_0 S_{ins} + \Theta_c(1 - \alpha\Delta p_0 S_{ins}) \quad (4)$$

where $q_{s.dir}$ is flux of direct solar radiation on the surface perpendicular to the sun's rays; $q_{s.dif}$ is flux density of diffuse solar radiation, averaged over all directions; k_{sh} is reduction factor, which takes into account the shaded portion of the conductor (shading coefficient); φ_s is the angle between the axis of the conductor and the direction of the sun's rays; I is current in the conductor; r_0 is the pursuit of resistance at 0 °C; Δp_0 is heat (active power losses), calculated from the resistance, defined at temperature of 0 °C; Θ_c is conductor temperature; S_{ins} is thermal insulation resistance per length unit, defined by equation (5).

$$S_{ins} = \frac{ln(r_2 / r_1)}{2\pi\lambda_{ins}} \tag{5}$$

where λ_{ins} is the thermal conductivity of the insulation; r_2 is the conductor outer radius; r_1 is the radius of the conductor without insulation.

The dependence of heat (active losses) in the overhead line conductor from the conductor temperature Θ_c as shown in equation (6)

$$\Delta p = \Delta p_0(1 + \alpha\Theta_c) \tag{6}$$

The conductor temperature was determined by equation (7):

$$\Theta_c = \frac{\Theta_{sur} + \Delta p_0 S_{ins}}{1 - \alpha\Delta p_0 S_{ins}} \tag{7}$$

Equation (4) is obtained by solving the heat equation in partial derivatives.

The equations (1-7) are a mathematical model of steady-state thermal regime of overhead insulation-covered conductor

3 The temperature gradient in the conductor

3.1. In the bare overhead conductor

One considers a cylindrical conductor of infinite length of diameter d_1. One assumes that the temperature in all points of the same surface of the conductor and the bulk density of the heat release is q_v = const. Under these conditions, the conductor temperature Θ is a function of distance from the conductor center r, and the heat equation in the steady-state, and its solution have the form as shown in equations (8), (9):

$$\frac{d^2\Theta}{dr^2} + \frac{1}{r}\frac{d\Theta}{dr} + \frac{q_v}{\lambda} = 0 \tag{8}$$

$$\Theta(r) = \Theta_{center} - \frac{q_v r^2}{4\lambda} \tag{9}$$

where q_v is volume density of heat; λ is a coefficient of thermal conductivity, Θ_{center} is the temperature in the conductor center.

In accordance with equation (9) temperature change from the center to the surface of the conductor and the average temperature gradient are shown in equations (10), (11)

$$\Delta\Theta = \Theta_{center} - \Theta_r = \frac{q_v r_1^2}{4\lambda} \tag{10}$$

$$\left(\frac{d\Theta}{dr}\right)_{av} = \frac{\Delta\Theta}{r_1} = \frac{q_v r_1}{4\lambda} \tag{11}$$

3.2 In the overhead insulation-covered conductor without dielectric losses

If there is no dielectric losses in the insulation the heat conduction equation to insulate has the form as shown in equation (12)

$$\frac{d^2\Theta}{dr^2} + \frac{1}{r}\frac{d\Theta}{dr} = 0 \tag{12}$$

As a result of integration of equation (5) obtain equation (13) [11]:

$$\Theta = C_1 \ln r + C_2 \tag{13}$$

where C_1, C_2 are integration constants determined from boundary conditions.

The final solution of equation (13) is presented in equation (14)

$$\Theta = \Theta_{sur} + \frac{q_v r_1^2}{2\lambda_{ins}} \ln\frac{r_2}{r} \tag{14}$$

If you set the temperature on the internal surface of the insulation, about equal to the temperature in the conductor center Θ_{center} then is obtained equation (15)

$$\Theta \approx \Theta_{center} + \frac{q_v r_1^2}{2\lambda_{ins}} \ln\frac{r_1}{r} \tag{15}$$

The temperature change in the insulation is presented by equation (16)

$$\Delta\Theta_{ins} = \Theta_{center} - \Theta_{sur} = \frac{q_v r_1^2}{2\lambda_{ins}} \ln\frac{r_2}{r_1} \tag{16}$$

3.3 In the overhead insulation-covered conductor with accounting of dielectric losses

If there are dielectric losses of insulation the heat equation has the form as shown in equation (17)

$$\frac{d^2\Theta}{dr^2} + \frac{1}{r}\frac{d\Theta}{dr} + \frac{q_{v,\text{ins}}}{\lambda_{ins}} = 0 \tag{17}$$

where $q_{v,ins}$ is the bulk density of heat insulation.

The value of $q_{v,ins}$, is proportional to the square of the electric field. If we consider the radial electric field of a single conductor, the tension is inversely proportional to r [12] and can be written by equation (18)

$$q_{v,\text{ins}} = \frac{K_1}{r^2} \quad (18)$$

Equation (17) takes the form as shown in equation (19)

$$\frac{d^2\Theta}{dr^2} + \frac{1}{r}\frac{d\Theta}{dr} + \frac{K_1}{r^2\lambda_{ins}} = 0 \quad (19)$$

The general solution is defined by the equation (20):

$$\Theta = C_1 \ln r - \frac{K_1 \ln^2 r}{2\lambda_{ins}} + C_2 \quad (20)$$

where C_1, C_2 are integration constants, K_1 is the coefficient of proportionality.

The final solution and the temperature change in the insulation take the form as shown in equations (21), (22)

$$\Theta = \Theta_{sur} + \frac{q_v r_1^2 - 2K_1 \ln r_1}{2\lambda_{ins}} \ln\frac{r_2}{r} + \frac{K_1}{2\lambda_{ins}}\left(\ln^2 r_2 - \ln^2 r\right) \quad (21)$$

$$\Delta\Theta_{ins} = \frac{q_v r_1^2 - 2K_1 \ln r_1}{2\lambda_{ins}} \ln\frac{r_2}{r_1} + \frac{K_1}{2\lambda_{ins}}\left(\ln^2 r_2 - \ln^2 r_1\right) \quad (22)$$

The nature of the temperature distribution inside the conductor is presented graphically in Figure1.

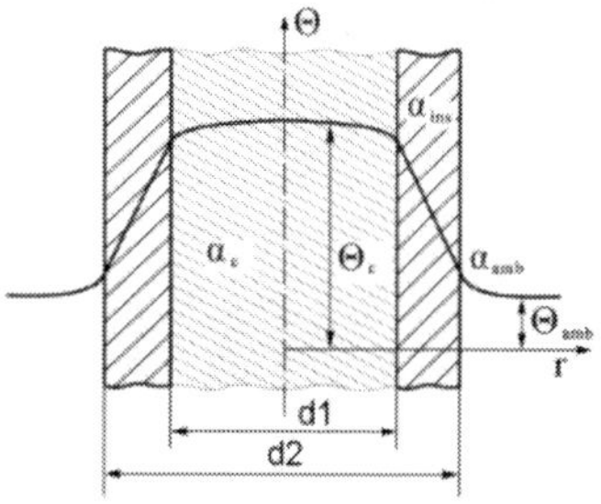

Figure 1. The temperature distribution in the conductor insulation.

Table 1. Conductor parameters

Sectional area	F	mm²	240
Chase resistance at 300 °K	r_{300}	Ohm/m	$2.8\cdot10^{-8}$
The temperature of the conductor	T_c	°C	90
Temperature coefficient of resistance of aluminum	α	°C⁻¹	0.004
Thermal conductivity of wires	λ	W/(m·K)	209
Thermal conductivity of insulation	λ_{ins}	W/(m·K)	0.4
The radius of the conductor	r_1	mm	8.74
The radius of the conductor assuming the insulation	r_2	mm	12
Permissible current	I	A	625

4 Simulation and results

4.1. Calculation of temperature and temperature gradient in the conductor

One considers wire brand SAX-240 with XLPE insulation with the parameters given in Table 1 [13] loaded with the permissible current at an ambient temperature of 300 K.

Heat in the conductor is

$$q_v = \frac{1}{l\cdot F} I^2 r \frac{l}{F} = \frac{I^2 r_{300}(1+\alpha(T_c - T_{amb}))}{F^2}$$
$$= \frac{625^2\cdot 2.8\cdot10^{-8}\left(1+0.004\left(273+90-300\right)\right)}{\left(240\cdot10^{-6}\right)^2} = 237739\,\text{W/m}^3$$

The temperature difference and a temperature gradient in the conductor is defined with equations (10), (11)

$$\Delta\Theta = \frac{q_v r_1^2}{4\lambda} = \frac{237739\cdot0.00874^2}{4\cdot209} = 0.022\ ^\circ\text{C}$$

$$\left(\frac{d\Theta}{dr}\right)_{av} = \frac{\Delta\Theta}{r_1} = \frac{0.022}{0.00874} = 2.52\ ^\circ\text{C/м}$$

The calculation results on equations (10) and (11) show that all points of the temperature of the conductor section can be considered equal.

4.2. Calculation of the change in temperature in the insulation excluding dielectric losses

One determines the change in temperature according to equation (16) for a conductor cross section of 240 mm², loaded to the permissible current of 625 A (Table 1).

$$\Delta\Theta_{ins} = \frac{q_v r_1^2}{2\lambda_{ins}} \ln\frac{r_2}{r_1}$$
$$= \frac{237739\cdot0.00874^2}{2\cdot0.4}\ln\frac{0.012}{0.00874} = 7.196^\circ\text{C}$$

4.3. Calculation of the temperature change in the insulation based on the dielectric losses

Assume clearly elevated levels of linear dielectric losses $\Delta p_{ins} \approx 10$ W/km, which corresponds to 35 kV cable with XLPE (Table 1). The value of K_1 is determined from the following equation:

$$\Delta p_{ins} = \int_{r_1}^{r_2} q_{v,\text{ins}} 2\pi r dr = 2\pi K_1 \int_{r_1}^{r_2} \frac{dr}{r} = 2\pi K_1 \ln\frac{r_2}{r_1}$$

Under these conditions for the given conductor obtain

$$K_1 = \frac{\Delta p_{ins}}{2\pi \ln \frac{r_2}{r_1}} = \frac{0.01}{2\pi \ln \frac{0.012}{0.00874}}$$

$$= \frac{0.01}{2\pi \cdot 0.316997} = 0.0050207\,\text{W/m}$$

$$\Delta\Theta_{ins} = \frac{237739 \cdot 0.00874^2 - 2 \cdot 0.0050207 \ln 0.00874}{2 \cdot 0.4}$$

$$\cdot \ln \frac{0.012}{0.00874} + \frac{0.0050207}{2 \cdot 0.4}\left(\ln^2 0.012 - \ln^2 0.00874\right)$$

$$= 7.21482 - 0.01823 = 7.197\text{癈}$$

The temperature difference do not show changes from the obtained without the dielectric losses. Therefore, in the thermal calculations dielectric losses can be ignored.

Table 2. Conductor parameters

Chase resistance at 20°C (Ohm/m)	Temperature coefficient of resistance (°C^{-1})	The diameter of the conductor (m)	Degree of black surface of the conductor	The absorption capacity of the conductor surface
$6.74 \cdot 10^{-5}$	0.0039	0.0285	0.5	0.5

Table 3. Practical cases considered

Case	1	2	3	4	5
Current (A)	600	650	600	970	600
Ambient temperature (°C)	40	40	40	40	40
Wind speed (m/s)	0,2	0,4	2	2	2
Wind direction (°)	90	90	90	90	90
The total solar radiation (W/m$^{2)}$	980	980	980	980	980
Altitude (m)	1600	1600	1600	1600	300

Table 4. The calculation results

Case		1	2	3	4	5
CIGRE	Θ_c (°C)	78.8	74	56.1	75	55.7
	ΔP (W/m)	29.36	33.95	27.37	75.85	27.33
IEEE	Θ_c (°C)	78.2	73.6	54.6	72.3	53.3
	ΔP (W/m)	29.3	33.92	27.24	75.23	27.12
FEM	Θ_c (°C)	80	76	55	73	54
	ΔP (W/m)	29.46	34.16	27.27	75.4	27.18
Proposed model	Θ_c (°C)	77.5	76	54.4	73.1	53
	ΔP (W/m)	29.24	34.16	27.22	75.42	27.1

4.4. Assessment of the temperature rating accuracy and active power losses on the proposed model

For the validation of the simulation according to Equations (1), (7) there were carried out numerical calculations of power losses and heat conductor of 429 brand-AL1/56-ST1. Initial data calculations are shown in Tables 2 and 3 [10], the results of calculations are shown in Tables 4 and 5. The results of calculations of the proposed model were compared to studies in [12] using the methods of IEEE [8], CIGRE [9], FEM [10].

Table 5. Errors in calculations of temperature and active power losses proposed model in comparison with CIGRE, IEEE, FEM methods

Case	1	2	3	4	5
$\Delta\Theta_{CIGRE}$ (°C)	0.7	2.4	0.2	0.8	0.3
$\Delta\Theta_{IEEE}$ (°C)	1.3	2	1.7	1.9	2.7
$\Delta\Theta_{FEM}$ (°C)	2.5	0	0.6	0.1	1
$\delta(\Delta P_{CIGRE})$ (%)	0.21	0.71	0.07	0.25	0.07
$\delta(\Delta P_{IEEE})$ (%)	0.41	0.62	0.55	0.57	0.84
$\delta(\Delta P_{FEM})$ (%)	0.75	0	0.18	0.03	0.29

The calculation of direct and diffuse components of solar radiation for the proposed model was carried out determining the highlands, for which we can assume approximately the ratio of forward and reverse components of solar radiation 1/10. Determining all this we take the value of the direct component $q_{s.dir}$ = 900 W/m^{2,} and reverse $q_{s.dif}$ = 80 W/m$^{2.}$ The shading coefficient is adopted $k_{sh} = 0.6$, and the angle between the axis of the conductor and the direction of the solar rays $\varphi_s = \pi/4$. At low wind speeds (0.2 and 0.4 m/s) the calculation is performed with the model proposed determining the natural convection, and with a wind speed of 2 m/s - determining the forced convection.

Active power losses is calculated according to equation (6).

To ease the analysis use the notation as shown in equations (23-28)

$$\Delta\Theta_{CIGRE} = \left|\Theta_c - \Theta^c_{CIGRE}\right| \tag{23}$$

$$\Delta\Theta_{IEEE} = \left|\Theta_c - \Theta^c_{IEEE}\right| \tag{24}$$

$$\Delta\Theta_{FEM} = \left|\Theta_c - \Theta^c_{FEM}\right| \tag{25}$$

$$\delta(\Delta P_{IEEE}) = \left|\frac{\Delta P - \Delta P_{IEEE}}{\Delta P_{IEEE}}\right| 100\% \quad (26)$$

$$\delta(\Delta P_{CIGRE}) = \left|\frac{\Delta P - \Delta P_{CIGRE}}{\Delta P_{CIGRE}}\right| 100\% \quad (27)$$

$$\delta(\Delta P_{FEM}) = \left|\frac{\Delta P - \Delta P_{FEM}}{\Delta P_{FEM}}\right| 100\% \quad (28)$$

where $\Delta\Theta_{CIGRE}$, $\Delta\Theta_{IEEE}$, $\Delta\Theta_{FEM}$ are absolute error of calculation proposed method by the temperature of the conductor with respect to the standards of CIGRE, IEEE, FEM; Θ_c is the temperature of the conductor calculated with the proposed method; Θ^c_{IEEE}, Θ^c_{CIGRE}, Θ^c_{FEM} are the temperature of the conductor calculated with the IEEE, CIGRE, FEM, respectively; $\delta(\Delta P_{IEEE})$, $\delta(\Delta P_{CIGRE})$, $\delta(\Delta P_{FEM})$ are the relative errors of calculation of power losses with the proposed method in comparison with the calculations of the losses of power with the standards of IEEE, CIGRE, FEM, respectively; ΔP is active power losses calculated with the proposed method; ΔP_{IEEE}, ΔP_{CIGRE}, ΔP_{FEM} are active power losses calculated using IEEE, CIGRE, FEM, respectively.

5 Conclusion

Table 5 shows that the absolute error in the calculation of temperature conductor on the proposed model does not exceed 2.7 °C, compared with IEEE and CIGRE, FEM. The relative error in the calculation of active power losses is less than 1%. The maximum deviation of the temperature of IEEE and CIGRE corresponding to the wind speed is v = 0.4 m/s. Perhaps this fact is due to the fact that the calculations for the proposed model for v = 0.4 m/s were carried out on the assumption of natural convection. However, calculations made for v = 0.4 m/s with the Finite Element Method gives the temperature 76 °C, which coincides with the result obtained with the proposed model. Calculations were carried out for the bare overhead conductor. Indirectly, it confirms the validity of the proposed theory, as the equation for the bare overhead conductor is a special case of conductors with insulation at $\lambda_{ins} = 0$.

Acknowledgment

The paper was prepared in Omsk State Technical University with financial support from Ministry of Education and Science in the framework of the agreement № 14.577.21.0097 on August 22, 2014. A unique identifier of applied research RFMEFI57714X 0097.

References

1. W.Q. Sun, Y. Zhang, C.M. Wang, and P. Song, IET Generat., Transmiss., Distrib., vol. **7**, no. 2, pp. 130-137, 2013
2. J. Heckenbergerova, P. Musilek, and K. Filimonenkov, *in Proc. IEEE Power Energy Soc. Gen. Meeting*, pp. 1-8, 2011.
3. H. T. Yip, C. An, J.G. Lloyd, P. Taylor, A. Michiorri, S. Jupe, M. Barlett, *10th IET International Conference on, IET*, pp. 1-5, 2010.
4. D. M. Greenwood, J.P. Gentle, K.S. Myers, P.J. Davison, I.J. West, J.W. Bush, G.L. Ingram, and M.C. Troffaes, IEEE Trans. Power Del., vol. **29**, no. 4, pp. 1849-1858, 2014.
5. T.O. Seppa, *IEEE Power Eng. Soc.Winter Meeting*, vol. **2**, pp. 1208-1211, 2002.
6. I. Albizu, E. Fernandez, P. Eguia, E. Torres, and A.J. Mazón, IEEE Trans. Power Del., vol. **28**, no. 1, pp. 3-10, 2013.
7. V.T. Morgan, Electric power systems research, vol. **6**, no.4, pp. 287-300, 1983.
8. IEEE, "Standard for calculating the current temperature of bare overhead conductors, Std 738," 2006.
9. Cigr´e Working Group 22.12, "Thermal behaviour of overhead conductors," Cigr´e Brochure 207, Aug. 2002.
10. F.A. Gomez, J. M. Garcia De Maria, D. Garcia Puertas, A. Bairi, R. Granizo Arrabe, *Proceedings of the 10th WSEAS international conference on communications, electrical & computer engineering, and 9th WSEAS international conference on Applied electromagnetics, wireless and optical communications. World Scientific and Engineering Academy and Society*, pp. 149-153, 2011.
11. A. In. Bulgarian, *Thermodynamics and Heat Transfer, Ripol Klassik*, 1975, p. 495.
12. K.S.Demirchian, L.R. Neumann, N.V. Korovkin, V.L. Chechurin, *Theoretical Foundations of Electrical Engineering*: *SPb: Peter,* vol. **3**, p. 377, 2006.
13. V.G. Gerasimov, *Electrical Engineering Handbook, MEI Publishing,* vol. 1, p. 440, 1995

Electromechanical-Traffic Model of Compression-Based Piezoelectric Energy Harvesting

B.C. Kok[1], Saleh Gareh[2], H.H. Goh[1] and C. Uttraphan[3]

[1]Department of Electrical Power Engineering, Faculty of Electrical & Electronic Engineering, UTHM 86400, Johor, Malaysia.
[2]Ph.D candidate in Electrical Engineering at Faculty of Electrical & Electronic Engineering, UTHM 86400, Johor, Malaysia.
[3]Department of Computer Engineering, Faculty of Electrical & Electronic Engineering, UTHM 86400, Johor, Malaysia.

Abstract. Piezoelectric energy harvesting has advantages over other alternative sources due to its large power density, ease of applications, and capability to be fabricated at different scales: macro, micro, and nano. This paper presents an electromechanical-traffic model for roadway compression-based piezoelectric energy harvesting system. A two-degree-of-freedom (2-DOF) electromechanical model has been developed for the piezoelectric energy harvesting unit to define its performance in power generation under a number of external excitations on road surface. Lead Zirconate Titanate (PZT-5H) is selected as the piezoelectric material to be used in this paper due to its high Piezoelectric Charge Constant (*d*) and Piezoelectric Voltage Constant (*g*) values. The main source of vibration energy that has been considered in this paper is the moving vehicle on the road. The effect of various frequencies on possible generated power caused by different vibration characteristics of moving vehicle has been studied. A single unit of circle-shape Piezoelectric Cymbal Transducer (PCT) with diameter of 32 mm and thickness of 0.3 mm be able to generate about 0.12 mW and 13 mW of electric power under 4 Hz and 20 Hz of excitation, respectively. The estimated power to be generated for multiple arrays of PCT is approximately 150 kW/ km. Thus, the developed electromechanical-traffic model has enormous potential to be used in estimating the macro scale of roadway power generation system.

1 Introduction

For the past ten years, many researchers have trained their sights on harvesting the vibration energy. This is due to its promising capacity to promote independently-powered wireless sensors [1]. Electromagnetic [2], electrostatic [3] and piezoelectric [4], [5] are the three kinds of electromechanical transducers that have been examined as vibration-operated generators. The underlying function of these electromechanical transducers is the alteration of mechanical energy into electrical energy. The vibration and stresses generated by vehicles along public roads are among the major sources of unexploited ambient energy. The harvesting of this type of energy can be realized with the utilization of a piezoelectric transducer (PZT) [6], [7]. The generation of electric voltage by PZTs is achieved through the exploitation of loading stress and vibration.

As depicted in Fig. 1, generator effect and sensor effect are the two terms that related to the direct piezoelectric effect of the piezoelectric materials, while motor effect and actuator effect are the terms related to the converse piezoelectric effect of the piezoelectric materials. In general, direct piezoelectric effect is employed for energy harvesting. Energy harvesting is a procedure whereby energy is taken out from the environment, altered into electrical energy, and stockpiled [8].

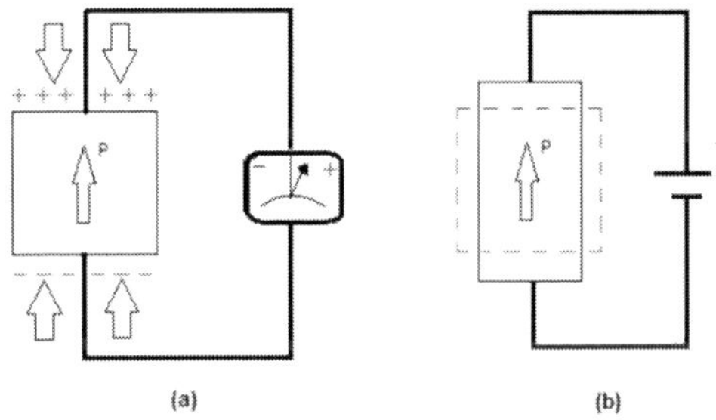

Figure 1. (a) Direct piezoelectric effect, (b) Converse piezoelectric effect.

The paper intents to present the two-degree-of-freedom (2-DOF) electromechanical model of compression-based piezoelectric harvesting unit for roadway application. A Cymbal transducer will be used to demonstrate the potential generated power and voltage of the developed model in various persuasive road vibration excitation frequencies as well as in the condition of variable external electrical loads.

2 Related works

The extraction of energy from vibration, temperature differentials and light are appeared to be the most effective means for micro energy harvesting. Recently, a compression-based energy harvester for roadways application has been introduced in [9], [10]. The harvester can be implanted into roadways for the purpose of exploiting electrical energy stemming from vibrations generated by traffic movement. Investigations revealed that a solitary harvesting module of 36 layers, under excitation of 1360 N of force amplitude and 6 Hz of frequency able to produce 85 mW of DC power.

A separate venture [11] employed a cantilever piezoelectric harvester to exploit the energy from vibrations produced by traffic movement over bridges. It has been observed that by maintaining a voltage between 1.8 and 3.6 V under a frequency lower than 15 Hz, mean power of as much as 0.03 mW can be generated.

An evaluation on the effectiveness of well-accepted piezoelectric transducers used for the harvesting of energy from tarmac walkways was carried out by [12], [13]. Eventually, they came up with an energy harvesting cymbal that could harvest output power of approximately 1.2 mW from tarmac walkways at a vehicle load frequency of 20 Hz. The velocity and types of the vehicle determines the frequency of the load on the implanted piezoelectric transducer. The conversion of unused acoustical energy into functional electrical energy was achieved by [14] with the development of the acoustic micro-energy harvester. At 60 Hz and 400 kΩ resistive load, the harvester successfully generated 0.9 V/ (ms^{-2}) and 1.79 μW/ (m^2s^{-4}).

Table 1 gives the comparison of three research groups who working on piezoelectric roadways applications. Obviously, the recent developed technologies in piezoelectric have demonstrated the possibility to harvest huge amount of electric power from roadways [7].

Table 1. Generated energy from different research groups.

Parameter	Genziko	ODOT	Innowattech
Generated power per km (single lane)	13 – 51 MW	486 kW	100 – 200 kW
Vehicles per hour (single lane)	600 – 2250	600	600

Table 2 illustrates the approximated amount of per unit possible harvested energy from two vibration harvesting sources i.e. from human and industry.

Table 2. Generated energy from vibration sources.

Source of energy (Vibration/ motion)	Harvested power
Human	4 μWcm^{-2}
Industry	100 μWcm^{-2}

3 Methodology

A two-degree-of-freedom (2-DOF) electromechanical model of the piezoelectric harvesting unit under the exciting force is portrayed in Fig. 2. This piezoelectric model receives its input from the force produced by traffic movement. This force is denoted in the model as *F(t)*.

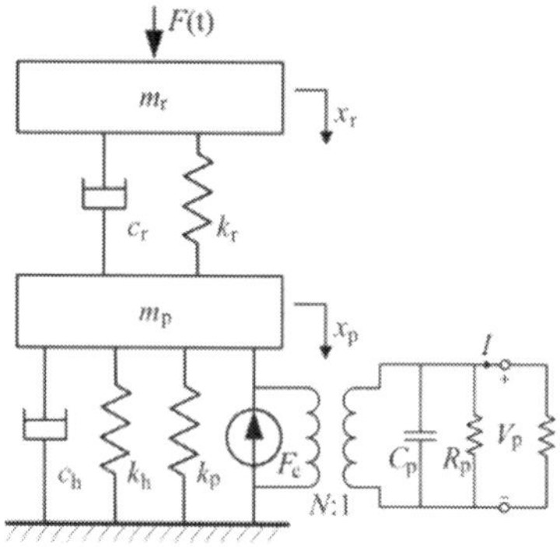

Figure 2. The electromechanical model of piezoelectric energy harvesting [10]

The symbols represented in the electromechanical model as well as the following equations are defined as follows:

m_r, c_r and k are the mass, damping and elastic coefficient of the rubber layer, respectively.

m_p and k_p are the mass and elastic coefficients of the piezoelectric stack, correspondingly.

The mechanized damping and elastic coefficients of the harvester's mechanical structure are symbolized as the parameters of c_h and k_h, respectively.

The deformation of the rubber layer and piezoelectric wafer-stack attributed to the external force are symbolized as the parameters X_r and X_p, respectively.

The clamped capacitance of the piezoelectric wafer-stack is symbolized as C_p

The piezoelectric leakage resistance and the electric loss property of the piezoelectric wafer-stack are symbolized as R_p and R_l, respectively.

The angular frequency of the exciting vibration is symbolized as ω and the frequency equivalent fuction of $F(t)$ is symbolized as $F(\omega)$.

The external electrical load is symbolized as R.

The normalized frequencies are symbolized as α and φ. The normalized electrical resistance is symbolized as β, and the electromechanical coupling alternative is symbolized as k_{em}.

The interrelation parameters that can be computed are given as follows:

$$\omega_1 = \sqrt{\frac{k_r}{m_r}}, \quad \omega_2 = \sqrt{\frac{k_a}{m_p}}$$

$$\xi_1 = \frac{c_r}{2\sqrt{m_r k_r}}, \quad \xi_2 = \frac{c_h}{2\sqrt{m_p k_a}}$$

$$\mu = \frac{m_r}{m_p}$$

$$k_a = k_p + k_h$$

$$\alpha = \frac{\omega_1}{\omega_2}$$

$$\varphi = \frac{\omega}{\omega_2}$$

$$\beta = \omega_2 R C_p$$

$$k_{em}^2 = \frac{N^2}{C_p k_a}$$

The generated voltage from the harvester is defined as:

$$V = \frac{F(\omega)k_{em}^2 \beta\varphi\alpha)}{N(A_1^2 + A_2^2)}[(A_1\alpha + 2A_2\xi_1\varphi) + (2A_1\xi_1\varphi - A_2\alpha)j] \quad (1)$$

The generated electric power by the harvester is written as:

$$P = \frac{F(\omega)^2 k_{em}^4 \beta^2\varphi^2\alpha^2)\{(A_1\alpha + 2A_2\xi_1\varphi)^2 + (2A_1\xi_1\varphi - A_2\alpha)^2\}}{2N^2(A_1^2 + A_2^2)R} \quad (2)$$

where,

$$A_1 = A_3\beta\varphi + (\alpha^2 - \varphi^2)\beta k_{em}^2\varphi + A_4$$

$$A_2 = A_4\beta\varphi + 2\xi_1\alpha\beta k_{em}^2\varphi^2 - A_3$$

$$A_3 = \alpha^2 - \varphi^2 - (1+\mu)\alpha^2\varphi^2 + \varphi^4 - 4\xi_1\xi_2\varphi^2\alpha$$

$$A_4 = 2\varphi[\alpha(1-\varphi^2 - \mu\varphi^2)\xi_1 + (\alpha^2 - \varphi^2)\xi_2]$$

The electromechanical model is derived based on the compression-based piezoelectric cymbal transducer [13]. This transducer enables the endurance of hefty loads present in the domain of public infrastructure schemes. A harvester equipped with this transducer can be implanted into public buildings, roads and highways for electric energy harvesting. The common properties for the piezoceramics are given in Table 3.

Table 3. The piezoelectric material properties.

Property	PZT-5H	PZT-8	PZT-5A
$d_{33}[pm/V]$	650	225	390
$d_{33}^{eff}[nm/V]$	110	38	66
$\varepsilon_{33}^T/\varepsilon_0$	3800	1000	1800
$C_p[nF]$	1745	459	826

The *d*-coefficient is the proportionality constant between dielectric displacement and stress or strain and electric field strength and their relationships are given in Equations (3) and (4). High *d*-coefficient is desirable in materials utilized as actuators, such as in motional and vibrational applications.

$$D = \varepsilon^T E + dT \quad (3)$$

$$S = s^E T + dT \quad (4)$$

where,
D is the dielectric displacement,
ε is the permittivity,
T is the stress,
E is the electric field strength,
S is the strain,
s is the elastic compliance,
d is the piezoelectric constant.

Lead Zirconate Titanate (PZT) is the most favored material in the field of ceramic engineering in which it comes in a variety of forms.

It is frequently used in the application of transducers, sensors and actuators. Presently, the most frequently utilized in engineeringfield of piezoceramics are PZT-5A and PZT-5H [15]. The piezoelectric model used in this study is PZT-5H which is a soft ceramic material that comes with a greater piezoelectric coefficient (direct charge coefficient), d_{33} as well as larger dielectric permittivity, ε_{33}^T.

A compression-based piezoelectric model is proposed to be utilized in the roadway energy harvester. Under the dynamic forces of compression, every single compression cycle produces energy in the form of pulses [10]. The model revealed that under harmonic excitation, the recommended stack harvester of 36 layers can produce as much as 200 mW of electrical energy [9]. This level of electrical energy is enough for the operation of nearly all forms of wireless sensors. Fig. 3 displays the electricity harvesting during the passage of a car (or any other vehicle) across the harvesters implanted in the roadway.

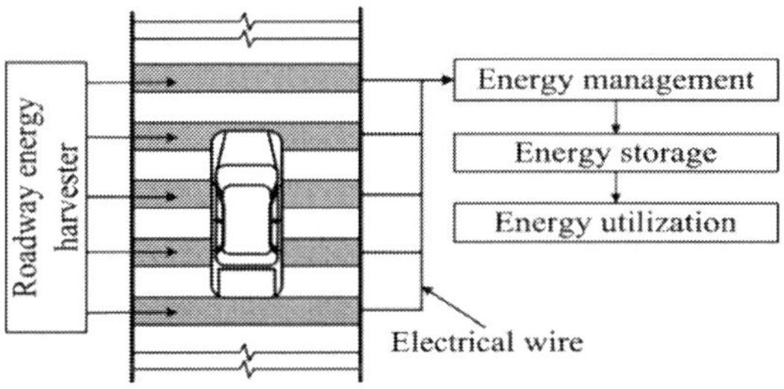

Figure 3. Roadway energy harvester on the road [9].

In order to demonstrate the potential of energy harvesting, a piezoelectric harvester unit is simulated so that the movement of a vehicle on the roadway is exploited. The electromechanical-traffic model is developed in Matlab environment in which the single cell of the piezoelectric consist of a poled piezoelectric disk (fully electrodes on top and bottom surface) which is sandwich with both truncated cone-shaped metal end caps, each containing a shallow air-filled cavity on their inner surface as shown in Fig. 4.

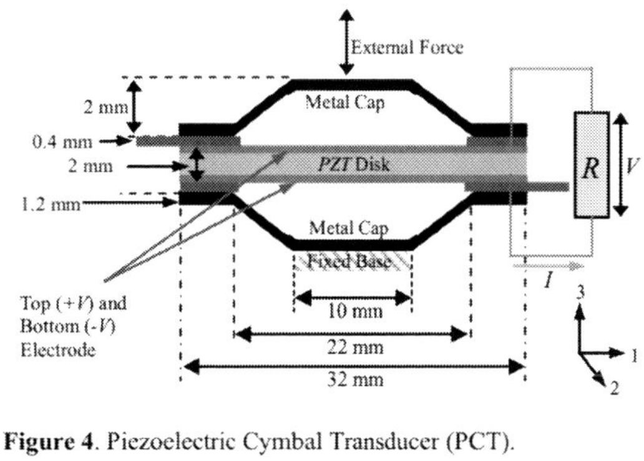

Figure 4. Piezoelectric Cymbal Transducer (PCT).

4 Simulation results

The passage of a vehicle across a pavement produces minor vibrations in the pavement. The conversion of these vibrations into electricity is achievable through the implanting of piezoelectric sensors into the road. The vibrations in fact depend on the road condition and type of vehicle passing through in terms of speed and mass of the vehicle.

The relationships between the amplitudes of the generated power and voltage with varying external electrical loads under the vibration excitation of 4 and 20 Hz, respectively is displayed in Figures 5 and 6.

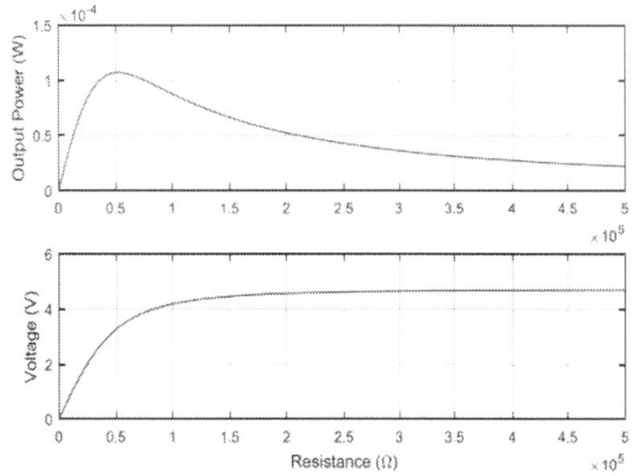

Figure 5. Output power and voltage vs. electrical loads under vibration excitation of 4 Hz.

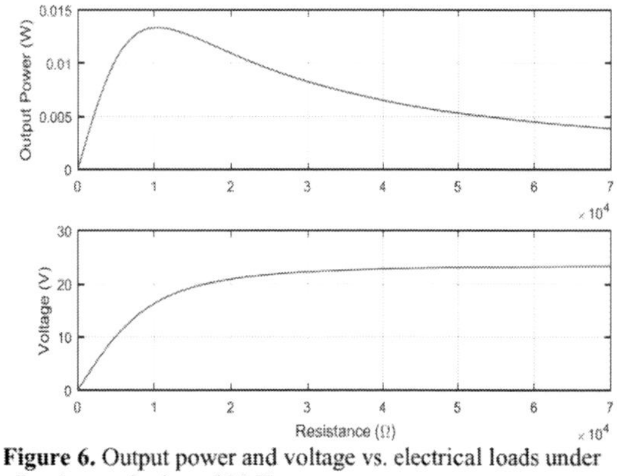

Figure 6. Output power and voltage vs. electrical loads under vibration excitation of 20 Hz.

The highest amplitude of output power as recorded in Fig. 5 with the vibration excitation of 4 Hz is 0.12 mW with an external load resistance of 50 kΩ. Within a particular external resistance span, the voltage is raised together with the value of the external loads. Upon the external resistance surpassing a particular value, the voltage becomes constant. The maximum voltage level is remained constant at about 4.8 V with the external load resistance of 200 kΩ onwards. Fig. 6 exhibits a significant maximum voltage increased to about 23 V with the vibration excitation of 20 Hz. Likewise, the output power is raised to about 13 mW at 10 kΩ of load.

The increase of excitation frequency leads to the raise of harvested power as well as the voltage as shown in Fig. 7.

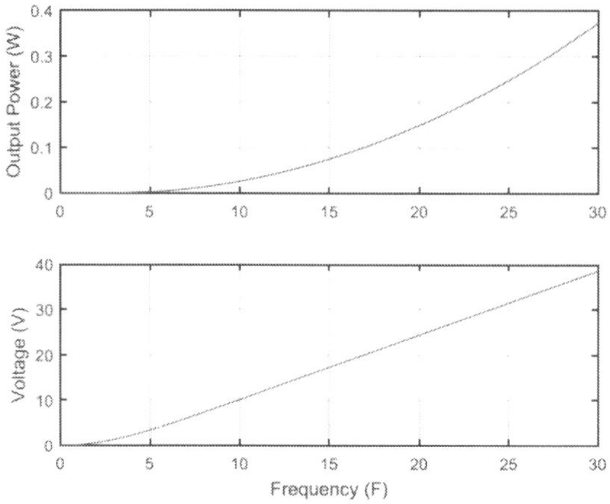

Figure 7. Harvested power and voltage vs. frequency.

Fig. 8 demonstrates the utilization of Cellular Automata (CA) to realize the simulation results of harvested power and voltage in real-time for a single lane traffic model.

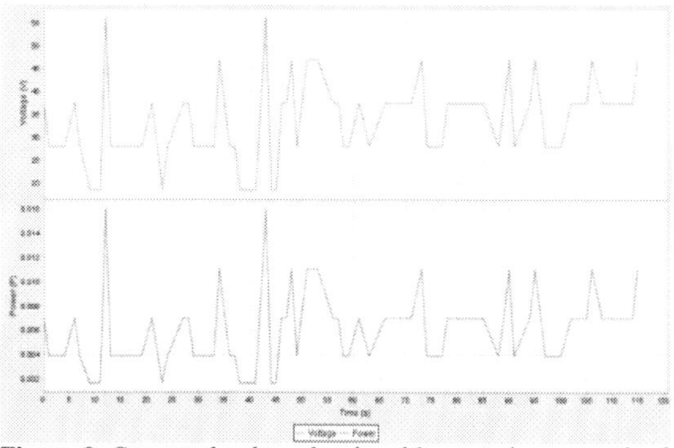

Figure 8. Generated voltage level and harvested power at load resistance of 100 kΩ.

The passage of 70 two-axle vehicles with an average speed of 70 km/h within 2 minutes across a single PCT resulted in an estimated overall power generation of 417 mW and average voltage of 34 V. The PCT arrays can be placed along one kilometer stretch of road to elevate the amount of power harvested.

5 Conclusion

This investigation delved into energy harvesting with a compression-based piezoelectric harvester that exploits vibrations derived from traffic movement. The energy acquired through this process can be used to supply electricity for the operation of remote control tools utilized in transportation systems. The outcomes from simulation revealed that the generation of as much as 13 mW of power under the excitation frequency of 20 Hz is achievable with a single PCT harvester. Other than the piezoelectric harvester, the external electrical load resistance also plays a role in determining the level of electrical power realized through compression-based piezoelectric harvesting system.

As an elevated velocity is indicative of an elevated exciting frequency, the output power rises in tandem with the velocity of a passing vehicle. With several series' of PCT, it is well within the means of the recommended roadway harvester to harvest about 150 kW per kilometer from a highway with a traffic volume of 600 vehicles per hour. The amount of power harvested is sufficient for the powering of common roadway appliances that include lighting schemes, emergency communication equipment and roadside billboards.

Acknowledgement

The authors gratefully acknowledged the financial support by the Ministry of Higher Education (MOHE) Malaysia under ERGS research grant with the vote number of E035 for this research works. The authors wish to express their heartfelt appreciativeness to Office for Research, Innovation, Commercialization and Consultancy Management (ORICC), UTHM for their continuous support and assistance in managing the research grant.

References

1. H.Y. Wang, , L.H. Tang, Y. Guo, X.B. Shan, and T. Xie, Journal of Zhejiang University SCIENCE A, **15**, 711-722 (2014)
2. M. El-Hami, P. Glynne-Jones, N.M. White, M. Hill, S. Beeby, E. James, A.D. Brown, and J.N. Ross, Sensors and Actuators A: Physical, **92**, 335-342 (2001)
3. M. Miyazaki, , H. Tanaka, G. Ono, T. Nagano, N. Ohkubo, T. Kawahara, and K. Yano, Low Power Electronics and Design, 2003. ISLPED'03. Proceedings of the 2003 International Symposium on, 193-198 (2003)
4. C. Keawboonchuay and T.G. Engel, Pulsed Power Conference, 2003. Digest of Technical Papers. PPC-2003. 14th IEEE International, **1**, 327-330 (2003)
5. J. Yang, Z. Chen, and Y. Hu, Ultrasonics, Ferroelectrics, and Frequency Control, IEEE Transactions on, **54**, 190-195 (2007)
6. S.F. Ali, M.I. Friswell, and S. Adhikari, Journal of Intelligent Material Systems and Structures, 1929 (2011)
7. D. Hill, A. Agarwal and N. Tong, *Assessment of Piezoelectric Materials for Roadway Energy Harvesting : Cost of Energy and Demonstration Roadmap: Final Project Report*, (2014)
8. S. Vassiliadis, *Advances in Modern Woven Fabrics Technology*, (2011)
9. X. Jiang, Y. Li, J. Li, J. Wang, and J. Yao, Journal of Renewable and Sustainable Energy, **6**, 043110 (2014)
10. X. Jiang, Y. Li, J. Wang, and J. Li, International Journal of Smart and Nano Materials, **5**, 152-168 (2014)
11. M. Peigney and D. Siegert, Smart Materials and Structures, **22**, 095019 (2013)
12. H. Zhao, J. Ling, and J. Yu, Journal of the Ceramic Society of Japan, **120**, 317-323 (2012)
13. H.G. Chua, B.C. Kok, and H.H. Goh, Energy and Sustainability V, **186**, 103 (2014)
14. K.F. Chen, J.H. Ho, and E.H. Yap, World Academy of Science, Engineering and Technology, International Journal of Electrical, Computer, Energetic, Electronic and Communication Engineering, **9**, 708-714 (2015)
15. K.A. Cook-Chennault, N. Thambi, and A.M. Sastry, Smart Materials and Structures, **17**, 043001 (2008)

Analysis of Asymmetrical Modes in Medium Voltage Electrical Grids with Compensated Neutral

Stanislav Girshin, Vladimir Goryunov, Evgenii Kuznetsov, Dmitrii Safonov and Elena Petrova

Omsk State Technical University, Energy Department, 644050 pr. Mira 11, Omsk, Russian Federation

Abstract. A mathematical model of steady-state radial medium voltage electrical network with earthed neutral via arc-suppression coils has been carried out. A detailed equivalent circuit of the examined network using the completed equivalent circuit of phase coordinates methods and Kirchhoff's laws has been developed. The system of mathematical models equations has been completed. The advantage of the created mathematical model is the ability to increase the accuracy of the simulation by a detailed account of features of a network with arc-suppression coils. The mathematical model considers the load currents, the non-linear nature of consumers, the impact of the power lines asymmetry. Using the developed mathematical model there are presented results of the carried out researches of three power lines with different cross sections with neutral grounding via arc-suppression coils. The analysis of asymmetry parameters impact of power lines on the value of the neutral displacement voltage and on the voltage value at the end of the lines at different values of the reactor inductance has been completed. The conclusion about the model possibility to perform validation of the reactor tuning, to evaluate voltage quality parameters, to carry out the formation of reliable initial conditions for calculating the overload in the single-phase earth faults. In conclusion the model performs validation of the reactor tuning and it evaluates voltage quality parameters in order to carry out the formation of reliable initial conditions. It also calculates the overload in the single-phase earth faults.

1 Introduction

Electrical medium voltage networks of a large number of countries (Germany, Austria, France, Italy, Finland, Sweden, Norway, Poland, Czech Republic, Russia, and others) use the mode of neutral grounding via arc-suppression coils. The popularity of this mode is due to the benefits of eliminating negative results common for electrical networks of single-phase earth fault. When single-phase ground fault the arc-suppression coils creates at the point of fault inductive current equal to the capacitative current. As a result, the total current of the fault is virtually zero, so there is no need for a fast tripping of the damaged power lines. The effectiveness of arc-suppression coils when faults in the network can be improved with further development of automated control systems of arc-suppression coils. This task can be achieved when using mathematical models of the electrical network with arc-suppression coils adequately reflecting the processes. One of the important features of the medium voltage electrical networks is asymmetry of operating modes, due to the presence of sufficiently long and untransposed lines.

Especially important asymmetry is in networks with compensation of capacitative currents and for solving problems connected with operation modes of neutral [1]-[4]. The presence of natural asymmetry can lead to abnormally large voltage of neutral displacement in normal modes [5]-[8] and requires detailed consideration. Moreover, the mode asymmetry leads to increased energy losses in the power network compared to a symmetric mode in the same consumer loads. Therefore, it is of interest to analyze such modes from energy efficiency points. Different approaches to the calculation of electric networks modes are presented in [9], [10]. Carried out and presented in this paper researches are to improve the accuracy of asymmetrical modes simulation considering not only the cross but also longitudinal equivalent circuits of the power line, as well as other characteristics of electrical networks with compensated neutral. The advanced mathematical model which allows to improve the quality of the research has been developed. This approach makes it possible to find the actual voltage drop in the examined electrical networks, the real voltage distribution considering asymmetry and determine the magnitude of the neutral displacement.

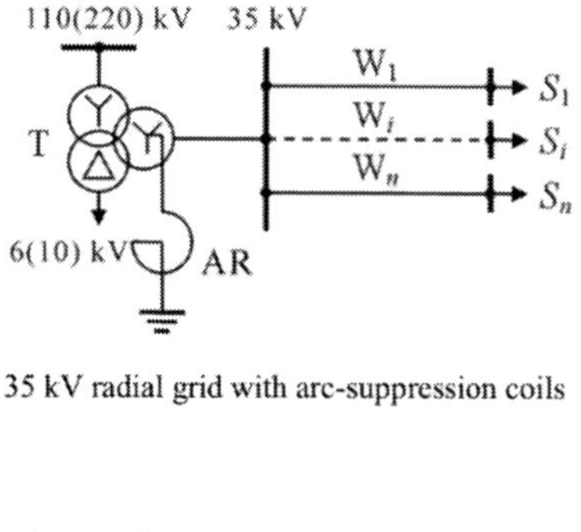

Figure 1. 35 kV radial grid with arc-suppression coils

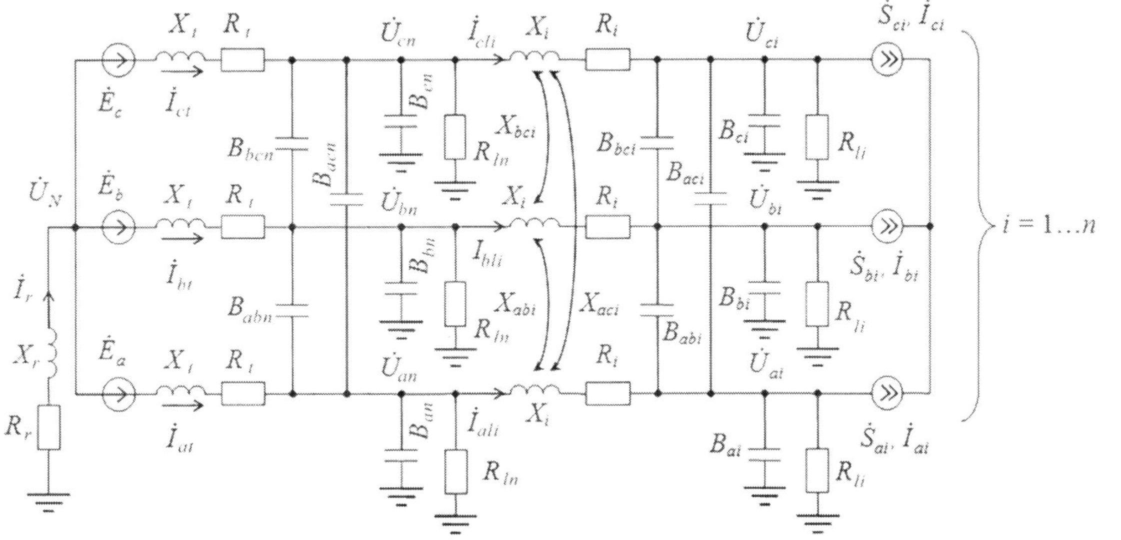

Figure 2. Equivalent circuit of 35 kV radial network with arc-suppression coils

2 Mathematical model

Radial network of 35 kV with arc-suppression coils are shown in Fig. 1 and its equivalent circuit are shown in Fig. 2.

The reason of the mode asymmetry is in nonequivalence of the geometric position of the overhead lines phase wiring. The result is in asymmetric own capacities, phase to phase (reciprocal) capacities [11] as well as the mutual inductance of each line.

The mathematic model is based on the method of phase coordinates. The allowances are:

1. The medium voltage winding of the transformer is considered as a three-phase source of EMF ($\dot{E}_a = const,\ \dot{E}_b = const,\ \dot{E}_c = const$). The EMF can be either symmetrical or asymmetrical.

2. Load is given with constant power, which corresponds to the actual source data and considers non-linear (in the inertial sense) character of consumers. When calculating the load currents one eliminates first voltage component of zero-sequence $\dot{U}_{a12i}$, $\dot{U}_{b12i}$, $\dot{U}_{c12i}$ and then current component of the zero sequence $\dot{I}_{0i}$, $\dot{I}_{ai}$, $\dot{I}_{bi}$, $\dot{I}_{ci}$.

The system of mode equations on a phase has the form by equation (1)-(27):

$$\dot{U}_{a12i} = \frac{1}{3}\left(2\dot{U}_{ai} - \dot{U}_{bi} - \dot{U}_{ci}\right) \quad (1)$$

$$\dot{U}_{an} = \dot{U}_N + \dot{E}_a - \dot{I}_{at}\left(R_t + jX_t\right) \quad (2)$$

$$\dot{I}_{0i} = \frac{1}{3}\left(\frac{\overset{*}{S}_{ai}}{\overset{*}{U}_{a12i}} + \frac{\overset{*}{S}_{bi}}{\overset{*}{U}_{b12i}} + \frac{\overset{*}{S}_{ci}}{\overset{*}{U}_{c12i}}\right) \quad (3)$$

$$\dot{I}_{ai} = \frac{\overset{*}{S}_{ai}}{\overset{*}{U}_{a12i}} - \dot{I}_{0i} \quad (4)$$

$$\dot{I}_{ali} = \dot{I}_{ai} + \dot{U}_{ai}\left[\frac{1}{R_{li}} + j\left(B_{ai} + B_{abi} + B_{aci}\right)\right] - j\dot{U}_{bi}B_{abi} - j\dot{U}_{ci}B_{aci} \quad (5)$$

$$\dot{I}_{at} = \sum_{i=1}^{n}\dot{I}_{ali} + \dot{U}_{an}\left[\frac{1}{R_{ln}} + j\left(B_{an} + B_{abn} + B_{acn}\right)\right] - j\dot{U}_{bn}B_{abn} - j\dot{U}_{cn}B_{acn} \quad (6)$$

$$\dot{I}_r = \dot{I}_{at} + \dot{I}_{bt} + \dot{I}_{ct} \quad (7)$$

$$\dot{U}_N = -\dot{I}_r\left(R_r + jX_r\right) \quad (8)$$

$$\dot{U}_{ai} = \dot{U}_{an} - \dot{I}_{ali}\left(R_i + jX_i\right) - j\dot{I}_{bli}X_{abi} - j\dot{I}_{cli}X_{aci} \quad (9)$$

$$B_{abn} = 0.5\omega\sum_{i=1}^{n}C_{0,ab,i}l_i \quad (10)$$

$$B_{abi} = 0.5\omega C_{0,ab,i}l_i \quad (11)$$

$$B_{aci} = 0.5\omega C_{0,ac,i}l_i \quad (12)$$

$$B_{ai} = 0.5\omega C_{0,a,i}l_i \quad (13)$$

$$C_{0,a} = \beta_a + \beta_{ab} + \beta_{ac} \quad (14)$$

$$C_{0,ab} = -\beta_{ab}, \quad C_{0,ac} = -\beta_{ac} \quad (15)$$

$$\boldsymbol{\beta} = \begin{pmatrix} \beta_a & \beta_{ab} & \beta_{ac} \\ \beta_{ba} & \beta_b & \beta_{bc} \\ \beta_{ca} & \beta_{cb} & \beta_c \end{pmatrix} = \boldsymbol{\alpha}^{-1} \quad (16)$$

$$\boldsymbol{\alpha} = \begin{pmatrix} \alpha_a & \alpha_{ab} & \alpha_{ac} \\ \alpha_{ba} & \alpha_b & \alpha_{bc} \\ \alpha_{ca} & \alpha_{cb} & \alpha_c \end{pmatrix} \quad (17)$$

$$\alpha_{i,i} = \frac{1}{2\pi\varepsilon_0} \ln\frac{2h_i}{r_i} \quad (18)$$

$$\alpha_{i,j} = \alpha_{j,i} = \frac{1}{2\pi\varepsilon_0} \ln\frac{r'_{i,j}}{r_{i,j}} \quad (19)$$

$$r'_{i,j} = \sqrt{r_{i,j}^2 + 4h_ih_j} \quad (20)$$

$$X_i = \omega\left(\frac{\mu_0 l_i}{2\pi}\left(\ln\frac{2l_i}{r_i} - 1\right) + \frac{\mu l_i}{8\pi}\right) \quad (21)$$

$$X_{abi} = \omega\frac{\mu_0 l_i}{2\pi}\left(\ln\frac{2l_i}{r_{ab,i}} - 1\right) \quad (22)$$

$$X_{bci} = 2\cdot 10^{-7}\omega l_i\left(\ln\frac{2l_i}{r_{bc,i}} - 1\right) \quad (23)$$

$$X_{aci} = 2\cdot 10^{-7}\omega l_i\left(\ln\frac{2l_i}{r_{ac,i}} - 1\right) \quad (24)$$

$$R_{ln} = \frac{2U_{nom}^2}{\sum_{i=1}^{n}\Delta P_{1,0,i} l_i}, \quad R_{li} = \frac{2U_{nom}^2}{\Delta P_{1,0,i} l_i} \quad (25)$$

$$X_r = \frac{k_e}{\omega\sum_{i=1}^{n}\left(C_{0,a,i} + C_{0,b,i} + C_{0,c,i}\right) l_i} \quad (26)$$

$$R_r = \Delta P_r \frac{U_{\mathrm{nom},r}^2}{S_{\mathrm{nom},r}^2} + R_g \quad (27)$$

where $i = 1 \ldots n$; n is a number of lines; $\dot{U}_{a12i}$, $\dot{U}_{b12i}$, $\dot{U}_{c12i}$ are voltage component of the zero sequence $\dot{U}_{ai}$, $\dot{U}_{bi}$, $\dot{U}_{ci}$ are phase voltages at the end of the line; $\dot{U}_{an}$, $\dot{U}_{bn}$, $\dot{U}_{cn}$ are a, b, c phase voltages respectively the neutral conductor; $\dot{U}_N$ is neutral displacement voltage; $\dot{E}_a$ is EMF of a phase of the power supply which is regarded as a booster; $\dot{I}_{at}$, $\dot{I}_{bt}$, $\dot{I}_{ct}$ are currents of a, b, c phases flowing through the booster; R_t, X_t are resistance and inductance of the low-voltage winding feeder transformer; $\dot{I}_{0i}$, $\dot{I}_{ai}$, $\dot{I}_{bi}$, $\dot{I}_{ci}$ are current component of the zero sequence; $\overset{*}{S}_{ai}$, $\overset{*}{S}_{bi}$, $\overset{*}{S}_{ci}$ are load power; $\dot{I}_{ali}$, $\dot{I}_{bli}$, $\dot{I}_{cli}$ are linear currents of a, b, c phases; R_{ln}, R_{li} are resistances to surface-leakage currents through the insulators at the start and in the end respectively; B_{ai}, B_{abi}, B_{aci}, B_{abn}, B_{acn}, B_{an} are internal and mutual capacity admittance of the line; $\dot{I}_r$ is reactor current; R_r, X_r are resistance and inductance of the reactor; X_i is internal inductive resistance of the line phase; X_{abi}, X_{aci} are mutual impedance of the line phases; R_i is resistance in the line; ω is angular frequency; $C_{0,a,i}$, $C_{0,b,i}$, $C_{0,c,i}$, $C_{0,ab,i}$, $C_{0,ac,i}$ are linear phase and phase-to-phase capacitances; ; l_i is the length of the line; β is a coefficient matrix of electrostatic induction; α is the matrix of potential coefficients; $\varepsilon_0 = 8.85\cdot 10^{-12}$ F/m is the dielectric constant; h_i is the height of the phase wire on the ground; r_i is the conductor radius; $r_{i,j}$ is the distance between the phase wires i and j; $r'_{i,j}$ is the distance from the wire i to j mirroring the wire relative to the surface of the earth; $\mu_0 = 4\pi\cdot 10^{-7}$ H/m is a magnetic constant; μ is absolute magnetic permeability of the conductor material taken equal to μ_0; $S_{nom,r}$, $U_{nom,r}$ are rated power and rated voltage of the reactor; $\Delta P_{l,0,i}$ is linear power loss in insulators; R_g is the equivalent earth resistance between the lines and the reactor capacities; k_e is a coefficient considering the inaccuracy of adjustment; ΔP_r is rated active power losses; U_{nom} is rated supply voltage.

Equations (1), (3) and (4) simulated electrical loads with the exclusion of the zero sequence currents. Equation (2) expresses the relation of phase voltage at the beginning of the line and neutral bias on the basis of the Kirchhoff's second law, and equation (9) expresses voltage relationships at the beginning and at the end of the line considering the mutual inductance. Equation (8) defines a neutral voltage displacement as the voltage drop at the reactor resistances. Equations (4), (5) and (6) represent the first law of Kirchhoff for the nodes at the end of the line, at the beginning of the line and the connection point of the reactor respectively. Capacitive admittances of the lines are determined with equations (10-20) based on the matrix inversion of potential coefficients (17). Self and mutual inductances of the lines are calculated with equations (21-24) via the geometric parameters. The mathematical model considers active resistances of the transverse lines of the branches, which are calculated according to equation (25) through the losses per unit length in insulators. Equation (27) allows to consider ground resistance together with the active impedance of the reactor. Equation (26) represents an equation of the reactor setting to the phase capacities of the lines. The complete system of the mode equations is obtained with adding to equations (1-27) of a phase equations describing the phase b and phase c, which are recorded similarly.

3 Simulation and results

The results of the numerical solution of the complete system of equations for the radial 35 kV network with three lines are presented. Initial data are given in Table 1. The MV winding of the transformer is considered as the source of three-phase EMF with internal resistance. Geometric parameters are defined for the lines, as well as active resistance per unit length and losses per unit length in insulators. Consumers capacities included at the end of each line are assumed to data on the lines. The reactor is set only to the resistance, because the inductive reactance is determined with the setting equation.

Table 1. Initial data for calculations

Element	Parameter	Value	Note
The booster TDTN-25000/110 (MV winding)	Resistance R_t	0.166 Ohms	Given to medium voltage
	Inductive impedance X_t	0	
	Modules of phase EMF E_a=E_b=E_c	$35/\sqrt{3}$ KV	Form symmetric system
Lines (general information)	The height of the phase wires A and C h_a=h_c	15m	Typical intermediate steel single circuit tower [12]
	The height of the phase wire B h_b	18m	
	Phase spacing r_{ab}	5m	
	Phase spacing r_{bc}	3.3 m	
	Phase spacing r_{ac}	5.3 m	
	Linear power loss in insulators ΔP_{L0}	0.3 kWh/km	On the basis of generalized data [13]
Line 1, AS-120/19	Length	8 kilometers	
	The radius of the wire	7.6 mm	
	Linear resistance r_0,	0.2492 ohms/km	
	Load power $\dot{S}$	6,6+ j3 MBA	Uniform distribution on phases
Line 2 AS-95/16	Length	8 kilometers	
	The radius of the wire	6.75 mm	
	Linear resistance r_0,	0.306 Ohm/km	
	Load power $\dot{S}$	4.5+ j3.6 MBA	Uniform distribution on phases
Line 3, AS 70/11	Length	10km	
	The radius of the wire	5.7 mm	
	Linear resistance r_0,	0.4286 ohms/km	
	Load power $\dot{S}$	3.6+ j 2.25 MBA	Uniform distribution on phases
Reactor	Resistance	10.6 ohms	

Table 2. The results of the set modes calculations

Reactor adjustment k_e		0.8	0.84	0.88	0.9	0.92	0.94	0.96	0.98
Voltage in the end of line l, kV	U_{a1}	21.87	22.19	22.46	22.54	22.56	22.49	22.34	22.11
	U_{b1}	19	19.2	19.61	19.92	20.28	20.68	21.1	21.5
	U_{c1}	18.64	18.18	17.58	17.24	16.9	16.59	16.34	16.18
Currents in the transformer , A	I_{at}	288.8	288.8	288.9	288.9	288.9	288.9	288.9	288.9
	I_{bt}	288.7	288.7	288.7	288.7	288.7	288.7	288.7	288.7
	I_{ct}	289	289	289	289	289	289	289	289
Neutral displacement voltage U_N, kW		2.106	2.484	2.895	3.097	3.285	3.446	3.568	3.644
The current in the reactor I_r, A		0.288	0.324	0.36	0.377	0.391	0.402	0.407	0.407
Total active power losses ΔP, kW		273.4	273.5	273.5	273.5	273.6	273.6	273.6	273.6

Calculations were made for the overcompensation modes, which correspond to the value k_e<1 in (26). This choice is due to the fact that the undercompensation mode and the more accurate adjustment can lead to resonance neutral displacement voltage and therefore are not recommended or even are unacceptable. The main results of the calculations are presented in Table 2. These results include: 1) the voltage at the load nodes (at the end of the lines), which determines the quality of electricity; 2) neutral voltage displacement, which determines the tolerance of this setting; 3) current in a reactor as an additional parameter which characterizes the degree of asymmetry of the mode; 4) the total active power losses in the network. It is evident that the load voltage is substantially non-symmetrical.

However, it has little effect on the quality of electricity, because the asymmetry is mainly due to zero sequence voltage, which is filtered when passing through the step-down transformers. Power losses in the network when approaching the resonance setting of the reactor increase, but a little. They are essentially the same in all modes as in the absence of the reactor (273.4 kW). It is due to the small asymmetry of the currents in the network and it indicates the energy efficiency of the considered modes.

4 Conclusions

1. In the table 2 is presented the voltages at the end of line 1 (the results for the other lines are similar). Real values of these voltages can be obtained only on the basis of models appropriate to the accuracy considered. Simplified models, not considering the load, will lead either to overrated voltage value at the end of the lines (if you consider the longitudinal branch), or to the voltage equation at the beginning and the end of lines (with neglect of the longitudinal branches).

2. If the voltage is noticeably asymmetrical, then for currents asymmetry is much less expressed (Table 2 shows the currents in the transformer, similar results are obtained for currents in the lines.) It should be expected, because the load is not sensitive to the residual voltage which is observed in practice. It proves the adequacy of the adopted model of the load.

3. Neutral displacement voltage increases monotonically with increasing reactor inductivity. The mode is expected as it is close to resonance. However, instead of the expected growth of voltage rise one observes some slowing. Thus, the neutral displacement voltage is more stable than it is obtained with the simplified model. Under these conditions, it is estimated around 10% from the nominal line voltage.

4. The presence of arcing reactor and changing its settings almost do not increase the energy losses in the network. Thus, the considered mode of the neutral operation is energy efficient.

The mathematical model can be used for a comprehensive analysis of steady-state modes of radial 35 kV networks with arc reactor, including the validation settings of the reactor and assessment of voltage quality. In addition, this model provides a realistic initial conditions for calculating the overvoltage in the single-phase earth faults.

Acknowledgment

The paper was prepared in Omsk State Technical University with financial support from Ministry of Education and Science in the framework of the agreement No. 14.577.21.0097 on August 22, 2014. A unique identifier of applied research RFMEFI57714X 0097.

References

1. X. T. Wang, J. M. Lin, L. G. Ban, Z. T. Xiang, and B. Han, Power Syst. Techn., **33**, 5, pp. 1-5, (2009)
2. K. Tao, and M.B.Liu, Power Syst. Protec. and Contr., **38**, 16, pp. 39-43, (2010)
3. S. Zhong and A. Abur, *in Proc. IEEE Power Eng. Soc.Winter Meeting*, pp. 975–979, (2002)
4. L. Zou, F. Lin, Z. Long, J. Xie, Power Syst. Techn., **32**, 2, pp. 283-286, (2008)
5. R. Burgess, and A. Ahfock, Australian Journ. of Electr. & Electron. Eng., 10, 1, pp. 127-136, (2013)
6. M. Qian, B. Zheng ; L. Ban, and Z. Xiang, IEEE Trans. Power Del., **30** , 3, pp. 1458 - 1466, (2015).
7. Z. Xu, X. Yan, X. Zhang, and A. Wen, IET Generat., Transmiss., Distrib., **7**, 1, pp. 1–8, (2013).
8. Zadeh M.R.D., Sanaye-Pasand M., Kadivar A., IEEE Trans. Power Del., **23**, pp. 2472-2479, (2008).
9. V. Leitloff, L. Pierrat, and R. Feuillet, Eur. Trans. Elect. Power Eng, **4**, pp. 145–153, (1994)
10. I.M. Sirota, S.N. Kislenko, and A.M. Mikhailov, *The neutral electrical networks*, Kiev: Nauk.Dumka, 264 p., (1985).
11. K.S.Demirchian, L.R.Neumann, N.V.Korovkin, and V.L.Chechurin, *Theory of Electrical Engineering: In 3 V,* The textbook for high schools, **3**, SPb: Peter, 377 p., (2006).
12. V.G.Gerasimov, *Electrical reference: In 4 V. V.3. Production, transfer and distribution of electric energy*, M .: MPEI Publishing, 964 p., (2002).
13. Yu.S. Zhelezko, A.V Artemyev, O.V. Savchenko *Calculation, analysis and valuation of losses in electric networks: a guide for practical calculations,* M .: Publishing House of NTs ENAS, 280 p., (2002).

Wavelet Transform - A New Tool for Analysis of Harmonics in Power Systems

D. S. Osipov, V. N. Gorunov, A.A. Bubenchikov and P.V. Katerov

Omsk State Technical University, Omsk, Russi

Abstract. This paper presents a review of application of discrete wavelet transform in power system transient analyses. Based on the discrete time domain approximation, the system components such as resistor and inductor are modeled respectively in discrete wavelet domain for the purpose of transient and steady state analyses. Numerical results for transient inductor model can be implemented by any kind of power system including normal and emergency operating modes.

1 Introduction

A wavelet transform (WT) is a mathematical tool for analyzing signals. Wavelet analysis is well suited to nonperiodic signals that may contain both stationary components and transient signals, such as the ones that can be found in power quality disturbances. The capacity of wavelets to adapt the window size improves the analysis of these types of nonstationary signals. Originally introduced in a group-theoretic setting, WT was soon realized to have powerful applications in various fields, which include analysing electric power systems [1], transient and steady state study electrical circuits [2] or analyzing power quality [3]. Harmonic, interharmonic, and subharmonic distortion can be studied using different WTs, in both stationary and nonstationary conditions, showing better performance in the measurement of time-varying waveform distortion [4].

The advantage of WT is that the band of analysis can be adjusted so that high frequency components and low frequency components can be detected precisely. Results from the wavelet transform are shown on both the time domain and the frequency domain.

Reference [4] provides extensive bibliographic review of applications of WTs in the measurement and analysis of harmonic distortion in power systems, discussing the performance of the different methods proposed in the technical literature.

2. Wavelet theory

Basically WT decomposes a given function f into its components on different scales or frequency bands. This is done by convoluting f with the translated and dilated wavelet ψ :

$$F_{\psi}(b,a)=\frac{1}{\sqrt{a}}f(t)\psi\left(\frac{t-b}{a}\right)dt \qquad (1)$$

In Equation (1) $f(t)$ is the signal to be analyzed, F_{ψ} is the wavelet transform of $f(t)$ based on mother wavelet $\psi(t)$, and b and a are the shift and dilation parameters, respectively. Each wavelet is created by scaling and translation operations in a special function called mother wavelet. A mother wavelet is a function that oscillates, has finite energy and zero mean value. In time-frequency domain the information regarding both time and frequency are preserved making it better than Fourier transform, i.e. a frequency domain approach. Also, the problem of fixed window size as in short time Fourier transform (STFT) is solved in WT.

A wavelet transform can be divided into a continuous wavelet transform (CWT) and a discrete wavelet transform (DWT). Fig. 1 represents CWT of the nonstationary signal under examination, the CWT giving a clear idea of amplitude, frequency and time variation in three-dimensional space.

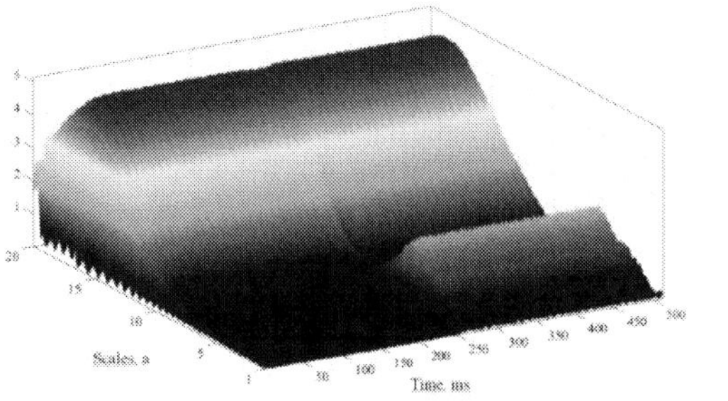

Figure 1. Continuous wavelet transform of nonstationary current signal under examination

The calculation of wavelet spectrum under CWT causes substantial computational difficulties at a continuous variation of parameters j and k.

The set of functions ψj,k is redundant, therefore to preserve a possibility of desampling out of its conversion, the sampling of these parameters is necessary. Sampling is often calculated by powers of two.

For the time being Haar wavelets have been replaced by other types of wavelets, but still Haar wavelets are rather comfortable to form theoretical basis of signals examination using wavelet transforms.

For the purpose of numerical analysis, only DWT is briefly introduced here. More detailed information on wavelet theory is available in [5], [6].

3 Transient inductor model

Since we know that the inductor current cannot change its value instantaneously, as otherwise the inductor voltage $v_L = L \cdot di_L/dt$ would become infinite, it follows that the inductor current immediately after closing the switch, must be the same as the inductor current just before closing the switch:

$$i_L(0_-) = i_L(0_+) \tag{2}$$

Consider the RL circuit of Fig. 2 with $R = 6.35\Omega$, $L = 0.127H$, and $v(t) = \sqrt{2} \cdot 220\sin(2\pi f + \varphi)$ (AC voltage source $f = 50Hz$, $\varphi = 30^0$). We may find the expressions for the inductor current $i_L(t)$ for $t > 0$, and plot them. Determination of the optimal parameters of RLC models is carried out by processing the instantaneous values of non-sinusoidal voltage and current signals of electric devices measured in the frequency range, limited with conditions of the used model adequately [7].

Using Kirchhoff's voltage law around the loop (Fig. 2) in the clockwise direction, we have ordinary differential equation:

$$v = Ri + L\frac{di}{dt} \tag{3}$$

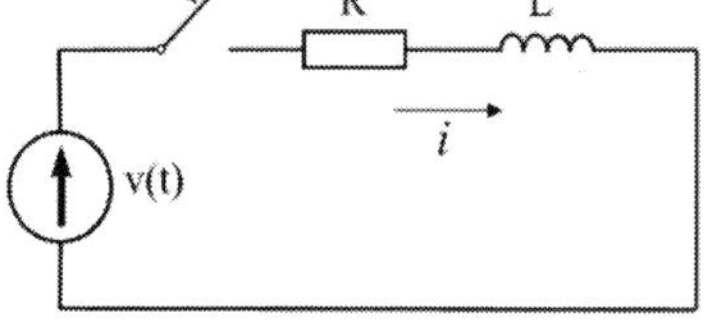

Figure 2. Analyzed RL-circuit transient

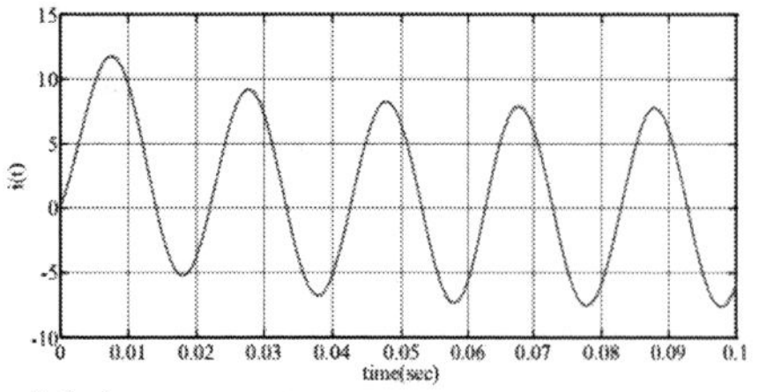

Figure 3. Inductor current

The solution is given by:

$$i_L(t) = 7.68\sin(2\pi f - 51^0) + 5.97e^{-t/0.02}$$

Reference [7] proposes to solve linear ordinary differential equations with wavelets – the wavelet recursion method.

Let $v(t)$ and $i_L(t)$ represent, respectively, the response and the forcing function vectors denoted at time $t = n\Delta t$, where Δt is the sampling interval. We may assume that the DWT representations of the vectors are:

$$v(t) = k\sum_j \sum_k v_{j,k}\psi_{j,k}(t) \tag{4}$$

$$i_L(t) = k\sum_j \sum_k i_{j,k}\psi_{j,k}(t) \tag{5}$$

The Right Formula Backward Difference (RFBD) equations begin with the backward difference approximation of the derivative in discrete time as follows [8]:

$$\frac{di_L}{dt} = \frac{i_L(t) - i_L(t - \Delta t)}{\Delta t} \tag{6}$$

Given the equation (6), the equation (3) can be rewritten:

$$\frac{1}{\Delta t}\left[i_L(t) - i_L(t - \Delta t)\right] + \frac{R}{L}i_L(t) = \frac{1}{L}v(t) \tag{7}$$

When the discrete wavelet series for $v(t)$ and $i_L(t)$ are substituted into a scalar version of Equation (7) using Equations (4-5) for the derivative approximation, one obtains (for small Δt)

$$\begin{gathered}\frac{1}{\Delta t}\left[k\sum_j \sum_k i_{Ljk}\psi_{jk}(t) - k\sum_j \sum_k i_{Ljk}\psi_{jk}(t - \Delta t)\right] + \\ + \frac{R}{L}\left[k\sum_j \sum_k i_{Ljk}\psi_{jk}(t)\right] = \frac{1}{L}\left[k\sum_j \sum_k v_{jk}\psi_{jk}(t)\right]\end{gathered} \tag{8}$$

where

$$\psi_{jk}(t) = \frac{1}{\sqrt{a_0}}\psi\left(\frac{t}{a} - b_0 k\right) \tag{9}$$

and

$$\psi_{jk}(t - \Delta t) = \frac{1}{\sqrt{a_0}}\psi\left(\frac{t - \Delta t}{a} - b_0 k\right) \tag{10}$$

The discrete wavelet transform has a fast recursive form that is most easily implemented using bank theory [9]. This algorithm requires only $O(N)$ steps instead of $O(N \cdot \ln(N))$ steps required by the FFT. Fig. 4 shows a simplistic way of looking at a wavelet decomposition of signal.

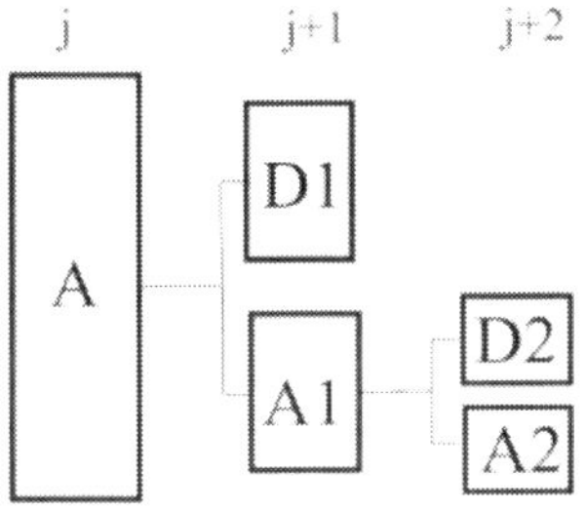

Figure 4. Simplistic view of a wavelet decomposition

Wavelet transform algorithm produces DWT coefficients starting from separating the original signals $v(t)$ and $i_L(t)$ of length N (number of simples) to 2 set of coefficients: approximate coefficients A1 by low pass filter and detail coefficients D1 by high pass filter. The length of each filter is equal to half of original length by down sampling function. The next step splits the approximate coefficients A1 in two parts again by the same process but replaces by A1 and produces A2 and D2 and so on. Thus, if the decomposition stopped at this point, the approximation of the signal at level j could be reconstructed by the sum of the approximation at level $j+1$ and the details at level $j+1$. That is

$$A_j = A_{j+1} + D_{j+1} \quad (11)$$

Since the wavelet and scaling functions have the aforementioned properties of orthonormality, one may view the Equation (8) as a linear combination of the detail coefficients and the approximation coefficients.

In the Multiresolution analysis (MRA) theory described by Mallat [5], a wavelet expansion of function (in this case function of inductor current and voltage) will have the form:

$$i_L(t) = \sum_{k=-\infty}^{\infty} i_{Lj,k}^{A} \varphi_{j,k}(t) + \sum_{j=J}^{\infty} \sum_{k=-\infty}^{\infty} i_{Lj,k}^{D} \psi_{j,k}(t) \quad (12)$$

$$v(t) = \sum_{k=-\infty}^{\infty} v_{j,k}^{A} \varphi_{j,k}(t) + \sum_{j=J}^{\infty} \sum_{k=-\infty}^{\infty} v_{j,k}^{D} \psi_{j,k}(t) \quad (13)$$

In equation (12-13) the superscripts D and A represent the detail and approximation coefficients.

Based on Parseval equality one can calculate the effective current within the required range of frequencies. In this paper the effective elementary frequency (50 Hz) can be obtained by approximation and detailed wavelet coefficients of the third level decomposition (in this case a maximum one):

$$I_{(50)} = N^{-\frac{1}{2}} \sqrt{\sum_{k=0}^{k_m} \left(i_{3,k}^{A}\right)^2 + \sum_{k=0}^{k_m} \left(i_{3,k}^{D}\right)^2} \quad (14)$$

The effective currents of higher harmonics $I_{(\nu)}$ are defined by detailed coefficients of the first two decomposition levels:

$$I_{(\nu)} = N^{-\frac{1}{2}} \sqrt{\sum_{j=1}^{2} \sum_{k=0}^{k_m} \left(i_{j,k}^{D}\right)^2} \quad (15)$$

where, N – the number of readings of the discrete signal examined

Thus the effective values obtained can be substituted in the formula of waveform distortion factor:

$$K_u = \frac{I_{(\nu)}}{I_{(50)}} = \sqrt{\frac{\sum_{j=1}^{2} \sum_{k=0}^{k_m} \left(i_{j,k}^{D}\right)^2}{\sum_{k=0}^{k_m} \left(i_{3,k}^{A}\right)^2 + \sum_{k=0}^{k_m} \left(i_{3,k}^{D}\right)^2}} \quad (16)$$

The task of wavelet analysis implementation for the voltage and current sine - wave signals distortion identification in industrial power systems with the wavelet carrier choice being substantiated is solved in [5]. The author examines the applicability of more than ten types of wavelets in power quality indices determination tasks. The four types of wavelets are presented in this paper (Table 1). With the wavelets specified we may define the effective currents of elementary frequency and higher harmonics with the subsequent waveform distortion factor being determined.

Table 1. Comparison of calculation accuracy of some wavelets

	Analytical calculation	Calculation on wavelet coefficients for wavelets			
		Haar	Daubechies 10	Daubechies 24	Daubechies 42
Effective value of elementary frequency $I_{(50)}$, A	0,707	0,6927	0,7176	0,7106	0,7095
Effective value of higher harmonics $I_{(\nu)}$, A	0,4302	0,4529	0,4124	0,4243	0,4261
Effective value of current I , A	0,8276	0,8276	0,8276	0,8276	0,8276
Total harmonic distortion (THD)	0,6085	0,6538	0,5747	0,5972	0,6005
Distortion coefficient error, %		7,4	5,6	1,9	1,3

Table 1 illustrates that the type of the wavelet used defines the distortion coefficient calculation accuracy. Daubechies wavelets of orders 24 and 42 (db24, db42), demonstrate the most acceptable results, the error not exceeding 2%. The proposed techniques of defining harmonic composition of current or voltage and the modern methods of active filtering of higher harmonics [6], [8] may help solve the tasks of power quality using the advantages of wavelet transform.

The decomposition of the signal into different frequency bands is simply obtained by successive high pass and low pass filtering of the time domain signal. The original signal $v(t)$ (or $i_L(t)$) is first passed through a half-band high pass filter g[n] and a half-band low pass filter h[n]. The half-band low pass filter removes all frequencies that are above half of the highest frequency, while the half-band high pass filter removes all frequencies that are below half of the highest frequency of the signal. The low pass filtering halves the resolution, but leaves the scale unchanged. The signal is then sub-sampled by two since half of the number of samples is redundant, according to the Nyquist's rule [10]. There are many perfect reconstruction filters . We may notice that Haar filter is the first Daubechies filter.

4. Nunerical experiment results

In this section a simple differential equation (3) is solved using proposed in [8] Right Formula Backward Difference method.

4.1 Steps of DWT analysis procedure

Step 1. Chose the type of wavelet. One reason for the popularity of wavelet technique is the vast choices of wavelets as the basis function, compared to the fixed choice of sine and cosine functions for Fourier analysis. This enables wavelet analysis to be adapted according to the expected characteristics of the signal. A coarse (irregular) wavelet is more suitable for a signal with sharp or abrupt transitions, while those smooth ones are better analyzed using regular wavelets. A mother wavelet is chosen by the user for the analysis. The selection of the mother wavelet is not a trivial issue. Different mother wavelets have different attributes which may or may not benefit in the analysis. Therefore the selection of the appropriate mother wavelet may be crucial [8]. We shall take Haar wavelet for the numerical experiment, which can be described as:

$$\psi(t)=\begin{cases}1 & 0\le t\le 1/2\\ -1 & 1/2\le t\le 1\\ 0 & otherwise\end{cases} \tag{17}$$

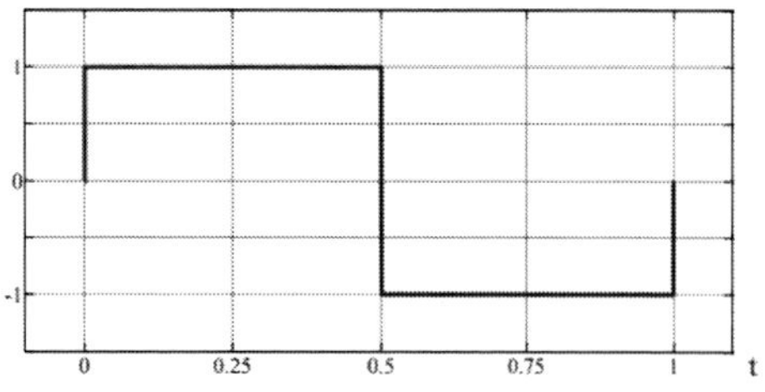

Figure 5. Haar wavelet

Its scaling function $\varphi(t)$ can be described as:

$$\varphi(t)=\begin{cases}1 & 0\le t\le 1\\ 0 & otherwise\end{cases} \quad (18$$

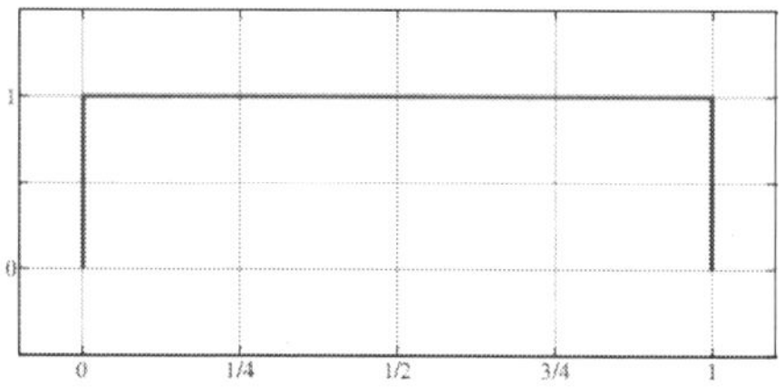

Figure 6. Haar scaling function

Step 2. Forward DWT with the selected wavelet filter, MRA decomposition scheme is performed on the system input signals to get their wavelet transform coefficients (Equations 4-5).

Step 3. Solve system equivalent circuit in wavelet domain.

$$\frac{1}{a_0^j\Delta t}i_{Ljk}^A-\frac{1}{a_0^j\Delta t}i_{Ljk-1}^A+\frac{R}{L}i_{Ljk}^A=\frac{1}{L}v_{jk}^A \tag{19}$$

$$\frac{1}{a_0^j\Delta t}i_{Ljk}^D-\frac{1}{a_0^j\Delta t}i_{Ljk-1}^D+\frac{R}{L}i_{Ljk}^D=\frac{1}{L}v_{jk}^D \tag{20}$$

In this numerical experiment $j=1$, and so $a_0^j=2^1=2$, The choosen step time $\Delta t=10^{-4}(s)=100\mu s$.

$$i_{Ljk}^A=\frac{\left(v_{jk}^A/L\right)+5\cdot 10^3\cdot i_{Ljk}^A}{5\cdot 10^3+\left(R/L\right)} \tag{21}$$

$$i_{Ljk}^D=\frac{\left(v_{jk}^D/L\right)+5\cdot 10^3\cdot i_{Ljk}^D}{5\cdot 10^3+\left(R/L\right)} \tag{22}$$

Step 4. Time-frequency response analysis: with the equivalent circuit, system transfer function in wavelet domain can also be set up. thus the characteristic of system time-frequency response can be easily achieved.

Step 5. Inverse DWT The MRA reconstruction scheme is performed on the wavelet transform coefficients of output signal to obtain their waveform in discrete time domain.

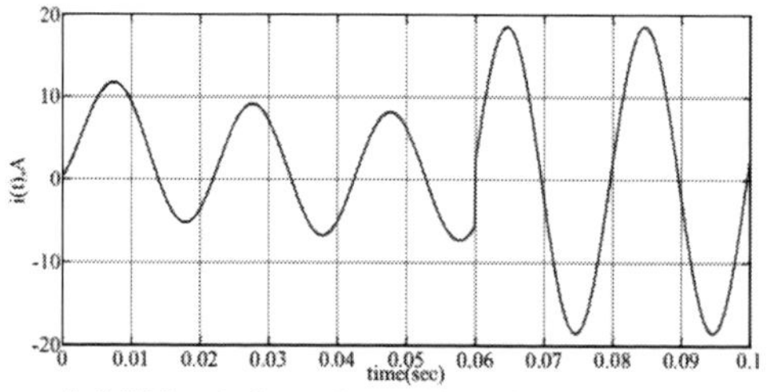

Figure 7. RFBD solutions after 300 iterations

4.2 Example analysis

Fig. 7, Fig. 8 illustrate how the recursion progress toward the solution (solution after 300 and 600 iterations).

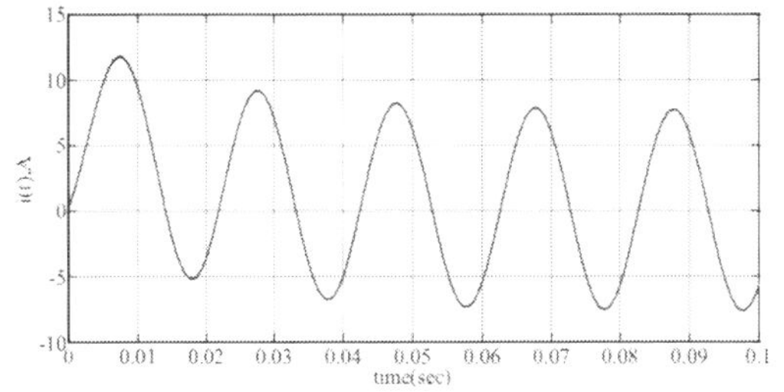

Figure 8. RFBD solutions after 600 iterations

5 Conclusion

This paper has presented a method for transient analysis of electric power systems. This method, referred to as the wavelet recursion method, makes use of a relatively new mathematical tool, the wavelet transform, which has certain benefits over techniques such as Fourier analysis when analyzing transient and nonstationary signals. Through the application of operator representation theory, the system component models in the wavelet domain are set up. Proposed in [2,8] models can be implemented by any type of orthogonal wavelet transform. Once the discrete time domain models are obtained, wavelet domain models can be easily calculated by MRA decomposition scheme.

An assessment of the advantages and disadvantages of the wavelet recursion method are as follows [8]:

- The DWT produces sparse representations of signals, thus sparse matrix techniques could be applied to this method to improve speed.
- In many cases, the wavelet recursion method converges relatively fast.
- Wavelet analysis software is readily available and adaptable to a variety of applications, including power engineering.
- Matrix inversions (at most, one inversion per level of decomposition) may be required.

In this paper the resistor model is considered as a scalar, the inductor model is based on the differential operator. The proposed scheme is suitable for transients and time varying analysis. Along with the proposed procedure, a switching of inductor is used as an example in this paper.

The paper was prepared in Omsk Srate Technical University with financial support from Ministry of Education and Science in the framework of the agreement № 14.577.21.0097 on August 22, 2014. A unique identifier of applied research RFMEFI57714X 0097.

References

1. P. F. Ribeiro, "Wavelets: A new approach to analyze power system distortions" in Proc. EPRI-Power Quality Steering Committee, Baltimore, MD, (1993).
2. T. Zheng, E. Makram, and A. A. Girgis, "Power system transient and harmonic studies using wavelet transform," IEEE Trans. Power Del., **14**, no. 4, pp. 1461–1468, Oct. (1999)
3. S. Santoso, E.J. Powers, and W.M. Grady, " Wavelet-based Power Quality Event Identification System, Part 1The Wavelet Transform and Feature Extraction of Power Quality Disturbances and Part 2: Design and Implementation," ERPI Report, August (1997)
4. J. Barros, R. I. Diego and M. de Apraiz. "Applications of Wavelet Transform for Analysis of Harmonic Distortion in Power Systems: A Review". IEEE Transactions on instrumentation and measurement, **61**, no. 10. pp. 2604-2611, October (2012)
5. Mallat, S. "A wavelet tour of signal processing" Academic Press, New York, (1998)
6. Dwight F. Mix and Kraig J. Olejniczak "Elements of wavelets for engineers and scientists". John Wiley & Sons, Inc. New York, NY, USA (2003)
7. A.N. Kirichenko, V.N. Gorunov, A.V. Schekochikhin, V.V. Barskov. "Algorithms of optimal parametric synthesis of RLC models in the problem of parameter identification of electrical devices". Dynamics of Systems, Mechanisms and Machines (Dynamics), (2014)
8. A. W. Calli, "Analysis of Electrical Transients In Power System Via A Novel Wavelet Recursion Method", Ph.D. Dissertation, Purdue University (1997)
9. G. Strang and T. Nguyen. Wavelets and Filter Banks. Wellesly-Cambridge Press. Wellesly. MA. (1996).
10. I.D. Nicolae, P.M. Nicolae , M.S. Nicolae. "Real-time Analysis Using Discrete Wavelet Transform in Power Systems" 15th International Power Electronics and Motion Control Conference, EPE-PEMC 2012 ECCE Europe, Novi Sad, Serbia

Enhancing the Performance of Building Load Forecasting Using Hybrid of GLSSVM – ABC Model

Mohammad Azhar Mat Daut[1, 2], Ahmad Sukri Ahmad[1, 2], Mohammad Yusri Hassan[1, 2, a], Hayati Abdullah[1, 3], Md Pauzi Abdullah[1,2] and Faridah Husin[1,2]

[1]Centre of Electrical Energy Systems (CEES), Institute of Future Energy, Universiti Teknologi Malaysia, 81310 Skudai, Johor Darul Ta'zim.
[2]Faculty of Electrical Engineering, Universiti Teknologi Malaysia (UTM), 81310 Johor Bahru, Johor, Malaysia
[3]Faculty of Mechanical Engineering, Universiti Teknologi Malaysia (UTM), 81310 Johor Bahru, Johor, Malaysia

Abstract. In conducting load forecasting, the accuracy of forecasting is an important aspect in planning and managing electricity. Thus, a new hybrid model is presented in this paper, which combines the Group Method of Data Handling, Least Square Support Vector Machine and Artificial Bee Colony (GLSSVM- ABC) for building load forecasting. Its performance accuracy has been compared with other methods by using the Mean Absolute Percentage Error (MAPE) and Root Means Square Error (RMSE). It was found that the proposed method has resulted in better performance accuracy in terms of both MAPE and RMSE. The MAPE analysis showed an increase in performance accuracy of more than 7 percent when compared to other methods. The RMSE analysis showed an increase in performance accuracy of more than 5 percent when compared to other methods. The results in this study showed that the proposed method is proven to be effective and has great potential for accurate building load forecasting.

1 Introduction

The development in electrical load forecasting is becoming interesting and load forecasting methods are constantly being improved. However, forecasting the electrical load is very difficult as electrical loads are frequently affected by several factors such as irregular behaviours, social, time and other non-linear factors [1].

Various methods have been used in modelling and analysing the data for forecasting purposes. Most of the improvements are for the purpose of increasing load forecasting accuracy. The accuracy of load forecasting can affect both users and suppliers. In the past few years, researchers have developed numerous forecasting model to increase the accuracy of load forecasting [1].

In general, methods for forecasting electrical load include using engineering methods, statistical methods and artificial intelligence methods. Among these methods, artificial intelligence method is the most frequently applied to conduct analysis [2]. There are also other models used in forecasting analysis such as decomposition and econometric models [3].

Artificial Neural Network (ANN) and Support Vector Machine (SVM) are also used widely in forecasting [2]. In the last few years, ANN is the most preferred method [4] due to its ability to deal with non-linear factors, and the accuracy of continuous function mapping can be achieved by a three layer neural network [1]. However, ANN requires a lot of training sample data and the selected initial weights can get the local optimal easily [5].

SVM is widely used in the research area and the industries due to its efficiency in solving non-linear problems even with small quantities of training data [2]. Bing Dong applied SVM method to forecast building energy consumption, and found that it performed better than other related models using neural network and genetic programming [6]. Hou and Lian proposed an application of SVM for cooling load forecasting. They concluded that the SVM could provide a promising alternative for cooling load forecasting [7].

Least Square SVM (LSSVM) is an improvement of SVM, proven better than ANN [8]. In the standard SVM model, the solution is addressed by quadratic programming. However, in the LSSVM model, the solution is addressed by a set of linear equations [9]. This improvement reduces the LSSVM complexity and requires less time. The important parts that play an important role in the LSSVM regression system are the regularization parameter and kernel parameter. Nevertheless, both parameters need to be selected properly by establishing a proper methodology to select the parameters [10].

Another sub-model of ANN is known as the Group Method of Data Handling (GMDH). This model has been used effectively with uncertainty. Linear and non-linear systems consist of a broad range of fields such as engineering, science and medicine [11]. The main idea of

[a] Corresponding author: yusrihutm@gmail.com

GMDH is to build an analytical function in a feed forward network. It is based on a quadratic node transfer function, and the coefficient of the model is obtained through the regression technique. Hongya et al [12] applied the GMDH to forecast electric load demand in Australia; the results showed that the GMDH performed better than the ARIMA model in the experiment. Tsado et. al. [13] also used GMDH for energy consumption forecasting, and the results showed the efficiency of GMDH in forecasting over the regression method.

Artificial Bee Colony (ABC) recently is seen as a competitor to the other existing optimization algorithm [8]. This method can conduct both global and local searches in each iteration, which is its main advantage from other method [14]. Furthermore, the parameters in ABC to control are less and require only simple mathematical equations [15]. These benefits can be applied further based on the optimization problems.

This paper proposes the combination of three algorithms; GMDH, LSSVM and ABC to improve the performance of forecasting. GLSSVM had been used to train the actual data with the other input, while the ABC was used to carry out the global and local searches for the best forecast. The forecast was then validated with the actual data as well as comparison with the other methods.

The remaining section of this paper is organized as follows; Section 2 introduces the fundamental of GLSSVM-ABC, while Section 3 explains the methodology of this study, and Section 4 discusses the result. The last section, Section 5 concludes this study.

2 The Fundamental of GLSSVM-ABC

This section introduces the fundamental of Group Method Data Handling with Least Square Support Vector Machine (GLSSVM) and Artificial Bee Colony (ABC0 in terms of theory and concepts.

2.1 GMDH and LSSVM (GLSSVM)

GLSSVM is the combination of GMDH and LSSVM models. This GMDH was introduced by Ivakhnenko [1] in early 1970s, while LSSVM was initiated by Vapnik [2]. The objective of the combination of these two models is to enhance their capability. The basic procedure of this hybrid model is carried out as follows:

Step 1: The training and testing data are separated from the normalized data.

Step 2: A combination of two input variable (x_i, x_j) is generated in each layer. The number of input variable is identified by using Equation (1). Regression of polynomial for this layer is created by establishing the quadratic expression which approximates the output in Equation (2).

$$^{M}C_2 = \frac{M!}{(M-2)!2!} \quad (1)$$

where M is the number of observations in the training set.

$$\frac{\partial L}{\partial w} = 0 \rightarrow w = \sum_{i=1}^{N} \alpha_i \varphi(x_i)$$

$$\frac{\partial L}{\partial b} = 0 \rightarrow \sum_{i=1}^{N} \alpha_i = 0$$

$$\frac{\partial L}{\partial e_i} = 0 \rightarrow \alpha_i = \lambda e_i$$

$$\frac{\partial L}{\partial \alpha_i} = 0 \rightarrow w^T \varphi(x_i) + b + e_i - y_{i=0} \quad (2)$$

$$i = 1,2,\ldots,N$$

Step 3: In the next layer, the new input is determined. The smallest root means square (RMSE) represents the output and variable for training data set, which are then combined as the input variable $\{x_1, x_2, \ldots, x_M, x'\}$ with $M = M + 1$. This new variable will be used as the input for the LSSVM model.

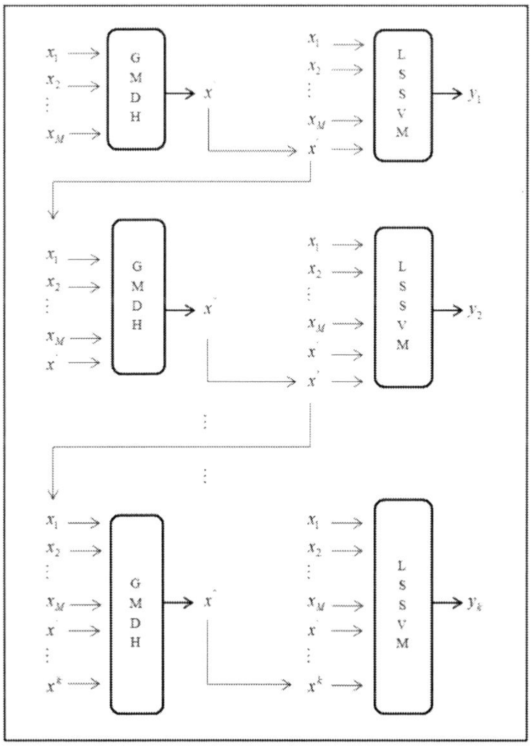

Figure 1. Structure of GLSSVM

Step 4: Steps 2 to 4 are repeated based on the number of iterations by using GLSSVM algorithm. The minimum value of RMSE from the GLSSVM algorithm is selected as the output model. The flow of GLSSVM process is illustrated in Fig. 1.

2.2 Artificial Bee Colony (ABC)

Karaboga and Basturk [16] developed the Artificial Bee Colony (ABC) algorithm inspired from the behaviour of honey bees searching for food. The main focus on this algorithm is to find the best amount of nectar (fitness) by finding the best position of food sources (solutions).

This algorithm can be divided into three phase which are employed bees (EB), onlooker bees (OB) and scout bees (SB). All phases have different task to solve. Generally, the EB phase will solve the position of the food sources. With the memory of the food source position, this bee will spread the information to the OB phase. Thus, the decision making needs to be done by OB to select the best of food source information given by EB. The last phase of this algorithm is SB. This SB is formed from of a few of employed bees, which leave the food sources and to find a new one. All the basic steps about this algorithm are enlightened in the following manner:

Step 1: Initializing the food source. In this stage, the solution will be generated randomly between the ranges of parameter by using Equation (3).

$$x_{ij} = x_j^{\min} + rand(0,1)\left(x_j^{\max} - x_j^{\min}\right) \quad (3)$$

where i represents i_{th}, the number of food source and j is the number of optimization variables associated with the i_{th} food source. Afterward, this process continues to evaluate the quality of the solution (fitness) by using Equation (4).

$$fitness_i = \begin{bmatrix} \dfrac{1}{(1+f_i)} & \text{if } f_i \geq 0 \\ 1+abs(f_i) & \text{if } f_i < 0 \end{bmatrix} \quad (4)$$

where the cost, f_i is cooperated with x_i.

Step 2: Assigning the food source to employee bees. Bees are employed to search for new food source in the memory by applying the Equation (5).

$$v_{ij} = x_{ij} + \varphi_{ij}(x_{ij} - x_{kj}) \quad (5)$$

where φ_i are consistently distributed number within a range of $[-1,1]$. A random optimization variable in the range of $[1,D]$ is represented as j ; where D is not a negative number. The randomly selected food source, k is different from i. Solution weight for higher probability is found by onlooker bees by using Equation (6).

$$p_i = \frac{fit_i}{\sum_i^{SN} fit_i} \quad (6)$$

where fit_i is the fitness solution and SN is the number of food source position.

Step 3: Onlooker bees will be in charge of the selection of the quality solution. Again, using Equation (5), the best solution based on the better solution weight is found, then evaluated by using Equation (6).

Step 4: Deciding the food source to be abandoned and assigning it to scout bees. After all the processes related to exploitation have been done and the number of food cannot be improved, the employed bees will become the scout bees and will process a random search by using Equation (7).

$$x_{id} = x_d^{\min} + rand(0,1)(x_d^{\max} - x_d^{\min}) \quad (7)$$

where $d = 1,2,...,n$.

Step 5: Memorizing the solution.

Step 6: Obtaining the output. If needed, steps 1 – 5 will be repeated.

3 Methodology

The description of data and evaluation of accuracy for building load forecasting are discussed in this section.

3.1 Description of Data

The set of data input which includes the dew point, dry bulb, humidity and pressure had been implemented in this study in order to evaluate the performance of the proposed method. From that, 70 % had been used as the training data while the remaining which is 439 sets of data was used for testing and represented as 'N'.

3.2 Evaluation of Accuracy

Three different types of evaluation had been used to evaluate the performance of accuracy, which include the Means Absolute Percentage Error (MAPE) and Root Means Square Error (RMSE). All these parameters are a major part in a forecasting study in order to differentiate the capability of the model. All their definitions are expressed as follows:

$$MAPE = \frac{1}{N}\left[\sum_{t=1}^{N}\left|\frac{F_t - A_t}{F_t}\right|\right] \times 100 \quad (8)$$

$$RMSE = \sqrt{\frac{1}{N}\sum_{t=1}^{N}(A_t - F_t)^2} \quad (9)$$

where t=1,2,...,x

A_t = Actual Load

F_t = Forecast Load

N = Number of test data

4 Results and Analysis

The performance accuracy results of the proposed GLSSVM- ABC method has been compared with the results of the GMDH, LSSVM, GLSSVM and GABC methods. The performance accuracy results as well as the actual and forecasted loads from all the methods are shown in Table 1. The results achieved by the proposed method were compared using MAPE and RMSE. From the table, it can be seen that all the models have different performance accuracy.

In terms of MAPE, it can be seen that, the proposed method has resulted in an improvement of the performance accuracy when compared to GLSSVM, GABC, LSSVM and GMDH. The improvement in accuracy when compared to GLSSVM is approximately 7 percent. The improvement in accuracy when compared to GABC and LSSVM are 30 percent and 49 percent respectively. The improvement in accuracy when compared to GMDH is 60 percent which is also the highest improvement achieved by the GLSSVM-ABC.

In terms of RMSE, the proposed method again showed the lowest error compared to the other methods. It can be seen that the proposed method has improved performance accuracy when compared to GLSSVM by almost 5 percent, GABC by 31 percent, LSSVM by 49 percent and the highest improvement achieved is 55 percent when compared to GMDH.

From the analysis, it shows that the proposed method has performed better for both MAPE and RMSE compared to the other methods. This is because the incorporation of the ABC with the GLSSVM has resulted in a balanced exploration, where it does not get stuck in a local minimum. This is important to avoid the over-fitting problem between the actual and forecasted results.

Table 1. Performance results of load forecasting

Algorithm	Actual Data (MW)	Forecasted Data (MW)	MAPE	RMSE
GMDH	30.059	29.067	0.00752	0.0473
LSSVM	30.059	30.846	0.00581	0.0421
GABC	30.059	30.632	0.00426	0.0307
GLSSVM	30.059	29.638	0.00319	0.0225
GLSSVM - ABC	30.059	30.457	0.00298	0.0213

5 Conclusions

This paper has described a proposed method for GLSSVM which incorporates the ABC algorithm based on the short term load forecasting, covering aspects such as dry bulb, dew point, humidity, pressure and historical load as input. The simulation result has shown that the proposed method has outperformed the other methods. In terms of MAPE, the proposed method has improved the accuracy of forecasted load in the range of 7% to 60%. The RMSE analysis showed that the accuracy of the forecasted load has been improved in the range of 5% to 55%. From these results, it can be concluded that the GLSSVM-ABC gave better results than the other methods in terms of MAPE and RMSE. It is envisaged that the proposed method has great potential for accurate load forecasting and can be very useful for the purpose of energy management in buildings.

Acknowledgment

This work was supported by the Ministry of Education Malaysia and Universiti Teknologi Malaysia through the Research University Grant (GUP) vot 07H57.

References

1. J. Wang, L. Li, D. Niu, and Z. Tan, An annual load forecasting model based on support vector regression with differential evolution algorithm, *Applied Energy*, **vol. 94**, (2012).
2. H. X. Zhao and F. Magoulès, A review on the prediction of building energy consumption, *Renewable and Sustainable Energy Reviews*, **vol. 16**, pp. 3586 - 3592, (2012).
3. L. Suganthi and Anand A. Samuel, Energy Models for Demand Forecasting-A Review, *Renewable and Sustainable Energy Reviews*, **vol. 16**, pp. 1223-1420, (2012).
4. K. Metaxiotis, A. Kagiannas, D. Askounis, and J. Psarras, Artificial intelligence in short term electric load forecasting: a state-of-the-art survey for the researcher, *Energy Conversion and Management*, **vol. 44**, pp. 1525–1534, (2003).
5. E. Avci, Selecting of the optimal feature subset and kernel parameters in digital modulation classification by using hybrid genetic algorithm–support vector machines, *Expert Systems with Applications*, **vol. 36**, pp. 1391-1402, (2009).
6. B. Dong, C. Cao, and S. E. Lee, Applying support vector machines to predict building energy consumption in tropical region, *Energy and Buildings*, **vol. 37**, pp. 545-553, 5//(2005).
7. Zhijian Hou and Z. Lian, An Application of Support Vector Machines in Cooling Load Prediction, presented at the International Systems and Application 2009, Wuhan, (2009).
8. Z. Mustaffa, Y. Yusof, and S. S. Kamaruddin, Application of LSSVM by ABC in Energy Commodity Price Forecasting, in *Power Engineering and Optimization Conference (PEOCO), 2014 IEEE 8th International*, Langkawi, (2014), pp. 94 - 98.
9. X. Yang, Comparison of the LS-SVM Based Load Forecasting Models, in *IEEE*, ed. Harbin, Heilongjiang, China: IEEE, (2011), pp. 2942 - 2945.
10. O. Hegazy, O. S. Soliman, and M. A. Salam, LSSVM-ABC Algorithm for Stock Price prediction,

International Journal of Computer Trends and Technology (IJCTT), **vol. 7**, pp. 81-92, (2014).
11. R. Samsudin, P.Saad, and A. Shabri, River flow time series using least squares support vector machines, *Hydrology and Earth System Sciences,* vol. 15, pp. 1835-1852, (2011).
12. Hongya Xu, Yao Dong, Jie Wu, and Weigang Zhao, Application of GMDH to Short Term Load Forecasting, *Advanced in Intelligent System,* vol. 138, pp. 27-32, (2012).
13. T. Jacob, U. A. Usman, S. Bemdoo, and A. A. Susan, Short-term Electrical Energy Consumption Forecasting Using GMDH-type Neural Network, *Journal of Electrical and Electronic Engineering,* vol. 3, pp. 42-47, (2015).
14. W.-C. Hong, Electric load forecasting by seasonal recurrent SVR (support vector regression) with chaotic artificial bee colony algorithm, *Energy,* vol. 36, pp. 5568-5578, 9// (2011).
15. Z. Mustaffa, Y. Yusof, and S. S. Kamaruddin, Application of LSSVM by ABC in energy commodity price forecasting, in *Power Engineering and Optimization Conference (PEOCO), 2014 IEEE 8th International*, (2014), pp. 94-98.
16. D. Karaboga and B. Basturk, A powerful and efficient algorithm for numerical function optimization: artificial bee colony (ABC) algorithm, *Journal of Global Optimization,* vol. 39, pp. 459-471, 2007/11/01 (2007).

Rule-based Energy Management System in an Experimental Microgrid with the Presence of Time of Use Tariffs

Mojtaba Moghimi[1], Domagoj Leskarac[1], Christopher Bennett[1], Junwei Lu[1] and Sascha Stegen[1]

[1]*Griffith School of Engineering, Griffith University, Nathan Campus, 4111, Brisbane, Australia*

Abstract. This paper aims to investigate a method of peak load shaving through the utilization of solar PV and battery energy storage whilst creating a cost effective Energy Management System (EMS). This is achieved by utilizing a rule-sets to manage and optimize a scheduling system with a forecasting algorithm. As Time of Use (ToU) tariffs change throughout the day, a cost benefit can be achieved when a smart energy storage system is appropriately employed. The EMS operation is tested on an experimental microgrid with commercial load considering payback period calculation.

1 Introduction

Microgrids are integrated energy systems comprising loads, storage systems, Distributed Energy Resources (DERs) and Renewable Energy Sources (RESs) [1]. The wide global adoption of renewable energy sources (RES) as distributed generation has led to an improved efficiency of microgrids and provided a reduction in carbon dioxide emissions on a global scale. One of the most common forms of RES is solar photovoltaics (PV), with a larger number of installations occurring on commercial and residential buildings [2]. Even though solar PV produces power with a normalized curve similar to that of the majority of commercial buildings, the intermittent nature of solar energy production makes grid management harder.

The utilization of a battery energy storage system (BESS) has the potential to remove the intermittency of solar PV energy production and can give the capability to shave the peak loads [3]. Due to high cost of BESS, the most effective size of energy storage and solar PV can be ascertained to minimize capital costs but provide enough storage to benefit the building load [4]. Different software and algorithms are used to increase the stability and the interconnection of microgrids [5].

Thus, to attain the full benefits of BESS, a forecasting and scheduling algorithm is a necessity. The optimal operating schedule of the storage device is obtained by maximizing an objective function which corresponds to the maximum benefit for the storage owner [6]. A forecasting algorithm for a BESS is used to determine when a peak load is likely to occur, how long that peak load will be occurring for and what the power requirement of that peak load is [7]. This is important information for a scheduling system that is required to best utilize the energy stored within the BESS. In order to create an optimized energy storage schedule, peak shaving, financial effects of peak and off peak tariffs and solar PV production have to be taken into account. Looking at the market, there are two main policies to charge the customers, Real Time Pricing (RTP) and Time of Use (ToU) tariffs. ToU tariffs provide two or three price levels for the customer (peak, shoulder and off peak prices). On the other hand, there is RTP which deals with retail electricity prices and whole sale electricity prices [8].

In the literature, different methods for optimal energy management in a microgrid has been proposed. In [9], a smart energy management system is proposed which is similar to the method used in this paper. However, the authors look at market prices and use complex optimization algorithms. Papers [2] and [3] present optimal energy management systems for PV and BESS systems. Paper [2] looks at the market price s and employs dynamic programming to optimize the objective function. The authors in paper [3] utilize model predictive control method to optimize the PV and BESS system operation.

The objective of this paper is to shave the demand peak in a commercial load and obtain the maximum benefit out of the PV and BESS scheduling in presence of ToU tariffs. In order to achieve this goal, an optimized cost effective Energy Management System (EMS) is proposed. This EMS reduces the peak demand and volume charges while optimally charges and discharges the BESS to get more benefits.

The rest of the paper is organized as follows; in section 2 the Energy Management System with forecast algorithm is explained. Section 3 represents the scheduling system and the objective function. In section 4, the case study of the experimental microgrid is investigated and the results of peak shaving and cost benefit analysis are represented. Section 5 concludes the paper.

2 Energy management system

The objectives of Energy Management Systems (EMS) are to shave the peak demand charges and optimally schedule the BESS to charge and discharge economically in the presence of time of use tariffs. Due to intermittency of renewable energy sources, a forecasting algorithm is required to compensate for uncertainties and provide high accuracy data for load and generation characteristics. As it is illustrated in Fig. 1, the EMS employs PV as the main generation source. Because of variation in PV generation, EMS cannot guarantee effective peak shaving with PV alone. Thus, a BESS is utilized in order to aid the system in shaving the peak demand in conjunction with solar PV. The optimal use of BESS requires a scheduling system which defines the time and amount of charging and discharging of BESS through the grid supply and solar PV where applicable. In the presence of ToU tariffs, the scheduling system charges the BESS when the tariffs are low and discharges when tariffs are high. By doing so, a cost beneficial system which accomplishes the objectives is created.

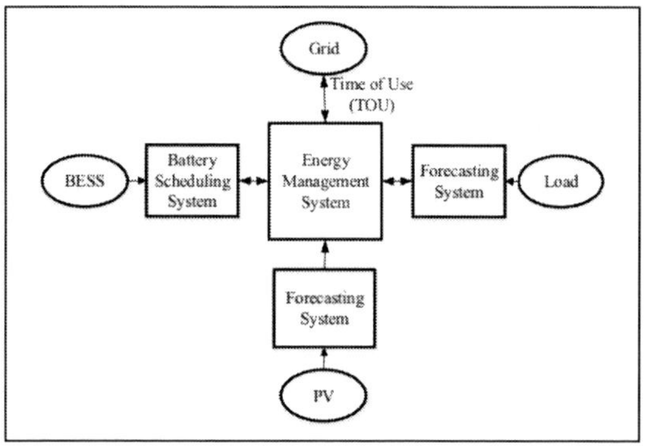

Figure 1. Energy Management System

2.1 Forecasting algorithm

The operation of the scheduling is aligned with uncertainties. In order to reduce the uncertainties, the forecasting system plays an important role. Load profile forecasts are used as input information in order to calculate an initial charge and discharge schedule for the current day in the scheduling algorithm. A load profile is the shape of the consumption of electricity or load throughout a 24 hour period.

The load profile forecasting approach used in the test site University building N44 control system follows the general method of [7] with forecast models modified to suit commercial building scenarios.

The average load forecast model is constructed according to the autoregressive moving average (ARMA) approach. The ARMA modelling approach incorporates variables that are lags of the time series being forecasted and lags of the historical forecast error [10]. The average load forecast model is depicted in Equations 1 and 2:

$$\hat{y}_t = \beta_0 + \sum_{i=1}^{n} \beta_i y_{t-i} + \hat{\varepsilon}_t + \varepsilon_t \quad (1)$$

where y_t is load at time t, $\hat{y}_t$ is the load forecast, β_0 is the y-intercept, β_i is the coefficient for time lag i, there are n number of time lags and ε_t is the forecast error and $\hat{\varepsilon}_t$ is the error forecast for time t and is expanded upon in (2):

$$\hat{\varepsilon}_d = \gamma_0 + \sum_{j=1}^{m} \gamma_j \varepsilon_{d-j} \quad (2)$$

where γ_0 is the y-intercept of the model, γ_j is the coefficient for the error time lag ε_{t-j} and m is the number of lags. After forecasting the peak load, the first stage in forecasting the load profile for the day is to forecast the load profile $LPF_{d,t}$ for the current day. The next stage in load profile forecast algorithm is the manipulation of $LPF_{d,t}$ to adjust it to be in accordance to the peak load forecast. The final stage in the forecasting the load profile for current day is to add the historical error forecast to the load profile forecast:

$$\widehat{LPF}_{d,t} = \widehat{LPF}_{d,t} + \hat{\varepsilon}_{d,t}, \quad for\ 1 \le t \le 96 \quad (3)$$

$$\hat{\varepsilon}_{d,t} = \gamma_0 + \gamma\varepsilon_{d-1,t}, \ for\ 1 \le t \le 96 \quad (4)$$

where γ_0 is the constant, γ is a vector of coefficients, j is the lag number and there are m number of lags. For more detailed explanation, papers [11, 12] are recommended.

3 Scheduling system

The main challenge in EMS is to develop a scheduling system for BESS to determine the optimal charge/discharge of battery considering ToU tariffs and peak demand shaving targets. A rule-base scheduling algorithm is designed to determine the charge/discharge of BESS. In order to apply the scheduling system to the battery and obtain a cost effective scheduling system, a model of BESS, an objective function and the constraints are defined. The main purpose of the objective function is to maximize the benefits of scheduling system by charging the BESS at low ToU tariffs and discharging at high ToU tariffs. Thus, a peak shaving algorithm is proposed to gain highest possible benefits from BESS scheduling.

The rule sets for scheduling are defined in Table 1. These rules define the charge/discharge characteristics of the BESS. As can be seen, the scheduling depends on the battery State of Charge (SOC), load demand and ToU tariffs.

Table 1. Scheduler Rule-Set
a) SOC Low

Scheduler	ToU (Off-peak)	ToU (Peak)
Load demand low	Charge	Charge
Load demand high	Standby	Discharge

b) SOC High

Scheduler	ToU (Off-peak)	ToU (Peak)
Load demand low	Charge	Standby
Load demand high	Discharge	Discharge

3.1 BESS modelling

The BESS is modelled as follows:

- BESS charging mode ($P_{BESS}(t) < 0$):

$$E(t) = E(t - \Delta T) - \eta_c P_{BESS}(t - \Delta T)\Delta T \quad (5)$$

- BESS discharging mode ($P_{BESS}(t) > 0$):

$$E(t) = E(t - \Delta T) - \frac{P_{BESS}(t - \Delta T)}{\eta_d}\Delta T \quad (6)$$

where $P_{BESS}(t)$ is the battery power (kW), E(t) is battery energy (kWh), η_c is battery charging efficiency, η_d is battery discharge efficiency, ΔT is the scheduling time step which is set at 15 min intervals in this study.
The BESS has some constraints which needs to be meet:

$$SOC_{min} \le SOC(t) \le SOC_{max} \quad (7)$$

$$P_{BESS}^{min} \le P_{BESS}(t) \le P_{BESS}^{max} \quad (8)$$

where SOC(t) is battery state of charge, SOC_{min} is minimum SOC of battery which is equal to *DOD* Depth of Discharge of battery, SOC_{max} is maximum SOC of battery which is upper bound of battery capacity. In this study, DOD is 80% and SOC_{max} is 100%.

3.2 Objective function

In order to maximize the benefits of the scheduling system, the objective function is expressed as:

$$\max f(t) = \max \sum_{t=t_0}^{T} \{P_{BESS}(t) \times ToU(t) \times \Delta t - P_{Load}(t) \times ToU(t) \times \Delta t\} \quad (9)$$

The goal of the objective function is to maximize the savings through utilizing the optimally scheduled BESS and minimize the costs by shaving the peak demand and volume.

3.3 Constraints

In this study the focus is in peak shaving and battery scheduling. While buying energy from grid, the policy is to charge when $ToU(t)$ is low and discharge when $ToU(t)$ is high. To achieve this, there are some constraints that should be meet. The other requirement for scheduling is the constraints definition which is as below;

Power balance:

$$P_{Load}(t) = P_{PV}(t) + P_{BESS}(t) + P_{Grid}(t) \quad (10)$$

BESS constraints:

$$SOC_{min} \le SOC(t) \le SOC_{max} \quad (11)$$

$$P_{BESS}^{min} \le P_{BESS}(t) \le P_{BESS}^{max} \quad (12)$$

$$E_{min} \le E(t) \le E_{max} \quad (13)$$

$$\eta_c P(t) \le C_{max} \quad (14)$$

$$\frac{P(t)}{\eta_d} \le D_{max} \quad (15)$$

Peak shaving:

$$P_{Grid}(t) \le P_{Grid}^{max} \quad (16)$$

where C_{max} is the max charge rate and D_{max} is the max discharge rate of battery which in this study is the same as the Statcom maximum charge and discharge rate (30kVA).

3.4 Peak shaving algorithm

In order to shave the peak demand, the BESS scheduling system requires a discharge algorithm which optimally discharges the battery and shaves the peak demand. The algorithm implemented in this study is as follows:

Step 1: The scheduling system sets the discharge level at the highest rate which is Statcom max discharge rate (30 KVA) in this case.

Step 2: If the BESS reduces the peak during the peak period in a way that the new peak is the same in the whole peak period, the discharge level is appropriate, otherwise;

Step 3: The discharge level is reduced one kW each time to reach to a level that the peak in peak period is created by battery, i.e. the battery has enough capacity to discharge during the peak period thoroughly not discharging the complete battery capacity for just a portion of peak period.

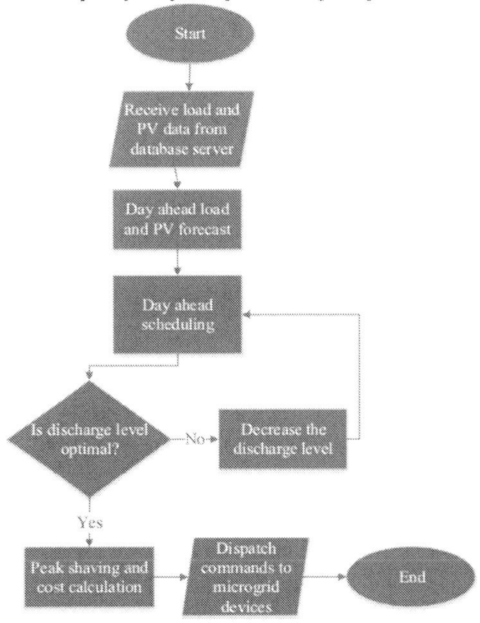

Figure 2. Scheduling algorithm

By going through the peak shaving algorithm, the optimal discharge level and consequently the maximum benefit out of peak shaving is achieved. The complete scheduling system is displayed in Fig. 2. As it can be seen, the forecasting algorithm is employed to do the initial scheduling and the peak shaving and cost calculation is derived out from real time scheduling.

4 Case study

4.1 Experimental Microgrid

To test the scheduling system performance, it is implemented on a Microgrid testing facility with commercial load characteristics. The Microgrid Testing Facility (MGTF) has been designed and implemented at Griffith University's Nathan Campus as a unique test bed for investigating all aspects of a Microgrid and Smartgrid network. One aspect of the MGTF incorporates distributed generation in the form of solar PV cells connected to a DC bus via parallel connected DC/DC converters. This DC bus connects the common DC components together to reduce the need for multiple power conversion stages; the energy storage tank and the Statcom inverter. 60kWh of Lithium-ion based energy storage is utilised to create the backbone of the bus. A 30kVA Distributed-Statcom provides the AC connection to the building and grid which also serves as the control mechanism for the energy storage solution.

The testing facility is equipped with a communication, monitoring and data management system. The main purpose of communication system is to facilitate data collection and measurement and also dispatch control commands to Microgrid devices. The system installation and communication details is explained in [13] and [14]. The complete experimental system in this study is shown in Fig. 3 with the BESS parameters specified in Table 2.

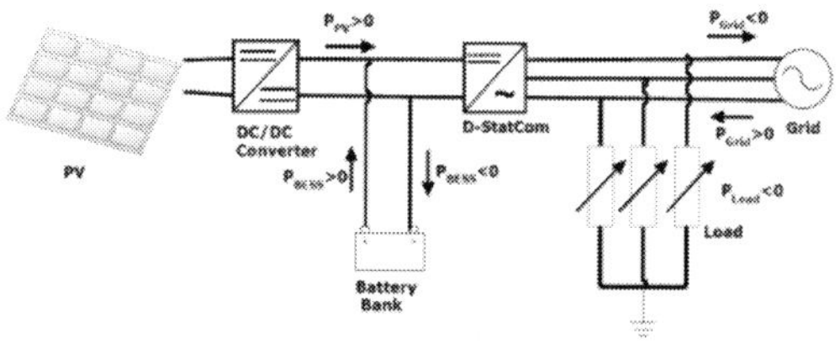

Figure 3. Microgrid structure with PV and BESS

Table 2. Battery parameters

BESS parameters	characteristics
BESS technology	Li-Ion
Rated capacity	60 kWh
Cycles (80% DOD)	7500
$C_{Capital}$	$60000

4.2 EMS operation

The Microgrid EMS performance is investigated observing PV, BESS and BESS scheduling system influence on peak load shaving and system cost reduction. The ToU tariffs for commercial loads are shown in Table 3 and 4. The tariffs are defined based on Australian energy tariffs in Queensland State. There are two tariffs (peak and off peak) for volume charges and one tariff for demand charges.

Table 3. Energy charges ToU

ToU tariffs	Price	Time
Peak	9.7 c/kWh	7 am - 8 pm
Off-peak	6.6 c/kWh	8 pm – 7 am

Table 4. Demand charges ToU

ToU tariffs	Price	Time
Peak	24.14 $/kW	Whole day

The results for the day ahead scheduling for a day in January 2016 is illustrated in Fig. 4 and 5, with a scheduling time step of 15 min. It can be seen in Fig. 4 that PV can reduce the peak demand and how BESS optimal scheduling increases the peak demand shaving and reduces the volume charges costs.

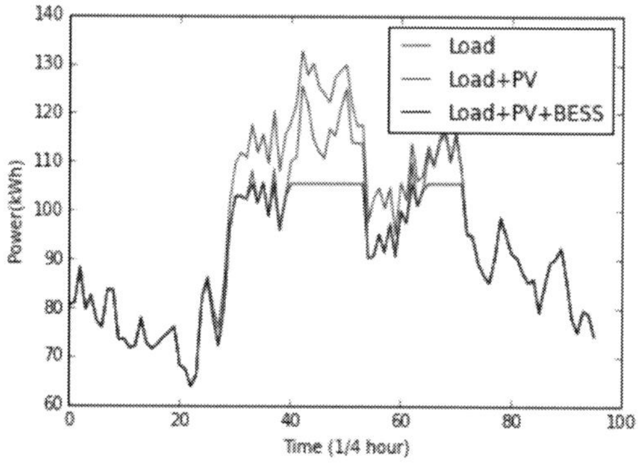

Figure 4. a) Load profile b) load profile with PV c) load profile with PV and BESS

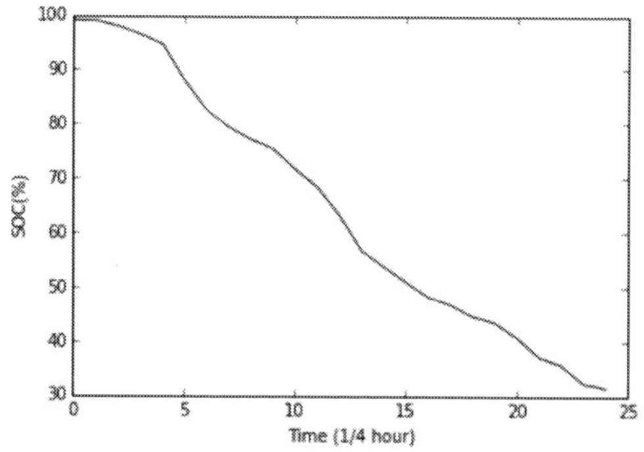

Figure 5. Battery SOC during peak period

In Fig. 5, the SOC of BESS during peak period is shown. The SOC is supposed to be high from 100 to 30 and low from 30 to 20. The BESS is not fully discharging during the peak period as it depends to the discharge level defined by scheduling system.

4.3 Cost benefit analysis

In order to analyse the economics, the payback period (PBP) of investment is investigated. Using PBP, the minimum time to recover the amount of investment of battery in estimated.

The battery investment and installation cost is calculated as:

$$C_{capital} = C_E E_{max} \quad (17)$$

E_{max} is the BESS energy capacity and C_E is BESS installation and production cost per kWh which is $1000 for the chosen battery technology. The PBP is calculated as follows:

$$\sum_{t=t_0}^{PBP} S_{annual} * (1+\mathrm{i})^{n-1} \geq C_{capital} \quad (18)$$

where S_{annual} is the annual savings from peak shaving, n is the year, $C_{capital}$ is the battery investment and installation cost, i is the interest rate per year which is 15% in this study.

The PBP of PV and BESS is calculated for the experimental Microgrid. As it can be seen in table 5 and 6, the BESS has a higher PBP but it brings more benefits to the system as well. Furthermore, PV and BESS provide a cost effective system for peak shaving applications.

Table 5. PBP for PV

Annual revenue	$6132
Capital cost	$40000
PBP	4.9 years

Table 6. PBP for BESS

Annual revenue	$8227
Capital cost	$61100
PBP	5.4 years

5 Conclusion

In this paper, an energy management system employing PV and BESS is proposed. The EMS aims to shave the peak load and achieve a cost beneficial solution for the experimental Microgrid. The rule-base scheduler is operating in a way that tries to get the most of energy form battery energy storage system and purchase the least form the grid. On the other hand, the forecasting module is employed in order to strengthen the accuracy of results of the energy management system.

Furthermore, the operation of EMS has been investigated for the following cases: load, load and PV as well as load and PV including BESS. The payback period for the PV and BESS system has been calculated considering the annual saving of the system and total system price. The performance of the scheduling system has been optimized for commercial operation which resulted in maximized benefits.

References

1. Parisio, A., et al., *Use of model predictive control for experimental microgrid optimization.* Applied Energy, **115**: p. 37-46, (2014)
2. Riffonneau, Y., et al., *Optimal Power Flow* Management *for Grid Connected PV Systems With Batteries.* Sustainable Energy, IEEE Transactions on. **2**(3): p. 309-320, (2011)
3. Perez, E., et al., *Predictive Power Control for PV* Plants *With Energy Storage.* Sustainable Energy, IEEE Transactions on. **4**(2): p. 482-490, (2013)
4. Teleke, S., et al., *Rule-Based Control of Battery Energy Storage for Dispatching Intermittent Renewable Sources.* Sustainable Energy, IEEE Transactions on, **1**(3): p. 117-124, (2010)
5. Hossain, M.J., et al., *Design of Robust Distributed Control for Interconnected Microgrids.* Smart Grid, IEEE Transactions on, (99): p. 1-12, (2015)
6. Telaretti, E., et al., *A Novel Operating Strategy for Customer-Side Energy Storages in Presence of Dynamic Electricity Prices.* Intelligent Industrial Systems, **1**(3): p. 233-244, (2015)
7. Bennett, C. J., R. A. Stewart, and J. W. Lu, *Forecasting low voltage distribution network demand profiles using a pattern recognition based expert system.* Energy, **67**: p. 200-212, (2014)
8. Sanseverino, E., et al., *A Self-Adapting* Approach *for Forecast-Less Scheduling of Electrical Energy Storage Systems in a Liberalized Energy Market.* Energies, **6**(11): p. 5738, (2013)
9. Chen, C., et al., *Smart energy management system for optimal microgrid economic* operation. Renewable Power Generation, IET, **5**(3): p. 258-267, (2011)
10. Bennett, C., et al. *Applicability of load* forecasting *techniques for customer energy storage control systems.* in *Power and Energy Engineering Conference (APPEEC), 2015 IEEE PES Asia-Pacific.* (2015)
11. Bennett, C. J., R. A. Stewart, and J. W. Lu, *Development of a three-phase battery energy storage scheduling and operation system for low voltage distribution networks.* Applied Energy, **146**(0): p. 122-134 (2015)
12. Bennett, C., R. Stewart, and J. Lu, *Autoregressive with Exogenous Variables and Neural Network Short-Term Load Forecast Models for Residential Low Voltage Distribution Networks.* Energies, **7**(5): p. 2938, (2014)
13. Leskarac, D., et al. *Testing facility for research and development of Smart-MicroGrid technologies.* in

Power and Energy Engineering Conference (APPEEC), 2015 IEEE PES Asia-Pacific. (2015)
14. Moghimi, M., et al. *Communication architecture and data acquisition for experimental MicroGrid installations*, in *Power and Energy Engineering Conference (APPEEC), 2015 IEEE PES Asia-Pacific.* (2015)

Out-of-step Prediction for Power System Using Improved Prony Algorithm

Fang SHI

Key Laboratory of Power System Intelligent Dispatch and Control of the Ministry of Education, Jinan 250061, China

Abstract. The development of the phasor measurement units (PMUs) and the wide-area measurement systems (WAMS) in power system provide abundant synchronous data for the new application. This paper proposes an online out-of-step monitoring and predicting scheme for the interconnected power systems based on the rotor angle measurement. The Prony algorithm is improved to predict whether and when a power swing will lead to out-of-step of the generator or the power systems. An energy-like weight coefficient approximation of the decomposed signals is introduced to maintain the accuracy and reliability of the model free dynamic prediction method. The effectiveness and applicability of the proposed scheme is illustrated using a 2-area-4-machine power system.

1 Introduction

The angle stability is essential for the security of both generators and the whole power system. Despite the development of the protection and control of power systems, blackouts still happened from time to time all over the world [1]. The unstable oscillation between a generator and the rest system (lead to generator out-of-step) or between two interconnected sub-systems (lead to islanding) are the main form of the power system instability.

The final report of the North American blackout on august 14, 2003 points out that 290 units (about 52,745 MW) tripped during the cascading failures [2]. Among these tripping cases, a large portion occurred after the grid lost synchronism. The study shows that many generators shut the unit down early in the cascade for the pre-designed protection points, which exacerbated the problem. The generator protections should be set tight enough to protect the unit but also wide enough to assure it remains connected to the grid as long as possible. A reliable electric system requires proper protection and control coordination between power plants and the transmission system. Therefore, to evaluate the severity and the evolution trend of the oscillation is important for setting the protection relays. In [3], NERC (American Electric Reliability Corporation) suggests that the generator owner and transmission owner shall work together to complete the coordination process. The SPS (Special Protection Schemes) has been proposed for prevent the cascading failures of the power system, which take advantage of all the available protection and control devices to adjust the operation state of the power system to a more secure level.

When out-of-step occurs for one generator in the power system, the referring generator should be tripped off immediately to prevent the shaft from suffering mechanical vibration torque. Otherwise, the separation of the two interconnected sub-systems should be carried out as soon as possible when the out-of-step exists between the two sub-systems. The traditional protection schemes are based on mho/impedance relay analysis for the SMIB or two equivalent machine model, which may can't appropriately reflect the physical characteristics of the power system in real-time especially when the system suffers cascading failures. It is of great importance that the stability estimation act accurately since the generator trip or power grid islanding will worsen the instability of the whole system. The operators also prefer some alarm information before the generator tripping or system islanding so they can take some measures to prevent it from happening, such as load control or capacitor/reactor switching, et. al. Therefore, the analysis of the power system dynamics and the prediction of the out-of-step conditions as well as the discriminative criterion are all need to be studied.

2 Related works

The stability of the power system has been studied for several decades. Overall, these methods can be twofold: (1) Model-based method; (2) Signal-based method. The classical method normally analyse the power system based on a simplified model. It needs a trade-off between the accurate equivalence and the available mathematic tools. The signal based method further considers that the dynamic trajectory of the system can imply its physical feature in a certain extent. These two methods can be used for decreasing the possibility of the out-of-step conditions and detecting the out-of-step scenarios as soon as they occurred.

Currently, all the major generators with a capacity larger than 300MW need to set the out-of-step protection. And the interconnecting transmission lines need to install the out-of-step separation devices. Classical power swing detection methods based on concentric characteristic or blinders, which need a sophisticated study for the power system in order to determine the appropriate settings. But the settings are fixed and may lead to mal-operation as it can't adapt to the changing operation conditions of the system. In addition, the above mentioned methods are computationally complex and require a number of settings which must coordinate with each other. In general, it is too late for the operators' intervention to prevent the instability using the conventional out-of-step detection schemes. In recent years, the up-to-date intelligent algorithm, such as like SVM, FUZZY, Neural Network and energy like methods based on the synchronous measurement are being studied in a quantitative way [4, 5].

Reference [6] proposes a prediction method for power system out-of-step. The singular points of the power system are calculated before plotting the state plane trajectory. The eigenvalue analysis method is used for determining the stability margins of the system. The plane trajectory is derived from SIMB model and the time information is extracted from discretization with small intervals. In [7], the active power (P), reactive power (Q) and the slip of frequency (f) at two different times are measured and synthesized after clearing the faults (t1 and t2). Then the critical phase angle are determined and used to check whether the slip will be greater than zero using an equation set. In [8], the angle difference, slip frequency, and acceleration are then used to detect an out-of-step (OOS) condition in the electrical power system. It is an accurate discriminative criterion but can not predict the out-of-step conditions beforehand.

In [9], a predictive out-of-step tripping scheme is used for the national power system of by monitoring slip (rate of change of their relative phase) and acceleration (rate of change of slip). It is based on comprehensive transient stability studies using actual dynamical models. In [10], a predictive, transient stability monitoring scheme for generator out-of-step protection is presented based on a Lyapunov like method. The state of the simplified model is monitored and the predefined energy value is chosen as the discriminative criterion for the generator out-of-step. But it can't tell when it will lose synchronism. In [11], the coordination of power swing detection, distance protection and out of step protection is studied. The characteristics during the out-of-step are summarized and the basic principles for tripping the generator are discussed.

Some practical discriminative criterions have been proposed for estimating the out-of-step potentials of the static power system in an off-line manner. However, the pre-defined fault analysis, such as N-1 or N-2, can not meet the numerous or cascading disturbance circumstances in the power system. The parameter and disturbance independent online used methods are urgently needed to be studied both in the literature and industry.

With the more and more applications of PMUs and WAMS, the rotor angle of the generators and the phasor of the bus voltage both can be accurately measured with the asynchronous time step. So the angle stability can be more precisely analysed for the large area interconnected power systems.

There are different ways to measure the position of the spinning generator shaft so as to obtain the rotor angle directly. The most efficient way is as shown in Fig. 1. It only needs to apply equally-spaced and aligned reflective tape on existing generator shafts, then a laser emitter and optimal pickup are mounted on the generator and pointed at the shaft. The rising and falling edges are detected and transferred into PMUs in which the data process is done and the time-stamped rotor angle of the generator can be obtained with high accuracy. The retrofit is very simple and costless, which make it possible to obtain the synchronous rotor angle of all the generators in the power system with the help of WAMS.

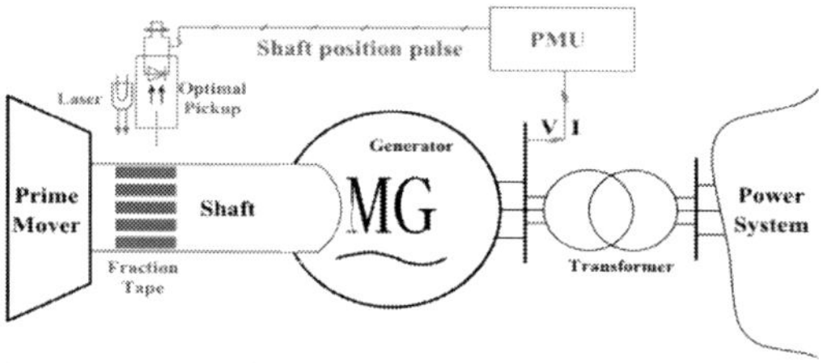

Figure 1. Rotor angle measurement

3 Basic theory and problem formulation

Mathematically, for a discrete signal F(k) ($k = 0, 1, 2, \ldots N-1$, and N is the number of the sampling points), it can be expressed as a linear summation of M distinct modes :

$$y(k) = \sum_{i=1}^{M} B_i e^{\lambda_i k \Delta t} \tag{1}$$

via the following three procedures [12].

1) Solve a linear prediction model (LPM)

If the signal consist of damped sinusoids and the number of the sampling data is no less than twice of the mode number, i.e., $N \geq 2M$, then part of the sampled data can be represented as linear summation of the rest data. The above process can be illustrated as follows:

$$\begin{bmatrix} F_M \\ F_{M+1} \\ \cdot \\ \cdot \\ F_{N-1} \end{bmatrix} = - \begin{bmatrix} F_{M-1} & F_{M-2} \cdots & F_0 \\ F_M & F_{M-1} \cdots & F_1 \\ & \cdot & \\ & \cdot & \\ F_{N-2} & F_{M-2} \cdots & F_{N-M-1} \end{bmatrix} \begin{bmatrix} P_1 \\ P_2 \\ \cdot \\ \cdot \\ P_M \end{bmatrix} \tag{2}$$

It can be rewritten in a compact form:

$$\mathbf{F} = \mathbf{\Phi}(F)\mathbf{P} \tag{3}$$

If N=2M, equation (3) is directly solvable; otherwise the least-square estimate of P can be obtained using the generalized inverse of matrix Φ as follows:

$$\mathbf{P} = (\mathbf{\Phi}^T\mathbf{\Phi})^{-1}\mathbf{\Phi}^T\mathbf{F} \tag{4}$$

2) Find the roots of the polynomial with coefficient of **P**

Solve for the roots (eigenvalues) of the following polynomial:

$$x^M + P_1x^{M-1} + P_2x^{M-2} + ... + P_{M-1}x + P_M = 0 \tag{5}$$

where the *mth* root is equal to e^{λ_m}.

3) Estimate the amplitude

Part of the linear equations representing F(k) with e^{λ_m} can be used for solving the amplitude B_M:

$$\begin{bmatrix} F_{k_1} \\ F_{k_2} \\ \vdots \\ F_{k_N} \end{bmatrix} = \begin{bmatrix} (e^{\lambda_1})^{k_1} & (e^{\lambda_2})^{k_1} ... (e^{\lambda_M})^{k_1} \\ (e^{\lambda_1})^{k_2} & (e^{\lambda_2})^{k_2} ... (e^{\lambda_M})^{k_2} \\ \vdots & \ddots \quad \vdots \\ (e^{\lambda_1})^{k_N} & (e^{\lambda_2})^{k_N} ... (e^{\lambda_M})^{k_N} \end{bmatrix} \begin{bmatrix} B_1 \\ B_2 \\ \vdots \\ B_M \end{bmatrix} \tag{6}$$

The compact form in matrix can be written as:

$$\mathbf{F} = \mathbf{AB} \tag{7}$$

If more than M samples are used, i.e. N > M, the least-square estimation of B can be solved by:

$$\mathbf{B} = (\mathbf{A}^T\mathbf{A})^{-1}\mathbf{A}^T\mathbf{F} \tag{8}$$

After the three procedures, the discrete signal F(k) can be approximately analysed using the analysis formula of (1). The fitness of formula (1) to the original signal data can be appraised by signal-to-noise-ratio (SNR) [13], which is defined as:

$$SNR = 20\log\left(\|y(k)\| / \|y(k) - \hat{y}(k)\|\right) \tag{9}$$

where $\|y(k)\|$ and $\|y(k) - \|\hat{y}(k)\|\|$ are the second norm of the measured signal and the norm of the error signal between the measured and estimated signals, respectively. Normally, it required that SNR ≥ 20 for a good fitting.

The advantages lie in the above transformation is that the characteristic of the discrete signal F(k) and the corresponding physical system can be evaluated in a more analytical way. Referring to the out-of-step problem, which initially reflects with the exponential separation or divergent oscillation of the rotor angle between one generator and the rest power system or between two sub-systems, the Prony algorithm can be potentially used for analysing the rotor angle trajectory.

During the dynamic process of the interconnected power system after a disturbance, the generators can be divided into two groups according to their movement characteristics. The generators with similar acceleration rotation speed in a certain time can be seen as coherent group. Assume generators i to n are identified as one coherent group, then accordingly, the centre of inertia (COI) reference can be calculated as:

$$\begin{cases} M_{sum} = \sum_{i=1}^{n} M_i \\ \omega_{COI} = \sum_{i=1}^{n} \frac{M_i\omega_i}{M_{sum}} \\ \delta_{COI} = \sum_{i=1}^{n} \frac{M_i\delta_i}{M_{sum}} \end{cases} \tag{10}$$

where M_{sum}, ω_{COI} and δ_{COI} are the inertial, equivalent speed and equivalent rotor angle of the coherency clusters, respectively. The rotor angle between the generator and the COI of the rest generators or between the COIs of the two groups is the direct discrimination criterion for the out-of-step conditions. Generally speaking, if the rotor angle difference is greater than 180° implies that out-of-step condition occurs.

Furthermore, the trajectory of the rotor angle difference in the future can be used for predicting the out-of-step conditions. In the power system, as the inertia characteristics of the power system, the oscillation dynamics of the rotor angle will not change swiftly. The rotor angle signal sampled in the past can be potentially used for predicting the movement trend of the rotor angle using Prony algorithm after some improvement.

4 Improvement of the Prony algorithm

4.1 The parameterization for Prony algorithm

There are three parameters in Prony analysis: Sampling frequency, the number of the data points and the order of the system. The basic principles for defining the parameters in the application of out-of-step detection and prediction are illustrated in the sequel.

Low frequency oscillation (0.1Hz~2.5Hz) is the potential danger for out-of-step of the power system. For interconnected power systems, the extra-low frequency oscillation needs to be concerned more particularly. Therefore, a sampling frequency of 25Hz is enough for out-of-step monitoring.

The number of the data points should cover at least one cycle of the oscillation with the possible lowest frequency. In practical use, several cycles of oscillation should be covered in order to derive the changing trend information. However, the rapid attenuation of the signal will be omitted if the time window is too long. Trade off must be made between the accuracy and the rapidity when choosing the length of the time window. Generally, 8s can be chosen as the time window for Prony analysis and there are total 200 data points during the data process.

The pre-assumed order of the system model is critical to the performance and accuracy of the Prony analysis. In the past, trier and error method is used for choosing the best order from 1 to half of the dada point number. The one with the least *SNR* is seen as the proper order. However, the prediction of the trajectory is not only relative to the accuracy of the curve fitting for the current signal wave but also to the stability or smoothness of the trajectory in the future. Define the diverging index of the Prony process as:

$$\mathrm{DI_m} = |\, e^{\lambda_m} | \tag{11}$$

which represents the divergence characteristic of the Prony curve fitting and the prediction. A supplementary stability index is defined as:

$$SSNR = SNR / \max(\mathrm{DI}_m) \quad (12)$$

It is used for evaluating the numerical stability of the prediction performance of the Prony algorithm, along with the SNR. So the order of the model can be chosen as the one with the optimal SSNR as shown in Fig. 2.

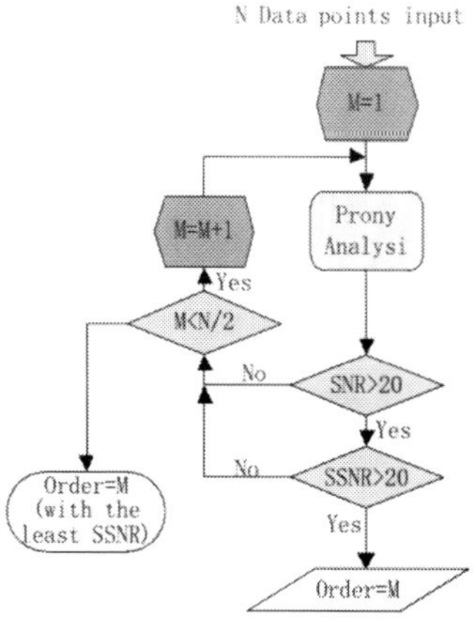

Figure 2. The order determination procedure

4.2 Energy index based order reduction

The order obtained via the above process could be a very large number. The order number of the Prony algorithm in the oscillation trajectory analysis is corresponding to the oscillation mode. But actually not each mode is equally important for the analysis of the physical nature. The out-of-step will happen when the oscillation energy keeps increasing. According to the physical meaning of the oscillation, define generalized oscillation energy as:

$$Energy_j = \sum_{i=1}^{N} |y(i)|^2 \quad (13)$$

The dominant oscillation modes will contribute a big proportion of the oscillation energy of the interconnected power system. The omission of the trivial energy components of the non-dominant oscillation modes will not significantly affect the dynamic movement characteristics of the power system. But it will reduce numerical instability in the prediction as the exponential expansion of the diverging index could exist in the omitted modes.

Suppose the generalized energy of the kth oscillation mode is:

$$Energy(k) = |B_k|^2 \cdot \sum_{i=1}^{N} |e^{\lambda_k i\Delta t}|^2, k = 1,2,...,M \quad (14)$$

Then sort the energy array in descending order, and denote K' is the data array after sorting. Generally, the sum of the energy of the first a few number (H) of oscillation modes will contribute a big proportion of the total energy of the signal, denote it as:

$$E_H = \sum_{j=1}^{H} Energy[K'(j)], j = 1,2,...H \quad (15)$$

If the following inequality holds:

$$E_H / Energy \geq 99\% \quad (16)$$

Then the curve fitting and prediction of the input data can be represented in a smaller number of the oscillation modes, as:

$$\tilde{y}(k) = \sum_{i=1}^{H} B_{K'(i)} e^{\lambda_{K'(i)} k\Delta t}, k = 1,2,...N, N+1,..., N+P \quad (17)$$

In this case, the order of the input signal is reduced and the numerical instability potential will be significantly decreased. The first N points data of $\tilde{y}(k)$ can be seen as the curve fitting to the original data points. The rest P points of data are the prediction of the trajectory signal which can be used for predicting the out-of-step condition.

5 Simulation results

The two-area power system shown in Fig. 3 is chosen as the example to verify the prediction scheme proposed in this paper. The parameters of the system can be found in [14]. Assume that a permanent three phase circuit short fault happens on bus 8 at 1.8s, after 100ms, part of the sectionalized bus is disconnected from the power grid by tripping one of the parallel transmission line between bus 7 and 9. Then the power system undergoes a dynamic oscillating process and after about 20.8s the system losses synchronization.

The PMUs are installed for every generator and the synchronous shaft position data will be sent to the control centre. The coherent grouping is carried out and the rotor angle between the two groups is calculated. The rotor angle between the two equivalent sub-systems is recorded and predicted via the improved Prony algorithm, the out-of-step conditions is monitored and predicted. As shown in Fig. 4, the rotor angle difference between every two generators is obtained via simulation. It can be seen that after 20.5s the angle difference is larger than 180°, which means the out-of-step occurs.

The rotor angle between the COI centres of sending sub-system (comprising of G1 and G2) and receiving sub-system (comprising of G3 and G4) is calculated, monitored and predicted using the proposed improved Prony algorithm. The Prony analysis is conducted once in every 0.5s, as it is computationally cost and not necessary to be re-conducted for every new data point. Fig. 5 shows the measured and the predicted value of the rotor angle between the COI of the sending and receiving groups. We can see that the predicted rotor angle trajectory is similar to the recorded one, which means that improved Prony algorithm is feasible in predicting the movement trend of the rotor angle different after decomposing the oscillation curve using the improved Prony algorithm.

The out-of-step detection is more straightforward as the rotor angle difference is directly used for judging the loss of synchronization. The alarm signal, which is multiplied by 180 in order to give a direct impression, is also shown in Fig. 5 with dashed line. It can be seen that the out-of-step condition is predicted about 5s before it truly happens, which gives the operators with opportunities to prevent the out-of-step and the islanding of the power system using some corrective measures, such as reducing power generation, controlling the power flow and supplementary damping control (HVDC or FACTS if possible).

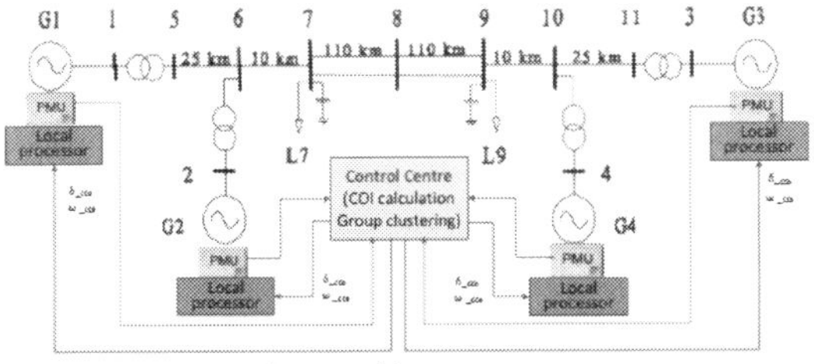

Figure 3. Two-area four-machine power system

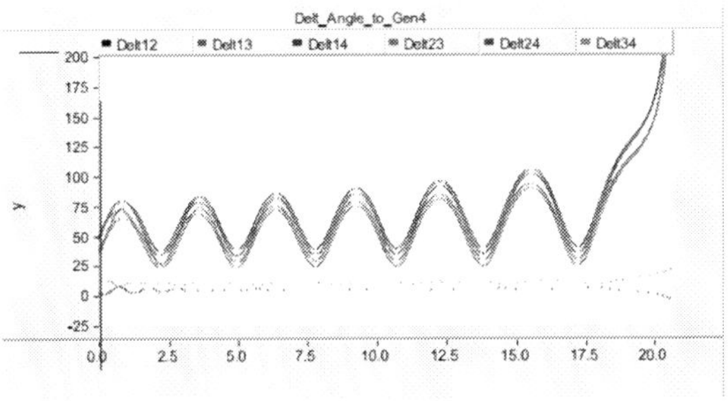

Figure 4. Rotor angle between the generators

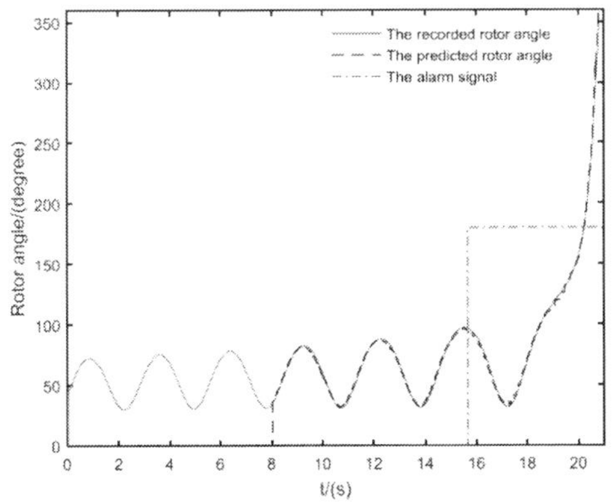

Figure 5. Predicted rotor angle via improved Prony algorithm.

6 Conclusion

With the increasing application of the PMU and WAMS in power system, an improved Prony algorithm is proposed for the out-of-step detection and prediction of interconnected power grids. The energy approximation method is used to reduce the order of the decomposed signals in order to improve the numerical stability of the prediction. The effectiveness and accuracy of the scheme is verified on a 2-area 4-machine power system.

Acknowledgement

This work is supported by National Natural Science Foundation of China (No: 51407109), China Postdoctoral Science Foundation (2014M551904), and the Fundamental Research Funds of Shandong University (No. 2082014GN034).

References

1. G. Zheng. Power system disturbance detection and classification based on wide area phasor measurements, Thesis, Tennessee Technological University, (2012)
2. U.S.-Canada Power System Outage Task Force, Final report on the august 14, 2003 blackout in the united states and canada: causes and recommendations, (2004)
3. NERC System Protection and Control Subcommittee, Power plant and transmission system protection coordination, (2012)
4. X. Wun, J. Zhao, A. Xu, H. Deng. Review on transient stability prediction methods based on real time wide-area phasor measurements, DRPT, Weihai, China (2011)
5. A. Sun, K. Hur, P. Zhang. A new unified scheme for controlled power system separation using synchronized phasor measurements. IEEE T. POWER SYST, **26**, 3 (2011)
6. B. Shrestha, R. Gokaraju. Apparatus and method for out-of-step protection using the analysis of trajectories of electrical measurements in state plane, Canada Patent, (2014)
7. W. Andrzej. Method of predicting transient stability of a synchronous generator and associated device, EU Patent, (2013)
8. A. Guzman-Casillas, Systems and methods for power swing and out-of-step detection using time stamped data, USA Patent, (2011)
9. R. Franco, C. Sena, N. Glauco, A. Giusto. Using synchrophasors for controlled islanding a prospective application for the uruguayan power system, IEEE T. POWER SYST, **28**, 2 (2013)
10. F. Evangelos ,A predictive out-of-step protection scheme based on PMU enabled distributed dynamic state estimation, Thesis, Georgia Institute of Technology, (2012)
11. J. Blumschein, Y. Yelgin, M. Kereit. Blackout prevention by power swing detection and out-of-step protection, JPEE, **2**, 4 (2014)
12. K. Altallaq, E. Feilat. Online detection of out-of-step operation based on prony analysis-impedance relaying, WSEAS Conference, (2005)
13. J. Hauer. Application of Prony analysis to the determination of modal content and equivalent models for measured power system response, IEEE T. POWER SYST, **6**, 3 (1991)
14. P. Kundur. Power system stability and control. New York: McGraw-Hill, (1993)

A Survey on Particle Swarm Optimization for Use in Distributed Generation Placement and Sizing

Syed Muhammad Arif[1,a], Akhtar Hussain[2] and Dong Ryeol Shin[1]

[1]*College of Information and Communication Engineering Sungkyunkwan University, Suwon, South Korea*
[2]*Department of Electrical Engineering, Incheon National University, Incheon, South Korea*

Abstract. This paper surveys the research and development of Particle Swarm Optimization (PSO) algorithm for use in selecting a suitable position for Distributed Generation (DG) units within a distribution network. Our discussion first covers the algorithm development of PSO and its use in neural networks. After establishing the foundations of PSO, we then explore its use in sizing and sitting of DG units in distribution network. Combining PSO with other optimization techniques for attaining better results is also discussed in this paper.

1 Introduction

In the conventional power system, the generating units are kept far away from the consumers. While it has certain advantages, the biggest disadvantage is the occurrence of heaving line losses when power is transported to the consumers. There can be many ways of reducing power loss in the distribution network. For example, the use of superconductors in transmission lines can reduce line losses. However, the superconductor technology is still in its early days and still quite expensive. More recently, the concept of Distributed Generation (DG) has come up which is environmentally friendly and economical. The sizing and placement of DG units can significantly affect the system losses. This has initiated considerable interest in finding an optimum position of the DG unit. To find a suitable place for a DG unit in Smart Grids, the authors in [1] use the Multi Criteria Decision Analysis. The optimization variables include Voltage Stability Index (VSI), Frequency Deviation in Substation (FDSS) and Frequency Deviation in DG (FDDG). The same authors in [2] have used circuit analysis to detemine the voltage stability index (VSI) for distribution networks connected in loop configuration.In this paper, we are more concerned with exploring PSO for optimizing the sizing and siting of DG units. Such an optimization problem canbe solved by different optimization algorithm [3], for instance Tuba Search (TS). However, is a time consuming algorithm and has a tendency to get trapped into local optimal value. Search Annealing (SA) is another possible algorithm which also consumes considerable time to get the optimal solution. As compared to TS and SA, Genetic Algorithm (GA) has been widely used for finding the global (or near global) optimal solution of the problem. PSO has been proven to be better than GA in terms of computational time and convergence speed [3]. Hence, this paper surveys the use of PSO in placing DG units.

2 Original PSO algorithm

The original PSO algorithm, which is a non-linear optimization algorithm, was discovered by James Kennedy and Russell Eberhart in 1995 [4], [5] while observing the social behaviour of animals and birds. The authors proposed the velocity and position equations for each particle travelling in a group. The prevailing velocity can be calculated by using the previous velocity and the distance between P_{best} and G_{best} as expressed by following formula [4], [5]. All particles adjust their position according to the personal flying experience called P best (personal best) as well as the flying experience of the other particles in the group called G best (Global best).

$$v_i^{k+1} = v_i^k + c_1 r_1 (Pbest_i^k - x_i^k) + c_2 r_2 (gbest^k - x_i^k) \quad (1a)$$

The current position can be calculated by adding the current velocity in previous position.

$$x_i^{k+1} = x_i^k + v_i^{k+1} \quad (1b)$$

V_i^k : Velocity of particle i at iteration k
$C1$: Cognitive factor; $C2$: Social factor
r1, r2: Uniformly distributed random number between 0-1
x_i^k : Position of particle i at iteration k
$Pbest_i^k$: Personal best of agent i at iteration k
$Gbest^k$: Best among all in the group

Kennedy and Eberhart [4], [5] have proposed the value of c1=c2= 2. If c1 is much greater than c2, each

[a] Corresponding author.

particle is remains attracted to its own personal best causing excessive wandering. Consequently, particles can not find the optimal solution and remain trapped in the local value. On the other hand, if c2 is much greater than c1, the particles are strongly attracted towards the global best position. This is also undesirable because the particles may move away from the optimal value.
All the particles in a search space should obey these three rules [6].

1. Avoid collision with the neighbours
2. Match their velocity with that of the neighbours
3. Stay in close vicinity with the neighbours

2.1 Modified PSO algorithm

The author in [7] modified the velocity equation by introducing the construction factor 'K' to insure the convergence of PSO. The modified expression is as follows:

$$v_i^{k+1} = K[v_i^k + c_1 r_1 (Pbest_i^k - x_i^k) + c_2 r_2 (gbest^k - x_i^k)] \quad (2)$$

$$\text{where } K = \frac{2}{\left|2-\varphi-\sqrt{\varphi^2-4\varphi}\right|} \quad where \varphi = C1 + C2, \varphi > 4$$

The authors in [8] introduced a new parameter called the inertia weight in the standard PSO form:

$$v_i^{k+1} = w v_i^k + c_1 r_1 (Pbest_i^k - x_i^k) + c_2 r_2 (gbest^k - x_i^k) \quad (3a)$$

$$v_i^{k+1} = w v_i^k + c_1 r_1 (Pbest_i^k - x_i^k) + c_2 r_2 (gbest^k - x_i^k) \quad (3b)$$

Refer to right side of equation (3a), which consists of three parts: the first part shows the previous velocity, the second and third parts are called the cognitive and social part. Without the second and third parts the particles keep on flying with the current velocity. Cognitive part increase the local search ability and social part increase the global search ability. The inertia weight w in Equation (3a) balances the local and global search abilities. The main advantage of introducing w is that it eliminates the requirement of careful setting of velocity. Also, increasing the inertia weight helps the search for a global solution and vice versa.

The authors in [9] have used w=0.7 with the hit and trial method to balance the global and local search abilities.

The authors in [10] have used variable weight factor for the optimal design of small renewable energy systems. In a search space particles are flying to find the optimal solution. The process of reaching the particles towards a common point is called convergence. The proposed method keeps on increasing the weight until convergence is achieved. This modified PSO technique shows faster convergence speed and needs less computational time as compared to the standard PSO technique. Mathmatically, the variable inertia weight is represented as follows:

$$w = \frac{(w_{max} - w_{min})}{\#iteration} * \text{current iteration no.} \quad (4)$$

The PSO flow chart shown in Figure 1, detail discussion of flow chart is beyond the scope of this paper. Details are given in [31]

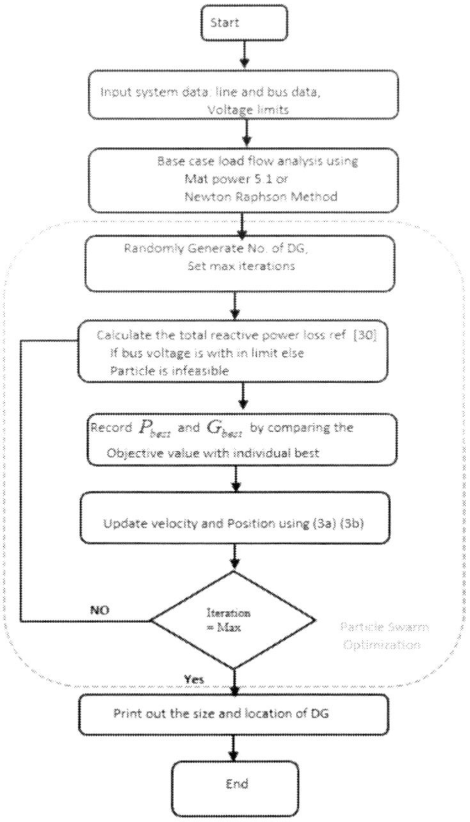

Figure 1. PSO flow chart.

3 Training a neural network using PSO

Data processing, classification and pattern recognition are widely used applications of Artificial Neural Network (ANN). ANN is a learning method motivated by the learning ability of human brain's neural systems which comprises of three layers structures: input, output and hidden layers. There are several learning algorithms, such as Back Propagation (BP) [11], Gradient Descendant [12], etc. One of the main disadvantages of these techniques is the slow convergence and tendency to get trapped in the local minima [13]. Since PSO is a population based search algorithm it can produce better results. One of the biggest issue with PSO is due to the strong competition between the flying particles in a search space the PSO get trap into local minima. To avoid getting trapped into the local minima and achieving fast convergence, the authors in [14] proposed Gaussian PSO with fuzzy reasoning based structural learning. Instead of uniformly distributed random variable, Gaussian random variable is proposed to escape from the local minima and the structural learning with fuzzy reasoning is proposed to improve the computational time by modifying the network structure. The authors in [15] propose optimization of type-2 fuzzy weights in back propagation learning for NN using PSO and GA. The type-2 fuzzy logic enhance the uncertainty in the learning while PSO and GA increases the convergence speed. It has been emphasized in [16], [17] that the weight affects the performance of the learning process. Another inspiring application of PSO in Neural Network is prostate cancer prediction system. The authors in [18] have proposed two-stage fuzzy neural network for prognosis system which provide more accurate prediction (due to fuzzy system) of prostate cancer results and translate it into information for doctor. [19] Shows the experimental study for pixel classification in satellite imaginary, the author did 1050 numbers of experiments and proposed the solution of two critical issues i.e. determination of hidden layers neurons and most discriminative spectral band. The multi-objective (MPSO) improve the classification accuracy and computation time.

4 Using PSO for distributed generation

Recent research shows that installing a DG unit in a distribution network leads to various technical, economical and enviromental benefits. However, in order to reap these benefits, it is very important to find an optimal location and size of a DG unit.There are a number of constraints that need to be considered before placing a DG unit [20], [37].

Equality constrains: Power flow equations represent the equality constrains (total power injected at any bus should be equal to total generation minus demand). For each bus the power flow equation should be satisfied, which can be handle by load flow such as Newton Raphson.

Inequality constrains: The thermal constrains and equiptment rating are called inequality constrains. These constrains can be handled by proper sizing the DG unit which can be done by using PSO. Inequality constrains are as follow [32]

1) Each segment in the distribution network has a thermal limits which should not exceed than its maximum rating.
2) The output of DG should not exceed then the power supplied by the substation.
3) The total number of DG connected in distribution network should be bounded by a certain number. By definition, only one DG unit can place at a given bus.

Boundary conditions: While placing the DG at a given bus the voltage magnitude and angle should be kept at a certain acceptable level.

4.1 Optimal sizing

The authors in [21] have used PSO for optimal sizing of the stand-alone hybrid power system. This system includes wind unit, electrolysers, a reformer, anaerobic reactor, fuel cell and some hydro tanks. The objective is to minimize the total cost of the system and the components have optimal sizes, such that the power demand by the system meets. It has been argued in [22]-[24] that PSO provides excellent convergence and feasible optimal solution for sizing the micro-grid system consisting of wind/PV/battery banks/ fuel cells and reformers. The comparative analysis in [25] focuses on the size of the off-grid Renewable Hybrid Energy System (micro hydro plant, fuel cell, wind turbine and solar panel etc.) using GA and PSO. The analysis shows that PSO is much better then GA in terms of CPU utilization and the number of iterations to find the optimal value. The authors in [26] proposed improved PSO which utilize the benefits of both the GA and PSO that can enhance global searching capability and convergence of particles. To improve the convergence rate the convergence factor K is introduce and the global search ability is enhance by introducing the idea of mutation of genetic algorithm this can escape the local trapping. In [27], the authors have proposed the structure of a micro grid consisting of Wind PV hybrid system with battery, Hydro tank and fuel cell. The author used PSO algorithm to achieve the best sizing of the DG unit for the proposed micro grid. The authors in [28] have proposed an approach for sizing the hybrid systems (Wind/PV/Diesel Generator/Fuel Cell/Battery and Hydro Tank). The ε-constrain method [29] is used to minimize the overall system cost and CO_2 emission produced by the diesel generators. The author in [30] proposed an approach for sitting and sizing of four types of DG units to enhance the loadability of primary distribution systems. The results show that there is a strong correlation between reactive power loss and loadability of the system

4.2 Optimal siting

The author in [31] have proposed multi-objective based approach to find the best location and size of different DG units with different load model having non-unity power factor. A simple PSO algorithm is used to obtain the best solution for the multi objective problem. The authors in [32] have proposed improved PSO algorithm for the placement and sizing of multiple DG units. The problem is divided into two sub problems: DG unit's optimal location (discrete optimization) and its optimal size (continuous optimization). In the standard PSO, a factor is multiplied with inequality constrains and added into the objective function called penalty factor. The proper selection of penalty factor is very difficult which differ form problem to problem. The author eliminates the needs of proper selection of penalty factor by using a unique PSO feature. If the particle flies away from the search space then restore its violated position to P_{best} this keeps the infeasible particles alive as a candidate. Author in [33] use the application of PSO for DG sizing and siting in the radial distribution network. The objective is to improve the voltage profile, and active power compensation by reducing the real power loss. By using the exact loss formula, the authors calculate the size of DG unit at every bus. The location of the DG unit is calculated by using the loss sensitivity factor. The bus having lowest power loss and lowest sensitivity factor will be the best location for DG placement. The authors in [34] use PSO for optimal placement of different types of DG units which can supply both real and reactive power in the distribution networks. The proposed PSO approach is tested on IEEE-69 bus distribution system and the result shows that as compared to other heuristic approaches, the proposed algorithm gives smaller DG size. The authors in [35] have used the dynamic sensitivity analysis method for the placement of capacitors which will reduce the search space problem. The results show that for more than one location, dynamic sensitivity is good because it helps in deciding other positions considering the previous locations and the value of capacitor. The comparison results show that the proposed PSO approach provides better global solution with greater saving. The authors in [36] have used PSO for optimal allocation of DG units for multi-phase unbalanced distribution network. The results show that optimal size and allocation of DG unit can reduce the total power loss and improve the voltage profile in the distribution network. The proposed method can place DG unit more effectively and in less computation time. The power loss reduction using PSO is better than power loss reduction using RPF (repeated power flow).

4.3 Hybrid PSO

Researchers are focusing on the hybridization of PSO with other techniques for getting better results. The authors in [37] have proposed hybrid PSO/GA algorithm to minimize the network power loss, better voltage regulation, and improve voltage stability index. The result shows that hybrid PSO/GA has better power minimization as compared to PSO and GA while GA has better voltage profile improvement as compared to PSO and Hybrid PSO/GA. The authors concluded that hybrid method offer a higher ability of finding the optimal solution. The authors in [38] have proposed hybrid PSO with analytical method (AM) to reduce the power loss and to improve the voltage profile. The sizing is done by AM and the placement is done by PSO. The result shows that loss reduction becomes more effective by increasing the number of DG units in the network. It also increases the voltage profile. In [39] the authors have proposed hybrid PSO which has a problem of being trapped into local optima and take a lot of time for getting the optimal solution. To balance the local and global searching ability, the same authors [40] have used fuzzy adaptive inference for tuning the inertia weight of PSO. The result shows that fuzzy Adaptive Particle Swarm Optimization (FAPSO) gives the optimal solution after 69 iterations while other PSO method need 190 cycles. The authors in [41] proposed a novel hybrid Rank Evolutionary PSO to improve the computation time (fastest) of multiple DG. The results shows that Rank Evolutionary PSO has stable performance and fast result in moving the particle to global optimal solution, less number of iteration, lowest standard deviation and fastest computing time as compared to other PSOs.

5 Conclusion

This paper presents a summary of the research and development in Particle Swarm optimization algorithm especially for DG placement and sizing. It highlights the standard and modified algorithm and constraints needed to be handled when the DG unit is placed in the distribution network. Several applications of PSO in training a neural network, distribution network for DG sizing and siting and hybrid PSO are also discussed in the literature.

Acknowledgement

The author would like to express out appreciation to Prof. Faraz Hasan, Massey University, New Zealand for sharing their pearls of wisdom during the course of this research. This research was supported by the MISP (Ministry of Science, ICT & Future Planning), Korea, under the National Program for Excellence in SW) (R2215-16-1005) Supervised by the IITP (Institute for Information & communications Technology Promotion).

References

1. Kazmi, SAA., Hasan, SF., & Shin, DR., "Multi Criteria Decision Analysis for Optimum DG Placement in Smart Grids", IEEE Innovative Smart Grids Technologies Conference (Asia) (2015)
2. Kazmi, SA., Hasan, SF., & Shin, DR.,"Circuit Analysis Approach for Calculating Voltage Stability Index", IEEE International Conference on Power Electronics and Drive Systems (2015)
3. A.M. El-Zonkoly, "Optimal placement of multi-distributed generation units including different load models using particle swarm optimisation" IET Gener.Transm. Distrib., **5**, 7, pp. 760–771, (2011)
4. J. Kennedy and R. Eberhart, "Particle swarm optimization," in Proc. IEEE Int. Conf. Neural Netw., **4**, pp. 1942–1948 (1995)
5. R. Eberhart and J. Kennedy, "A new optimizer using particle swarm theory," in Proc. 6th Int. Symp. Micro Machine Human Science, pp. 39–43 (1995)
6. Reynolds, C. W. Flocks, Herds, and Schools "A Distributed Behavioral Model, in Computer Graphics", **21**(4) (SIGGRAPH) pages 25-34 (1987)
7. R. C. Eberhart and Y. Shi, "Comparing Inertia Weights and Constriction Factors in Particle Swarm Optimization" IEEE (2000)
8. Shi, Y. and Eberhart, R.C., "A modified particle swarm optimizer", Proceedings IEEE International Conference on Evolutionary Computation, 69-73. Piscataway, NJ: IEEE Press. (May 1998)
9. S.M. Hakimi and S.M. Moghaddas-Tafreshi, "Optimal sizing of a stand-alone hybrid power system via particle swarm Optimization for Kahnouj area in south-east of Iran", Renewable Energy, **34** 1855–1862 (2009)
10. Ahmed Hassan, Magdi Saadawi, Mahmoud Kandil, Mohammed Saeed, "Modified particle swarm optimization technique for optimal design of small renewable energy system supplying a specific load at Mansoura University", IET Renew. Power Genre., **9**, 5, pp. 474–483 (2015)

11. Marco Gori, Alberto Tesi, "On the problem of local minima in back-propagation algorithm", Neural Netw., **5**(4) 465–471 (1992)
12. M. F. Moller, "A Scaled Conjugate Gradient Algorithm for Fast Supervised Learning", PB-339 Reprint, Computer science Department, University of Aaur- hus, Denmark, (1990)
13. C.H.Chen, H.Lai, "Anempirical study of the Gradient descentand the conjugate gradient back propagation neural networks", in: Proceedings OCEANS' 92 Mastering the Oceans through Technology, **1**, 1991, pp. 132–135 (1992)
14. Haydee Melo, Junzo Watada, "Gaussian-PSO with fuzzy reasoning based on structural learning for traininga Neura lNetwork", Neurocomputing, 172-405–412 (2016)
15. Fernando Gaxiola, Patricia Melina, Fevrier Valdez, Juan R. Castro, Oscar Castillo, "Optimization of type-2 fuzzy weights in backpropagation learning forneural networks using GAs and PSO", Applied Soft Computing, 38-860–871 (2016)
16. Mohammad Ali Ahmadi, "Connectionist approach estimates gas–oil relative permeability in petroleum reservoirs: Application to reservoir simulation", Fuel, 140-429–439 (2015)
17. Mohammad-Ali Ahmadi, Mohammad Reza Ahmadi, Seyed Moein Hosseini, Mohammad Ebadi, "Connectionist model predicts the porosity and permeability of petroleum reservoirs by means of petro-physicallogs: Application of artificial intelligence", Journal of Petroleum Science and Engineering 123-183–200 (2014)
18. Ren-Jieh Kuoa, Man-Hsin Huangb, Wei-Che Chenga, Chih-Chieh Linc,d, Yung-Hung Wue, "Application of a two-stage fuzzy neural network to a prostate cancer prognosis system", Artificial Intelligence in Medicine 63 -119–133 (2015)
19. Rajesh K. Agrawal, Narendra G. Bawane, "Multiobjective PSO based adaption of neural network topology for pixel classification in satellite imagery", Applied Soft Computing 28217–225 (2015)
20. Florina Rotaru, Gianfranco Chicco, Gheorghe Grigoras ,Gheorghe Cartina,"Two-stage distributed generation optimal sizing with clustering-based node selection", Electrical Power and Energy Systems 40 120–129 (2012)
21. H. Hassanzadehfard, S.M. Moghaddas-Tafreshi, S.M. Hakimi, "Optimal Sizing of an Islanded Micro-Grid for an Area in North-West Iran Using Particle Swarm Optimization Based on Reliability Concept", World Renewable Energy Congress –Sweden (2011)
22. S.M. Moghaddas-Tafreshi and S.M.Hakimi, "Optimal Sizing of a stand-alone Hybrid Power System via Particle Swarm Optimization (PSO)", International Power Engineering Conference -IPEC (2007)
23. H. Valizadeh Haghi, S. M. Hakimi, and S. M. Moghaddas Tafreshi ,"Optimal Sizing of a Hybrid Power System Considering Wind Power Uncertainty Using PSO-Embedded Stochastic Simulation", PMAPS (2010)
24. M. Bashir and J. Sadeh, "Size Optimization of New Hybrid Stand-alone Renewable Energy System Considering a Reliability Index", IEEE (2012)
25. Bhimsen Tudu, Sibsankar Majumder, Kamal K. Mandal, and Niladri Chakraborty,"Comparative Performance Study of Genetic Algorithm and Particle Swarm Optimization Applied on Off-grid Renewable Hybrid Energy System",Springer-Verlag Berlin Heidelberg (2011)
26. Y. S. Zhao, J. Zhan, Y. Zhang, D. P. Wang and B. G. Zou,"The optimal capacity configuration of an independent wind/Pv hybrid power supply system based on improved pso algorithm" (2009)
27. Arash Navaeefard, S.M.Moghaddas Tafreshi, Mostafa Barzegari, Amir Jalali Shahrood, "Optimal Sizing of Distributed Energy Resources in Microgrid Considering Wind Energy Uncertainty with Respect to Reliability", International Energy Conference IEEE (2010)
28. Masoud Sharafi , Tarek Y. ELMekkawy,"Multi-objective optimal design of hybrid renewable energy systems using PSO-simulation based approach", Renewable Energy 68-67-79 (2014)
29. Kaveh Khalili-Damghani, Amir-Reza Abtahi, Madjid Tavana, "A new multi-objective particle swarm optimization method for solving reliability redundancy allocation problems", Reliability Engineering and System Safety 111 -58–75 (2013)
30. Nguyen Cong Hien, Nadarajah Mithulananthan, Senior Member, IEEE, and R. C. Bansal, Senior Member, IEEE, "Location and Sizing of Distributed Generation Units for Loadabilty Enhancement in Primary Feeder", IEEE systems journal, **7**, 4, (December 2013)
31. A. M. El-Zonkoly,"Optimal placement of multi-distributed generation units including different load models using particle swarm optimization", Swarm and Evolutionary Computation **1** 50–59 (2011)
32. M. R. AlRashidi , M. F. AlHajri, "Optimal planning of multiple distributed generation sources in distribution networks: A new approach", Energy Conversion and Management **52** 3301–3308 (2011)
33. Satish Kansal, B.B.R. Sai, Barjeev Tyagi, Vishal Kumar, "Optimal placement of distributed generation in distribution networks", International Journal of Engineering, Science and Technology, **3**, 3, pp. 47-55 (2011)
34. Satish Kansal, Vishal Kumar, Barjeev Tyagi, "Optimal placement of different type of DG sources in distribution Networks", Electrical Power and Energy Systems 53-752–760 (2013)
35. S.P. Singh and A.R. Rao, "Optimal allocation of capacitors in distribution systems using particle swarm optimization", Electrical Power and Energy Systems 43-1267–1275 (2012)
36. Samir Dahal, Hossein Salehfar, "Impact of distributed generators in the power loss and voltage profile of three phase unbalanced distribution network", Electrical Power and Energy Systems 77-256–262 (2016)
37. M.H. Moradi , M. Abedini, "A combination of genetic algorithm and particle swarm optimization

for optimal DG location and sizing in distribution systems", Electrical Power and Energy Systems 34-66–74 (2012)
38. Satish Kansal,Vishal Kumar, Barjeev Tyagi, "Hybrid approach for optimal placement of multiple DGs of multiple types in distribution networks", Electrical Power and Energy Systems 75-226–235 (2016)
39. Shuheng Chen, Weihao Hu, Chi Su, Xiaoxu Zhang, Zhe Chen, "Optimal reactive power and voltage control in distribution networks with distributed generators by fuzzy adaptive hybrid particle swarm optimisation method", IET Gener. Transm. Distrib., **9**, 11, pp. 1096–1103 (2015)
40. Wen Shan Tan, Mohammad Yusri Hassan, Hasimah Abdul Rahman, Md Pauzi Abdullah, Faridah Hussin,"Multi-distributed generation planning using hybrid particle swarm optimization- gravitational search algorithm including voltage rise issue",IET Gener. Transm. Distrib., **7**, 9, pp. 929–942 (2013)
41. J. J. Jamian, M. W. Mustafa, H. Mokhlis, "Optimal multiple distributed generation output through rank evolutionary particles warm optimization", Neurocomputing 152-190–198 (2015)

The Design and Optimization of Combined Rogowski Coil Based on Printed Circuit Board

Pongchit Saetang[1] and Amnart Suksri[1, 2]

[1]Department of Electrical Engineering, Faculty of Engineering, Khon Kaen University, Khon Kaen, 40002, Thailand, Tel. 043-347-031, Fax. 043-347-032
[2]Center for Alternative Energy Research and Development, Department of Mechanical Engineering, Faculty of Engineering, Khon Kaen University, Khon Kaen, 40002, Thailand, Tel. 0804614416

Abstract. Rogowski coil has been widely used for measurement of small magnitude with fast rise time signal such as partial discharge (PD), but Rogowski coil based on Printed circuit board (PCB Rogowski) has limitation for this application. This paper presents the prototype of Combined Rogowski coil that can overcome the drawback of PCB Rogowski coil. Experimental results showed that the proposed coil was able to measure a simulated of PD signal and its output signal indicated that the coil is suitable for a detection of small magnitude signal when compare with the other coils. The proposed Combined Rogowski coil in this research has upper frequency (f_h) of 53.59 MHz and mutual inductance (M) of 0.417 µH, while PCB Rogowski coil has f_h of 22.48 MHz and M of 0.026 µH. The oscillation of signal output that caused by non-uniform winding also has diminished. The electrical parameters that affected the performance of coil were controlled by the geometric variables and optimized by multi-objective genetic algorithm to achieve the optimum design of coil.

1 Introduction

Rogowski coil is a toroidal coil wound on a nonmagnetic core form. The coil has been widely used for the detection and measurement of electric currents in many applications [1]. Rogowski coil has many advantages such as linearity, has no effects from saturations, galvanic isolation and low cost. However, traditional coil is sometime hard to achieve on uniform winding that can cause errors to output measurement [2]. PCB Rogowski coil is designed by CAD methods, a perfect spacing is easy to achieve. Nevertheless, PCB Rogowski coil has a narrow bandwidth and very low sensitivity because inherent defect from its structure [3] meaning PCB Rogowski coil is not suitable for the detection of small and wide bandwidth signal such as partial discharge signal.

In this paper, a Combined Rogowski coil is proposed. The design coil was looking into an optimization of the bandwidth and sensitivity of the coil. It is sought by using genetic algorithm for multi-objective optimization technique.

2 Rogowski coil principle

Rogowski coil was developed by Rogowski and Steinhas in 1912. It has been made by placing the winding on a flexible or rigid toroidal core and then bending it round the conductor whose current is going to be measured. Rogowski coil works by sensing the magnetic field in space around the conductor, Ampere's Law and Faraday's Law provide the relationship between current in the conductor $i_p(t)$ and output of Rogowski coil $V_c(t)$ as

$$V_c(t) = -M\frac{di_p(t)}{dt} \quad (1)$$

where

$$M = \frac{\mu_0 Nh}{2\pi}\ln\frac{b}{a} \quad (2)$$

Here M is the mutual inductance between conductor and Rogowski coil, μ_0 is the magnetic permeability of air ($4\pi \times 10^{-7}$ H/m), a is inner radius, b is outer radius, h is height of core and N is number of turns. In order to cancels external fields, the winding can be formed by returning the wire through the center of winding [4]. As in Fig. 1, shows the dimension of Rogowski coil. The behavior of Rogowski coil can be represented by the equivalent circuit of lumped parameter model [5] as shown in Fig. 2.

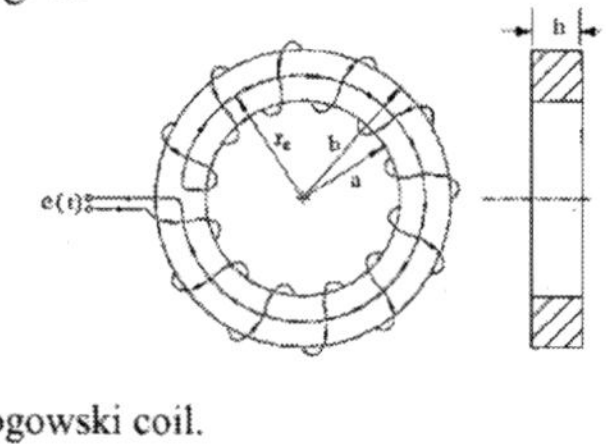

Figure 1. Rogowski coil.

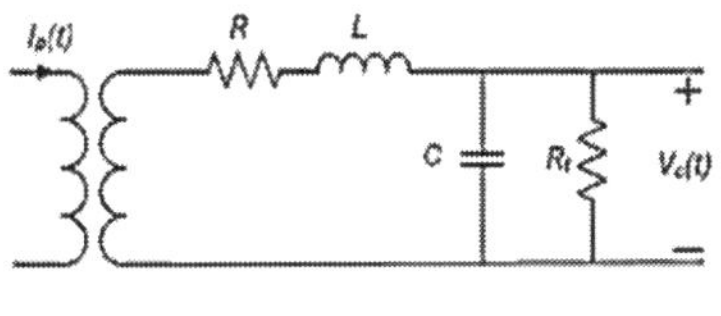

Figure 2. Lumped parameter of Rogowski coil.

Electrical parameters for rectangular cross-section of coil can be calculated as follows:

$$R = \frac{\rho l}{\pi (\frac{d}{2})^2} \quad (3)$$

$$L = \frac{\mu_0 N^2 h}{2\pi} \ln \frac{b}{a} \quad (4)$$

$$C_l = (4\pi^2 \varepsilon_0 k \frac{b+a}{2}) / (\log \frac{b+a}{2} / \frac{b-a}{2}) \quad (5)$$

$$C_t = \pi \varepsilon_0 k l / N \log(\frac{\pi(a+b)}{Nd} + \sqrt{\pi(a+b)^2/(\mathrm{Nd}-1)}) \quad (6)$$

where R, L, C are the resistance, inductance and capacitance of coil, ρ is electrical resistivity, l is length of wire or length of imprint, d is wire diameter, ε_0 is the electric permittivity of air $(8.854x10^{-12}$ F/m) and k is relative permittivity.

The coil's capacitance C , composed of two capacitances in Rogowski coil. The first one is the turn to return capacitance of coil, C_l can be calculated as acoaxial system [6] and the second one is the turn to turn capacitance of coil, C_t. The capacitance between turns would be a chain of series capacitors, C_t which is smaller than C_l and it can be neglected [7]. From the lumped parameters model, the transfer function $H(s)$ for this model is given by

$$H(s) = \frac{V_{out}(s)}{I_p(s)} = \frac{\frac{M}{LC}s}{s^2 + \left(\frac{L + RR_tC}{R_tLC}\right)s + \left(\frac{R + R_t}{R_tLC}\right)} \quad (7)$$

where R_t is damping resistor.

The variation of electrical parameters that affect the frequency response of the coil such as inductance and capacitance of coil are shown in Fig. 3,4.

When L is increased by adding more number of turn or increase the height of core, a decrease of the sensitivity in the high frequency band is found. So, the inductance should be suitable so that it will result in high sensitivity while maintaining the sensitivity in the high frequency band. Increasing of C will lead to the reduction of upper frequency response while maintaining sensitivity. The capacitance has to be kept as low as possible to avoid the drop of high frequency.

As for now, there are many kinds of PCB Rogowski coil, but most of them are made by replacing the windings by imprint of PCB then connected them into a loop, and the height of core is the thickness of baseboard. Since baseboard is very thin and it has high relative permittivity, resulting to the PCB Rogowski coil has low sensitivity and large capacitance. Therefore PCB Rogowski coil is not suitable for the high frequency application.

We propose a Combined Rogowski coil which is quite similar to the PCB Rogowski coil but the Combined Rogowski coil uses copper wire as a half loop to be connected to the baseboard that has the imprint as half loop for a complete coil. While PCB Rogowski coil uses only imprint on PCB as a loop. There are three advantages of this design. Firstly, the oscillation of output signal caused by reflection of electromagnetic wave due tonon-uniform winding is diminishing. Second, using a copper wire as a half loop makes the coil has low capacitance because core is an air instead of baseboard material which has a low value of permittivity. (As a result would improve the upper frequency of coil) Lastly, copper wire made coil's cross-section larger than PCB Rogowski coil, which is a mean of improving the sensitivity of the coil.

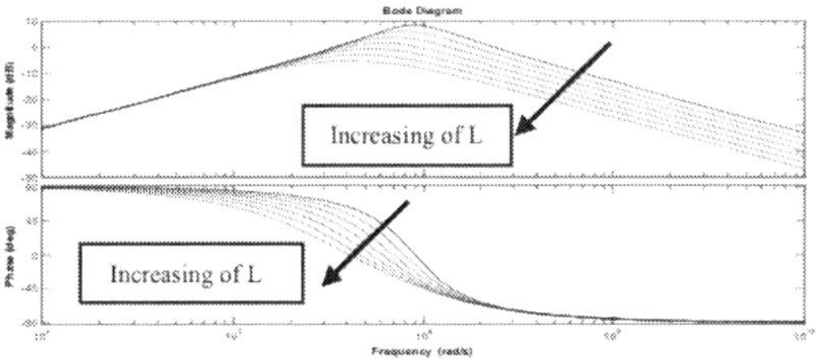

Figure 3. Inductance L changes.

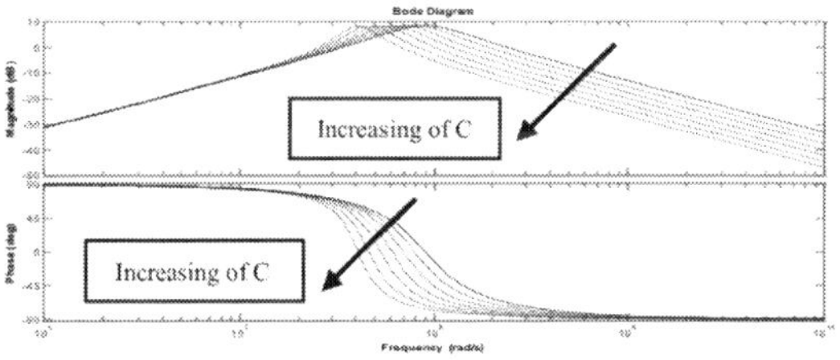

Figure 4. Capacitance C changes.

3 Experimental measurement

Two Rogowski coils, PCB Rogowski coil (RC1), Combined Rogowski coil (RC2) have been constructed. The parameters of constructed coil are present in Table 1 and bode plot of two coils are shown in Fig.5. It can be proved that Combined Rogowski coil (RC2) has higher upper frequency and sensitivity than PCB Rogowski coil (RC1) when considering with the same size (inner, outer radius) and number of turns. Combined Rogowski coil may be more feasible for signal of small amplitude and fast rise time current signal measurement than PCB Rogowski coil.

The measuring were made using RC1, RC2 and non-uniform Rogowski coil are shown in Fig. 6, from left to right. These coil were clamped on XLPE cable to detect the signal from PD generator. The output of coils are measured by Tektronix TDS 1001B oscilloscope.

The model of PD generation is a needle-plate electrode configuration. It consisted of needle electrode which connects to high voltage and plate electrode as a return that connected to ground. The needle-plate gap was setup to 1mm. The current pulse of PD model and its frequency spectrum are shown in Fig. 7, Fig. 8.

Table 1. Parameters of constructed coils.

Parameters of coils	RC1	RC2
Inner radius$a(m)$	0.03	0.03
Outer radius$b(m)$	0.09	0.09
Height of core $h(m)$	0.0032	0.05
Inductance $L(\mu H)$	1.0153	15.864
Mutual inductance$M(\mu H)$	0.02671	0.41747
Capacitance $C(pF)$	121.03	30.257
Resistance$R(Ohm)$	0.6695	0.3928
Damping resistance$R_t(Ohm)$	100	100
Number of turns $N(Turn)$	38	38

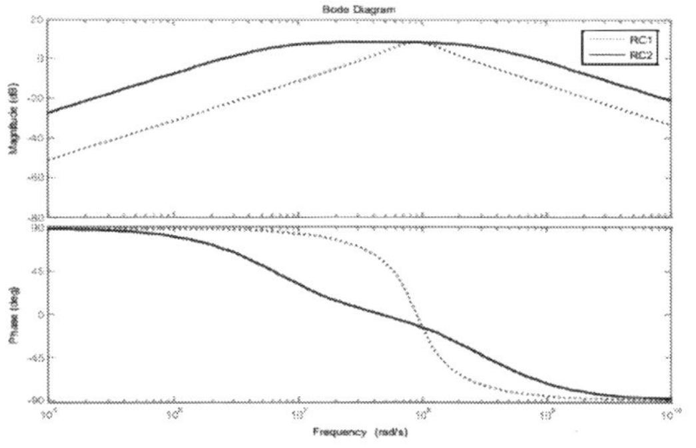

Figure 5. Frequency response of two coils.

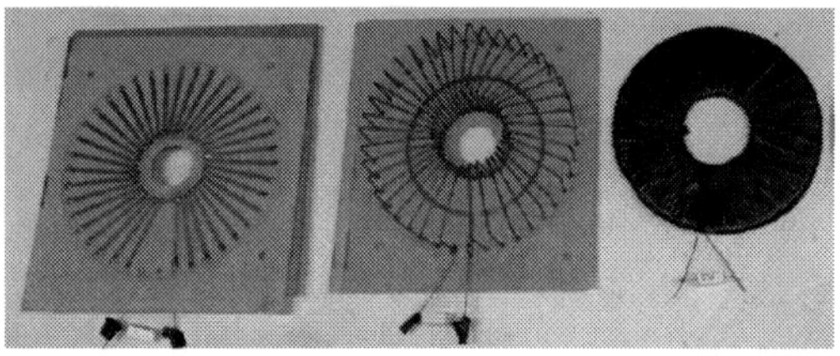

Figure 6. Photo of three coils.

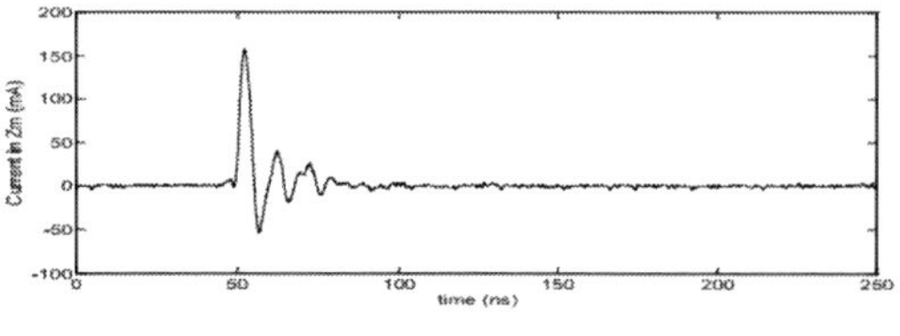

Figure 7. Current pulse.

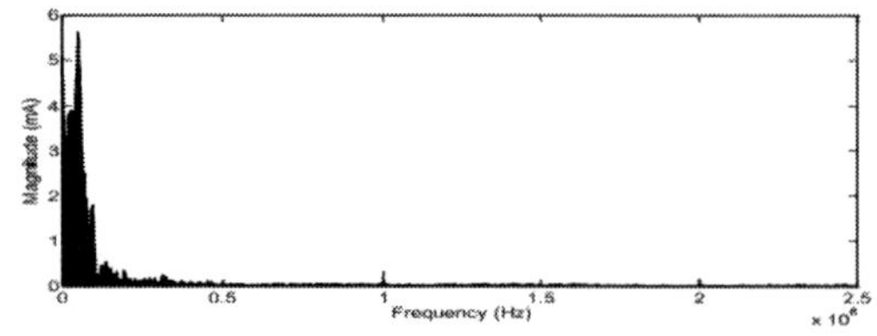

Figure 8. Spectrum of the current pulse.

The measured waveforms are shown in Fig. 9, 10, 11.The output signal from Combine Rogowski coil has higher amplitude than that of the other coils because of high mutual inductance and also its output doesn't oscillate which is the good characteristic of PCB Rogowski coil.

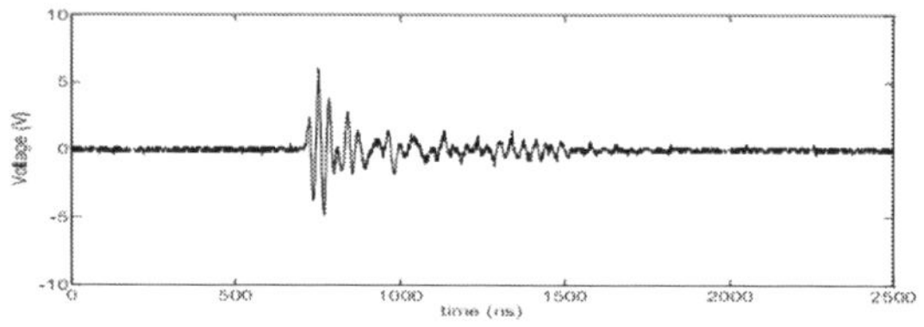

Figure 9. Pulse detected by RC1.

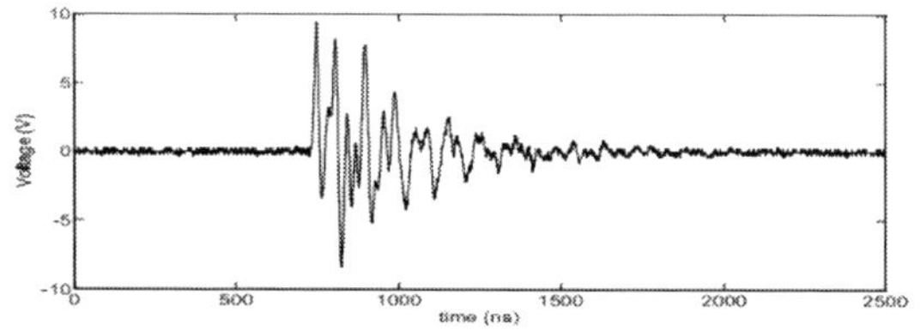

Figure 10. Pulse detected by RC2.

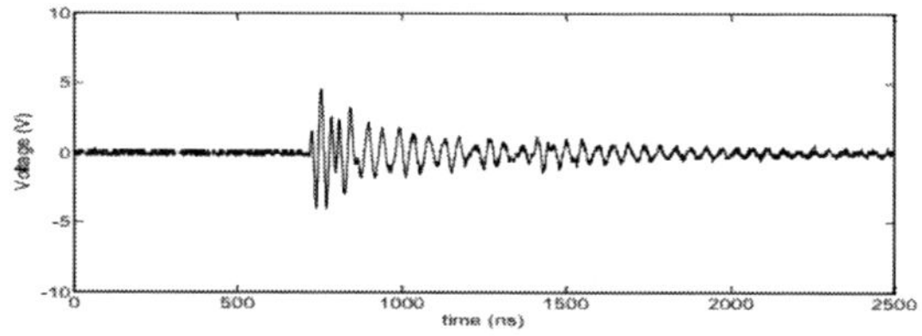

Figure 11. Pulse detected by non-uniform RC.

4 Optimization

Genetic algorithm is a method for finding an optimum values of problem based on a natural selection process which mimics biological evolution, when considering the problem in real life there are more than one objectives, optimizing for a particular objective may result to the unacceptable solution to the other objective. So that a multi-objective optimization problem was investigated, a set of solution where each set of solution satisfies the objectives at acceptable point without dominated by any other solution as call Pareto front [8].

For tuning Rogowski coil to be able to measure a small current, we must improve the mutual inductance by increasing the height of the core or number of turns. Nevertheless, increasing of these parameters will also improve inductance of coil which means the sensitivity of coil at high frequency might be diminished.

According to the equation (7), upper frequency limitation can be found

$$f_h = \frac{(L + RR_tC) + \sqrt{(L + RR_tC)^2 + 4(R_tLC)(R + R_t)}}{4\pi R_t LC} \quad (8)$$

And the value of suitable damping resistance can be determined as [9].

$$R_t = \frac{1}{2}\sqrt{\frac{L}{C}} \tag{9}$$

From the equation (8), (9), the upper frequency limitation can be rewritten as

$$f_h = \frac{\left(L + \frac{R}{2}\sqrt{LC}\right) + \sqrt{2L^2 + +3RL\sqrt{LC} + \frac{1}{4}R^2LC}}{2\pi L\sqrt{LC}} \tag{10}$$

Using Rogowski coil for measuring a very low current, but with a very high frequency such as PD signals have been very attracted moreattention. The challenge is the requirement of a Rogowski coil with high sensitivity and wide bandwidth that can cope up with the PD signals.

Objective functions for maximization are mutual inductance from (2) which directly related with sensitivity of Rogowski coil and upper frequency limitation from (10). These *R, L, C* parameters are a function of geometric parameters *a, b, h, N* and its constraints according to the inside diameter based design method (IDDM) [10] are shown in Table 2.

Table 2. Geometric constraints.

Parameters of coil	Constraints value
Inner radius$a(m)$	0.035-0.05
Outer radius$b(m)$	0.057 - 0.137
Height of core $h(m)$	0.01 - 0.05
Number of turns $N(turns)$	50 - 82

When analyze multi objective genetic algorithm tool in Matlab by setting population size of 100 groups, cross over fraction 0.7 and pareto fraction 0.1 then, we shall get a group of solution as shown in Table 3 and Fig. 12, shows Pareto front(Local pareto solution).

Table 3. Set of pareto optimal solutions.

No.	Parameters					M	f_h
	a (m)	b (m)	h (m)	N (Turns)	R_t (Ohm)	(μH)	(MHz)
1	0.0477	0.1369	0.049	79	595.33	0.8226	7.204
2	0.0499	0.0801	0.010	82	318.21	0.0818	37.70
3	0.0498	0.0841	0.010	82	318.05	0.0912	34.02
4	0.0477	0.1369	0.049	79	595.33	0.8226	7.204
5	0.0446	0.0826	0.014	73	353.20	0.1331	28.76
6	0.0467	0.0826	0.013	77	350.17	0.1146	30.92
7	0.0443	0.0895	0.022	73	434.13	0.2312	20.24
8	0.0474	0.0832	0.011	78	324.73	0.0992	32.99
9	0.0464	0.1190	0.049	76	615.25	0.7068	8.856
10	0.0438	0.0811	0.025	72	467.75	0.2226	22.63

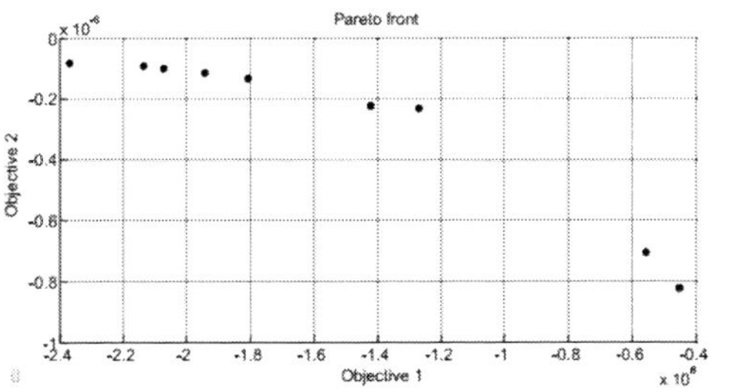

Figure 12. Pareto front.

5 Conclusion

The proposed Combined Rogowski coil has been constructed. There are many advantages on small signal detection due to low value of coil's capacitance that will improve the upper frequency limitation. Also, the coil has more cross-section area leads to high mutual inductance and uniform winding to diminish the oscillation from output signal. The optimization technique using multi-objective genetic algorithm is applied for improving the sensitivity and the bandwidth of coil in order to achieve the best optimum permissible value.

References

1. D. A. Ward, J.L.T. Exon, Engineering Science and Education Journal, *Using Rogowski coils for transient current measurements*, **2**, pp. 105-113, (1993)
2. W. Chengyuan, ICMA,*Design of Printed-Circuit Board Rogowski Coil for Highly Accurate Current Measurement*, pp. 3801-3806, (2007)
3. Y. Zhang, Power System Technology Conference, *The reliable design of PCB Rogowski Coil current transformer*, pp. 1-4,(2006)
4. L. Kojovic, IEEE Computer Applications in Power, *PCB Rogowski Coils Benefit Relay Protection*, **15**, pp. 50-53, (2002)
5. M. Argueso, Scientific Instruments, implementation of Rogowski coil for the measurement of partial discharges, **76**, (2005)
6. M. Shafiq,Elsevier Measurement, *Effect of geometrical parameters on high frequency performance of Rogowski coil for partial discharge measurements*, **49**, pp.126–137, (2014)
7. G. Robles, Instrumentation and Measurement Technology Conference Proceedings, *Identification of parameters in a Rogowski coil used for the measurement of partial discharges*, pp. 1-4, (2007)
8. K. Deb, Chichester Wiley, *Multi-Objective Optimization Using Evolutionary Algorithms*, (2001)
9. M. Shafiq, Sensors Journal IEEE, *Parameters Identification and Modeling of High-Frequency Current Transducer for Partial Discharge Measurements*, **13**, pp. 1081-1091, (2012)
10. D. Porto, 3th IAS, *Design of a new air-cored current transformer part analytical modeling and experimental validation*, **1**, (2004)

DOI: 10.1051/matecconf/20167010015

Hotspots Detection in Photovoltaic Modules Using Infrared Thermography

April M. Salazar[1,2] and Erees Queen B. Macabebe[1]

[1]Ateneo de Manila University, Department of Electronics, Computer, and Communications Engineering, Quezon City, Philippines
[2]Ateneo de Davao University, ECE Department, School of Engineering and Architecture, Davao City, Philippines

Abstract. An increased interest on generating power from renewable sources has led to an increase in solar photovoltaic (PV) system installations worldwide. Power generation of such systems is affected by factors that can be identified early on through efficient monitoring techniques. This study developed a non-invasive technique that can detect localized heating and quantify the area of the hotspots, a potential cause of degradation in photovoltaic systems. This is done by the use of infrared thermography, a well-accepted non-destructive evaluation technique that allows contactless, real-time inspection. In this approach, thermal images or thermograms of an operating PV module were taken using an infrared camera. These thermograms were analyzed by a Hotspot Detection algorithm implemented in MATLAB. Prior to image processing, images were converted to CIE L*a*b color space making k-means clustering implementation computationally efficient. K-means clustering is an iterative technique that segments data into *k* clusters which was used to isolate hotspots. The devised algorithm detected hotspots in the modules being observed. In addition, average temperature and relative area is provided to quantify the hotspot. Various features and conditions leading to hotspots such as crack, junction box and shading were investigated in this study.

1 Introduction

There has been a rising awareness worldwide on the use of solar photovoltaic energy. Reasons behind are not only for addressing climate change, but also for generating new financial opportunities and most importantly for giving energy access to countless people still deprived of modern energy facilities. This global clamor to harness energy from the sun has grown rapidly over the past decade as stated in the Renewables 2015 Global Status Report which underscored that solar PV global capacity increased from 3.7 Gigawatts in 2004 to 177 Gigawatts in 2014. Of this, 40 gigawatts was added in 2014 alone [1].

Even in the Philippines, solar PV installations have grown rapidly as evidenced by the numerous projects implemented across the country including installations in shopping malls, residences and commercial establishments. The largest installation to date is the 100MW solar power plant in Cadiz City, Negros Occidental.

The concern now however is that, after these solar modules have been installed, less attention is given to evaluating whether or not these modules are operating at its optimal rated conditions after five or ten years of use. An ideal solar PV module is made to last for 20 to 25 years. However due to various factors, this lifespan is reduced and without proper monitoring, one will never know if a solar PV module has already become unproductive. Usually, the cause of these damages starts out as a hotspot.

A hotspot is a defect found in a solar photovoltaic module that negatively affects the normal operation of the PV module. There are several conditions that may lead to a hotspot. One simple cause is shading. Shading can be produced by a branch of a tree, an antenna, a nearby building, dirt or anything that physically blocks the solar PV module. What happens is that, when one cell is shaded, it reduces the current through the good cells. The good cells that produce higher voltages usually reverse bias the bad cell. When this occurs, there is large dissipation of power in the poor cell. Large dissipation of power in a small area results to local overheating or hotspots. This can lead to continual yield losses and in worst case scenarios, hotspots can even catch fire [2].

Another potential cause of hotspot is an existing defect in the module like a crack or a dent. This defect becomes a load thus current also concentrate to this area, causing an enormous dissipation of power in the damaged cell, hence, creating a hotspot. As a result, the output power decreases, thereby decreasing the efficiency of the PV module.

Current-voltage (*I-V*) characterization is used to evaluate the performance of the solar PV modules and obtain parameters such as maximum power output and conversion efficiency. However, it is impractical to do in large-scale PV systems because modules need to be individually disconnected from the array and attached to the *I-V* curve tracers. It is, therefore, time-consuming and inconvenient if large solar PV power plants are to be considered.

With the said considerations in mind, a thermal imaging system that can be used to evaluate the operating conditions of a solar PV module is introduced in this study. Thus, the objectives of this study are as follows: (1) Develop a non-invasive technique to locate hotspots in the solar PV module using an infrared camera and a hotspot detection algorithm, (2) Isolate areas with higher temperature intensities using k-means clustering and quantify these hotspots by determining the average temperature and identifying the relative size (area), and (3) Determine possible factors or conditions contributing to hotspots.

This technique eliminates the need to detach each solar PV module from the array for testing since it can identify damaged or problematic PV modules through the thermograms and relative size of the hotspots, which can then be verified through *I-V* characterization.

2 Background

2.1 Infrared thermography

Amongst existing condition monitoring techniques and non-destructive testing methods today, infrared thermography is considered a promising tool for fast and reliable fault detection. Infrared thermography uses mid-wave (MWIR, from ~3 to 5μm) or long-wave (LWIR, from ~7 to 14μm) infrared sensors to obtain thermal images or thermograms of objects under inspection. Based on Planck's black body radiation law, all objects emit infrared radiation proportional to their temperatures. For instance, a study lamp which is turned off will emit low power radiation at wavelengths mostly higher than 1 μm, outside the human visible range. If the lamp is turned on, the temperature will rise and it will glow red due to the higher energies emitted and enters the visible spectrum. If it continues to rise in temperature, radiation may change from red to violet and then the light looks white [3]. Hence, IR thermography is a feasible tool to determine the surface temperature as well as any abnormalities to the temperature trend of a body, or a solar PV module under inspection.

2.2 K-means clustering

In this study, k-means clustering algorithm is taken as a segmentation tool to proceed with the processing of the infrared images. K-Means algorithm categorizes the input data points into multiple groups based on their intrinsic distance from each other. The algorithm aims to find groupings among them by gathering around a centroid. The points are clustered around centroids, *k*, which is achieved by

$$V = \sum_{i=1}^{k} \sum_{x_j \in S_i} (x_j - \mu_i)^2 \quad (1)$$

where S_i are the clusters created, $i = 1,2,\ldots k$ and μ_i is the mean of all the points $x_j \in S_i$ [4]. The cycle of recomputing and assigning data points to the centroid repeats until the centroids do not change or when a certain threshold or criteria is achieved. The most important parameter to consider in this type of clustering is the value of *k*. It is important that *k* is appropriate to the image being clustered. This is why some researchers prefer *k* to be unsupervised based on the color patterns of the image under study [5].

2.3 CIE *L*a*b** color space

Prior processing, the RGB values of the IR images were converted to CIE $L^*a^*b^*$. CIELAB, like the RGB, is a color space, specified by the International Commission on Illumination (*Commission internationale de l'éclairage*, hence its *CIE* acronym). The three coordinates of CIELAB represent the lightness of the color, with $L^* = 0$ yields black and $L^* = 100$ is white. The a^* represent its position between red and green, where negative values indicate green and positive values indicate red. Its position between yellow and blue is represented as b^* where negative values indicate blue and positive values indicate yellow. Distinct from the RGB and CMYK color space, $L^*a^*b^*$ color is intended to approximate human vision and seeks perceptual uniformity. L^* component also closely matches human interpretation of lightness [6].

3 Experimental methods

FLIR C2 infrared camera was utilized in this study. It is a portable infrared camera with an infrared sensor resolution of 80x60 and a thermal sensitivity of <0.10°C. The camera, having a range of 7.5-14μm, can display the object temperature within the range of -10°C to +150°C. Two camera setting conditions had to be considered when taking infrared images of the PV module under inspection. Emissivity *e* is the ratio of the energy radiated from a material's surface to that radiated from a blackbody under the same temperature conditions and wavelength [7]. Objects normally have emissivity ranging from 0.1 to 0.95. In this case, $e = 0.90$ was used, within the typical value range for glass. Another setting condition is the reflective temperature which was set at 10°C, as recommended by the camera manufacturer. This parameter compensates for the radiation reflected on the object and those between the camera and the PV module.

Distance setting between the camera and the PV module was set at 1m. Furthermore, the images should be taken at camera angle between 5° and 60° to avoid casting shadows on the PV module [8]. In this study, images were taken at angles between 20°to 35°, estimated with respect to a line normal to the surface of the PV module. Several images were taken at several time intervals to eliminate the effects of glare or temporary shading factors like clouds in the data. With these settings, infrared images of the solar PV module were collected.

Irradiance, ambient temperature and module temperature were also recorded during the image

 DOI: 10.1051/matecconf/20167010015

collection stage. A temperature data logging circuit using DS18B20 digital temperature sensors and an Arduino Uno microcontroller was devised to measure the temperature of the PV module at four points. A DHT22 temperature and humidity sensor was used to measure the ambient temperature as well as the relative humidity of the environment. For the irradiance measurement, Daystar DS-05 Solar meter was used. For the image processing component of the system, the Image Processing Toolbox in MATLAB was utilized. Fig. 1 illustrates the hotspot detection process.

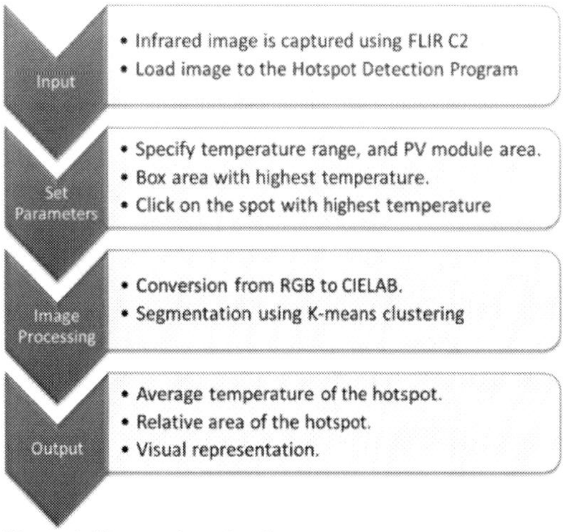

Figure 1. Hotspot detection flow

An infrared image of the PV module is obtained and loaded to the Hotspot Detection Program. The minimum and maximum temperature is then specified by the user for internal computations of the average temperature of the hotspot. These values are displayed in the color bar of the infrared image. The user also specifies the solar PV module area by clicking the four corners of the PV module. Due to camera limitations, it is possible that only parts of the module are captured and not the entire solar PV module. The relative percent area of the potential hotspot is computed based on the specified PV module area. To further specify the spot with the highest temperature, it is necessary to box in the area with the highest temperature.

To lessen the color into 2 dimensions, the RGB values of the image are converted to CIE $L^*a^*b^*$. The conversion from RGB to CIE $L^*a^*b^*$ reduces the computational complexity of the algorithm since the color scale was reduced from 3 to 2 dimensions. This minimizes scattering or large variations of clusters and reduces computation time.

Image segmentation utilizes k-means clustering. K-means is used to group pixels that are in the same temperature range. Data points are grouped to the nearest cluster index using Euclidean distance. After the image is clustered into k groups, the user must select the cluster with the highest temperature range then, finally, the isolated hotspot is displayed including the average temperature value and relative percent area affected by the hotspot.

4 Results and discussions

4.1. Hotspots detection program

Fig. 2 displays an implementation of the Hotspot Detection algorithm developed in the study with $k = 3$. In this run, the potential hotspot has an area of 13.08% and a temperature of about 51°C. As seen in the figure, the Hotspot Detection algorithm developed was able to isolate the area with localized heating.

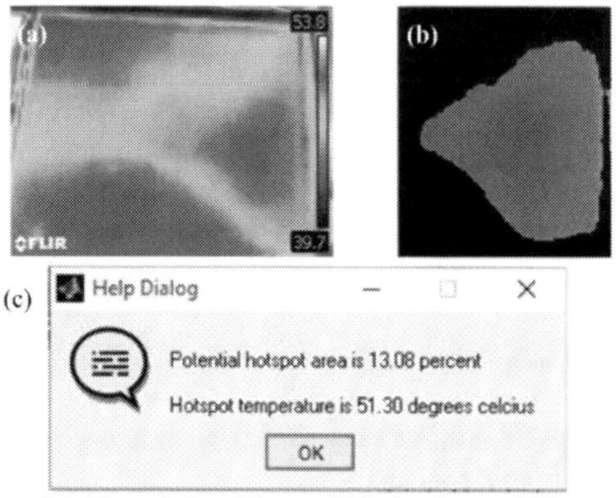

Figure 2. (a) Input Infrared Image (b) Isolated potential hotspot, (c) Summary of area and temperature of the hotspot

One PV module considered for the study is a polycrystalline silicon PV module rated at 50W. Fig. 3 shows a sample run on the IR image of the 50W PV module using the hotspot detection algorithm developed.

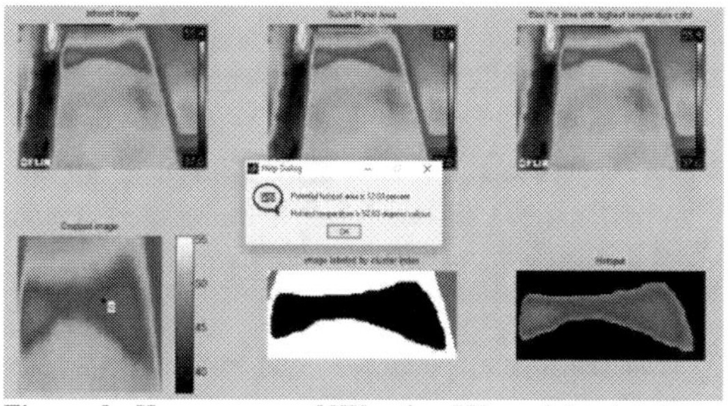

Figure 3. Hotspot on a 50W solar PV module showing a relative area percentage of 12.03 % at 52°C, Irr = 749 W/m^2.

4.2. Cracks cause hotspots

The IR image of the 50W PV module in Fig. 3 show a relatively large hotspot. Due to its age, it is possible that PV module has accumulated defects but was not diagnosed since its deployment. Fig. 4 show that an area on the upper part of the 50W PV module consistently measured the highest temperature. Images were taken at different days to ensure that the hotspot was not caused by clouds or any other temporary obstruction that may cause shading. Despite changing the capture angle, as seen in Fig. 4 (c) and (d), a consistent white spot is observed. Using the algorithm on one of its IR image, this particular hotspot measured an average temperture of 52°C.

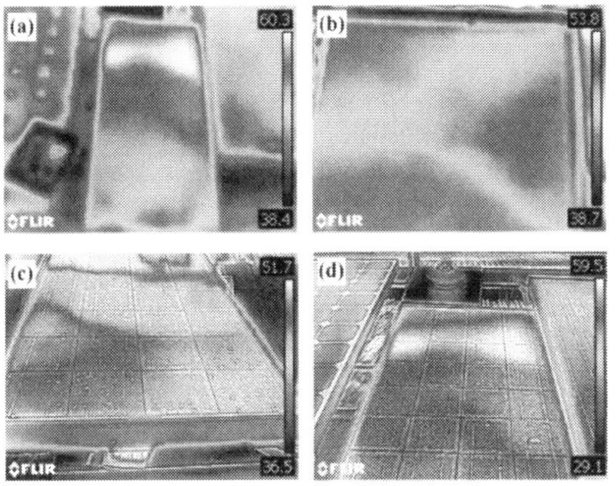

Figure 4. Infrared images of the 50W Module taken as (a) whole module, (b) half module, (c) and (d) at different angles.

By inspecting the PV module, the area with the potential hotspot contains a crack. This defect caused this region to dissipate more power. Hence, produced colors corresponding to higher temperatures. To further validate this observation, temperature at varied points of the panel was recorded using the temperature data logging circuit created.

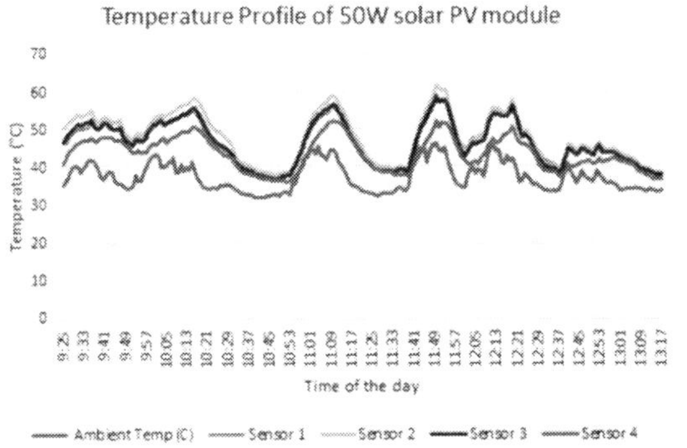

Temperature Sensors Legend:

1 At glass side, where the junction box is
2 At glass side, where the crack is
3 At glass side, center part of module
4 At the back of the module, center

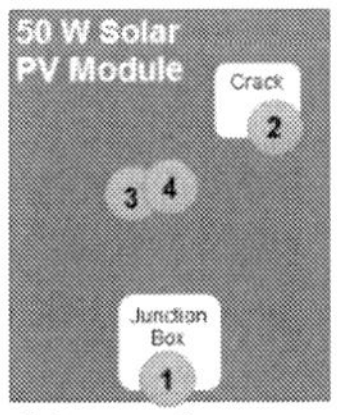

Figure 5. Temperature readings at 4 points of the panel

Sensor 2, placed on the cracked area of the PV module, has recorded higher temperature readings when compared with sensors 1 and 3. Sensor 2 has an average temperature difference of 1.6°C to sensor 1 and 1.3°C to sensor 3. In addition, sensor 2 has an average temperature difference of 4.5°C when compared to sensor 4 readings which is expected since sensor 4 is located at the back of the PV module. This supports the initial observation that the crack caused localized heating in the module.

4.3. Junction box as a source of hotspots

Common among the PV modules investigated in this study is that localized heating was observed in the areas where the junction box is located. Figure 6 shows IR images of another aged 50W polycrystalline PV module. It did not manifest any other hotspot aside from the area at the bottom of the image where the junction box is located. Using the Hotspot Detection algorithm, the hotspot was identified to be about 1.9% at around 54 °C.

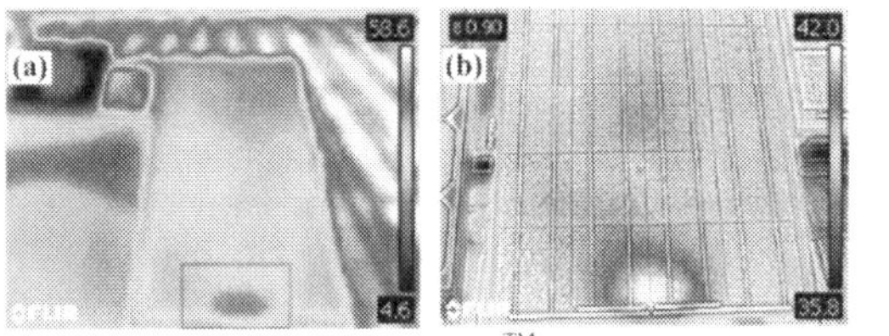

Figure 6. (a) Infrared and (b) MSX™ images of the 50W solar PV module with junction box, indicated in (a), as cause of hotspot.

To confirm this observation, different solar PV modules were investigated. Fig. 7 presents the IR image of (a) a 10W polycrystalline PV module, and two monocrystalline PV modules with peak power ratings at (b) 280W and (c) 250W. They were found to exhibit the same condition, that the area where the junction box is located created thermal stress on the PV module and thus lead to hotspots. The junction box contains the bypass diodes of the solar module.

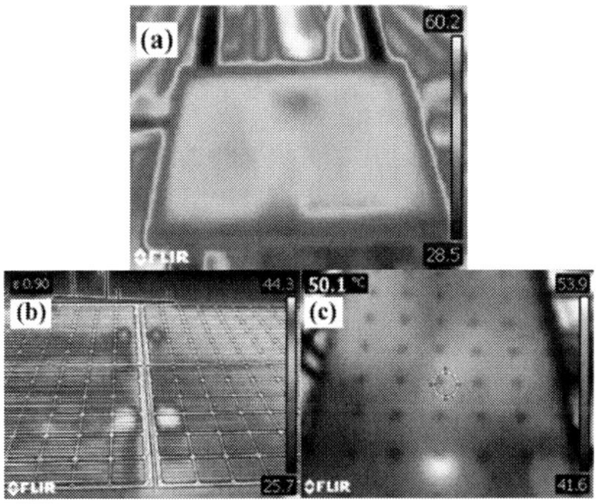

Figure 7. Different solar PV modules with junction box as source of hotspot: (a) 10W Polycrystalline, (b) 280W Monocrystalline, and (c) 250W Monocrystalline PV modules.

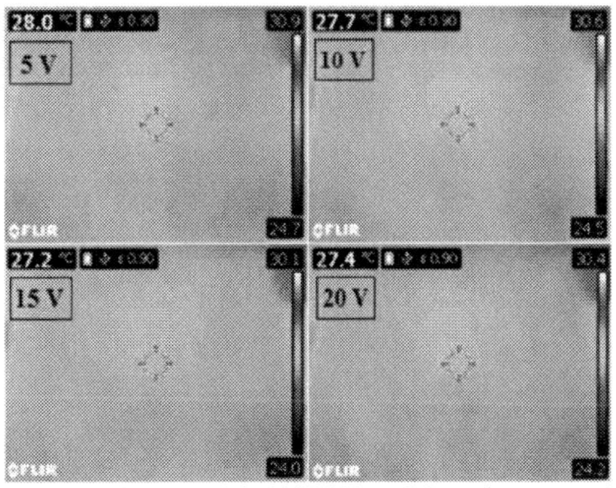

Figure 8. IR images of the10W PV module under reverse-bias conditions: 5 V, 10 V, 15 V and 20 V.

4.4. Shading factors lead to hotspots

A reverse-bias test was conducted on a 10W polycrystalline silicon PV module. Under reverse-bias conditions, external current is fed to the solar PV module and the solar cells begin to emit light in the infrared range. Defects in the material fail to produce light and so faults become apparent. Therefore, it is possible to detect material defects using the reverse-bias test. Fig. 8 shows the IR images of the 10 W PV module under reverse-bias conditions where no defects were detected.

To confirm this defect-free condition, infrared images were taken with the module under operating conditions. Fig. 9 shows the IR images and the corresponding photos of the PV module taken under operating conditions. Initial infrared images, however, yielded a particular hotspot pattern shown in Fig. 9(a). There were observable red spots on the PV module that were not evident during the reverse-bias test. Upon checking, the glass surface of the PV module contained adhesive remains, the patterns of which correspond to the pattern of the red spots.

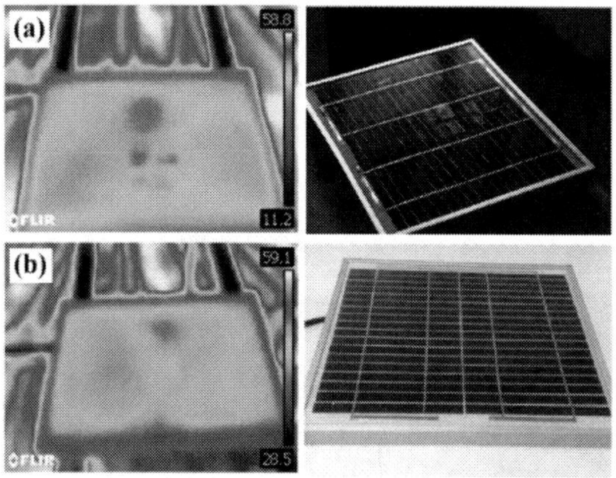

Figure 9. Infrared images and photos of the 10W solar PV module taken (a) with the adhesive and (b) without the adhesive.

When the adhesive was removed and infrared images were taken again, only one small hotspot remained on the PV module, shown in Fig. 9(b). This corresponds to the localized heating at the junction box. The hotspot area which initially affected 4.54% of the module was reduced to 1.61% when the shading factor was removed. This results support the observations that anything that physically blocks the solar PV module from direct sun exposure, causing a partial shading effect, can lead to hotspots.

5 Conclusions

The thermal imaging system developed in this study evaluated different solar PV modules and detected potential hotspots. Through infrared thermography and k-means clustering, local heating areas were isolated. Relative percent area affected by the hotspot as well as the temperature were obtained using the Hotspot Detection algorithm developed.

There are many factors that contribute to the existence of hotspots on solar PV modules. Among the possible causes of hotspots are underlying defects such as a crack. Junction boxes, found at the back of the PV modules also contribute to hotspots. Even adhesives, which caused partial shading on the solar PV module contributed to hotspot formation. Anything, no matter the size, that directly covers the solar PV module from the sun contributes to shading and can lead to hotspots.

Acknowledgement

The researchers would like to express their gratitude to the Department of Science and Technology – Engineering Research and Development for Technology (DOST-ERDT) for supporting this study, and Ateneo de Davao University School of Engineering and Architecture for supporting A.M. Salazar.

References

1. *Renewables 2015 Global Status Report – Key Findings*, **8** (2015) REN21
2. Q Cells, Retrieved from https://www.q-cells.com/products/technology.html
3. C. Ibarra-Castanedo, A. Bendada, and X. P. V Maldague, *5th Pan American Conference for NDT*, (2011).
4. M. C. Chandhok, S. Chaturvedi, and A A Khurshid, IJIT **1** 11 (2012).
5. Mathworks Retrieved from http://www.mathworks.com/help/stats/k-means-clustering.html#brah7f_-1.
6. R. Mente, B. Dhandra and G. Mukarambi, IJCSE **3** 51 (2014).
7. NPL. Retrived from http://www.npl.co.uk/reference/faqs/what-is-emissivity-and-why-is-it-important-(faq-thermal)
8. FLIR. Retrived from http://www.flir.co.uk/instruments/building/display/?id=41872

Developing the City-Level Climate Change Master Plan: The Case of Energy Sector in Bangkok

Wongkot Wongsapai[1], Ariyah Mekhakul[2], Manaswee Arayasiri[2], Kazuhito Yamada[3], Mariko Fujimori[3] and Hisaaki Kato[3]

[1]Faculty of Engineering, Chiang Mai University, Chiang Mai, 50200, Thailand
[2]Bangkok Metropolitan Administration Office, Mitrmaitree Road, Din Dang, Bangkok, 10400, Thailand
[3]Pacific Consultants Co. Ltd., 3-22 Kanda-Nishikicho, Chiyoda-ku, Tokyo, 101-8462, Japan

Abstract. This study focuses in the developing energy sector plan of city-level climate change master plan in Bangkok, the capital city of Thailand. The energy consumption, which consists of electricity and fossil fuels, are investigated and forecasted to be Business-as-Usual (BaU) scenario covering year 2013 to 2023 by using trend analysis. The Greenhouse Gas (GHG) emission mitigation scenario was then analyzed by applying the energy reduction from the national energy efficiency development plan as well as the own reduction activities by Bangkok Metropolitan Administration (BMA). We can conclude that if the national energy efficiency plan can fully implemented, the GHG reduction potential of the city from energy sector would be at around 18% from BAU in 2023 from all energy conservation and renewable energy measures.

1 Introduction

Climate change is one of the largest challenges to the current and future development of human society. The Intergovernmental Panel on Climate Change (IPCC) issued its 5th Assessment Report [1] and warned that warming of the climate system is unequivocal, and it is extremely likely that human influence has been the dominant cause of the observed warming since the mid-20th century. Hence, the long-term climate change policies and plans which focus mainly in mega-city level are very important to each country due to the very high impact to overall GHG emission of the country.

The World Bank Study[2] pointed out that Bangkok Metropolitan Region (consists of Bangkok, Nonthaburi province, Samut Prakarn province, and Pathum Thani province) might face a serious challenge by climate-induced large scale flooding within the next 50 years and urged actions to increase the readiness to cope with such extreme events [3], [4]. Bangkok area was hit by a large scale flooding in 2011 with huge economics and social loss were recorded.

BMA also implemented the Bangkok Action Plan on Global Warming Mitigation 2007-2012 [5], aiming at the GHG reduction of 15% by undertaking five initiatives as follows: (i) expansion of mass transit and improvement of traffic system, (ii) promotion of the use of renewable energy, (iii) improvement of building electricity consumption efficiency, (iv) improvement of solid waste management and wastewater treatment efficiency, and (v) expansion of park areas. After ending of the plan on 2012, BMA policymakers had efforts to develop the next ten-year climate change master plan covering 2013 to 2023, by related sectors. This new master plan would support to the national climate change policy which Thailand will put her efforts to reduce the GHG in the range of 7 to 20% below the BaU in energy and transport sectors in 2020 through her NAMAs [6]-[8].

2 Scope of this study

2.1 Economic sector classification

BMA develops the Bangkok Master Plan on Climate Change 2013-2023, which is the city level long-term climate change plan. It consists of five main sectors; i.e. (i) energy efficiency and renewable energy, (ii) environmental sustainable transport, (iii) efficient solid waste management and wastewater treatment, (iv) green urban planning, and (v) adaptation planning.

This paper focuses only on the development of the energy sector, the biggest emission source of Bangkok, which includes the mitigation actions in energy efficiency and renewable development activities in Bangkok and exclude the energy consumption in transport and waste sectors. The energy sector in this study then consists of BMA governmental category and other civil category (industry, commercial, and residential parts). It should be noted here that there are only few activities of agriculture sector in Bangkok, hence we neglected the agricultural sector of this city in this study.

DOI: 10.1051/matecconf/20167011001

2.2 Greenhouse gas emission type

The energy sector covers only CO_2 emissions from energy utilization.

2.3 Level of this study

In this study, we focus mainly in the entire Bangkok area. This level (hereafter, Bkk) would focus all energy consumption and reduction potential which occur in Bangkok city, as a whole area and focus all government and non-government categories.

2.4 Energy sector data collection

We investigated and developed our energy data structure into two main types; (i) electricity, and (ii) thermal energy. All data are covering the year 2009-2012.

In Bangkok area, secondary data is main source of collecting energy data which is divided into the data from all sectors responsible to Bangkok entire area. All of collected data can be divided into 4 energy types which are electricity, oil, natural gas and coal as followings:

2.4.1 Electricity

Electricity consumption data in Bangkok is collected in secondary data from Metropolitan Electricity Authority directly requested data, not publish) in the year 2009-2012, covering residential, small and large general service, specific business service, government institute, non-profit organization, and temporary user and public lighting.

2.4.2 Fuel and petroleum products

Oil consumption data in Bangkok is collected in secondary data from Department of Energy Business (DOEB) [9]and classified in gasohol (all types), gasoline, diesel, fuel oil, LPG and other types of oil.

2.4.3 Natural gas

NG data is collected in secondary data from PTT Plc. (directly requested data, not published).

2.4.4 Coal

Coal data is collected from Department of Alternative Energy Development and Efficiency (DEDE).

Figure 1 illustrates the Structure and sources of energy data in this study.

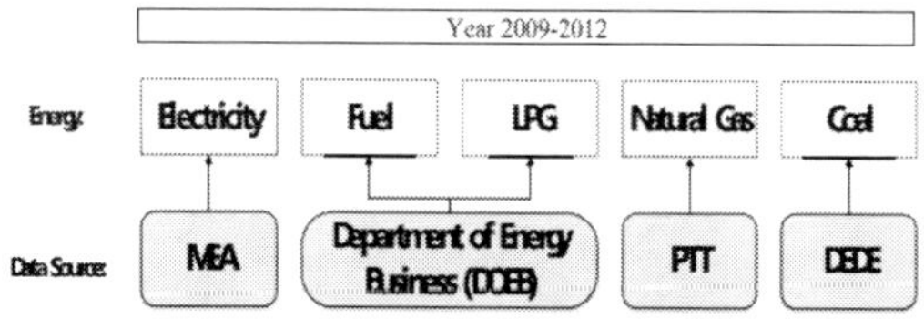

Note: Fuel means oil based petroleum products

Figure 1. Structure and sources of energy data.

3 The BAU scenario

Like other climate change and energy plans, it is important to see GHG emission amount by comparing cases without mitigation actions (business-as-usual or BAU) and with actions. In other words, how much GHG is reduced in quantity is one crucial approach to assessing the degree of success in mitigation actions in an objective way. In this Master Plan, GHG emissions have also been quantified for the two emission scenarios, namely the case of BAU and the case with emission reduction by taking mitigation actions.

Figure 2 illustrates the concept of how to develop our BAU of Bangkok climate change model in this study.

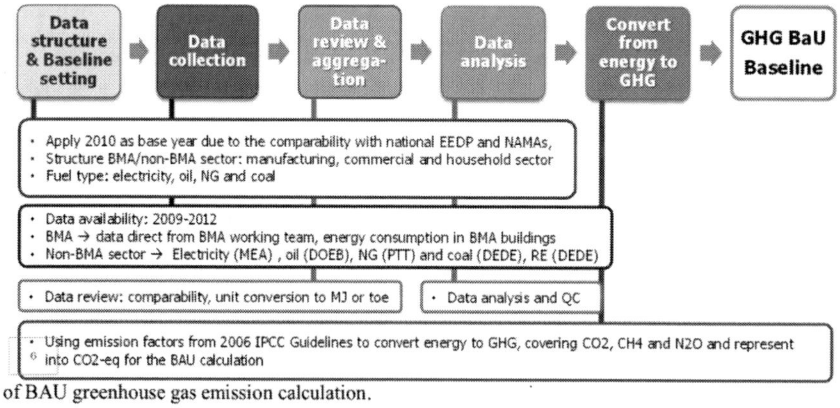

Figure 2. Flow of BAU greenhouse gas emission calculation.

By using the trend analysis from equation (1), the energy consumption in Bangkok area was then forecasted separately by each fuel type.

$$y_t = \sum_{i=1}^{t-1} \frac{y_i}{t-1} \qquad (1)$$

while y = forecast of y, in energy unit

t = time, annually

The results of the forecasted energy consumption in BAU, presented in Table 1, would then calculate into GHG value by using equation (2). Finally, BAU in Bangkok from the year 2013 to 2023 presents in Table 2.

Table 1. Energy consumption in BAU in Bangkok.

Type (Million MJ)	2013	2018	2023
Electricity	111,960.00	131,040.00	150,120.00
Thermal energy	85,057.11	91,665.53	102,699.59
Total	197,017.11	222,705.53	252,819.59

Table 2. GHG emission in BAU in Bangkok.

Energy type	BAU emissions (Million tCO_2-eq/year)							
	2009	2010	2011	2012	2013	2014	2015	2016
Electricity	15.76	16.70	16.76	17.83	18.55	19.19	19.82	20.45
Thermal	7.83	6.81	7.09	7.20	7.05	7.11	7.17	7.23
Total	23.59	23.58	23.86	25.02	25.60	26.29	26.99	27.68

Energy type	BAU emissions (Million tCO_2-eq/year)						
	2017	2018	2019	2020	2021	2022	2023
Electricity	21.08	21.71	22.35	22.98	23.61	24.24	24.88
Thermal	7.41	7.60	7.78	7.96	8.14	8.33	8.51
Total	28.49	29.31	30.12	30.94	31.75	32.57	33.38

$$\textit{Emissions estimate} = AD_t \times EF \quad (2)$$

while AD = Activity data, in energy unit per year

EF = Emission factor of each fuel type

In this study, we applied the emission factor from Revised 1996 IPCC Guidelines for National Greenhouse Gas Inventories [9] in each thermal energy in BAU.

4 The GHG mitigation scenario

In GHG mitigation scenario, we focus to the GHG reduction potential from the energy efficiency and renewable energy in Bangkok, exclude transport and waste sector, covering the year 2013-2023. All steps present in figure 3.

In this Master Plan, possible mitigation actions are divided into two major categories; (a) actions which BMA can implement by themselves, and (b) actions which are implemented by other stakeholders within Bangkok area. It should be noted that the GHG emission from activities by BMA itself is less than 1% of the total emission from whole Bangkok area. Therefore, it is important to consider the effect of mitigation actions of category (b). In addition, because it is difficult for BMA to monitor and control the mitigation actions of category (b), we have selected related national plans to calculate GHG mitigation potential in entire Bangkok area.

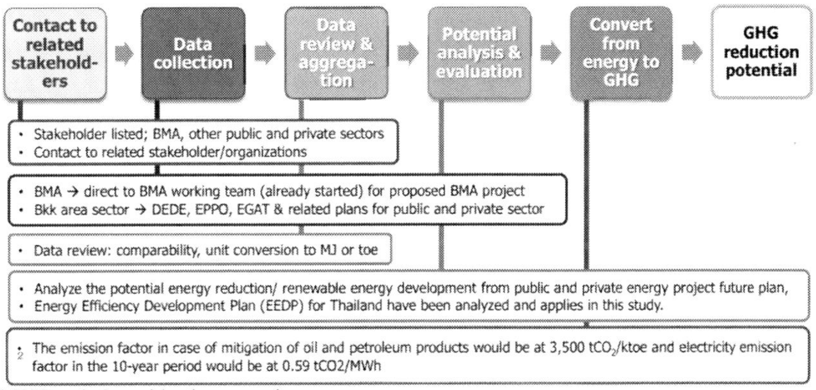

Figure 3. Flow of GHG emission mitigation scenario.

By using the national energy efficiency development plan (EEP) [10] and alternative energy development plan, we assume that the energy reduction potential in Bangkok, in percentage, is as same as the energy reduction of the country. This means that all calculation are based on the national plan but focus in city level.

In case of energy conservation, we can conclude the annual energy reduction which exclude transport and waste sector, in Bangkok as follows;

For electricity conservation, based on EEP

- Residential sector would be at 0.25% per year,
- Commercial building would be at 1.5% per year,
- Industry sector would be at 0.81 % per year

For thermal energy conservation, based on EEP

- Commercial building would be at 0.3% per year,
- Industry sector would be at 0.5 % per year

For emission actor, due to this study focus in the long-term period, hence the future energy consumption proportion would be very flexible. We then applied one universal figure of emission factor, average from various fuel types in Thai energy market. The emission factor in case of mitigation of oil and petroleum products would be at 3,500 tCO_2/ktoe and electricity emission factor in the 10-year period would be at 0.59 tCO_2/MWh (or 7,000 tCO_2/ktoe). The results in this scenario in Bangkok illustrate in figure 4 and table 3 and the amount of GHG mitigation in Bangkok presents in Table 4.

Table 3. Energy consumption in GHG mitigation scenario.

Type (Million MJ)	2013	2018	2023
Electricity	985.79	13,940.89	29,209.39
Thermal energy	106.93	5,869.96	15,869.22
Total	1,092.72	19,810.85	45,078.61

Table 4. GHG emission in GHG mitigation scenario.

Type (Thousand t-CO2eq)	2013	2018	2023
Electricity	152.17	2,236.24	4,692.47
Thermal energy	8.86	486.34	1,314.80
Total	161.02	2,722.58	6,007.27

DOI: 10.1051/matecconf/20167011001

Table 5. Selected countermeasures in GHG mitigation scenario of Bangkok master plan.

Sector	Measure		Possible mitigation actions	2013		2018		2023	
				Energy reduction (Million MJ)	GHG mitigation (Thousand tCO2-eq)	Energy reduction (Million MJ)	GHG mitigation (Thousand tCO2-eq)	Energy reduction (Million MJ)	GHG mitigation (Thousand tCO2-eq)
Residential	Promotion of energy saving house	a	Promotion of low-carbon/energy saving detached house (Publicity of cost benefit from the viewpoint of LCC, backup exhibition)	13.50	2.23	16.20	2.67	18.90	3.12
		b	Facility equipment introduction promotion of energy saving house (LED, energy saving air conditioning, etc.)						
	Promotion of energy saving repair work	a	Publicity of cost benefit by repair work for energy saving	9.00	1.49	10.80	1.79	12.60	2.09
		b	Promotion of repair work for energy saving (heat barrier film, renew air conditioning device, etc.)						
	Promotion of energy saving home appliances	a	Purchase promotion of energy saving home electric appliances (air conditioning, fridge, TV etc.)	9.00	1.49	10.80	1.79	12.60	2.09
	Promotion of energy saving action	a	Promote better understanding of air conditioner maintenance	27.00	4.47	32.40	5.37	37.80	6.26
	Others	a	Promote solar panel installation （subsidy system or mediating installable roof）	9.00	1.49	10.80	1.79	12.60	2.09
Commercial	Promotion of energy saving building	a	Incentive for constructing/repairing saving energy factory (tax reduction, subsidy, zero-interest finance etc.)	55.94	9.23	66.53	10.97	77.11	12.72
	Promotion of energy saving repair work for existing building	a	Conduct energy saving inspection of public buildings	55.94	9.23	66.53	10.97	77.11	12.72
		b	Promotion of ESCO business for existing buildings						
			(Educate ESCO business, advertisement promotion support, subsidy system for energy saving diagnostic)						
		c	Promotion of repair work for energy saving: insulation upgrade by double glazing, heat barrier film, renew air conditioning device (subsidy system etc.)						
		d	Publicity of cost benefit by Electricity Peak-Cut (automatic control facility of Electricity Peak-Cut, etc.)						
	Promotion of energy saving action	a	Promote saving energy activity (publicity of cost benefit etc)	167.83	27.68	199.58	32.92	231.34	38.16
		b	Raise preset cooling temperature at public buildings						
			Turn off lightings during lunch break						
		c	Thorough power saving setting on PC or OA equipment						
		d	Commendation for saving energy activity						
	Others	a	Promote solar panel installation （subsidy system or mediating installable roof）	167.83	27.68	199.58	32.92	224.99	37.11
Industry	Promotion of energy saving factory	a	Incentive for constructing/retrofitting saving energy factory (tax reduction, subsidy, etc.)	145.08	23.93	160.56	26.48	176.04	29.04
	Promotion of energy saving repair work for existing factory	a	Conduct energy saving inspection of public factories	145.08	23.93	160.56	26.48	176.04	29.04
		b	Promotion of repair work for energy saving (subsidy system etc.)						
		c	Publicity of cost benefit by Electricity Peak-Cut						
	Promotion of energy saving action	a	Promotion activity for factory's energy saving technique (for SMEs)	377.21	62.22	417.46	68.85	457.70	75.49
		b	Commendation for saving energy activity						
	Others	a	Promote Solar Energy （subsidy system or mediating installable roof）	33.18	2.75	430.63	35.68	1199.11	99.35
		b	Promote beneficial use of factory exhaust heat						

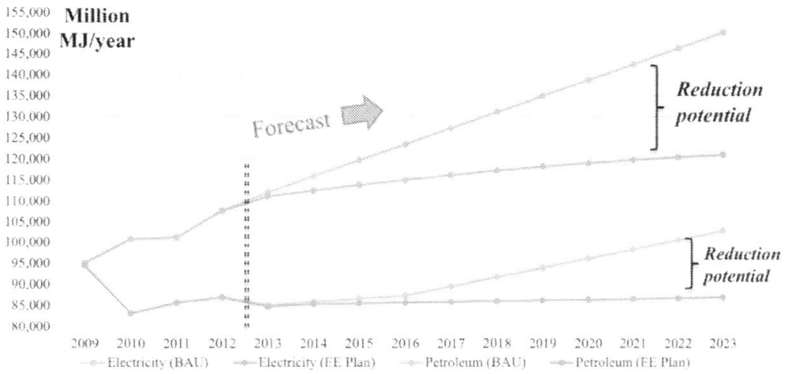

Figure 4. Flow of GHG emission mitigation scenario from 2009 to 2023.

From this scenario, we can conclude that if the EEP can fully implemented, it would reduce the energy by 17.83% while there is also potential of GHG emission reduction at around 18.00% from the BAU in the year 2023. However, it must be noticed that it is a potential but not the emission reduction target of the Master Plan.

Selected energy countermeasures including the GHG mitigation is presented in table 5. Most energy measures cover in three economic sectors while there is small potential of renewable energy in Bangkok except the solar roof top. Wind and biomass power plant are not easy to construct. On-ground solar farm is not economic-feasible because of the very expensive land. The energy conservation measures consist of the housekeeping, process improvement, major change equipment, and public awareness concept. The potential of energy conservation and renewable energy measures are adjusted from the national energy conservation potential.

5 Conclusions

This study focuses in the developing of city-level GHG mitigation master plan from energy sector in Bangkok. The energy consumption in all economic sectors except transport and waste sector are also investigated and forecasted to be BAU scenario covering year 2013 to 2023 by using trend analysis.

The GHG emission mitigation scenario was then analyzed by applying the energy reduction from the national energy efficiency plan. We can conclude that the energy reduction potential from energy sector would be at 17.83% while the potential of GHG emission reduction of the city from energy sector would be at around 18.0% from BAU in 2023 from all energy measures.

Acknowledgement

The authors would like to thank the Japan International Cooperation Agency (JICA) for all support. We remain culpable for any remaining errors.

References

1. IPCC *AR5*, WG1 Report (2014)
2. World bank, *Climate change impact and adaptation study for Bangkok Metropolitan region, Final report*, World Bank, Washington, DC. (2009)
3. Sanit Wongsa, 2011 Thailand Flood," *Journal of Disaster Research*, **8**, 3, pp. 380-385, (2013)
4. Poaponsakorn, N. and Meethom, P., "Impact of the 2011 Floods, and Flood Management in Thailand," *ERIA Discussion Paper Series*, ERIA-DP-2013-24, Economic Research Institute for ASEAN and East Asia, (November 2013)
5. BMA, *The Bangkok Global Warming Mitigation Action Plan 2007-2012*. Bangkok Metropolitan Area Management office, Bangkok, Thailand, (2007)
6. UNFCCC, *Communication on Thailand's NAMAs*, United Nations Framework Convention on Climate Change. (Nov. 2014). Available: http://unfccc.int/files/meetings/cop_15/copenhagen_accord/application/pdf/thailandcphaccord_app2.pdf
7. Communication on *Thailand's Nationally Appropriate Mitigation Actions (NAMA)*, Official letter to the UNFCCC. (2014). Available: http://unfccc.int/files/meetings/cop_15/copenhagen_accord/application/pdf/thailandcphaccord_app2.pdf
8. Promjiraprawat, K and Limmeechokchai, B. "Assessment of Thailand's energy policies and CO_2 emissions: analyses of energy efficiency measures and renewable power generation", Energies, **5**, 8, pp. 3074-3093, (August 2012)
9. DOEB, *Energy Business Statistics of Thailand*, Department of Energy Business, Thailand. (November 2014). Available: http://www.doeb.go.th/v5/en/service_stat.php
10. IPCC, *Revised 1996 IPCC Guidelines for National Greenhouse Gas Inventories*, Intergovernmental Panel on Climate Change, 2006. 1st ed. Chichester, U.K.: Wiley, ch. 2, pp. 45-47 (1993)
11. EPPO, *Thailand 20-Year Energy Efficiency Plan 2011-2030*, Energy Policy and Planning Office, Thailand. (2011). Available: http://www.eppo.go.th/encon/ee-20yrs/EEDP_Eng.pdf

Removal of High C and N Contents in Synthetic Wastewater Using Internal Circulation of Anaerobic and Anoxic/Oxic Activated Sludge Processes

Nittaya Boontian

School of Environmental Engineering, Institute of Engineering, Suranaree University of Technology

Abstract. Internal circulation (IC) of activated sludge anaerobic and anoxic/oxic processes was used to treat high carbon (C) and nitrogen (N) synthetic wastewater in order to achieve an effluent standard. The 1st & 2nd internal circulation anoxic (ICA) has shown a great advantage. Because under anoxic conditions heterotrophs utilize organic matter (OM). OM and NO_3^--N are an electron donor and an electron donor, respectively. As a result, under conditions high carbon and NO_3^--N, N_2 production can generate higher than with 1st ICA BNR system only. Aerobic IC to anoxic, 1st ICA, and effluent IC to anoxic, 2nd ICA, could remove 100% of COD, 99.60% of TKN, and 99.90% of NH_4^+-N. The organic loading rate and NH_4^+-N loading rate are 68.38 kg-COD m^{-3} d^{-1} and 9.86 kg NH_4^+-N m^{-3} d^{-1}, respectively in synthetic wastewater. Performance of these ICA activated sludge treatment was achieved the discharge standard with effluent COD less than 0 mg L^{-1}, effluent NH_4^+-N less than 4 mg NH_4^+-N L^{-1}, effluent NO_2^--N less than 0.1 mg NO_2^--N L^{-1}, and NO_3^--N less than 0.02 mg NO_3^--N L^{-1}.

1 Introduction

Biological nitrification-denitrification is the most commonly used process for nitrogen removal from industries handling protein-rich materials or other nitrogen compounds generate, effluents with very high loads of ammonia. The degree of treatment required and the selection of suitable treatment technique are usually dependent upon the wastewater composition. A post treatment may be required mainly to remove ammonia before discharge and biological nitrification–denitrification being the most extensively used process. As the C/N ratio after anaerobic digestion is generally insufficient for denitrification, the main challenge in combining the anaerobic digestion process with an activated sludge nitrification/denitrification process lies in the management of organic matter (OM). Optimize the management of OM between the anaerobic and anoxic/oxic reactors. For that, proposed a combined process based on an anaerobic phase followed by an anoxic/oxic reactor in order to removal nitrogen by nitrification/denitrification. To provide enough effluent denitrification in the anoxic/aerobic reactor, a fraction of the effluent was fed directly to the anoxic/oxic reactor. This study focused on the use of recycled effluent. Operating reactor with pumping the water from effluent to anoxic zone could significantly increase the effluent quality. The purpose of this work was to present the design and functionality of a lab-scale dynamic effluent activated processes for biological removal of organic substrate and nitrogen applying anoxic-oxic process.

2 Material and methods

2.1 Process configuration

The system (Fig. 1) and the operating design for each instance of influent flow rate, reactor volumes, return sludge ratio, internal recirculation flow rate ratios, hydraulic retention time (HRT), solid retention time (SRT) and O_2 concentration were obtained from lab scale results. These are given in Table 1. This lab scale processes consisted of four reactors in series. They included anaerobic, anoxic, aerobic phase and clarifier stages with effective volumes 5, 5, 20 and 30 L, respectively. In addition, the total HRT of anaerobic, anoxic and aerobic conditions, and SRT were 12 h and 20 days, respectively. The operating conditions of this system were: influent wastewater flow rate, return activated sludge (RAS) flow rate, 1st and 2nd internal circulation anoxic (ICA) flowrate. O_2 concentration was maintained between 2-4 mg L^{-1} in the aerobic zone. Average process temperature was maintained at 37 °C.

Table 1. The operation condition for lab-scale biological nutrient removal (BNR) treatment plant.

Flow rate	Unit (L h^{-1})	HRT	Unit (h)
Influent	2.5	Anaerobic	2
RAS	2.5	anoxic	2
1st and 2nd ICA	7.5 and 0.75	Aerobic	8

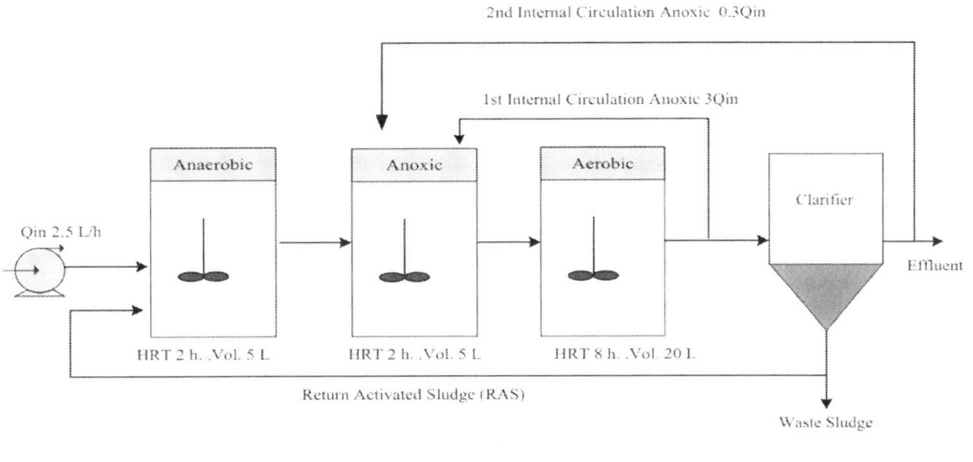

Figure 1. Schematic diagram of lab-scale BNR processes.

2.2 Wastewater feed

Synthetic wastewater used in this study was prepared according to Coelho et.al [1], with some modification. Wastewater contained ammonium chloride and urea as the nitrogen sources. The average concentrations of TKN and ammonia nitrogen in the fluent were 967.28 ± 109.41 mg L^{-1} and 786.33 ± 210.95 mg L^{-1}. The COD in influent averaged 5,465.44 ± 232.91 mg L^{-1}. The C/N ratio is 6.95±0.47.

2.3 Analytical methods

The experiment carried on about 41 days and was considered complete when the range of changes of particular parameters in the effluent (soluble chemical oxygen demand (SCOD), total kjeldahl nitrogen (TKN), ammonia-nitrogen (NH_4^+-N), nitrite-nitrogen (NO_2^--N), nitrate-nitrogen (NO_3^--N) within 7 days did not exceed 5%-10%. Daily measurements of all reactors include SCOD, TKN, NH_4^+-N, NO_2^--N, NO_3^--N), pH, dissolved oxygen (DO), mixed liquor suspended solids (MLSS), mixed liquor volatile suspended solids (MLVSS). The analyses were performed according to APHA [2].

3 Results and discussion

3.1 Mixed liquor suspended solids (MLSS), mixed liquor volatile suspended solids (MLVSS) and SCOD

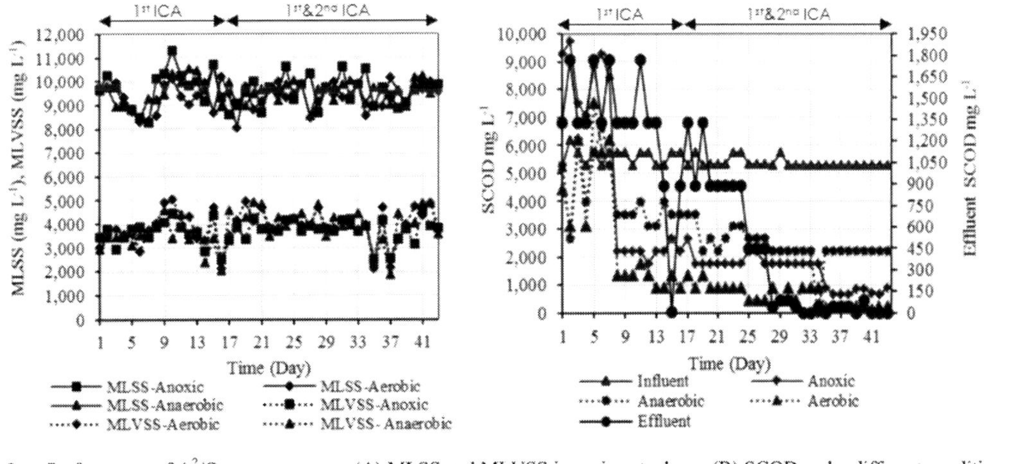

Figure 2. Performance of A^2/O processes: (A) MLSS and MLVSS in various tanks (B) SCOD under different conditions

As shown in Fig. 2, MLSS concentration in various tanks fluctuated similarly for all conditions from day until day 41, whereas under anoxic phase MLSS was slightly higher than others. Under conditions 1st and 2nd ICA, from day 17 up to day 41 effluent SCOD removal was very high (Fig. 2) in comparison to under condition 1st ICA only. Operation of 1st ICA was between day 1 and day 16. The overall combined process removed 100% of SCOD. Considering that all biodegradable SCOD was then possible to calculate the biodegradable SCOD at the influent and the SCOD removal in each reactor. A quantity of SCOD removal in anaerobic, anoxic and aerobic reactors were 58.34%, 60.01%, and 75.03 respective under 1st & 2nd ICA. However, the MLSS trend is similar in both situations, 1st ICA only and 1st & 2nd ICA. The 2nd ICA did not affect the MLSS. MLSS and MLVSS concentration in all the three chambers were same during the operation and remains almost constant during the whole operation period. This might be due to the equal HRTs and SRTs of all the three chambers. During 20-day SRT period the average MLSS and MLVSS concentration for all the three chamber remained constant. Within the chambers there was no noticeable effect of circulation change but on the day 35 and 37, decrease in the average MLVSS concentration in the three chambers, where it rapidly drop to 2,345 mg L^{-1} and 2,195 mg L^{-1}, respectively. After that both MLSS and MLVSS was almost stable for the three chamber at 9,620 mg L^{-1} and 4,080 mg L^{-1}, respectively. COD removal rate for both conditions was different through the operation, affected by the changes in the circulation. Based on the average incoming SCOD concentration of 5,465.44 ± 232.91 mg L^{-1}, the removal efficiency was between 69.23% and 84.62% with the average 75.39% (Std. Dev. = 4.27) for 1st conditions, and between 99% and 100% with average 99.38% (Std. Dev. = 0.61) for 2nd condition, respectively.

3.2 Dissolved Oxygen (DO) and Total Kjeldahl Nitrogen (TKN)

Aeration to all the aerobic chamber was supplied between 2.5-5 mg-O_2 L^{-1}. DO concentration decreased with the increase in concentration of influent COD and ammonia. Fig. 3, the DO concentration in the average, and increase to 0.65 mg-O_2 L^{-1} for 2nd chamber, anoxic phase, the last chamber, aerobic first chamber, anaerobic phase, was 0.37 mg-O_2 L^{-1} on phase, and increase to 3.92 mg-O_2 L^{-1}. DO concentration influenced TKN removal efficiency, which was about 62.47% - 73.35% and 93.06% - 100% at conditions of only 1st ICA and 1st & 2nd ICA as shown in Fig. 3. Higher organic carbon concentration in the discussed accelerated the activity of heterotrophic bacteria that utilize oxygen for COD oxidation. For this reason, high initial COD concentration may result in oxygen depletion zones in activated sludge that favor nitrogen removal. Zhu et al., [3] also observed a linear relationship between SND and DO concentration, but SND was obtained even under high DO concentration (4.5 mg L^{-1}). The average of maintained DO concentration in 1st & 2nd ICA is higher, 4.31 mg-O_2 L^{-1}, than in 1st ICA only, 3.34 mg-O_2 L^{-1}, as shown in Fig. 2. The median DO level, defined as the mid-value between the minimum and maximum DO values in given aeration period for only 1st ICA and 1st & 2nd ICA, were 3.34 mg-O_2 $L^{-1}$and 4.04 mg-O_2 L^{-1}, respectively. To obtain the true average DO concentration, integration must be performed for the DO curve during the aeration period and the division by the time; however, such integration was not performed in this experiment. Using the average DO concentration together with the minimum and maximum concentration would be the accurate way to describe the pattern of DO level.

3.3 Removal of Ammonia by nitrification

Nitrites was not observed at any reactor during the experiment. It was detect only in effluent. At specific circulation showed a significant impact on rate of nitrification. Working at 1st ICA and 1st & 2nd ICA, nitrification efficiencies were about 79.61% and 98.45%, respectively. When 1st ICA and 1st & 2nd ICA, the residual $NH4^+$-N the effluent was decrease to 151.76 mg- NH_4^+-N L^{-1}, and 8.87 mg-NH_4^+-N L^{-1}, influent 924.65 mg-NH_4^+-N L^{-1}.

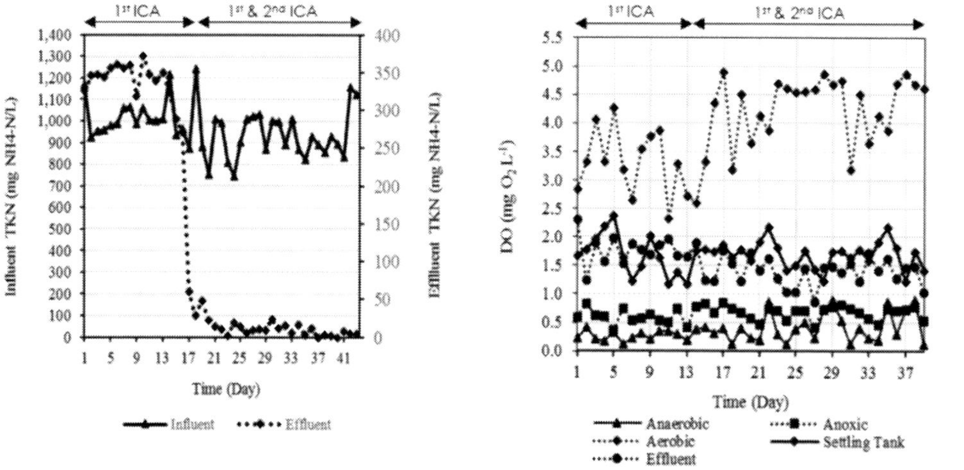

Figure 3. Performance of A^2/O processes: (A) TKN variation influent and effluent (B) DO under different conditions

Fig. 4 (A) depicts the influent, and effluent NH_4^+-N profiles with different conditions of operations. NH_4^+-N was not completely converted into NO_3^- and also NO_2^- was also detected in effluent. This is may be due to some ammonia gas losses favoured by the high pH in the reactor. The aerobic reactor can be operated at high pH value, higher than 7.5, at this condition without affecting the nitrification process. In some of the cases the dissolved oxygen concentration was lower than 1 mg- L^{-1} in the aerobic chamber because of the high COD and ammonia loading rates. Under this condition, there may be the creation of some anoxic zones which resulted in the loss of NO_3^- by simultaneous nitrification and denitrification process. Nitrogen could be removed by both assimilation into biomass and biological nitrification-denitrification process. It has been demonstrated that cell assimilation is another factor which contribute up to 15-20% of the influent TN concentration in a pre-denitrification process. The pre-denitrification was carried out by internal aerobic circulation nitrate rich both MLSS and liquid containing back to the anoxic reactor. The anoxic reactor was fed with the anaerobic wastewater with almost negligible NO_3^-. Average effluent NO_3^- and NO_2^- was 0.02 mg-NO_3^--N L^{-1} and 0.10 mg-NO_2^--N L^{-1}, respectively. Effluent profile of NO_3^- and NO_2^- are shown in Fig. 4 (B). During aerobic mode average NO_3^- concentrations for 1st ICA and 1st & 2nd ICA due to nitrification were 0.44 mg- NO_3^--N L^{-1}, 0.49 mg-NO_3^--N L^{-1}, respectively. During anoxic phase average NO_3^- concentrations for 1st ICA and 1st & 2nd ICA in the denitrifying chamber were 0.36 mg-NO_3^--N L^{-1}, 0.39 mg-NO_3^--N L^{-1}, respectively. Denitrification process requires organic carbon as electron donor. Insufficient availability of suitable organic substrate may result in poor denitrification. According to the literature, 3.3 – 5.0 g-COD g-NO_3^--N^{-1} ratio is required to achieve complete denitrification [4]. It can be noticed that in this study the supplied COD/NO_3^- was more than the required, so it could not be a reason for comparatively low denitrification efficiency. The denitrification efficiency was dependent on the NO_3^- and NO_2^- recycle rate. In this experiment, NO_3^- and NO_2^- was mainly dependent on the circulation rate of both MLSS and liquid from aerobic and effluent to anoxic chamber.

3.4 Biomass Characteristics and Variation in pH

Flocs with poorer settling properties are normally generated in an activated sludge treating organic matter, when the organic loading rate is high. The biomass concentration and organic loading rate (ALR) were 6.84 kg-NH_4^+-N m^{-3} $d^{-1}$and 10.97 kg-NH_4^+-N m^{-3} d^{-1} for 1st ICA and 1st & 2nd ICA, respectively. This index was between the 50 – 100 mL g-VSS^{-1} usually found for sludge with good settling characteristic in activated units treating organic matter [5]. A sludge with SVI of 80.79 mL g-VSS^{-1}, 85.31 mL g-VSS^{-1} for 1st ICA and 1st & 2nd ICA, respectively was obtained as shown in Fig. 5 (A). In this study, fed only with ammonia, a slightly compact sludge with high SVI was developed and, thus, it was possible to operate at slightly higher loads while maintaining higher biomass concentration in the system than those indicated for organic matter treatment. Fig. 5 (B) depicts pH profiles, along the length of reactor in all the three chamber (anaerobic, anoxic and aerobic), final clarifier, and effluent. The pH decreased in the first anaerobic chamber, and then remained almost constant from anoxic to the last aerobic chamber. Constant pH value from aerobic chamber suggested that the complete nitrification was achieved at this point. This was also confirmed by the NH_4^+-N residual concentration, which was almost zero in effluent (Fig. 4 (A)). The pH variations along the aerobic zones are believed to be jointly caused by a number of factors. The nitrification is the most pH influencing process, which causes the pH to decrease due to alkalinity consumption. It is very important to maintain pH to a certain level for efficient nitrification as researchers have reported that nitrification is inhibited at a low pH value (>6.5) [6]. However, in the present system pH remained above 6.5.

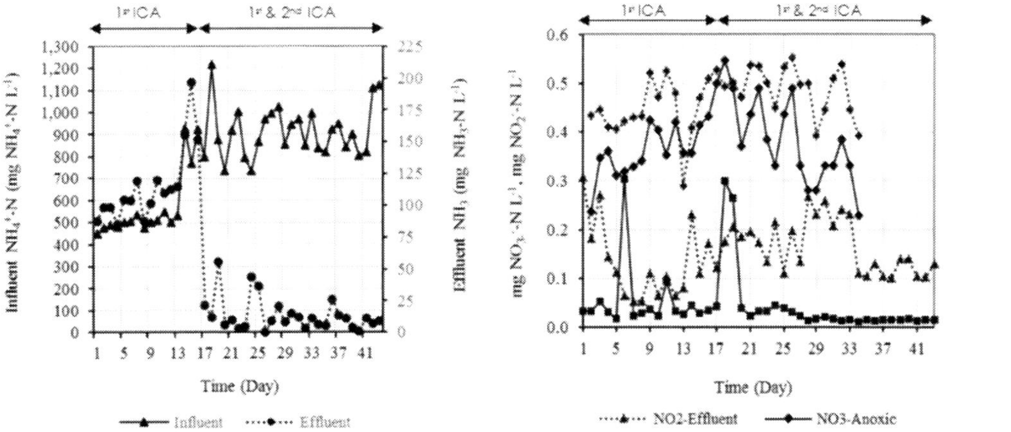

Figure 4. Performance of A^2/O processes: (A) NH_4^+-N concentration profiles (B) NO_3^- and NO_2^- under different conditions

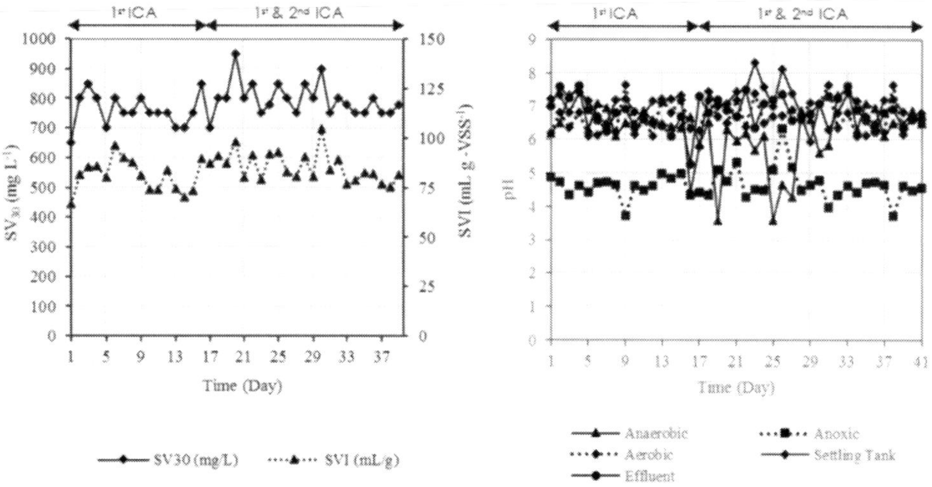

Figure 5. Performance of A^2/O processes: (A) Variation of SV_{30} and SVI (B) Variation in pH change

3.5 Performance of the 1st ICA versus 1st & 2nd ICA BNR system

The overall performance of the 1st ICA versus 1st & 2nd ICA BNR system is shown in Table 2. The pollutants were efficiently in 1st & 2nd ICA BNR system removed with SCOD 100%, TKN removal 99.60% and NH_4^+-N removal 99.90%, respectively. These parameters were better than 1st ICA only. The removal of SCOD, TKN and NH_4^+-N are 84.62%, 73.35% and 97.27%, respectively. The 1st & 2nd ICA BNR system is superior to the 1st ICA BNR system only using in coupling 2nd ICA. Because under anoxic conditions heterotrophs utilize organic matter as electron donor and NO_3^--N act as electron donor. As a result, under conditions high carbon and NO_3^--N, N_2 production can generate higher than with 1st ICA BNR system only.

Table 2. Overall effluent by 1st and 1st & 2nd ICA BNR system

Sys.	SCOD mg L^{-1}	TKN mg-NH_4^+-N L^{-1}	NH_4^+-N mg-NH_4^+-N L^{-1}	NO_3^--N mg-NO_3^--N L^{-1}
1st ICA	1,328	60.48	21.75	0.04
1st&2nd ICA	0	0	1.12	0.01

4. Conclusion

Internal circulation (IC) of activated sludge anaerobic and anoxic/oxic processes was used to treat high carbon and nitrogen synthetic wastewater. The research proved that return effluent to anoxic phase can achieve discharge standard. Treatment only with 1st ICA BNR system, the discharge cannot meet the standard. The 2nd ICA with 0.3Qin has shown a great advantage. The 1st & 2nd ICA BNR system could remove 100% of SCOD, 99.60% of TKN, and 99.90% of NH_4^+-N. The organic loading rate and NH_4^+-N loading rate are 68.38 kg-COD m^{-3} d^{-1} and 9.86 kg NH_4^+-N m^{-3} d^{-1}, respectively.

References

1. Coelho M. A. Z., Russo C., Araújo O. Q. F. Optimization of a sequencing batch reactor for biological nitrogen removal. Water Res. **34**, 10 (2000)
2. APHA, AWWA, WEF, Standard methods for the examination of water and wastewater (18th ed.). APHA, AWWA. WEF, Washington DC. (1992)
3. Zhu G. B., Peng Y. Z., Wu S. Y., Wang S. Y., Xu S. X., Simultaneous nitrification and denitrication in step feeding biological nitrogen removal process, J Environ Sci , **19**, 9 (2007)
4. K. Karim, S. K. Gupta. Continuous biotransformation and removal of nitrophenols under denitrifying conditions. Water Res. **37**, 12 (2003)
5. D. H. Eikelboom and H. J. J, van Buijsen, TNO Report A94A, Microscopic sludge investigation manual TNO Research Institute for Environmental Hygiene. Water and Soil Division, Delft, The Netherlands, (1981)
6. Bae W, Baek S., Chung J., Lee Y. Optimal operational factors for nitrite accumulation in batch reactors. Biodegradation **12**, 5 (2002)

The Novel Design and Manufacturing Technology of Densified RDF from Reclaimed Landfill without a Mixing Binding Agent Using a Hydraulic Hot Pressing Machine

Somrat Kerdsuwan, Pubet Meenaroch and Thitaaus Chalermcharoenrat

The Waste Incineration Research Center, Department of Mechanical and Aerospace Engineering, Faculty of Engineering, Science and Technology Research Institute, King Mongkut's University of Technology North Bangkok, Bangkok, Thailand

Abstract. The manufacturing of RDF (Refuse Derived Fuel) from a conventional cold press extrusion machine is not suitable for producing RDF from reclaimed landfill since it is not identical in shape and form after production due to the swelling of the plastic fraction contained inside the reclaimed landfill and hence needs a very high compression force. Moreover, a binder agent is needed in order to keep the RDF in a similar shape and form. A novel design and manufacturing technology for a hydraulic hot pressing machine has been established and can produce high-quality RDF without any binder. The two electrical heaters are installed at the inner core and on the surface of the mold. The compression force on the mold is performed by a hydraulic jack. In addition, a newly-designed locking plate system which is designed by a slider to open and close along the paired horizontal slots, can reduce the cycle time of the manufacturing process and yield higher productivity. The testing properties of the RDF produced by the novel hydraulic hot pressing machine include the examination of size, shape, weight, unit density, bulk density, compression strength, moisture content, and heating value. The results showed that the RDF is suitable to be used as feedstock in an incinerator or gasifier to produce green and clean energy from reclaimed landfill.

1 Introduction

Nowadays, landfill, which is regraded as a simplified technology for the disposal of municipal solid waste (MSW) in many countries, is widely used [1]. However, due to a huge amount of MSW, most of it has been disposed by open dumps, which causes subsequent problems, for instance, health problems, leachate, requirement of land space, etc. Consequently, it is necessary to reclaim land for reuse. The residual left after the urban mining process is called reclaim landfill, which needs a suitable disposal technology that can recover green and clean energy effectively and sustainablility. Incineration or gasification technology, which is a thermal treatment option, can be considered as an alternative choice for reducing the mass and volume of reclaimed landfill by burning and can convert it into alternative green and clean energy. Considering the combustible fraction of reclaimed landfill, which is mostly plastic and paper, it can be used as feedstock with an incinerator or gasifier. However, in order to improve its efficiency, reclaimed landfill should be converted into RDF (Refuse Derived Fuel), and consequently, its quality will improve, including an increase in bulk density, low moisture, high heating value, etc. [2]. Normally, the RDF can be prepared with three main technologies, e.g. a piston press, a screw press, and a pelletizing method. There can be in several forms, such as briquettes and pellets, which depend on the mold used. Previously, the technology of compressed RDF used the hydraulic hot pressing system while the sample was compressed, the mold cover or locking plate was locked by a bolt. The heater was only installed at the surface of the mold. Hence, the RDF that is produced from this machine easily cracks because it has not been heated at the inner core [3]. The aim of this work is to redesign and manufacture the hydraulic hot pressing machine to overcome this difficulty and manufacture of RDF from reclaimed landfill without a mixing biding agent.

2 Review of RDF production

The classification of RDFs can be divided into 7 groups as shown in Table 1. Each of these RDFs has different properties. RDF-1 is MSW that has been sorted out to remove oversized bulky fraction only. To use MSW as fuel for a fluidized-bed combustion system, the combustible waste is ground to coarse particle size of approximately 6 cubic inches or less and it is classified as RDF-2. RDF-3 are the removal of the incombustible fraction (metal, glass, and other entrained inorganics), then will be cut in size until the particle size is smaller than 2 inches. RDF-4 is combustible waste in the form of powder. RDF-5 is highly-densified fuel (more than 600 kg/m^3 in density) by compression and appears in several forms such as pellets, slugs, cubettes, briquettes or similar forms. The benefits of compression are increased bulk density, low moisture, high heating value, and being easy to transport. The combustible waste processed into a liquid or in the form of slurry is called RDF-6. Finally, RDF-7 is the combustible fuel in the form of gaseous which is produced by the gasification process [2].

The next consideration is the high operating cost due to the use of an additional fuel burners in the primary combustion chamber. The purpose of this burner is to sustain the desired combustion temperature in the chamber when burning high-moisture waste content, especially with wet fraction MSW (food waste, vegetation waste, etc.) The more that this additional burner is used, the more additional fuel must be used (diesel oil, for example), which results in high operating costs. In order to overcome this, the less use or even no use of the burner muse be taken into account, even by improving the combustion process by using a novel design of combustion air distribution or by mixing the characteristics of waste and air.

Table 1. ASTM classification of RDFs [2]

Class	Form	Description
1. RDF-1 (MSW)	Raw	Municipal solid waste with minimal processing to remove oversized bulky waste
2. RDF-2 (C-RDF)	Coarse	MSW processed to coarse particle size with or without ferrous metal separation such that 95% by weight passes through a 6 in. square mesh screen
3. RDF-3 (f-RDF)	Fluff	Shredded fuel derived from MSW processed for the removal of metal, glass and other entrained inorganics. The particle size of this material is such that 95% by weight passes through a 2 in. square mesh screen.
4. RDF-4 (p-RDF)	Powder	Combustible waste fraction processed into powdered form such that 95% by weight passes through a 10 mesh screen (0.035 in. square)
5. RDF-5 (d-RDF)	Densified	Combustible waste fraction densified (compressed) into pellets, slugs, cubettes, briquettes or similar forms
6. RDF-6	Liquid	Combustible waste fraction processed into a liquid fuel
7. RDF-7	Gas	Combustible waste fraction processed into gaseous fuel

Conventionally, the extruded machine can be used to prepare RDF-5. Reclaimed landfill from old dumpsite landfill urban mining mainly contsisting of paper and plastic is compressed using a screw extrusion machine [4], as shown in Fig. 1. The length, height, and width of the screw extrusion body used in this study were 800, 1000, and 500 mm, respectively. The other parts were a feeding hopper, a 7500 watt motor, a 300 watt heater coil, a feeding screw, and a die (inner diameter of 50 mm). To produce RDF, the MSW, which is fed at the rate 200-300 kg/hr into the hopper, moves down into the feeding screw connected to the motor and compresses the MSW in the die. At the same time, the heater is operated at 300°C in order to melt the fraction of plasstic. Finally, the extruded RDF is manually cut into 50 – 60 mm lengths.

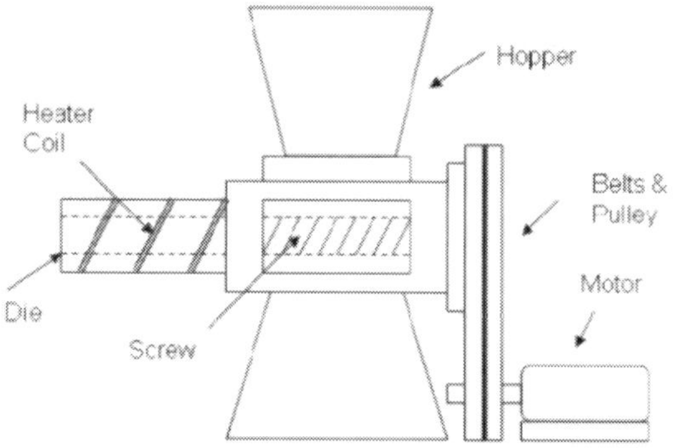

Figure 1. Screw extrusion machine with a heater at the die [4]

Nevertheless, the products produced by the screw extrusion machine are not of the same size or shape because the compressed RDF swells at the end of the extrusion die. Morover, the screw extrusion system cannot control the mixing ratio of RDF composition accurately and additionally the heat generated in the extrusion machine causes the plastic to melt; thus, some volatile matter is released, resulting in lowering the heating value of the fuel [5]. Therefore, the hydraulic hot pressing machine can be considered a way to overcome this disadvantage. This technique helps with the production of RDF-5 from reclaimed landfill that is more stable in size and shape.

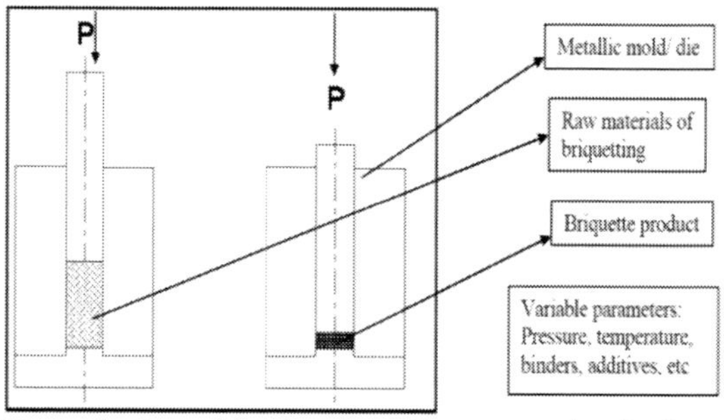

Figure 2. Assembly drawing of a conventional piston press densification machine [7]

3 Innovative design of a hydraulic hot pressing machine without using mixing binder

The different densification technologies of pressing machines can be classified according to several determining factors, including *applied compression pressure*, *operating condition*, and the *feeding mode of operation*. The applied compression pressure consists of low pressure (up to 5 MPa), intermediate pressure (5-100 MPa), and high pressure (above 100 MPa). The operating conditions are hot and high pressure densification, and cold and low pressure densification where the feeding mode of the operation is batch and continuous [6]. The conventional piston press densification machine (Fig. 2) has been designed to produce RDF-5 in a cylindrical briquette shape without a central hole. The operating condition is a batch operation with *the cold compress* of *low pressure force* applied to loosen the raw materials together *with a binder agent* into a mold in order to achieve a completely solid uniform shape of the mold after removal of the compression force.

The conceptual design of the novel hydraulic hot pressing machine presented in this study is given in Fig. 3 and Fig. 4. This new design can produce RDF-5 from reclaimed landfill in cylindrical briquette shape with or without a central hole. The operating condition is *the hot compress force* applied to loosen the raw materials of the reclaimed landfill *without a binder agent* in the mold to get a completely solid uniform shape of the mold with a central hole after removal of the compression force based on the batch operation mode. The RDF-5 in the cylindrical briquette shape with the central hole provides better combustion characteristics due to the larger specific area and because of the high combustion rate. In addition, the newly-designed locking plate system which is designed by the slider to open and closed along the paired horizontal slots can reduce the cycle time of the manufacturing process and yield higher productivity.

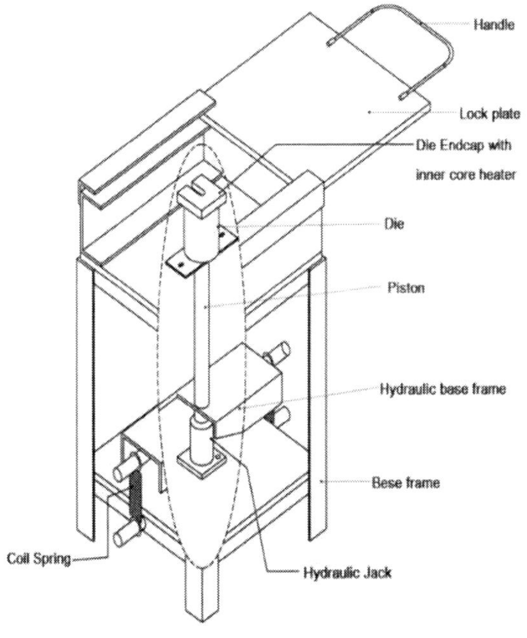

Figure 3. Conceptual design of a novel hydraulic hot pressing machine

The hydraulic jack can generate a compression force of approximately 11 MPa inside the mold cavity. The heater was installed at the inner core in order to apply heat to the surface of the central hole area and the entire surface of the mold. The machine is made of steel with a dimension of 300 mm in width, 300 mm in length, and 580 mm in height. The mold to produce the RDF-5 is of a cylindrical shape with a central hole inner diameter of 48 mm and a length of 100 mm. The pressing unit has a hydraulic piston stoke of 750 mm. In addition, the elctrical power of the heater capacity applied to the entire outer mold surface is 300 watts and can sustain an operating temperature of 120°C. The electrical heater power at the inner core of the mold is 200 watts with an equal operating temperature on the outer surface. The compressing time holds for 30 seconds per batch. Reclaimed landfill is put inside the mold, then a piston is compressed with the hydraulic jack. While it is compressed, the elctric heater switches on to melt the plastic fraction inside the reclaimed landfill at the outer surface and in the central hole area. The prototype machine can produce molding RDF-5 from reclaimed landfill which contains plastic of more than 50 %-wt. and bulk density higher than 600 kg/m^3 [8]. The production capacity of this machine is 1.2 kg/hr. The results of the physical and chemical properties of RDF-5 produced from the prototype of the hydraulic piston hot press densification machine are shown in [9].

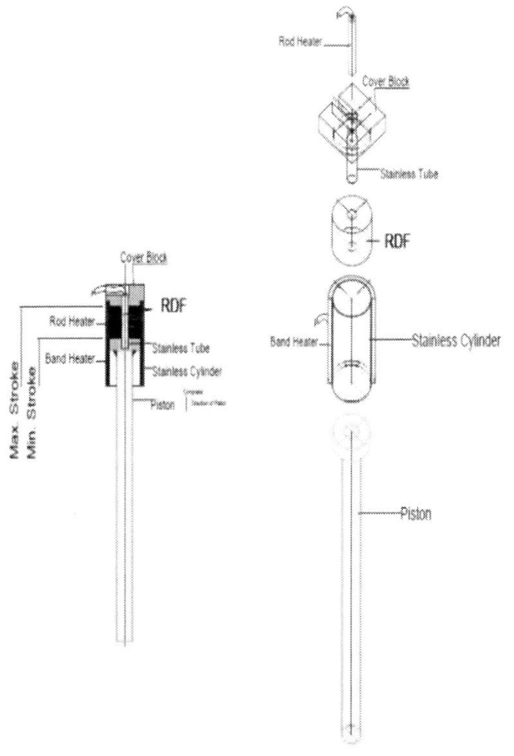

Figure 4. Assembly drawing of a novel hydraulic hot pressing machine

4 Summary

The novel hydraulic hot pressing machine in the present study was designed and manufactured in order to overcome the difficulty in producing RDF from reclaimed landfill, which cannot be done in using a conventional cold press densification machine. The RDF produced using the novel machine is similar in shape and form, with high bulk density and no use of any binder agent. The test of the physical and chemical properties of the RDF demonstrated that it is suitable to be used as feedstock in an incinerator or gasifier to produce green and clean energy from reclaimed landfill.

References

1. Pollution Control Department (PCD), *Municipal Solid Waste Composition Survey and Analysis Project in Thailand: Final Report*, Thailand: Pollution Control Department, (2015)
2. Robert E. Sommerlad, W. Randall Seeker, Abraham Finkelstein, James D. Kilgroe, *Environmental Characterization of Refuse Derived Fuel Incinerator Technology*, 1988 National Waste Processing Conference, Pennsylvania, (1998)
3. Tawach Prechthai, Maddala Padmasri, Chettiyappan, *Quality Assessment of Mined MSW from and Open Dumpsite for Recycling Potential*, Conservation and Recycling, vol. 53, pp. 70-78, (2008)
4. A.B. Nasrin, A.N. Ma, Y.M. Choo, S. Mohamad, M.H. Rohaya, A. Azali and Z.Zainal, *Oil Palm Biomass as Potential Substitution Raw Materials for Commercial Biomass Briquettes Production*, American Journal of Applied Sciences, vol. 5, pp 179-183, (2008)
5. Kejian Wang, *Description of Extrude Swell for Polymer Nanocomposites*, Materials, vol. 3, pp. 386-400, (2010)
6. United Nations Environmental Programme, *Technologies for Converting Waste Agricultural Biomass to Energy*, Division of Technology, Industry and Economics, International Environmental Technology Centre Osaka, (June 2013)
7. Ramesh M. Singh, *Biobriquetting in Nepal – Scope and Potentials : a Review*, Engineering and Technology, vol. 9, pp. 104-120, (2013)
8. Aatonio C. Caputo, Pacifico M. Pelagagge, *RDF Production Plants: I Design and Costs*, Applied Thermal Engineering, vol. 22, pp. 423-437, (2002)
9. Somboon Chalermcharoenrat, Nuth Sirirermrux, Krongkaew Laohalidanond, Somrat Kerdsuwan, *Optimization of the Production of Densified RDF from Reclaimed Landfill without Mixing Binding Agent Using Hydraulic Hot Pressing Machine*, 2014 the 5th TSME International Conference on Mechanical Engineering, Chiang Mai, (2014)

Developing System Thinking Approach for Sustainable Destination Management in Lake Toba

Ricardo Situmeang

Engineering Management, Faculty of Industrial Technology. Del Institute of Technology Jalan Sisingamangaraja Sitoluama Laguboti Toba Samosir North Sumatera Indonesia

Abstract. Lake Toba as one of the largest natural lake in the world occupying the caldera of super-volcano features diverse topography, remarkable historical heritage and archeological value. Because of its unique and rich features, lake toba has become a popular destination for tourists. Considering the significance and the potential of the tourism industry for economic development of the society both locally and nationally, there is an urgency to establish competitiveness and attractiveness the destinations using effective and sustainable strategies based on the market conditions, and the other to balance of interests of stakeholders. The aim of this study is to propose the important of developing system thinking for destination management to be sustainable, as well as the opportunities to pursue strategic policies and conditions for constituting different types of management structures. The process includes the development of a systems thinking approach that represents a holistic understanding of the interconnectedness and relationships between the various components that impact on sustainable development of tourism in Lake Toba. The paper is intended to use this as a framework for decisions and capacity building by government and private stakeholders who share the responsibility in developing, managing and sustaining the system.

1 Introduction

Destination development is a dynamic and complex system which contains numerous components that involves a diverse array of stakeholders. These components are interconnected through non-linear relationships while each stakeholder holds different objective, agendas and interests. It requires management of all tourist resources. This is a holistic process that ensures tourism adds value to our economy, social fabric and ecology of our communities. Tourism can be an economic driver, generating jobs and contributing vibrant lifestyle benefits to our communities. At the same time it should maintain cultural integrity, essential ecological processes, biological diversity and life supporting ecosystems.

This often causes conflict that necessitates compromises based on a better understanding of each other's needs and requirements. The tourism system is also strongly influenced by various internal and external factors resulting in an extremely complex and dynamic system with high levels of uncertainty and unpredictability. In order to manage the system in a sustainable way, it is necessary to provide stakeholders with effective learning mechanisms to align their views, to help them make more informed decisions, and to adapt to the continually changing environments.

At the same time tourism needs to be managed to ensure that it leaves a positive legacy for current and future generations. Importantly the tourism sector needs to ensure that it is considered in the broader context of regional development and that it is recognized for its overall contribution and economic value to the region. To be effective, planning, development and marketing activity must be based on research and the needs of the consumer. To compete effectively, destinations have to deliver wonderful experiences and excellent value to visitors. (Sterman, 2002)

Destination Management Organization (DMO) in Indonesia has been implemented since 2011 by the Ministry of Tourism in 16 destination clusters. One of the characteristics of DMO is the utilization of network approach consisted of key agents of tourism in developing communication on tourism development in the area. Ownership on sustainable management in destination area has become one of the important aspects in competitiveness improvement. Such improvement is hoped to minimize the local and sectorial ego appeared from the implementation of regional autonomy. The power of governance development collaboration creates the helpful communication, partnership, and coordination in destination area i.e. also region. DMO is considerably new in Indonesia. If management is divided into three dimensions; structural (public), nonstructural (sociocultural, community) and market (business), then DMO is a part of the nonstructural dimension. The role of

 DOI: 10.1051/matecconf/20167011004

DMO is to reinforce and empower the existing management pattern, thus it complement the structural governance in public and private sector. (Ministry of Tourism, 2016)

Lake Toba development focused on the preservation of natural and cultural creativity of communities around Lake Toba. Lake Toba is now being promoted as geopark tourist area. There are priority tourist destinations that are become government focus such as to provide added value to the sustainability, development of tourist destinations, developing infrastructures and accessibility. (Lumbanraja, 2012).

DMO program in Lake Toba is one of priority for Indonesian government that is to design strategy and action plan for tourism development in Lake Toba. Developing destinations in Lake Toba require spatial integration, support facilities and infrastructure, connectivity and accessibility as well as amenities with various support facilities. To create multi sector development required an integrated process that aims to build commitment of stakeholders in developing DMO program in Lake Toba. (Ministry of Tourism, 2015)

Before developing tourism plan in Lake Toba, it is important to consider market potential that appeal to conditions, infrastructure, accessibility, and management, regulatory, economic and social value. (Teguh, 2015)

2. Tourism system models

The competitiveness and attractiveness of the destinations suggest the use of effective, sustainable strategies and systems that are based on market conditions and preferences, as well as on the balancing of interests inside the tourist destination. Currently, the call for systemic change is becoming increasingly strident. Unfortunately, the word *system* has been popularized without a fundamental understanding of its implications, to the point where everything is a system but nothing really is treated as one. Many people say they are using a systems approach, but almost no one really is. Furthermore, popular interpretations of systems tend to use inappropriate mechanical models and metaphors. Decision makers need to fully understand why our current approaches won't work and what is different about the systems approach. (Senge, 2010)

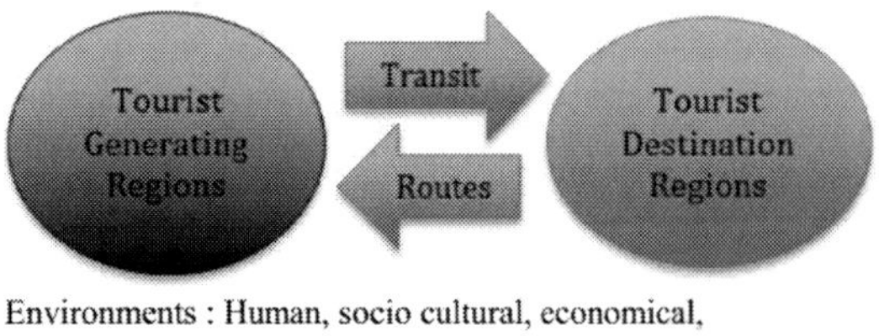

Environments : Human, socio cultural, economical, technological, physical, political, legal etc

Figure 1. The geographical components of tourism. Source :Leiper (1979)

2.1 Tourism system

For the last decade, we are introduced with various models of tourism development. Most of them are focused on destination areas instead of defining them as "system" (Gunn, 1994; Leiper, 1990; Mill & Morisson, 1998). Generally, a system has a structure and the environments (Bertalanffy, 1971).

2.2 The destination life cycle model

Destination development is about creating and maintaining visitor experiences, saleable products and services that meet visitors' needs and expectations and facilitate their travel and stay within a destination. Traditionally this focus has been on assets and products. Increasingly visitors are looking to enjoy unique and fulfilling experiences that bring a destination to life and make it memorable. The most authentic visitor experiences may be developed outside the regular tourism offering. These will provide opportunities for visitors to experience the landscape, people, way of life, culture and food enjoyed by the local community. Make sure to invest in soft infrastructure rather than just hard infrastructure such as transport, accommodation and transport.

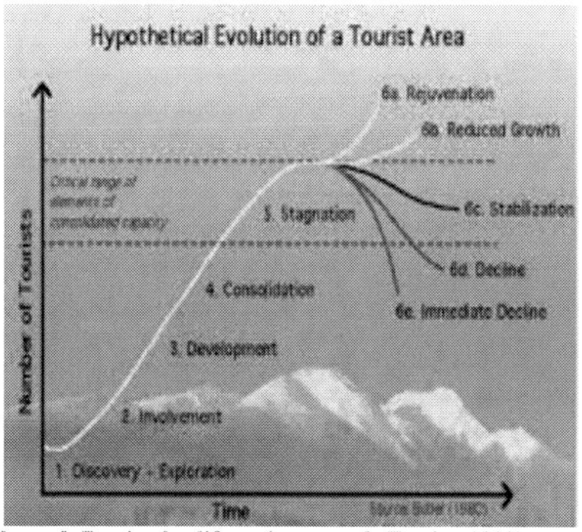

Figure 2. Destination life cycle approach (Buttler, 1980)

Buttlers model (1980), has become of the best known theories of tourism evolution at destinations (Weaver & Lawton, 2002). This model represents the relationship between an increasing number of tourists and the development of a tourist over time, as a life cycle. The cycle includes six stages: exploration, involvement development, consolidation, stagnation and post-stagnation. This model also explains the involvement process of the local community in those stages

3 System thinking approach

System thinking allows people to make their understanding of social systems explicit and improve them in the same way that people can use engineering principles to make something explicit. Systems thinking will help us to understand the value of other opinions, and ability to see things from different perspective. A mental models which is ingrained assumptions that ultimately influence how we see things and what we do, can be a good place to start. (Meadows, 2008)

The habits of systems thinkers include: considering long and short-term consequences of actions such as, identifying the circular nature of complex cause and effect relationships (the bee buzzing around the flower is a system, where the bee needs the flower and the flower needs the bee); and looking at things from different angles and perspectives.

Systems thinking approach is fundamentally different from traditional form of analysis. Traditional analysis focused on the separating the individuals pieces of what is being studied. Systems thinking is not just about the tools to help us see the world with a better lens; it also can give us greater grasp of why things happen a certain way. Traditional thinking often see things in linear ways without considering others perspectives and its relationships. Things are circular in systems thinking, and recognizing the complex nature of cause-and-effect relationships can help us to understand why things happen.(Senge, 2010

Linear Thinking

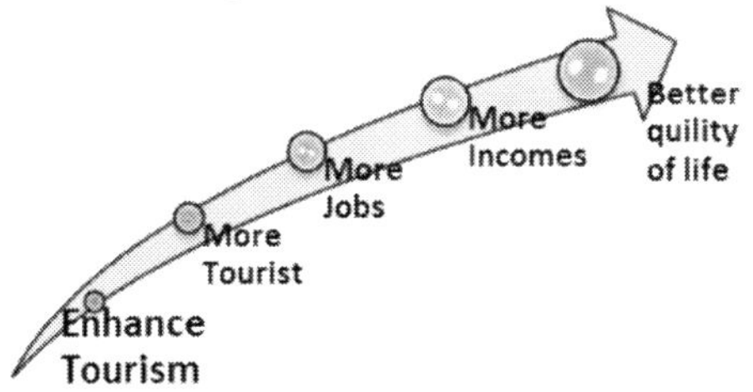

Figure 3. Linear thinking approach source: Modification ricardo, 2016

4. System thinking for sustainable tourism development in Lake Toba

4.1 Backgrounds

The application of systems thinking has grown extensively and encompassed work in many diverse fields such as business (Sterman, 2000), health (Cavana, Davies, Robson, & Wilson, 1999), commodity systems (Sawin, et al., 2003), agricultural production systems (J. Wilson, 2004), natural resource management (Allison & Hobbs, 2006), environmental conflict management (Elias, 2008), decision making and consensus building (Kambiz E Maani, 2002; Kambiz E Maani & Maharaj, 2004), and human resource management (Quatro, Waldman, & Galvin, 2007) Nevertheless, this is the first study using a systems thinking approach to apply in the sustainable development of tourism in a biosphere reserve context. It would therefore have high potential to be applied globally, considering the importance of tourism in many economies and a comprehensive network of Lake Toba (UNESCO, 2009)

Using systems thinking approaches for developing integrated destination management organization will enable us to see from another perspective and look deeper to why some events play out in certain ways. If we develop those habits of thinking systemically and look at any tourisms issue, we will ask different questions. This means that instead of isolating smaller parts of system being studied, we can develop something strikingly different conclusions than those generated by traditional forms of analysis, especially when what is being studied is dynamically complex or has a great deal of feedback from other sources, internal and external. This approach has been developed as an emerging discipline to understand complexity and dynamic changes (K. Maani, E. & Cavana, 2007). In addition, it offers new powerful paradigm, a specialized language and methodology.

The character of system thinking makes it extremely effective on the most difficult types of problems to solve: those involving complex issues, those that depend a great deal dependence on the past or on the actions of others, and those stemming from ineffective coordination among those who involved. Many important problems that plague us today are complexes, involve multiple actors, and are at least partly the result of past actions that were taken to alleviate them. Dealing with such problems is notoriously difficult and the results of conventional solutions are often poor enough to create discouragement about the prospects of ever effectively addressing them. One of the key benefits of systems thinking is its ability to deal effectively with just these types problems and to riase our thinking to the level at which we create the results we want as individuals and organizations even in those difficult situations marked by complexity, great numbers of interactions and the absence or ineffectiveness (Sterman, 2002)

4.2 System Thinking Approach : Process and Results

Maani and Cavana (2007) simplified the approach by using the analogy of an iceberg to visualize four levels of thinking (Fig. 6). The events level represents the tip of the iceberg which describes symptoms of reality. However, most of the decisions and interventions take place at this level. This is because events are the most visible part and often require immediate attention and action (Maani and Cavana 2007).Therefore, it often receives quick fixes and leads to "fire fighting" or "band aid" behavior (Bosch, et al., 2007). The next level of thinking is patterns where a larger set of data points are linked together to create the trends or patterns behaviour of events. The following level of thinking is systemic structures which presents the causal relationship within different parts of the system

There is yet another deeper level of thinking located at the bottom of the iceberg which hardly ever comes to the surface. This is the mental model of individuals and organisations that influence why things work the way they do. Mental models reflect the beliefs, values and assumptions that we personally hold, and how they underlie our reasons for doing things the way we do (K. Maani, E. & Cavana, 2007). The events represent the shallowest and most visible level of reality while mental models reflect the deepest and most profound assumptions, norms and motivation. Leverage from

 DOI: 10.1051/matecconf/20167011004

events to mental models is increasing. Hence, the deeper the level we apply, the stronger is the leverage that can be perceived. The tools of systems thinking focus on the four levels of thinking, thus they provide a systemic framework to deal with the most complex or stubborn problems that we are facing in everyday life and work. Those tools include Increasing leveragecausal loop diagram, stock an flow diagram, computer simulation, learning laboratory and group modelling (K. Maani, E. & Cavana, 2007)

4.2.1 Problem structuring

In this stage, 60 participants from 7 different groups of relevant stakeholders (Table 1) participated in focus group discussion (FGD) or key information interviews (KI) during the period of September to November 2016 to identify key issues, potential, and challenges for sustainable tourism development in lake toba.

Table 1. Stakeholders participation in the interview

No	Stakeholder Groups	No. of Interviews	No. of People involve	Interview	Male	Female
1	Local community	4	20	FGD	14	6
2	Local Authority	15	15	FGD & KI	7	8
3	Private sector	16	16	FGD	10	6
4	Academic	2	2	KI	1	1
5	Non- government organization	1	1	KI	1	0
6	Hotel/ Restaurant manager	5	5	KI	1	4
7	Tourism Advisor	1	1	KI	1	

Table 1 Stakeholders participation in the interview . The identification of key issues is slightly different amongst groups based on their different backgrounds (see Fig. 2). The main issues identified were: (i) lack of law enforcement (ii) no waste prevention (iii) lack of coordination between stakeholders (v) Poor infrastructures and tourism facilities. These main issues are grouped into four broad categories namely politic, environmental conservation, economic development, social development as summarized in Fig. 4 as follow

Politic	Environmental Conservation	Economic Development	Social Development
Lack of law enforcement	No waste prevention	Lack of job opportunity	Lack of human capacity
Lack of coordination between stakeholders	No pollution control program	Poor infrastructure	Lack of hospitality
	Illegal forest exploitation	Poor accessibility	Poor friendliness
		Poor amenities	
		Insufficient tourist facilities	

Figure 4. Key identified issues for sustainable development of tourism in Lake Toba

4.2.2 Developing causal loop model

UNWTO (2007) in a Practical Guide to Tourism Destination Management explained that DMOS have typically undertaken marketing activities, reviews their remit is becoming far broader, to become a strategic leader in destination development. In this study, the participants consist of local government planning committee (including tourism, agriculture and forestry, industry, natural resources, education, transportation, national park) practitioners (hotel manager, travel agent, and tourist information center), community leaders, and NGOs. About 30% of participants believes system thinking is an important approach by actors and stakeholders as an adhesive factor of cooperation (collaboration). Hardly there will be parties objected to benefit of system thinking. As a result, the concept of mutually beneficial (win-win) between related parties can be formed trough consensus. When the aspect of 'benefits' has been obtained, it is expected that each stakeholders will develop ownership which is characterized by commitment-related programs. This commitment can be seen from how much participation and empowerment can be realized. The collaborative management also can coordinate all the components to develop synergy that led to strengthening the competitiveness of Lake Toba.

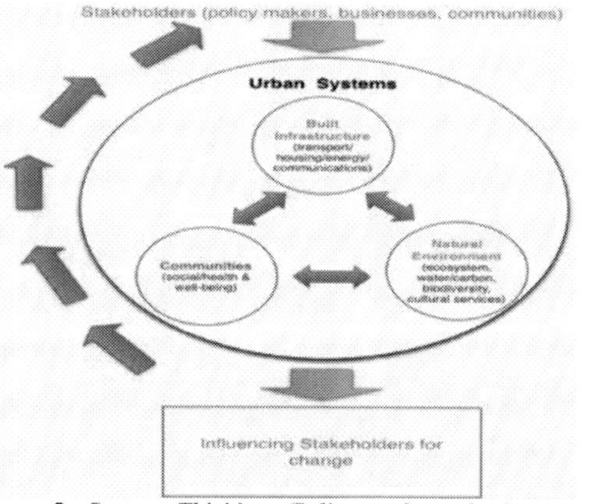

Figure 5. System Thinking (Policy makers, business and communities) with stakeholders. Modification Ricardo, 2016

Systemic thinking provides an organizing means of understanding the interconnectedness of life events in the world. Perhaps it is the lack of systemic thinking examples in destination management that impedes a comprehensive and wide-ranging source of applications from which one could draw. Applying systemic thinking in destination management is a difficult task largely because of systemic factors in education that compound its application. Although some resources exist that attempt to apply systemic thinking in education (see Case, 1992; Isaacson & Bamberg, 1992), the paucity of such examples raises some possible concerns, including the misunderstanding of what is meant by systemic thinking. Some leaders need to understand more than just behavioral incidents and institutional operations. There are factors associated with each of these that may shed light on responses and solutions.

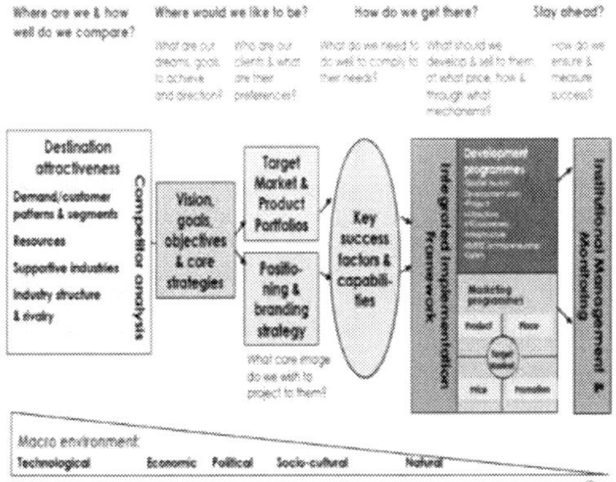

Figure 6. Integrated destination framework.

5. Conclusion

Destination management involves various factors includes environmental, social, health, educational, economic and political dimensions. Lake toba shows that all these aspects were brought light clearly and rational by using system thinking approach. When issuing tourism development policies, policy makers need to understand the nature of local tourism and the role of participation in policy making. In addition, they should study the results of previous tourism policies and apply policy practically (Bramwekk & Meyer, 2007; Stevenson, Airey, & Miller, 2008; Liu & Wall, 2003). Policy-makers should also embrace economic, social and environmental tools to boost the development of local tourism (Panayotou, 1994; Collins, 199; Palmer & Riera, 2003)

This study has addressed tourism as a complex and dynamic system. The system includes many interacting components. The development of tourism in a sustainable way impinges on and is subject to many different factors. This is especially true in Lake Toba context where the system includes three interacting sub-systems of economy, environment and socio-demography, and involves diverse stakeholders, each of whom holds different objectives. The process of developing a casual loop model or qualitative modeling would help governmental agencies, hoteliers and planners to understand complex relationships. This system thinking is very useful to simplify, clarify and integrate complex problems and the key leverage points of the tourism system in Lake Toba.

References

1. DeBrine, Peter. UNESCO *World Heritage and Sustainable Tourism.* UNESCO World Heritage Center. Paris France. UNESCO
2. Sterman D, John (2002) *System Dynamic: Systems Thinking and Modeling for a Complex World.* MIT Sloan School of Management Cambridge MA
3. Moejodo et al, 2006. *Lake Toba: Experience and Lessons Learned Brief.* Lake Toba Heritage Foundation
4. Parlagutan Siahaan. *Pesan Danau Toba (Message from Lake Toba Bulletin)*, p. 9, September (1999)
5. Robinson, M. and Picard, D. (2006) *Tourism, Culture and Sustainable Development*, Division of Cultural Policies and Intercultural Dialogue, Culture Sector, UNESCO.
6. Lumbanraja,Victor (2012) *Tourism Area Life Cycle in Lake Toba.* IJG, **44**, no. 2, December 2012 (150-160) Faculty of Geography UGM and The Indonesian Geographers Association.
7. Scheyvens, Regina (2011) *The challenge of sustainable development in the Maldives: Understanding the social and political dimensions of sustainability.* Asia Pacific Viewpoint, **52**, no. 2, , pp148-164, August (2011)

Biogas Initiative from Swine Farm in Southern Thailand

Det Damrongsak and Wongkot Wongsapai

Energy Technology for Environment Research Center, Faculty of Engineering, Chiang Mai University, Chiang Mai, 50200, Thailand

Abstract. First biogas pipeline network has been well established in southern Thailand. About 1,273 households, accountable for about 87% of the total of 1,466 households in the district, get the benefits from biogas energy in many ways. Key success to this initiative is the collaboration between all parties, i.e., swine farm owners, households, and government officials. Swine farm owners are responsible for the design and construction of the biogas plants. Households pay some contributions regarding labor work and maintenance cost on biogas system and its pipeline network. Government officials are responsible for financial and technical supports to both parties. Indeed biogas energy offers an alternative source of heat energy for cooking fuel in this region.

1 Introduction

Thailand alternative energy consumption has increased continuously due to the national policy of alternative energy development that targeted to increase more alternative energy consumption in all sectors, especially alternative energy which can be produced in the country, to decrease the energy consumption from fossil fuel and the energy imports. In 2014, alternative energy consumption in Thailand was 9,025 ktoe, an increase of 9.6% from the previous year and shared 11.9% of the total final energy consumption. This affected decreasing of energy imports, as amounts of $6,000 million, and also decreasing of CO_2 emission as amounts of 27.68 million tons [1], [2]. Recently, Department of Alternative Energy Development and Efficiency (DEDE) has presented the new Alternative Energy Development Plan 2015 (AEDP 2015), aiming to promote alternative energy usage at 30% of the total energy consumption by 2036 and reduce dependency on fossil fuel and energy import [3].

Prakha district is located in Phatthalung province in southern Thailand. Many people in the district own swine farms. Swine manure creates the environmental concerns to people in the district. These concerns include bad smell, wastewater, flies, and insect caused by the swine manure. These problems make it difficult for people in the district to live their lives well under these conditions. Therefore, the group of swine farm owners tried to find a suitable solution to reduce the pollutions produced from their farms in the district. Finally, they came up with the idea of constructing the biogas system from swine manure to produce biogas for energy use and for reduction of pollutions in the district. In Thailand, biogas technology has been a successful renewable energy technology developed and generally distributed. Livestock farms commonly utilize biogas for heat and electricity generation [4]-[7].

Phatthalung Provincial Energy Office, who was responsible for energy issue in the district area, recognized the problems and what solution they must take to resolve this environmental concerns, and make the local people and farm owners live together pleasantly. They acknowledged the important and development of the biogas system in Prakha district. They received a sufficient amounts of funds from several sources to build the network of biogas system from the swine manure to reduce the use of LPG in the households. Since the energy supply for households is predominantly LPG for cooking, the biogas would be fed to households to replace the use of LPG. They set up the rule and regulation in related to the use of biogas. Each household has to pay $1 per month for the right to use the biogas. This money would be used for maintenance of the biogas pipeline system and also for biogas safety system as well. This project not only helped the people in the community to reduce the living expenses, but also helped substantially in reducing the environmental impact from swine farms in long term. This biogas project in the district obviously follows the principle of the sufficiency economy.

Since 2008, over $770,000 funding from the government and others was secured to provide engineering design and construction of the biogas plants and their related equipment, pipeline to distribute biogas, and biogas stove burner. The project is ongoing and currently in its sixth year of implementation. The biogas network covers eight villages in the district providing biogas to 1,273 households. There are still some areas remaining to be developed under this project.

2 Biogas program initiative

Due to the high energy prices in the country, Thailand strategic plan for alternative energy development strongly advocates the production of alternative energy for national use and contains a key target to increase the alternative energy share of the total energy consumption to 30% by 2036. To achieve that target, Thai government encourages local communities and private sectors to produce and use alternative energy. For example, financial subsidy for electric power plant construction and the introduction of Feed-in Tariff (FIT) program have been provided to entrepreneurs who produce electricity from alternative energy such as biogas and solar energy.

In addition, pollutions from swine manure were of great concern to all people living in the district. Bad smell, wastewater and insects made it difficult for people to live. Biogas system is the key to solve all of these problems. Hence, the main focus of the biogas project is on community-based biogas production for local use, and on reduction of the effect of pollutions produced by swine manure.

As for the supporting groups to this initiative, Energy Conservation Promotion Fund (ENCON), the Department of Alternative Energy Development and Efficiency (DEDE), and PTT Public Company Limited provided high amounts of funds to start up this project. This funds were used for project management and construction of the biogas plants in the swine farms, pipeline network in the district, and its required equipment. Energy Technology for Environment Research Center (ETE) from Chiang Mai University (CMU) together with PTT provided training and technical supports such as main and branch pipelines maintenance, gas leakage inspection. The key to the success of this initiative program is the mutual understand and collaboration between the swine farm owners, households in the district, and all supports from government and private sectors.

The primary source of heat energy used for cooking in most households is Liquefied Petroleum Gas (LPG). To replace LPG use with biogas, it is imperative that the biogas system along with the network of pipeline be constructed in the district. Therefore, ECON, DEDE, PTT, and ETE provided $770,000 for construction of 12 sites of biogas plants, their required equipment, and all technical supports. The main biogas pipeline is 28 kilometers in distance. All main pipelines are placed about 50 centimeters beneath the soil surface with signposts for every 100 m apart to indicate where the pipelines are. Branches of pipelines to distribute biogas from the main pipeline to households are 70 kilometers in distance. This network of biogas pipeline in Prakha district is unprecedentedly the first biogas network in Thailand. In order to prolong the use of biogas system, PTT and ETE are responsible for technical supports and training course such as inspection of biogas leakage, biogas safety and all maintenance issues. With biogas as a cooking fuel, the new biogas stoves have to be given to all households to replace LPG stoves.

When the biogas was produced by swine farm owners, it was distributed through pipeline network to all households who took part in this program. The total biogas capacity of all sites is 2,822m^3/day, in which it helps in reducing the release of methane (CH_4) to atmosphere by 1,130kg/day equivalent to 28.25 ton CO_2. Because each household use LPG for cooking with the average of 0.23kg/day equivalent to 0.5m^3/day of biogas (1m^3 of biogas is equivalent to 0.46kg of LPG [8].), hence there is more than enough biogas to cover as many as 1,273 households in the district.

Prakha district consists of 1,466 households [9]. Figure 1 shows the numbers of households with and without biogas energy. There are 1,273 households (87%) joining the biogas program. Only remaining 193 households (13%) are not yet joined the program due to some difficulties such as unreachable location and lacking of budget.

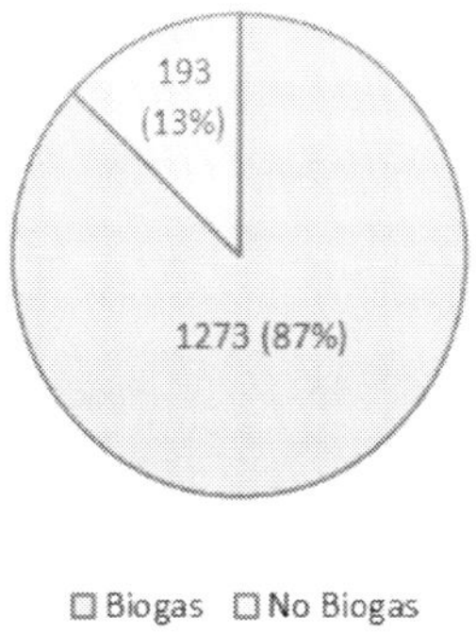

Figure 1. Numbers of households with and without biogas energy.

3 Analysis of outcomes

3.1 Relationships between stakeholders

Phatthalung Provincial Energy Office has been the main organization to run this biogas project since 2008. Therefore, it has good relationship with all parties involved in this project. Figure 2 shows relationships between participants and how they get involved in this biogas initiative.

The provincial energy office endorsed by Prakha district municipality successfully set the meeting between swine farm owners and households to talk about the development of the biogas system in the district. This project would help in reducing the energy expenses on LPG use for cooking, and reducing environmental pollutions created by swine farms in this area as well. Then the provincial energy office used the conclusions from this meeting to propose DEDE for the support for the development of biogas system from swine farms in Prakha district.

ETE is responsible for course training on biogas technology and its safety concerns and other technical supports, therefore, the relationship between ETE and households and swine farm owners is purely technical-based on biogas system. The swine farm owners have a good and informal relationship with households in the district because the biogas system helps in reducing pollutions in the areas caused by swine manure. And households who link up to biogas network is provided free biogas from the farm owners at the same time.

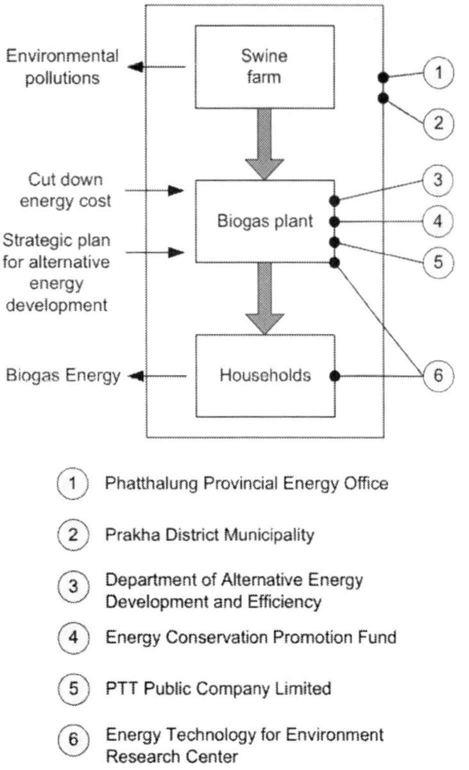

Figure 2. Relationships between participants and how they get involved in this biogas initiative.

3.2 Use of biogas energy

Prior to launching the biogas program in this area, the primary source of energy use for cooking in the district is LPG. Each household would use LPG about 0.23kg/day. This requires 293kg/day of LPG to cover 1,273 households who take part in this project. To achieve these amounts of LPG use per day, it requires 637 m^3/day of biogas. So with the existence of biogas system from swine farms in the district, all households are able to reduce the use of LPG by 106,945kg/year. This can be converted to the expense of approximately $80,000 energy payment saving for a year. With the presence of biogas, each household is able to cut down the expense on LPG by estimated $6 per month.

Even though the biogas plant owners distribute biogas regularly to households in the district for free, households need to pay $1 monthly to Prakha district municipality as they use these amounts of income for maintenance purpose of the biogas system and its pipeline. Some remaining biogas are generally used to produce heat and electricity for using in the farm itself.

3.3 Carbon dioxide emissions

In terms of reduction of carbon dioxide (CO_2) emissions to earth atmosphere [10], the biogas project plays an important role in reducing the use of fossil fuel. 1,273 households in the district change from fossil fuel to biogas. Recently, they only use LPG for backup source of energy, just in case of the biogas shortage that may occur from the biogas network from time to time.

With an amount of 106,945kg/year of LPG that had been used for cooking prior to the use of biogas fuel, an equivalent of 333 ton CO_2 per year has been released to atmosphere. After biogas system and pipeline network were completely constructed. Estimated 637m^3/day of biogas has been used in replacement of LPG in households that participated with this biogas initiative. These amounts of biogas equal 255kg/day of methane (CH_4). With these amounts of methane use as a cooking fuel, only 0.73 ton CO_2 equivalent per day or 265 ton CO_2 equivalent per year has been emitted to atmosphere. Biogas helps in decreasing equivalent of 68 ton CO_2 per year. Figure 3 shows amounts of equivalent of ton CO_2 emitted by LPG and biogas from 1,273 households.

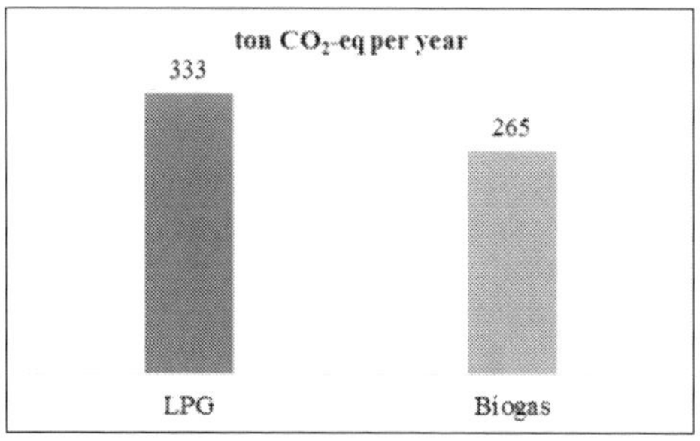

Figure 3. Amounts of equivalent of ton CO_2 emitted by LPG and biogas from 1,273 households.

3.4 More potential on biogas

There are still some rooms available to increase the use of biogas. There are total of 1,466 households in Prakha district. 193 households have not used biogas as an alternative source of energy yet. They still use LPG as a primary source of heat energy. They are accountable for 16,200 kg of LPG per year. With these amounts, equivalent of 50 ton CO_2 have been emitted to atmosphere. If all the remaining households change cooking fuel to biogas, it would require estimated 35,200 m^3 of biogas to do the same task. With these amounts of biogas, equivalent of 40 ton CO_2 have been emitted to atmosphere. This would reduce amounts of 10 ton CO_2 emitted to atmosphere if all of 193 households switch from LPG to biogas. Figure 4 shows amounts of equivalent of ton CO_2 emitted by LPG and biogas from 193 households.

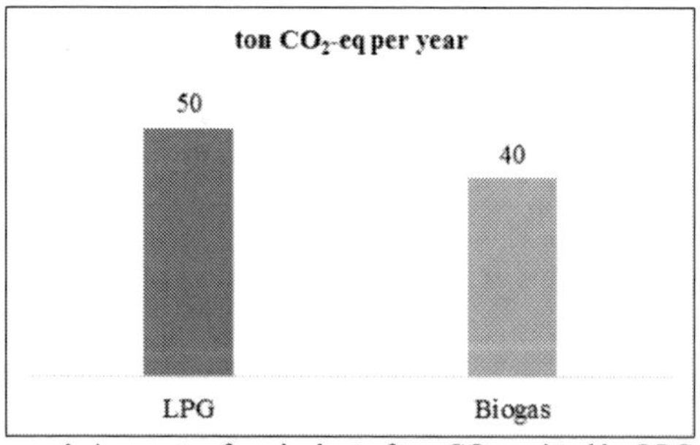

Figure 4. Amounts of equivalent of ton CO_2 emitted by LPG and biogas from 193 households.

3.5 Land use

Biogas system not only helps in developing the land use, but also helps in improving the environmental impact both inside swine farms and surrounding community area. Biogas plants are well constructed inside swine farm area. In addition, the biogas main pipelines are mostly put beneath the soil surface with signpost directing where the lines are. Therefore, the biogas plants and their pipelines do not affect the nearby natural resources and agricultural land of the people in the vicinity area. Since, biogas systems belong to the swine farm owners. The farm owners are just responsible for producing and supplying biogas to households through the network of pipelines. Hence, there is no issue on land tenure.

4 Obstacles and solutions

4.1 Technical issue

Technical issue related to biogas technology has been primary obstacle prior to the beginning of initiative program. This obstacle included design and construction of biogas plant and its accessory equipment. How to properly operate biogas system was another challenge needed to be concerned with. Many biogas plant owners still do not have adequate knowledge of biogas system and its technology. Likewise, common people are also not aware of technology and its benefit of biogas system as well. So it is imperative that basic trainings on biogas technology need to be done to owners of the biogas plants and end users (households).

Many workers in biogas plant do not fully understand the operation and safety of biogas system. Therefore, biogas safety is another important topic that needs to be paid attention. Training course on biogas safety needs to be carried out especially to all workers who work closely with the biogas system day in and day out. This is done to ensure that all workers have better knowledge to work safely on biogas system.

4.2 Capital investment

Capital Investment was another issue need to be confronted. Some biogas plant owners are able to deal with all design and construction costs. There are still some other owners who do not have sufficient budget to construct biogas plant. With substantial subsidy from the government agency or soft loan from bank, it would be of great for owners to make a prompt decision regarding the construction of a new biogas plant. Moreover, subsidy is also an important element to help plant owners to decrease the payback period of the investment on biogas system.

4.3 Replication and up-scaling

With collaboration between all involved parties, technical and financial supports from related organization, the network of biogas system in Prakha district becomes a successful program. It becomes good demonstration sites of biogas technologies where government officials and other communities can learn from experience of this community. Thus, the biogas program in Prakha district can definitely be upscaled or replicated to some other areas of the country. This would directly increase more alternative energy use that goes along well at the same time with recent alternative energy development plan from the government. To ensure the success and sustainability of this biogas program, it is essential to make biogas technology knowledge and its benefit available to biogas owners, end users, and general public.

5 Conclusion

First biogas pipeline network has been successfully constructed in southern Thailand. Most households in the district receive biogas energy using for cooking fuel. This helps each household in reducing the energy cost from LPG by about $1 per month. Not only biogas can be used as a source of heat energy in households, but also helps to reduce the environmental concerns caused by swine manure. Biogas plant owners and local communities are essential elements to the success of the program. As swine farm owners and workers are responsible for construction and operation of biogas system, local communities have better knowledge and understanding on biogas technology and its benefit. Their collaboration plays an important role to the success of this biogas initiative.

Acknowledgment

Technical data and other assistances from Energy Technology for Environment Research Center, Faculty of Engineering, Chiang Mai University are gratefully acknowledged.

References

1. DEDE, *Thailand Alternative Energy Situation*, Department of Alternative Energy Development and Efficiency, Ministry of Energy, Thailand (2014)
2. DEDE, *Energy Balance of Thailand*, Department of Alternative Energy Development and Efficiency, Ministry of Energy, Thailand (2014)
3. DEDE, *Alternative Energy Development Plan: AEDP 2015*, Department of Alternative Energy Development and Efficiency, Ministry of Energy, Thailand (2015)
4. N. Tippayawong, A. Promwangkwa, and P. Rerkkriangkrai, "Long term operation of a small biogas/diesel dual-fuel engine for on-farm electricity generation," *Biosystems Engineering*, **98**, pp. 26-32 (2007)
5. W. Wongsapai, P. Thienburanathum, and P. Rerkkriengkrai, "Biogas situation and development in Thai swine farm," *International Conference on Renewable Energy and Power Quality (ICREPQ'08)*, Santander, Spain, (12-14 March 2008)

6. D. Damrongsak and N. Tippayawong, "Experimental investigation of an automotive air conditioning system driven by a small biogas engine," *Applied Thermal Engineering*, **30**, pp. 400-405 (2010)
7. D. Damrongsak and N. Tippayawong, "Performance and thermoeconomic analysis of a biogas engine powered ventilation system for livestock building," *Engineering Journal*, **18**, no. 3, pp. 1-10 (2014)
8. S. Babel, J. Sae-Tang, and A. Pecharaply, "Anaerobic co-digestion of sewage and brewery sludge for biogas production and land application," *International Journal of Environmental Science and Technology*, **6**, 1, pp. 131-140 (2009)
9. PSAO, *General Information*, Prakha Subdistrict Administration Organization, Phatthalung, Thailand. Available: http://www.prakha.go.th
10. IPCC, *2006 IPCC Guidelines for National Greenhouse Gas Inventories*, Intergovernmental Panel on Climate Change (2006) Available: http://www.ipcc-nggip.iges.or.jp/public/2006gl

DOI: 10.1051/matecconf/20167012002

A review of community-based solar home system projects in the Philippines

Erees Queen B. Macabebe[1,2], Reynaldo C. Guerrero, Jr.[2], Aleta C. Domdom[3], Aison S. Garcia[2], Emma E. Porio[4], Samuel Matthew G. Dumlao[1] and Teresita R. Perez[5]

[1]*Ateneo de Manila University, Department of Electronics, Computer, and Communications Engineering, Quezon City 1108 Philippines*
[2]*Solarsolutions, Inc., Quezon City 1108 Philippines*
[3]*Ateneo de Manila University, Department of Economics, Quezon City 1108 Philippines*
[4]*Ateneo de Manila University, Department of Sociology and Anthropology, Quezon City 1108 Philippines*
[5]*Ateneo de Manila University, Department of Environmental Sciences, Quezon City 1108 Philippines*

Abstract. Solar Home Systems (SHS) are easy to deploy in island and in remote communities where grid connection is costly. However, issues related to maintenance of these systems emerge after they are deployed because of the remoteness and inaccessibility of the communities. This study looked into community-based programs in the Philippines and investigated the following: (1) social preparation, (2) role of the community in the project, and (3) sustainability of the program. In this paper, three communities under two government programs offering SHS are presented. These programs are the Solar Power Technology Support (SPOTS) program of the Department of Agrarian Reform (DAR) and the Household Electrification Program (HEP) of the Department of Energy (DOE). A focused group discussion and key informant interviews were conducted in two communities in Bukidnon province and in a community in Kalinga to obtain information from the project beneficiaries and SHS users on the preparation, implementation and maintenance of the projects. The results revealed that emphasis on the economic value of the technology, proper training of the locals on the technical and management aspects of the project, as well as the establishment of a supply chain for replacement parts are crucial factors for the sustainability of the programs.

1 Introduction

Solar energy is the least tapped of all the renewable energy sources in the Philippines. In the 2013 Energy Sector Accomplishment Report of the Department of Energy, the country's primary energy supply in 2012 reached 42.90 million tons of oil equivalent (MTOE). Figure 1 shows the primary supply mix with about 45.98% of the energy supply coming from renewable and natural gas sources. Geothermal power having the biggest share at 20.55% while solar/wind contributes a meager 0.02%.

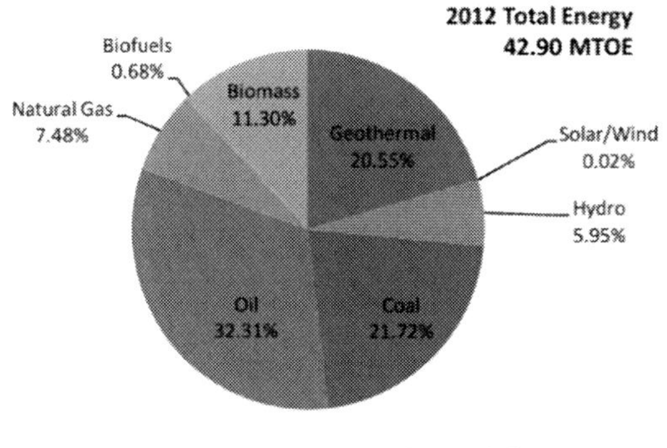

Figure 1. 2012 Primary Energy Supply Mix [1]

However, solar remains the most viable renewable energy source in the country. The use of solar photovoltaic systems in off-grid communities is most appropriate because the Philippines is near the equator, between 5° N and 20° N latitude, and the country receives an annual average of between 4.5 - 5.5 kWh/m^2/day of solar energy. Figure 2 shows the map of the solar energy potential of the country using the Climatological Solar Radiation Model in a resource assessment conducted by NREL [2].

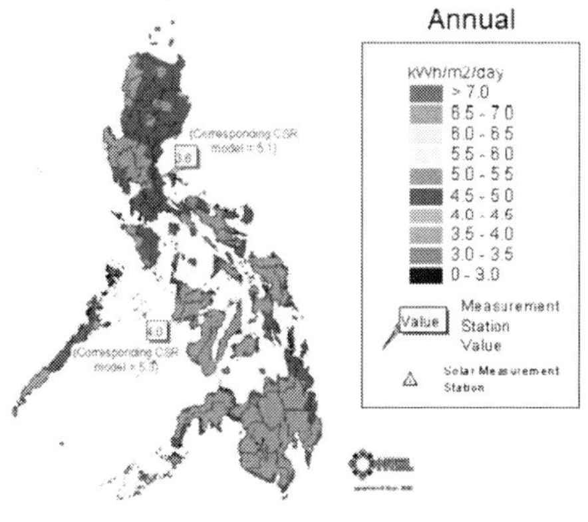

Figure 2. Solar energy potential of the Philippines

The DAR-SPOTS program is an integrated social and agricultural project [3]. The primary objectives of the program were economic and social in nature, which includes the following: (1) Address poverty in the un-energized and off-grid agrarian reform communities (ARCs) through the introduction of specific PV applications in agriculture and rural enterprise development, (2) Increase agricultural productivity and promote economic/livelihood diversification, (3) Capacitate and empower local communities and organization in the ARCs, (4) Apply solar power as enabling technology to promote sustainable development.

To achieve the objectives of the program, the following were undertaken: (1) solar electrification which includes installation, commissioning and hand-over of solar packages to recipient communities, (2) agricultural and rural enterprise development, and (3) institutional development such as formation of People's Organization (PO) in all beneficiary communities and PO capacity building. The target beneficiaries are Agrarian Reform Communities (ARC) in the country and one of the selection criteria is their non-inclusion in electrification projects of other agencies such as DOE and NEA.

THE DOE-HEP involves energizing off-grid households using mature renewable energy technologies such as photovoltaic in SHS and streetlights, and micro-hydro systems [4]. The aim is to address the electrification needs of households in areas where grid connection is currently not feasible. This program was designed to contribute to attaining the target of 90% household electrification nationwide by 2017. Through this program, DOE plans to contribute at least 2,000 households every year to achieve the target.

HEP was funded by the national government subsidies from the DOE's Special Account in the General Fund (Fund 151). Thus, DOE directly implemented HEP through its three field offices across the country. The DOE was responsible for hardware procurement, shipment and installation of the power systems. It provided technical training of users and technicians. DOE field personnel conducted social preparation and community organizing in collaboration with Team Energy Foundation, Inc. The DOE assigned field personnel took charge of all stages of project activities: development and identification of beneficiaries, inventory, implementation, and monitoring. The DOE coordinated with the local government units (LGU) and the electric cooperatives (EC) during the implementation of the project.

Once the projects were completed, the DOE turned over the renewable systems to community power associations, composed of household beneficiaries, for operation and maintenance. The community associations are expected to manage and maintain the systems using the membership and maintenance fees to ensure the sustainability of the project.

The goal of this study is two-fold: to review the implementation procedure of the SHS projects, and to find out what measures are undertaken by the community to maintain the SHS and sustain the program. Three communities were chosen as study sites: two beneficiaries of DAR-SPOTS in the Municipality of Kadingilan in Bukidnon Province and the third community is a beneficiary of DOE-HEP located near the boundary of Tabuk City, Kalinga and Paracelis, Mountain Province.

2 Methodology

Site visits were made for data collection and field study. In-depth interview and focused group discussions (FGD) were conducted in each case study sites with at least 10 participants from the community and members of the cooperatives. Also Key Informant Interviews with implementing agencies, NGOs and relevant local government units were conducted to gather information on the projects. Data triangulation was conducted by reviewing accomplishment reports and documents obtained from stakeholders, the government agencies responsible in the implementation of the program, as well as third party reports to verify the information from different sources. The case studies focused on four questions. (1) What was the role of the community in the project? (2) How were they prepared before the technology was introduced? (3) How did they adapt to the technology and what efforts were undertaken by the community to sustain the project? (4) What support did they get from the implementing agency to sustain the project?

The impact of the programs and projects were analysed by considering the following key aspects of sustainability: technology and the capacity of the community members to learn the skills required to operate and maintain the SHS, environmental, social and economic. These key aspects were analyzed following the Five Level Framework [5] which looks into (1) Systems level - the set-up or equipment in the selected area, (2) Success level - what the community or agency wants to achieve for the project, (3) Strategic level - guiding principles during the implementation, (4) Actions level - actions done to achieve the goal of the project, and (5) Tools level - tools used to achieve the goal of the project.

3 Results and Discussion

Figure 3. Project implementation flow followed by both government programs.

In both programs, the government agencies involved followed a similar procedure in implementing the projects. It followed the paradigm in Figure 3 where community consultations and trainings were conducted prior to system deployment. One key aspect in these projects is that to qualify for the program, a local cooperative or an association is required to operate the project and maintain the SHS in the community [3, 4].

3.1 Kadingilan Municipality: Cabadiangan and Pay-as Agrarian Reform Communities

Kadingilan is a third class municipality in the Province of Bukidnon in the Philippines. The topography of the municipality is mountainous and hilly characterized by rough terrain. According to the 2010 census [6], Kadingilan has 31, 756 residents distributed in 17 Barangays. Cabadiangan with 4, 521 residents and Pay-as with 1, 863 residents are two of the barangay beneficiaries of DAR-SPOTS. The two barangays are accessible from the town center via a dirt road. Figure 4 shows the participants of the focused group discussion that was conducted to gather information from the community for this project.

Figure 4. Community members from Pay-as gathered for the Focused Group Discussion.

Table 1 summarizes how the project was implemented in the two communities. Since SPOTS is a social and agricultural project, there were several consultations on the most appropriate rural enterprise that the community can engage in. Once the systems were turned over to the ARCs, they are responsible for the operations and maintenance of these systems.

To avail the SHS, one should be a member of the cooperative. To become a member, one has to undergo the Pre-membership education seminar (PMES) which includes an introduction on the role of cooperative members and their benefits. This is conducted by the officers of the cooperative and integrated into the seminar is the training on the maintenance of the SHS which includes battery care and cleaning of the solar PV module. This training is conducted by the technician.

A total of 187 availed of the SHS in Cabadiangan and Pay-as. Each household agreed to pay the cooperative a monthly subscription fee of Php 220.00 (USD 4.70) to cover the operations and maintenance costs of the solar PV systems. This amount is less than what the household would normally pay for in a month using kerosene lamps. A few years later, after consultation with DAR, the monthly fee was reduced to Php 110.00 (USD 2.35) in Cabadiangan and Php 100.00 (USD 2.13) in Pay-as on the condition that the cooperative will no longer be responsible for parts replacement like bulbs when they are damaged or no longer functional, and for the battery solution.

Table 1. DAR-SPOTS implementation in Cabadiangan and Pay-as.

Consultation	▪ In 2006, DAR conducted a survey and consultation with the communities. ▪ Pay-as formed the Mountain Hilltop Association and Cabadiangan under their existing Cabadiangan Agrarian Reform Beneficiaries Multipurpose Cooperative were chosen beneficiaries of the program. ▪ Several consultations were conducted to determine the most appropriate rural enterprise for each community.
Social Preparation	▪ 5 Members of the cooperative, some of them officers, attended the seminar about the SPOTS program. ▪ BP Solar trained two members of the cooperative on the maintenance and repair of the solar home system.
Deployment and Maintenance	▪ In 2008, 147 SHS in Cabadiangan and 40 SHS in Pay-as were installed. ▪ The trained members of the community provide the maintenance support and training for members of the cooperative who availed of the SHS.
Monitoring	▪ The cooperatives collect the monthly fees for the operations and maintenance of the SHS. ▪ A personnel from the Municipal Agrarian Reform Office is assigned to each ARC to monitor the implementation of the program and mediate if there are issues between the cooperative and its SHS subscribers.

In Cabadiangan, the technician checks the battery every 3 months, whereas in Pay-as, the technician makes his round once every month. Also, the effectiveness of the response to complaints depends on the location of the household and the type of complaint. Households that are far from the barangay do not receive immediate support because of the travel time. In one of the cooperatives where the role of the technician has become voluntary, the response mechanism can take a few days.

The cooperative can pull-out the SHS from households who could not keep up with the payments, and from those households who have already connected to the electricity grid and chose to forego the SHS. The SHS can then be transferred to another household. The cooperative charges an installation fee, Php 1,000 (USD 21.33) in Cabadiangan and Php 1,200 (USD 25.60) in Pay-as, to cover the cost of pulling-out the system and installing it to the new household. Since the project was implemented in 2008, the local electric cooperative finally managed to install grid connections in these barangays. However, the cost of connecting to the grid is still expensive for many of these households because they are spread out and they have to pay for the cost of the physical connection from the main post to their house. Thus, many still prefer to use the SHS. But the existence of the grid allowed the program to expand in its scope in the ARC. Households from nearby sitios that were once not prioritized by the program can have a chance to avail of the refurbished SHS, as long as they are members of the cooperative.

3.2 Tabuk City: Sitio Makilo in Brgy. Calaccad

Calacaad is a barangay of Tabuk City in the Province of Kalinga and remains rural with 1,339 residents [6]. Sitio Makilo is a farming community located in the southern-most part of Calaccad, close to Kalinga-Mountain Province border. Because of its very remote location, the route from Calaccad proper can only be traversed on a motorbike and requires several river crossings. Figure 5 is a typical home in the community.

Figure 5. Typical house with a 30 W solar PV system.

Table 2 summarizes the project implementation in Sitio Makilo. Identification and consultations with the community by DOE were conducted in coordination with the local electric cooperative, Kalinga Apayao Electric Cooperative (KAELCO). Also, Kalinga Apayao State College is the DOE-Accredited Renewable Energy Center (AREC) that led the implementation of the program.

The subsidized cost of the 30 Wp SHS was priced at Php 5,000 (USD 106.64). This amount was intended for the operations and maintenance costs of the SHS and was managed by the Makilo Solar Home System Association (MASOHOSA). It was decided by DOE that each household make an initial payment of Php 1,000 and the remaining amount be paid in an instalment basis during the harvest season, which occurs at least twice in a year.

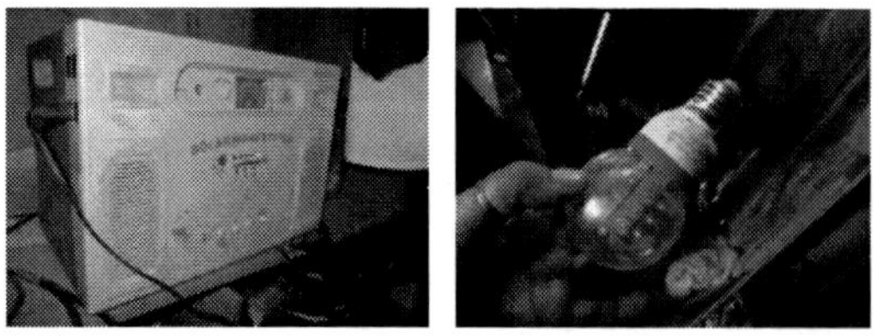

Figure 6. The SHS charge controller with radio feature, and the 12V, 1.2 W LED bulb that came with the SHS.

The DOE conducted seminars to introduce the technology to the households. However, in the focused group discussion (FGD) conducted for this study, the FGD participants all agreed that maintenance was not explained well during the seminar. They rely on the knowledge of the electrician in the community to troubleshoot the system when the need arises.

Currently, all 124 households continue to use their SHS even though many have already encountered problems since the SHS was installed. Despite the difficulty in acquiring replacement parts, the community managed to find substitutes though sometimes they resort to more expensive parts, like the use of truck batteries or the 13W LED bulbs. The SHS originally came with 3 pieces of 3W LED and 1 piece of 1W LED bulbs. Figure 6 shows the charge controller and the LED bulb that came with the PV system.

Table 2. DOE-HEP implementation in Sitio Makilo.

Consultation	▪ In 2010, DOE conducted a consultation attended by 75 residents in Sitio Makilo. ▪ In 2011, the Makilo Solar Home System Association (MASOHOSA) was established with 124 members to comply with one of the requirements of the HEP, which is the presence of an association or people's organization.
Social Preparation	▪ DOE conducted seminars to introduce the technology to the households. ▪ A technician was trained to maintain the SHS.
Deployment and Maintenance	▪ 124 SHS were installed in Sitio Makilo under the supervision of KASC-AREC. ▪ The trained technician provides the maintenance support. ▪ KASC-AREC also provides technical assistance to the MASOHOSA.
Monitoring	▪ MASOHOSA monitors the SHS and collects the Php 5,000 for the operations and maintenance of the SHS. This amount is payable under negotiated instalment terms. ▪ KASC-AREC liaise with DOE in monitoring the SHS since ownership of the systems remain with DOE.

Ownership of the SHS remains with the DOE, as of Feb 2015. The MOA is not clear whether the ownership will be transferred to the MASOHOSA. Also, now that the sitio is electrified, it is not clear to the MASOHOSA if the systems must be transferred to unelectrified households or they can remain with the original subscribers even if they are already connected to the grid.

Under the HEP, the solar home system consists of a solar panel, a charge controller with integrated radio, a sealed type battery and 4 bulbs (3 units of 3W bulb and 1 unit of 1W bulb). The SPOTS program has a similar type of system but instead of LED, fluorescent lamps were provided with the SHS and the charge controller or system regulator is operated through a pre-paid scheme. However, some SPOTS beneficiaries in Pay-as feel that the system should be bigger to power other appliances. In Cabadiangan, four lamps were reduced to three to control the usage of the system. This means that the system was not sized properly or that the load is more than what was designed. Also, the prepaid mechanism was ineffective because delinquent subscribers found a way to modify the regulator in order to by-pass the system.

For the maintenance of the SHS, households in Cabadiangan and Pay-as were taught basic maintenance such as wiping the panels and keeping batteries in a dry place. Households in Sitio Makilo however feel that the maintenance of the SHS was not explained well during the seminar. Also, there was no clear policy on the ownership and responsibility of the repair and maintenance of the solar home system.

The community association keeps track of the monitoring frequency of the trained technician and the number of functional solar home systems. The most common problem is the availability of replacement parts. Parts used to build the SHS are imported and batteries designed for solar applications are also not readily available to the community. Some prepaid regulators in the SPOTS program stopped working after 2-3 years and were not replaced. Also, there were no replacements for the fluorescent lamps/tubes. The size is different from those available in local hardware stores. This is also true in Sitio Makilo as there are no suppliers for LED bulbs and batteries that are compatible with their system. They had to resort to LED bulbs with higher wattage.

In all three communities, savings and potential income through the programs were the emphasis in the seminars conducted during the social preparation. For instance, the monthly fees should be the same or less than the amount spent on kerosene because the sustainability of the program relied on the households' capacity to pay. Also, the RE technology enabled the household to increase their hours of productivity, which directly or indirectly affected their livelihood. A commissioned study [4] reported that rural electrification, in general, increased the per capita income in beneficiary areas by 36% based on the 2010 Annual Poverty Indicator Survey. This is typical in developing countries where the economic impact of the RE technology being introduced is the primary consideration. Unlike in developed countries where community-based RE projects are undertaken as a step towards sustainable energy generation to mitigate climate change [7].

4 Conclusion and Recommendations

This study looked into three community-based SHS projects in the Philippines. Project implementation requires several key elements to achieve community adoption and sustainability : (1) Conduct consultation with the community, which includes resource and site assessment to properly design the SHS, and social and environmental impact assessment to make sure that the RE technology and its scale will not have an adverse impact on the environment and on the community, (2) Social preparation includes educating the community members for technology adaptation, especially where local community is ethnic minority group. The system should be simple so that the technology can be transferred to the locals. Also, highlight the benefits to convince the community members that the technology is viable and useful to them, (3) Training the community members as local personnel to manage the project, its implementation and its maintenance. Technology transfer must be ensured and safety must be emphasized in these trainings, (4) Financing can be obtained from government, foundations and international grant giving bodies. However, this should be limited to subsidies. The community or its members still need to have counter funding for the project so they have a sense of ownership of the system. This ensures that they will take care and make use of the system properly, (5) Operations and maintenance is the responsibility of the cooperative within the community. Upon turnover, the cooperative continues to manage and monitor the project. Thus, capacity building in all aspects of the projects must be conducted properly, (6) Local Materials/Locally available components must be used so that in case of damages and repairs, parts of the system that need to be replaced are easily available to the community. The implementing agency can also partner with a local hardware/electrical store who can guarantee that replacement parts are always on stock, and (7) Sustainability of the project is ensured if there is an income generating component to the project. In the cases presented, the beneficiaries are the community associations, and the monthly subscription fee for operations and maintenance of the SHS sustains the project. Also, the SHS subscription fees must be less than what the household used to pay for lighting so that the households can afford to pay the monthly fees.

Acknowledgement

This work was supported by the Toyota Foundation under the International Grant Program [D14-N-0082].

References

1. *Energy Sector Accomplishment Report*. Department of Energy, 2013. Web. 15 Oct. 2015.
2. D. Renné, P. Gray-Hann, R. George and L. Brady. *Assessment of Solar Resources in the Philippines: Internal Review Report*. National Renewable Energy Laboratory, 2000. Print.
3. *Solar Power Technology Support Project (SPOTS) Project Completion Report*. Department of Agrarian Reform, 2005. Print.
4. A.M. Navarro. *Cost Efficiency and Effectiveness of the Sitio and Household Electrification Program*. Department of Budget and Management, 2013. Web. 31 Dec. 2015.
5. D. Waldron, et.al. *Guide to the Framework for Strategic Sustainable Development*. Blekinge Institute of Technology, 2008. Web. 17 Feb. 2016.
6. C.N. Ericta. "The 2010 Census of Population and Housing Reveals the Philippine Population at 92.34 Million." Philippine Statistics Authority, 14 Apr. 2012. Web. 31 Dec. 2015.
7. L. Greenius, E. Jagniecki, K. Thompson. *Moving Towards Sustainable Community Renewable Energy: A Strategic Approach for Communities*. Blekinge Institute of Technology, 2010. Thesis.

Power and Frequency Fluctuations of MPPT Algorithms in Variable Irradiance Environments

Syed Ali A. Kazmi[1], Debraj Basu[2], Syed Faraz Hasan[2] and Dong-Ryeol Shin[1]

[1]*College of Information and Communication Engineering, Sungkyunkwan University, South Korea*
[2]*School of Engineering and Advanced Technology, Massey University, New Zealand*

Abstract. Solar cells use the natural sunlight for a variety of purposes including electrical power generation. In order to maximize the power obtained from a solar cell, it has to operate at what is known as the Maximum Power Point (MPP). Intelligent algorithms that can track MPP for a given cell are crucial for the success of a large scale deployment of solar panels. This paper investigates the performance of two well-known Maximum Power Point Tracking (MPPT) algorithms. A variable irradiance environment has been considered in this work, where sunlight varies significantly during the day. Towards the end, frequency stability analysis is conducted to evaluate the relevant impact of MPPT algorithms on frequency stability of system.

1 Introduction

One of the most important uses of solar energy technology has been in generating renewable or so called green electricity. The so-called photovoltaic (PV) cell (commonly referred to as solar cell) makes use of sun's energy to generate a voltage which can derive the load. Solar cells inherently produce DC power, which can be converted into AC power using appropriate inverters. A solar panel is a combination of a number of solar cells, each of which produces a modest amount of electrical power [1].

Recently, an increasing number of rooftop panels are being considered for connection to the main grid's distribution network. Since these small-scale solar-powered generators are connected directly onto the distribution network, they are referred to as distributed generation (DG) units or more specifically renewable energy generation (REG) [2]. The DG units are different from the traditional generators in several ways. First, they are connected to the grid at distribution network, which is closer to the load and operates at low voltage level. Second, these generators are connected to the grid through inverters because the main power grid is AC in nature. In the conventional power grids, the balance between the supply and the demand of the power is achieved by controlling rotating inertia (e.g. using torque, etc) [3]. In solar-powered DG units, inertial components do not exist so they cannot use the conventional droop control methods [4]. While there are a number of other open research issues in relation with using solar panels as DG units, we focus on optimizing their power output in this paper.

The power output of a solar panel depends on the amount of exposure it receives from the sun. Better the line of sight it maintains with the sun, the better its power output gets. A number of approaches have been proposed and adopted that physically change the orientation of solar panels so they remain aligned with the sun all the times [5]. This ensures higher received electromagnetic radiation from the sun, which is often expressed as irradiance [6]. Even while the panel is well-aligned with the sun, it cannot extract maximum output until it is operated at Maximum Power Point (MPP). MPP is an electrical operating point which enables all modules of a solar panel to produce the maximum output power. Since it is not a physical point in space, no inertial movement is required to reach MPP [7]. In order to understand the notion of MPP, let's consider the diagram shown in Figure 1. As shown in the figure, a solar panel (SP) is connected to a battery via a DC-DC controller. The battery ensures smooth and regulated delivery of power to the AC and DC loads.

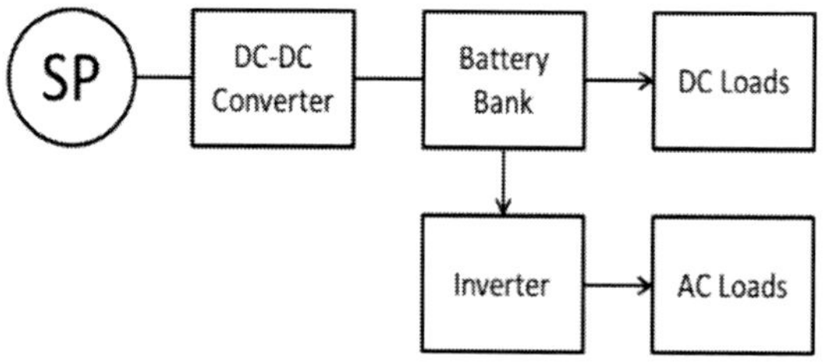

Figure 1. A general block diagram illustrating the use of solar panels for supporting AC and DC loads.

DOI: 10.1051/matecconf/20167012003

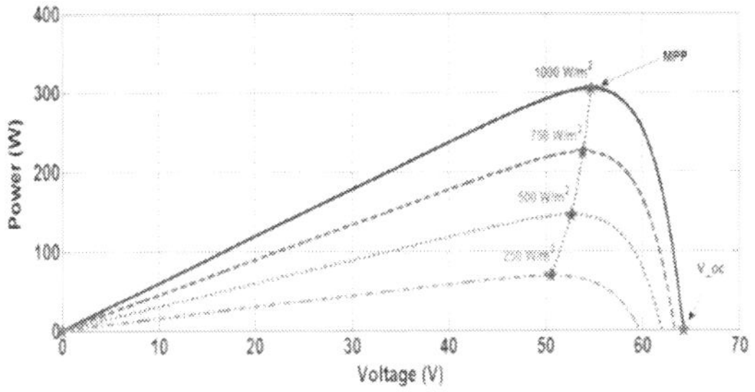

Figure 2. Typical PV curves for different values of irradiance.

The controller maintains the operating conditions so that the extracted power can be maximized in a 305W PV panel capable of supplying 3050Wh load. This PV system is further supported with 2×100 AH Li-ion battery of 12V each in series resulting in 24V. The depth of discharge is 70%. At inverter, PV system is capable of supplying 220V AC at 60Hz besides DC. This system can-be used in both off-grid and on-grid applications. The Power-Voltage graphs for different values of irradiance are shown in Figure 2. The variable irradiance pattern is taken from [8] and is shown in Figure 3. Note that in order to extract maximum power from the solar cells, voltage level is maximize and thus the optimum voltage is called MPP voltage.

A number of maximum power point tracking (MPPT) algorithms have been proposed to ensure that a solar panel operates at its MPP. The main objective of this research work is to examine the performance of two common MPPT algorithms in environments where the solar energy does not remain constant. This study is motivated by the fact that all countries across the world do not get a consistent source of power from the sun.

2 Legacy MPPT algorithms

The Perturb and Observe (P&O) algorithm is an MPPT method that searches for MPP by changing the battery voltage [9]. As shown in Figure 1, once connected, the load shall assume the same voltage level as that asserted by the battery. The current drawn by the load, or conversely power extracted from the solar panel, shall therefore depend on the battery voltage. P&O algorithm keeps increasing the battery voltage until the power starts to decrease (see Figure 2). Once a decrease in output power is noticed, the P&O algorithm reduces the battery voltage so as to reach MPP.

The other commonly used MPPT algorithm is the incremental Conductance (IC) algorithm [10]. The IC of a solar panel is defined by the ratio *dI/dV*. The IC makes changes in *dI/dV* and observes another ratio *dP/dV*. The ratio *dP/dV* represents the slope. The IC continues to increase *dI/dV* as long as a positive change in *dP/dV* is observed. It decreases *dI/dV* when the slope *dP/dV* becomes negative. Thus, the IC method determines the MPP using successive iterations. A comparative study on various MPPT techniques is done by Subudhi and Pradhan [11]. Among various techniques, P&O and IC methods are widely used in commercial sectors.

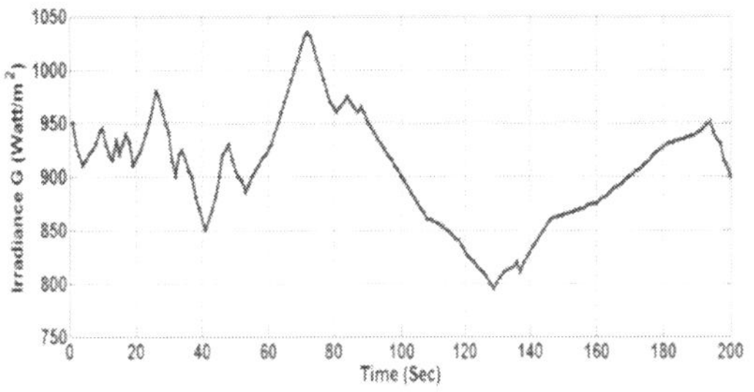

Figure 3. A 200-sec interval has been selected as irradiance pattern from [8].

However under highly variable and shady irradiation conditions, P&O do-not perform well due to causing oscillation between two voltage levels close to MPP.

Several notable works on MPPT algorithms have been done over the past some time. For example, Sera et al [12] have introduced the dP-P&O technique, which drives actual change in the power due to the perturbation. Likewise, Jiang et al [13] have proposed an MPPT method with an adaptive step-size and adaptive-perturbation-frequency algorithm. Their method continuously modulates the perturbation step size and the interval between any two perturbations while observing the load current. A faster convergence speed has been observed as a consequence.

As a new paradigm to track the maximum power points in multiple maxima conditions that arises due to variable irradiation, artificially intelligent techniques for MPPT have been proposed in literature [14]. Artificial neural network (ANN) has ability to recognize and estimate unknown parameters. Thus ANN is an ideal tool to use in MPP tracking especially in variable radiation conditions [15]. Lian et al [16] have proposed hybrid method based on particle swarm optimization (PSO) and P&O algorithms. An improved PSO method to reduce steady state oscillation in proposed in [17]. A novel PSO algorithm based on voltage window limits has been proposed in [18].

3 Analyses and discussion

This section first describes the PV models used in the simulations and then explains the performance of P&O and IC algorithms.

3.1. PV models

MATLAB based PV models for P&O and IC have been used in study and are given in [19, 20] for different system capacities. The irradiation pattern in Figure 3 is used to determine the fluctuations in power output of the two MPPT algorithms.

3.2 Output power

Figure 4 shows the power output from the solar panel of 100KW and 200KW capacity when IC and P&O algorithms are used.

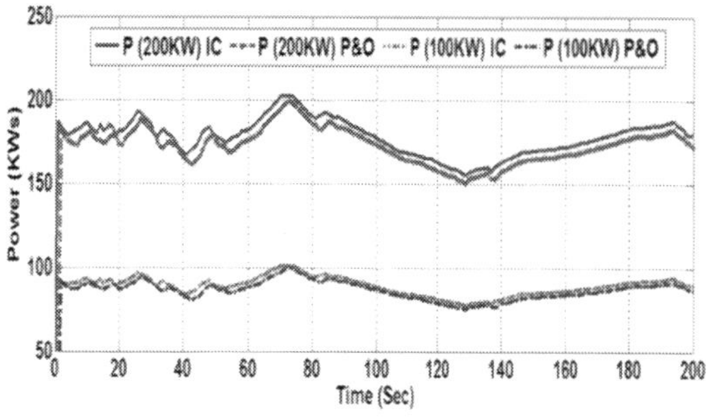

Figure 4. Fluctuations in power output for IC and P&O algorithms.

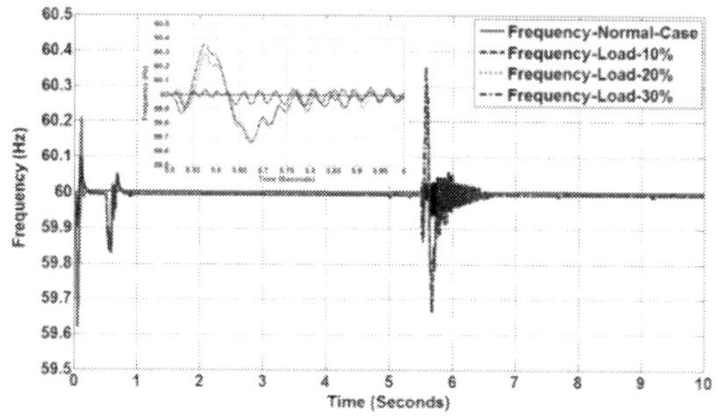

Figure 5. Frequency response of IC method when load is increased.

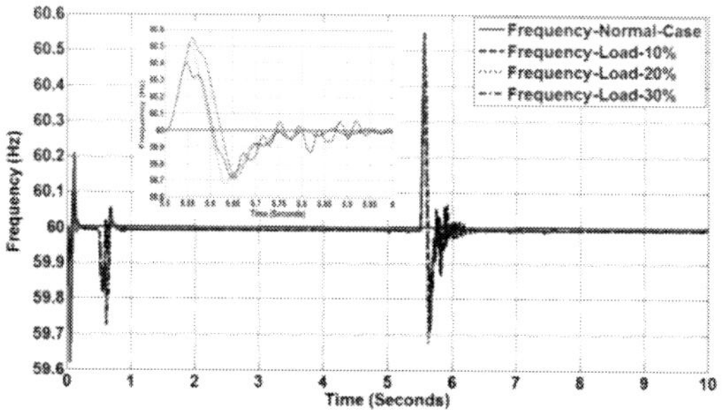

Figure 6. Frequency response of P&O method when load is increased.

It is obvious that the fluctuations in the power output relate directly with the variations in the irradiation pattern (see Figure 3). Note that the magnitude of 100KW power output from IC and P&O algorithms is not quite different. It is interesting to note from Figure 4 that the response time of IC is slightly quicker than P&O. Table 1 shows the standard deviation (Std) and variance (Var) of the power output fluctuations shown in Figure 4. The performance of P&O and IC in 200KW system is also tabulated in Table 1. Regardless of the system capacity, P&O exhibits a much larger Std and Var in the output power. More specifically, the Std in P&O is twice that in IC for both 100KW and 200KW systems. The difference between P&O and IC in-terms of Var is much larger. Table I suggests that the variance in power output in P&O is four times more than the same in IC under 100KW and 200KW systems respectively. Also the power difference for 200KW system is more in P&O than IC. This assessment makes IC more sensible choice in-terms of power regulation.

3.3 Frequency response

The frequency response of a distribution network has been analysed here at a node which houses a DG (REG) unit of 200KW. The DG unit connected here is PV generator that feeds power into the distribution network. This section examines the changes in system frequency (i) when load increases, and (ii) when the main supply is suspended momentarily. The changes in the load and the main supply are made at *t=5.5sec*. The frequency response of the network under these conditions is observed based on IC and P&O algorithms. Figures 5 and 6 show the changes in frequency when the load is increased by 10%, 20% and 30% using IC and P&O methods respectively.

Table 1. Std and Var values of PV systems (100KW / 200KW).

Algorithm	Capacity	Std	Var
IC	100KW	8.234	67.807
	200KW	10.068	101.366
P&O	100KW	16.469	271.225
	200KW	20.136	405.464

Both figures also show a magnified image of the time instants during which the load change is applied. It can be seen from the figures that both IC and P&O methods respond to increase in load in similar fashion. However note that unlike the IC method, P&O increases the grid frequency beyond the tolerable range *(±0.5Hz)* at the time the load is increased *(t=5.5sec)*. On the other hand, the fluctuation in frequency dies out slightly sooner in P&O in comparison with the IC method. Figures 7 and 8 show the changes in frequency when the main supply is suspended at t=5.5sec for IC and P&O methods respectively. The frequency spike in both methods is well beyond the tolerable range at the time a deliberate fault is introduced. Again, P&O algorithm exhibits a higher rise in the frequency. It is interesting to note that unlike figures 5 and 6, the frequency fluctuations in the IC method die out sooner than P&O method.

4 Conclusions

The paper examines two common MPPT algorithms in-terms of output power and operating frequency. Power and frequency are important parameters within a power system that can significantly affect its performance. It has been observed in the paper that the IC and P&O methods respond the same way to the fluctuations in the irradiance pattern. It has been shown that IC method can result in slightly increased power output.

We have also shown that the IC method performs better than the P&O method in keeping the grid frequency under the tolerable threshold.

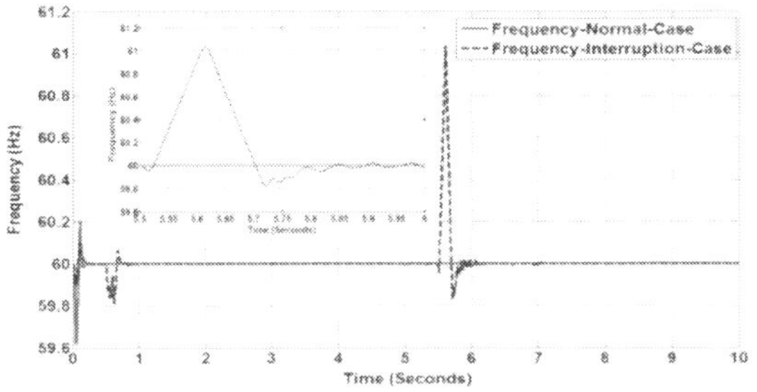

Figure 7. Frequency response of IC method under loss of mains condition.

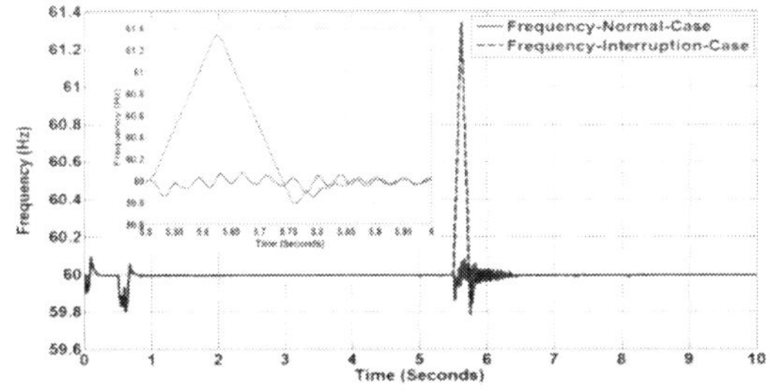

Figure 8. Frequency response of P&O method under loss of mains condition.

Acknowledgement

Funded through Massey University Research Fund 2016.

References

1. S. A. Rahman, R. K. Varma and T. Vanderheide, "Generalised model of a photovoltaic panel", IET Renewable Power Generation, 8(3), (2014)
2. S. A. Kazmi, S. T. Shah, D. R. Shin and S. F. Hasan, "Circuit Analysis Approach for determining Voltage Stability Index", IEEE PEDS, (2015)
3. T. L. Vandoorn, J. D. M. De Kooning, B. Meersman, J. M. Guerrero and L. Vandevelde, "Automatic Power-Sharing Modification of P/V Droop Controllers in Low-Voltage Resistive Microgrids", IEEE Transactions on Power Delivery, 27(4), (2012)
4. H. Han, Y. Liu, Y. Sun, M. Su and J. M. Guerrero, "An Improved Droop Control Strategy for Reactive Power Sharing in Islanded Microgrid", IEEE Transactions on Power Electronics, 30(6), (2015)
5. G. M. Dousoky, A. -H. M. El-Sayed and M. Shoyama, "Increasing energy efficiency in solar radiation trackers for photovoltaic arrays", IEEE Energy Conversion Congress and Exposition, (2012)
6. Y. S. Khoo, A. Nobre, R. Malhotra, Y. Dazhi, R. Ruther, T. Reindl and A. G. Aberle, "Optimal Orientation and Tilt Angle for Maximizing in- Plane Solar Irradiation for PV Applications in Singapore", IEEE Journal of Photovoltaics, 4(2), (2014)
7. C. Konstantopoulos and E. Koutroulis, "Global Maximum Power Point Tracking of Flexible Photovoltaic Modules", IEEE Transactions on Power Electronics, 29(6), (2014)
8. M.Datta, T.Senjyu, A.Yona, T.Funabashi and C.H Kim,"A Frequency-Control Approach by Photovoltaic Generator in a PV-Diesel Hybrid Power System", IEEE Trans on Energy Conversion, (2011)
9. B. Bendib, H. Belmili and F. Krim, "A survey of the most used MPPT methods: Conventional and advanced algorithms applied for photovoltaics systems", Elsevier, Renewable and Sustainable Energy Reviews 45(1), (2015)
10. T. Esram, and P. L. Chapman, "Comparison of Photovoltaic Array Maximum Power Point Tracking Techniques", IEEE Transactions on Energy Conversion, 22(2), (2007).
11. B. Subudhi and R. Pradhaan, "A Comparative study on Maximum Power Point Tracking Techniques for Photovoltaic Power Systems", IEEE Transactions on Sustainable Energy, 4(1), (2013)
12. D. Sera, T. Kerekes, R. Teodorescu and F. Blaabjerg, "Improved MPPT Algorithms for Rapidly Changing Environmental Conditions", 12th International on Power Electronics and Motion Control Conférence, (2006)
13. Y. Jiang and J. A. A. Qahouq, "Adaptive Step Size with Adaptive-Perturbation Frequency Digital MPPT Controller for a Single-Sensor Photovoltaic Solar System", IEEE Transactions on Power Electronics, 28(7), (2013)
14. L. M. Elobaid, A. K. Abdelsalam and E. E. Zakzouk, "Artificial neural network-based photovoltaic maximum power point tracking techniques: a survey", IET Renewable Power Generation, (2015).
15. K. Ishaque and Z. Salam, "A review of maximum power point tracking techniques of PV system for uniform insolation and partial shading condition", Renewable and Sustainable Energy, Elsevier, 19(1), (2013)
16. K. L. Lian, J. H. Jhang and I. S. Tian, "A Maximum Power Point Tracking Method Based on Perturb-and-Observe Combined With Particle Swarm Optimization", IEEE Journal Of Photovoltaics, 4(2), (2014)
17. K. Ishaque, Z. Salam, M. Amjad and S. Mekhilef, "An Improved Particle Swarm Optimization (PSO)–Based MPPT for PV With Reduced Steady-State Oscillation", IEEE Transactions On Power Electronics, 27(8), (2012)
18. Z. Qing, Z. Xing, L. Chun, L. S. Shou and C. Xiaojing, "An Improved Particle Swarm Optimization MPPT algorithm Based on Voltage Window Restrictions", 17th International Conference on Electrical Machines and Systems (ICEMS), (2014)
19. Simulink-based PV model in MATLAB. Available online:http://au.mathworks.com/help/physmod/sps/examples/average-model-of-a-100-kw-grid-connected-pv-array.html.
20. P. Giroux, "Grid-Connected PV Array", MATLAB Central", avalable online: http://kr.mathworks.com/matlabcentral/fileexchange/34752-grid-connected-pv-array.

DOI: 10.1051/matecconf/20167012004

Modelling of a CFD Microscale Model and Its Application in Wind Energy Resource Assessment

Jie-shun Yue[1,a], Song-ping Wu[1,2] and Fei-shi Xu[3]

[1]*School of Aeronautical Science and Engineering, Beihang University, 100083, Beijing, P. R. China*
[2]*National computational fluid dynamics laboratory of China, 100083, Beijing, P. R. China*
[3]*Sino-French Engineering School, Beihang University, 100083, Beijing, P. R. China*

Abstract. The prediction of a wind farm near the wind turbines has a significant effect on the safety as well as economy of wind power generation. To assess the wind resource distribution within a complex terrain, a computational fluid dynamics (CFD) based wind farm forecast microscale model is developed. The model uses the Reynolds Averaged Navier-Stokes (RANS) model to characterize the turbulence. By using the results of Weather Research and Forecasting (WRF) mesoscale weather forecast model as the input of the CFD model, a coupled model of CFD-WRF is established. A special method is used for the treatment of the information interchange on the lateral boundary between two models. This established coupled model is applied in predicting the wind farm near a wind turbine in Hong Gang-zi, Jilin, China. The results from this simulation are compared to real measured data. On this basis, the accuracy and efficiency of turbulence characterization schemes are discussed. It indicates that this coupling system is easy to implement and can make these two separate models work in parallel. The CFD model coupled with WRF has the advantage of high accuracy and fast speed, which makes it valid for the wind power generation.

1 Introduction

Energy is one of the most valuable consumables of human society. With energy shortage becoming obvious, the demand for renewable energy is rapidly increasing. Among these available new energy resources is wind power which has widely been used during recent years. However, drastic fluctuations of wind power generation being caused by random wind variation will result in a serious strike on the power grid [1]. The reliable and accurate prediction of wind farm near the wind turbines can effectively improve the safety and economy for wind power generation. Currently, the most general method for wind farm prediction is using the mesoscale weather forecast models, e.g. WRF, MM5 [2]. The demand for a more precise prediction of atmospheric processes calls for the improvement of the resolution of model and grid. But it will cause extreme increase of computation and time expense if the fine mesh model is used with higher resolution in the whole prediction area and consequently making the model unable to meet the requirements of fast prediction. Therefore, the solution is applying the microscale model in the concerned local region on the background of mesoscale meteorological model. This method is called the coupling model method. Models developed on computational fluid dynamics (CFD) method can be used to simulate the local wind field on a fine grid. Several meteorological CFD models have been founded, such as WindSim of Norway and Meteodyn WT of France [3][4].

The coupling method is derived from the idea of the regional model developed by Richardson [5]. The advantage of the coupling method is that the two models that are based on the coarse grid and the fine grid respectively can be independently programmed and launched. Between the two models it only needs to build a reasonable coupling scheme to achieve the exchange of information, thus helping to improve the efficiency and accuracy of prediction. In recent years, along with the air pollution problem becoming the international issue, the method coupling the mesoscale model and microscale CFD model has widely been applied in the simulation of urban atmospheric flow and environmental pollution flow. M. Tewari [6] calculated the wind speed and direction above the Salt Lake City via coupling WRF and CFD-Urban method. Wyszogrodzki [7] simulated the pollutant dispersion of Oklahoma City by coupling WRF model with a CFD-LES solver. Kwak [8] did the air quality simulation in a high-raise building area of Seoul by coupling a CFD model with mesoscale meteorological and chemistry-transport model. In the field of wind energy, Katurji [9] simulated the turbulent flow of complex terrain by coupling the WRF model and WindSim model. Gopalan [10] compared several different CFD models by coupling them with the WRF model. The results showed that the CFD model can refine

[a] Corresponding author: buaayjs@buaa.edu.cn

the flow structure and simulate the unsteady variation of the vortex in the wind field. Emeis [11] suggested that turbulence, thermal convection and surface-induced secondary circulations could be the three main challenges to mesoscale–microscale models.

The speed of wind is naturally much lower than the speed of sound. Assuming that the wind velocity does not influence the density of flow, the simulation will adopt the incompressible flow control equations, so that the pressure and velocity can be solved separately [12]. For the wind field prediction, it's more suitable to use the compressible flow control equations due to the severe changes in air density. Yet in order to avoid the apparent decrease of convergence rate and accuracy, i.e. 'stiffness' problems, while solving the compressible equations, the preconditioning for equations is necessary. The preconditioning method has been well developed and been widely used in solving a variety of low speed flow problems [13][14].

Taking the advantage of CFD's ability of accurately simulating the flow details and using the preconditioning method for the compressible Navier-Stokes equation of the low speed flow, this paper developed a fine mesh method for local wind field prediction. A coupling process of the CFD model and the mesoscale weather forecast model WRF is established. The coupled model carries out a refinement on the coarse mesh in WRF. A special method for information exchange on the lateral boundary of two models is adopted. The wind speed near Hong Gang-zi in Jilin, China wind power station is calculated as an application of this coupled model. The results are compared with the results generated by WRF and the measured data of wind towers.

2 Method: Forecast model and coupling process

2.1 Dynamics equation

WRF model uses the dynamics equations built in the p-σ coordinates, the output results will be transformed into spherical coordinates of the earth. Due to the small computational area of the regional wind field, the curvature of the earth can be neglected, and thus the CFD model can be established in local Cartesian coordinate. Considering the external force, the dynamics equations of the compressible NS equations is as follows:

$$\frac{\partial \mathbf{q}}{\partial t} + \nabla \cdot \mathbf{F} = \nabla \cdot \mathbf{F}_v + \mathbf{s} \tag{1}$$

where $\mathbf{q} = [\rho, \rho u, \rho v, \rho w, E]^T$ is the conserved variables, $\mathbf{s} = [0, f\rho v, -f\rho u, -\rho g, 0]^T$ is the source term, $f = 2\Omega \sin\varphi$ is the Coriolis force and g is gravitational acceleration.

If a reference pressure p_0 is specified, the difference between the pressure of a certain point and the reference pressure, is called the perturbation pressure, denoted as $p' = p - p_0$. At this time, the state equation can be written as:

$$p = p' + p_0 = \rho R T \tag{2}$$

Among dozens of parameterize schemes of turbulence, two Reynolds-averaged Navier–Stokes turbulence models are considered. One is B-L model. This model is an algebra model, which does not need to introduce extra control equations as well as costs small amount of calculation. It especially suits for the simulation of near wall turbulent flow [15]. The other approach to parameterize the turbulence is using the k-ω SST model. This model solves two dynamic equations, one is the equation for turbulence kinetic energy, k, the other is for turbulence dissipation rate, ω. This model is more complex than the B-L model, but it is suitable for all kinds of turbulence including near wall flow and shear flow [16].

2.2 CFD discretization scheme based on the preconditioning method

In the case of low speed flow, the NS equations will appear the 'stiffness' problem of slow convergence rate and poor accuracy. In order to overcome this problem, the preconditioning for the equations is necessary. For the change of wind speed within the 24h-72h, it needs to adopt a dual time step algorithm for calculating the unsteady time integration. A pseudo time τ is introduced and the preconditioning only plays role on the pseudo time term.

According to the preconditioning algorithms, the prime variable is $\mathbf{Q} = [p, u, v, w, T]^T$ and the preconditioning matrix is [17]:

$$\Gamma = \begin{pmatrix} \frac{1}{\varepsilon RT} & 0 & 0 & 0 & -\frac{\rho}{T} \\ \frac{u}{\varepsilon RT} & \rho & 0 & 0 & -\frac{\rho u}{T} \\ \frac{v}{\varepsilon RT} & 0 & \rho & 0 & -\frac{\rho v}{T} \\ \frac{w}{\varepsilon RT} & 0 & 0 & \rho & -\frac{\rho w}{T} \\ \frac{H}{\varepsilon RT} - 1 & \rho u & \rho v & \rho w & \rho c_p - \frac{\rho H}{T} \end{pmatrix} \tag{3}$$

where the preconditioning parameter is $\varepsilon = M_r^2$, $M_r = \min(1, \max(M, M_0))$, while M and M_0 are the local Mach number and reference Mach number, respectively.

The control equation after preconditioning is:

$$\Gamma \frac{\partial \mathbf{Q}}{\partial \tau} + \frac{\partial \mathbf{q}}{\partial t} + \nabla \cdot \mathbf{F} = \nabla \cdot \mathbf{F}_v + \mathbf{s} \tag{4}$$

During the numerical solution of the above equations, LUSGS algorithm is used for the inner iteration along the pseudo time. The backward Euler implicit scheme is used for the outer iteration of physical time, where the convection term using implicit processing and the viscous term and the source terms using explicit processing. Inviscid flux is discretized by the Roe scheme.

2.3 The coupling of coarse mesh and fine mesh model

The output file format of mesoscale prediction model WRF is NetCDF, which contains geographic coordinates of the coarse grid and physical quantities such as velocity, pressure and humidity. In this coupling model, the data of time, geographic coordinates, velocity, pressure and temperature are extracted as the required information for CFD computation. All the extracted data is stored in single files in time series. Along with the time advance, the program regularly read the file of each time, as the input to update the computational boundary condition. The working process of the coupling model is shown in Fig. 1:

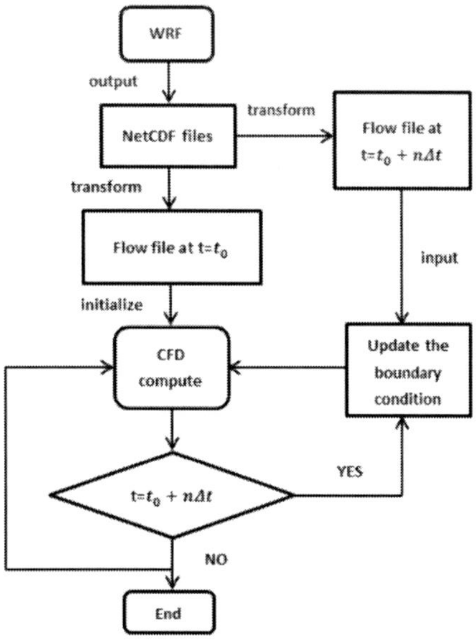

Figure 1. Working process of the coupling model.

The horizontal grid scale in the WRF model is thousands of meters, while vertical hundreds of meters. The coupling model chooses several pieces of coarse grid corresponding to the concerned local wind field to be the refining region. The refinement decreases the mesh size by one order of magnitude. Physical quantities on fine mesh point are obtained by interpolating the values on coarse mesh, of which the interpolated result in the initial time is regarded as the initial flow field for CFD simulation. The interpolated results in the follow-up moment are used as boundary conditions, to achieve the information exchange between the two models.

There are two methods to deal with boundary conditions. One is to directly use interpolated data to update the physical quantities on the boundary of the CFD calculation, which is called the fixed boundary condition. In specific, if the boundary is the flow entrance, by given the velocity and density, the temperature is obtained from interpolation of the inner field, while pressure is determined by the state equation; if the boundary is outlet, by given the outlet pressure (result from the interpolation), velocity and temperature are obtained from the interpolation of the inner field [18], while density is determined by the state equation.

To reduce the horizontal gradient of variable near the boundary, another method applied is called the transition zone method [19]. In this method, several layers of grid, i.e. the transition zone, should be added outside the computational area. In the outermost lateral of the transition zone, the fixed boundary condition is applied, while in the innermost lateral the result from CFD calculation is preserved. For the transition zone, a transition function is used for linking these two parts.

The time step of CFD is smaller than the time interval of WRF data. Therefore, during the calculation of the CFD model, at each physical time step, it needs to do the interpolation for WRF data between two adjacent moments.

3 Numerical forecast design

The wind field near a wind tower in Hong Gangzi of Jilin, China is selected for the wind velocity prediction. The average altitude of this area is about 130m. The horizontal resolution of WRF is 5km. 4 coarse grids around the wind tower are chosen with the total area for CFD model becoming 10km×10km. As is seen from the computational area for the coupling model in Figure 2, the red dot is the wind tower location, while the selected grid for refinement is within the dashed-line box. Each coarse grid is divided into 20 × 20 fine mesh, namely a horizontal resolution of 250m for CFD model. In the vertical direction, the upper boundary is taken as 1km and the height of the first layer from the ground is 2 meters. The coordinates of the wind tower are 123.892 east longitude and 45.549 north latitude, in which 4 measuring points arranged at the height of 10m, 30m, 50m and 70m, respectively.

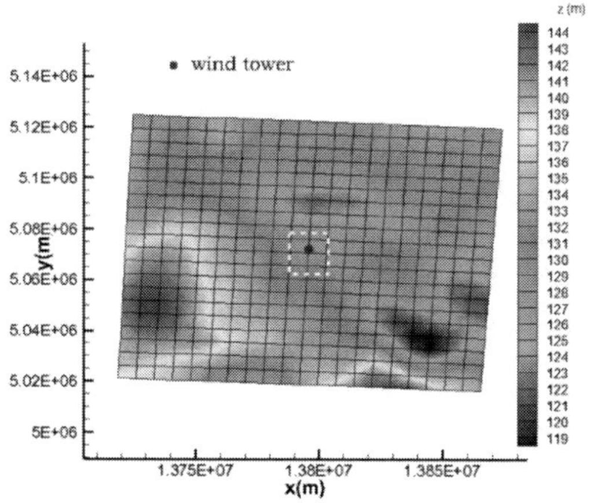

Figure 2. Computational area for the coupling model.

The whole period for a prediction is 72h, i.e. 3 days. WRF provides a data every 15 minutes. Hence the boundary values need to be updated every 15 minutes. A new forecast starts every 24h according to the data sequence of WRF. 12:00 is chosen as the starting time for a 72h forecast. The prediction process lasts from July 24th to July 29th, 2012, with a total of 6 time series (Table 1).

For the time integration scheme, the dual-time step method is adopted. The iterative process includes the inner iteration of pseudo time and the outer iteration of real physical time marching. In order to improve the efficiency and ensure the high time accuracy, the time step of outer iteration is 20s, and the maximum step of inner iteration is 20.

Table 1. Time series.

Serial number	Starting time	Serial number	Starting time
1	7/24 12:00	2	7/25 12:00
3	7/26 12:00	4	7/27 12:00
5	7/28 12:00	6	7/29 12:00

4 Results and analysis

4.1 Results

Fig. 3 and Fig. 4 show the instantaneous distribution of wind velocity component u in extracted fine mesh area at the height of about 50 meter. Comparing the WRF and the numerical results of this coupled model, it reveals that, because of the refinement of the grids and improvement of the model's accuracy, CFD model can be able to describe the flow details and characterize the gradient of the wind field variable. This feature makes it particularly suitable for processing in a region with a plurality of wind towers or wind turbines.

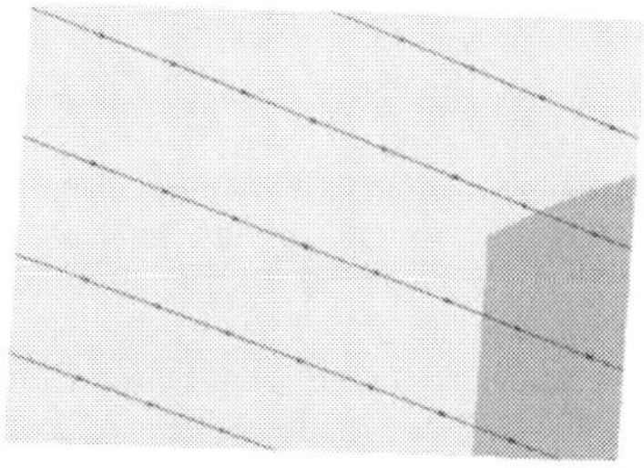

Figure 3. Counter of velocity component u of WRF model.

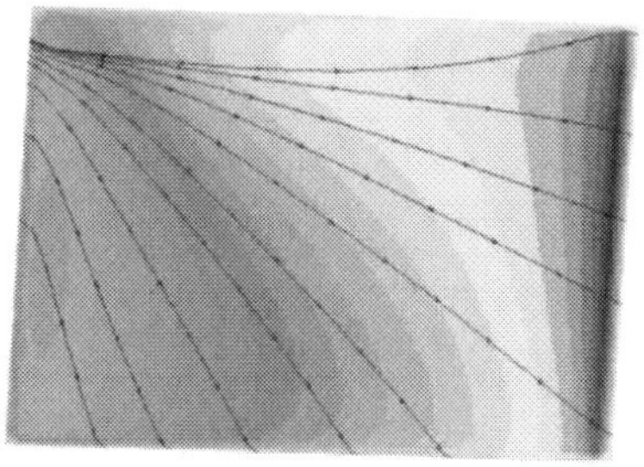

Figure 4. Counter of velocity component u of CFD result.

The result of a set of time series (July 26th 12:00) at the height of 70m is presented in Figure 5 and Figure 6. It can be roughly seen from the figures that, through enhancing the accuracy of the WRF model, the result of the coupling model is more close to the measured curve of wind velocity. On the other hand, the wind direction results provided by WRF and CFD model have almost the same trend.

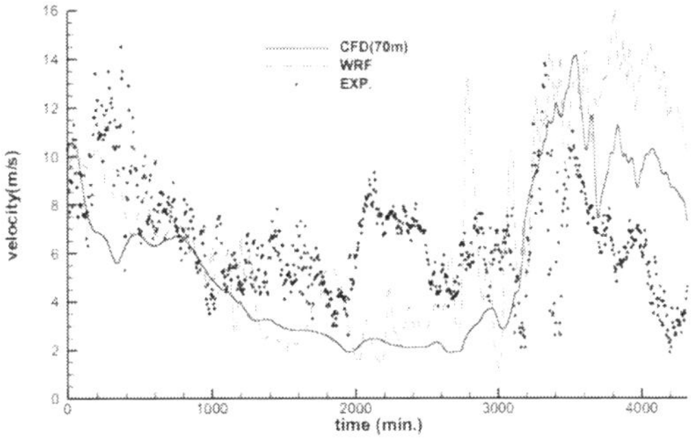

Figure 5. Wind velocity curve of 7/26 12:00 at 70m.

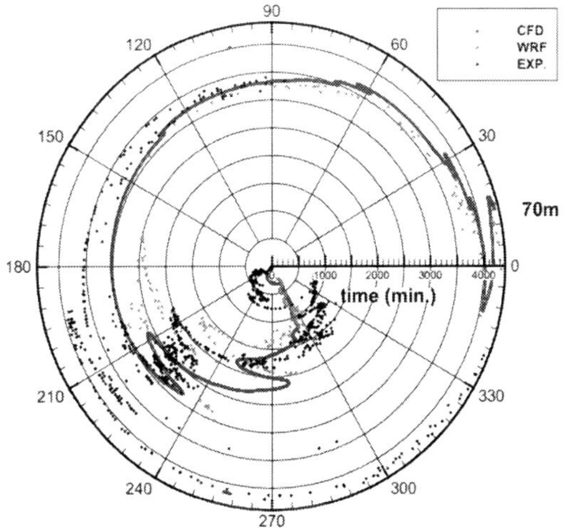

Figure 6. Wind direction curve of 7/26 12:00 at 70m.

Due to the results of WRF are used as the initial field and the boundary condition for CFD calculations. The results of CFD and WRF have strong similarity. It can be named as 'the following property' of the fine mesh model for the coarse mesh model. Because the WRF model is the only input of the CFD model, the wind velocity of two models increases or decreases similarly with the inflection point appearing in different positions. In fact, the CFD model corrects the position of the inflection point, so that the shape of the wind speed curve is closer to the measured data.

CFD simulation has a certain time lag effect relative to WRF data, namely 'the lag property' of fine mesh model to the coarse mesh model. That is because in each time step, the data of WRF is just input as the boundary condition for CFD calculation, which means the changes of the flow field will take time to spread from the boundary to interior of the wind field.

The CFD simulation has a certain 'smoothing' effect on the wind velocity curve, making the temporal variation of wind velocity not so intense. Results of WRF vibrate acutely at some point, seriously deviating from the measured data. CFD model can avoid wind velocity to appear large fluctuation, which improves the calculation results of WRF. By modifying the calculation time step, the 'smoothing' effect can be increased or decreased, to achieve the resolution adjustment according to the actual situation and the requirements.

4.2 Parallel programming and time consumption

Accuracy of CFD simulation requires a big amount of computation time. In order to increase the efficiency of the coupling model and meet the requirement of fast forecast, parallel programming is necessary. The OpenMP multi-thread parallel programming is adopted. By optimizing the loop calculation in the CFD simulation, this method is easy to program and can provide a considerable improvement in efficiency.

Table 2. Computation time of the coupling model.

	Time(s)
Serial	19617
Parallel	15163

The parallel model is running with 6 threads. Tab. 2 reveals the time consumption before and after the parallel is adopted. The parallel gives the program an acceleration of more than 20%.

4.3 Error analysis

In order to evaluate the optimization effect of CFD model, the average error and the correlation coefficient between the WRF and CFD results are compared with the measured data of each time series. The average error indicates the mean value of the error between computational results and measured data, while the correlation coefficient indicates the degree of the linear correlation between computational results and measured data. Computing the 6 sets of time series shown in Table 1 with B-L model, the error analysis at the height of 10m and 50m is carried out and the results are shown in Fig. 7 and Fig. 8, respectively.

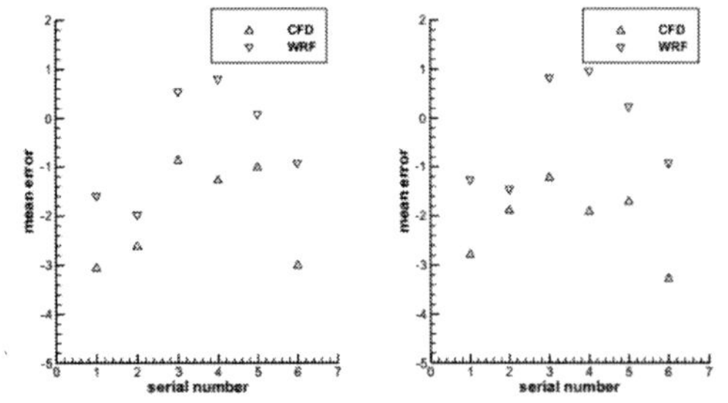

(a) (b)

Figure 7. Mean error (a) 10m (b) 50m.

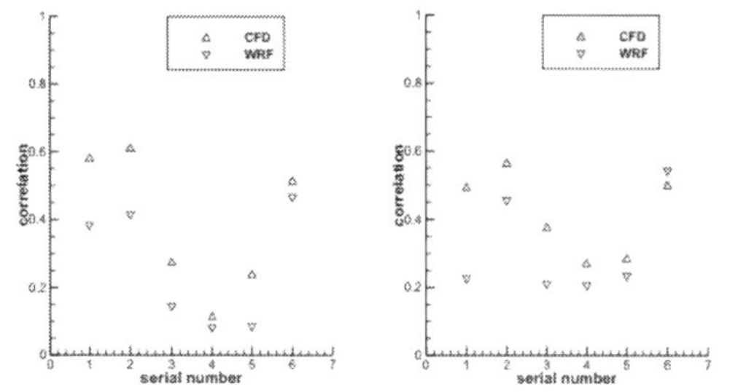

(a) (b)

Figure 8. Correlation coefficient (a) 10m (b) 50m.

From Fig. 7 it can be seen that, the coupled CFD model has a larger mean value than the WRF model, whereas the values are minus. It shows that in general, the wind velocity calculated by CFD is smaller than that by WRF. The near wall boundary layer effect decreases the wind speed magnitude. Therefore, the extremely high wind speed deriving from WRF can be avoided. Figure 8 shows that the coupled CFD model has lager correlation coefficient than the WRF model, which means the curve computed by the CFD model is closer to the measured data. Overall, the result of coupling model is more sophisticated, closer to the measured wind velocity and has optimized the forecast results of WRF.

4.4 Comparison of turbulence models

Two models are chosen for the parameterization of turbulence, one the algebraic B-L model, the other is two-equation k-ω SST model. Theoretically, the algebraic operation of B-L model is efficient and fast, and it is suitable for near wall flow. However, it is not suitable for the wake flow, the shear flow, etc. SST model is a relatively well-developed two-equation model, which more accurately calculates the adverse pressure gradient and separated flow. Yet solving two extra equations, computational efficiency of this model is lower than the algebraic model. The model has to take into account the efficiency and accuracy of wind prediction. Thus the evaluation of these two models is very important. Table 3 is the comparison of the two models in the aspect of calculation time.

Table 3. Computation time of two turbulent models.

	Time(s)
B-L	15163
k-ω SST	16386

The results show that for short-term forecasting, both models can achieve completing 72 hours' forecasts within 5 hours. Since the turbulence equations are not a major part of the calculation, and the parallel programming methods also have a significant impact on the calculation speed, B-L mode computing speed is not much faster than the k-ω SST model. From this point of view, by introducing of parallel computing, slow shortcomings of two-equation k-ω SST model calculations can be made up.

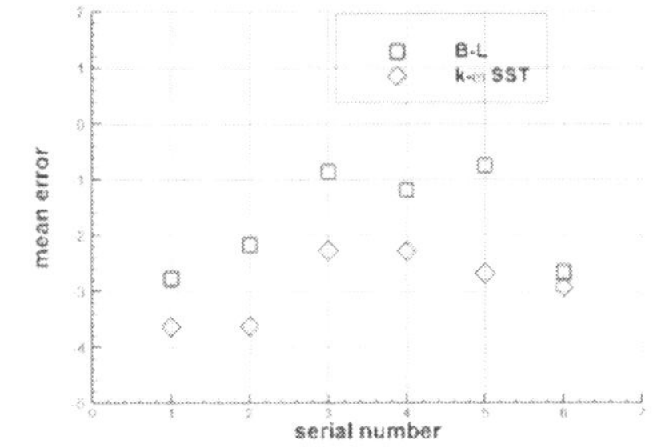

Figure 9. Mean error of different turbulent models.

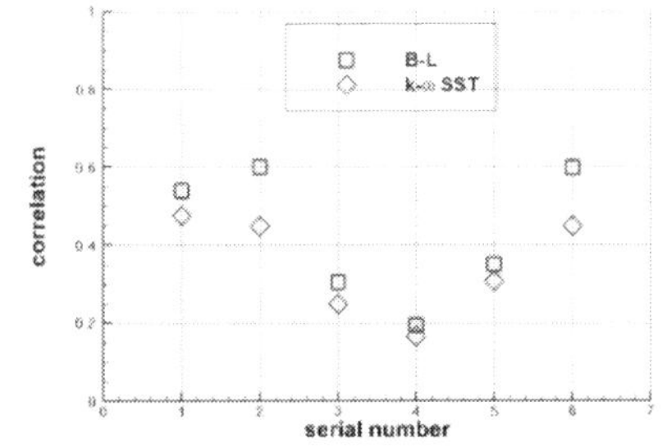

Figure 10. Correlation coefficient of different turbulent models.

Fig. 9 and Fig. 10 further compare the average error and the correlation coefficient of B-L model and k- ω SST model. It can be seen from the figures that, compared to the k-ω SST model, the B-L model is more accurate. B-L model is an algebraic model, so in each calculation step it directly gives the exact value of the turbulent viscosity coefficient. While in k-ω SST model the accurate obtaining of turbulent viscosity coefficient requires a lot of iteration, and the excessive number of iteration will reduce the efficiency. Thus, considering the computational efficiency and accuracy, for this coupling mode, B-L model is a better choice. However, the terrain of this forecast is relatively flat. Whether k-ω SST model will perform better in the undulating terrain with mountains still needs to be verified.

5 Conclusion

The author has established a CFD model suitable for microscale wind field simulation. This model uses the preconditioning technique to solve the compressible NS equations, parameterizes the turbulent viscosity coefficient with the BL model and k-ω SST model. On this basis, by coupling the CFD model and WRF model, a nested model which can realize the fine forecast for local wind field is developed. The coupled model is successfully applied in the wind field prediction near a wind tower in Hong Gang-zi, Jilin, China. The prediction results are compared with that of the WRF model and the measured data.

The CFD model can be quickly coupled with the WRF model, and can carry on the delicate simulation of the local wind field, to describe the wind field details. According to the results of a series of time sequence, as the fine grid model, the CFD model can effectively augment the prediction precision of the WRF model. The wind velocity curve of the coupled model is overall close to that of WRF, presenting 'the follow property', 'the lag property' and 'smoothing' effect. These characteristics of CFD model modify the inflection point and smoothness of WRF model curve, thus make it closer to the measured data. In addition, as can be seen from the error analysis, CFD model can significantly reduce the calculation error of the WRF model.

The established CFD and WRF coupling model can undertake delicate simulation for the wind field above local complex terrains and can be used for short-term wind velocity forecast. In order to improve the accuracy of CFD model and more accurately describe the velocity fluctuations, the next research direction will be to reduce the coupling degree, optimize the iterative scheme of time integration, to optimize the turbulence parameterization schemes, and to consider the land surface process.

References

1. M. Jannatia, S. H. Hosseiniana, B. Vahidi, Renew. Sust. Energ. Rev. **29** 158-127(2014)
2. A. Gsella, A. de Meij, A. Kerschbaumer et al, Atmos. Environ. **89** 797-806(2014)
3. S. Milashuk, W. Crane, Environ. Mod. Soft. **26** 429-433(2011)
4. J. Manning, P. Hancock, R. Whiting, Wind Eng. **5** 477-500(2010)
5. L. Richardson, *Weather Prediction by Numerical Process*(London: Cambridge University Press, 1922)
6. M. Tewari, H. Kusaka, F. Chen et al, Atmos. Res. **96** 656-664(2011)
7. A. Wyszogrodzki, S. Miao, F. Chen. Atmos. Res. **118** 324-345(2012)
8. K. Kwak, J. Baik, Y. Ryu et al, Atmos. Environ. **100** 167-177(2015)
9. M. Katurji, (Ph. D. thesis, University of Canterbury, (2011)
10. H. Gopalan, C. Gundling, K. Brown, J. Wind Eng. Ind. Aerod. **132** 13-26 (2014)
11. S. Emeis, J. Wind Eng. and Ind. Aerod. **144** 24-30(2015)
12. C. Bruneau, M. Saad, Comput. Fluids, **35** 326–348(2006)
13. E. Turkel, Appl. Numer. Math. **12** 257-284(1993)
14. Y. Colin, H. Deniau, J. Boussuge, Comput. Fluids, **47** 1-15(2011)
15. B. Baldwin, H. Lomax, Aerospace Sciences Meeting, 16th(1978)
16. F. Menter, I. Rumser, 25th AIAA Fluid Dynamics Conference, Colorado(1994)
17. C. Merkle, P. Buelow, J. Sullivan et al, AIAA J. **36** 515–521(1998)
18. E. Turkel, Ann. Rev. Fl. Mech. **31** 385-416(1999)
19. A. McDoland, PINSA, **65** 91-105(1999)

Biofuel Production by Fermentation of Water Plants and Agricultural Lignocellulosic by-Products

Yaakov Anker [1], Faina Nakonechny[2], Betty Niazov[2], Svetlana Lugovskoy[2] and Marina Nisnevitch[2]

[1] Environmental Research, Regional R&D Center, Ariel University, Israel
[2]Department of Chemical Engineering, Biotechnology and Materials, Ariel University, Israel

Abstract. While at present most energy crops are depriving human feedstock, fermentation of agricultural residues and fast growing water plants possesses a good prospect to become a significant source for bio-fuel; as both substrates are widely available and do not require agricultural areas. Water hyacinth for instance can be cultivated in fresh, brackish or wastewater and owing to its rapid growth and availability. Since owing to its natural abundance it is considered to be an invasive plant in most continents, its utilization and use as a renewable energy source may also contribute for its dilution and control. Agricultural lignocellulosic surplus by-products are also a promising fermentable substrate for bioethanol production, as it decreases both disposal expenses and greenhouse gases emissions. This paper describes a scheme and methodology for transformation of any lignocellulosic biomass into biofuel by simple cost effective operation scheme, integrating an innovative process of mechanochemical activation pre-treatment followed by fermentation of the herbal digest and ethanol production through differential distillation. Under this approach several complex and costly staged of conventional ethanol production scheme may be replaced and by genetic engineering of custom fermenting microorganisms the fermentation process becomes a fully continuous industrial process.

1 Introduction

Despite the development of new fossil fuels recovery techniques, reducing production costs; ethanol production is increasing, owing to its positive environmental impact and economic benefits. The notion of ethanol advantages over fossil fuels have led to implementation programs in 64 countries, whereas in the USA and Brazil its portion is already greater than 10% (Fig. 1) and in the EU the 10% bar is scheduled to 2020 [www.afdc.energy.gov/data/].

World production of ethanol as transportion fuel increased between 2000 and 2007 from 17 billion to more than 52 billion litres, and in 2011 its production reached 84.6 billion litres [1-2]. Most cars in the U.S.A. can consume blends of gasoline fuel with up to 10% ethanol, and in Brazil the fuel blend consists from 25% ethanol and 75% gasoline. By December 2011 in Brazil there already were 14.8 million dual-fuel vehicles and 1.5 million dual-fuel motorcycles, which regularly were using 100% ethanol fuel [3].

The most common source for biofuel today are sugar-rich feedstock's, including sugarcane, sugar beet and various fruits, which are converted into ethanol with relatively high yield. However, large-scale expansion of these sources is limited by the feedstock prices. The choice of such a feedstock for ethanol production is problematic since the biomass requirements either directly or indirectly compete with food crops for arable land resources. For this reason, production of ethanol from lignocellulosic surplus by-products (e.g. wood residue, sugarcane bagasse, corn fibres, rice straw, cassava waste etc.) has received widespread interest owing to their availability, abundance and relatively low cost [4-6]. A parallel trend is utilization of marginal lands for cultivation of CAM bioenergy crops, such as *Agave* and *Opuntia* [7]. Nevertheless, these of other CAM plants are also considered as possible human feed source and once cultivated these marginal lands are no longer marginal, but rather farmlands.

Aquatic plants belong to another group of promising renewable energy resource. The use of aquatic plants has benefits, such as, an absence of competition with grains and vegetables on arable farmland, low cultivation expenses and the potential of artificial water reservoirs and aeration pools purification. The use of free-floating aquatic plants for self-remediation of hazardous organic compounds and heavy metals from concomitant water is approved by several directives [8].

Water hyacinth (*Eichhornia crassipes*) as an example is a good potential candidate for ethanol feedstock. Its biomass has about 48% hemicellulose, 18% cellulose and 3.5% lignin [8] and regardless composition variability it is considered to be rich in hemicellulose and with low lignin content [10]. In addition, since it is a typical menace infesting the backwaters in many places around the world [11-12], effective technologies for ethanol production from biomass can serve the purpose of aquatic weed removal. Processing biomass into biofuel through the use of high technology and complicated instrumentations may limit their commercialization and industrial applicability, especially for developing countries. Moreover, surplus biomass processing in large scale plants may not be economical owing to the problems in collection, transportation and infrastructural requirements [13]. Localized ethanol production systems may be far more efficient, especially in countries where agriculture is decentralized and the biomass availability is distributed.

Such systems will allow the use of regionally available biomass types, such as agriculture by-products, water hyacinth and other aquatic species and also other types of organic feedstocks. This scenario may also include some economic and social benefits for rural community being less capital intensive and demanding less skilled workforce [14], enabling new mean of employment and income to rural communities.

At present, centralized ethanol production systems are economically obligatory owing to the complex biomass pre-treatment and the enzymatic saccharification (especially the sourcing of enzymes), which pose several technical challenges to distributed localized ethanol industry[15]. These challenges are being addressed in the current paper with the development of a new innovative mechanochemical pre-treatment and in integrative continuous ethanol production operation scheme.

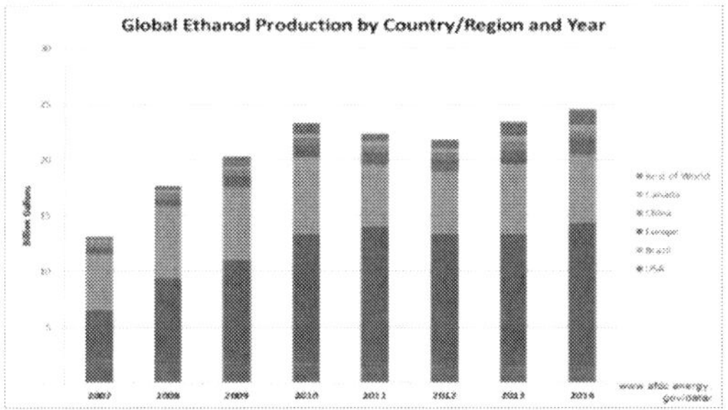

Figure 1. World ethanol production for the years 2004-2014

2 Methodology

An integrative and continuous process was developed for the optimal transformation of water plant biomass and agricultural wastes into ethanol. Following biomass chopping with a rotary blade device a novel mechanochemical fermentation pre-treatment is applied. This treatment replaces most of the costly stages of polysaccharides depletion into monosugars and optimizes the subsequent fermentation process owing to the increase in biomass active surface area [16]. The transformation of the post-treated biomass into ethanol is done by differential microbial fermentation with optimal conditions for fermentation of the monomer sugars into ethanol. At the last stage, an optimal scheme for ethanol recovery and concentration by direct continuous distillation is yet under development.

2.1 Mechanochemical pretreatment

Biomass pre-treatment methods may be classified into different categories, e.g. physical, chemical, biological, and combination of these methods [17]. Physical pre-treatment alters biomass structure by mechanical stress without additional chemical or biological reagent. Lignocellulosic materials are difficult to process in raw condition; owing to the celluloses, hemicelluloses and lignin, strong linked structure [18]. The pre-treatment process should increase biomass active surface area and pore sizes [19] and also reduce cellulose and hemicellulose degree of crystallinity and polymerization ([17], [20]).

Through the physical rendering of the cellulose and hemicellulose rigid structure, the mechanochemical process can increase their solubility and break them into monomers, making the biomass suitable for fermentation. The mechanochemical process includes pressure based water extraction and graining of the biomass, followed by an activation treatment, which converts the plant's polysaccharides into monosaccharides. The main distinction of polysaccharides from (mono-) oligosaccharides is the presence of the glycoside bond between (mono-) oligosaccharides (Fig. 2). Disruption of these bonds leads to production of monosaccharides: glucose in the case of cellulose (Fig. 2a) and, mainly, xylose in the case of hemicelluloses (Fig. 2b).

Agricultural wastes contain mostly, cellulose and aquatic plants biomass mostly, hemicellulose, making them both suitable for mechanochemical pre-treatment. Polysaccharides transformation into monosaccharides increases the sustenance of nutriment and decreases the biofuel production process fermentation time.

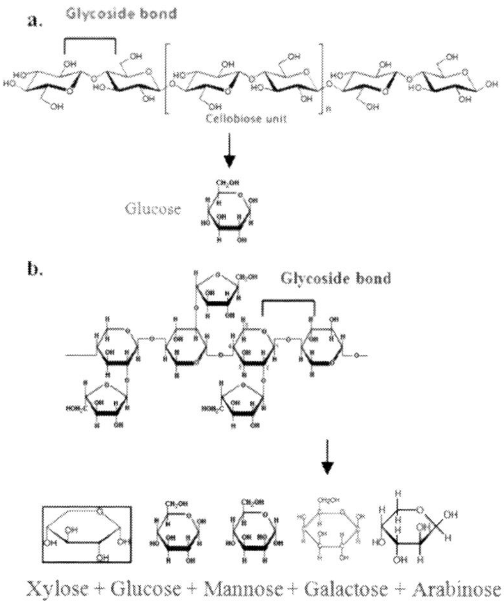

Figure 2. Conceptual scheme of polysaccharides' transformation into monosaccharides: a. cellulose and b. hemicellulose.

2.2 Fermentation process

While traditional fermenting yeast (*S. cerevisiae*) produce ethanol from hexoses, recent genetic engineering has produced microorganisms that ferments pentoses in addition to hexoses into ethanol [21-22]. Several fermenters were tested during the research including: Common baker's yeast *Saccharomyces cerevisiae*, *S. cerevisiae* NBRC 2346, recombinant bacterial strains *E. coli* KO11, *Klebsiella oxytoca*, *Zymomonas mobilis* or a yeast strain *Pichia stipitis* [23].

The fermentation scheme under development is differentially applying specific fermenters in a sequential fermentation process, allowing optimal ethanol recovery from various biomass substrates.

3 Results and discussion

The research consisted of two stages, at the beginning water hyacinth plants from a water plant, cultivation pond (Fig. 3), were processed into homogenous biomass. The plants were harvested and air-dried. The dried plants were divided into three groups: **a.** plants sliced with a knife, **b.** plants roughly shredded and the **c.** group was treated mechanochemically. After that, each group was subjected to two kinds of treatments: acid hydrolysis and heating in aqueous medium in autoclave (121°C, 1 atm, 15 min).

Figure 3. 5x6 m pond (free-flow wetland), used for wastewater sanitation stage of the Ariel University student dormitories, wastewater reclamation project

The plant biomass was separated from the liquids and xylose concentration was measured by spectrophotometric assay (Fig. 4). As can be seen, after both treatments, by acid (1% sulfuric acid) and by water, the highest xylose concentration was achieved after a short (1-5 minutes) mechanochemical treatment.

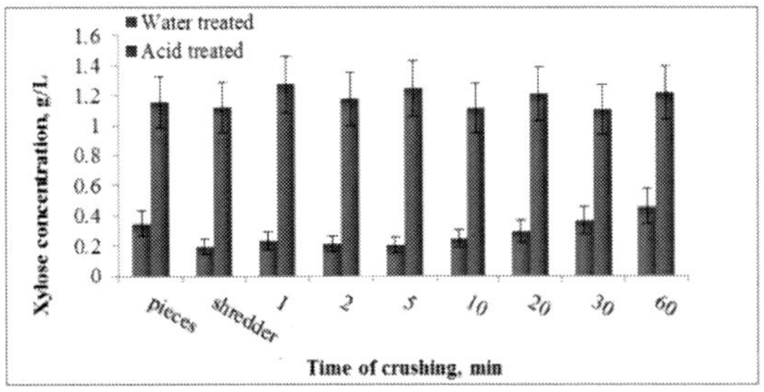

Figure 4. Xylose concentration extracted from **a**, **b** and **c** types water hyacinth biomasses, after acid and aqueous treatments and various stages of milling.

During process optimization, the biomass to milling balls ratio and milling speed were tested in various constellations (Fig. 5). This procedure determined the optimal milling speed (650 rpm), temperature (90°C optimal and 55°C nominal) and mass to balls ratio (1/40).

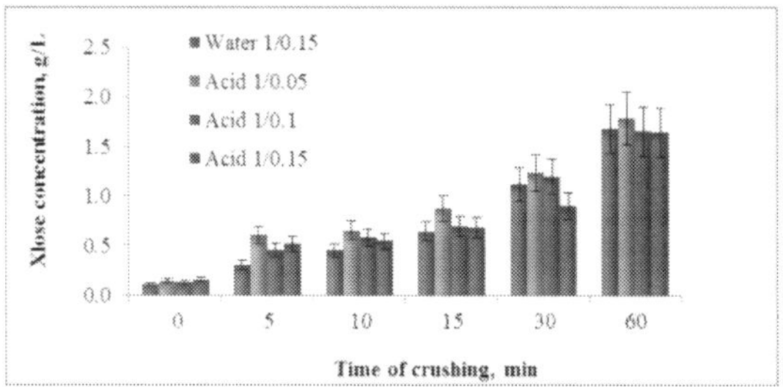

Figure 5. The results of xylose concentration after water and acid treatment (in several acid/biomas ratios) with 1/40 biomass to balls ratio and milling speed of 650 rpm.

Within this stage *Pichia stipitis* yeast were grown on a xylose containing medium. The growth curve of the yeast showed, that *Pichia stipitis* cells are capable of using xylose as organic substrate. In addition, other substrates in different combination were also evaluated. The substrates that were evaluated consisted of hydrolyzate synthetic fermentation substrate, Xylose in different concentrations and processed biomass with various xylose concentrations. The mechanochemical process was applied to the water hyacinth biomass and in the following stage to other organic materials. The treatment consisted of three hours duration in 650 RPM and wet biomass to milling balls ratio of 1/30, which was found nominal (vs. 1/40) in the previous research stage.

The fermentation productivity comparison was done both by monitoring the yeast growth curve and measurements of the ethanol concentration (Fig. 6). While the mixture of commercial synthetic reference substrate and water hyacinth biomass yielded the highest ethanol concentrations (18 g/L) after eight days the treated water hyacinth biomass alone demonstrated slightly lower concentrations (16 g/L) after ten days, which is with about the same economical balance owing to the material price of the commercial substrate.

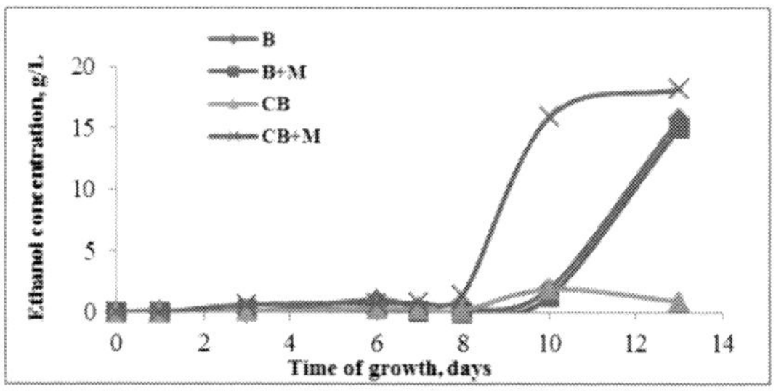

Figure 6. Ethanol production by fermentation with *Pichia stipitis* yeast in: B- Treated water hyacinth fermentation substrate with 14 g/L initial xylose concentration M- synthetic fermentation medium and CB- reference of 14 g/L xylose synthetic substrate.

The second research stage was focused on identification and evaluation of various agriculture products and aquatic plants as potential fermentation substrates. The suitability evaluation was done through measurements of their sugars combination and lipids and also protein contents; before and after nominal mechanochemical treatment. It was found that protein concentration (Fig. 7) is the most valid parameter for comparison. The results indicates that while protein levels in fruits are not affected by the mechanochemical treatment, grass, which is an equivalent to most surplus agriculture by-products, as well as aquatic plants such as water lettuce (*Pistia stratiotes*) are susceptible to the treatment, which improves their fermentation potential.

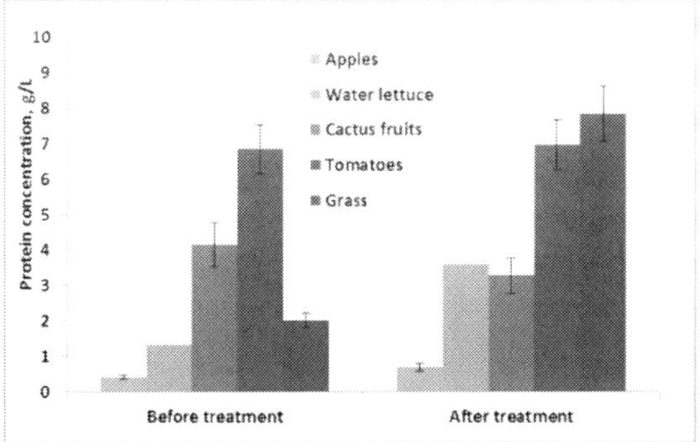

Figure 7. Dissolved protein concentrations, in various organic biomasses, before and after mechanochemical pre-treatment.

4 Conclusions

The present study is aimed to apply a simple and reliable process for the bioconversion of water hyacinth and agricultural wastes into ethanol. The proposed approach futile complex operations that are currently essential and as such, are expected to provide the cost efficiency allowing the technology implementation throughout developed and developing countries.

In general, increasing efforts and funds are being dedicated to the development and application of surplus biomass conversion into liquid bio-fuels and in particular ethanol, which provides several advantages over fossil fuels [23]-[24]:

1. Utilization of abundant and inexpensive sources of renewable resources
2. Reduction in greenhouse gas emission and toxic substances
3. Macroeconomic benefits for under-developed rural communities and their social aspect of sustainability
4. Improving national energy security

The possibility of converting aquatic plants and surplus agriculture products into ethanol is currently established in a number of developing countries [10-11]. While the approach advocated here favours distributed production units, such pilot facilities have not yet become industrial [26]. The methodology developed, may economically improve such systems by utilizing vast range of raw materials for ethanol production [27]. Moreover, the environmental impact in such cases will always be positive either through reduction in agriculture wastes or through mitigation of invasive aquatic flora growth [28-29]. In developed and developing countries this approach may be applied as wastewater treatment by aquatic flora grown in wastewater reservoirs and its transformation into biomass suitable for ethanol production [30]. The operation scheme developed throughout this research consists of several stages that may be utilized as basis for the development of continuous system for biomass transformation into ethanol:

1. Biomass chopping
2. Mechanochemical process
3. Differential fermentation
4. Ethanol refinement

At the present R&D stage, biomass chopping is done by commercial high-speed rotary blade cutter. The mechanochemical process is in an upscaling stage as well as the processed, biomass differential fermentation stage. The final stage of continuous ethanol distillation is also under development, which is being done directly in semi-industrial capacity.

Acknowledgement

The research is funded by the Israeli ministry of environmental protection.

References

1. Bringezu S., H. Schütz, M. O'Brien, L. Kauppi, R.W. Howarth and J. McNeely, Report of UNEP (United Nations Environment Programme), Towards Sustainable Production and Use of Resources: Assessing Biofuels (2009).
2. RFA (Renewable Fuels Association), Accelerating Industry Innovation - 2012. Ethanol Industry Outlook (2012).
3. Flavin C., J. L. Sawin, L. Mastny, M. H. Aeck, S. Hunt, A. MacEvitt and P. Stair, Report of American Energy, The renewable path to energy security, Worldwatch Institute and Center for American Progress (2006).
4. Gonzales G., Lopes-Santin J., Caminal G. and Sola C. Dilute acid hydrolysis of wheat straw hemicellulose at moderate temperature: a simplified kinetic model. Biotechnol Bioeng, 28:288-293, (1986)
5. Pessoa Jr. A., Mancilha I.M. and Sato S. Acid hydrolysis of hemicellulose from sugarcane bagasse. Braz. J. Chem. Eng. 14:291-297, (1997).
6. Hu G., J.A. Heitmann and O.J. Rojas, Feedstocks pretreatment strategies for producing ethanol from wood, bark and forest residues, BioResources, 3:270–294, (2008)
7. Yang, L., Lu, M., Carl, S., Mayer, J. A., Cushman, J. C., Tian, E., & Lin, H., Biomass characterization of Agave and Opuntia as potential biofuel feedstocks. Biomass and Bioenergy, 76, 43–53. doi:10.1016/j.biombioe.2015.03.004, (2015)

8. Mishima D., Kuniki M., Sei K., Soda S., Ike M. and Fujita M. Ethanol production from candidate energy crops: water hyacinth (Eichhornia crassipes) and water lettuce (Pistia stratiotes L.). Bioresour. Technol. 99:2495-2500, (2008).
9. Nigam J. N. Bioconversion of water-hyacinth (Eichhornia crassipes) hemicellulose acid hydrolysate to motor fuel ethanol by xylose-fermenting yeast, J. Biotechnol. 97:107-116, (2002)
10. Abraham M. and Kurup G.M. Pretreatment studies of cellulose wastes for optimization of cellulose enzyme activity. Appl. Biochem. Biotechnol.62:201-211, (1997).
11. EEA, 2012. The impacts of invasive alien species in Europe. EEA Technical Report No. 16/2012. Luxembourg: Publications Office of the European Union.
12. Gichuki J., Omondi R., Boera P., Tom Okorut T., Said Matano A., Jembe T. and Ofulla A. Water hyacinth Eichhornia crassipes (Mart.) Solms-Laubach dynamics and succession in the Nyanza Gulf of Lake Victoria (East Africa): implications forwater quality and biodiversity conservation. The Scientific World Journal 2012, 10 pp.
13. Ganguly A., Das S., Bhattacharya A., Singh P., Chatterjee P. K. and Dey A. Studies on the production of xylose from water hyacinth. Chemical Science, 2(1):1-7, (2013)
14. Aswathy U.S., Sukumaran R. K., Devi G. L., Rajasree K.P., Singhania R. R. and Pandey A. Bio-ethanol from water hyacinth biomass: An evaluation of enzymatic saccharification strategy. Bioresource Technology, 101:925-930, (2010)
15. Ganguly A., Das S., Dey A., Das R. and Chatterjee P. K. Optimization of xylose yield from water hyacinth for ethanol production using Taguchi Technique. IOSR Journal of Pharmacy and Biological Sciences (IOSR-JPBS), 3(5):1-9, (2012)
16. Harun M.Y., Radiaha A.B.D., Abidina Z.Z. and Yunus R. Effect of physical pretreatment on dilute acid hydrolysis of water hyacinth (Eichhornia crassipes). Bioresource Technology,102:5193-5199, (2011)
17. Silverstein R., Y. Chen, R. Sharma-Shivappa, M. Boyette and J. Osborne, A comparison of chemical pretreatment methods for improving saccharification of cotton stalks, Bioresour. Technol. 98:3000–3011, (2007)
18. Boudet A. M., S. Kajita, J. Grima-Pettenati and D. Goffner, Lignins and lignocellulosics: a better control of synthesis for new and improved uses, Trends Plant Sci, 8:576–581, (2003)
19. Da Costa Sousa L., S.P.S. Chundawat, V. Balan and B.E. Dale, 'Cradle-to-grave' assessment of existing lignocellulose pretreatment technologies, Curr. Opin. Biotechnol. 20:339–347, (2009)
20. Binod P., R. Sindhu, R. Singhania, S. Vikram, L. Devi, S. Nagalakshmi, N. Kurien, R. Sukumaran and A. Pandey, Bioethanol production from rice straw: an overview, Bioresour. Technol. 101:4767–4774, (2010)
21. Dien B. S., Cotta M. A. and Jeffries T.W. Bacteria engineered for fuel ethanol production: current status, Appl. Microbiol. Biotechnol. 63:258-266, (2003)
22. Wooley R., Ruth M., Glassner D. and Sheehan J. Process design and costing of bioethanol technology: a tool for determining the status and direction of research and development. Biotechnol. Prog. 15:794-803, (1999)
23. Krahulec S., Kratzer R., Longus K. and Nidetzky B. Comparison of Scheffersomyces stipitis strains CBS 5773 and CBS 6054 with regard to their xylose metabolism: implications for xylose fermentation. Microbiology Open,1:64-70, (2012)
24. Knauf M. and Moniruzzaman M. Lignocellulosic biomass processing: A perspective. Int Sugar J. 106:147-150, (2004)
25. Lin Y. and Tanaka S. Ethanol fermentation from biomass resources: current state and prospects. Appl Microbiol Biotechnol. 69:627-642, (2006)
26. Abdullah M. O., Lim S.F., Salleh S.F., Tai T.M., Leo S.L. and Umar A.K. Development of a small-scale reactor system for bioethanol production from agriculture waste geared towards small industries applications in Malaysia. Agriculture.12:12-15, (2011)
27. Hahn-Hägerdal B., Galbe M., Gorwa-Grauslund M. F., Lidén G. and Zacchi G. Bio-ethanol - the fuel of tomorrow from the residues of today. Trends Biotechnol.24:549-556, (2006)
28. Theuri M. Water hyacinth – Can its aggressive invasion be controlled? Environmental Development,7:139-154, (2013)
29. Zhang Y., Zhang D. and Barrett S. Genetic uniformity characterises the invasive spread of water hyacinth (Eichhornia crassipes), a clonal aquatic plant. Molecular Ecology,19:1774-1786, (2010)
30. Sarkar N., Ghosh S.K., Bannerjee S. and Aikat K. Bioethanol production from agricultural wastes: An overview. Renewable Energy, 37:19-27, (2012)

AUTHOR INDEX

AUTHOR INDEX

AUTHOR INDEX